Textile Materials for Lightweight Constructions

Chokri Cherif
Editor

Textile Materials for Lightweight Constructions

Technologies - Methods - Materials - Properties

Springer

Editor
Chokri Cherif
TU Dresden
Dresden
Germany

Translated by/Translators:
David Friemann, M.A., Dr. Mir Mohammad Badrul Hasan (Chapter 13)
Vignaesh Sankaran (Chapter 16)

ISBN 978-3-662-50199-3 ISBN 978-3-662-46341-3 (eBook)
DOI 10.1007/978-3-662-46341-3

Springer Heidelberg New York Dordrecht London

Softcover reprint of the hardcover 1st edition 2016

Printed on acid-free paper

Springer-Verlag GmbH Berlin Heidelberg is part of Springer Science+Business Media (www.springer.com)

Preface

The supply of natural resources for economic development is limited. In order to prevent the destruction of humanity's own livelihood, this limited supply requires a gentle and efficient handling of the available energy sources and materials by all the branches of industry, aimed at reducing emission of pollutants and the efficient use of materials. This has become particularly clear against the backdrop of globalization with the abrupt increase in global transportation and the steady growth of individual mobility. Particularly in branches of industry relying on the accelerated transportation of large masses, passenger and goods traffic, as well as mechanical and plant engineering, innovative lightweight construction technologies based on plastics are more important than ever before. These aspects of material and energy efficiency also apply for wooden and concrete reinforcements. In the construction industry, the number of applications increases for fiber-reinforced slender and filigree concrete components, for fiber-reinforced plastics, as well as for the reinforcement and restoration of existing structures. Furthermore, textile membranes are highly efficient and extremely lightweight construction materials with adjustable functionalities, making them relevant for a wide range of applications.

Textile materials and semifinished products have a versatile property potential and often act as carriers and driving forces behind innovative developments. They are distinguished by the use of high-performance fiber materials and advanced technologies. In the past decades, a unique interdisciplinary spectrum of knowledge has been evolved worldwide in the field of textile technology. The focus is placed on polymeric, mineral, and metallic fiber-based materials for use in high-tech applications. These textile materials will remain a crucial group of high-performance materials and will be established as a significant research priority in twenty-first-century material science. The fiber and textile technology research institutions will become the centers of an indispensable multidisciplinary research of innovative technologies and products.

The combination of material science, nanotechnology, microsystems technology, bionics, physics, and chemistry results in a new product range with properties adjustable to the individual demands in a wide range. The gamut and depth of the required processes and materials are immense and highly complex. Even products with unique characteristics and fundamental approaches toward intelligent and self-learning materials are realizable.

The aim of this first edition is to fully exploit the performance potential and the variety of textile materials and semifinished products. Experts of textile technology will share basic knowledge of textile and ready-made technology as well as future-oriented special knowledge for the manufacture and use of high-tech textiles. They show the possibilities for the application of textile structures in lightweight construction. Therefore, this book will concentrate on the detailed portrayal and description of the entire textile process chain from fiber material to the diverse yarn constructions and various textile semifinished products in two- and three-dimensional shapes, but will also touch upon preforming and interphase/interface design. Beyond those, tests according to valid norms and special, recently developed test methods for textile lightweight construction will be introduced. This reference book is rounded off by remarks concerning the modeling and simulation technology for the structure-mechanical calculations of highly anisotropic, flexible high-performance textiles, and exemplary applications from the fields of fiber-reinforced composites, textile concrete, and textile membranes. This aims to exemplify the potential of textile structures as innovative lightweight construction material by the specific selection and combination of textile processes for the realization of a nearly unlimited number of property profiles, and finding possibilities of functional integration as well as designing of near-net shape components. The aim is to create a conscious motivation for a wider use of textile high-performance materials in lightweight construction applications on a large scale, which will soon start their triumph in the field of fiber-reinforced composite materials.

The deliberations included in this book are based on long years of interdisciplinary research and development work, including special research areas and research clusters in the fields of fiber-reinforced composites, textile-reinforced concrete, and textile membranes. These research projects along the entire textile process chain are promoted at the Institute of Textile Machinery and High-Performance Material Technology of TU Dresden. Extensive teaching material could be gathered from engineering education and doctoral studies in textile and assembly technology as well as lightweight construction, all of which contributed to the successful creation of this textbook.

Dresden, Germany
June 2014

Chokri Cherif

Acknowledgment

This textbook is the result of a combined effort of scientists of the Institute of Textile Machinery and High-Performance Material Technology (ITM) at Dresden University of Technology and external experts from research and teaching. The majority of them are experienced researchers with extensive knowledge in the areas of textile and assembly technology as well as of textile-reinforced lightweight construction gathered from working and teaching in these fields. Other contributions were provided from junior researchers, whose graduate studies yielded specialized insights into their respective fields. The intense cooperation of the editor with the individual authors and the resulting coherence of the content and interlinking of the chapters have created a consistent work on the subject of "Textile Materials for Lightweight Construction" from fiber materials and semifinished products to preforms, along the textile supply chain.

This textbook was first published in a German-language edition in 2011, which laid the groundwork for this translated version.

First of all, all staff members of the ITM involved in the creation of this book have to be acknowledged. Apart from conducting their activities in research and teaching, their personal dedication and commitment were crucial for the success of this project. This book is based on years of experience of the scientists of the ITM designated as authors. My special thanks are due to all authors of individual chapters for their contributions and understanding collaboration in ensuring the success of this book.

I would like to express my personal gratitude to the other experts: Dr. Beata Lehmann (former research fellow at ITM, currently research fellow at the Institute of Surface and Manufacturing Engineering at TU Dresden), Dr. Georg Haaseman (Institute of Solid Mechanics at TU Dresden), and Dr. Silvio Weiland (TUDALIT Markenverband e.V., Dresden), who contributed to some of the special fields. Prof. Dr.-Ing. Hilmar Fuchs of the Sächsisches Textilforschungsinstitut e.V. Chemnitz, honorary professor for technical textiles at TU Dresden, has kindly contributed to this textbook.

My gratitude is also due to all authors for their support in the translation of this book and for their invaluable advice. Many of them have left the ITM or the respective institutions since the original publication of the book to take over highly responsible managerial positions in the industry or in public institutions.

I would like to cordially thank Annett Dörfel for editorial coordination and design.

I would also like to mention the contribution of David Friemann, M.A., who translated the textbook into English in close cooperation with the authors. Equally, I would like to thank M.Sc. Moniruddoza Ashir and Dr.-Ing. Mir Mohammad Badrul Hasan for translating the graphical elements and figures and proofreading.

More thanks are due to Dr. Ezzedine Laourine, Janine Kleinow, Dipl.-Designer (FH) Aram Haydeyan, Stefanie Fiedler, Anja Wenzel, Moritz Egger, as well as Richard Müller, as representative of many students of the ITM responsible for the creation of illustrations, graphics, and complementary research work contained in the German edition of the book.

I do not want to omit friends and experts, who have contributed to the successful creation of the German edition of this textbook by correcting and advising professionally wherever required. My personal thanks are due to Prof. Dr.-Ing. habil. Dr. h. c. Peter Offermann (former Institute Director of ITM), Prof. Dr. rer. nat. Volker Rossbach (former Chairman of Textile Finishing), Prof. Dr. Frank Ficker (Hof University of Applied Sciences, Institute of Material Science, Department Münchberg), Dr. Harald Brünig (Leibniz-Institute for Polymer Research Dresden e. V.), Dr. Adnan Wahhoud (Lindauer Dornier GmbH, Lindau), Peter Maier, Peter Rotter and Martin Leidel (LIBA Maschinenbaufabrik, Naila), and Dr. Christian Callhoff (Mehler Texnologies GmbH, Hückelhoven).

We would also like to extend our gratitude to our partners in research institutions and industry for the rewarding dialogue and lively exchange of experiences and for the supply of large number of current picture materials, which are provided with the corresponding sources in the respective chapters.

I would finally like to thank the publisher for the extraordinary cooperation, for helpful advice during the compilation of the master copy, and for designing the textbook.

Contents

List of Contributors

Chokri Cherif Institute of Textile Machinery and High Performance Material Technology, TU Dresden, Dresden, Germany

Olaf Diestel Institute of Textile Machinery and High Performance Material Technology, TU Dresden, Dresden, Germany

Thomas Engler Formerly: Institute of Textile Machinery and High Performance Material Technology, TU Dresden, Dresden, Germany

Christiane Freudenberg Institute of Textile Machinery and High Performance Material Technology, TU Dresden, Dresden, Germany

Hilmar Fuchs Sächsisches Textilforschungsinstitut e. V., Chemnitz, Germany

Lina Girdauskaite Institute of Textile Machinery and High Performance Material Technology, TU Dresden, Dresden, Germany

Georg Haasemann Formerly: Institute of Solid Mechanics, TU Dresden, Dresden, Germany

Jan Hausding Formerly: Institute of Textile Machinery and High Performance Material Technology, TU Dresden, Dresden, Germany

Claudia Herzberg Formerly: Institute of Textile Machinery and High Performance Material Technology, TU Dresden, Dresden, Germany

Gerald Hoffmann Institute of Textile Machinery and High Performance Material Technology, TU Dresden, Dresden, Germany

Evelin Hufnagl Formerly: Institute of Textile Machinery and High Performance Material Technology, TU Dresden, Dresden, Germany

Heike Hund Institute of Textile Machinery and High Performance Material Technology, TU Dresden, Dresden, Germany

Rolf-Dieter Hund Institute of Textile Machinery and High Performance Material Technology, TU Dresden, Dresden, Germany

Roland Kleicke KLEICKE ingenieure, Demitz-Thumitz, Germany

Formerly: Institute of Textile Machinery and High Performance Material Technology, TU Dresden, Dresden, Germany

Cornelia Sennewald née Kowtsch Institute of Textile Machinery and High Performance Material Technology, TU Dresden, Dresden, Germany

Sybille Krzywinski Institute of Textile Machinery and High Performance Material Technology, TU Dresden, Dresden, Germany

Ezzeddine Laourine Albstadt-Sigmaringen University, Albstadt, Germany

Formerly: Institute of Textile Machinery and High Performance Material Technology, TU Dresden, Dresden, Germany

Beata Lehmann Institute of Manufacturing Technology, TU Dresden, Dresden, Germany

Jan Märtin Formerly: Institute of Textile Machinery and High Performance Material Technology, TU Dresden, Dresden, Germany

Kathrin Pietsch Institute of Textile Machinery and High Performance Material Technology, TU Dresden, Dresden, Germany

Thomas Pusch Institute of Textile Machinery and High Performance Material Technology, TU Dresden, Dresden, Germany

Hartmut Rödel Institute of Textile Machinery and High Performance Material Technology, TU Dresden, Dresden, Germany

Mirko Schade Formerly: Institute of Textile Machinery and High Performance Material Technology, TU Dresden, Dresden, Germany

Wolfgang Trümper Institute of Textile Machinery and High Performance Material Technology, TU Dresden, Dresden, Germany

Silvio Weiland TUDALIT Markenverband e.V., Dresden, Germany

Chapter 1
Introduction

Chokri Cherif

The recent concepts and trends in lightweight construction applications and the development of suitable fiber-based materials and matrix systems, as well as thoroughly automated manufacturing concepts contribute to an increased use of high-performance fibers and make fiber-reinforced composites (FRCs) the defining material class of the twentyfirst century. Textile materials and semi-finished products act as carriers and drivers of these innovative developments and are an important basis of quantum leaps in resource efficiency, CO_2 emission reduction and development of products able to meet consumer demands with wholly new ideas. In the future, energy consumption coverage in all civil and economic sectors will require wider use of renewable and CO_2 neutral energy sources and concepts, leading to innovations and changes within the energy sector. As a result, a paradigm shift in the use of materials driven by energy and resource scarcity will put fiber-based high-performance materials and the products based on them in high demand by a variety of industries [1].

Continuous fiber-reinforced composites, as a relatively new material class, consist of a tensile load-absorbing textile reinforcement structure and a shaping, compressive-load-absorbing matrix material. Composite materials also include textile membranes consisting of a coated or foil-laminated textile surface as reinforcement.

The excellent properties of fiber-reinforced composites, such as their high specific strength and stiffness, good damping properties, chemical resistance and low thermal expansion lead to the increased use of fiber-based lightweight construction products frequently classified as high-grade construction materials. In comparison to conventional materials, particularly based on metals, they are distinguished by outstanding corrosion resistance, ductility and weight reduction. In

C. Cherif (✉)
Institute of Textile Machinery and High Performance Material Technology, TU Dresden, Dresden, Germany
e-mail: chokri.cherif@tu-dresden.de

C. Cherif (ed.), *Textile Materials for Lightweight Constructions*,
DOI 10.1007/978-3-662-46341-3_1

order to fully exploit the potential of the fiber reinforcement in the composite, the fibers have to be arranged stretched along the main load directions and suitably embedded into the matrix. The textile materials and semi-finished products providing reinforcement are often transferred into near-net shape component geometries by means of a large range of different manufacturing technologies.

Continuous fiber-reinforced composite components based on different matrix systems also display flexible adaptability of their structure and thus adjustability of material properties and property anisotropy to the individual requirements of processing and construction. Therefore, they possess high potential for a cost-efficient production of tailor-made composite components with broad use in lightweight construction for conventional and new application areas. In this regard, the setup of efficient, non-stop process chains for the development of textile-based lightweight construction structures is crucial. A requirement-suited designing of FRC components based on carbon fibers can attain weight reductions of 30 % compared to aluminum or even 70 % compared to steel. In construction, weight reductions of up to 80 % over the conventional ferroconcrete structures are feasible.

Due to their structural variety, textile-based lightweight construction structures resulting from the combination of textile reinforcement components with plastic and mineral-based matrix systems have already been established as an innovative, economically viable and suitably designable material indispensable to lightweight construction applications. The potential of FRCs in fields of application requiring a great lightweight construction advantage is widely recognized. The market for fiber-reinforced plastic composites (FRPCs) has registered above-average growth rates in recent years [2]. Particularly, FRPCs based on short-fiber reinforcements, as manufactured in *sheet molding compound* (SMC) or the various long-fiber-reinforced thermoplastic (*LFT*) methods, have been used in the serial production of secondary components in passenger and utility vehicles [3]. A similar situation can be observed in construction, regarding market-proven short fiber concrete components such as facade elements and carbon-fiber-reinforced plastic composite lamellas for the strengthening of buildings. Numerous recent lightweight construction developments are aimed at a dedicated enhancement and acceleration of the use of continuous fiber-reinforced composite components. Apart from the expensive *Pre-preg technology* (pre-impregnated textile structures with a resin system), science and industry are closely tracking alternative production methods based on dry (i.e. non-pre-impregnated) reinforcement structures in combination with thermosetting matrix systems or through the use of thermoplastic-based hybrid constructions, aspiring to establish possibilities of a highly productive processing into complex, function-integrated composite components.

FRPCs are preferentially produced from a basis of thermosetting matrix systems. The broad use of such FRPC solutions, especially in aeronautics, wind power and sports equipment technology, has contributed to this technology's special potential concerning material, ecological and energy efficiency under economical aspects. They are increasingly being used in structural components and applications with challenging requirements. Despite the excellent properties and the considerable increase in the use of FRPCs particularly in aeronautics, the currently realizable cycle times are a limiting factor for application in large series. So far, their industrial applications have been largely limited to "prestige object" in the form of spoilers, side panels, tailgates, doors or roofs for vehicles [4]. The consistent automation of steps for the reproducible FRPC component manufacture and the development of cost-efficient textile semi-finished products, highly reactive resin systems and fast impregnation and consolidation technologies, which are at the center of attention of recent research are laying the groundwork for a promising industrial implementation in large-scale production areas.

The high potential of continuous fiber-reinforced thermoplastic composite materials for serial use in complex, highly loaded and recyclable vehicle and machine engineering parts have put them in the center of intensive developments in science and industry aspiring to realize dry structures. The possible specific strengths and stiffnesses that can be achieved are considerably higher and can be ten times as high as that of metallic materials, while requiring equal amounts of material. Load cases caused by alterations in component geometry due to thermal expansion can be avoided or generally defined by purposefully combining carbon fibers and matrix materials.

The various existing applications in the field of textile-reinforced concrete demonstrate the effectiveness of textile reinforcements as well as their practicality. A large number of applications of textile concrete have already been implemented. Most of them are related to the free shaping properties and low material consumption leading to the reduction of CO_2 emissions by conserving energy and material, in particular cement. Especially for the strengthening of old ferroconcrete constructions or the maintenance of existing structures, the use of textile reinforcements is becoming a technology with good prospects [5, 6]. The small geometrical changes and the great capabilities of the high-performance fibers, the correlating low own-weight loads and relatively easy applicability of the textile-reinforced fine-grained concrete layers lead to completely innovative usage and structural design possibilities. The almost unlimited formability and load-adapted arrangement of rovings in textile structures, their free designing and dimensioning, often inspired by nature, result in complex, aesthetically pleasing, and delicate architecture with economic market relevance.

Apart from FRPCs and textile concrete, textile membranes, usually thin ones mainly stressed by tensile forces, are another innovative class of technical materials adaptable to material choice and construction to fulfill a variety of different functions: for example, separating, delineating, encasing, filtering, load bearing and distribution; or protecting against weather, sound or heat. This results in a wide application spectrum ranging from membranes for textile construction to sails in high-performance yachting. Furthermore, it also includes sun protection textiles, advertising surfaces, tents, geotextiles and truck tarpaulins. Developments so far have been aimed at achieving high tensile strengths and Young's moduli at small membrane mass. Thus, textile membranes are the basis of unlimited innovative technical lightweight construction solutions in the high-performance field.

The load-adapted alignment of textile reinforcement structures in complex-shape components adds high engineering-technological requirements on both the calculation and simulation software used as well as on the realization of appropriate machinery concepts and production methods. Compared to metallic materials, fiber-reinforced composites, at purposeful choice of reinforcement structures and low-fiber-damage production, can allow a more cost-efficient component designing. This requires the selection of suitable technology and the coordination of the chosen material with the geometry, regarding design and construction aspects. If this is taken into account, composite materials help realize complex component geometries that would be costly to realize with metallic materials.

However, innovative and efficient lightweight construction means designing each individual part with the global concept of the application in mind. In product development, economical and ecological aspects (e.g. end-of-life vehicle laws, emission and fuel consumption regulations) are understood not as antagonism but as supplement and symbiosis [7]. Efficient use of available and the development of new fiber materials, construction methods, and technologies for lightweight construction products can contribute to sustainable climate protection. The lower the weight of cars, airplanes or machines, the smaller the amount of energy required to operate them and the fewer greenhouse gasses and air pollutants are emitted. While load-adapted and recyclable designing and near-net shape manufacturing methods can help attain high material efficiency in production, energy efficiency in the daily use of the products or parts is a decisive competitive criterion.

The technologies currently used for the production of textile fiber materials, structures and *preforms* (near-net shape and dry fiber structure) and their subsequent exact positioning within the impregnation tool are usually suitable for a number of different application areas, regardless of the component manufacturing methods and the matrix material used. A sustainable breakthrough for the wider use of textile-based lightweight constructions in large series can only be achieved by thoroughly exploiting the different textile constructions' potential for the respective applications [8]. The extremely wide range of textile materials, production technologies, and their combinations complicate the user's efficient design of manufacturing processes and component-related choice of suitable materials. The following complexes and challenges are relevant to ensure an efficient and reproducible manufacture of processable and load-adapted textile semi-finished products and preforms from a multitude of textile constructions with near-net shape geometries, suitable for the economical processing into complex and high-loaded fiber composite parts for automotive and machine engineering, construction, membrane technology as well as the elastomer and timber industries [9]:

- reduction of fiber and semi finished product cost by correct selection of textile fiber materials and manufacturing technologies, waste minimization, e.g. by employing near-net shape production methods and damage free processing of high-performance fiber materials aimed at attaining highest possible material efficiency,
- development and selection of cost-effective technologies and machine types for the realization of dry structures or their hybridization/(partial) impregnation, temporary fixing of textile semi-finished products for the production of complex,

suitable and easy-to-handle preforms with component-adjusted reinforcement as well as near-net shape/thickness-adjusted dimensions for medium and large series,
- selection of reproducible and automated preforming processes and concepts, with subsequent impregnation and consolidation at short cycle times,
- cutting of production costs and process times of components by correctly selecting textile reinforcement structures and related technologies dependent on the matrix systems, impregnating and consolidating quickly and flawlessly, as well as choosing adjusted machine concepts and impregnation strategies accordingly,
- tailor-made *interface design* by means of surface modification and realization of adapted interfaces for the application-related adjustment of composite properties between reinforcement and matrix component,
- development of intelligent concepts for a function-integrated lightweight construction in multi-material design,
- development of powerful, non-destructive and fast testing methods and quality control systems for textile materials, semi-finished products and preforms, as well as
- development and implementation of effective and reliable simulation software for the design of flexible reinforcement structures to produce complex components

This textbook emphasizes the detailed introduction and interpretation of definitions as well as the distinct delineation of textile production methods. These will help avoid misinterpretations and enable the most efficient use of the near-unlimited possibilities of textile and ready-made technology for energy efficient fiber-reinforced plastic composites in the areas of lightweight construction. Specialized know-how is conveyed, taking into account exact textile terminology and interconnections and contexts of the multifaceted textile process steps. Furthermore, special methods that are currently being developed are presented and research trends are illustrated. The potential, properties and versatility of textile materials and semi-finished products as well as their combinations into completely new possibilities and product geometries are to be extracted in particular in the course of this book.

The focus of this book will be on an illustration of the importance of textile constructions based on fiber-based high-performance materials for function-integrated lightweight construction applications. This includes applications of fiber-reinforced plastic composites on thermoplastic and thermosetting bases, of textile membranes and of lightweight constructions with concrete. Furthermore, reinforcements for wooden and elastomer materials are other target groups. For the first time, manufacturers of composite materials will obtain a comprehensive overview of the possibilities and the potential of fiber and textile technologies and the corresponding machine technologies for an economical production of suitable 2D and 3D textile structures for lightweight construction in complex architectures, as previous relevant textbooks only touch upon selected fields.

The book is to be a "classic" as a modern teaching and learning package and is targeted at students, engineers, designers and developers as well as research institutions in the areas of textile and ready-made technology, plastics, elastomer and wood engineering, lightweight construction, fiber-reinforced plastic composites, materials science, civil engineering and architecture. The book concentrates mainly on textile materials and semi-finished products. The selected application examples in the textbook (FRPCs, textile concrete and textile membranes) are limited to a short introduction of the requirements for the components and are followed by explanations of the required textile process chains and the component production technologies necessary to fulfill them. For purposes of additional information on the subject areas of fiber-reinforced plastic composites, textile concrete and textile membrane technology, relevant textbooks and publications are listed.

References

1. SGL Group: Textilbewehrter Beton - ein neuer Verbundwerkstoff für die Bauindustrie (Fachpresse-Information). Meitingen, Deutschland, 28. November 2008
2. ANONYM: Composite Materials Price Trends, Forecast and Analysis, Marktanalyse. Dallas, USA, 2007
3. NASSAUER, J.: Automobil + Innovation - Internationale Märkte, technische Entwicklung. Nürnberg, 2007
4. NEITZEL, M.; MITSCHANG, P.: Handbuch Verbundwerkstoffe. München, Wien : Carl Hanser Verlag, 2004
5. CURBACH, M.; MICHLER, H.; WEILAND, S.; JESSE, D.: Textilbewehrter Beton - Innovativ! Leicht! Formbar! In: BetonWerk International 11 (2008), Nr. 5, S. 62–72
6. DUBEY, A. (Hrsg.): Textile-reinforced concrete. Farmington Hills, USA : American Concrete Institute, 2008
7. HUFENBACH, W.: Materialeffizienz durch Systemleichtbau. In: Proceedings. 11. Dresdner Leichtbausymposium. Dresden, Deutschland, 2007
8. CHERIF, Ch.; DIESTEL, O.; GRIES, Th.: Textile Verstärkungen, Halbzeuge und deren textiltechnische Fertigung. In: HUFENBACH, W. (Hrsg.): Textile Verbundbauweisen und Fertigungstechnologien für Leichtbaustrukturen des Maschinen- und Fahrzeugbaus. Dresden: SDV - Die Medien AG, 2007
9. CHERIF, Ch.: Trends bei textilbasiertem Leichtbau / Trends in textile based lightweight design. In: Technische Textilien/Technical Textiles 51 (2008), S. 22–23, E22–E23

Additional Literature

LÄSSIG, R.; EISENHUT, M.; MATHIAS, A.; SCHULTE, R.T.; PETERS, F.; KÜHMANN, T.; WALDMANN, T.; BEGEMANN, W.: Series production of high-strength composites: perspectives for the German engineering industry. VDMA study, 2012

BAI, J.: Advanced fibre-reinforced polymer (FRP) composites for structural applications. Woodhead Publishing, 2013, ISBN: 978-0-85709-418-6

VEIT, D.: Simulation in textile technology: theory and application. Woodhead Publishing, 2012, ISBN 0857097083

CHERIF, Ch.; KRZYWINSKI, S.; LIN, H.; SCHULZ, Ch.; HAASEMANN, G.: New Process Chain for Realisation of Complex 2D/3D Weft Knitted Fabrics for Thermoplastic Composite Applications. In: Procedia Materials Science 2 (2013), pp. 111–129

GAND, A. K.; CHAN, T.-M.; MOTTRAM, J. T.: Civil and structural engineering applications, recent trends, research and developments on pultruded fiber reinforced polymer closed sections: a review. In: Frontiers of Structural and Civil Engineering 7 (2013), No. 3, pp. 227-244

Chapter 2
The Textile Process Chain and Classification of Textile Semi-finished Products

Chokri Cherif

This chapter gives a general overview of the most important steps of the textile process chain and will thus facilitate a deeper understanding of the material group of functional textiles. The introductory material- and process-related definitions concerning fibers, yarns, fabrics and their further processing are explained in depth in the following chapters. The scope of technical textiles has been extended far beyond the original technical application areas. The steady and intense use of micro system and nanotechnology, measurement and sensor technology, plasma technology and modern finishing techniques are suitable to equip textiles with specific, adjustable properties and functions. One main characteristic of functional textiles is their orientation toward functionality, performance and an added value in comparison with conventional textiles.

2.1 Introduction

For decades, the European textile industry has been experiencing a structural change focused on the development of innovative high-quality products. Current trends and the know-how transferred into practice reveal the great potential of textile innovation. This does not only affect the textile industries, but also other branches of industry and products. Apart from the classical uses in garment and home textiles, technical applications are present in nearly all areas of everyday life. The production of technical textiles is a new, innovative, and promising field. Technical textiles are often distinguished by their functional diversity, and specific know-how is required for their design and production.

C. Cherif (✉)
Institute of Textile Machinery and High Performance Material Technology, TU Dresden, Dresden, Germany
e-mail: chokri.cherif@tu-dresden.de

C. Cherif (ed.), *Textile Materials for Lightweight Constructions*,
DOI 10.1007/978-3-662-46341-3_2

The use of these technical textiles is multi-faceted and not limited to clothing or home textiles, but extends to a variety of disciplines like automobile construction, aeronautics, construction engineering, and architecture, as well as healthcare, and security services.

Technical textiles feature the use of high-performance fibers, highly developed technologies and the incorporation of other, often non-textile materials. Their properties display an extremely versatile potential and turn both textile materials and the related production methods into mainstays and driving forces for the development of innovative products. The fiber materials and textiles with their unique properties are the best precursors for new products and technologies, e.g. in the fields of materials science and microsystems engineering, and for intelligent and adaptive materials.

Technical textiles are characterized by diversity, compatibility, functionality, flexibility and interactivity. These properties have broadened the range of applications and allow the development and opening of entirely new product groups. The range of variation and functionality of technical textiles is extremely large because of the near-unlimited multitude of property profiles resulting from fiber type and mixture, yarn formation, fabric production, as well as surface modifications and functionalizations on various production levels. These possibilities create perfect conditions for compatibility and connection with other, non-textile materials like plastics, metals and concretes. The combination of technical textiles with micro systems technology creates interactive data and information media [1] and integrated sensor and actuator networks, used for instance for structural monitoring and oscillation dampening in composite components. This allows the flexible use and customization of textile materials and semi-finished products with their adjustable properties.

The use of technical textiles as an independent product group is by now well-established in nearly all disciplines, beyond applications in technical areas. This requires an intense analysis of terminology and distinction of technical and functional products from conventional garment, home and household textiles. With the steady and intense use in micro systems technology, nanotechnology, metrological and sensor technology, modern finishing technology and bionics sees textiles being fitted with specific adjustable properties and functions are going far beyond the requirements of technical applications. Therefore, the term functional textiles will be favored within the framework of this textbook. This product class is distinguished by its focus on functionality, performance and the obvious additional usefulness in comparison with conventional textiles.

Textile technology and its variety of production methods offer outstanding possibilities for the development of bionics-based solutions. The most popular bionic product in garment production is probably the hook-and-loop fastener, fashioned after the natural seed distribution mechanism of the burdock plants [2]. Other inspirations in plants include bamboo, horsetail or arundo characterized by extreme stability with long stems and thin-walled culms. These construction principles are exploited in the development of structurally optimized fiber composite materials with a similar combination of stability and small mass [3–5]. Leaf

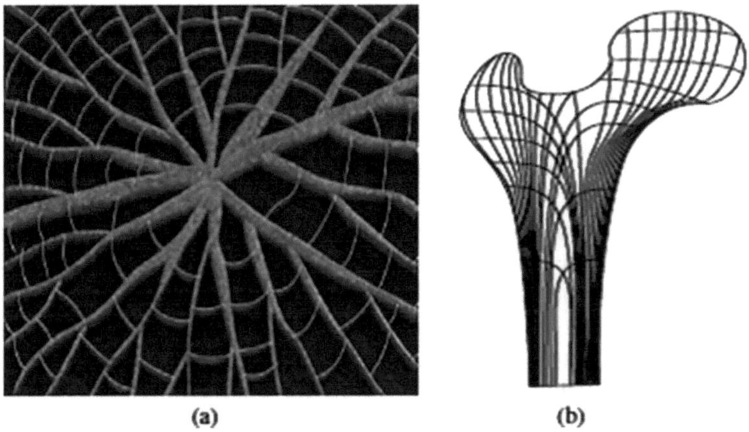

Fig 2.1 Lily pad (**a**) and bone structure (**b**)

structures also serve as inspiration for the design of light, yet highly rigid, fiber composite component structures like shells with stiff reinforcement ribs. Complex 3D geometries with lightweight construction characteristics, for instance based on lily pads (Fig. 2.1a), can be produced by means of textile construction. *Bionics* are potentially suited for the production of complex and three-dimensionally loaded lightweight constructions constituting an optimized construction with force-flow-adapted design and special force application systems, analogous to the human hip joint (Fig. 2.1b) and the corresponding articular cartilage tissue. Flexibly customizable fiber and textile technology offers an ideal foundation for the emulation of biological solutions in all their complexity and range.

The spectrum and depth of the required textile materials and processes are immense and highly complex. Therefore, this book concentrates on the portrayal and description of the textile process chain from fiber material to different yarn construction and 2D or 3D textile semi-finished products, preforming, interface and interface layer design, their testing according to current norms as well as newly developed testing methods for lightweight construction. This includes the fields of fiber-reinforced plastic composites (FRPC), textile-reinforced concrete and textile membranes. Altogether, this chapter conveys basic knowledge on the representation of the textile process chain, its links and relations, and the correct classification of textile materials and semi-finished products.

2.2 Textile Process Chain

2.2.1 Representation

In order to give a clear representation of the enormous range of textile processes and establish an overview of the versatility of the possible combinations, it is necessary to abstract the textile processes and limit their depiction to most important process steps. Incidentally, it has to be noted that textile materials and processes are virtually unlimited in their combination possibilities. This distinguishes fiber and textile technology with regard to the design of variable, anisotropic structural properties of fiber or textile compound components.

Figure 2.2 gives a general overview of the most important steps of the textile process chain: *primary* and *secondary spinning*, *winding, twisting*, warp yarn preparation, the processes for creating plane and three-dimensional textile construction, *finishing* and *ready-made manufacturing.*

Primary spinning includes the production of continuous man-made fiber materials from natural and synthetic polymers as well as non-polymer raw materials. For high-performance applications, these materials are often processed from their unaltered state into textile semi-finished products and finished products. To improve processing conditions, the man-made fibers are finished with sizing and

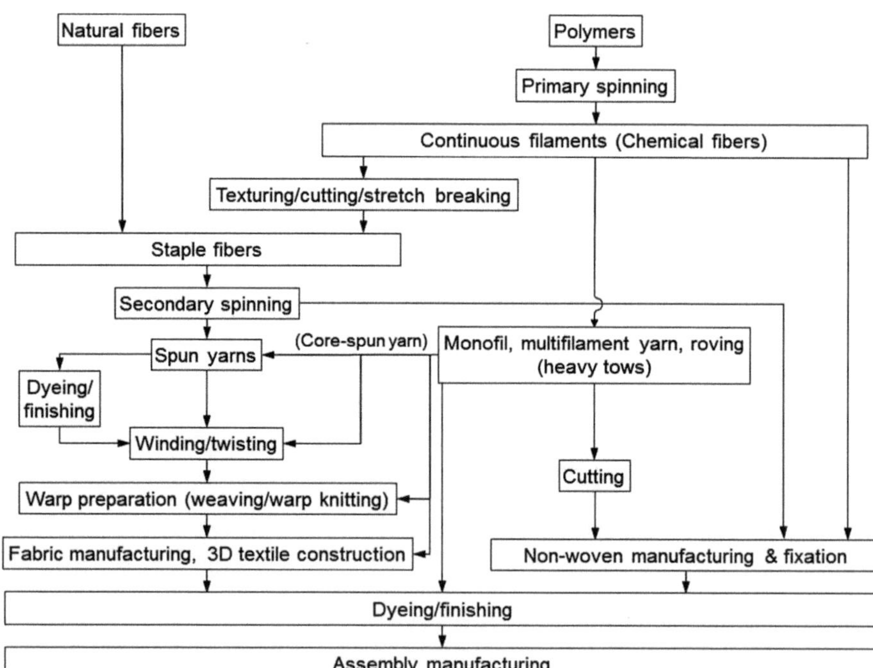

Fig. 2.2 Overview of the stages of the textile process chain

finishing materials during spinning. This allows a low-damage, unimpeded further processing of often shear-force sensitive fiber materials and prevents electrostatic charging.

Natural and synthetically manufactured (man-made) fibers of non-continuous lengths (cut or stretch broken) are spun into staple fiber yarns during *secondary spinning* processes. Often, cut or continuous man-made fibers are mixed with natural fibers and processed into hybrid yarns or fabrics. The use of the term *secondary spinning* is limited to fields in which natural fibers and cut or stretch broken (i.e. non-continuous) man-made fibers are processed. It serves to differentiate between the actual fiber production (*primary spinning* for the manufacture of continuous fibers) and the subsequent process steps required for spun yarn production.

Winding, twisting and warp yarn preparation all serve to convert yarns into suitable forms for the subsequent process steps. This includes fabric manufacturing, finishing and assembly techniques. The processes and their combinations portrayed in Fig. 2.2 can be expanded and customized arbitrarily for the corresponding applications and products.

2.2.2 Definition

In order to bring important basic concepts to the readers' minds and ease them into the subject of fiber and textile technology, the following sections will clarify the most important terms and notions. This will provide readers with a basis to understanding and more profoundly engaging themselves in the textile materials, constructions and technologies described below. The focus is on the portrayal and interpretation of definitions based on the current national and international standards and on the unambiguous distinction of textile materials, semi-finished products, finished products and necessary production methods. This will help prevent misinterpretations and fully tap the potential of Textile and ready-made technology's near-unlimited possibilities in energy-efficient lightweight construction designs and composite materials. The latest expert knowledge, with respect to exact textile terminology, and the relations and interconnections of the various textile process steps, will be imparted to the reader. The standards and norms associated with the most important textile structures will be included in the respective sections.

2.2.2.1 Textile Fiber Materials

Textile Fiber Materials can be classified into natural and man-made fibers. Because of the extremely high industrial demand for tailor-made fiber-based materials in a number of applications, and of the continuously growing world population, the global consumption of fibers is largely met with man-made fibers, which will be

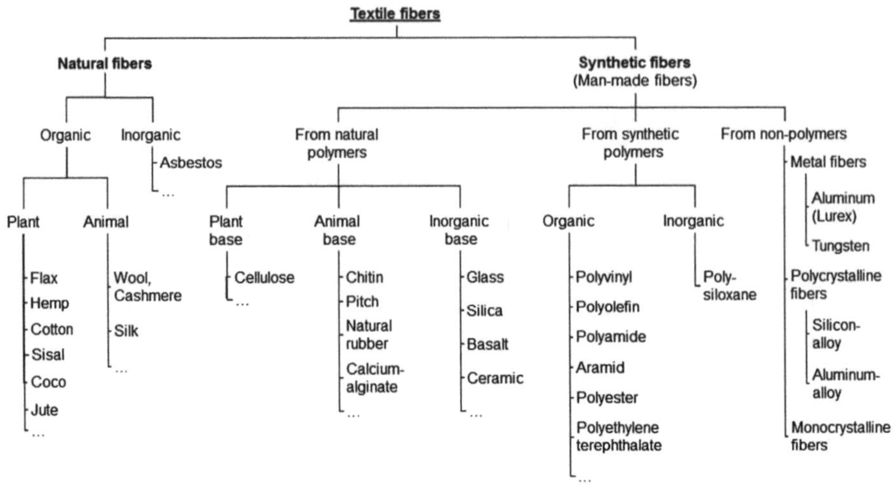

Fig. 2.3 Classification of textile fibers

used more and more frequently in the future. The various textile fiber materials are shown in Fig. 2.3. The overview illustrate the range of textile fiber raw materials and their classification. Details of the inner and outer fiber structures and the resulting properties will be clarified in Chap. 3.

Natural fibers is the umbrella term for all textile fibers and fiber materials processed from plant or animal origin. They have to be differentiated from man-made fibers, which are produced synthetically. Regenerated fibers, such as bamboo viscose or Lyocell are not classed with natural fibers.

Natural fibers are classified into two main groups: organic and inorganic fibers. Organic fibers are subdivided into plant (or cellulosic) fibers, such as cotton, jute, hemp, sisal and kapok, and animal fibers, such as wool or silk. Mineral asbestos fibers are one example of inorganic natural fibers. Natural fibers are often limited in their length, with silk fibers being an exception at lengths often exceeding several hundreds of meters.

Man-made fibers are industrially produced textile fiber material and can be produced synthetically in infinite lengths. Fiber-forming polymers are macromolecules with a relative molecular weight of at least 10,000. More than 1,000 atoms are involved in the formation of a macromolecule. Usually, these polymers originate from the covalent bond of monomers resulting from polyaddition, polycondensation and polymerization reactions [6, 7]. Man-made fibers are classified into three categories:

– Man-made fibers from natural polymers: these textile fiber materials can originate from plants (e.g. viscose, acetate), animals (e.g. chitin, bitumen as a precursor for carbon fiber production, and alginate) or from inorganic sources. Fiber materials made from natural polymers or inorganic origin play a crucial part in lightweight construction applications. Some examples are different types of glass fibers, silica glass, basalt and ceramics.

- Man-made fibers from synthetic polymers: the macromolecules of synthetic fiber materials result from stringing together monomers based on single atoms or molecules. The formation mechanisms of macromolecules will be explained in detail in Chap. 3. This group of fibers contains the largest number of fiber types, which are also the most common in practical applications. Some of the most important synthetic fibers are polyester, polyamide, aramid, polyimide, polyurethane, polyethylene, polypropylene and fiber materials of the polyvinyl group. Some of these synthetically produced fibers are used, for instance, as reinforcement components, thermoplastic matrixes for fiber-reinforced plastic composites or for the stabilization of non-rigid textile structures or crack minimization in concrete applications
- Man-made fibers from non-polymer materials: this includes monocrystalline and polycrystalline fibers, as well as metal fibers, for examples those based on steel, aluminum and tungsten.

The chemical fibers can be found in different forms in practice (see Chap. 3). Their properties can be purposefully adjusted during production. In addition, their application range can be broadened systematically by chemically or physically modifying or functionalizing the fiber surface (surface or interface design), for example increasing temperature resistance or adapting composite properties to the matrix.

2.2.2.2 Fibers, Filaments and Staple Fibers

According to DIN 60000, textile fiber materials are textile-technologically processable, linear structures. They are very slim and flexible, and display sufficient strength for both textile processing and use. Textile fiber materials are the elementary construct for the formation of yarns, non-woven and other fabrics. They are mainly loadable by tension.

Textile fibers can be classified into *staple fibers* and *continuous fibers*.

Staple fibers are limited in their length. Sometimes called spun fibers, they bear this name, even though are they are not necessarily spun in every case. They can also be processed into non-woven fabrics for the production of mats and felt. Staple fibers include natural fibers as well as continuous fibers cut or stretch-broken to the desired staple length. Non-spinnable, very short fibers are called *flock fibers* or *linters*. Fibers of great, practically unlimited length are called *filaments* or *capillaries*. In practice, filaments are also defined as fibers with a length of at least 1,000 mm. However, this limit is not fixed absolutely and depends on a number of conditions and circumstances, such as the size of the intended component. Filaments include:

- All synthetically produced man-made fibers, except for products which are cut or stretch broken to a staple length or a staple length distribution, and
- *Natural silk*, although not being labeled as a filament in common usage

2.2.2.3 Fiber and Yarn Fineness as Textile Physical Reference Values

Textile fiber materials are extremely versatile and the cross-section varies depending on fiber type. Natural fibers are of course subject to natural variation in their geometric dimensions. Their length is naturally limited and they are highly inhomogeneous. The fiber cross-sections are mostly unevenly round and irregular, sometimes even vary throughout fiber length. Some of them display undefined cavities. Among natural fibers of the same type, fiber lengths may also vary. As a result, it is extremely complex and impractical to use the fiber cross-section as a reference value for the establishment of fiber and yarn fineness. Therefore, weight and length are used as reference values for the establishment of fiber and *yarn fineness* in all linear textile structures. In this context, various fineness systems (*numbering systems*), which are often country- or material-specific, have been used in the past. Measuring the fineness in "*tex*" (*mass numbering*) has become one of the most prevalent systems. This *fineness* (symbolic abbreviation: Tt) is a textile-specific term and designates the ratio of mass to length. It is expressed in "*tex*" (1 tex = 1 g/1,000 m). Both geometrical variables (mass and length) can be measured precisely. Apart from *mass numbering*, there are other fineness definitions in use. They include *length numbering* [Metric number Nm (m/g)], the (*Titre-*) *Denier* system (den: 1 g/9,000 m), and the *English cotton yarn number* (Ne: 840 yards/1 lb). In the designation of fiber materials on carbon basis, the number of filaments in the yarn cross-section is commonly used. 50 K signifies a number of 50,000 individual filaments within the roving or the heavy tow. The carbon filaments have round cross-sections with diameters of typically 7 μm.

Table 2.1 gives an overview of the conversions between the different fineness systems.

To determine the fiber and yarn fineness, the quotient of breaking load and fiber or yarn fineness is calculated. This fineness-related force (N/tex) for linear fiber materials is used as a substitute for stress (force/area), which is used for non-fiber-based materials, such as metals and plastics. In order to illustrate the lightweight construction potential of textile high-performance fiber materials, specific strength or rigidity are commonly used, as they represent the relation between fiber strength or elastic modulus and fiber density.

Table 2.1 Conversion between the fineness systems

	tex	Nm	Ne_B	Td
tex	–	1,000/*tex*	590.541/*tex*	9 *tex*
Nm	1,000/*Nm*	–	0.590 *Nm*	9,000/*Nm*
Ne	590.541/Ne_B	1.693 Ne_B	–	5,341.87/Ne_B
den	0.111 *Td*	9,000/*Td*	5,314.87/*Td*	–

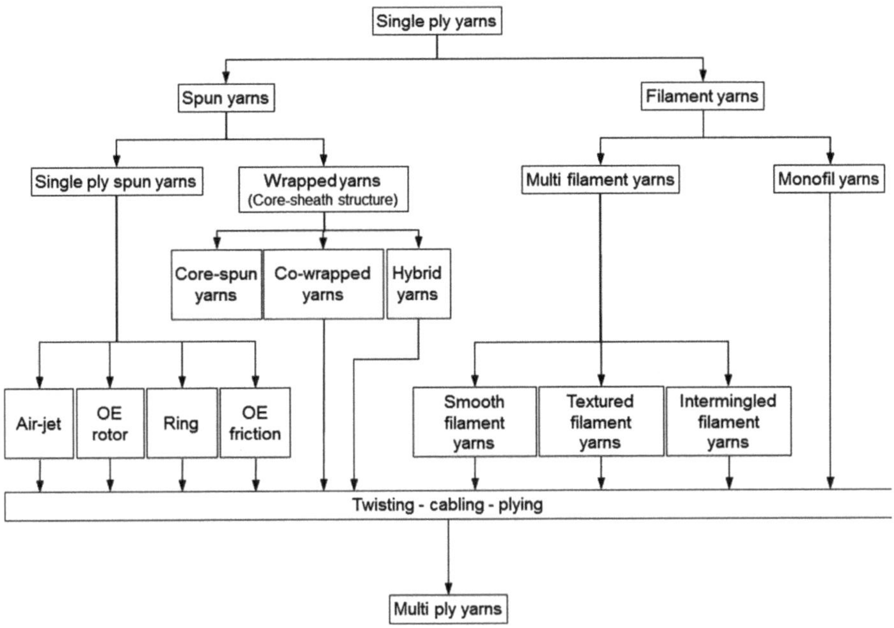

Fig. 2.4 Technological classification of yarns (single ply and multi ply yarns)

2.2.2.4 Yarns, Rovings, Heavy Tows and Plied Yarns

According to DIN 60900-1, *yarns* include all linear textile structures consisting of textile fibers. Single yarns are an elementary component for further yarn constructions like *ply yarns*, *twisted yarns*, or even textile surface structures. Yarns are made of textile fibers (spun fibers, filaments or tapes), which are usually form-fitted by twisting or bonded with special auxiliary materials. Yarns are often called *threads* in the context of a certain application or a technological explanation, e.g. weft thread or sewing thread. Yarns are categorized into spun yarns and filament yarns (Fig. 2.4).

A *spun fiber yarn* consists of spun fibers and is formed by continuous elongation of the particular fiber material and twisting of all or part of the fibers among themselves by means of various operating principles (mechanical or pneumatic). The twisting of the yarn is ideally performed by real twisting of all fibers around the yarn's longitudinal axis. These twistings ensure a stabilization of the yarn and thus allow full utilization of the fibers' substantial strength by form-fit or force-fit transfer of forces between the fibers. Additionally, twistings enable the purposeful attainment of effects and properties.

The most important spinning methods are *ring spinning*, *open end (OE) rotor spinning*, *air jet spinning*, and *OE friction spinning*. These methods are (machine-) technologically designed according to purpose, material and fiber length. The corresponding spun yarns from organic (jute, hemp) and inorganic natural fibers (e.g. basalt), or cut man-made fibers, are also interesting for fiber composite applications.

In special and modified spinning procedures (e.g. OE friction spinning, OE rotor spinning and ring spinning), filament yarns constitute the core component which is wrapped with spun fibers acting as the sheath component. Alternatively, parallelized spun fibers or spun yarns are wrapped with a filament yarn. The characteristic of this yarn construction is the distinctive core-sheath-structure, often called *core-spun yarn*, *co-wrapped yarn* or *hybrid yarn*.

Spun fiber yarns can also be produced in non-twist processes like adhesive bonding.

The properties of *spun fiber yarns* depend strongly on the fiber material's characteristics, the yarn structure and spinning method. The orientation, level of bonding, arrangement, length, number of fibers in the yarn cross section, and twist rate (number of twists per meter) of the fibers are decisive process and fiber parameters. Textile physical properties of single spun yarns are relatively low compared to those of filament yarns with stretched fiber alignment. *Spun yarns* are often designated according to production methods. Because of the variety of yarn construction types, an in-depth description will be foregone in favor of a reference to DIN 60900-1.

Filament yarns are classified into *monofil* and *multifilament yarns*. *Monofils* consist of a single filament with a diameter of >0.1 mm and are used for technical applications. In conventional textiles, e.g. in clothing, diameters of monofils on the scale of 20 μm are feasible. In contrast to the monofils, a *multifilament yarn* contains a number of individual filaments with or without twist. The term filament yarn is clearly defined in DIN 60900-1. *Multifilament yarns* is a generic term for derived notions used in practice and covers the entire fineness and material range. For textile lightweight construction applications based on high-performance fiber materials, the term roving has gained acceptance. Multifilament yarns made from carbon and displaying extremely high yarn fineness above 2,400 tex are usually called Heavy Tows. Since the term is specific to the use of carbon, a roving with a fineness ranging from 300 tex to 2,400 tex is commonly called a Low Tow, although the fineness range is not defined universally. Details for the various yarn constructions are given in Chap. 4.

2.2.2.5 Textile Fabrics and Three-Dimensional Textile Constructions

In fiber composite components, tensile forces are primarily absorbed by the rovings embedded in the matrix. Therefore, the fiber material has to be aligned in load direction. Apart from the direct use of cut or continuous reinforcement fibers, textile fabrics with filament yarns are used as reinforcements for complex components. This allows not only the realization of customized fiber arrangements, but also an efficient component manufacture. The application potential and acceptance of reinforcement semi-finished products depend mainly on the state of textile

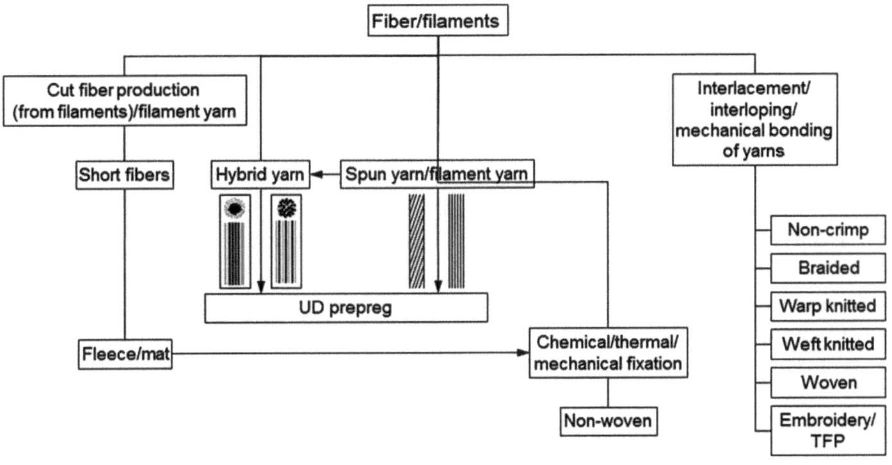

Fig. 2.5 Classification of interlooping textile structures

processing technology. The term textile fabrics, which is used for a number of constructions differing from each other with regard to the connection of individual basic elements (unit cell) and in terms of the type of arrangement of reinforcement yarn systems. On a more basic level, textile semi-finished products can be differentiated into planar and three-dimensional textiles.

Figure 2.5 gives an overview of the variety of feasible geometries and the versatility of planar and three-dimensional textile structures with different degrees of complexity inherent in the respective technology. Some of the most common textile fabric structures are woven, warp-knitted or weft-knitted non-wovens and braids. From these base structures, a large number of special designs have been derived by means of suitable enhancements or combinations of different textile technologies which have become state of the art in the meantime. This includes multiaxial non-crimped fabrics and semi-finished products manufactured using the *Tailored Fiber Placement* (TFP) method. These production methods are very advanced and have seen use in various areas of applications.

Apart from textile geometry, other distinctions can be made based on the arrangement of reinforcement yarns. Fundamentally, the current stage of development of textile fabric formation methods allows three-dimensional reinforcement yarn arrangements with open or closed fabrics. Following DIN 60000, the most important textile manufacturing methods will be presented in short as source technologies for the development of complex textile constructions. This norms and chapters of this book associated with the respective segments will be given in brackets.

Woven fabrics (DIN ISO 9354, DIN 61100-1, DIN 61100-2, see Chap. 5): woven fabrics are the oldest man-made textile surface structures. The oldest woven fabric fragments, manufactured on so-called weight looms, date from 7,000 years ago. Today, high-end, computer-controlled weaving machines are used to produce load-adapted woven fabrics with high strength, rigidity and energy absorption capability. Conventional 2D weaves consist of at least two yarn systems crossing each other perpendicularly. The yarns running lengthwise (longitudinal or manufacturing direction) are called warp yarns, while the yarns running crosswise (lateral) are referred to as weft yarns. The type of interlacing of warp yarns with weft yarns, i.e. the alternating underpass and overpass, is referred to as a weave. It has an influence on the product's appearance, mechanical properties and drapability. Each of the two yarn systems can consist of several warp or weft yarns. Weaving with several yarn systems, i.e. ground and stuffer yarns in warp and weft direction and additional weft yarns as pile or binding yarns, enables the manufacture of multilayer and three-dimensional structures.

Weft-knitted and *warp-knitted fabrics* (DIN 4921, ISO 7839, DIN 62050, DIN 8388, DIN 8640, DIN 61211, see Chaps. 6 and 7): In contrast to woven structures, textile fabrics in weft-knitting and warp-knitting are created by forming yarns into stitch loops which are then connected with each other. Warp-knitting has the special distinction of simultaneously forming one or several yarn sheets, which are also called knitting yarn systems, into stitches. A warp-knitting yarn system, therefore, is a multitude of parallel running yarns sharing the same function in the formation of the warp-knitted fabric [8]. In weft-knitted fabrics, the stitches are formed successively across the production direction (single-yarn knitted fabric) [9]. Knitted fabrics are used in the creation of complex geometries, as these textile semi-finished products are distinguished in terms of high stretchability and drapability, and are highly versatile for a variety of applications due to the combination of various types of interlacement. Thus, they allow for complex near-net shape geometries. However, the stitch-like arrangement of yarns in knitted fabric makes it highly elastic, which is a disadvantage in highly-strained composite components. To fully exploit the potential of knitted structures in composite applications, stretched yarn systems, which are responsible for force transmission, are integrated into the stitch system to realize non-crimp semi-finished products.

Stitch-bonded fabrics (DIN 61211, see Chap. 7): Stitch-bonding, as a variation of warp-knitting, is a method for the manufacture of textile fabric structures. It is based on the principle of connecting yarn sheets or fabric structures, using the stitches of one or more warp-knitting yarn systems. In stitch-bonding, yarn sheets are inserted into two parallel transport devices at one or more consecutive lay-up stations. The stacked yarn sheets are then guided to the warp-knitting unit, and connected by the stitches of the knitting yarn system to form a stable bi- or multiaxial non-crimp fabric [10].

Braids (DIN 60000, see Chap. 8): Braids are formed by the continuous crossing of at least three yarns, usually running diagonally to the direction of production. Additional axial yarns, so-called 0°-yarns or pillar yarns can be integrated into the braid for axial reinforcement. Braids can be realized as plane or three-dimensional structures.

Nonwoven fabrics (DIN EN 29092, see Chap. 9): Nonwoven fabrics are fabric structures in the form of mats or webs from directionally or randomly orientated fibers, with form-fit, force-fit or bonding connections. In contrast to woven and knitted fabrics, nonwoven fabric formation is performed without the process step of yarn production. All fibers can be processed into nonwoven fabrics, which are sometimes referred to as fiber mats and are frequently used for lightly loaded components without or in combination with a plastic matrix, as in spun-laid nonwovens and glass mats, and *SMC* or *GMT* semi-finished products.

Embroidered fabrics (DIN 60000, see Chap. 10): These surface structures are characterized by embroidery yarns being drawn through embroidery grounds such as woven or warp-knitted fabrics. In some procedures, the embroidery ground can later be removed entirely or partially. One technology derived from embroidery is *Tailored Fiber Placement* (TFP), which offers the possibility of realizing various textile structures with fiber orientation and locally adjustable fiber count adapted to the direction of load [11].

Three-dimensional textile constructions: In practice, complex component geometries requiring the development of suitable textile 3D-semi-finished products (*preforms*) are often indispensable. A number of innovative textile production methods for the manufacture of 3D-textiles rely largely on the further development of existing technologies for textile fabric structures. One method often used in practice is based on the combination of different individual structures into complex preforms by means of textile joining methods (see Chap. 12). This approach is referred to as *differential construction*. *Integral construction*, on the other hand, is characterized by the manufacture of as many structural elements of a complete preform in a single step as possible. The number of individual elements and the resulting number of joinings are considerably reduced by this construction. However, the degree of complexity of such integral 3D-structures and geometries is limited. Special textile structures and the required technologies as well as development directions and possibilities will be treated in detail in Chaps. 5–10 and 12.

2.2.2.6 Finishing

Materials with customized surface properties are of great interest for a large number of applications in lightweight construction and in connection with the realization of integrated sensor networks. In the interdisciplinary field of nanotechnological material synthesis, it has been observed that surfaces and interfaces in nature are often nanostructured systems with several components consisting of polymers and inorganic constituents and display technically relevant and desirable properties [12–16]. In this context, much of the current development work in modern material science is aimed at hybrid systems of reinforcement materials, matrices and nanostructures to attain customized surface characteristics and functionalities.

Apart from the influence of the properties of reinforcement fibers and matrix, the character of the interaction of these two components is a decisive factor in the performance of composite materials. In addition to mechanical characteristics of the individual components, it is their adhesion that determines load transmission between the components and crack propagation. Strength and toughness of a fiber composite material can be altered significantly by the interface of fiber and matrix.

The design of interface and surface modification defines how tensions are transmitted from the matrix to the fibers, and as a result, the chemical, thermal and/or mechanical properties of the FRPC itself [17]. Finishing of the surfaces and interfaces allows the integration of sensory and actuator functions into the composite components for purposes of structural monitoring, self-diagnosis and self-regulation.

Enhancing adhesion quality by means of surface modification is highly relevant, as it ensures industrial usability and development of suitable textile structures. The bonding between fiber and matrix is the decisive criterion for the quality assessment of textile-reinforced composite materials. While a mechanical fixation is usually sufficient for lightly loaded composite components, dynamically and mechanically heavily loaded components definitely require a chemical bonding of the fibers to the matrix. Adhesion in the fiber-matrix interface is important for load transfer and force transmission. To guarantee reliable composites with high mechanical characteristics, the interaction between matrix and reinforcement component has to be adjusted purposefully with customized *interface design*.

For the processing of the materials, all process steps from the fiber, the roving, and the fabric structure to the ready-made 3D-reinforcement semi-finished products in fixed form (*Preform*) have to be employed. The finishing of fiber and yarn materials allows an extraordinary combination of diversely functionalized fibers and yarns for the manufacturing of textile structures. Wet chemical processes, plasma treatments, sol-gel procedures and functional coatings are available methods for the finishing of textile materials. Intelligent functional coatings enable the realization of innovative functional systems. Combined use of novel methods of physical self-organization, surface chemistry and surface structuring makes specific multi-scale interface architectures feasible, which fully exploit the performance potential of the individual composite components and achieve their full functionalization [17]. Details on the finishing of textile structures can be found in Chap. 13.

2.2.2.7 Assembly and Preforming

In assembling technological processes, textile semi-finished products (e.g. non-crimp fabrics, woven fabrics, braids) are converted into near-net shape preforms and assembled either individually or in combination. The assembly process begins with the construction of individual parts of the preform, where the correct choice of semi-finished products and their directional integration into the preform structure have to be considered to ensure mechanical functionality of the textile-reinforced composite component to be produced. Moreover, it should be noted that, in the process of preform production, the shaping of individual parts is performed without creasing and with a defined alteration of the originally created yarn orientation during draping.

Using textile-adapted cutting methods, the specific individual parts, commonly provided meter-wise, are cut out from the semi-finished products. Before, during, or after the process, protection of the cutting edge has to be ensured to prevent the loss of peripheral yarns.

Familiar sewing technology is suited for the assembly of the individual parts into a near-net shape *preform*, while large-format and complex preforms are more easily accessible by using the novel principle of unilateral sewing. Any kind of sewing technique requires sewing yarns by virtue of their fiber material composition, yarn structure, and preparation able to withstand the sewing process, hold together the preform in its textile form, and, if applicable, contribute to the reinforcement in *out-of-plane* direction as so-called z-reinforcements in the finished textile-reinforced component. Sewing, in any case, also causes punctures and perforations, which reduce *in-plane* properties when applied during preform construction.

Alternative connection technologies for textiles are offered by welding, adhesive bonding and the use of binders. Welding requires thermoplastic fiber materials, although fiber material mixtures with a proportion of thermoplastic behavior can also be used. Adhesive bonding in preform production requires compatibility with the matrix material. Thermally activated adhesives, often called binders, can be used in preform assembly to ensure the shaping of the preform parts well into composite component manufacturing process. This does not constitute a load-bearing function within the composite component. A local application of binder or adhesive allows a defined manipulation of the draping behavior of individual parts before or immediately after cutting.

The handling of individual textile parts is closely connected to the assembly process, beginning with the removal from the cutting table, followed by the defined conveyance to the assembly workstations, up to the delivery of the textile preform to the component manufacturing process. For reasons of reproducibility, CNC-controlled machines are preferred for cutting and various assembly steps. While cross table systems are state of the art for any work on the plane, robot-guided joining techniques are used most commonly for three-dimensional assembly work. Details on the assembly of textile preforms will be given in Chap. 12.

2.2.2.8 Universal Textile-Physical Parameters for the Characterization of Textile Fiber Materials, Yarns, Textile Surface Structures and Components

The assessment of quality and usability of textile structures depends on their textile-physical parameters. For the determination of these parameters, regulations have been devised, which are usually set down in test standards. These standards are internationally authoritative and applied in particular in international commercial movement of goods. The use of test standards includes the strict compliance of testing conditions like testing speed, testing climate, sample assembly and test procedure, and guarantees the comparability of the specific values detected by different test centers. This procedure is necessary, since textile-physical parameters of textile structures depend on the testing conditions and methods.

Instruments for the testing of textile materials and structures require a mechanical layout suited to the dimensions of the fiber materials and the textile structure. Therefore, four distinctions are often made in overviews of the field of textile testing. They are:

- Fiber/Filament Testing
- Yarn Testing
- Fabric Testing
- Component Testing

Within each of these areas, respective typical textile-physical parameters are established, which are often formally identical but require highly specific testing conditions and measuring principles. The characterization is generally performed according to the following principles, for which some examples are given in brackets:

- Material (type, composition, fiber fineness, humidity/sizing/matrix ratio)
- Material properties (thermal expansion coefficient, dielectric number, thermal conductivity)
- Geometrical properties (diameter, length, regularity, homogeneity)
- Constructional properties (weave type, yarn twists, yarn density of textile fabrics, layer buildup of composite components)
- Mass (linear mass of filaments and yarns, areic mass of textile semi-finished products)
- Dimensional change at low speeds (Young's modulus, ultimate stress and ultimate strain, flexural rigidity, shear stiffness, torsional stiffness, creep, crack formation and delamination in composites)
- Dimensional change at high speeds (ultimate stress and ultimate strain, impact resistance, fatigue endurance limits)
- Crash behavior (crash energies, residual strength after impact, crack formation, delamination)
- Interactions with partners (friction, abrasion, air permeability)
- Characterization of surface and interface properties (surface energy, wettability)

The possibilities of a metrological characterization of textile structures are extensive, which makes a comprehensive overview in this book impossible. This is due to the variety of materials, the geometrical dimensions ranging from macromolecule to finished component, the distinct measuring principles and application possibilities. The textile-physical and chemical parameters noted in brackets are typical values. However, the listing is not complete. The development of complex tailor-made textile structures for special applications sometimes requires the conception of new measuring technologies. A selection of relevant testing methods used in the characterization of textile structures and the composite components manufactured from them in technical applications is included in Chap. 14.

2.3 Textile Semi-finished Products and Preforms for Lightweight Construction

2.3.1 Classification, Distinction, and Definitions

For a better representation of textile semi-finished products and preforms for lightweight construction and to ensure a better understanding of the selection of suitable technologies, structures and geometries, important terminology and their distinctions have to be discussed first. Figures 2.6 and 2.7 give relevant characteristics for the differentiation of geometries and reinforcement structures in textile semi-finished products.

2.3.1.1 Geometry Versus Structure

When assessing existing textile fabric formation methods with regard to their suitability for the realization of contour-close 2D and 3D reinforcement textiles, the construction of the textile is the first priority. The array arrangement of the yarn and the appearance of the fabric are important for both the specification of textile reinforcement semi-finished products and for the structure-mechanical properties of the composite components manufactured from them. In the context of textile products, the terms one-/two-/2.5-/three-dimensional are used on the one hand for characterizations of reinforcement yarn positions and on the other hand for the determination of semi-finished product geometry. For an unambiguous classification of terminology, a distinction will be made hereafter, between the structure and

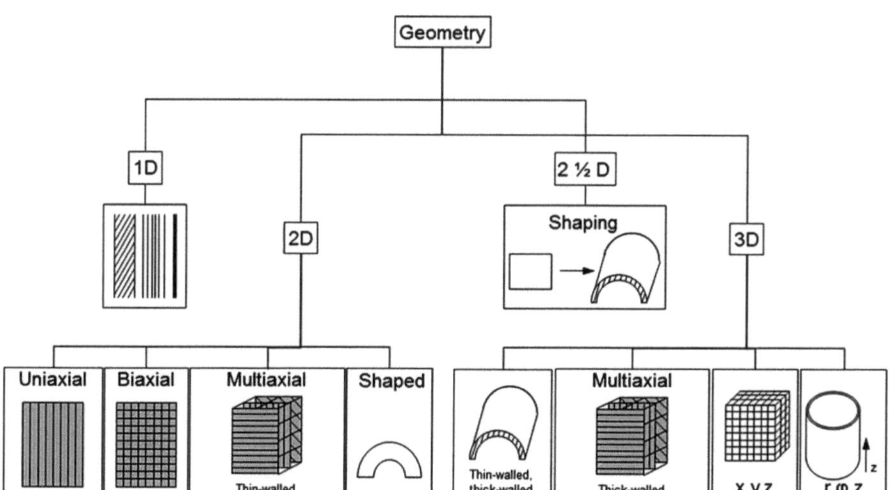

Fig. 2.6 Shape of textiles-geometry

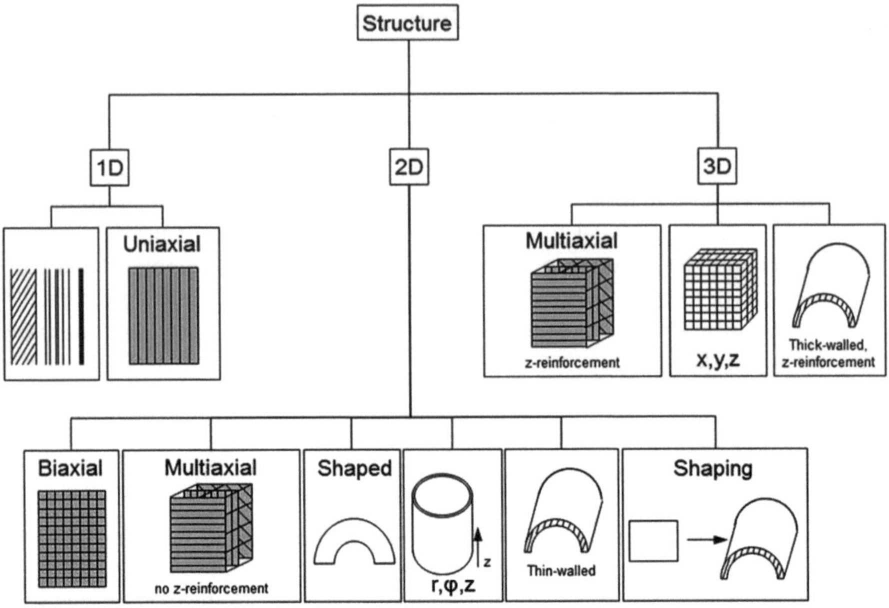

Fig. 2.7 Shape of textiles-reinforcement structure

the *geometry*. The *structure* itself can be subdivided into *reinforcement structure* and *weave-related structure*.

2.3.1.2 Geometry of a Textile Semi-finished Product

Geometry denotes the linear, flat or three-dimensional appearance of the textile semi-finished product with and without influence of additional processes for the production of the net shape, where the type of arrangement of the reinforcement yarn systems plays only a minor role (Fig. 2.6).

One-dimensional geometry: This includes linear structures with a high slenderness ratio (length to cross-section ratio), for example in the form of monofiles, spun fiber yarns, rovings or twisted yarns.

Two-dimensional geometry: A two-dimensional geometry of a textile signifies a plane textile fabric with a thickness that is negligible in comparison to its surface area.

2.5-D geometry: Textile fabrics with a "two-and-a-half-dimensional" geometry are defined as having a thickness that is negligible in comparison to their surface area and they can be shaped into three-dimensional constructs or net shapes by means of forming, draping or assembly processes.

Three-dimensional geometry: This includes volume-forming or thin, spatially designed, shell-like textile architectures manufactured within in a single process step without additional influence from subsequent steps. Volume-forming textiles subjected to retroactive forming processes are therefore also included in constructs with three-dimensional geometry.

2.3.1.3 Reinforcement Structure of a Textile Semi-finished Product

In contrast to the geometry of a textile semi-finished product, a design of the reinforcement structure focuses on the orientation of the yarn systems for reinforcement, which is the main task of textile semi-finished products in mechanically strained fiber composite components.

One-dimensional reinforcement structure: the reinforcement of the textile semi-finished products is oriented primarily in one preferential direction (*unidirectional*, UD). This concerns stretched yarns and unidirectionally reinforced textile fabrics. Any UD-reinforced fabric is a one-dimensional structure with 2D geometry.

Two-dimensional reinforcement structure: A textile construction features a two-dimensional reinforcement structure, if the reinforcement components are primarily planar and oriented in at least two different directions. Bi-, tri- and multi-axial planar reinforced structures (2D geometries) are also considered among the two-dimensional reinforcement structures. Thin-walled 3D geometries which can be produced in a single textile-technical process step without additional effect from subsequent steps, and which contain at least two preferred directions of the reinforcement components in the plane, are 2D reinforcement structures, as they do not contain reinforcements in thickness direction. This becomes clear from the phaseout of the shell-shaped 3D geometry. Thick-walled multilayered reinforcement structures without reinforcement system in thickness direction are 2D reinforcement structures with a 3D geometry.

Three-dimensional reinforcement structure: A three-dimensional reinforcement structure features reinforcement yarn systems oriented in all three spatial directions and thus ensures the corresponding reinforcement effect within the composite. In general, 3D structures require a 3D geometry based on volume-forming textile architecture. The advantages of this integral 3D structure include the significant improvement in mechanical properties of the component in z-direction and impact behavior as well as in the reduction of delamination risk.

2.3.1.4 Weave-Related Structure of a Textile Semi-finished Product

In contrast to the reinforcement structure, which takes into account the orientation of the reinforcement yarns, the *weave-related structure* describes the local or global orientation of the various reinforcement yarn systems to each other. This reflects the type of yarn crossing or interlacing within the textile semi-finished product. In the specific case, it concerns the type of weave.

2.3.1.5 Textile Semi-finished Products with Open and Closed Appearance

Apart from the type of the textile structure based on various reinforcement yarn systems, the appearance of the textile semi-finished products plays a decisive part in the application in composite components. According to the matrix system to be used (plastic, coating or mineral basis) and the component's degree of strain, reinforcement structures are classified into those with compact (high fiber volume content) and lattice-like reinforcement structures. Textiles with an open appearance are used for the reinforcement of matrices with solid aggregates (e.g. concrete) or for the strengthening of non-highly loaded fiber-reinforced plastic composite components. Highly loaded composite component can only be realized with closed structures with high fiber volume content.

2.3.2 Preform and Preforming

Preform is the term used for a single- or multi-layered dry textile structure which is impregnated with a suitable matrix system in an individual, subsequent process. The geometry of the textile reinforcement structure (*preform*) is largely congruent with the eventual component geometry and ensures a suitable yarn orientation according to the load direction. The methods for manufacturing preforms can be categorized into *direct* and *sequential preforming* (Fig. 2.8).

Direct preforming denotes manufacturing processes in which a usually three-dimensional, integral preform is produced in a single step. Its geometric complexity

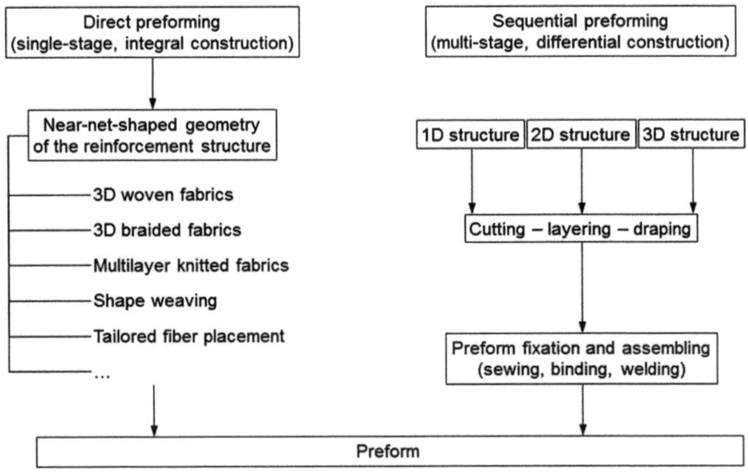

Fig. 2.8 Method of preform production

and the attainable fiber volume content depend on the manufacturing method. While short fibers (secondary structures) are used for conventional methods, direct processes of textile 2D and 3D preform manufacture use continuous fibers. For the production of complex high-performance components (primary structures), the existing standard methods are further developed. This makes the desired lightweight construction effect with spatial stiffnesses and strengths feasible at smaller mass.

Sequential methods for the manufacture of preforms in *differential design* are classified into two classes. Binder forming technology uses binders to fix the filament yarn layers of the preform relative to one another in the net shape to be modeled. Beyond that, the binder can be used for structural fixation. It is crucial to select a binder that is compatible with the respective polymer matrix and thus allows fixation outside of the machine at low curing temperatures. Binders can be applied in their solid or liquid form but must not cause limitations to the required forming behavior and permeability. Furthermore, binders can be inserted as yarns during the manufacture of the textile fabric. A suitable binder design is a prerequisite for the increase in efficiency of composite forming technologies.

The production of preforms by means of conventional methods of textile assembly technology has been popular since the 1980s. Here, the *cut-and-sew technique* is used primarily for the assembly of simple preforms. To produce complex geometries from textile reinforcement structures, several technological developments in stitching and sewing technology have resulted, among others, in a range of unilateral sewing methods, most of which are robot-guided [18–20]. Their use for the insertion of fixing seams can be accompanied by fiber damage resulting in the degradation of mechanical *in-plane properties* of the component. The advantages of these textile joining techniques include the improvement of the components mechanical properties in z-direction by sewing and stitching yarns, which in turn lowers the risk of delamination and improves impact behavior. Details are given in Chap. 12.

2.3.3 Advantages of the Integration of the Matrix as Continuous Fiber

For thermoplastic fiber composite materials, more and more applications are being opened up. One crucial advantage over composite materials with thermoset matrices is the shorter attainable processing times of the original materials during component manufacture. Thus, thermoplastic components can be produced cost-effectively and efficiently in highly productive processes, such as injection molding, deep-drawing and pressing. Further advantages of thermoplastics are their thermal ductility, the possibility of repeated melting and re-shaping, their superior impact resistance and damage tolerance, as well as their greater repair-friendliness.

While initially short-fiber-reinforced fiber composite materials were realized principally, the recent focus of research has been shifted to the development of long-fiber-reinforced thermoplastics (LFT), including the use of continuous-fiber-reinforcements [21–24]. In order to advance into the field of highly loaded fiber composite materials, continuous-fiber-reinforcements with a fiber length matching the component dimensions are necessary. The challenge in the application of thermoplastic matrix materials is posed by their high melting viscosity, as the injection of highly viscous thermoplastic material at high processing pressures causes structural distortions of the preforms in the machine cavity. Therefore, the flow paths of the molten thermoplastic during component impregnation have to be minimized. One approach to a solution is the previous impregnation or hybridization of textile materials and semi-finished products. The fiber volume ratio of the matrix and reinforcement components required for the respective component can be customized and preadjusted by means of the mixing ratio.

One possible approach to manufacture hybrid textile semi-finished products is the combined processing of reinforcement yarns and thermoplastic yarns. A much improved mixing of the reinforcement and matrix components can be attained by the use of *hybrid yarns* in the textile manufacturing process. A preferably homogeneous mixing of both components allows short flow paths for the highly viscous molten matrix during composite consolidation and it is a basic prerequisite for high composite quality. Hybrid yarns with continuous reinforcement filaments can be produced by winding, wrapping, twisting or impregnating reinforcement yarns with matrix powders, in which both components are aligned side-by-side or in a core/sheath structure in the yarn cross-section. Hybrid yarns with a substantially homogeneous fiber/matrix distribution and a low bending stiffness suitable for textile processing can be produced by integrating thermoplastic filaments like polypropylene, in glass rovings (Twintex®) during glass filament production or by a commingling process [25]. In practice, numerous hybrid yarn constructions are available, which can be differentiated regarding the forms of individual components, hybrid yarn type and structure. The most important hybrid yarn constructions are given in Fig. 2.9.

Other methods for preliminary impregnation of the reinforcement component with matrix material in non-textile form include different approaches, such as *Film-Stacking, powder impregnation, hot melt extrusion* and wetting in polymer solutions [27]. With the use of one of these methods, yarn-like prepregs with widely varying properties can be produced. The resulting pre-impregnated yarns exhibit high bending stiffness and their usability for further textile processing, in particular for the manufacture of complex reinforcement semi-finished products, is limited [28]. These problems and possibilities are dealt in detail in Chap. 11.

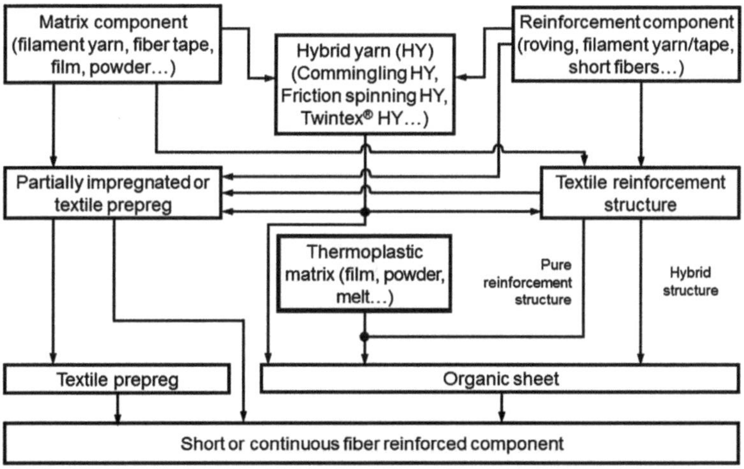

Fig. 2.9 Hybrid yarn construction from filament or spun yarns according to [26]

2.4 Application and Performance Potential of Textile Semi-finished Products and Preforms in Lightweight Construction

In the development of material- and cost-efficient structural components, lightweight construction with textile-reinforced composite materials has many advantages over conventional constructions. Here, functionally integrated lightweight design in mixed textile construction forms plays an important role. In particular their high strength and stiffness in combination with their small weight, the adjustable short-time-dynamic properties (*Impact*), the great variety of textile methods and structures as well as the economic manufacture with high reproducibility, suitability for series production and recyclability make the young continuous-fiber-reinforced composite group of materials in fiber-reinforced plastic composites, textile concrete and textile membrane technology interesting and promising for future lightweight construction applications in various industries. Textile-reinforced composite materials have the highest flexibility in comparison to other material groups, and can be considered almost predestinated for the use in optimal material mixtures for the combined design approach necessary for complex requirements in lightweight construction [29].

Textile materials and semi-finished products as innovative materials display an extremely versatile property potential and are one of the most important high-tech material groups for the present and the future. They are a basic prerequisite for the creation of innovative products with new scalable properties. Continuous-fiber-reinforced composite materials have an especially high potential for the series use in complex highly loaded lightweight components in vehicle and machine engineering, for the reinforcement of slender, filigree concrete parts, and for the redevelopment and maintenance of existing structures. They contribute to the significant reduction of masses and to energy conservation. They are distinguished by their flexible customizability to material structure and the resulting adjustability of material properties and property anisotropy to the existing processing and component requirements.

The versatile textile processes and their combinations for the manufacture of 2D and 3D structures and preforms offer various possibilities and parameters, the variations of which allow a far-reaching manipulation and suitable adjustment of the characteristics of the products to be manufactured. The following textile constructions for lightweight structures with customized properties can be realized by actively and purposefully forming the textile material:

- any spatial alignment of the load-bearing yarn systems (1D, 2D, and 3D structures),
- force fit orientation of the yarns and quantitative determination of load-bearing yarn systems by load case, e.g. biaxial, multiaxial or polar,
- matching to the component geometry and design, for instance in freeform surfaces, complex profiles, tubular and *spacer* structures, and
- hybridization and functional integration.

The wide range of existing and available textile materials, structures and methods, economic manufacture and suitability for series production give this group of materials a promising future [30]. The increased use and growing importance of textile-reinforced composite materials for mass markets also raises performance requirements of textile semi-finished products for structural components. The suitable choice of materials and proper manufacture of textile semi-finished products offers numerous advantages over conventional materials, such as an improved performance to weight ratio by anisotropic fiber alignment for greater lightweight construction benefit. Contrary to the expensive prepreg technology, dry reinforcement structures and preforms are particularly advantageous for the economic production of lightweight construction components in large series applications.

The potential of textile semi-finished products as an innovative lightweight construction material, the purposeful selection and combination of textile materials and processes, and the customized, force-fit-suitable alignment of rovings in the textile structures, results in a nearly unlimited variety of property profiles and design possibilities, up to function-integrated *near-net shape* components. Assembly technology offers a maximum of flexibility regarding the connection of textile fabrics to form suitable preforms. The assembly of the individual, load-adapted reinforcement textiles into integral preforms is performed by means of modern joining technologies, such as sewing, welding or adhesive bonding. This approach has proven its effectiveness, as assembly and reinforcement seams can be inserted variably into the preform. The use of assembly technology has made it possible to produce "*spacer preforms*" of larger sizes and more complex geometry as well as with integrated functions.

In general, it can be stated that fiber-based semi-finished products and preforms for lightweight construction applications are not only relevant economically in aeronautics, but that they are also viable on mass markets in automotive and transport engineering, mechanical engineering, the building industry, and in textile membrane technology. A paradigm shift is to be expected, in which the material industry will increasingly substitute traditional, monolithic materials like aluminum and titanium by fiber composite materials [31]. This development process is already well advanced and will revolutionize material science through the extreme flexibility of the textile materials and their processing into nearly unlimitedly complex constructions with scalable properties. Textile materials and semi-finished products offer a broad range of variation and an enormous diversity of possibilities, including the suitable customization of load-bearing structures with regard to strength, stiffness, impact behavior and energy absorption capabilities.

References

1. MEYER-STOCK, L.: Forschungsaufgabe Technische Textilien. In: KNECHT, P. (Ed.): *Technische Textilien*. Frankfurt am Main : Deutscher Fachverlag GmbH, 2006
2. OERTEL, D.; GRUNEWALD, A.: *Potenziale und Anwendungsperspektiven der Bionik*. http://www.tab-beim-bundestag.de/de/pdf/publikationen/berichte/TAB-Arbeitsberichtab108.pdf (06-04-2011)
3. LING, Z.; JIANFENG, M.; TING, W.; DENGHAI, X.: Lightweight Design of Mechanical Structures Based on Structural Bionic Methodology. In: *Journal of Bionic Engineering* 7 (2010), pp. 224–231. DOI 10.1016/S1672–6529(09)60239–0
4. HONGJIE, J.; YIDU, Z.; WUYI, C.: The Lightweight Design of Low RCS Pylon Based on Structural Bionics. In: *Journal of Bionic Engineering* 7 (2010), pp. 182–190. DOI 10.1016/S1672–6529(09)60207–9
5. MILWICH, M.; SPECK, T.: Solving engineering problems with the help of nature's wisdom. In: *American Journal of Botany* 93 (2006), pp. 1455–1465. DOI 10.3732/ajb.93.10.1455
6. BOBETH, W.: *Textile Faserstoffe - Beschaffenheit und Eigenschaften*. Berlin : Springer Verlag, 1993
7. WULFHORST, B.: *Textile Fertigungsverfahren - Eine Einführung*. München, Wien : Carl Hanser Verlag, 1998
8. RENZ, R.; FLECKEISEN, M.: *Bindungslehre der Ketten- und Nähwirkerei*. Leipzig : VEB Fachbuchverlag Leipzig, 1980
9. WEBER, K. P.; WEBER, M.: *Wirkerei und Strickerei - Technologische und bindungstechnische Grundlagen*. Frankfurt am Main : Deutscher Fachverlag, 1997
10. HAUSDING, J.: *Multiaxiale Gelege auf Basis der Kettenwirktechnik - Technologie für Mehrschichtverbunde mit variabler Lagenanordnung*. Dresden, Technische Universität Dresden, Fakultät Maschinenwesen, Dissertation, 2010
11. FELTIN, D.: *Entwicklung von textilen Halbzeugen für Faserverbundwerkstoffe unter Verwendung von Stickautomaten*. Dresden, Technische Universität Dresden, Fakultät Maschinenwesen, Dissertation, 1997
12. PUKANSZKY, B.: Interfaces and interphases in multicomponent materials: past, present, future. In: *European Polymer Journal* 41 (2005), No. 4, pp. 645–662. DOI 10.1016/j.eurpolymj.2004.10.035
13. WAGNER, H.D.; VAIA, R.A.: Nanocomposites: issues at the interface. In: *Materials Today* (2004), pp. 38–42

14. SANCHEZ, C.; LEBEAU, B.; CHAPUT, F.; BOILOT, J. P.: Optical properties of functional hybrid organic-inorganic nanocomposites. In: *Advanced Materials Journal* 15 (2003), No. 23, pp. 1969–1994. DOI 10.1002/adma200300389
15. KING, W. P.; SAXENA, S.; NELSON, B. A.; WEEKS, B. L.; PITCHIMANI, R.: Nanoscale thermal analysis of an energetic material. In: *Nano Letters* 6 (2006), No. 9, pp. 2145–2149. DOI 10.1021/nl061196p
16. CHO, K.; KIM, D.; YOON, S.: Effect of substrate surface energy on transcrystalline growth and its effect on interfacial adhesion of semicrystalline polymers. In: *Macromolecules* 36 (2003), pp. 7652–7660. DOI 10.1021/ma034597p
17. MÄDER, E.: *Grenzflächen, Grenzschichten und mechanische Eigenschaften faserverstärkter Polymerwerkstoffe*. Dresden, Technische Universität Dresden, Fakultät Maschinenwesen, Habilitation, 2001
18. GRUNDMANN, T. C.: *Automatisiertes Preforming für schalenförmige komplexe Faserverbundbauteile*. Aachen, RWTH Aachen, Fakultät Maschinenwesen, Dissertation, 2009
19. NEITZEL, M.; MITSCHANG, P.: *Handbuch Verbundwerkstoffe. Werkstoffe, Verarbeitung, Anwendung*. München, Wien : Carl Hanser Verlag, 2004
20. HERZBERG, C.; KRZYWINSKI, S.; RÖDEL, H.: Konfektionstechnische Fertigung mehrschichtiger 3D-Preforms für Composites. In: *Verbundwerkstoffe und Werkstoffverbunde*. Weinheim, New York : WILEY-VCH Verlag GmbH, 2001, pp. 49–55
21. BRÜSSEL, R.; ERNST, H.; GEIGER, O.; HENNING, F.; KRAUSE, W.: LFT - mit Technologieinnovationen zu neuen Anwendungen. LFT - Equipment modifications enable the realisation of new applications. In: *Proceedings. 7. Internationale AVK-Tagung für verstärkte Kunststoffe und duroplastische Formmassen*. Baden-Baden, Germany, 2004, pp. A8/1–A8/11
22. MAISON, S.; THIBOUT, C.; GARCIN, J.-L.; PAYEN, H.; SIBOIS, H.; COIFFER-COLAS, C.; VAUTEY, P.; BOULNOIS, P.: Technical Developments in Thermoplastic Composite Fuselages. In: *SAMPE Journal* 34 (1998), No. 5, pp. 33–43
23. BÜRKLE, E.; SIEVERDING, M.; MITZLER, J.: Spritzgießverarbeitung von langfaserverstärktem Polypropylen. In: *Kunststoffe* 93 (2003), pp. 47–50
24. SIGL, K. P.: Einarbeitung von Endlos-Glasfaserrovings in Thermoplaste im Zuge eines Einstufenprozesses. Incorperation of endless glassfiber-rovings in thermoplastic materials in the course of a single-stage process. In: *Proceedings. 15. Stuttgarter Kunststoff-Kolloquium*. Stuttgart, Germany, 1997, pp. 1–10
25. CHOI, B.-D.; OFFERMANN, P.; DIESTEL, O.: Development of Multidirectional Reinforced Weft-Knitted Fabrics Made from Comingling Hybrid Yarns for Complex Light Weight Constructions. In: *Proceedings. 13. Internationales Techtextil Symposium*. Frankfurt, Germany, 2005
26. CHOI, B.-D.: *Entwicklung von Commingling-Hybridgarnen für langfaserverstärkte thermoplastische Verbundwerkstoffe*. Dresden, Technische Universität Dresden, Fakultät Maschinenwesen, Dissertation, 2005
27. EHRLER, P.: FVW aus Trägerfaser-Filamentgarnen und Thermoplast. In: *Chemiefaser/ Textilindustrie* 41/93 (1991), pp. T121–T127
28. RAMANI, K.; HOYLE, C.: Processing of thermoplastic composites using a powder slurry technique, II. Coating and consolidation. In: *Materials and Manufacturing Processes* 10 (1995), No. 6, pp. 1183–1200. DOI 10.1080/10426919508935101
29. *DFG-Sonderforschungsbereich SFB 639 "Textilverstärkte Verbundkomponenten für funktionsintegrierende Mischbauweisen bei komplexen Leichtbauanwendungen"*. http://www.tudresden.de/mw/ilk/sfb639/ (06-04-2011)
30. CHERIF, Ch.; RÖDEL, H.; HOFFMANN, G.; DIESTEL, O.; HERZBERG, C.; PAUL, C.; SCHULZ, C.; GROSSMANN, K.; MÜHL, A.; MÄDER, E.; BRÜNIG, H.: Textile Verarbeitungstechnologien für hybridgarnbasierte komplexe Preformstrukturen. In: *Zeitschrift Kunststofftechnik/ Journal of Plastics Technology* 5 (2009), No. 2, pp. 103–129
31. FLEMMING, M.; ZIEGMANN, G.; ROTH, S.: *Faserverbundbauweisen - Fasern und Matrizes*. Berlin, Heidelberg, New York : Springer Verlag, 1995

Additional Literature

ABOUNAIM, M.; CHERIF, Ch.: Flat-knitted innovative three-dimensional spacer fabrics: a competitive solution for lightweight composite applications. In: *Textile Research Journal* 82 (2012), No. 3, pp. 288-298

CHEN, X.; TAYLOR, L. W.; TSAI, L.: An overview on fabrication of three-dimensional woven textile preforms for composites. In: *Textile Research Journal* 9 (2011), No. 81, pp. 932-944

DENKENA, B.; HORST, P.; SCHMIDT, C.; BEHR, M.; KRIEGLSTEINER, J.: Efficient Production of CFRP Lightweight Structures on the Basis of Manufacturing Considerations at an Early Design Stage. In. Denkena, B. (Edt.): *New Production Technologies in Aerospace Industry*. Springer International Publishing, 2014

HUFENBACH, W.; BÖHM, R.; THIEME, M.; WINKLER, A.; MÄDER, E.; RAUSCH, J.; SCHADE, M.: Polypropylene/glass fibre 3D-textile reinforced composites for automotive applications. In: *Materials & Design* 3 (2011), No. 32, pp. 1468-1476

UDDIN, N.: *Developments in Fiber-Reinforced Polymer (FRP) Composites for Civil Engineering*. Woodhead Publishing, 2013, ISBN-13: 978-0-85709-234-2

Chapter 3
Textile Fiber Materials

Christiane Freudenberg

Textile fiber materials are the basic element for textile semi-finished products and the products manufactured from them. Thus, they are crucial for the product properties. Based on molecular and supramolecular structures, and ensuring optimum synthesis and fiber formation processes, high-quality fiber materials with tailor-made characteristics with the ability to function as reinforcement fibers or thermoplastic matrix fibers are created. The following chapter explains the general, complex relationships between initial materials, production, structure and properties. Detailed attention will be paid to the common commercially available reinforcement fiber materials, such as glass, carbon, and aramid fiber materials. Examples of other types of reinforcement fiber materials and thermoplastic fiber materials acting as matrix fibers will be given. A short introduction will be provided for the property optimization by means of surface modifications and material combinations.

3.1 Introduction

Textile fiber materials are classified into natural and man-made fibers, according to their origin. The group of organic man-made fibers made from synthetic polymers is the most extensive. Further details regarding the classification of textile fiber materials are given in Sect. 2.3.2.

Textile fiber materials are the basic element of textile semi-finished products and the products manufactured from them. The specific fiber material characteristics have a significant influence on the product characteristics. Building on the structure

C. Freudenberg (✉)
Institute of Textile Machinery and High Performance Material Technology, TU Dresden, Dresden, Germany
e-mail: christiane.freudenberg@tu-dresden.de

C. Cherif (ed.), *Textile Materials for Lightweight Constructions*,
DOI 10.1007/978-3-662-46341-3_3

and characteristics of the natural fiber material, bionics are employed to modify man-made fiber materials and develop new fiber materials with tailor-made properties, for instance for technical purposes. The steadily increasing number of applications, as well as the demand for processability on high-performance machines along the entire textile value-added chain and adapted to optimized practical use characteristics, in combination with the growing respective market, the development and modification requirements of man-made fiber materials are complex. Equally, knowledge regarding the characteristics stemming from production, fiber geometry and topography (which can be influenced by surface modifications), and supramolecular and molecular structure and their interaction are required for a targeted and requirement-adapted fiber material selection, due to the large variety of fiber materials. These complex relations are illustrated in Fig. 3.1 The individual aspects will be treated generally in this chapter, with selected fiber materials serving to exemplify certain aspects.

The remarks regarding the fiber production of man-made fibers are concerned with the significant methods: melt spinning and wet spinning. Both processes create filaments with adjustable fiber material parameters and the ability to be directly provided for further steps in the textile process chain as monofil and multi-filament yarns. Filament yarns, therefore, can be crimped, surface-modified/surface-functionalized or combined with one another, depending on the product requirements. Cutting or converting the continuous fibers to requirement-adjusted lengths creates staple fibers, which are specifically post-treated before their processing in further process steps, e.g. in yarn production. Details concerning yarn production can be found in Chap. 4.

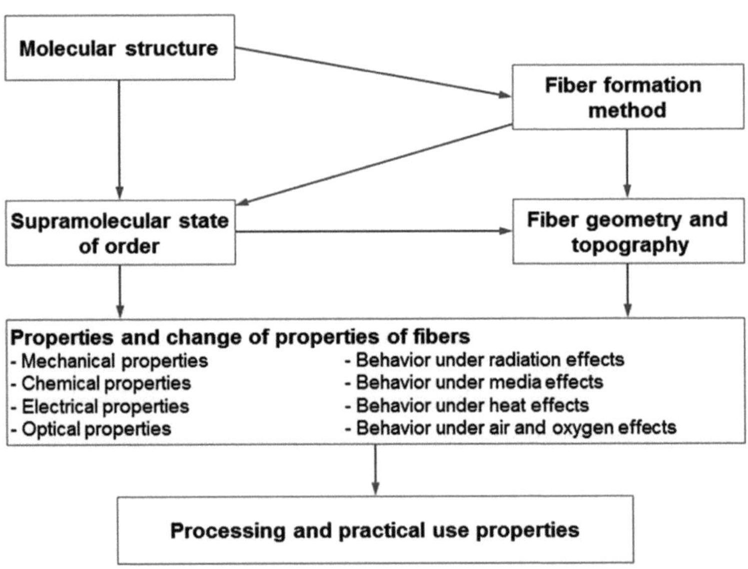

Fig. 3.1 Relation between structure, fiber formation method, and properties of textile fiber materials (according to [1])

Table 3.1 Reinforcement and matrix fiber materials

Reinforcement fibers	
Glass fiber (GF)	Ceramic fibers
Carbon fiber (CF)	-Aluminum oxide Al_2O_3
Aramid fiber (AR)	-Silicon carbide SiC
Ultra-high-molecular-weight polyethylene UHMWPE	Basalt fibers Basalt
Liquid crystal polymers LCP	Metal fibers
High-tenacity polyester PES, ht	-Steel (Steel)
High-tenacity polyamide PA, ht	-Aluminum (Al)
High-tenacity polypropylene PP, ht	Flax LI
	Sisal SI
Matrix fibers	
Polypropylene PP	Polyetherimide PEI
Polyamide PA	Polyether ether ketone PEEK
Polyester PES	Polybenzimidazole PBI_M
Polyethersulfone PSU	Polybenzoxazole PBO_M
Polyphenylenesulfide PPS	Polytetrafluorethylene PTFE

The representation of molecular and supramolecular structures contributes to a basic chemical understanding, which helps comprehend the abovementioned relations and use them specifically.

After an introduction to the basics, details will be given concerning reinforcement fibers. *Reinforcement fibers* are characterized by their high tensile strengths and Young's modulus at low densities and low maximum elongations at break. Due to targeted structural influencing, different fiber materials offer various fiber material types with special characteristics for different applications (Table 3.1).

Reinforcement fibers in the composite component act as force-absorbing and therefore load-bearing parts. In contrast, the tasks of the *matrix* primarily include fiber fixation, force introduction and distribution, and the absorption of compression loads. Due to the manifold types and the possibility to specifically generate fiber material properties, thermoplastic fiber materials can serve as matrix systems under certain conditions and for special applications. The reinforcement and matrix materials made from thermoplastic polymers will be introduced briefly below.

3.2 Technological Basics of Synthetic-Fiber Production

3.2.1 Principles of Synthetic-Fiber Production

The production of *man-made* fibers based on polymer materials includes the physical or physical and chemical conversion of linear polymers of high molecular weight into thin, continuous fibers or filaments. The base material is made fluid during the spinning of the synthetic fibers, and pressed through small openings into a medium, where the fibers solidify. The randomly arranged macromolecules in the filaments then have to be aligned in the direction of the fiber axis (orientation, e.g. by drawing/stretching). Finally, a thermal post-processing (*heatsetting*) is necessary to lower internal tension. The three most important steps of synthetic fiber production are spinning, drawing, and subsequent post-treatment. The following sections will provide details regarding these steps, mainly based on [1, 2].

3.2.1.1 Spinning/Synthetic Fiber Production

The viscous flow condition of the base materials required for *spinning*, depending on the polymer properties, is usually achieved by melting or solution. By means of spinning pumps, the highly viscous spinning dope is transferred through pipes to the nozzle holes, which serve as shaping elements. The material is pressed through the holes and relieved by taking off while still fluid. During this step, a pre-alignment of the chain molecules can be achieved. Crucial fiber parameters (e.g. fiber fineness) and the supramolecular structure are created during the solidification of the spun filament by cooling or solvent removal by chemical reaction or coagulation. *Coagulation* refers to the transition from the fluid into a gelular state. For the manufacture of most established synthetic fiber materials from high-molecular-weight linear polymers, the following production principles are available (Fig. 3.2):

- Spinning from the melt: *melt spinning*, or
- Spinning from the solution: *solvent spinning* (wet and dry spinning methods)

Other specialized spinning methods include gel spinning, reactive spinning, matrix spinning, and electro-spinning.

Spinning from the melt is used for polymers which can be melted into highly viscous fluids in extruders, and whose properties ensure a homogeneous melt for a sufficient duration. Melt spinning has the advantage over solvent spinning by not requiring the production, use, and recycling of a solvent. Furthermore, cleaning sub-processes, such as filtering and ventilation are redundant. Filaments made from polyester (PES), polyamide (PA), polypropylene (PP), and polyethylene (PE) are produced in melt spinning processes.

Spinning from the solution (solvent spinning) is used for polymers with a melting temperature above their decomposition range and requires the preparation of a high-quality spinning solution by providing suitable solvents. The solvents are required to meet certain general standards like low toxicity, eco-friendliness, non-flammability, and explosion safety, as well as low costs. Furthermore, solvent recovery is subject to high ecological and economical requirements. The characteristics of the spinning solution are defined by the parameters viscosity, polymer concentration, and solution stability. After shaping, which is preceded by the cleaning of the spinning solution by filtering (ensuring purity) and degassing (removal of air and gas bubbles), the solvent has to be removed to solidify the polymer. For this, the spun filaments are sent through a hot air chamber in the dry

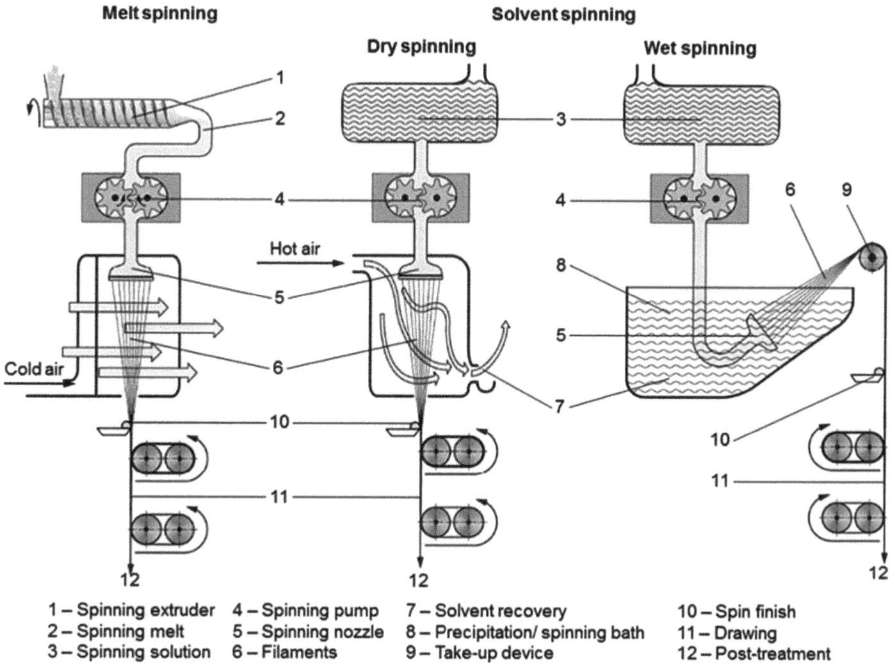

Fig. 3.2 Production principles of synthetic fiber production

spinning process. Utilizing concurrent or countercurrent principles, the conditioned air causes the solvent to diffuse from the solidifying polymer, in relation to the filament orientation direction and at adjusted flow rates. Due to the volume decrease, a negative pressure is created within the fiber material structure, leading to a deformation of the cross-section. The flow rate of the hot air depends on the delivery speed and the physical properties of the solvent, such as explosion limit or evaporation relation. In wet spinning, the filaments pressed from the spinning nozzle coagulate in a chemical or precipitation bath. The diffusion of the solvent, accompanied by osmotic processes, is typical of this process. The diffusion of the solvent in the precipitating agent increases the polymer concentration and, by extension, causes the coagulation and solidification of the filament. On the other hand, some of the solvent is displaced by the precipitating agent. Especially in the interphases a boundary structure is created, which strives to achieve a round cross-sectional form. The targeted variation of the spinning solution composition allows the design of a variety of different fiber cross-sections. Repeated washing and drying are necessary post-treatments in wet spinning to remove any solvent remaining in the filament.

Dry spinning is primarily used to produce filaments from polyacrylonitrile (PAN), acetate (CA), and polyvinyl chloride (CLF). Wet spinning processes are utilized to manufacture filaments from viscose (CV), cupro (CUP), but also polyacrylonitrile (PAN) and polyvinyl chloride (CLF).

3.2.1.2 Drawing/Stretching

In the un-drawn state, the low orientation of the macromolecules gives the spun filament a high deformability and low strength. The alignment of the chain molecules in the longitudinal direction of the fiber begins during the drafting part of the spinning process. A specific orientation of the macromolecules in relation to the fiber axis contributes to achieving the required properties. This structural influencing is realized by *drawing* and *stretching* processes. Figure 3.3 shows the basic plot of the yarn count-related stress and strain in dependence on the draw ratio, exemplified by PA filaments.

The *draw ratio*, also referred to as degree of stretching, is determined by the speed ratio of take-up and delivery devices. In contrast, the preceding spinning draft is defined as the relation of take-up speed and the average filament speed upon leaving the nozzle. Drawing, which usually takes place above the glass temperature, consists of the following processes:

- Fiber length increase at simultaneous fiber cross-section decrease,
- Alignment of the structural fiber elements in the direction of the force,
- increase of the number of crystalline areas by means of post-crystallization,
- unwinding of polymer chain loops and decrease of distance between structural fiber elements,
- reduction of inhomogeneities, and
- increase of interfibrillar, amorphous polymer chains (according to model by Prevorsek, Fig. 3.7 in Sect. 3.2.3.2).

During these processes, a bottleneck effect, similar to tensile testing, can occur (Fig. 3.4).

The increasing demand for higher productivity and high quality of filament yarns inspired the process-stage integration of drawing into the melt spinning process. Depending on the draw ratio, the macromolecules display various degrees of orientation, from which a classification of filament yarns can be derived, as in Table 3.2. In low oriented yarns (LOY) and partially oriented yarns (POY), the desired remaining orientation is achieved in subsequent processes. In LOY yarns,

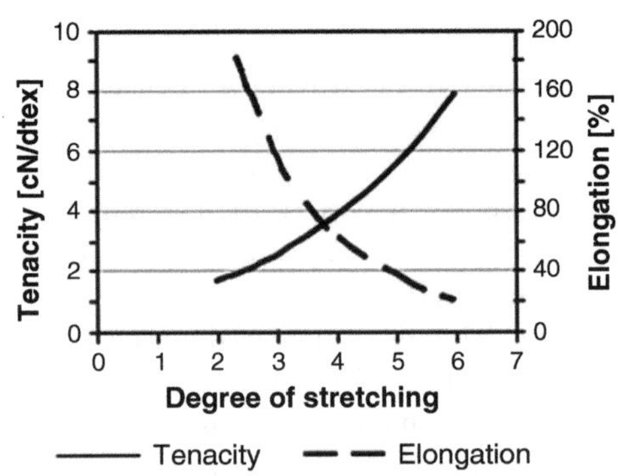

Fig. 3.3 Yarn count-related stress and strain in dependence on the draw ratio of PA 6.6 filament (according to [2])

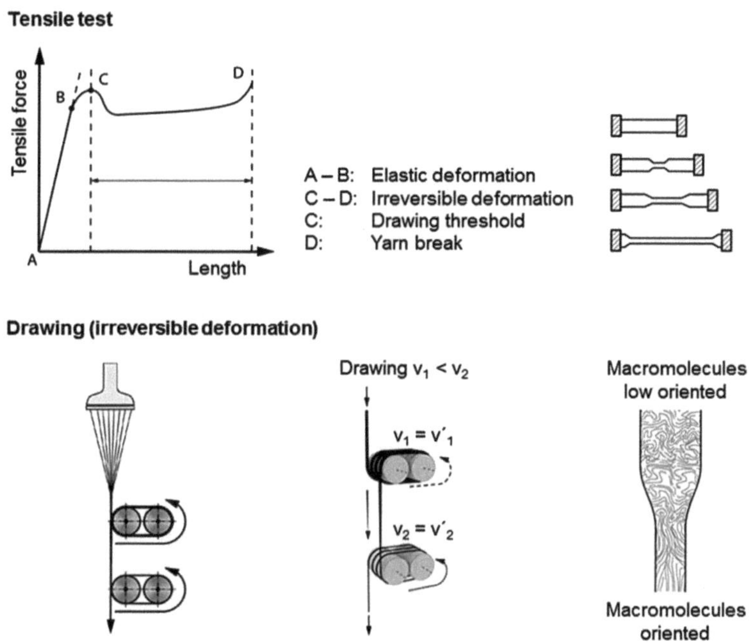

Fig. 3.4 Stretching sequence with neckings

Table 3.2 Classification of melt-spun filament yarns by their degrees of orientation [3, 4]

Designation	Yarn take-up speeds (m/min)	Yarn elongation (%)
LOY low oriented yarn	Up to 1,800	200–300
MOY medium oriented yarn	1,800–2,800	200–300
POY pre(partially) oriented yarn	2,800–4,200	70–220
HOY high oriented yarn	4,000–6,000	50–70
FOY full oriented yarn	>6,000	20–50

drawing is followed by plying or texturizing, while POY yarns are treated with combined processes, such as drawn plying, drawn spooling, drawn texturizing, and warp-yarn drawing. Beyond these, other material-specific combinations are possible.

3.2.1.3 Post-treatment

To create requirement-adapted properties, the spun filaments can be subjected to various post-treatment processes. This creates a desire fiber length or crimp, which is stabilized, for instance, by heatsetting. For these processes, various methods are available, which are particularly relevant for the manufacture of spun yarns. Here, only post-treatment methods relevant for the production of fiber-reinforced plastic composite semi-finished products are discussed.

To achieve optimal operation characteristics of filament yarns made of synthetic fibers during production and unimpeded processing on textile machines, certain auxiliary agents from aqueous solution or emulsions are applied to the filament yarns. If the agents are applied to un-drawn filaments, they are referred to as *sizings* or *spinning preparations*, while *avivages* (post-sizings) are applied after drawing.

The filaments touch a roller covered with a sizing coating, or a finger furnished with a sizing agent coat. There, they take the spinning preparation, which has to meet requirements regarding the realization of yarn cohesion, near-temperature-independent and consistent friction behavior during drawing and, if applicable, texturizing, resistance to corrosion and depositions on the yarn guides, anti-static effects, possibility of thorough washing, and chemical neutrality. Another method used for high-speed spinning due to the homogeneous sizing application is to prepare individual filaments using a fine dosage pump. To improve homogeneity, the sizing agent can be foamed up for multi-yarn systems.

The requirements to be fulfilled by the avivage are similar to those of sizing agents. Avivages are supposed to smoothen the fiber surface, realize a certain fiber-fiber adhesion, prevent electrostatic charging, and ensure an adjustable adhesion with the matrix in the case of fiber-reinforced composites. For avivage application, a number of methods are available, roller and revolving-screen application, dipping application, or spraying among them.

The use of filament yarns with a more or less strongly crimped structure is of secondary importance for the manufacture of semi-finished products for fiber-reinforced composite materials. Hybrid-yarn production by air texturization is a special case. Here, the aim is to homogeneously intermingle two or more fiber material components, one of which, as a thermoplastic component, assumes the matrix function in the finished composite material. In contrast, *texturization* in general serves to create a permanent crimping and impart a voluminous character similar to that of wool on the originally straight and smooth filaments of a filament yarn. Usually, this involves mechanical/thermal processes (e.g. false-twist, torsion, stuffing-box crimping procedures) or chemical/thermal processes (e.g. bi-component processes).

3.2.2 Fiber Parameters

The characterization of textile fiber materials relies on geometrical parameters as well as properties closely related to fiber geometry and of crucial importance regarding processing and application characteristics of these materials. The use of parameters like fiber length, fiber cross-section, fiber fineness (yarn count, see Sect. 2.2.2.3), fiber crimp or twist, and topography or surface design, allow a comprehensive description of fiber geometries, which are random in natural fibers, but specifically adjustable in synthetic fibers.

To determine the geometric parameters, a number of methods are available and the most significant and relevant ones are mentioned in this section and, in more detail, in Sect. 14.3.

3.2.2.1 Fiber Length

The extension of an individual textile fiber in a fully stretched position without strain loading represents the fiber length. As changes in humidity cause changes in length in some fiber materials (e.g. wool, PA), this parameter always has to be determined in experiments under atmospheric conditions. Furthermore, previous mechanical strains (e.g. tearing, pressing) can release latent, time-dependent tensions in connection with moisture effects, causing changes in length. For further processing, especially in secondary spinning, knowledge of fiber length distribution is crucial to achieve high-quality fiber cohesion in the spun yarn [5].

Natural fibers have a finite length for natural growth reasons, with a classification being necessary between short fibers (especially cotton) and long fibers (stem fibers, leaf fibers, fruit fibers, e.g. flax, jute, coconut, sisal). Most long fibers are "technical fibers" distinguished by fiber bundles formed from several elementary fibers by pectins.

As mentioned in Sect. 2.2, silk (e.g. made by the Bombyx mori silkworm moth) holds a special rank regarding fiber length. The entire fiber length of a single silk cocoon can be up 4,000 m, of which 1,500 m can be fiber material fit to be reeled for processing in reel silk spinning. In natural fibers, the crucial factors influencing fiber length are, for instance, the kind of plant (including genus, type, species, provenance), climatic growth conditions, and the manner of harvesting and preparation, are also the reasons for great inhomogeneities and differences in quality.

Due to the technologies used in their production, synthetic fibers are theoretically of unlimited length. The filaments produced in a range of spinning processes are shortened to the lengths and length distributions required by their respective applications, using various means of cutting or converting. The fiber length to be produced, including the fiber length distribution of the spun fibers conform to the required combination with natural fibers or other types of synthetic fibers, the respective yarn production technology, or other processing methods used during composite component production, e.g. fiber-reinforced concrete, fiber-sprayed concrete.

The basic separation principles listed in Table 3.3 are the groundwork for various technologies to produce fibers of requirement-adapted length (spun fibers).

Table 3.3 Separation principles for spun-fiber production [5]

Separation principle	Method	Technology	Fiber lengths
Force effects in fiber cross-section direction (radial)	Cutting method	Knife or spiral cutting drum in a cutting converter	defined
Force effect in longitudinal fiber direction (axial)	Stretch-breaking method	Difference of radial speed between delivery and take-up roller pairs in the stretch-breaking machine	Broad random distribution
Radial and axial force effects	Air-blowing process	Flowing medium inclined toward the fiber axis	randomly distributed

3.2.2.2 Fiber Cross-Section

The cross-section created by a perpendicular cut through the fiber in cross-wise direction represents the fiber cross-section and illustrates structural traits of the fiber interior and surface. Fiber cross-sections are preferably created with a microtome and viewed microscopically under polarized light. The cross-section shapes of textile fibers are extremely varied and influence the processing behavior (primarily in secondary spinning) and application characteristics [2, 5].

Natural fibers, with the distinctive cross-sections, which are characteristic for the respective fiber material and which, like fiber length, depend on a number of conditions, are easily identified by their fiber cross-section and other special traits. Cotton fibers, for instance, display kidney-shapes cross-sections, while those of the elementary flax fibers are polygonal in cross-section, with central lumina (cavities). Sheep's wool has an oval to round cross-section with an external scale structure, clearly visible in a longitudinal view. Boiled silk, as an elementary rope (hank), has irregular, rounded, triangular or multiangular cross-sections.

The cross-sections of synthetic fibers are adjustable to the respective requirements and applications. Usually, melt-spun fibers from PA, PES, PE, PP, and other polymers, as well as fibers made from glass, quartz and metal, have round cross-sections. The realization of special fiber cross-sections (e.g. star-shape, ribbon-shape, corner-shape profiles) as solid or hollow profiles is achieved by using specially-shape nozzle holes in the spinning nozzles. Special attention has to be paid to wear effects on the nozzle holes, as these can cause cross-section deformations, resulting in significant quality losses. Synthetic fibers produced with solvent spinning processes change their shape after exiting the round spinning nozzle. Fibers created by dry spinning usually form dumbbell-, kidney-, x-, or y-shapes cross-sections, with a less pronounced corrugation or overlap distinguishing them from wet-spun fibers. In the latter, the selected composition of the spinning bath and the timing of the individual sub-processes (coagulation and regeneration) allow an objective influence on the fiber cross-sections. If the two sub-processes are running parallelly, filaments with a core-sheath structure are created, as the regeneration in the sheath section takes place at a point when the coagulation in the core is not yet completed. This is especially true for viscose production with strongly overlapped fiber cross-sections. If the regeneration is performed after coagulation, more rounded and less overlapped fiber cross-sections, as well as fibers (e.g. modal fibers) with a relatively homogeneous supramolecular structure across the fiber cross-section are created [5].

In some individual production methods of solvent spinning, the round cross-section is retained after coagulation. In a few rare cases, profile nozzles are used [2].

Apart from requirement- and process-related microcavities (gas volume completely surrounded by fiber substance) and pores (gas-filled indentations in the fiber surface), as well as the purposeful creation of air-filled channels running axially through the fiber (accessible from outside on one or both sides), synthetic fibers usually are compactly structured on the inside [6].

Multi-component spinning methods allow the production of synthetic fibers consisting of several components. The polymers are provided in *side-by-side*, *core/sheath,* or *islands-in-a-sea*, *separation-type*, or *multilayer-type* arrangements, and can generate special fiber material properties and, by extension, special fiber types, e.g. fiber crimping and material combination for reinforcement and matrix fiber [2]. Multi-component fibers in a *separation*-type matrix/fibril arrangement offer a possibility to produce microfibers with yarn counts smaller than a decitex. This is equivalent to fibers whose cross-section dimension is smaller than 10 μm. For this, a component is either dissolved in a solvent, or a splitting after a targeted force effect is induced by cooling after the spinning process, or by the introduction of tensile forces, tearing the adhesive points.

3.2.2.3 Fiber Yarn Count

Apart from the cross-sectional form, the diameter is also used for the characterization of textiles fibers with round cross-sections, e.g. for glass filaments. For further characterization of textile fiber materials, the textile-physical variable of yarn count is used. Yarn count, according to the standard Tex system, is the quotient of mass and length. Due to the small cross-section dimensions of fibers and filaments, the mass is usually related to a fiber length of 10,000 m. From this, the statement of fiber yarn count in decitex (1 dtex = 1 g/10,000 m) is derived. In Sect. 2.2.2, the reasons for introducing this textile-specific variable depending on length and mass, influenced by cross-section area and density, as the various yarn count systems are detailed.

3.2.2.4 Fiber Crimp/Torsion

Fiber crimp refers to the in-plane or spatial course of the fiber [7]. To assess the crimp, quantitative descriptions of the crimping form (e.g. crimping arch count per unit of length, arch length, arch height, degree of curl, crimp) and statements regarding the stability under mechanical and thermal loads (e.g. crimp rigidity) are made. Torsions are rotations around the fiber axis in Z or S direction occurring naturally in cotton fibers. Crimp and torsion are important for the processing of fibers into yarns, as well as for the application characteristics like strength, handling, and bulkiness. Therefore, a specific texturization of the fibers or filaments is necessary [6, 8].

Without specific efforts before or during production, synthetic fibers do not exhibit crimp. The basic principles for deflecting the fiber axis described below allow the purposeful setting of desired crimp effects (*Texturizing*) of a fiber [9]:

1. Stress build-up within the previously balanced fiber by means of external force effects, usually under higher temperature effects, followed by positional stabilization
   - by means of stress relief (heatsetting), or
   - by maintaining the deflection by means of additional force effects,
2. Relief of existing internal stresses
   - within one fiber or filament with different structural units due to molecular mobility and positional stabilization by establishing balance (e.g. bicomponent method), or
   - between fibers or filaments with different structural properties, caused by external force effects due to the shrinkage of a component, and subsequent positional stabilization by force effects (e.g. differential shrinkage process).

Regarding (1) The basic principle of mechanical/thermal processes is the sequence of deformation, fixation, and reforming, with a possible inclusion of a previous heating step. Depending on the manner of mechanical deformation, a classification is made into twist-, stuffing-, blade-, gear-, and knit-fixing processes. False twisting is the most common method in practical application.

Regarding (2) Due to inhomogeneous stresses caused by molecular and supramolecular structures in the cross-section, sheep's wool fibers, as well as bicomponent fibers specifically synthesized during production, and viscose fibers with differently pronounced core-sheath structures exhibit structure-inherent crimp.

3.2.2.5 Topography

Topography describes the surface design (micro-morphological structure) of the fiber materials, which can be characterized by the form of the indentations due to pores, breaks, tears and fibrils. Topography analyses are based on microscopic and physical-chemical examinations aiming for qualitative or quantitative statements [10]. The topography of textile fiber materials influences the properties of the textile semi-finished product with regard to gloss, handling, and especially to the interactions with preparation agents and matrix systems.

The topography of textile fiber materials is very diverse. In natural fibers, it varies primarily because of respective growth conditions and preparation methods. Their topography can be modified specifically by special finishing measures. Before drying, cotton has a round fiber cross-section and an almost smooth and structureless surface. Drying processes change these fiber parameters into a kidney-shaped cross-section and a surface showing pronounced grooves. A treatment with sodium hydroxide solution (mercerization) causes the fiber to swell, creating a round cross-section and a smooth surface. Wool has a distinct flake structure, with imbricately arranged flakes pointing at the tip of the hair. This surface can also be modified in order to reduce felting.

For solvent-spun fiber materials, electron-microscopic examinations show noticeable grooves and micropores in the fiber surface. Furthermore, PAN, for instance, displays a fibril structure consisting of rough fibrils (fibril bundles) with cross-section sizes of ca. 1 μm, and capillary fibrils (cross-section dimensions of ca. 10 nm) integrated into them, leading to a very distinctive topography. This is caused by coagulation, solvent discharge, and drawing processes [6].

Melt-spun fiber materials are mostly smooth and barely structured, unless non-identical substances like delustring agents are integrated. Electron-microscopically, fine fibril-like longitudinal stripes are visible on the surface of PA. As in solvent-spinning, drawing influences the formation of such fibril-like structures [6].

Changing the topography of organic and inorganic synthetic fibers can be done by means of a variety of methods. Especially for the utilization of textile fiber materials in composites, research and development offer a wide field of activity, with priority on the prevention of fiber damages, the adjustable fiber/matrix adhesion (see Chap. 13), and the economy of the processes.

3.2.3 Molecular and Supramolecular Structure of Textile Fiber Materials

The quality and properties of textile fibers result from the chemical composition and the formation/production process, which is followed by voluntary or involuntary modifications and specific finishing treatments. The primary initial materials for textile fibers are natural or synthetic organic polymers, but also inorganic substances like glass, rocks, or metal. Natural organic polymers (biopolymers) consist of the main components polysaccharides, polypeptides, or polyisoprenes, and are "pre-synthesized". Based on plant cellulose or plant/animal proteins, synthetic fiber materials are created from natural organic polymers by solution and subsequent regeneration of the biopolymers. Synthetic organic polymers are created by methods of organic chemistry, which are used to connect individual components. Fiber-forming synthetic polymers for the production of textile fiber materials are distinguished by the following characteristic traits [1]:

- optimal (usually high) molecular weight at narrow molecular weight distribution,
- linear macromolecule shape, preferably without branching or cross-linking
- bonding types of varying energy:
  - homopolar: covalent bonds in chain direction of the macromolecules,
  - inter- and intramolecular interactions,
  - partially crystalline structure: all states between "non-oriented" (amorphous) and "highest state of orientation" (crystalline) are possible,

- possibility to create concentrated solutions or resistant melts as fiber spinning base materials, and
- dyeability and finishing ability

The formation of natural fibers is more or less strongly influenced by growth conditions, which interact with environmental conditions, impacting supramolecular (physical) structure and fiber geometry. The production processes of synthetic fibers, on the other hand, are based on the specific transfer of a polymer solution or melt into a fiber form, followed by drawing and fixation. The design of these processes offers the possibility to achieve required properties with regard to supramolecular structure, yarn count, and topography, as well as adjustable mechanical fiber parameters.

Building on the knowledge of molecular and supramolecular structure, fiber materials with tailor-made properties can be produced, especially for technical purposes. To correctly interpret these properties and utilize them optimally, sufficient knowledge of the structural design is indispensable.

3.2.3.1 Molecular Structure

More than 1,000 atoms contribute to the structure of fiber-forming polymers. The macromolecules resulting from the covalent bonding of the monomers as a result of the three reaction mechanisms have a relative molecular weight of at least 10,000. The reaction mechanisms are given in simplified form in Fig. 3.5 and explained below.

- In polymerization, monomers containing double bonds are turned into long polymer chains in a chain reaction without discernible stages. This happens without rearrangements or elimination of molecular constituents. Fiber materials

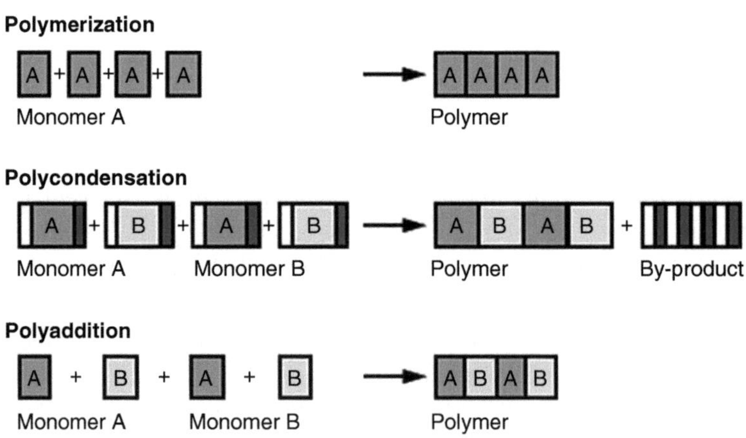

Fig. 3.5 Principle illustration of reaction mechanisms

made from PAN, CLF, polytetrafluorethylene, PE, and PP are synthesized by polymerization.

- Polycondensation progressively connects hetero- and homogeneous molecules into a polycondensate, while dissociating side products (e.g. water, short-chain alcohols). This requires at least two reactive groups per monomer. Polycondensates are primarily produced by ester formation (e.g. PES) and carboxylic acid amide formation [PA, Aramid (AR)].
- Polyaddition, which works progressively and without dissociations, first creates short molecule chains of a few monomers (oligomeres), which can react with one another or with longer chains. The monomers have at least two reactive groups each and react by shifting atoms and electron pairs. This reaction mechanism is used to produce fiber materials from polyurethane elastomere (EL).

The *molecular structure* describes the chemical structure of macromolecules and is characterized by the parameters constitution, configuration, and conformation. A comprehensive representation from [1, 6] will give an overview for the sake of basic chemical understanding.

The *constitution* describes type and arrangement of chain-atoms (of the monomer units) as well as the type of bond connecting them, the type, position, and polarity of the functional groups, and the molecular weight or degree of polymerization as well as the molecular weight distribution.

The monomers introduce "repeating units" or monomer units into the macromolecules. The smallest, ever-repeating unit is referred to as "structural element". The structural element can be larger, smaller or of equal size as the repeating unit. If the polymer is formed from a single type of monomer, a homopolymer is created. If several different monomers are used, heteropolymers (copolymers) are formed. If the build-up reaction involves bifunctional monomers, the primary valence linkage is made to a linear, chain macromolecule. Branched or crosslinked polymers are created by using tri- or polyfunctional monomers, which usually are unsuitable for fiber formation. Depending on the initial monomers and the build-up reaction, specific binding links are created in the macromolecules, influencing the supramolecular structure (intermolecular interactions) and fiber material properties like hydrolytic stability and strength. Functional groups and their accessibility in the binding links and on the side groups of the polymer backbone, as well as on the end groups on main and side groups are of crucial importance for chemical reactivity. These are acid carboxyl groups (–COOH), basic amino groups ($-NH_2$), and chemically relatively inert but polar hydroxyl (–OH) and nitrile (–CN) groups. The length of the macromolecules is an important parameter of the fiber-forming polymers for the processing and application properties and is designated with the molecular weight or the degree of polymerizations (Eq. 3.1).

$$P = \frac{M}{M_0} \tag{3.1}$$

P (–) Degree of polymerization
M (g/mol) Polymer molecular weight
M_0 (g/mol) Monomer molecular weight

Due to different molecular weights and lengths of the macromolecules, resulting from the formation processes, detailed examinations regarding the spinnability usually rely not only on the most frequently used parameter of average degree of polymerization, but also on molecular weight and the distribution of polymerization degrees. The various measuring methods result in different average values (e.g. averages of number and weight). The number average is determined by Eq. 3.2, and the weight average by Eq. 3.3 [11]:

$$M_n = \frac{\sum n_i \times M_i}{\sum n_i} \tag{3.2}$$

$$M_w = \frac{\sum m_i \times M_i}{\sum m_i} = \frac{\sum n_i \times M_i^2}{\sum n_i \times M_i} \tag{3.3}$$

M_n (g/mol) Number average of molecular weight average values
M_w (g/mol) Weight average of molecular weight average values
M_i (g/mol) Relative molecular mass of a narrow molecular fraction
n_i (mol) Number of molecules of the molecular fraction
m_i (g) Total mass of all molecules of the molecular fraction

In practice, molecular inconsistency is crucial for all statements regarding the length of the chain molecules and their length distribution (Eq. 3.4) [11]:

$$U = \frac{M_w}{M_n} - 1 \tag{3.4}$$

U (–) Molecular inconsistency

Configuration (tacticity or stereoregularity) refers to the spatial orientation of substituents around a certain atom. The linkage of the monomers during the synthesis reaction defines the spatial structure, without considering rotations of single bonds. A transformation of different configurations is therefore only possible by breaking and re-forming chemical bonds. If the substituents are statistically distributed spatially, polymers are referred to as atactic and in the presence of an ordered structure, they are termed tactic or stereoregular. If the substituents are regularly present on the same side of the plane defined by the C–C skeleton, isotactic polymers are created. A regular change of the spatial direction of the substituents leads to syndiotactic polymers (Fig. 3.6a).

Conformation describes the preferred spatial position of the groups of atoms possible due to the free rotation of C–C bonds, e.g. in helical conformation

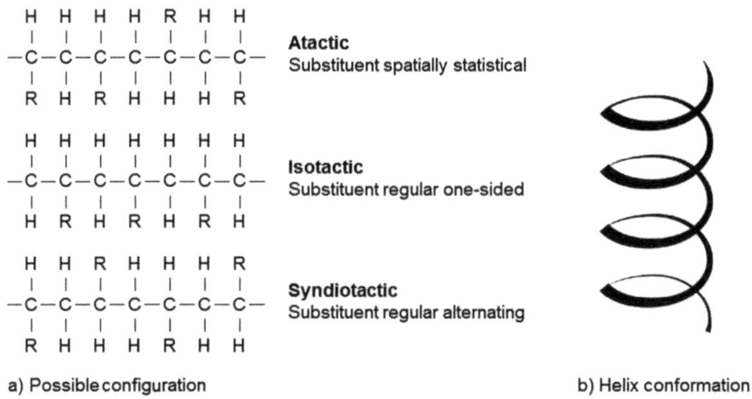

Fig. 3.6 Possibilities of configuration and conformation in fiber materials (according to [1])

(Fig. 3.6b). The spatial position is expressed by the angle of rotation and the energy level depending on it. Different conformations of a certain energy level can be transferred into one another without dissociating chemical bonds. An easy transfer is realizable if small potential barriers (difference between the highest maximum and the lowest minimum) exist. This facilitates mobile and flexible macromolecules, great bond lengths between the chain-atoms, a high number of competing layers at identical substituents, as well as low or small substituents. The flexibility of the polymers and the mobility of the chain segments influence melting temperature, melting viscosity and crystallization behavior, and therefore, indirectly, all properties (e.g. strength, rigidity) related to the crystalline states.

3.2.3.2 Supramolecular Structure

The *supramolecular structure* of fiber-forming polymers is a term for the organization and spatial position of the individual molecules to each other, which are influenced by the conditions during fiber formation processes and subsequent processing (e.g. drawing). Only the combined effects of several macromolecules lead to the specific polymer properties of the textile fiber materials.

Entropy, which causes the entanglement tendency of the macromolecules and strives to achieve maximum disorder, and the energy caused by the interactions between the molecule chains are equally responsible for the formation of macromolecular solid bodies. Higher energy leads to the formation of strong intermolecular interactions, resulting in the creation of polymers with high degrees of orientation, which are usually suitable for fiber formation. The supramolecular structure is influenced by the intermolecular interaction, the packing of the macromolecules, the degree of orientation of the crystalline domains and molecular segments, individual chains and aggregates (loose, weakly bonded connection of ions, molecules and other particles) in the non-crystalline areas.

The supramolecular states of orientation of fiber-forming polymers, therefore, contain the crystalline and amorphous fraction as well as the colloidal structure. The crystalline fraction is determined not only by the packing of the chain structure, but also by crystallite quantity (depending on the crystallize dimensions, lattice structure, and thermal stability), crystal system, and degree of crystallinity. The glass temperature characterizes the amorphous fraction as much as density and orientation or stretched position, as well as the related chain refolding and concentration of the amorphous molecules. The colloidal structure, which is created by the spatial arrangement of crystalline and amorphous section, is characterized by fibrils or paracrystalline layer lattices, the interphases between amorphous and crystalline areas, and by the creation of microvacuoles and cavities [1, 6].

The following explanation of the parameters aims to help the understanding of the structure/property relation of fiber materials [1, 6].

Intermolecular interactions are effective forces of attraction and repulsion, created by the reciprocal approximation of macromolecules. Depending on the mobility and possibly extant side groups, as well as on external conditions like temperature and shear forces, the macromolecules strive to achieve a position characterized by the energy minimum, preferably the crystalline state. The determining intermolecular interactions for the structure and properties of textile fiber materials include dispersion forces (PES, PE, PP), directional forces (PES, PAN), π-electron interactions (PES), and hydrogen bonds (CO, CV, SE, PA, EL). The intensity of the interactive forces between the molecules is responsible for their cohesion and, thus, for fiber-material-dependent mechanical parameters like strength and rigidity.

Fiber density is the quotient of mass in a certain volume and depends largely on the conformation, the volume of extant side groups, and the degree of orientation. Furthermore, it is influenced by the process conditions, as these affect the distances between the macromolecules. Macromolecules without side groups and in all-trans conformation are packed more densely than α-helix and macromolecules with large substituents or with atactic configuration. The fiber density increases within a chemically defined (equal) polymer with increasing degree of crystallinity. Due to structural differences, average and extreme values for crystalline and amorphous sections are usually mentioned in relevant tables. The fiber density affects properties like mechanical parameters (e.g. strength), swelling capability, or moisture and dye absorption, but also fiber volume count in fiber composites for lightweight construction purposes.

The crystalline structure, which is important for the *degree of crystallinity* (fraction of crystalline sections), is the strict spatial orientation of chain-atoms or chain links of several linearly stretched or parallelly folded macromolecules at maximum effective inter- and intramolecular interactions and formation of the highest-possible degree or orientation. Here, the prerequisites for the formation of stable crystalline sections are a densely packed, three-dimensional orientation of the chains as well as a change of the internal energy to an energy minimum. The smallest, periodically repeating crystal unit is the elementary cell, which is defined by its limiting surfaces and their angles to each other.

The temperature and orientation conditions during the manufacturing influence the usually imperfect and two-phase process of crystallization of fiber-forming polymers. In general, thermodynamic balance is not achieved during crystallization, and kinetic inhibitions occur, which prevent a complete crystallization, as does the presence of different chain lengths and chain loopings. The crystallization can be separated into the phases of primary crystallization (nucleation growth into morphological structures) and secondary crystallization (increase of the degree of order).

Due to the complexity of the crystallization process and the resulting supramolecular structures, various authors have developed structural models contributing to better understanding of the structure and derived properties. The three-phase model for synthetic fiber materials by Prevorsek [12] (Fig. 3.7) is based on elements of previous models and describes not only a crystalline and an amorphous phase which represents molecules primarily connecting crystalline sections, but also a second amorphous phase. This phase consists of stretched, unoriented chains in the interfibrillary space.

In summary, it can be said that the crystalline state of fiber-forming polymers in dependence on the molecular (chemical) structure and the crystallization conditions

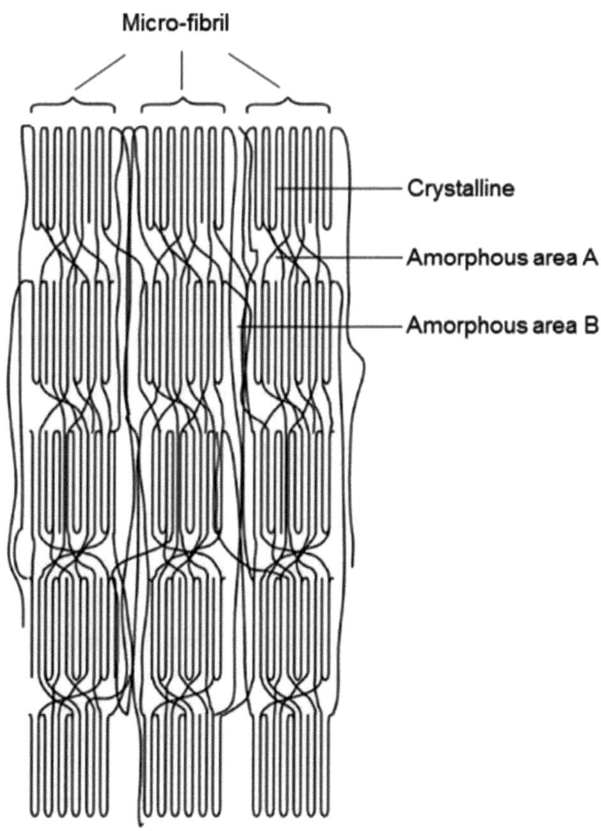

Fig. 3.7 Structural model for synthetic fiber materials, exemplified by PA 6, according to Prevorsek (cited in [1] based on [12])

(temperature, orientation) is greatly influenceable and adjustable to product requirements. The degree of crystallinity, which can be characterized by various means and can reach more than 90 %, determines the mechanical properties, like strength, resistance, deformation behavior, and sorption behavior.

The orientation is the alignment of macromolecules or macromolecule segments in non-crystalline (amorphous) sections, of crystallites or crystalline sections, and of fibrils in a preferred direction, i.e. in the direction of the acting forces (here, this is the fiber axis direction). To determine the *degree of orientation*, methods are available which state the degree of alignment of the fiber construction elements in fiber axis direction as average value. In practice, statements are made regarding the degree of drawing or the draw ratio.

In non-crystallizing polymers, the orientation is continuous and incomplete, in which stabilization is ensured by external stresses, cooling under glass transition temperature, interlinking, or the removal of solvent. In crystallizing polymers, the orientation effect is largely fixated by the crystallites, and orientation is realized suddenly and completely. As fiber-forming polymers feature amorphous and crystalline sections, the following effects are assumed during orientation: the ratio of crystalline and amorphous sections in the micro-fibrils decreases in favor of non-crystalline macromolecules filling the interfibrillary space (amorphous sections B in Fig. 3.7). The growth of the interfibrillary space in width primarily increases strength properties.

The reasons for the monaxial orientation of the macromolecules during fiber production are:

- primarily elastic stretching deformation of the polymer melt or solution in the entry area of the nozzle
- shear deformation during passage through the nozzle channel in spinning, resulting in a partial neutralization due to relaxation processes,
- stretch deformation as a result of the drawing during solidification
- monaxial stretching in the solidified state during the post-spinning processes

The specifically adjustable process of monaxial stretching can be used to modify fibers as required by the applications allowing the tailor-made realization of component-related fiber properties.

3.2.4 Surface Preparation for Textile Processing

At the surface between the textile fibers and the substances coming into contact with them, interactions take place which depend on the respective materials and the distance between them. The occurring interactions, e.g. covalent bonds, coulombic interaction, dispersion forces, and hydrogen bonds are varied and of different strength. The forces and interaction energies related to an area unit result from multiplying the forces or energies occurring between the individual molecule groups by the number of the interacting molecule groups present in the area unit.

To determine these, a number of methods for the characterization of the individual force components are used, e.g. by making statements regarding the type and surface density of functional groups by means of the Fourier transform infrared spectroscopy (FTIR), photoelectron spectroscopy (ESCA, XPS), or zeta potential measurements. The free surface energy determined by contact angle measurement, as well as the *zeta potential* (fiber charge) are considerably influenced by surface modifications of the fiber materials, e.g. by introducing functional groups or applying preparation agents and textile auxiliaries [10].

The application of spinning preparations or avivages during or after fiber production is absolutely necessary for the further textile processing (see Sect. 3.2.1.3). The absorption of these textile auxiliaries dissolved or dispersed in water is determined by the zeta potential of the fiber material. Knowledge of the fiber charge allows an optimized selection and addition of the textile auxiliaries. The usually negatively charged fibers adsorb the cation-active additives much better than anion-active additives.

The application-adjusted use of textile fiber materials in the form of filaments, yarns or fabrics, as reinforcement fiber in fiber-reinforced composites requires knowledge of the properties and behavior of the matrix systems in relation with the fiber materials. On the other hand, knowledge regarding the interface created between fiber surface and matrix is indispensible. The mechanical strength of a component is defined by the adhesion between reinforcement fiber and matrix. In a large number of applications (e.g. in heavy-duty components), a strong bonding of the reinforcement fibers to the matrix is absolutely vital. For other applications (e.g. crash and ballistics protection), a less strong adhesion is advantageous, as a greater energy conversion is achieved by a detaching interface than by "too excellent adhesion" [13]. The targeted modification of the fiber surface helps realize chemically utilizable anchoring points.

The following approaches to surface modification for high-performance fiber material are the current state of the art and research (see Chap. 13):

- functionalization during spinning:
  - coating with special sizing consisting of film formers to protect against mechanical damages, adhesion agents to specifically encourage fiber/matrix adhesion, and processing agents (e.g. wetting agents, softening agents, static inhibitors) in aqueous solution or more rarely in organic solvents [13–15], and
  - surface oxidation on carbon fiber materials (wet, dry, and anodic oxidation) followed immediately by coating with special sizing [16]
- functionalization by chemical processes after spinning:
  - impregnation with aqueous resin systems, and
  - impregnation with bifunctional adhesive solutions

- chemical/physical methods after spinning, primarily for synthetic fibers:
  - plasma treatment,
  - fluorination and oxyfluorination (also relevant for carbon fibers), and
  - electron beam treatment

Apart from high-performance fiber materials, natural fibers are also used in the fiber-reinforced composite industry. Here, too, interface issues play an important role, as the natural plant fibers are strongly hygroscopic. Possibilities to design the interface include, for instance, the finishing with silanes and the coating with functionalized polypropylenes [15].

3.3 Reinforcement Fibers

3.3.1 Introduction

In composite materials, at least two components with various functions are combined. By selecting the material combinations, the advantageous properties are combined with each other. The fiber materials, which display better tensile strengths, higher Young's moduli and lower strains than the matrix materials, are the reinforcing, load-bearing components in the composite, determining to a large extent, the mechanical properties of the composite.

A number of textile fiber materials is available for the use as reinforcement fiber. Reinforcement fibers display not only good mechanical properties at low density (usually <3 g/cm^3), they also offer great functional characteristics and can be classified into *high-performance fibers* and high-strength fiber. High-performance fibers are fibers with extremely high levels of physical and chemical properties. High-performance fibers include glass fibers (GF), carbon fibers (CF), and aramid (AR) fibers. For applications with moderate mechanical requirements, as in membranes, high-strength polymer fiber materials like PES and PA can be used. These serve as coating substrates, for example for Teflon coatings. In sections with relatively low mechanical requirements, synthetic and natural standard fibers can be used for reinforcement. Flax and sisal are used to exemplify this. Here, it has to be noted that the mentioned synthetic fibers (e.g. PES, PA) in fiber-reinforced composites in which the high performance fibers preferentially make up the reinforcement component can also be used as thermoplastic matrix system. This is explained in detail in Sect. 3.4.

The most important reinforcement fibers in composite production are glass (especially E-glass), carbon, and aramid fibers. Therefore, the focus of the following remarks will be on the production, structure, and functional properties of these fiber materials. The details concerning other reinforcement fibers not yet used widely in practice will finish this section. As the further textile processing of very stiff and sometimes extremely brittle fiber materials is a special challenge, exemplary technological properties are highlighted.

The gathered specific values are from various sources. Research has shown that there are no clear boundaries for the classification between carbon and aramid fiber materials. Therefore, deviations within the sources and from other sources can occur, despite diligent selection.

3.3.2 Glass Fiber Materials

Glass is a member of the group of inorganic-non-metal materials. It consists primarily of finely powdered, iron-free, and highly pure quartz sand and soda (sodium carbonate) and/or potassium carbonate as flow agent, and other additives depending on the type of glass and the respective application. Fiberglass is a group of technical glasses with tailor-made chemical and physical properties, consisting of the raw materials such as quartz powder (silicon dioxide), limestone (calcium carbonate), colemantite (borate mineral), kaolin (kaolinite), and other finely dosed oxides (e.g. boron and aluminum oxides, calcium and magnesium oxides). It optimally matches requirements for the processing into glass fibers. Glass fibers, whose melt composition differs significantly from the melts of solid glasses, are the general term for glass fibers, micro-glass fibers, insulating glass fibers, silica fibers, and hollow glass fibers, with the focus here being on glass fibers producible as filaments or spun fibers [17].

3.3.2.1 Production of Glass Filaments and Fibers

A large number of special melt spinning methods for the production of textile glass fibers is available. The classification of the production can be based on the initial materials or with regard to the technology used in fiber formation.

In 90 % of all production cases (97 % for E-glass) glass filaments are produced by *nozzle drawing* (Fig. 3.8), which is a direct spinning procedure. One exception is the production of glass filaments with diameters of 7 μm (maximum). Here, the two-stage process is used, in which the melt is produced from pre-produced glass pellets. In direct spinning, the initial materials are dosed and mixed depending on the respective glass fiber type. The blend is melted in a special melting system (e.g. a unit melter) at ca. 1,400 °C and kept at temperature range of 1,250–1,350 °C for further processing. A feeding system allows the melt to flow to the drawing nozzles, which work at temperatures of 1,200 °C. The melt exits the bores in the nozzle bottom, which consists of platinum/rhodium alloys. Depending on the desired filament diameter and the realizable take-up speed, the diameters of the nozzle bores vary between 1 and 2 mm. Drawing nozzles with 400–2,400 bores are used, test runs have been made with 4,800 bores. After leaving the nozzle, the filaments are mechanically taken up continuously at high speeds, e.g. by means of

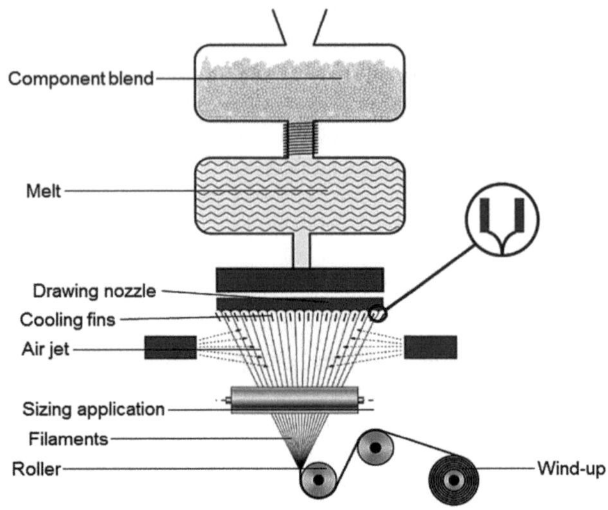

Fig. 3.8 Schematic representation of the nozzle drawing process

quickly rotating spools. The take-up speed for filament diameters >14 μm is about 1,200–1,500 m/min. At smaller diameters (<10 μm), 3,000–3,600 m/min are feasible. By setting the take-up speed, the desired filaments are tapered to the desired diameter (5–24 μm). To prevent broken filaments due to active take-up forces, the filaments also pass through a cooling unit (cooling fins made from copper or silver). To ensure a safe further processing, the filaments solidified in the short cooling face are coated with a sizing by a roller system (sizing rollers). The sizing material has to be selected with regard to the further processing steps or the projected application of the textile semi-finished product. Textile auxiliaries are available for the processing of glass fiber materials and their later use in textile applications, for instance in protective gear against heat and fire, for filters or home textiles. This includes adhesive-free textile sizing, e.g. based on starch, plant oils or cationic wetting agents, lowering fiber-fiber friction by ensuring good yarn cohesion or improving the sliding properties between fiber and yarn guide during textile processing. For the use of textile semi-finished products on the basis of glass fibers in fiber-reinforced composite applications, the applied textile sizing has to be removed, usually by heat treatment, as it would otherwise negatively impact the fiber/matrix composite formation. As a rule, adhesive-containing layers are used for composite materials, since these meet the process-engineering requirements and ensure optimal fiber/matrix adhesion. The adhesion promoter selection (e.g. organosilane compounds) considerably influences fiber/matrix adhesion and must therefore be adjusted to the respective matrix system. Before winding on suitable bobbins, the filaments are gathered and may be provided with a low twist. Roving production immediately after spinning is an exception in this case [15, 17–19].

95 % of all glass staple fibers are produced using the *drum drawing process* (Fig. 3.9), which is a two-step process, the first step of which consists of processing the blend of initial materials into pellets. These are the base material for the second

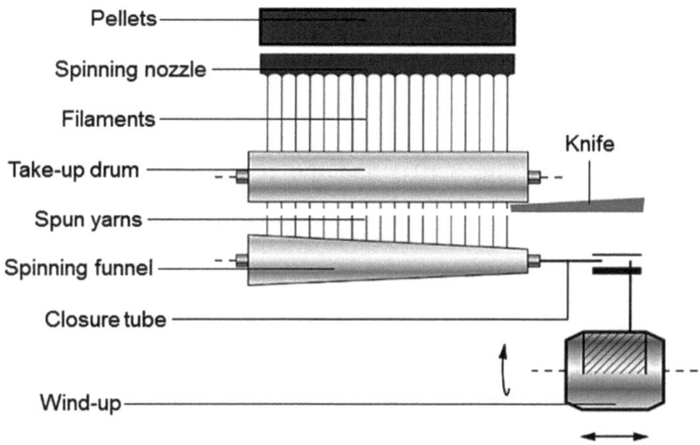

Fig. 3.9 Schematic representation of drum drawing

phase, which contains the actual spun glass fiber production. In the spinning nozzles, the pellets are melted, using an electrical resistance heating operating at temperatures between 1,000 and 1,200 °C. The melt escapes the openings of the platinum nozzle bar, which can have between 250 and 1,000 bores per bar depending on the requirement. A rotating drum draws the glass droplets created at the openings into filaments of 8–11 μm, at variable speeds of up 3,500 m/min. The sizings, which are adapted to the later application, can be applied immediately after the nozzle or by means of the take-up drum. For the continuous production of spun fibers, cooled-down glass filaments (the produced fibers have a length of 2 cm–1 m) are removed mechanically from the drum by means of scrapers or knifes. The fibers are gathered by a longitudinal opening in the fixed-position spinning funnel, where they are vortexed by a rotating closed tube. Afterwards, they are processed into a fiber yarn by a yarn production process. The mechanical and pneumatic protective twisting is followed by the winding of the spun glass fiber roving onto cylindrical cross-wound packages, which is then delivered to the subsequent processing stages [17, 18, 20, 21].

The drawn-rod method, in which precisely calibrated glass rods are the base material, allows the production of glass filaments and spun glass fibers.

In the manufacture of glass fiber materials, especially of glass filament yarns for use in composite applications, the selected sizing plays a special part. To influence the properties of the interface between fiber and matrix in accordance with the product requirements, established sizing systems are modified, tailor-made systems are newly developed, and possibilities are created to integrate additional functions into the sizing, e.g. by means of carbon nanotubes [22–24].

3.3.2.2 Structure of Glass Fiber Materials

Since the quick cooling of the melt during glass filament production prevents crystallization, the fiber-forming components cannot form regular crystal lattices, resulting in an isotropy, i.e. a directional independency with regard to the arrangement of the molecules (Fig. 3.10a) and the properties. The formation of the amorphous state is essentially dependent on the cooling speed and the temperature range of the melt, and influences the properties of the glass fiber types as much as the chemical composition does.

With regard to the textile glass fibers for use in fiber composites, only silicate glasses will be discussed in detail. These glasses consist of the two components silicone dioxide (SiO_2) as network former and the network-filling silicates or borates. The basic building element of the network, the SiO_4 tetrahedron, is shown in Fig. 3.10c. Due to the lack of oxygen atoms, the formation of Si–O–Si bridges causes polymerization resulting in the formation of high-molecular silicates, in which several SiO_4 tetrahedrons share a single oxygen atom. In multi-component glasses, such as textile glass fibers, the quartz glass network is modified. The suitable cations to be integrated according to glass fiber type, have to change the network by cleaving the Si–O–Si bridges. The resulting structure is given schematically in Fig. 3.10b and is determined by the metallic oxides used as network modifiers. A distinctive characteristic and the reason for the high strength and elasticity module values are the three-dimensional covalent bonds between silicon and oxygen. The network modifiers (e.g. Na_2O, K_2O, CaO, BaO, PbO) are not glass-forming, but property-affecting oxides [15, 17].

Due to the versatility of the textile glass fibers, chemical compositions based on the DIN 1259 (page 1) and ISR R 2078 standards have been summarized in the literature. These compositions are given in Table 3.4 for selected reinforcement fibers.

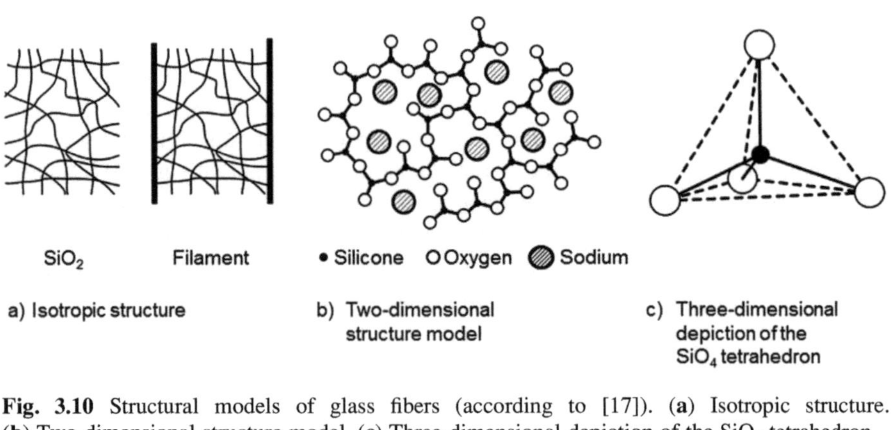

Fig. 3.10 Structural models of glass fibers (according to [17]). (**a**) Isotropic structure. (**b**) Two-dimensional structure model. (**c**) Three-dimensional depiction of the SiO_4 tetrahedron

Table 3.4 Chemical composition of selected glass fiber types [12, 17, 18, 25, 26]

Components (mass %)	Type E	Type AR	Type R	Type S	Silica	Type M
SiO_2	50.0 … 56.0	60.9 … 62.0	60	62.0 … 65.0	99.9	53.5
Al_2O_3	12.0 … 16.0	–	24.0 … 25.0	20.0 … 26.0	–	–
CaO	16.0 … 25.0	4.8	6.0 … 9.0	–	–	13.0
MgO	≤6.0	0.1	6.0 … 9.0	10.0 … 15.0	–	9.0
B_2O_3	6.0 … 13.0	–	–	≤1.2	–	–
F	≤0.7	–	–	–	–	–
Na_2O	0.3 … 2.0	14.3	0.4	≤1.1	–	–
ZrO_2	–	10.2	–	–	–	2.0
K_2O	0.2 … 0.5	2.7	0.1	–	–	–
Fe_2O_3	0.3	–	0.3	–	–	0.5
TiO_2	–	6.5	0.2	–	–	8.0
ZnO	–	–	–	–	–	–
CaF_2	–	–	–	–	–	–
LiO_2	–	–	–	–	–	3.0
SO_3	–	0.2	–	–	–	–
BeO	–	–	–	–	–	8.0
CeO	–	–	–	–	–	3.0

The following overview primarily contains main components present in the individual glass fiber types besides SiO_2, and the outstanding derived applications [13, 18, 27]:

E-glass — Aluminum borosilicate glass with less than 2 % alkali oxides for general plastic reinforcement and electrical applications, most common glass fiber type

AR glass — Alkali-containing glass with an increased level of zirconium oxide, alkali resistance for use in concrete reinforcement

R glass — Aluminosilicate glass with added calcium and magnesium oxide, high mechanical requirements even at high temperatures

S glass — Aluminosilicate glass with added magnesium oxide, high mechanical requirements even at high temperatures

Silicate — High SiO_2 mass content ratio (>99.9), very high temperature resistance

M glass — Beryllium—containing glass, high Young's modulus, used for the highest mechanical requirements

3.3.2.3 Functional Properties of Glass Filaments

Textile glass fibers for composite applications are characterized by excellent mechanical properties, such as high tensile strength at low elongation and low density, which are almost identical in both transverse and longitudinal direction of the fibers due to isotropy. It is noteworthy that the strength of textile glass is higher than that of a monolith. In addition, the strength of the textile glass increases with decreasing fiber diameter. The reason is suspected to be the type of fracture initiation. Occurring mechanical damages (e.g. cracks, notches, splittings, stages) create defects inside and on the surface of the fiber. With reduced fiber diameter, the occurrence of continuity disturbances is lessened, resulting in a strength increase [15, 28, 29].

If the tensile strength of a glass filament is determined immediately after solidification, it can be observed that it will be ca. 20 % above the tensile strength of filaments taken off a bobbin. The elongation is fully elastic, and the stress-strain behavior to failure is linear. In comparison to other high-performance fibers, especially carbon fibers, glass fiber have a lower Young's modulus (ca. 70–90 GPa). For the use in composite materials, which require extremely high rigidities, conventional glass fiber types are unsuitable. For this special area of applications, a high-modulus fiber (M glass) was developed, particularly for military utilizations [15].

For the application of glass filament yarns, it has to be kept in mind that the high strength of the individual filaments does not directly equal the strength of yarns or rovings. The experimentally determined tensile strength of a yarn or roving is smaller than the calculated strength of the sum of the included individual filaments, as these are not loaded simultaneously and evenly during testing, which causes a premature damage of single filaments. This behavior is usually countered by the targeted utilization of preparation agents. Therefore, the strength in the composite can be approximated to the summary strength of the individual filaments, provided that the processing does not excessively damage the fibers.

Apart from the parameters given in Table 3.5 for many technical applications of lightweight construction, knowledge of the long-term behavior under stresses and loads matching the life time of the component is essential. This is required to estimate relaxation effects and creep behavior over a multi-year life time of the fiber composite materials. In tests on glass filament yarns under endurance stresses,

Table 3.5 Ranges of mechanical properties of glass filaments made from selected glass types [17, 27, 30]

Parameter	Type E	Type AR	Type R
Density (g/cm^3)	2.52 … 2.60	2.70	2.50 … 2.53
Young's modulus (GPa)	72 … 77	76	83 … 87
Tensile strength (MPa)	3,400 … 3,700	2,000	4,400 … 4,750
Elongation at break (%)	3.3 … 4.8	2.6	4.1 … 5.4
Parameter	Type S	Silica	Type M
Density (g/cm^3)	2.45 … 2.55	2.00	2.89
Young's modulus (GPa)	75 … 88	56 … 66	87 … 115
Tensile strength (MPa)	4,300 … 4,900	800	4,750 … 4,900
Elongation at break (%)	4.2 … 5.4	1.5	4.0

hour- and day-long observations revealed relaxation occurrences under constant tensile forces as well as retardation appearances under constant elongations [31].

Based on these findings, recent research efforts include extensive fundamental experiments over the course of months to examine the long-term behavior of AR glass filaments for use in textile concrete. During the pre-stressing phase, an irreversible length increase of the filament yarn takes place. The reasons are to be found in the orientation of the individual filaments along the tensile force direction, and in the resulting local stress peaks, which can cause the local failure of individual filaments by exceeding the tensile strength. The ensuing creep of the filament yarn leads to changes in length. These elongation growths occur in addition to the generally occurring elongations and come to circa 0.3 ‰ during the examination period of several months at permanent loads of ca. 50 % of the maximum tensile strength. In relation to the occurring elastic short-time elongation from the loading, this is equal to a temporal increase (ε_{creep}) of the order of 3–5 %. This time-dependence of the elongation of AR glass filaments of 640 tex is shown exemplarily in Fig. 3.11 [32].

The mechanical properties of glass fibers change under environmental and chemical influence. Storing R and S glass under standard conditions and in distilled water, for instance, decreases their tensile strength by 10–25 %, in E glass by 20–35 % [15, 17, 25]. The influence of the duration of exposure to water or humid air on tensile strength is shown in Fig. 3.12. Alkali-free glass fibers or glass fibers with low alkali content, i.e. glass fibers without or with low content of alkali oxides, are more resistant to water. The effect of water damages the fiber surface. The existing Si–O and Si–O–Si bonds are polar and as a result capable to form secondary and, possibly primary valence bonds, which derives from the bonding of water and cationic compounds to glass surface. Here, not only molecular water, but also free or bonded OH groups can be integrated into the glass structure. These reactive hydroxyl groups can react with the water on the fiber surface, and leads to a further

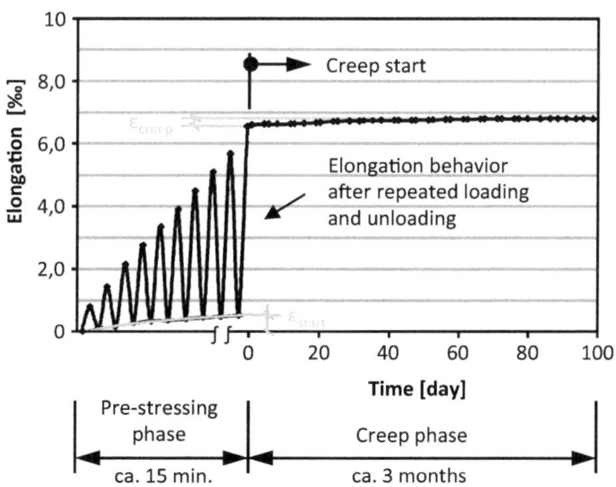

Fig. 3.11 Temporal development of yarn elongation under long-term stress in AR glass filament yarn of 640 tex (according to [32])

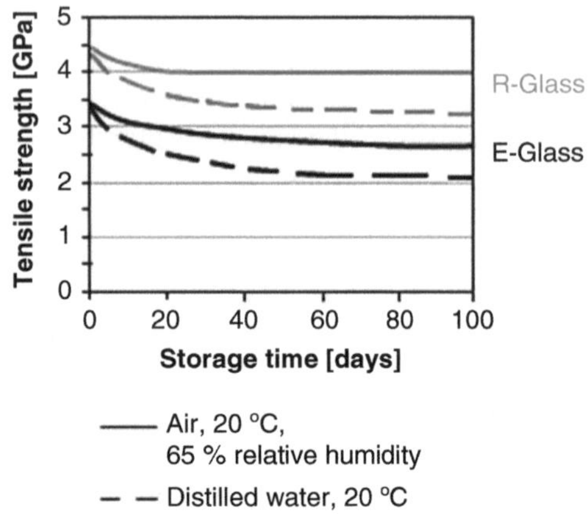

Fig. 3.12 Tensile strength of glass filaments under storage in humid air and water (according to [25], cited in [15])

decrease in strength. However, they can also be used for the chemical bonding of adhesives, to improve composite properties [15].

The influence of acidic (e.g. sulphuric acid, hydrochloric acid) or basic media (e.g. sodium hydroxide solution) can also reduce strength dramatically (by up to 30 %), depending on the glass fiber type. This applies especially to alkali-free and low-alkali types. Under long-term exposure to basic media, the fiber substance is ablated until it dissolves [15, 17, 33–35].

High requirements with regard to thermal properties and combustibility are set for the use of textile glass fibers in the fiber composite field. Glass fibers have the following relevant parameters: linear coefficient of thermal expansion α ($\alpha = 5 \times 10^{-6}\ K^{-1}$), specific heat capacity c [$c = 840$ J/(kg K)], and thermal conductivity λ [$\lambda = 0.85$ to 1.0 W/(m/k)] [17].

Other significant factors are the changes of strength under extreme short-term temperature loads, and the permanent load limit. Depending on the glass type, glass fibers do not lose strength under temperatures up to 200 °C (E glass) or above (250 °C for R and S glasses). At higher temperatures, strength will deteriorate even at short exposure durations, due to the occurring structural changes. This goes hand in hand with significantly decreased loading capacity in later applications [36–38]. This is shown in an example in Fig. 3.13, left-hand side. Silica fiber materials are an exception to this, as they are thermally loadable to 1,000 °C without considerable strength deterioration. The influence of temperature on the Young's modulus is given exemplarily in Fig. 3.13, right-hand side.

The strength losses and Young's modulus deteriorations originate from the high temperatures causing a desizing, which leads to water absorption and hydrolytic degradation in glass fibers [15]. On the other hand, the "frozen" (solidified) structure approaches that of compact glass, which also results in strength losses.

Due to consolidation of the individual filament in the yarn, coating the glass filament yarns demonstrably improves the mechanical properties. Coated yarns have a significantly higher strength than uncoated yarns, while their Young's moduli are almost identical. Research efforts [39] aim to specify the material

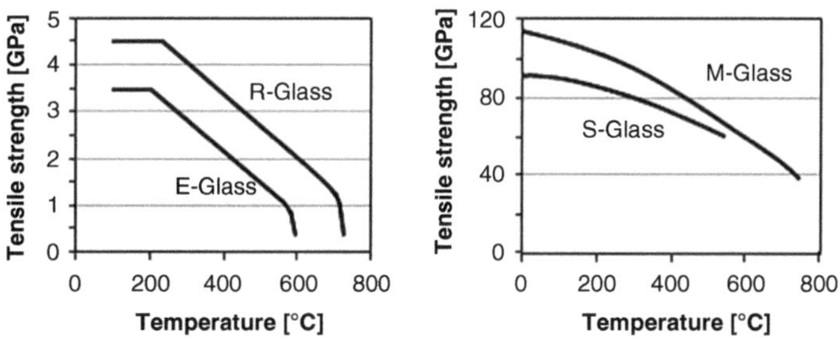

Fig. 3.13 Tensile strength and Young's modulus of glass filaments under temperature loads (according to [25], cited in [15])

behavior resulting from coating under various conditions. Among other things, the behavior under thermal stress is analyzed, in order to investigate the effects resulting from the used coating systems and to develop further suitable coating recipes. The effects on the mechanical and thermal characteristics were extensively examined on AR glass filament yarns with an organic coating from carboxylated styrene-butadiene copolymers (SBR) and compared to the uncoated reference material. In the uncoated material, the strength remains nearly constant up to 500 °C. At a further temperature increase, continuous strength deterioration occurs, caused by the beginning phase transition of the glass into the fusion state. The SBR-coated yarns show much higher strengths up to a temperature of ca. 300 °C, after which the thermal degradation of the applied coating begins. Due to the resulting heat release, further exposure to temperature accelerates the strength decrease, which is still on a higher level than in the uncoated yarns. Only at a temperature above 700 °C, the yarn is almost completely damaged. In all cases, the yarns exhibit a near-constant modulus of elasticity at temperatures under ca. 500 °C. This modulus only starts to decrease slowly at further temperature increases. This is due to the phase transition of the material, and is an indirect measurement for the damage to the yarn cross-section area [40].

The electrical properties of glass fibers are characterized by a high specific electrical resistance ρ in dependence of the temperature [$\rho = 10^{15}\ \Omega$ cm (T = 20 °C), $\rho = 10^{13}\ \Omega$ cm (T = 250 °C), $\rho = 10^{11}\ \Omega$ cm (T = 450 °C), $\rho = 10^{7}\ \Omega$ cm (T = 700 °C), low permittivity ε (ε = 5.8 to 6.7 (at 10^6 Hz)], and a dissipation factor tan δ at 10^6 Hz with (20–35) × 10^{-4} [17].

3.3.2.4 Textile Glass Products and Applications

Glass fiber materials in the form of glass filament or spun glass fiber yarns offer a wide range of economic applications. Far more than 80 % of glass filament or spun glass fiber yarns are used to reinforce plastics. In Europe, the production amount of glass-fiber-reinforced plastic in 2010 was ca. one million tons [41]. Due to the advantageous cost-benefit ratio, E-glass filament yarns are the primary reinforcement material, which makes them the most common high-performance reinforcement fibers.

In the following, only a few application examples for textile glass will be given. Different glass filaments in combination with textile production methods, type and geometry of fiber reinforcement, and selection and modification of the matrix systems offers a multitude of possibilities to optimize the functional value of glass-fiber-reinforced composite materials (such as glass-fiber-reinforced plastics), for instance for force-absorbing components in machine and plant engineering or in textile concrete.

Apart from the application in high-performance composite materials, textile glass fibers are used as reinforcements for industrial functional parts (e.g. grinding wheels, belts, clutch linings), as filters in dust removal technology, as insulation for thermal and electrical protection, as textile membranes, impermeable membranes and plaster reinforcement in construction, and as decoration and wallpaper material.

3.3.3 Carbon Fiber Materials

Fibrous carbon materials are pyrolytically produced from organic carbon compounds in the form of whiskers (monocrystals) or fibers. Based on the fiber structure and orientation of the crystallites, a distinction is made between isotropic fibers without identifiable preferred orientation, and anisotropic fibers with distinctive layer planes oriented parallel to the fiber axis. The strength of isotropic carbon fiber materials is relatively low, which makes them suitable for fillers, strings, and packings for thermal insulations, instead of composite materials. All further remarks below apply to anisotropic fibers consisting of at least 90 % carbon.

3.3.3.1 Production of Carbon Filaments

The production of carbon fibers by controlled pyrolysis requires initial materials that meet high requirements regarding spinnability, infusibility, low carbon loss during thermal degradation, and easy restructuring of the carbon skeleton into a graphite structure, as well as retention of the fiber form. Therefore, PAN, pitches, and, to a lesser degree, viscose are available as resources [15, 18]. The production of carbon fibers is based mainly on two methods, which rely on similar processes for the thermal degradation of the initial materials, which are referred to as precursors, to realize a very high carbon content (Fig. 3.14).

The spinning of the precursor is followed by the process stages drawing, stabilization (oxidation), carbonization, and graphitization, all of which influence the properties and aim to achieve maximum alignment of the synthesized graphite layers in longitudinal direction of the fibers to ensure extremely high mechanical characteristics. Here, the quality of the precursor materials and the selected process

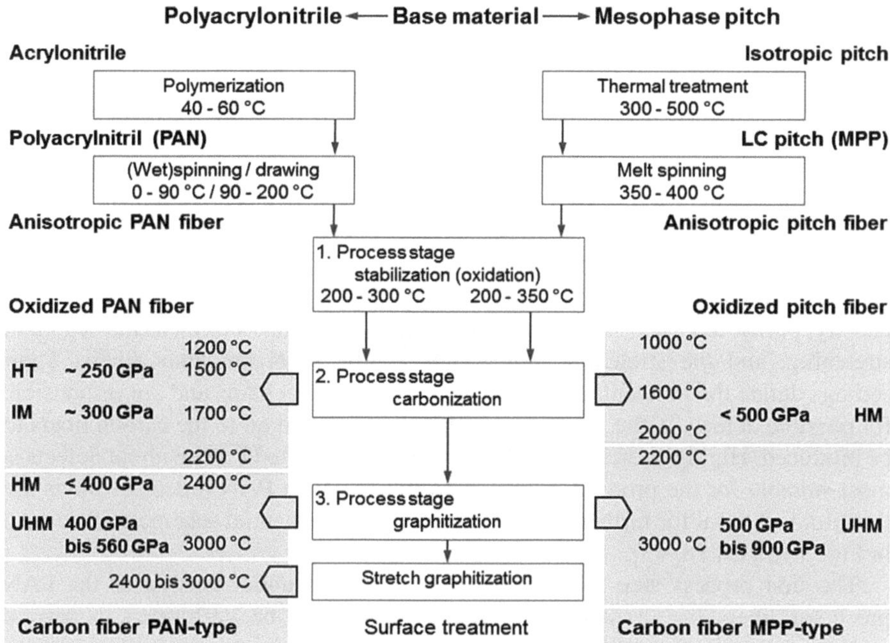

Fig. 3.14 Carbon fiber production (according to [42], cited in [43])

temperatures considerably influence the extremely high strengths or extremely high Young's modulus. For the further processing and the successful application, surface treatment is indispensible.

To produce carbon fiber materials based on mesophase pitch (MPP), residues from petroleum refining, as they accrue in the primary distillation of coal tar, are used. The useable pitch has to be thoroughly refined before being processed into MPP, to comply with the extremely high purity requirements. During thermal treatments at 300–500 °C, their density makes mesophase spherulites sink and allows their separation from the isotropic matters and other solid material particles. The pitch forms liquid crystals and begins to polymerize. This state is referred to as mesophase pitch, which is melted by a spinning extruder and pushed through the spinning nozzles for fiber production. The stretching aligns the macromolecules along the fiber axis and adjusts the required filament diameters and yarn count. To ensure the retention of the fiber structure, the anisotropic pitch fiber is stabilized by air oxidation at 200–350 °C, making it incombustible. Depending on the respective process parameters, carbonization and graphitization are performed under a protective gas atmosphere (nitrogen or argon, for instance) at temperatures ranging from 1,000 to 2,000 °C, or up to 3,000 °C. The objective selection of process temperatures allows the realization of tailor-made mechanical properties. Thus, fibers can be produced specifically with high strength or extremely high Young's moduli. As MPP-based carbon fibers only lose a small part of their weight and retain ca. 80 % of their substance, the process stages can be shorter than in the carbon fiber manufacture on the basis of PAN. However, MPP-based carbon fibers are particularly sensitive to compression forces [15, 18, 43].

For the production of carbon fiber materials on a PAN basis, a polymerization reaction is used to synthesize the precursor polyacrylonitrile (Fig. 3.15, left-hand side) from acrylonitrile and up to 15 % comonomer. The use of comonomers is required due to the strong exothermic reaction, although they contaminate the fiber material by impacting the structure negatively. Therefore, the aim is the utilization of pure PAN (Fig. 3.15, right-hand side) as a precursor fiber. The following factors are the dominant influences on quality: the spinning solution, regarding homogeneity, purity, and used solvent, the coagulation conditions as well as the degree of stretching, and the stretching temperature of the PAN precursor fibers. Their settings define the molecular orientation of the polymer chains and are responsible for possible defects in the filament, which are transferred on to the carbon fibers to be produced. Highly stretched and therefore highly oriented PAN without defects is most suitable for the production of carbon fibers. These PAN precursor fibers are the initial material for further process stages, whose sequential scheme is illustrated in Fig. 3.16 [15, 18, 43].

The first process step is the thermal stabilization under tension of the PAN precursor fibers by an oxidation process, which can be performed in several

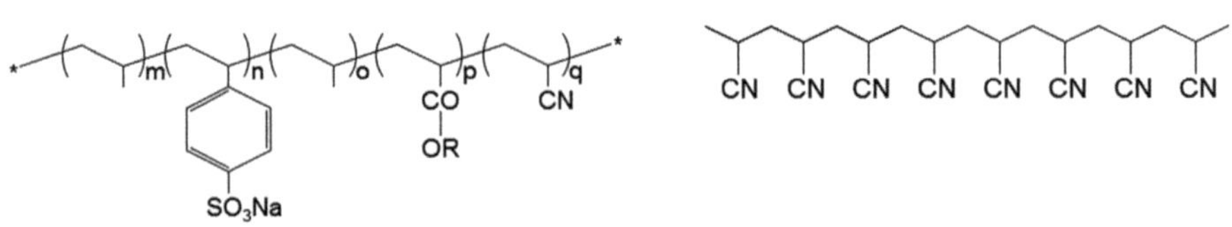

Fig. 3.15 Chemical structure of PAN (*left*: PAN with comonomers, *right*: pure PAN as precursor fibers)

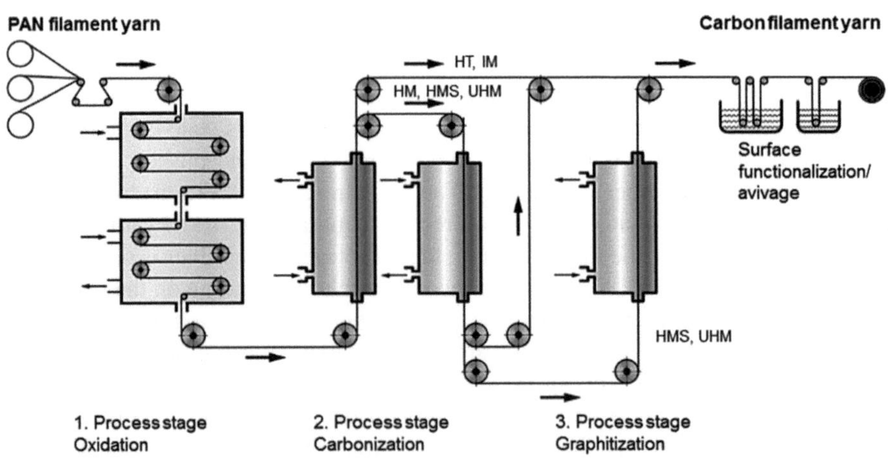

Fig. 3.16 Production sequence of PAN-based carbon filaments (according to [15, 18])

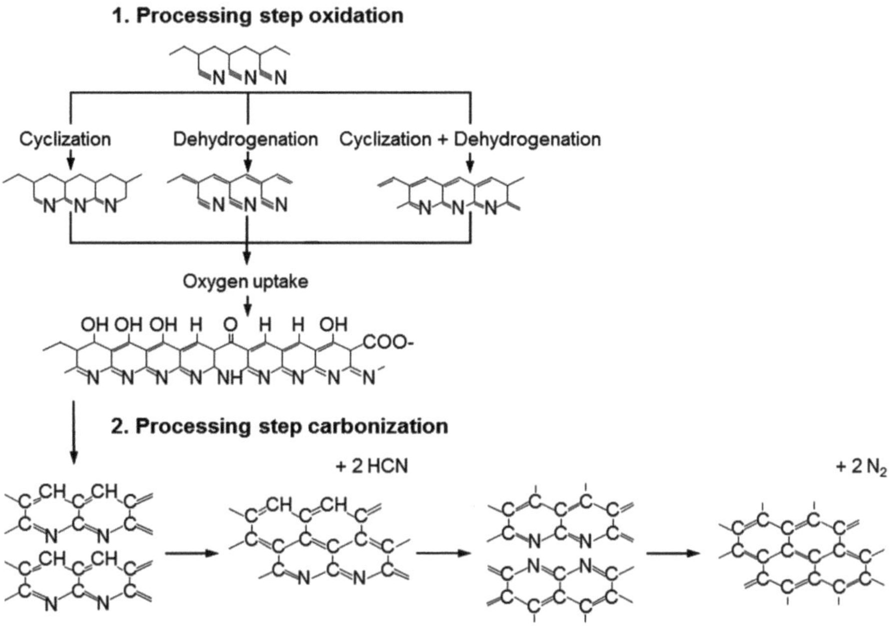

Fig. 3.17 Schematic representation of stabilization and carbonization (according to [15, 18])

multi-stage furnaces at temperature from 200 to 300 °C. Color changes from white to black and density increases caused by the formation of volatile by-products characterize the highly exothermic process. By cyclization of the nitrile groups and dehydrogenation of the C/C chain by oxygen, the linear polymer chains are transformed into more thermally stable hexagonal ring structures (Fig. 3.17). During the second processing stage, carbonization, which is also realized in several carbonization furnaces, the carbon content ratio is increased from ca. 60 % to more than 90 or even 95 %. The exact composition depending on the pre-treatment, reaction time, and carbonization temperature can range from 1,200 to 2,000 °C. The thermal degradation, performed under an inert gas atmosphere (e.g. nitrogen), is virtually finished at 1,700 °C and leads to the elimination of HCN, NH_3, H_2O, and CO_2, loss of additional mass, increase in density, and decrease in fiber cross-section. For the production of high-modulus fibers, the carbonized fibers are subjected to a third processing stage, referred to as graphitization. The thermal treatment of the fibers at temperatures of 2,400 to 3,000 °C under protective gas atmosphere (e.g. argon), increases the degree of orientation of the graphite layers created during carbonization (Figs. 3.17 and 3.19). An additional stretching of the carbon fibers in the plastic range of the carbon above 2,400 °C, the so-called stretch graphitization, contributes positively to the mechanical properties [18, 43–45].

To realize a low-fiber-damage textile processing, the very brittle and flexurally sensitive carbon fiber materials are provided with a surface functionalization and preparation (avivage). For the utilization of the outstanding mechanical properties

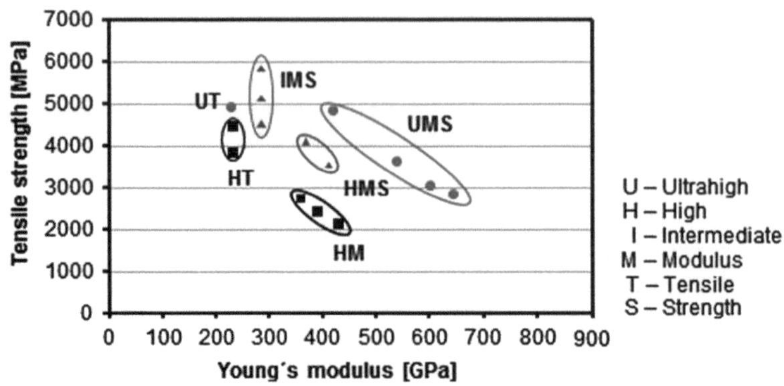

Fig. 3.18 Different types of carbon fibers (according to [13])

of carbon fibers in composite materials, an adjusted fiber-matrix adhesion is required. As carbon/plastic composite use mainly epoxy resin as matrix systems, textile auxiliaries based on epoxy resin blends are common, which also serve as protection during textile processing and as adhesion promoters between fiber and epoxy resin [13]. Other methods to realize the bonding possibilities for a matrix include the oxidative surface treatment in wet, dry, and anodic oxidation processes [43].

Corresponding to the properties of the precursor materials and process parameters during the production of carbon fibers, the specific material values of carbon fibers are set specifically, from which the individual developments in several groups of carbon fiber types are derived:

- High-tensile (HT) carbon fibers
- High-modulus (HM) carbon fibers
- Intermediate modulus (IM) carbon fibers
- High-modulus/High strength (HMS) carbon fibers
- High strain and tenacity (HST) carbon fibers
- Ultrahigh-modulus (UHM) carbon fibers

For the commercially common carbon fiber types, the groups are represented in Fig. 3.18 with the parameters Young's modulus and strength.

The production of carbon fibers is usually realized in filament form with diameters of 5–10 μm, with the filament number selected according to the final packaging form (roving, heavy tow). In rare cases, plied yarns are produced (see Sect. 4.3).

3.3.3.2 Structure of Carbon Fiber Materials

Graphite layers are the elementary structural elements of carbon fibers and are oriented in the direction of the fiber axis. The carbon atoms within these layers are

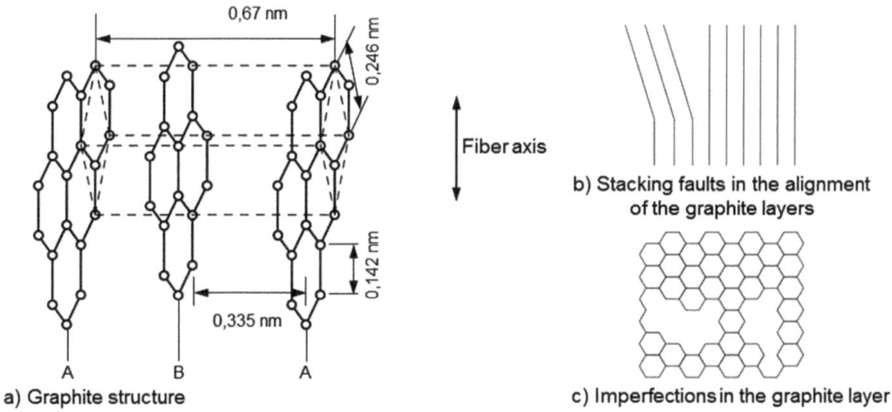

Fig. 3.19 (**a**) Structure of the hexagonal graphite (according to [43]), (**b** and **c**) lattice defects in the graphite structure (according to [42], cited in [43])

connected by strong covalent bonds. The perpendicularly acting, weak van der Waals bonding forces realize the cohesion of the individual graphite layers. These chemical bonds result in extreme tensile strengths and low transverse strength, while the orientation of the layers is decisive for the value of the Young's modulus.

The great variety of carbon fiber types is related to the present carbon content ratio and the residual contents of nitrogen and hydrogen, and at the same time on the more or less pronounced structural deviations from the ideal graphite lattice (Fig. 3.19a). The 100 % paracrystalline structure of the carbon fibers is characterized by increased layer plane distances, in which lattice defects (e.g. stacking faults or imperfections) can occur (Fig. 3.19b, c). This increase in layer plane distances is caused by the torsion of the parallel layers around the crystalline C-axis. The result is a turbostratic structure, i.e. layers on top of each other are parallel, but do not assume a preferred direction to each other [43, 44].

3.3.3.3 Functional Properties of Carbon Filaments

Due to their high anisotropic structure, carbon fibers are characterized by outstanding mechanical properties at low density as well as very high stiffness and strength in fiber axis direction. An overview is provided in Tables 3.6 and 3.7. The low transverse strength and the related compression sensitivity perpendicular to the fiber axis are caused by the low van der Waals forces between the graphite layers.

The relation between process temperature at carbonization or graphitization and the mechanical properties tensile strength and Young's modulus is sketched in Fig. 3.20. Higher Young's moduli are achievable at higher process temperatures, while high tensile strengths are realized at temperature of ca. 1,300 °C. This means that the development of new carbon fiber types is made with the focus on one of the parameters, always with regard to the planned application and aim. Furthermore, it

Table 3.6 Ranges of mechanical filament properties of selected PAN-based carbon fiber types [15, 16, 19, 43, 46]

Parameter	HT	IM	HM	HST
Density (g/cm^3)	1.74 ... 1.80	1.73 ... 1.80	1.76 ... 1.96	1.78 ... 1.83
Young's modulus (GPa)	200 ... 250	250 ... 400	300 ... 500	230 ... 270
Tensile strength (MPa)	2,700 ... 3,750	3,400 ... 5,900	1,750 ... 3,200	3,900 ... 7,000
Elongation at break (%)	1.20 ... 1.60	1.10 ... 1.93	0.35 ... 1.00	1.70 ... 2.40

Table 3.7 Ranges of mechanical filament properties of extremely high-modulus carbon fiber types [15, 19, 43]

Parameter	PAN-HMS	PAN-UHM	MPP-HM
Density (g/cm^3)	1.85	2.00	2.15
Young's modulus (GPa)	550	560	900
Tensile strength (MPa)	3,600	1,850	3,500
Elongation at break (%)	0.65	0.40	0.40

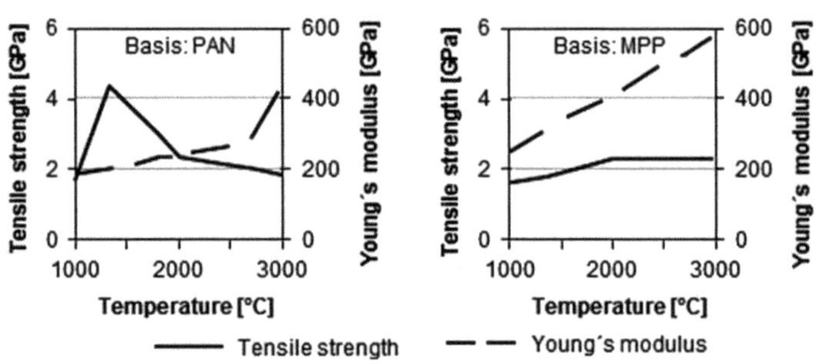

Fig. 3.20 Dependence of the tensile strength and Young's modulus on the process temperature (according to [47], cited in [15])

becomes clear that higher Young's moduli are achievable with MPP-based carbon fibers at lower process temperatures than with PAN-based carbon fibers. Beyond that, a correlation exists between the tensile modulus of the fiber and the compressive strength of the fiber, as shown in Fig. 3.21.

The filament diameter is an important variable for the formation of mechanical properties. A greater filament diameter improves the compressive strength of the carbon-based composite materials. In contrast, the number of surface defects is reduced, which improves strength and elongation at break in filaments of small diameters. Small bending radii are easier formed with filaments of small diameters, which is an advantage in the realization of complex-shape components [15].

As in the application of glass filament yarns, knowledge of the long-term behavior of carbon fibers under practical conditions is required to use them in composite applications. Research has shown that an initial elongation of 2 ‰ is not followed by further changes in length under loading with 50 % of the maximum tensile strength (Fig. 3.22).

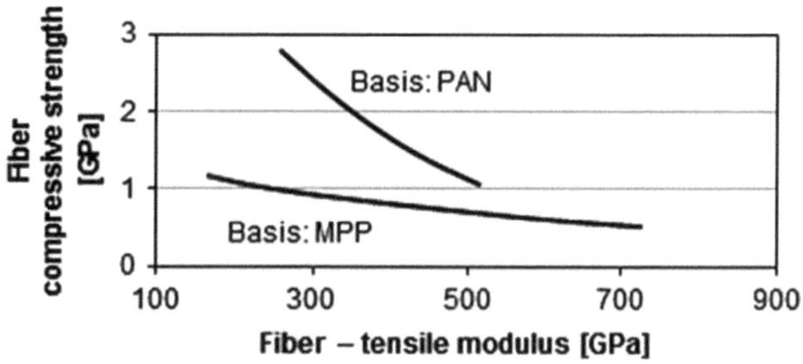

Fig. 3.21 Relation of the compressive strength of the fiber and tensile modulus for various carbon fibers (according to [48], cited in [15])

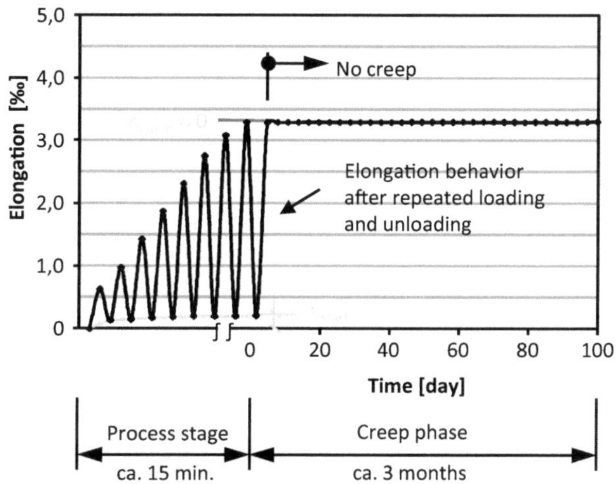

Fig. 3.22 Change in yarn elongation over time under long-term loading of a carbon filament of 800 tex (according to [32])

Carbon fiber materials are corrosion-resistant and more stable under media influence than other fiber materials. They possess good chemical resistance to most acids, alkalis, and solvents.

The thermal properties of carbon fibers depend on the fiber type and are direction-dependent due to the high anisotropy of the fiber structure. The linear coefficient of thermal expansion α is very low, lies in the negative range and has a value of ca. $1.6 \times 10^{-6}\ K^{-1}$ for types with a higher Young's modulus up to 700 GPa. Other parameters are the specific heat capacity c [c = 710 J/(kg K)] and a high, fiber-type-related thermal conductivity λ [Normal type: λ = 5 W/(m/K), High-modulus type: λ = 115 W/(m/K)] [43].

For the application, knowledge of the strength and Young's modulus behaviors under temperature exposure is essential. Under ambient conditions, carbon filament yarns do not experience noteworthy strength losses at temperatures below 400 °C [49, 50]. Oxidation effects occurring at temperatures from 300 °C upward create carbon dioxides and monoxides, which change the yarn structure and cause strength losses. In an inert atmosphere, the strength of carbon filament yarns will only gradually decrease at temperatures from ca. 1,000 °C upward [51, 52].

As in glass filament yarns, carbon filament yarns benefit greatly from a coating, especially with regard to the mechanical properties of yarn and composite (see

Sect. 3.3.2.3). In a standard atmosphere, uncoated carbon filament yarns of an 800 tex yarn count display tensile strengths of ca. 1,500 MPa. An applied organic SBR coating creates an adhesive connection, which allows it to transfer loads homogeneously, increasing tensile strength to ca. 2,500 MPa. To determine the influence of coating on the behavior of carbon filament yarns under effects of high temperatures, experiments were performed, similar to the tests on AR glass filament yarns. Due to the present organically-based coating which oxidizes at temperatures of ca. 300 °C and releases energy, carbon fiber materials exhibit structure-changing and strength-reducing processes at temperatures as low as 300–400 °C [53]. Temperature-resistant coating recipes are needed to prevent the oxidation of the glass fibers by means of creating a film around the fibers.

The electrical properties of carbon fibers are distinguished by a low specific electrical resistance, which ranges from 10^{-3} to 10^{-5} Ω cm depending on fiber type, and by the related electrical conductivity [43, 46].

3.3.3.4 Products from Carbon Fiber Materials and Applications

Carbon fibers are "young" fibers, which have been produced on an industrial scale since 1971. Due to their property range and the existing potential for various industries, their production has increased steadily, but is still marginal in comparison to conventional materials. The global production capacity in 2008 was ca. 65,000 tons. Due to the development of new carbon-fiber-reinforced composite materials, the global use of carbon fibers increased from 26,000 tons in 2006 to ca. 44,000 tons in 2010. Applications in industrial areas and aeronautics have played considerable parts in this increase. In these industries, the use of carbon fibers has increased by ca. 80 % from 2006 to 2010. The demand is predicted to reach 50,000 tons in 2012 [54, 55].

Due to the excellent mechanical properties at low density, carbon fiber materials are primarily used in composite material industries. The most important examples include heavy-duty components in aircraft, container, and mechanical engineering, as well as elements in automobile and sports equipment construction. Textile concrete is emerging as another application area of carbon fiber materials. Due to the electrical conductivity, carbon fibers are also used as conductive elements in components or protective garments with monitoring functions.

Carbon fibers in the form of felts can be used as thermal insulation material for high operating temperatures or as packing strings for temperature-resistant and corrosion-resistant shaft packings, or as nonwoven fabrics for fuel cells.

3.3.4 Aramid Fiber Materials

The term aramid refers to all aromatic polyamides. Aramids are synthetic fibers made from synthetic polymers, constructed from aromatic long-chain polyamide, in which at least 85 % of the mass is linked directly with two aromatic rings, where up to 60 % of the amide bonds can be replaced by imide bonds. To produce aramid fiber materials, three basic polymers are currently available, all of which are produced in polycondensation processes:

- poly-m-phenylene isophtalamide (MPIA) for the production of meta-type fibers (m-AR)
- poly-m-phenylene terephthalamide (PPTA) for the production of para-type fibers (p-AR)
- poly-p-phenylene 3,4-diphenylene terephthalamide for the production of p-type copolymer fibers

As the production of composite materials uses para-aramid fibers, the following remarks will focus on this type of fibers.

3.3.4.1 Production of Aramid Filament Fibers

From the monomers p-phenylenediamine (PPD) and terephthaloyl dichloride (TDC), the poly-p-phenylene terephthalamide (PPTA) is synthesized after solvent condensation (Fig. 3.23). Organic solvents are used, such as *N*-methylpyrrolidone (NMP) and Hexamethyl phosphoramide with added calcium chloride, which does not contribute to the chemical reaction. The resulting polymer is washed with a sodium hydroxide solution, and the formed hydrochloric acid is neutralized.

As the melting temperature of PPTA is above the decomposition temperature, filaments are formed in solvent spinning processes, with the wet spinning method established for para-aramid fibers. The process sequence is schematically illustrated in Fig. 3.24.

PPTS is dissolved in concentrated sulfuric acid at temperature from 80 to 100 °C, to achieve low viscosities and an anisotropic character of the spinning solution. To realize high fiber strengths, the production of the spinning solution has to include due regard to high molecular masses of the polymer and a high concentration of the spinning solution. The fluid-crystalline spinning solution is heated above the melting point of 80 °C, pressed through the spinning nozzles, guided

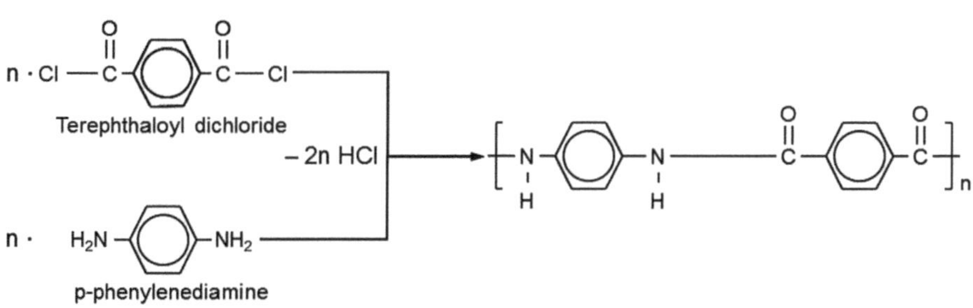

Fig. 3.23 Synthesis of poly-p-phenylene terephthalamide (according to [56])

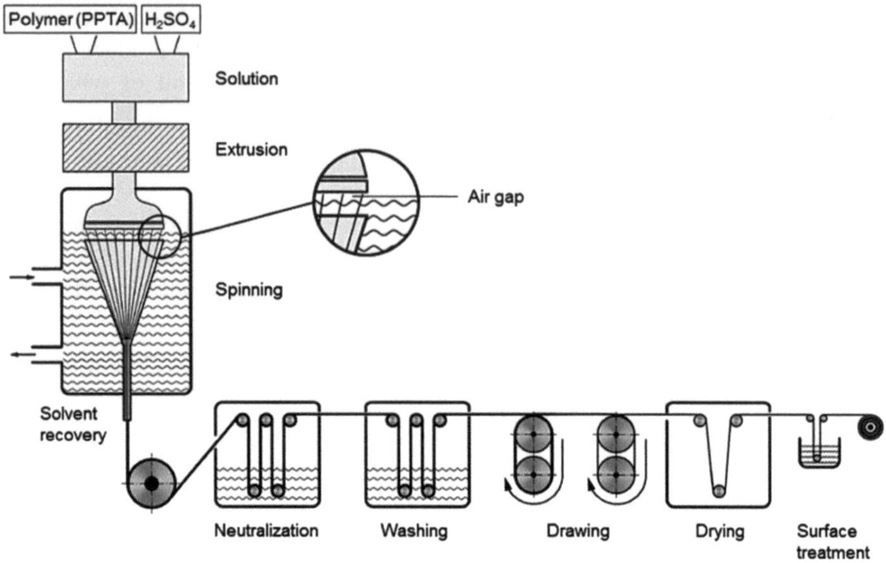

Fig. 3.24 Production of aramid fibers

through an *air gap* and spun at 0–20 °C in an aqueous sulfuric acid spinning bath, which is funneled downward from the surrounding container. This solvent-"dry-jet-wet" spinning process pulls down the spun filaments by the downward stream of the spinning bath, with the usual take-up speeds ranging between 100 and 200 m/min and possibly up to 600 m/min with additives. The non-oriented crystalline fractions of the spinning solution are given a pre-orientation by shear forces in the spinning nozzle, which is removed immediately below the nozzles, before they are oriented again by drawing in the air gap due to high take-up speeds. This orientation of the macromolecules is frozen during coagulation in the spinning bath [15, 56–58].

The actual spinning is followed by the post-treatment processes of neutralization, washing, drawing, and drying. To modify the properties, especially to increase crystallinity and Young's modulus, para-aramid fibers can be subjected to a one- or two-stage hot stretching at temperatures of 300–600 °C [56, 57]. The filament diameter tapers from ca. 25.0 μm to ca. 12.5 μm [15]. Another modification of the filament structure and properties is offered by fixation on rollers with temperatures from 400 to 420 °C [57]. For subsequent processing of the filament yarns, an avivage has to be applied.

3.3.4.2 Structure of Aramid Fiber Materials

The structure of para-aramid fibers (Fig. 3.25) in longitudinal fiber direction is characterized by rigid macromolecules consisting of amide bridges and aromatic rings and affected by strong one-dimensional, covalent bonding forces. These

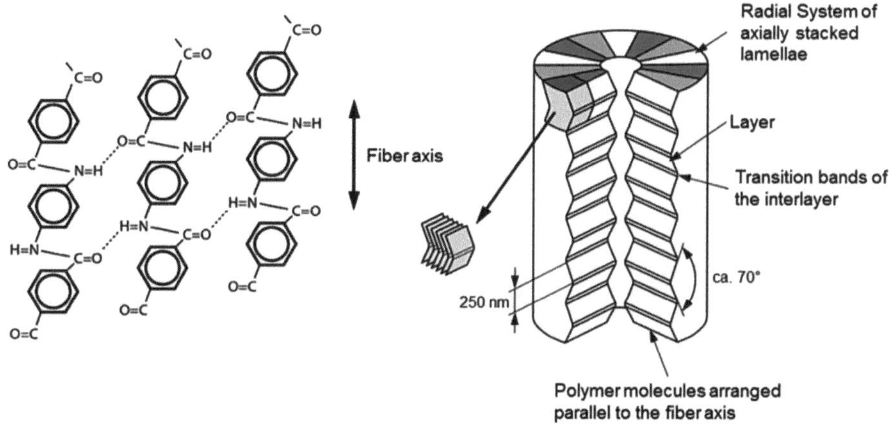

Fig. 3.25 Structure of para-aramid fibers (according to [56, 59], cited in [15])

stretched molecule chains are connected laterally by weaker hydrogen bonds and form layers which are regularly folded in longitudinal fiber direction. These axially arranged layers are connected by the even weaker van der Waals bonds [15, 56].

The anisotropic, fibrillar, highly crystalline structure with highly oriented and strong macromolecules in longitudinal fiber direction can be influenced specifically by spinning conditions, in order to create desired property profiles of the fiber types. For instance, drawing further increases crystallinity and results in the synthesis of high-modulus fibers.

3.3.4.3 Functional Properties of Aramid Filaments

The strongly pronounced structure of para-aramid fibers results in characteristic mechanical properties like high strength and high rigidity at low density (Table 3.8), which are optimized corresponding to product requirements. In comparison to brittle reinforcement fibers like glass and carbon, aramid fibers are highly tenacious, i.e. energy absorption will cause plastic deformation until failure. The rotatability around the longitudinal axis is limited by the fibrillar and layered structure. Apart from high ductility or tenacity, the high energy absorption capacity and outstanding impact load resistance have to be emphasized. Their structure makes aramid fibers sensitive to compressive loads in axial direction. This becomes apparent in filament knot tests, which reveal damages of the compression-loaded fiber side from the inside of the knot by bulging and buckling. This is primarily caused by the fibrillation of the fiber [15, 16, 56].

The stress-strain behavior and the creep behavior for various para-aramid fibers in are shown Fig. 3.26. Para-aramid fibers, depending on the respective type, are dimensionally stable under tensile loads. The elongation over a period of 1 year at a

Table 3.8 Ranges of mechanical properties of selected para-aramid filaments [15, 16, 19, 46, 56]

Parameter					
Density (g/cm³)	1.39 ... 1.44	1.44	1.45 ... 1.47	1.45	1.45
Young's modulus (GPa)	58 ... 80	100	120 ... 135	186	80.3
Tensile strength (MPa)	2,760 ... 3,000	3,150	2,800 ... 3,620	3,400	3,600
Elongation at break (%)	3.3 ... 4.4	2.0	1.9 ... 2.9	2.0	4.0

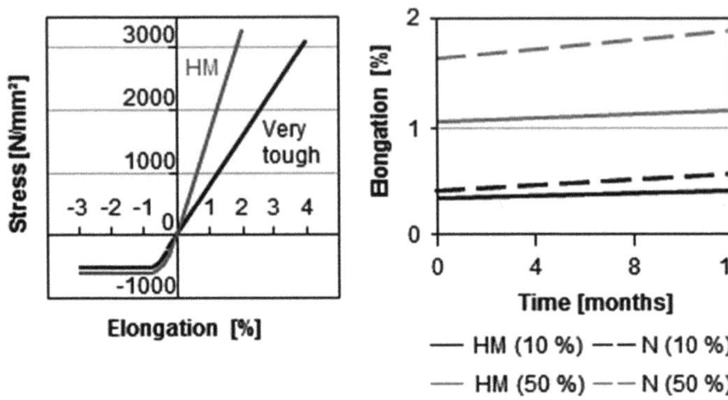

Fig. 3.26 *Left*: stress-strain curves [according to [46]), *right*: creep behavior at loading with 10 and 50 % of the maximum tensile strength of para-aramid fibers (according to [60])

Table 3.9 Residual tensile strength (in %) of para-aramid after storage in aggressive media [15]

	Exposure duration		
Media	12 days	40 days	120 days
Nitric acid 10 %	28.8	23.1	16.9
Hydrochloric acid 10 %	33.6	25.3	16.8
Sulfuric acid 10 %	87.7	85.0	49.7
Sodium hydroxide solution 10 %	59.6	38.4	27.8

load of 50 % of the maximum tensile strength is 0.13 % for high-modulus fibers, at 10 % of the maximum tensile strength it is 0.04 % [60].

Aramid fiber materials are stable under media influences. They display good chemical resistance to most acids, alkalis, and solvents, except very strong acids and alkalis (Table 3.9).

In comparison to glass and carbon fibers, aramid fibers exhibit hygroscopic behavior. Under standard atmosphere (temperature: 20 °C, relative humidity: 65 %), aramid fibers, depending on the fiber type, absorb 5–7 % of moisture [13, 58]. As this moisture absorption does not significantly reduce the maximum tensile strength of para-aramid fibers, an adsorption of the moisture in the micropores of the interfibrillar areas of the fiber is assumed [15, 56].

Aramid fibers, just like other organic fibers, suffer from macromolecule degradation under UV radiation. The resulting loss of tensile strength depends on the

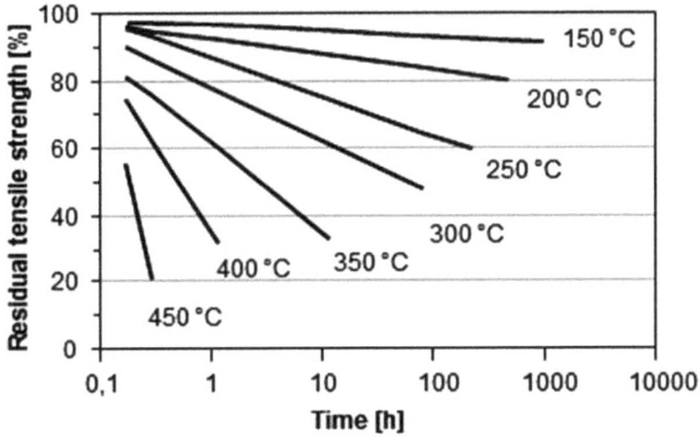

Fig. 3.27 Residual tensile strength ratio of para-aramid fibers under thermal load (according to [60])

duration of the exposure. The residual strength after 16 weeks is only 65–80 % of the original tensile strength, and only ca. 30 % after 120 weeks [56]. The strength losses occur under direct sun light or in tests under normal glass [15]. Therefore, UV protection should be included in the processing. The light resistance can be increased by the inclusion of pigments or UV absorption agents, which are applied to the fiber surface by means of resins [58].

The thermal properties of aramid fibers depend on the fiber type, and are direction-dependent due to the high anisotropy of the fiber structure. The linear coefficient of thermal expansion is very low and it is in the range from -2 to $-6 \times 10^{-6}\,K^{-1}$, with the pronounced reduction in length at higher temperatures due to shrinkage behavior. The maximum radial coefficient of thermal expansion is $70 \times 10^{-6}\,K^{-1}$, which is high compared to glass and carbon fibers (Table 3.16). Further parameters include the specific heat capacity c [$c = 1.4$ J/(g K)] and thermal conductivity λ [$\lambda = 0.05$ W/(m/k)] [45, 46].

Up to a temperature load of 160 °C, the mechanical properties like tensile strength and Young's modulus decrease by less than 10 % compared to the values registered under standard atmosphere. At higher temperatures, the fibers experience loss of strength (Fig. 3.27), and decrease in the Young's modulus. At 200 °C, the Young's modulus decreases by ca. 20 %, and tensile strength drops by up to 40 %. The decomposition temperature of para-aramid is 500 °C. In the low-temperature range, they retain their properties up to −70 °C [15, 56].

The electrical properties of para-aramid fibers are characterized by a low permittivity ε ($\varepsilon = 2.5$–4.1 at 10^6 Hz) and a high specific electrical resistance of $10^{15}\,\Omega$ cm [46, 56].

3.3.4.4 Product Examples for the Application of Aramid Fiber Materials

The use range of para-aramid fibers is very versatile. In the field of protective garments, para-aramid fibers are used as protective textiles or embedded in thermoplastic elastomers, primarily to prevent ballistic injuries. Apart from that, fiber composite materials with aramid reinforcement fibers are used for components designed to withstand impact loads. This utilizes the high energy absorption provided by the fibrillation. Further applications are in object and personal protection, especially in cutting and stabbing protection. This includes cut-resistant and tear-propagation-protective membranes, e.g. for truck tarpaulins. Apart from thermally more stable meta-aramid fibers, para-aramids can also be used as asbestos replacement for heat-protective work wear. Other applications are: clutch and friction linings, seals, cables, nets, or flame-retardant system for airplane seats.

Para-aramid fibers have very good mechanical properties, which are utilized in the application as reinforcement fibers for elastomers, for instance in tires, belt, and transport belt reinforcement. Analogous to meta-aramids, these properties are used as coating substrate for the production of membranes (e.g. foldable containers, roofing tarpaulins, and rubber boats) [56, 61].

Due to the low density, aramid fibers can be used to produce extremely lightweight FRPCs, with strengths comparable to carbon-fiber-reinforced FRPCs. Due to the significantly lower compressive strength in fiber direction compared to the tensile strength of aramid fibers, this compression sensitivity in fiber direction has to be kept in mind for all applications. This makes AR-reinforced components suitable as reinforcement for tensile-loaded components, while they are unsuitable for applications designed for flexural and compression loads [46]. Due to their tenaciousness, para-aramid fibers are also suitable for components which have to be lightweight and are subjected to dynamic and shock-like loads [16].

3.3.5 Other High-Strength Synthetic Fibers made from Organic-Origin Synthetic Polymers

3.3.5.1 Polyethylene Fiber Material

Polyethylene fibers are organic synthetic fibers polymerized from ethylene (C_2H_4) and classified into standard and highly-stretched/high-molecular fibers. “Extended Chain” polyethylene fibers (ECPE) or “Ultra High Molecular Weight” polyethylene fibers (UHMWPE) are used as reinforcement fibers. ECPE and UHMWPE are produced in a gel spinning process, which is an adapted solvent spinning method (wet technology), with the polymer being dissolved at increased temperatures. The first step of the high stretching is performed at ca. 120 °C. This dissolves the microcrystallites formed in the spinning bath (water bath), allowing the solvent to vanish. The macromolecules are further oriented along the fiber axis by post-stretching the solvent-free filaments at temperatures of ca. 140 °C, where they are stretched up to the 100-fold of their original length.

Polyethylene has a one-dimensional, highly anisotropic structure, which is characterized by a high degree of polymerization and a very high orientation of the macromolecules at small chain folding.

Table 3.10 Ranges of mechanical properties of selected filaments made from high-strength synthetic fiber materials [15, 27]

Parameter	UHMWPE	PA, hf	PES, hf	PP, hf	LCP
Density (g/cm^3)	0.97	1.13 ... 1.16	1.10 ... 1.39	0.90	1.41
Young's modulus (GPa)	87 ... 170	4.0 ... 8.3	10 ... 15	0.5 ... 5.0	100
Tensile strength (MPa)	2,800 ... 3,100	780 ... 930	820 ... 1,200	455 ... 670	2,700
Elongation at break (%)	2.7 ... 3.5	14.0 ... 22.0	8.0 ... 22.0	8.0 ... 10.0	3.0

Characteristic properties of the high-molecular polyethylene fibers include their density of $\rho = 0.97$ g/cm^3, which is very low for textile fabric materials, and their very high strength and rigidity, which is very high compared to other thermoplastic fiber materials (Table 3.10). The disadvantages of these fibers are their low temperature resistance and the low axial compression strength, which is similar to that of aramid fibers. Under standard atmosphere conditions, polyethylene fiber materials are resistant to aggressive media and to acids and alkalis (as long as temperatures are below 80 °C) [15].

Due to the high specific work at impact, ballistic protection systems are main application areas of these fibers. The combination of different reinforcement materials, e.g. hybrid yarns made from carbon and UHMWPE fibers can purposefully broaden the property profile of the components. With regard to the required fiber-matrix adhesion in composite applications, the surface of high-molecular polyethylene fibers have to be functionalized.

3.3.5.2 Technical Thermoplastic Fiber Materials

As shown in Sect. 3.2.3, the molecular and supramolecular structure of synthetic fiber materials made from synthetic polymers can be modified according to the requirements. Increasing the degrees of polymerization and crystallinity as well as orienting the macromolecules increases the Young's modulus and strength. This allows the synthesis of high-strength fiber materials, which are also used in technical areas, e.g. in composite materials and as membranes. The integration of ductile technical filaments (e.g. PA, PES, PP) in carbon- or glass-reinforced composites improves the technical properties, such as impact characteristics, fatigue resistance, splintering behavior, weight, and cost effectiveness.

Further developments in the field of synthetic fiber materials, include meltable liquid crystal polymers. Important fibers for this material are aromatic polyester fibers. These liquid crystalline polymers (LCP) have certain advantages, such as very low water absorption, excellent fire behavior, and high thermal stability.

Examples of the mechanical properties of high-strength filaments made from UHMWPE, PA, PES, PP, and LCP are given in Table 3.10.

3.3.6 Other High-Strength Synthetic Fibers Made from Natural Polymers of Inorganic Origin

3.3.6.1 Ceramic Fibers

Ceramic fibers, like glass fibers, are inorganic, non-metallic materials, produced from ceramic base materials like quartz, zircon, kaolin, or aluminum oxide. Apart from oxidic (e.g. aluminum oxide) and non-oxidic (e.g. silicon carbide) ceramic fiber types, silicon carbide fibers with a core yarn made from wolfram or carbon are available. The single-component ceramic fibers allow a broader application range than multi-component ceramic fibers, due to their small fiber diameter (3–20 μm). Therefore, these fiber groups will be in the focus of the following remarks.

Oxidic ceramic fibers—aluminum oxide fibers—can be produced in dispersion spinning and solvent spinning processes. In dispersion spinning, an aqueous suspension is made from the fiber-forming components (main ingredient Al_2O_3, with a fraction of SiO_2), the spinning agent, and, if necessary, from other additives needed for compact sintering. The fiber is formed using different methods by pneumatically stowing, drying, and burning the suspension at low temperatures. The so-called green fiber is then flame-burned in a sintering process stage. In solvent spinning, a viscous, concentrated solution of aluminum compounds (e.g. basic aluminum compounds) with the added auxiliary materials (e.g. water-soluble polymers and SiO_2) is used for spinning. The spinning solution is extruded into dry air through the spinning nozzle. The created hydrochloric acid evaporates from the spun filaments under the influence of increasing temperatures. After that, sintering is performed, creating a microscopically roughened surface resulting in a good mechanical connection to the matrix [1, 15].

Aluminum oxide fibers are polycrystalline and contain three-dimensional co-valent bonds. The extreme brittleness of the fibers can be improved regarding elongation at break by adding small amounts of silicone oxide (SiO_2). This, however, also reduces temperature resistance. The melting temperature is 2,045 °C. The addition of SiO_2 leads to creep of the fiber at temperatures over 1,000 °C, and to a loss of strength and reduction of the Young's modulus at temperatures from 1,100 °C to 1,500 °C [15].

Non-oxidic ceramic fibers—silicon carbide fibers (of various types)—are melt-spun from an initial polymer called polycarbosilane, which is created by a multi-stage process. The air-solidified filaments are incombustible and are integrated into the silicon carbide fiber by a heat treatment under inert atmosphere at temperatures from 1,200 to 1,400 °C. As the fiber consists of very small, cohesively connected particles, it has an almost isotropic structure and a smooth surface [15]. Silicon carbide fibers, which also contain high amounts of nitrogen and boron, are referred to SiCN or SiBCN fibers [62] respectively.

Ceramic fibers have a high tensile strength and high Young's modulus, high chemical resistance and very high temperature resistance. The specific electrical resistance varies, depending on the temperature and fiber type, and ranges between

Table 3.11 Ranges of the mechanical properties of selected ceramic, basalt, and metal fibers [15, 18, 27, 46, 62, 68]

Parameter	Al_2O_3	SiC	Basalt	MTF (steel)	MTF (Al)
Density (g/cm^3)	2.7 ... 4.1	2.35 ... 3.14	2.75	7.8 ... 7.9	2.8
Young's modulus (GPa)	150 ... 380	170 ... 420	89	210	72
Tensile strength (MPa)	1,700 ... 2,930	1,500 ... 3,600	2,000 ... 4,840	200 ... 2,500	460
Elongation at break (%)	0.4 ... 1.1	0.4 ... 1.1	3.15	1.0 ... 2.0	Not available

10^2 and 10^4 Ω cm, which places it in the range of semi-conductors [15]. Selected properties are given in Table 3.11.

3.3.6.2 Basalt Fibers

Basalt is a lava rock with glassy characteristics and consists primarily of the oxides of silicon, aluminum, iron, calcium and magnesium. Only basalt rocks with a silicon dioxide content above 46 % are suitable for the production of filaments [63]. The requirements on a reproducible melt spinning process regarding viscosity, crystallization, surface tension, and chemical and thermal homogeneity of the melt are very high. In addition, constant process parameters (e.g. temperature) are vital. Ensuring a homogeneous melt is complicated due to the natural variations of basalt rocks, which makes the abovementioned parameters unstable [64–66].

The realizable filament diameters range from 9 to 12 μm, depending on the working temperature of the nozzles, the size of the bore in the nozzle bottom, and the speed of the winding process [67]. The basalt filaments exhibit very good physical, chemical, and mechanical properties (Table 3.11), which are comparable to those of glass filaments. Furthermore, basalt fiber materials display high thermal stability [65].

3.3.7 *Metal Fibers*

Metal fibers include fiber produced from pure metals, alloys, and metalloids using various mechanical or thermal methods. Furthermore, synthetic fiber materials and metal fibers can be specifically metalized to improve properties. The production of the metals, the manufacture of alloys, and metallization will not be included in the following.

The mechanical wire or metal fiber production is based on the conventional wire drawing and nozzle drawing process, and the subsequent bundle wire drawing method. In wire drawing, the wire is subjected to multiple drawing stages reducing the diameter (depending on the material) from 8 to 2 mm. Steel wire is then annealed at 600–900 °C before being quenched. To achieve small filament diameters, conventionally drawn thin wires are embedded in a ductile and chemically more instable matrix (e.g. copper). This composite is subjected to a drawing process, tapering the total diameter. After chemically removing the matrix, metal multifilaments with very small diameters of 4–25 μm are left. By breaking these filaments into staple fibers of 50–150 mm in length and subsequent spinning, metal spun fiber yarns are produced. Other variations of mechanical metal fiber production are based on cutting, e.g. by so-called chasing or by vibrating cutter heads, resulting in fiber dimensions of 10–250 μm. However, these mechanical methods can only process certain, e.g. some iron- and copper-based materials [69].

Thermal methods (rapid solidification processes) are based on producing the metal fibers and wires by direction extrusion from the hotmelt phase, with a subsequent quenching process. The Taylor method achieves fiber diameters of 50 μm, melt spinning in rotating fluids creates fiber diameters of 50–500 μm. Especially brittle fibers like aluminum, copper, and zinc, or fibers made from aluminum and copper alloys, are produced by melt extraction [18, 69, 70].

Depending on the production method, fiber cross-sections and surfaces differ. Fibers produced by cutting methods have a rough surface and a lower fineness-related maximum tensile force, due to the notch effect. Crucial properties of metal fibers are summarized in Table 3.11. It is to be noted that the melting point, depending on the material, is ca. 1,400 °C for high-grade steel fibers [69].

The combination of certain metals exhibits the shape memory effect (SME) in their alloys. Metals with this property are referred to as *shape memory alloys* (SMA). These include nickel/titanium alloys, alloys made from copper, zinc, aluminum, or tin, as well as alloys made from copper, aluminum, and nickel. Regarding the memory effect, a distinction is made between thermal and mechanical shape memory. Applications of these materials in the composite field are promising [69, 71].

3.3.8 Natural Fibers

As stated in Chap. 2, a wide range of natural fibers is available. For composite applications, plant fibers from stems (e.g. flax) or leaves (e.g. sisal) are relevant. The structure of plant fibers is hierarchical and very complex. The geometrical and mechanical properties of plant fibers differ due to growth conditions and between species of plants. Significant properties are summarized in Table 3.12.

The technical flax fiber, which is a fiber bundle, is embedded in the bark layer between cambium and epidermis. The fiber bundle consists of several elementary fibers (length: 20–40 mm) connected by pectins, a vegetable glue. The bundle has a length between 200 and 1,400 mm, and a fineness of 1–4 tex. This technical fiber is extracted by means of a number of subsequent process steps removing vegetable

Table 3.12 Ranges of mechanical properties of fiber bundles (of technical fibers) of selected plant fibers [15]

Parameter	Flax	Sisal
Density (g/cm^3)	1.50	1.30 ... 1.45
Young's modulus (GPa)	80 ... 100	9.4 ... 38
Tensile strength (MPa)	1,100	568 ... 850
Elongation at break (%)	2.0 ... 3.0	2.2

glue, ligneous components, short fibers, and the marrow from inside the stem [1, 3, 5, 7, 57, 72].

Sisal fibers, whose elementary fibers have a length of 3 mm, form fiber strands in the long (1–2 m) leaves of the Agava sisilana. The fibers are extracted by squashing and scraping the leaf tissue [1, 3, 7, 57].

Apart from flax and sisal, other technical fibers like jute and hemp are used in fiber-reinforced composite material applications. They are applied with thermoplastic and thermoset matrices at low mechanical loads of the components. As plant fibers absorb moisture, measures have to be taken to ensure fiber-matrix adhesion and to limit the resulting tensions in the composite [15, 19, 73, 74].

3.3.9 Technological Properties of High-Performance Fibers

Due to the excellent mechanical properties of high-performance fibers, e.g. glass, carbon, and aramid, like high strength and high Young's modulus, knowledge regarding the significant technological properties are required for their optimized application. Apart from that, the brittleness of glass and carbon fiber materials, and the UV sensitivity of aramid fiber materials as well as the electrical conductivity of carbon have to be included in processing decisions. To do this, measures regarding UV protection of the aramid fiber materials have to be taken, and on the other hand regarding enclosures and seals to prevent short circuits in the control and drive units of the machinery used for the processing of carbon fibers. In addition, functional units based on electrical contacting have to be replaced with units relying on other effective mechanisms (e.g. optical warp-stop motions on weaving machines).

The further processing of the high-performance fiber materials, especially of carbon and glass filament yarns, causes damages to these materials to some degree and can (significantly) decrease the excellent strength values [16]. Extensive research proves the efforts to optimize a low-fiber-damage processing of such yarns on textile machines. Summarized remarks are included below regarding the friction and bending behavior in the processing of these brittle fiber materials, as they are distinguished by their very high strengths and rigidities.

The friction behavior between the filament yarns and the yarn-guiding elements is determined by yarn-related factors like yarn materials (fineness, twist, structure), degree of stretching, dynamometric properties, surface structure and preparation,

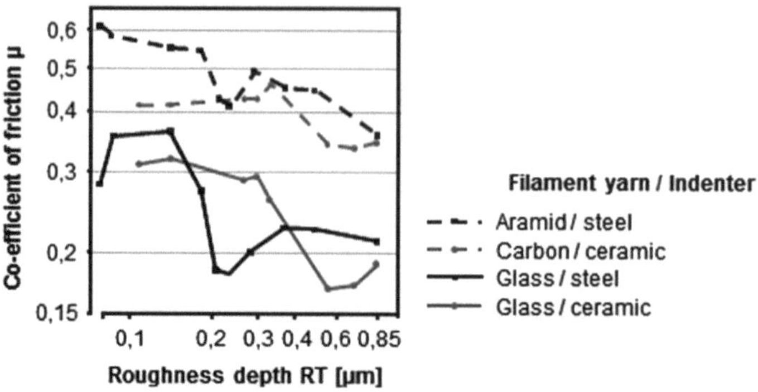

Fig. 3.28 Influence of roughness depth of the indenter surface on the friction coefficient in various yarn-indenter pairings (according to [28])

and by indenter-related factors like material, surface design, and diameter alike. In addition, process-related parameters like yarn speed, wrap-around angle, and atmospheric conditions also influence the friction behavior. The optimization of all possible parameters reduces friction and benefits processing with low yarn damages. Research has shown exemplarily that the roughness depth of the indenter, i.e. the yarn guides in the processing sequence, are the dominating influence on the friction coefficient (Fig. 3.28). Skillful material selection and combination, as well as surface treatment of the filament yarns contribute to a minimization of friction, and to a damage-free processing of the high-performance fibers [28].

The bending behavior of textile fiber materials is described by various textile-physical tests, and partly determines the processing properties of yarns and textile semi-finished products. Loop strength and filament knot tests are complex strain tests providing qualitative insights into the brittleness and processing behavior. The percental loop strength of glass fibers in relation to strength (1.2 %) is much lower than in conventional high-strength materials (e.g. high-strength PA: 70 %). The filament knot tests on 9 µm-diameter EC glass filaments reveal a specific radius of curvature of 23. The specific radius of curvature reflects the smallest possible radius of curvature at the moment of knot rupture, in relation to the filament diameter. With increasing filament diameters, the specific radius of curvature decreases (specific radius of curvature of ca. 18, at filament diameters of 13 µm) [17, 28]. This shows that loop strength and filament knot tests are generally feasible. However, the results show that filament and fiber yarns made from textile glass cannot be formed into knot for all practical intents and purposes [18]. The high brittleness of the carbon filament yarns is reflected by the greater specific radius of curvature, which, depending on fiber type, ranges from 25 to 60. The fracture behavior is assessed by means of scanning electron microscopic fracture tests and shows characteristic differences between axial and combined fractures [28].

The minimization of yarn strains during processing aims to prevent filament fractures. Guiding and deflection that do not damage the yarns can be achieved by reducing the number of guiding and deflection points, realizing smaller looping angles and higher curvature radii, and using suitable yarn preparations. Selecting the optimal processing speed is another crucial measure. Reducing the processing speed results in decreased tensile and friction loading, but an insufficient number of revolutions in the machine can cause an unstable running of the yarn and an uneven fabric appearance. Furthermore, the textile machines have to be adapted constructively to the technological conditions, which result in part from the technological and functional, but primarily from the mechanical properties of the fiber materials. Details are included in Chaps. 5, 6, 7, and 10.

3.3.10 Overview of Reinforcement Fibers

The properties of reinforcement fibers are defined considerably by their molecular and supramolecular structure, which depends on the molecular structure itself and on the production conditions (growth conditions in natural fibers).

Structural parameters like chemical bonding or orientation of the fiber construction elements are crucial for the formation of tensile strength and Young's modulus. For instance, three-dimensional covalent bonds, as they occur in glass fiber materials, are the cause of highly direction-independent tensile strengths. The two-dimensional bond in carbon fiber materials, and the distinct orientation of their macromolecules, as well as a degree of crystallinity of almost 100 % are the cause of their extremely high tensile strengths and Young's moduli, but also for their transverse compression sensitivity. In contrast to the brittle carbon and glass fiber materials, tenacious synthetic fiber materials like para-aramid or high-molecular polyethylenes have covalent bonds only in fiber direction, which makes them extremely durable under loads in fiber direction. However, they display a high degree of crystallinity and a strict spatial orientation of the macromolecules, giving them, too, a near 100 % crystalline lattice structure. Like aramid, plant and high-strength synthetic fibers are characterized by a 1d structure, but have a low degree of crystallinity and low orientation. An overview for the production and structure of high-performance fiber materials is given in Table 3.13.

Ceramic and basalt fibers are comparable to glass fibers, regarding their structure, although ceramics are a poly-crystalline material, instead of an amorphous one.

When comparing the individual fiber materials, not only the individual mechanical parameters should be considered, but also the anisotropy of the properties (except for glass fiber and ceramic materials). The lightweight construction potential of the textile reinforcement fibers, especially of high-performance fibers becomes apparent when the mechanical parameters are related to the density or fineness, creating comparable parameters like specific strength or specific Young's modulus.

In Tables 3.14, 3.15, 3.16 and 3.17, an overview is provided of selected reinforcement fibers and the significant, composite-relevant specific values. This

Table 3.13 Overview regarding the production and structure of high-performance fiber materials

Parameter	Glass	Carbon	Aramid
Raw materials	Quartz sand, additives	Precursor: PAN or pitch	Monomers: PPD and TDC, solvents
Process	Melt from mixture or pellets, melt spinning including sizing application GF filament	Stabilization Carbonization Graphitization Surface treatment CF filament	Polymerization Solving Extrusion Solvent spinning Stretching Avivage AR filament
Structure	Three-dimensional, isotropic Amorphous	Two-dimensional, layer-shaped, anisotropic Crystalline	One-dimensional, fiber-shaped/fibrillar, highly anisotropic Crystalline
Bonding	3d-covalent	2d-covalent van der Waals bond	1d-covalent van der Waals bond Hydrogen bonds
Crystallinity	None	~100 % (para-crystalline)	~100 % (para-crystalline)
Orientation	None	High	Very high

Table 3.14 Overview of the specific values of high-performance fiber materials: glass fibers

Parameter	Type E	Type AR	Type R/S
Filament diameter (μm)	3 ... 25	3 ... 25	3 ... 15
Density (g/cm^3)	2.52 ... 2.60	2.70	2.45 ... 2.55
Tensile strength, axial (MPa)	3,400 ... 3,700	3,000	4,300 ... 4,900
Spec. tensile strength, axial (MPa cm^3/g)	1,300 ... 1,470	1,110	1,690 ... 2,000
Young's modulus, axial (GPa)	72 ... 77	73	75 ... 88
Spec. Young's modulus (GPa cm^3/g)	27.7 ... 30.6	27.1	29.4 ... 35.9
Young's modulus, radial (GPa)	72 ... 77	73	75 ... 88
Elongation at break (%)	3.3 ... 4.8	4.3	4.2 ... 5.4
Thermal expansion coefficient			
Axial (10^{-6}/K)	5.0	Not available	4.0
Radial (10^{-6}/K)	5.0	Not available	4.0
Melting temperature (°C)	840 ... 1,500	1,300 ... 1,500	1,000
Decomposition temperature (°C)	Not available	Not available	Not available
Permanent temperature range (°C)	250 ... 350	400	300
Moisture absorption under standard atmosphere (%)	$\leq$0.1	$\leq$0.1	$\leq$0.1

overview has an orientation character and should help any pre-selection of the fiber materials.

Special relevance for high-performance products is attributed to high-performance filament yarns made from glass, carbon, and aramid fiber materials.

Table 3.15 Overview of the specific values of high-performance fiber materials: carbon fiber materials

Parameter	HT[i]	HST[i]	IM[i]
Filament diameter (μm)	7 ... 8	5 ... 7	5 ... 7
Density (g/cm³)	1.74 ... 1.80	1.78 ... 1.83	1.73 ... 1.80
Tensile strength, axial (MPa)	2,700 ... 3,750	3,900 ... 7,000	3,400 ... 5,900
Spec. tensile strength, axial (MPa cm³/g)	1,500 ... 2,160	2,140 ... 3,930	1,970 ... 3,410
Young's modulus, axial (GPa)	200 ... 250	230 ... 270	250 ... 400
Spec. Young's modulus (GPa cm³/g)	111.1 ... 144.0	126.0 ... 152.0	138.9 ... 231.2
Young's modulus, radial (GPa)	15	Not available	Not available
Elongation at break (%)	1.2 ... 1.6	1.7 ... 2.4	1.1 ... 1.93
Thermal expansion coefficient			
Axial (10^{-6}/K)	−0.1 ... −0.7	−1.0	−1.2
Radial (10^{-6}/K)	10	10	12
Melting temperature (°C)			
Decomposition temperature (°C)	3,650*	3,650*	3,650*
Permanent temperature range (°C)	400 ... 500		500
Moisture absorption under standard atmosphere (%)	≤0.1	≤0.1	≤0.1
Parameter	HM[i]	HMS[i]	MPP-HM[ii]
Filament diameter (μm)	4 ... 8	5 ... 7	Not available
Density (g/cm³)	1.76 ... 1.96	1.85	2.15
Tensile strength, axial (MPa)	1,750 ... 3,200	3,600	3,500
Spec. tensile strength, axial (MPa cm³/g)	890 ... 1,820	1,950	1,630
Young's modulus, axial (GPa)	300 ... 500	550	900
Spec. Young's modulus (GPa cm³/g)	153 ... 284	297.30	481.60
Young's modulus, radial (GPa)	5.7	Not available	Not available
Elongation at break (%)	0.35 ... 1.0	0.65	0.4
Thermal expansion coefficient			
Axial (10^{-6}/K)	−0.1	−1.3	Not available
Radial (10^{-6}/K)	Up to 30	Up to 30	Not available
Melting temperature (°C)			
Decomposition temperature (°C)	3,650*	3,650*	3,650*
Permanent temperature range (°C)	500 ... 600	500	600
Moisture absorption under standard atmosphere (%)	≤0.1	≤0.1	≤0.1

[i]PAN-based
[ii]MPP-based
*General specification for CF

Table 3.16 Overview of specific values of high-performance fiber materials: synthetic high-performance fiber materials

Parameter	Para-aramid N-type	HM-type	UHMWPE
Filament diameter (μm)	12	12	27 ... 38
Density (g/cm³)	1.39 ... 1.44	1.45 ... 1.47	0.97
Tensile strength, axial (MPa)	2,760 ... 3,000	2,800 ... 3,620	2,800 ... 3,100
Spec. tensile strength, axial (MPa cm³/g)	1,920 ... 2,160	1,900 ... 2,500	2,890 ... 3,200
Young's modulus, axial (GPa)	58 ... 80	120 ... 186	87 ... 170
Spec. Young's modulus (GPa cm³/g)	40 ... 58	82 ... 128	80 ... 175
Young's modulus, radial (GPa)	Not available	Not available	Not available
Elongation at break (%)	3.3 ... 4.4	1.9 ... 2.9	2.7 ... 3.5
Thermal expansion coefficient			
Axial (10^{-6}/K)	−2.0 ... −6.6	−2.0 ... −6.6	Not available
Radial (10^{-6}/K)	40	52	Not available
Melting temperature (°C)	>500	>500	140
Decomposition temperature (°C)	~550	~550	Not available
Permanent temperature range (°C)	180	180 ... 250	60 ... 121
Moisture absorption under standard atmosphere (%)	~7.0	~3.5	Not available

Table 3.17 Overview of specific values of high-performance fiber materials: ceramic and metal fiber materials

Parameter	Ceramic Al_2O_3	SiC	Metal Steel
Filament diameter (μm)	15 ... 20	10 ... 15	1 ... 100
Density (g/cm³)	2.7 ... 4.1	2.35 ... 3.14	7.8 ... 7.9
Tensile strength, axial (MPa)	1,700 ... 2,930	1,500 ... 3,600	200 ... 1,600
Spec. tensile strength, axial (MPa cm³/g)	410 ... 1,090	480 ... 1,530	30 ... 210
Young's modulus, axial (GPa)	150 ... 380	170 ... 420	210
Spec. Young's modulus (GPa cm³/g)	14 ... 93	54 ... 179	25.5 ... 27
Young's modulus, radial (GPa)	Not available	Not available	Not available
Elongation at break (%)	0.4 ... 1.1	0.4 ... 1.1	1.0 ... 2.0
Thermal expansion coefficient			
Axial (10^{-6}/K)	6.5 ... 8.9	3.1	11 ... 25
Radial (10^{-6}/K)	6.5 ... 8.9	Not available	Not available
Melting temperature (°C)	1,815 (2,045)	1,815[a]	1,400
Decomposition temperature (°C)	Not available	Not available	Not available
Permanent temperature range (°C)	Up to 1,430	Up to 1,430	Up to 1,370
Moisture absorption under standard atmosphere (%)	0	0	0

[a]General specification for ceramic fiber

Due to their characteristic molecular and supramolecular structure, these high-performance materials are distinguished by their extremely high tensile strengths and Young's moduli. The specific Young's modulus of glass fiber materials is in the range of 27–36 GPa cm^3/g, and reaches 110–480 GPa cm^3/g for carbon fiber materials. The specific tensile strength in axial direction reaches 1.1–2.0 GPa cm^3/g for glass fiber materials, and 1.5–3.9 GPa cm^3/g for carbon fiber materials. In comparison, steel fibers have a specific Young's modulus of ca. 26 GPa cm^3/g, and a specific tensile strengths of 0.03–0.2 GPa cm^3/g.

The lightweight construction potential becomes most apparent for fiber-reinforced plastic component applications in which large masses are moved or quickly accelerated, e.g. in aeronautics and aerospace industries as well as in mechanical and plant engineering and construction. On the other hands, textile reinforcements made from carbon and glass are used in textile-reinforced concrete. The use of high-performance fiber materials saves resources and materials during production as well as energy during active use. Apart from the abovementioned high-performance fiber materials, fiber materials made from synthetic materials or natural fibers can be used to reinforce lightweight construction components, for instance for natural-fiber-reinforced plastics and as reinforcement of wooden constructions.

The development of new and improved textile semi-finished products and their more widespread application in lightweight construction require the provision of efficient high-quality reinforcement fibers at sufficient amount. From this, the following future development trends can be derived:

- eco-friendly technology and plant development to optimize the individual production methods, aiming to increase energy, material, and cost efficiency
- development of requirement-adapted and cost-effective carbon fiber material types with small numbers of defects (increase of strength and Young's modulus), and
- development/optimization of cost-effective synthetic fiber materials for use as reinforcement or matrix fibers

In the future, systematic lightweight construction in multi-material design with high material and energy efficiency will be focused on. High-performance fibers like glass or carbon, and synthetic fiber materials like para-aramid and UHMWPE contribute greatly to this development. In particular, the anisotropy of the textile reinforcement materials and their advantageous mass-performance ratio, as well as the possibility to design complex and heavy-duty components allow the flexible adjustability of the material structure, and a targeted adjustability of tailor-made, anisotropic component properties.

3.4 Matrix Fibers from Thermoplastic Polymers

3.4.1 Tasks and General Characteristics of Matrix Fibers

The matrix is the second component of any fiber-reinforced composite material. The materials surrounding the reinforcement fibers are supposed to realize an effective composite and assume the following tasks [13]:

- fixation of the reinforcement fibers in the desired geometric arrangement, and protection of the external shape
- introduction of forces into the reinforcement fibers, and force distribution
- support of the reinforcement fiber under compression loads, and
- protection of the reinforcement fiber against external influences

Polymer matrix materials are high-molecular organic compounds created by different bonding mechanisms (see Sect. 3.2.3.1). In general, all known thermosets and thermoplastics are suitable matrix systems for fiber-reinforced plastics. In comparison to thermosets, thermoplastics have advantages like higher ductility, higher elongation at break, unlimited storage time, and better ecological processability, making them a more attractive option. To improve impregnation by reducing the flow paths, thermoplastic materials are available as fibers or filament, whose specific use opens new application possibilities.

3.4.2 Major Matrix Fibers

The stress-strain behavior of the matrix, and the adhesion between matrix and reinforcement fiber, with due regard to the fiber surface modifications, determine composite properties to a large extent. Thermoplastic fiber materials contain amorphous and crystalline section within the supramolecular structure. A high degree of crystallinity increases properties like melting range, tensile strength, Young's modulus, toughness and solvent resistance. Impact resistance and resistance to tear formation, however, decrease. Thermoplastic materials exhibit a visco-elastic behavior, i.e. the specific values have to be regarded in relation to temperature, loading speed and loading time. One special trait of partially crystalline polymers is the viscoplastic range occurring during heating, followed by the melting range at continued heating. The temperature dependence of the viscosity of various polymer melts is given in Fig. 3.29. In the molten state, thermoplastics are moldable and formable, and can be transferred from the respective fiber form into a space-filling form, which secures the composite [15].

Apart from the abovementioned aspects, surface energy has to be taken into account in the selection of the thermoplastic fibers to be used as matrix fibers [75]. Table 3.18 gives an overview of the synthetic fiber materials utilizable as matrix fibers, including relevant properties.

Fig. 3.29 Viscosity behavior in thermoplastic polymers in comparison to a thermoset (according to [15]

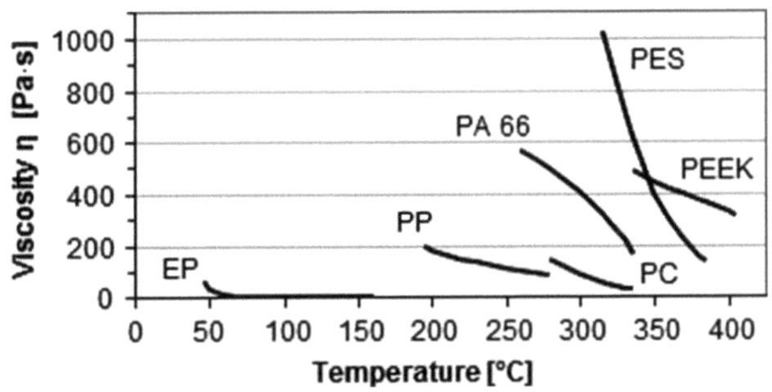

Table 3.18 Standard values of selected thermoplastic fiber materials (according to [27, 36, 57, 61, 75])

Parameter	PP	PA 6.6	PES	PPS
Density (g/cm^3)	0.91	1.14	1.39	1.24
Young's modulus (GPa)	0.5 ... 5.0	1.1 ... 3.0	3 ... 15	2.48 ... 6.2
Tensile strength (MPa)	210 ... 660	360 ... 720	350 ... 830	322 ... 496
Elongation at break (%)	15 ... 90	26 ... 75	30 ... 44	20 ... 30
Melting temperature (°C)	175	255 ... 260	250 ... 260	285
Softening range (°C)	150 ... 155	220 ... 235	230 ... 250	Not available
Permanent temperature range (°C)	Not available	75 ... 85	140 ... 160	190
Freezing temperature (°C)	−12 ... −20	45 ... 65	70 ... 80	Not available
Decomposition temperature (°C)	328 ... 410	310 ... 380	283 ... 306	Not available
Surface energy (10^{-3} j/m^2)	26	37	Not available	

Parameter	PEI	PEEK	PTFE
Density (g/cm^3)	1.28	1.3	2.1 ... 2.3
Young's modulus (GPa)	Not available	3.8	Not available
Tensile strength (MPa)	243 ... 346	80 ... 90	176 ... 308
Elongation at break (%)	38 ... 80	16 ... 80	19
Melting temperature (°C)	225	355	Not available
Softening range (°C)	Not available	Not available	327
Permanent temperature range (°C)	170	220 ... 260	280
Freezing temperature (°C)	Not available	144	Not available
Decomposition temperature (°C)	Not available	Not available	Not available
Surface energy (10^{-3} j/m^2)	Not available	Not available	22

Polypropylene (*PP*) has been established as matrix system for GMTs over the past years, and is also an important bulk plastic material of Europe-wide importance. The fiber material polypropylene makes up 25 % of the European synthetic fiber production, with the numbers steadily rising. Due to the increased use of fiber-form polypropylene as matrix material, the application potential can be improved significantly.

Polyamides (*PA*), in comparison to PP, are distinguished by their higher strength, rigidity, toughness, form stability under thermal influences, and higher abrasion resistance. However, they display higher viscosity at higher temperatures. Depending on the requirements, matrix materials produced from PA 6, PA 6.6, PA 11, or PA 12 can be used. The modification of the molecular structure, e.g. by inserting aromatic groups, improves properties and extends the application of PA fibers as matrix systems in FRPCs [15, 57].

Polyester (*PES*) possesses interesting potential for automobile construction applications due to the higher permanent temperature resistance in comparison to standard polymers and polypropylene. So far, no noteworthy applications in practice are known [15, 57].

Newly developed polymers with improved properties, which are distinguished by the integration of sulfur in the backbone in the form of sulfide or sulfone, can also be used as matrix fibers in FRPCs. Materials like *polyethersulfone* (*PSU*) *and polyphenylene sulfide* (*PPS*) are characterized by high form stability under heat (distinguished by the shear modulus in dependence of the temperature), and high flame retardance, especially in comparison to abovementioned polymers. Their UV resistance, however, is low [15, 57, 61].

Because of the imide and aromatic groups, *polyetherimide* (*PEI*) is characterized by its high long-term heat resistance in comparison to PA and PES, high flame retardance and minimal flue gas development . Despite a high chemical resistance, PEI is soluble in different solvents (e.g. Methyl ethyl ketone) [15, 57, 61].

Poly ether ether ketone (PEEK) is distinguished by apart from their excellent mechanical and good electrical properties, by a high continuous operating temperature, high flame retardance at low gas emission as well as a good friction and abrasion resistance even at high temperatures. These properties are the benchmark for the use of thermoplastic polymers in the fiber composite material [15, 57, 61].

Other thermoplastic fiber materials can be used as matrix systems in composite materials, for instance fiber materials based on *polybenzimidazolene* (PBI_M) and *polybenzoxazolene* (PBO_M).

Polytetrafluorethylene fibers (*PTFE*) are either spun directly, using suspension or matrix wet spinning, or indirectly by fibrillation of a PTFE foil. High-quality PTFE fibers are transferred into a gel-like state at 327 °C, which enables them to work as a matrix system with effective properties [57, 61].

3.5 Requirement-Adapted Further Processing in the Textile Process Chain

3.5.1 Hybrid Filament Yarns

Depending on the respective application, the packaging of textile fiber materials differs. Usually, the reinforcement fibers are used as continuous fibers/filaments in a yarn or roving/heavy tow. To improve, for instance, the mechanical properties of the semi-finished product or component, filaments of different fiber materials can be combined. This results in the combination of the advantageous properties of different reinforcement fibers. Thus, for example, the energy-absorbing properties of aramid fibers can be specifically combined with the high-modulus characteristics of carbon fibers, resulting in tailor-made component properties. Additionally, fiber material combinations allow the homogeneous blending of reinforcement and matrix fibers at adjustable ratios in the filament yarn. Such hybrid yarns can be processed by textile technology into 2D and 3D geometries and structures. By local partial melting, it is possible to fix the flexible textile semi-finished product. Furthermore, the use of hybrid filament yarns consisting of reinforcement and matrix fibers is an advantage for the composite properties, as the short flow paths of the molten matrix allow the realization of an even and nearly complete impregnation of the reinforcement filaments. Details regarding hybrid yarns are given in Sect. 4.3.

3.5.2 Finishing of Fibers

To improve the properties of textile fiber materials corresponding to the requirements, or to generate new, additional functionalities in the fiber materials, textile materials are subjected to finishing processes. These finishings have various purposes and the types can be of mechanical, physical, or chemical nature. Conventional textile finishing include roughening for a bulky surface, mercerizing to improve the properties of cotton (e.g. dyeing behavior), the anti-felt and moth-resistance finishing to improve the performance of wool, dyeing/printing, impregnation, and laminating. The finishing can be performed on the fiber or the filament, the yarn, the textile fabric, or the assembled product, depending on the respective method.

In technical applications of textile fibers, the focus is placed on the creation of extended functionalities by fiber finishing. This goes far beyond the requirement of processability without fiber damage, which is realized by spinning preparation and avivage. The additional surface modification (see Sect. 3.2.4) of reinforcement fibers in the form of fibers, yarns, or fabrics, for instance, creates a tailor-made fiber-matrix adhesion in fiber-reinforced composite materials and realizes highly dense structures in membranes. A pioneering approach is the realization of sensor and actuator networks by permanent or temporary functional integration into the fiber materials or their surface (e.g. integration of additional function in the sizing by means of carbon nanotubes), and by the increase of fracture energy by nano-structured surfaces via the use of carbon nanotubes. Details regarding the finishing of textile fiber materials and the structures produced from them are given in Chap. 13.

References

1. BERGER, W.; FISCHER, P.; MALLY, A.: Struktur der textilen Faserstoffe. In: BOBETH, W. (Hrsg.): *Textile Faserstoffe: Beschaffenheit und Eigenschaften.* 1st Edition. Berlin, Heidelberg : Springer Verlag, 1993
2. FALKAI, v. B.: Fasern, Herstellungsverfahren. In: FALKAI, v. B. (Hrsg.): *Synthesefasern: Grundlagen, Technologie, Verarbeitung und Anwendung.* Weinheim : Verlag Chemie, 1981
3. WULFHORST, B.: *Textile Fertigungsverfahren: Eine Einführung.* München, Wien : Carl Hanser Verlag, 1989
4. KUMAR, S.; AGRAWAL, A.K.: How to produce PET POY at higher speeds? In: *Chemical Fibers International* (2002), No. 6, pp. 418
5. BOBETH, W.: Fasergeometrie. In: BOBETH, W. (Hrsg.): *Textile Faserstoffe: Beschaffenheit und Eigenschaften.* 1st Edition. Berlin, Heidelberg : Springer Verlag, 1993
6. BONART, R.; ORTH, H.: Struktur. In: FALKAI, v. B. (Hrsg.): *Synthesefasern: Grundlagen, Technologie, Verarbeitung und Anwendung.* 1st Edition. Weinheim : Verlag Chemie, 1981
7. HAUDEK, H.W.; VITI, E.: *Textilfasern.* Wien-Perchtoldsdorf : Verlag Johann L. Bondi Sohn, 1980
8. GALL, H.: Texturierung synthetischer Filamentgarne. In: FALKAI, v. B. (Hrsg.): *Synthesefasern: Grundlagen, Technologie, Verarbeitung und Anwendung.* 1st Edition. Weinheim : Verlag Chemie, 1981
9. MIKUT, I.: *Gestaltungsmerkmale textiler Faserstoffe-Kennzeichnung, Einflussgrößen, Wirkung.* Dresden, Technische Universität Dresden, Dissertation, 1940
10. BOBETH, W.; JACOBASCH, H.-J.: Topographie und Oberflächeneigenschaften. In: BOBETH, W. (Hrsg.): *Textile Faserstoffe: Beschaffenheit und Eigenschaften.* 1st Edition. Berlin, Heidelberg: Springer Verlag, 1993
11. BAUR, E.; BRINKMANN, S.; OSSWALD, T. A.; SCHMACHTENBERG, E.: *Saechtling Kunststoff Taschenbuch.* 30th Edition. München : Carl Hanser Verlag, 2007
12. GEIL, P. H.; BEAR, E.; WADA, Y.: *The Solid State of Polymers.* New York : Marcel Dekker, 1974
13. EHRENSTEIN, G. W.: *Faserverbund-Kunststoffe: Werkstoffe, Verarbeitung, Eigenschaften.* München, Wien : Carl Hanser Verlag, 2006
14. LANGE, P. J.; AKKER, P. G.; MÄDER, E.; GAO, S.-L.; PRASITHPHOL, W.; YOUNG, R. J.: Controlled interfacial adhesion of Twaron aramid fibres in composites by the finish formulation. In: *Composites Science and Technology* 67 (2007), pp. 2027–2035
15. FLEMMING, M.; ZIEGMANN, G.; ROTH, S.: *Faserverbundbauweisen: Fasern und Matrices.* Berlin, Heidelberg : Springer Verlag, 1995
16. MICHAELI, W.; WEGENER, M.: *Einführung in die Technologie der Faserverbundwerkstoffe.* München, Wien : Carl Hanser Verlag, 1989
17. WULFHORST, B.; KALDENHOF, R.; HÖRSTING, K.: Faserstofftabelle nach P.-A. Koch: Glasfasern. In: *Technische Textilien* 36 (1993), pp. T68–T86
18. MANSMANN, M.; KLINGHOLZ, R.; WIEDEMANN, K.; KURT, A. F.; GÖLDEN, D.; OVERHOFF, D.: Anorganische Fasern. In: FALKAI, v. B. (Hrsg.): *Synthesefasern: Grundlagen, Technologie, Verarbeitung und Anwendung.* 1st Edition. Weinheim : Verlag Chemie, 1981
19. NEITZEL, M.; MITSCHANG, P.: *Handbuch der Verbundwerkstoffe.* München, Wien : Carl Hanser Verlag, 2004
20. KIENER, R.: Glas-Stapelfaserprodukte: Verfahrenstechniken und Anwendungsgebiete. In: *Chemiefasern/Textilindustrie* 37/98 (1987), pp. T11–T14
21. KIENER, R.: Glas-Stapelfasergewebe - Herstellung der Garne und Verarbeitung auf Greiferwebmaschinen. In: LOY, W. (Hrsg.): *Taschenbuch für die Textil-Industrie.* 1st Edition. Berlin : Fachverlag Schiele Schön GmbH, 1990
22. CHERIF, Ch.; RÖDEL, H.; HOFFMANNN, G.; DIESTEL, O.; HERZBERG, C.; PAUL, Ch.; SCHULZ, Ch.; GROSSMANN, K.; MÜHL, A.; MÄDER, E.; BRÜNIG, H.: Textile

Verarbeitungstechnologien für hybridgarnbasierte komplexe Preformstrukturen / Textile manufacturing technologies for hybrid based complex preform structures. In: *Kunststofftechnik / Journal of Plastics Technology* 5 (2009), No. 2, pp. 103–129
23. MÄDER, E.: *Grenzflächen, Grenzschichten und mechanische Eigenschaften faserverstärkter Polymerwerkstoffe*. Dresden, Technische Universität Dresden, Habilitation, 2001
24. SCHEFFLER, C.; GAO, S.-L.; PLONKA, R.; MÄDER, E.; HEMPEL, S.; BUTLER, M.: Interphase modification of alkali-resistant glass fibres and carbon fibres for textile reinforced concrete II:Water adsorption and composite interphases. In: *Composites Science and Technology* 69 (2009), pp. 905–912
25. KLEINHOLZ, R.: Neue Erkenntnisse bei Textilglasfasern zum Verstärken von Kunststoffen. In: *Proceedings. 22. Internationale Chemiefasertagung Dornbirn*. Dornbirn, Österreich, 1983
26. SCHMIDT, K. A.: *Textilglas für die Kunststoffverstärkung*. 2nd Edition. Speyer : Zechner Hüthig Verlag GmbH, 1972
27. BOBETH, W.; FAULSTICH, H.; MALLY, A.: Mechanische Eigenschaften. In: BOBETH, W. (Hrsg.): *Textile Faserstoffe: Beschaffenheit und Eigenschaften*. Berlin, Heidelberg : Springer Verlag, 1993
28. BECKER, G.: *Struktur und mechanisch-technologische Eigenschaften von Filamentgarnen für Faserverbundwerkstoffe*. Aachen, RWTH Aachen, Dissertation, 1991
29. KREVELEN, D. W.: Verbundwerkstoffe. In: *Proceedings. 22. Internationale Chemiefasertagung Dornbirn*. Dornbirn, Österreich, 1983
30. BUTLER, M.; HEMPEL, S.; MECHTCHERINE, V.: Zeitliche Entwicklung des Verbundes von AR-Glas- und Kohlenstofffaser-Multifilamentgarnen in zementgebundenen Matrices. In: CURBACH, F. (Hrsg.): *Textilbeton - Theorie und Praxis: Tagungsband zum 4. Kolloquium zu Textilbewehrten Tragwerken (CTRS4) und zur 1. Anwendertagung, Dresden, 3.-5.6.2009*. Dresden : Technische Universität Dresden, 2009, pp. 213–226
31. ABDKADER, A.: *Mechanische Eigenschaften unter Berücksichtigung des Verbundverhaltens zwischen den Filamenten und Dauerhaftigkeit*. Dresden, Technische Universität Dresden, Fakultät Maschinenwesen, Dissertation, 2004
32. SEIDEL, A.; YOUNES, A.; ENGLER, Th.; CHERIF, Ch.: On the mechanical behavior of carbon and glass fiber filament yarns under long-term load. In: *Proceedings. ACI 2010 Spring Convention*. Chicago (Illinois), USA, 2010
33. MEYER, O.: Glasfasern für das Verstärken von Kunststoffen. In: *Kunststoff-Rundschau* 2 (1955), No. 4
34. SCHEFFLER, C.: *Zur Beurteilung von AR-Glasfasern in alkalischer Umgebung*. Dresden, Technische Universität Dresden, Fakultät Maschinenwesen, Dissertation, 2009
35. BUTLER, M.: *Zur Dauerhaftigkeit von Verbundwerkstoffen aus zementgebundenen Matrices und alkaliresistenten Glasfaser-Multifilamentgarnen*. Dresden, Technische Universität Dresden, Dissertation, 2009
36. BOBETH, W.; MALLY, A.: Thermisches Verhalten. In: BOBETH, W. (Hrsg.): *Textile Faserstoffe: Beschaffenheit und Eigenschaften*. 1st Edition. Berlin, Heidelberg : Springer Verlag, 1993
37. *Basalt, Fasern und Gewebe*. http://www.basfiber.com/en/basfiber.shtml (10.02.2011)
38. NOELLE, G.: *Technik der Glasherstellung*. 3rd Edition. Stuttgart : Deutscher Verlag für Grundstoffindustrie, 1997
39. KÖCKRITZ, U.: *In-situ Polymerbeschichtung zur Strukturstabilisierung offener nähgewirkter Gelege*. Dresden, Technische Universität Dresden, Fakultät Maschinenwesen, Dissertation, 2007
40. YOUNES, A.; SEIDEL, A.; ENGLER, Th.; CHERIF, Ch.: Effects of high temperature and long term stress on the material behaviour of high performance fibres for composites. In: *World Journal of Engineering* 7 (2010), No. 4, pp. 309–315
41. WITTEN, E.; SCHUSTER, A.: *Composites-Marktbericht: Marktentwicklungen, Herausforderungen und Chancen*. http://www.avk-tv.de/news.php?id=134 (12.05.2011). Version: 2010

42. FITZER, E.; HEINE, M.; JACOBSEN, G.: Kohlenstoffasern. In: HESSLER, H. (Hrsg.): *Verstärkte Kunststoffe in der Luft- und Raumfahrt.* Stuttgart : Verlag W. Kohlhammer, 1986
43. WULFHORST, B.; BECKER, G.: Faserstofftabelle nach P.-A. Koch: Carbonfasern. In: *Chemiefasern/Textilindustrie* 39/91, pp. 1277–1284
44. GLAWION, E.: Behutsame Herstellung von Carbonfasern. In: *Technische Textilien* 53 (2010), No. 5, pp. 182
45. KIRK, R. E.; OTHMER, D. F.; GRAYSON, M.: *Encyclopedia of chemical technology - 5 : Carbon and graphite fibers to chlorocarbons and chlorohydrocarbons-C1.* 4th Edition. John Wiley Sons, 1993
46. EHRENSTEIN, G. W.: *Faserverbundkunststoffe: Werkstoffe-Verarbeitung-Eigenschaften.* München, Wien : Carl Hanser Verlag, 1992
47. FITZER, E.; WEISS, R.: *Oberflächenbehandlung von Kohlenstoffasern; Verarbeiten und Anwenden kohlenstoffaserverstärkter Kunststoffe.* Düsseldorf : VDI-Verlag GmbH, 1989
48. EDIE, D.; FITZER, E.; RHEE, B.: *Present and Future Reinforcing Fibres - From Solid to Hollow.* Wiesbaden : Verbundwerk, 1991
49. PAPAKONSTATINOU, C.; BALAGURU, P.; LYON, R.: Comparative study of high temperature composites. In: *Composites Part B* 32 (2001), pp. 637–649
50. LONG, G. T.: Influence of Boron Treatment on Oxidation of Carbon Fibre in Air. In: *Journal of Applied Polymer Science* 59 (1996), pp. 915–921
51. SAUDER, C.; LAMON, J.; PAILLER, R.: Thermomechanical properties of carbon fibres at high temperatures (up to 2000 °C). In: *Composites Science and Technology* 62 (2002), pp. 499–504
52. SUMIDA, A.; FUJISAKI, T.; WATANABE, K.; KATO, T.: Heat resistance of Continuous Fibre Reinforced Plastic Rods. In: BURGONE, C. (Hrsg.): 5th Symposium on Fibrereinforced-Plastic Reinforcement of Concrete Structures (FRPRCS-5). London : Elsevier, 2001, pp. 791–802
53. YOUNES, A.; SEIDEL, A.; ENGLER, T.; CHERIF, Ch.: Materialverhalten von AR-Glas- und Carbonfilamentgarnen unter Dauerlast- sowie unter Hochtemperatureinwirkung. In: CURBACH, M. (Hrsg.); JESSE, F. (Hrsg.): Textilbeton - Theorie und Praxis: Tagungsband zum 4. Kolloquium zu Textilbewehrten Tragwerken (CTRS4) und zur 1. Anwendertagung, Dresden, 3.-5.6.2009. Dresden : Technische Universität Dresden, 2009, pp. 1–16
54. SCHNEIDER, M.: Carbon fibre products for mechanical engineering applications - Kohlenstofffaser-Produkte für den Maschinenbau. In: Proceedings. 2. Aachen-Dresden International Textile Conference. Dresden, Germany, 2008
55. WARNECKE, M.; WILMS, Chr.; SEIDE, G.; GRIES, Th.: Der Carbonfasermarkt - ein aktueller Überblick. In: Technische Textilien 53 (2010), No. 6, pp. 216
56. WULFHORST, B.; BÜSGEN, A.: Faserstofftabelle nach P.-A. Koch: Aramidfasern. In: Chemiefasern/Textilindustrie 39/91 (1989), pp. 1263–1270
57. FOURNÉ, F.: Synthetische Fasern: Herstellung, Maschinen und Apparate, Eigenschaften; Handbuch für Anlagenplanung, Maschinenkonstruktion und Betrieb. München, Wien : Carl Hanser Verlag, 1995
58. ISTEL, E.; PELOUSEK, H.; OERTEL, H.; MOORWESSEL, D.; FALKAI, B. et al.: Organische Fasern. In: FALKAI, B. (Hrsg.): Synthesefasern: Grundlagen, Technologie, Verarbeitung und Anwendung. 1st Edition. Weinheim : Verlag Chemie, 1981
59. MORGAN, R. J.; ALLRED, E. A.: Aramid Fiber Composites. In: LEE, S. M. (Hrsg.): Handbook of Composites Reinforcements. Weinheim : VCH-Verlagsgesellschaft mbH, 1993
60. ANONYMOUS: Prospekt Aramid Products: Twaron® Spinnfasern und deren Anwendungen. Wuppertal, 1997
61. LOY, W.: Chemiefasern für technische Textilprodukte. Frankfurt am Main : Deutscher Fachverlag, 2001
62. CLAUSS, B.: Fasern und Preformtechniken zur Herstellung keramischer Verbundwerkstoffe. In: Proceedings. DGM-Fortbildungsseminar"Keramische Verbundwerkstoffe". Würzburg, Germany, 2004

63. XIE, E.; LI, Z.: Application Prospect of Basalt Fiber. In: Fibercomposites (2003), No. 3, pp. 17–20
64. LIU, J.: Untersuchung von Verbundwerkstoffen mit Basalt- und PBO-Faser-Verstärkung. Dresden, Technische Universität Dresden, Dissertation, 2007
65. MILITKÝ, J.; KOVACIC, V.; RUNEROVÁ, J.: Influence of thermal treatment on tensile failure of basalt fibers. In: Engineering Fracture Mechanics (2002), pp. 1025–1033
66. MILITKÝ, J.; KOVACIC, V.: Ultimate mechanical properties of basalt filaments. In: Textile Research Journal 66 (1996), pp. 225–229
67. SARAVANAN, D.: Spinning the rocks - basalt fibres. In: IE (I) Journal-TX 86 (2006), pp. 39–45
68. Suter Kunststoffe AG: Fasern und Gewebe - Basalt. http://www.swiss-composite.ch (01.06.2011)
69. MAC, T.; HOUIS, S.; GRIES, T.: Faserstofftabelle nach P.-A. Koch: Metallfasern. In: Technische Textilien 47 (2004), No. 1, pp. 17–32
70. HOFF, H. G.; MÄGEL, M.; OFFERMANN, P.: Textile Verarbeitung von Stahlfasern und Stahlfäden. In: Technische Textilien 46 (2003), No. 3, pp. 219–221
71. PAUL, C.: Funktionalisierung von duroplastischen Faserverbundwerkstoffen durch Hybridgarne. Dresden, Technische Universität Dresden, Fakultät Maschinenwesen, Dissertation, 2010
72. SATLOW, G.; ZAREMBA, S.; WULFHORST, B.: Faserstofftabelle nach P.-A. Koch: Flachs sowie andere Bast- und Hartfasern. In: Chemiefasern/Textilindustie 44/96 (1994), pp. 765–785
73. PHILIPP, K.: Naturfaserverbundwerkstoffe im automobilen Innenraum. In: Technische Textilien 47 (2004), No. 1
74. KARUS, M.; KAUP, M.: Naturfasereinsatz in der europäischen Automobilindustrie. In: Technische Textilien 44 (2001), No. 4
75. HORNBOGEN, E.: Werkstoffe: Aufbau und Eigenschaften von Keramik-, Metall-, Polymer- und Verbundwerkstoffen. 6th Edition. Berlin, Heidelberg : Springer Verlag, 1994

Additional Literature

EHRENSTEIN G. W.: Polymerwerkstoffe – Struktur und mechanisches Verhalten, 1st Edition Carl Hanser Verlag München Wien (1978)

FLEMMING M.; MUSCH G.: Werkstoffkennwerte, ETH Zürich, Institut für Konstruktion und Bauwesen

HORNBOGEN E.: Werkstoffe – Aufbau und Eigenschaften von Keramik-, Metall-, Polymer- und Verbundwerkstoffen, 6th Edition, Springer Verlag Berlin Heidelberg (1994)

IAFL Publications. "Textile Fibers: Developments an Innovations". New Delhi, India. 2000, pp. 642-677

International Fiber Science and Technology Series /15, Handbook of Fiber Chemistry (1998), Lewi, Pearce (ed), 2. ed. Macel Dekker, Inc. New York (ISBN 0-8247-9471-0)

McINTYRE J. E.; RICHARDS A. F.; EAST A. J.; COX R.; MATHER R. R.: Synthetic fibres: nylon, polyester, acrylic, polyolefin. Meclntyre (edited) 1 ed. Woodhead Publishing Ltd and CRC Press LLC (ISBN 1 85573 5881) (2005)

MOTRON W. E.; HEARLE J. W. S.: Physical Properties of Textile Fibres, 3rd. edition by The Textile Institute (ISBN 1 870812 41 7) (1993)

NIEDERSTADT G.: Ökonomischer und Ökologischer Leichtbau mit faserverstärkten Polymeren, Expert-Verlag Renningen-Malmsheim (1997)

Progress in Textiles: Science & Technology, Volume 2 Textile Fibres: Development and Innovations. (2000) Kothari (Editor), 1. Ed., IAFL Publications New Delhi (ISBN 81-901033-1-8)

ZERONIAN S. H.; XIE Q.; BUSCHLE-DILLER G.; HOLMES S.; IINGLESBY M. K.: Relationships between the Mechanical Properties of Synthetic Fibers. In: Journal of Textile Institute, 85 No. 3, pp. 293-300. (1994)

Chapter 4
Yarn Constructions and Yarn Formation Techniques

Beata Lehmann and Claudia Herzberg*

Yarns are an important basic element for both the production and the assembly of textile reinforcement structures. They consist of either a 100 % reinforcement fibers or a blend of reinforcement and matrix fibers. They are made from filaments and/or staple fibers by means of different technologies, which allow the customization of structure and properties of the yarns according to the respective functional requirements. These chapters give an overview of the yarns currently used in lightweight construction and shows that their design has considerably influence of further processing and on the characteristics of composite materials. During textile processing, the yarns have to be processed easily at high speeds and must be formable force- or form-fit. Defined, anisotropic characteristics are achieved by the preferred orientation of the fibers in the yarn and the 2D or 3D yarn orientation during the production of textile semi-finished products. In the composite material itself, the yarn structures offer mechanic fixations.

4.1 Introduction and Overview

4.1.1 Introduction

The production of textile semi-finished products (see Chaps. 5–8) is based on the processing of single and/or multifold yarns (see Sects. 4.2 and 4.3) into textile

*Author of Sect. 4.5

B. Lehmann
Institute of Manufacturing Technology, TU Dresden, Dresden, Germany
e-mail: beata.lehmann@iws.fraunhofer.de

C. Herzberg (✉)
Formerly: Institute of Textile Machinery and High Performance Material Technology, TU Dresden, Dresden, Germany
e-mail: i.textilmaschinen@tu-dresden.de

C. Cherif (ed.), *Textile Materials for Lightweight Constructions*,
DOI 10.1007/978-3-662-46341-3_4

fabrics, and their subsequent local reinforcement and/or insert and functional integration (see Chap. 10) or their use in the assembly of the textile fabrics into complex preforms (see Chap. 12). The latter includes the use of special yarn constructions as sewing yarn (see Sect. 4.5). Furthermore, yarns can be processed directly by winding, pultrusion (see Chap. 11) or by nonwoven reinforcement with yarns (see Chap. 9).

The mechanical properties (e.g. strength, elasticity, stiffness) of the, preferably non-rigid, easily drapeable, and homogenous yarns are highly anisotropic due to the preferentially stretched position of the fibers in the yarn, and the yarn orientation at a high length/diameter ratio. The course of the yarn within the textile semi-finished product has to be adapted to the expected loads to fully exploit the potential offered by their properties. In general, textile semi-finished products made from yarns show better mechanical properties compared with semi-finished products made from non-woven materials (e.g. higher strengths and stiffnesses, higher energy absorption capacity).

Depending on the requirements (weaving, knitting, sewing, and technical yarns) and the precursor fiber materials, yarns are produced with different technologies. Single yarns can contain one or multiple filaments, consist of staple fibers or be constructed from a blend of filaments and staple fibers. Depending on the mechanical or processing requirements, single yarns are processed into plied yarns, a process usually requiring additional process steps after mechanical yarn processing.

The following requirements are put on yarn production technologies:

- Low fiber damage during production processes,
- Use of suitable yarn carriers (e.g. paper, plastic, or metal sleeves) or packagings without assistive equipment (e.g. bumps—compressed and bound contents of a sliver can) for the storage of yarns or intermediate products, especially of slivers (e.g. packaging without assistive equipment, sliver cans),
- Securing of an economic production.

4.1.2 Yarn Parameters and Yarn Structure

Apart from the common macroscopic parameters (yarn testing parameters, see Sect. 2.2.2.8) quantified by testing (see Chap. 14), the *yarn structure*, i.e. the orientation of the fibers within the yarn itself, has to be described. This structure depends on the respective yarn construction. Knowledge of the yarn structure allows an assessment of the yarn parameters, as well as their modification during further textile processing, and of the delamination behavior in the final composite material.

The yarn structure results from fiber geometry (length, fineness, cross-section), straightness (plain, undulated), arrangement (degree of parallelism), orientation (in yarn axis direction or at a gradient angle with the yarn axis), and the degree of homogeneity of the fiber distribution across the yarn cross-section and length. Here, the similarity of the yarn cross-section to a circle increases with improved twisting or wrapping of the fibers. Otherwise, the yarn cross-section is either oval or rectangular.

The degree of dimensional stability under mechanical loads depends on yarn construction-related compressive forces between the fibers and a finishing or sizing agent (see Sect. 3.2.4) possibly interacting with the fibers. If there are no physical (no twisting or wrapping of the fibers) and/or chemical forces at interplay between the fibers within a yarn, the yarn cross-section can be formed as desired by small tensile and/or compressive forces, which makes porosity adjustable. In absence of compressive forces between the fibers, as they are created by, for instance, twisting-together of the fibers, an impregnation of the yarn in polymer solution is often required to enhance yarn cohesion.

Under mechanical loads, relative displacement between the fibers occurs, because the fiber bundle transfers from static friction to dynamic friction under increasing loads, such as high yarn tensile loads during textile semi-finished product manufacture. When exceeding a critical tensile load value, different fiber elongations (caused by disparate fiber orientation or dissimilar mechanical properties) can cause individual fibers to break, and/or result in relative movements between the fibers (fiber migration). This can be facilitated by interior yarn faults (inhomogeneous pore size distribution).

A distinction has to be made between internal and external yarn structure. Both are described in an unloaded state in which they are connected elastically.

Internal yarn structure (Fig. 4.1, Table 4.1): This structure is defined by the porosity of the yarn. Porosity describes the proportion of cavities (pores) present in the yarn volume A_{Pore}, determined by packing density of the fibers within the yarn ρ_{Yarn}. At an ideal hexagonal orientation of circular fibers the minimum percentage of porosity in the yarn is 93 %, independent of the individual fiber diameter:

$$A_{\text{Pore}} = 1 - \rho_{\text{Yarn}} = 1 - \frac{\pi}{2\sqrt{3}} = 0.93 \tag{4.1}$$

The real percentage of porosity is usually much higher and depends on the yarn construction. Various pore geometries can occur. The pores can be closed (without connection to other pores) or open (connected to other pores, forming a pore labyrinth). Open pores can be connected to the external yarn structure. This outward opening of the internal yarn structure increases the interface of the yarn, which is

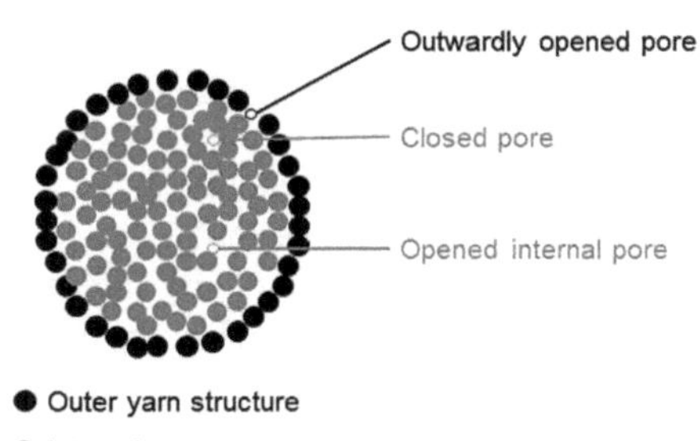

Fig. 4.1 Schematic of the internal and external yarn structure

Table 4.1 Important yarn test parameters of the internal yarn structure and their significance in the process chain

Yarn Test Parameter	Significant for
Stress-strain curve with maximum tensile force, maximum tensile elongation, tensile modulus of elasticity	Textile processing and assembly, e.g. in weaving and sewing
Flexural modulus of elasticity	Forming (behavior under tensile and bending load), especially during stitch formation processes and draping
Compressive strength proportion and size of pores opening to the external yarn structure	Depth of impregnation during composite production and composite strength

Table 4.2 Important yarn test parameters for the external yarn structure, and their significance for the process chain

Yarn test parameter	Significant for
Coefficient of friction against other materials[a]	Textile processing and assembly (smooth running over or through the yarn guiding elements)
Contact angle for the assessment of the wetting behavour[a, b]	Mechanical/chemical interface design Reinforcement/matrix and mechanical composite properties

[a]In connection with sizing/finishing agents (see Sect. 13.5.1)
[b]Necessary, but insufficient requirement for optimum fiber-matrix adhesion

crucial for its permeability, e.g. for the impregnation with a thermoset matrix system. Especially at low packing densities, the yarn porosity can decrease outwardly, creating a core-sheath pore structure. The roughness of the fiber surface and the porosity determine the contact area between the fibers and thus the fiber-fiber friction.

External yarn structure (Fig. 4.1): It is determined by the exterior fibers in the yarn surface. They define the yarn cross-section and its homogeneity along the length of the yarn (fluctuations of the yarn cross-section cause optical thick and thin places, neps) and the microporous roughness of the yarn surface. The external yarn structure is of special importance for the yarn test parameters given in Table 4.2. The microporous roughness, in combination with the sizing or a suitable yarn finishing, determines the wetting behavior of the reinforcement fibers by the matrix and the resulting adhesion characteristics. The connection to the internal yarn structure is established via open pores. The higher the ratio of matrix contact area and cross-section area in the yarn, the more beneficent the more favorable will the effects on matrix permeability be, considering that the capillary forces stemming from the yarn pore sizes enable the permeation of matrix particles.

Targeted adaptations of the internal and external yarn structure allow adjustments to the yarn behavior during yarn production by way of the process parameters.

4.1.3 Yarns Made from Fiber Material Blends (Hybrid Yarn)

Yarns can be made up of one or several fiber materials. If the yarn contains only one fiber material, it is a reinforcement fiber yarn. Yarns consisting of two or more fiber materials are classified as *hybrid yarns*. The combination of reinforcement fibers with thermoplastic fibers is on common example of this.

Furthermore, yarns can be combined with non-textile thermoplastic components in the form of solutions, molten masses, powders or foil coating during finishing processes. The typical textile properties (good draping behavior, flexibility) are largely lost during these processes.

All textile reinforcement fibers and most spinnable polymer fibers can be blended with each other [1].

The following essential fiber material combinations for an optimization of yarn properties are possible:

1. Different reinforcement fibers for:
    - Achieving a greater variety of physical (including mechanical) properties, such as balanced relations of stiffness, strength, tensile elongation at break, higher impact resistance, improved damping characteristics and
    - Cost reduction,
2. Reinforcement fibers and thermoplastic fibers for:
    - The creation of dried, textile semi-finished products for thermoplastic composites with preferably stretched reinforcement fibers and adjustable reinforcement fiber volume content as well as easy handling and storage,
    - An ensured textile processability of the yarns with nearly all known textile technologies for the production of textile 2D and 3D semi-finished products, and
    - Solidification of textile semi-finished products, especially of nonwovens, by melting of the thermoplastic component, or their fixation of further processing,
3. Combination of 1 and 2,
4. Combination of 1 (and 2) with functional yarns (see Sect. 4.6), or
5. Combination of, for instance, 1 with auxiliary fibers, such as PVA fibers, which can later be removed from the yarn, for example for improvements to porosity [2].

The blending itself can be integrated into chemical fiber production for economical improvements, but is usually performed in a separate process step. As there is no standard chemical fiber production method, the topic is included in this chapter for the sake of a consistent overview of fiber blends in yarns.

The composition and homogeneity of the *fiber blend* (Fig. 4.2) can be adjusted technologically according to the requirements by the selection of yarn finenesses to be used and of suitable yarn construction methods:

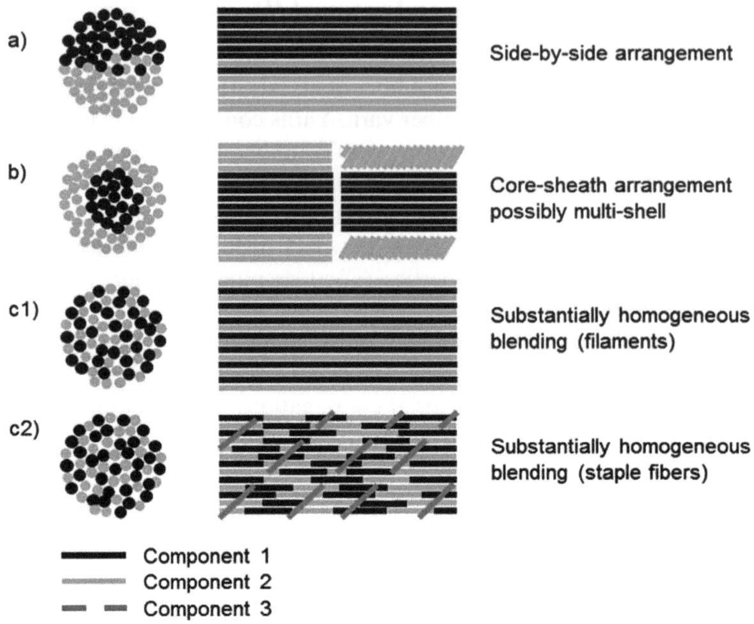

Fig. 4.2 Basic possibilities of blending two fiber materials

(a) The arrangement of the components along the length of the yarn can, for instance, be effected by twisting one or multiple yarns.
(b) The yarn surface can contain one or more components, either homogeneously distributed or segmented along the yarn length.

(c1/c2) A component can also serve as a placeholder. It is removed by the addition of a suitable matrix, thus increasing porosity.

Here, the individual components (filaments and/or staple fibers) and their arrangement (fiber orientation in yarn axis direction or at angles to it) can be modified in a variety of ways. This allows the production of hybrid yarns varying greatly with regard to their mechanical and function properties.

The homogeneity of blending in longitudinal and transverse direction depends on both the choice of textile method and on the geometrical yarn parameters (fiber cross-section, fineness, and number) as well as expansion coefficient) has to be considered.

The blend of reinforcement and thermoplastic fibers will be considered here mainly in its relation to the composite manufacturing process.

Due to the high viscosity of thermoplastics (100–5,000 Pas) in comparison to thermoset materials (usually <1 Pas), any impregnation of the reinforcement fibers during the manufacturing process is generally complicated [3]. This is to be alleviated by means of optimized homogeneity of the fiber blend in the yarn and the resulting short flow paths of the matrix, which also shortens the required process times in composite production. After the solidification of the melt, the components adhere to one another. When the resulting composite is exposed to loads, the matrix, which fixates the reinforcement fibers in the defined position, transfers the forces into the reinforcement fibers via the interface, which is the third component of every composite material.

Alternatively, the fiber materials can be mixed in the textile semi-finished fabric, for example by blending the yarns in one or several yarn systems, by using different yarns in the yarn systems or by stacking textile fabrics made from different fiber materials. All of these possibilities result in decreased homogeneity of the blend, making the hybridization of yarns a crucial topic.

In blends of reinforcement and thermoplastic fibers in a hybrid yarn, the *fiber mass percentage* and the *fiber volume percentage* are stated. The fiber mass percentage Γ is calculated from the relation of reinforcement fiber mass m_{Fiber} and total mass $m_{Composite}$ (including matrix fiber mass m_{Matrix}):

$$\Gamma = \frac{m_{\text{Fiber}}}{m_{\text{Composite}}} \times 100\% = \frac{m_{\text{Fiber}}}{m_{\text{Fiber}} + m_{\text{Matrix}}} \times 100\%$$
$$\Gamma = \frac{n_{\text{Fiber}} \times Tt_{\text{Fiber}}}{n_{\text{Fiber}} \times Tt_{\text{Fiber}} + n_{\text{Matrix}} \times Tt_{\text{Matrix}}} \times 100\% \tag{4.2}$$

n_{Fiber}, n_{Matrix} (–)	Number of filaments
Tt_{Fiber}, Tt_{Matrix} (tex)	Filament fineness.

This is important for the calculation of blend prices.

In the composite, the *fiber volume percentage* ϕ is calculated from the relation of reinforcement fiber volume V_{Fiber} and total volume $V_{Composite}$ (including V_{Matrix}: *matrix fiber volume*), neglecting pores:

$$\varphi = \frac{V_{\text{Fiber}}}{V_{\text{Composite}}} \times 100\,\%$$
$$\varphi = \frac{V_{\text{Fiber}}}{V_{\text{Fiber}} + V_{\text{Matrix}}} \times 100\,\% \tag{4.3}$$

considering

$$V = \frac{m}{\rho} \tag{4.4}$$

V (m^3)	volume
M (kg)	mass
P (kg/m^3)	density.

The following is valid after applying Eqs. 4.3 and 4.4 after rearranging and including Eq. 4.2:

$$\varphi = \frac{m_{\text{Fiber}}}{\rho Fiber\left(\frac{m_{\text{Fiber}}}{\rho_{\text{Fiber}}} + \frac{m_{\text{Matrix}}}{\rho_{\text{Matrix}}}\right)} \times 100\% \tag{4.5}$$

$$\varphi = \frac{n_{\text{Fiber}} \times Tt_{\text{Fiber}} \times \rho_{\text{Matrix}}}{n_{\text{Fiber}} \times Tt_{\text{Fiber}} \times \rho_{\text{Matrix}} + n_{\text{Matrix}} \times Tt_{\text{Matrix}} \times \rho_{\text{Fiber}}} \tag{4.6}$$

Fiber volume percentage is an important parameter for the composite material and its mechanical design.

For a homogenous distribution of both components, hybridization and filament diameter are important. Here, the components are supposed to show equal or nearly equal filament diameter [4, 5]. The use of thermoplastic filaments with small diameters as reinforcement fibers is recommended for a more homogeneous impregnation [6].

This recommendation does not apply to the production of staple fiber blends. It is a challenge to retain the homogenous blend of staple fibers. For example, fine, long, and even fibers are preferably located in the yarn core, while coarse, short, and crimped fibers are present primarily in the yarn surface. It is even more complicated to retain the blend across all process steps to the formation of the yarn. This results from the different fiber properties, such as different surface structure of the fibers or their varying energy of deformation, both of which cause different static and dynamic coefficient of friction. Under tensile loads, for instance, this can cause different fiber movements resulting in changes of the fiber arrangement and eventually in separation of the different fiber types.

Beyond that, depending on the yarn structure (especially in filament yarns), mechanical loads during further textile processing can cause fiber migration resulting in a separation of reinforcement and thermoplastic fibers. Ye et al. attribute this to the different fiber stiffnesses [7]. Long et al. confirm this assessment in the example of weaving, by determining that the compressive and tensile forces affecting the yarn are higher on the top and bottom sides of the yarn, causing a migration of the stiffer reinforcement fibers [1].

An overview of hybrid yarn production is offered by [3, 8].

4.2 Single Yarn for Semi-finished Yarn Products

4.2.1 Classification

Single yarns are manufactured from natural staple fibers or from flat filament yarns/direct rovings and the staple fibers cut or stretch-broken from them in synthetic fiber production, as described in Chap. 3 (Fig. 4.3). Depending on the length of the fibers contained in the yarn, the following yarn formation and processing methods are available.

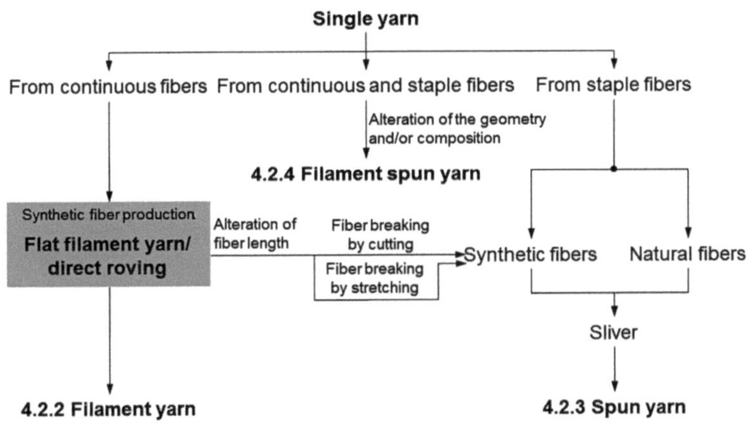

Fig. 4.3 Classification of single yarns according to length of the fibers contained in the yarn

Apart from that, the following simple yarns can be used in textile semi-finished products:

- Flat tape yarns: flat foil strips cut from a foil without surface structure
- Fibrillated flat tape yarns: reticular structure made from a flat foil tape, mechanically fibrillated

4.2.2 Filament Yarn

4.2.2.1 Introduction

The synthetic fibers of theoretically infinite length produced in chemical fiber production consist either of a single filament (monofilament length: >0.1 mm) or of many stretched, parallel filaments (flat multi-filament yarn: mostly <300 tex, direct roving (continuous filament strand): mostly 300 tex and more). As the individual filaments in the yarn are initially flat, parallel, and twistless, and do not have any filament cohesion, the exclusively mechanical yarn processing aims to alter yarn geometry and/or fiber material composition in order to achieve an improved textile processability (Fig. 4.4).

The filaments of various fiber materials can be blended either during chemical fiber production or in an additional process step, where the geometrical yarn parameters are changed simultaneously.

The following specific details supplement the generally applicable yarn parameters (see Sect. 2.2.2.8) for the specific characterization of *filament yarns*:

- Filament fineness, filament cross-section,
- Number of filaments,
- Filament course (flat, wavy, loops),
- Filament orientation (parallel, random), and
- Possibly number of twists and twist direction (compact yarns)

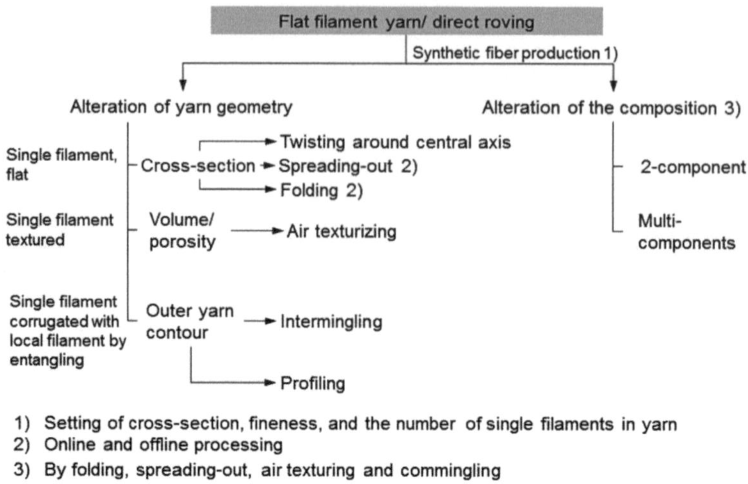

Fig. 4.4 Classification of filament yarns by manner of modification

The mechanically processed filament yarns are post-treated, with the sizing/ finishing agents and heat treatment (see Sect. 3.2.4) near-universally being integrated in the yarn processing machines.

4.2.2.2 Flat Filament Yarn and Filament Yarn Twisted Around the Central Axis

By means of twisting around the central yarn axis, the yarn geometry is changed. The yarn is given a round cross-section, while the packing density is increased, resulting in a reduction of the cross-sectional area of the yarn. Twisting around the central axis facilitates the improved filament cohesion of the brittle reinforcement filaments (preventing the spreading-out of filaments by electrostatic charging, reduction of breaking individual filaments and lint formation). At the same time, spools connected by twisting around the central axis allow the elimination of yarn faults caused by broken filaments.

Flat filament yarn (usually <300 tex): The number of filaments in the yarn follows the definition of yarn fineness, e.g. 14 tex f 40.

Filament yarn twisted around the central axis (usually from 300 tex upward): A single filament yarn is twisted around its axis in S or Z direction, the number of twists in the yarn is given as follows: 40 Z 250. Twisting machines can be used for this purpose (see Sect. 4.3).

4.2.2.3 Roving and Heavy Tow

From 100 % Reinforcement Fibers

Direct roving or single-end roving are filament yarns produced in a single step in synthetic fiber production, usually with finenesses from 300 tex upward, containing flat reinforcement filaments without twist, and formed immediately after the application of a sizing material adjusted to further processing. Therefore, direct rovings are also known as parallel rovings. They feature a ribbon-shaped, i.e. elliptical to rectangular cross-section, and are generally deformable by mechanical forces. Direct rovings are available as packages without supporting element (internal doffing) or on spools (external doffing) (Fig. 4.5).

Assembled Roving or multi-end roving: Assembling means the folding of reinforcement fibers. The assembled roving is delivered from a defined number of rovings of equal tension (feed package with multi-filament yarn or direct rovings) treated with a sizing adapted to further processing. It is available on roving spools with external doffing or as package without supporting element with internal doffing. The assembled roving usually does not feature producer twists, but can contain a small number (no more than 15 producer twists/m) of them. During production, the tensioned individual roving should always be of equal length to avoid a spreading-out of filaments between individual rovings of the assembled roving.

Rovings are usually designated by the statement of their number of individual filaments and/or the statement of yarn fineness. For historical reasons the designation depends on the fiber material. The following applies as a general rule:

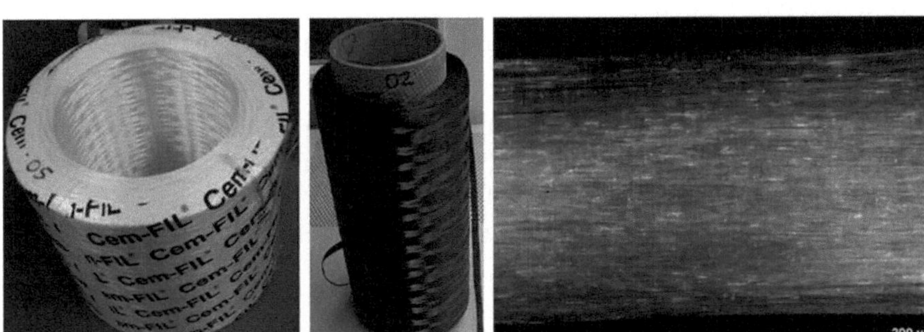

a) Direct roving from AR-GF (Cem-Fil®, 2400 tex, OCV™Reinforcements)

b) Direct roving from CF (TENAX® STS40 F13 24k 1600 tex 5S, TohoTenax Europe GmbH)
F13: Type with ca. 1.0 % sizing based on polyurethane
5S: 5S rotation/m

c) Direct roving from AR-GF (Cem-Fil®, 640 tex, OCV™Reinforcements)

Fig. 4.5 Direct roving as (**a**) package without supporting element and (**b**) spool, as well as (**c**) detailed view

- CF: Number of individual filaments (number) in steps of 1,000 filaments (K) (1 K ... 24 K),
- GF: Fineness in tex ≥300 tex, and
- AR: Fineness in tex ≥300 tex, or denier ≥2,700 denier
  (The non-SI unit denier is usually used for AR fibers in business)

Heavy Tow: A very robust assembled roving with a high number of filaments and the resulting larger cross-sectional area is called heavy tow. Currently, they are exclusively produced from CF, and the following rule applies:

- CF: >24 K (customary finenesses, e.g.: 48 K, 50 K, 100 K).

Heavy tows are significantly lower priced than rovings [9]. Before further processing in a textile process, they have to be spread out as evenly as possible to reduce the distance between individual filaments of the inner yarn structure. This enables the creation of homogeneous masses per unit area in the textile semi-finished product, a good impregnation of the reinforcement fibers by the matrix, resulting in the ability to manufacture homogenous and cost-effective composites. To fully utilize these economic advantages, a compacting during further textile processing has to be preempted, for instance by modifying the knitting yarn tension settings.

From Several Fiber Material Components

Hybrid roving/Hybrid heavy tow: These types contain filaments or flat tape yarns made from various fiber materials, which can be arranged as follows:

- Parallel to one another: Parallel-hybrid (SBS: side-by-side) roving or heavy town (see Fig. 4.2a), or
- Nearly ideally blended: In-situ hybrid (COM in-situ commingled) roving or heavy tow (see Fig. 4.2c1)

Rovings are either directly processed further by pultrusion or filament winding, or they are used for the manufacture of textile semi-finished products. Apart from these, chopped rovings are available, for example for the production of GMTs (glass mat-reinforced thermoplastics) and SMCs (sheet-molded compound).

Production in a Separate Process

Folding/assembling and winding are performed with customary machines adapted to the respective reinforcement fibers (GF, CF, AR, or basalt fibers).

During folding, the tension of the individual hanks has to be monitored. Because deviations from the rectilinear orientation and the related different filament lengths cause lower maximum tensile forces of the rovings at slightly increased strain at maximum load.

Spreading-out (one fiber material component): This is primarily used for heavy tows and can be realized with the same processes as the spreading-out of multiple fiber material components.

Spreading-out (two or more fiber material components): The usually uncoated fiber components to be mixed are spread-out to a defined width my means of various methods: electrostatically [10–12], mechanically with spreader bars, spreader rollers, spreader combs or knives [13–16], pneumatically by means of jets operated with pressurized air, e.g. fan nozzles or slot nozzles [17–19]; other possibilities are spreading by means of a jet of liquid [13, 20] or acoustic spreading [21]. The components are then joined via rollers or rods and either fixed in their structure (with sizing materials [20] or by wrapping with matrix filaments [22]). The filaments in a hybrid roving are mostly aligned parallel to the yarn axis.

Due to the spread-out width, both the yarn structure and the yarn properties are influenced significantly, where the filaments are homogenously blended, resulting in higher adhesive and frictional forces [23]. Spreading-out can also be employed in combination with the insertion of yarns in fabric formation processes.

Spread-blended hybrid yarn: In this application example [24, 25], the reinforcement filaments are spread out mechanically by means of a spreading wheel consisting of disks frontally fitted with needles embedded in a fan shape on a central axis. The needles penetrate between the filaments of the reinforcement yarn and distribute the filaments on a larger width, as the needle distance increases in take-up direction. With special modules (fold rollers with corresponding web roller), a wider spreading of the reinforcement filaments becomes possible. Subsequently, the reinforcement filaments are mixed with the matrix filaments pneumatically inserted with advancement from a warp beam, and the state of blending is fixated before winding by a coating based on aqueous polymer dispersions.

CF/PA and CF/PEEK rovings are offered by Cytec Industries Inc. [26].

Production During Yarn Formation in Chemical Fiber Production Processes

Online hybrid yarn spinning: This process was developed my Vetrotex and adopted by OCV™ Reinforcement [27]. The hybrid rovings consisting of blends of E-Glass and thermoplastic filaments [e.g. PP or copolyester (co-PBT)] are commercially available under the Twintex® label.

The Dresden-based Leibniz Institute of Polymer Research Dresden integrates the melt spinning plant into the glass fiber drawing (Fig. 4.6), and focuses its research on the development of glass fiber blends with other technical thermoplastics, such as PA and PET, while specifically adjusting fiber volume content ratios and filament finenesses. Wind-up speed and flow-rate determine the filament finenesses. For the thermoplastic component, e.g. PP, several nozzle plates with varying number of bores and geometries are in place, so that the number of PP filaments can by within the hybrid roving as well as the diameter/surface rations of PP can be variegated. The process parameters are being optimized [28–30]. However, the available machinery can only be used for the manufacture of small

Fig. 4.6 Online hybrid yarn spinning (*Source*: Leibniz Institute of Polymer Research Dresden)

amounts, and research potential remains with regard to the full utilization of hybridization.

In comparison to hybridization during air texturization (Sect. 4.2.2.4), the following advantages result at smaller variability of the possible fiber materials and filament numbers:

- Minimization of glass filament damage during processing, and thus improved mechanical yarn properties,
- More homogeneous blending of both filament components,
- No thermal shrinkage, and
- Increased economic efficiency owing to a reduction of process steps

A blend of, for instance, 52 % GF with 48 % PP results in the most favorable mechanical properties and is most suited for further textile processing [31].

4.2.2.4 Air-Texturized and Intermingled Filament Yarn

Air-texturized and intermingled filament yarns are manufactured by blow processes. These are mechanical processes in which a texture is imparted on the filament yarns by running them through a cold, gaseous, streaming medium (air) under overfeed [32]. This method increases the total volume, elastic elongation, and the porosity of the filament yarns/rovings.

The input yarns are usually fine to coarse POY (partially oriented yarns) (see Sect. 3.2.1) from synthetic fiber production, whose polymer chains have been fully oriented by heating and drawing. In contrast to other texturing methods, non-thermoplastic fiber materials and reinforcement fibers, e.g. glass filament

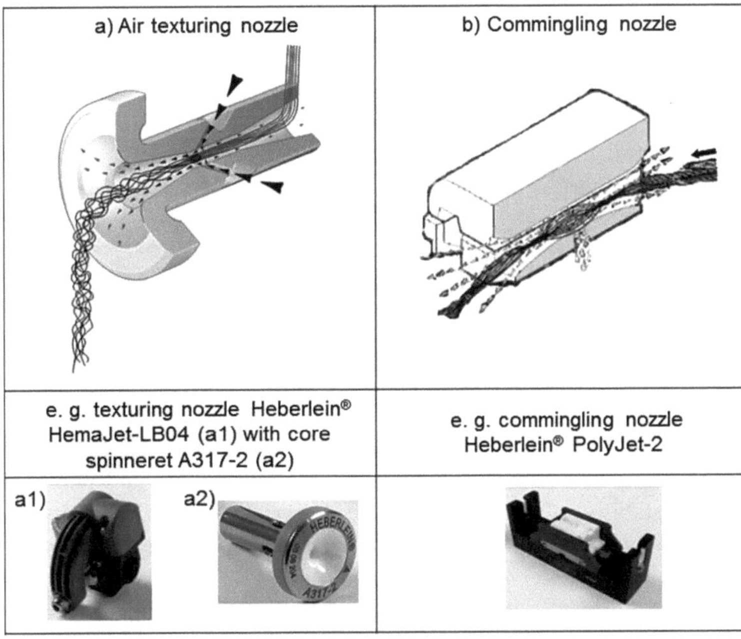

Fig. 4.7 Air nozzle for texturization purposes (**a**) and for intermingling (**b**) (*Source*: Oerlikon Heberlein Temco Wattwil AG)

yarns, can be processed. The process is very flexible with regard to suitable fiber materials and blends.

The air is applied via an air-texturizing nozzle, whose air canal can run at different angles to the yarn canal. Figure 4.7 shows the respective nozzle constructions for this area of application and the resulting yarn structures.

Intermingling

The yarn is mingled with itself, which is also known as intermingling.

Air-texturized filament yarn: The pressurized air flow (Fig. 4.7a) meets the filament yarn in the nozzle at a sharp angle. It is slowed by the widening of the yarn canal, which spreads the filaments of the inserted filament yarn. As the filament yarn is inserted at up to 80 % overfeeding ratio (=ratio of initial speed of the untexturized yarn and take-up speed of the texturized yarn), individual filaments in the yarn may be displaced lengthwise to the yarn axis. This creates loops, which interconnect or progress beyond the yarn surface, which can swell out the yarn. Thus, air-texturized yarns have a strong core and a yarn sheath containing outward-pointing loops of various sizes. The characteristics can be similar to those of staple fiber yarns, but another possibility lies in the production of compact yarns by a suitable selection of machine parameters.

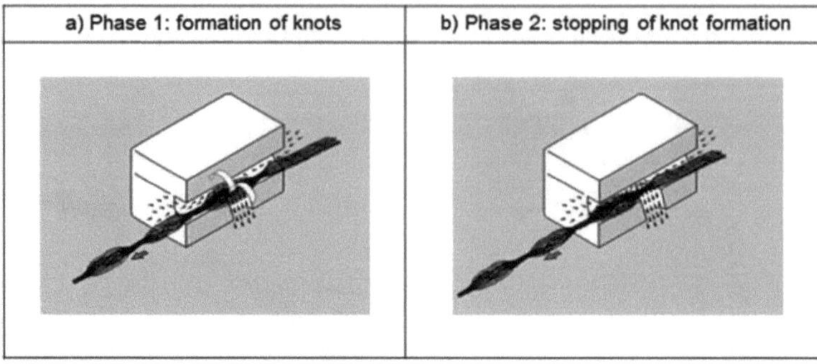

Fig. 4.8 Intermingling in two phases (*Source*: Oerlikon Heberlein Temco Wattwil AG). (**a**) Phase 1: Formation of nips; (**b**) Phase 2: Termination of nip formation

Behind the exit of the air nozzle, a impact element can be inserted to deflect the flow. Strong turbulences can result between impact element and nozzle surface, increasing the stability of the yarn structure. However, impact elements are rarely used for reinforcement filaments, as they are easily damaged by impacts caused by the turbulences [33].

Intermingled filament yarn: The intermingling air jet (Fig. 4.7b) meets the filament yarn at a right angle in the nozzle, where the filament yarn is intermingled by the pressurized air rushing in from the air duct.

The yarn structure (Fig. 4.8) is formed in two phases (Oerlikon Heberlein Temco Wattwil AG):

1. Formation of nips (Fig. 4.8a): The individual filaments are separated by the air flow and set into a rotating motion upon entry into the intermingling nozzle. This creates an accumulation of false twist at the entry and exit points of the nozzle. This accumulation is referred to as a nip.
2. Termination of nip formation (Fig. 4.8b): The yarn moves through the nozzle at delivery speed. Thus, the nip formed in phase 1 approached the air flow, stopping the twist of the individual filaments.

During intermingling, isolated intertwinings of the filaments are created at defined distances, taking the shape of a relatively high number of small loops, so-called *interlaces*. Between the intermingling points are yarn lengths containing primarily non-intermingled, i.e. open yarn points. The yarn is characterized by the following parameters:

1. Number of nips per meter, or intermingling density
2. Average length of nips
3. Degree of intermingling: indicates the ratio of mingling section length to yarn length in % [34]

The intermingling uniformity (i.e. the alternation of opened sections and nips), and the intermingling stability (loss of nips under certain loads the yarn is exposed to) are additional parameters for yarn characterization. The Oerlikon Heberlein Temco Wattwil AG tests for intermingling density, homogeneity, and stability.

During processing, the risk of filament damage by excessive air pressure is high [35].

The characteristics of the input yarns (number of filaments, fineness, cross-section, yarn fineness, stiffness in bending) and important process parameters like nozzle geometry (angle between inlet hole and yarn channel, yarn and air duct diameter, air canal profiles), yarn arrangement in front of the nozzle, air pressure, yarn overfeeding (air texturizing) or yarn tension (Intermingling), humidification, pre-heating temperature, take-up speed, and yarn tension at winding all contribute significantly to the characteristics of processing and finished yarn (yarn structure).

Texturization or intermingling can increase yarn cohesion as much as rotation around the central axis or twisting. This reduces the danger of broken filaments locking or building up fluff during further processing, which would cause production disruptions.

Commingling

The process of texturizing or mingling two or more yarns with one another is called commingling. The yarns produced in this manner are known as commingling yarns.

Air-texturized commingling yarn: The production is performed on modified air-texturizing machines using special nozzles, e.g. Heberlein® HemaJet-LB04 (Oerlikon Heberlein Temco Wattwil AG). The components to be mixed, e.g. reinforcement and thermoplastic filaments are inserted by separate, individually adjustable delivery mechanisms (Fig. 4.9) of the nozzle at different

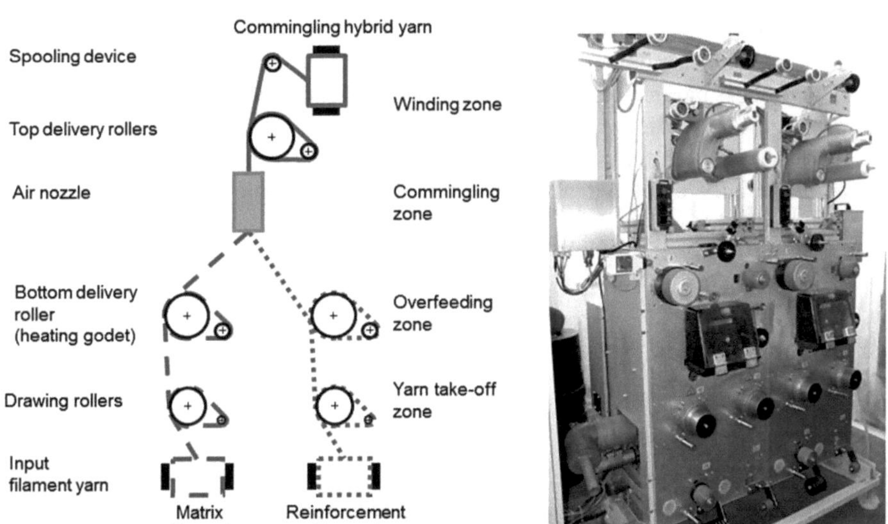

Fig. 4.9 Commingling yarn production. (**a**) Schematic representation of commingling yarn production (two components in the yarn). (**b**) Air-texturization machine by Stähle (three components in the yarn)

overfeeding. Afterwards, the components are thoroughly opened with cold or hot pressurized air, and then intermingled with one another in the air flow, aiming at a highly cohesive, homogeneously mixed yarn. These steps are performed in a core-effect process, i.e. the thermoplastic filament yarns have a greater overfeeding. Thus, the reinforcement filaments are kept stretched and arranged in the yarn core (protection against damages). Loops are created primarily by the matrix filaments (effects), which also provide a mechanical binding between both fiber components. For an optimum delivery behavior at high speeds, the following rule applies to yarn overfeeding: the sum of core and effect is not to exceed 45 % [36].

By heating the matrix yarn on the heated godets without drawing, the matrix filaments can be thermally fixed. This can prevent shrinking of the matrix filaments during consolidation, which helps avoid stress cracks in the composite [37].

The influence of yarn and process parameters on the yarn structure and the resulting yarn properties have been researched extensively [38]. To increase the blending homogeneity of the components and ensure lower filament damages, [6] recommends the selection of a smaller diameter of thermoplastic fibers than of the reinforcement fibers, while retaining the blending ratio. This also applies to online hybrid yarn spinning.

Air-texturized commingling yarn: These are manufactured on modified air texturizing machines fitted with special nozzles, e.g. SlideJet™-HFP15-2 (Oerlikon Heberlein Temco Wattwil AG), with which several yarns can be intermingled with one another. Apart from that, these nozzles can be used for the intermingling of filaments with elastane or staple fiber yarn, or of filaments with elastane and staple fiber yarn, if the filament component predominates.

Various intermingled commingling yarn structures (GF/PP, GF/PA, GF/PET) are characterized by Alagirusamy et al. [39]. They determine that the GF/PP yarns display the smallest number of nips/m and the lowest degree of intermingling at identical process parameters, making them the greatest challenge for intermingling, and placing them in the focus of later examinations [40, 41].

Hybrid yarns from GF or CF, in combination with various thermoplastic fibers, are commercially available under the brand names Comfil®-G and Comfil®-C (Comfil ApS) [42].

4.2.3 Slivers and Spun Yarn

4.2.3.1 Introduction

Spun yarns from carbon fiber rovings were first produced around the middle or late 1980s (among others, by Courtaulds's Heltra Division, ICI Fiberite (Tempe, Arizona), DuPont (Wilmington, Delaware). Over the past years, these yarns and their intermediate products in the form of fiber tapes made from oriented, non-continuous reinforcement fibers have made a comeback. This is caused by demands from the industry, aiming to replace manual laminating processes with faster and automated technologies like automated vacuum forming or diaphragm forming. Dry textile semi-finished products containing oriented staple fibers in yarns are well-suited to this task. In contrast to continuous fibers, they display superior drapability and formability without risk of losing their favorable mechanical characteristics, since the fibers in the yarn can be moved along one another relatively well (so-called intralaminary sliding occurring at local overelongation of the yarn during forming), allowing the creation of curved component parts. Owing to this, simple semi-finished product geometries can be processed into very complex components, thus reducing production costs and widening the available range of producible composite components. Furthermore, staple fiber tapes or spun yarns can be processed in a continuous direct extraction process.

Direct rovings produced in synthetic fiber production are processed into non-continuous-fiber slivers of cut or stretch-broken fibers. Natural fibers of plantal origin are delivered as fiber bales (pressed fiber tufts). They are processed in a multi-step secondary spinning process. Figure 4.10 shows that in fibers of finite length arise new possibilities for the composition of the yarn by blending fiber flocks (natural fibers) or slivers (natural fibers, synthetic fibers), in which a more homogeneous blending especially in longitudinal direction can be achieved in comparison to filament blending. The resulting yarn geometry of the spun yarns is created by mechanical processing of the sliver (in particular by drafting and doubling) and finally stabilized by twisting all or a fraction of the fibers around the longitudinal axis of the yarn, for which a number of final spinning processes are available. For the processing of reinforcement fibers, ring spinning and friction spinning are particularly important, as these processes also allow the processing of

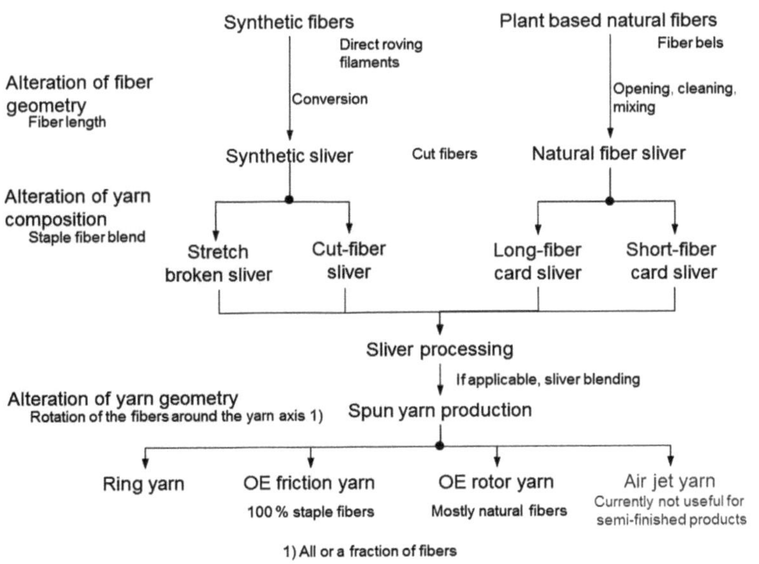

Fig. 4.10 Classification of the most important staple fiber yarns according to manner of modification

longer staple fibers (>65 mm). Therefore, synthetic reinforcement fibers are usually spun on machine trains used in long fiber spinning. Yarns containing plantal natural fibers can be produced with processes used in long and short fiber spinning, especially by OE rotor spinning.

The following parameters are used to supplement the general yarn parameters (see Sect. 2.2.2.8) in the specific characterization of spun yarns:

- Number of twists T and direction of twists, and
- Twist factor α_{tex}

α_{tex} is a tabulated value [43] and describes the number of twists of a yarn with a fineness of 1,000 tex. With T (number of twists in twists/m) and Tt (fineness in tex), the following applies:

$$\alpha tex = T \times \sqrt{Tt} \tag{4.7}$$

By plugging in the tabulated value α_{tex} into 4.7, the required number of twists for a yarn of defined fineness and equal strength as the 1,000 tex yarn can be determined.

A typical exterior structure of spun yarns is given in Fig. 4.11. The fibers are oriented at an average gradient angle γ to the yarn axis; a portion of the fiber ends is not incorporated and gives the yarn protruding fiber ends and fiber loops, which constitute the so-called hairiness. This creates a coarser and more voluminous surface, improving the incorporation of the reinforcement fibers into the matrix by positive locking, making it effectively an adhesion agent to the matrix. The hairiness depends particularly on the relation of fiber length and fiber fineness, on the spinning method, and the selected process parameters.

The fibers have to be fitted with a suitable finishing agent to enable an improved processing in later process steps.

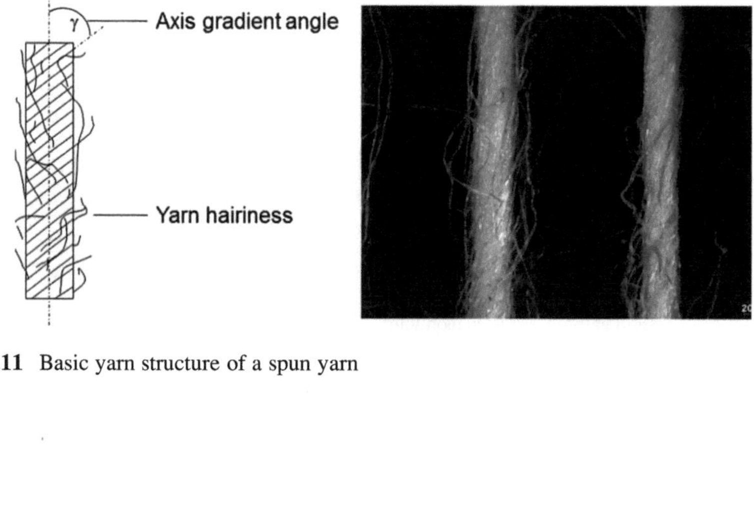

Fig. 4.11 Basic yarn structure of a spun yarn

4.2.3.2 Staple Fiber Sliver

A staple fiber sliver is a linear, nearly untwisted coarse structure, e.g. with a typical fineness of 5 ktex, with a high number of largely parallel, non-continuous fibers oriented in longitudinal direction. It has only a small strength, as the cohesion of the fibers by adhesion and force influences is determined by the dynamic friction between the fibers. It can be drawn easily and is equally suitable for the production of homogenous fiber blends from reinforcement (including natural) and thermoplastic synthetic fibers. The staple fiber slivers are usually stored in delivery cans (Fig. 4.12). They are referred to as continuous discontinuous tow or CD tow, and can be processed directly into pre-consolidated tapes [44], where the extrusion can be performed with or without imparting a twist.

In comparison to yarns and rovings, the following advantages result:

- Better draping behavior and thus improved suitability for deep-drawing
- Shorter process times during further processing, as well as
- Lower production costs of selected component parts

On the other hand, diminished mechanical characteristics are to be expected due to the finite fiber length.

Production by Stretch-Breaking (Stretch-Broken Sliver)

Hexcel (Dublin, California), Schappé Techniques (Charnoz, France) and Pharr Yarns have a variety of technologies available for the manufacture of slivers made from reinforcement filaments created by breaking fibers by stretching. These technologies will be explained in the following. Fundamentally, the filaments are always incrementally stretch-broken between the drafting roller pairs at the points of lowest local maximum load [45].

Hexcel: Stretch Broken Carbon Fiber (SBCF) in the form of 6 K or 12 K CF rovings consisting of unsized AS4 fibers (7 μm fiber diameter) or IM7 fibers (5.4 μm fiber diameter): The method is patent-protected [46] and discussed in detail in [47]. It is based on the drawing of a roving run through drafting roller pairs rotating

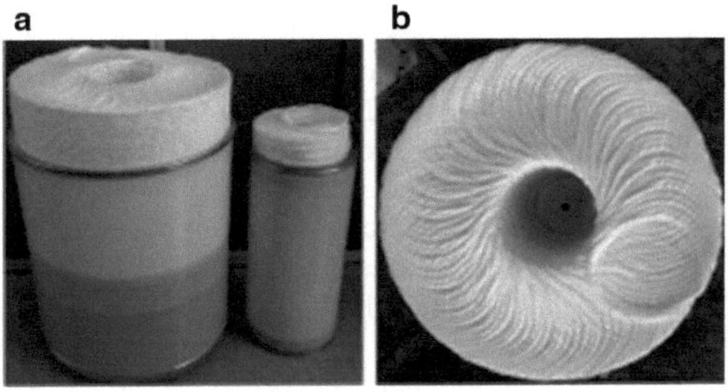

Fig. 4.12 (**a**) Delivery cans for sliver storage, (**b**) *top view* of a staple fiber sliver stored in *cycloid shape*

at increasing speed in the take-up direction, which are stretch-broken randomly at the weakest spots of the filaments at circa 10 % elongation. The main characteristics of the stretch-broken slivers are (see [48], with regard to [49]):

- The filament breaks are distributed randomly along the length of the sliver.
- The average fiber length (at normal distribution) is 10.2 cm (first generation) or 7.1 cm (first generation), while the fiber length distribution is less wide in the second generation.
- The sliver is sprayed with a water-soluble epoxy bath serving as an adhesion agent, then dried and wound up.

By means of a heat treatment, the bath can be solvated to minimize friction between the fibers.

Unidirectional prepreg tapes or textile semi-finished products can be produced from these slivers consisting of stretch-broken fibers. They have been stated to achieve 95 % of the strength of comparable continuous-fiber rovings at 2 % stretch [50].

Schappe techniques stretch-broken slivers from 12 to 24 K CF or AR, and hybrid stretch-broken fiber slivers (blends with PA, PPS, PEEK, LCP): the Schappe method [51] is applied on an in-house-developed machine and is also based on the drafting of the rovings to the tearing of the filaments at the points of lowest maximum load. The average fiber length is 80 mm, with the lengths of the individual fibers ranging from 40 to 200 mm.

The rovings are first spread and then stretch-broken in two steps [52]:

- 1st step: The continuous-fiber strand is elongated by 11 %, causing a random stretch-breaking of the fibers, whose maximum elongation is only circa 2 %
- 2nd step: The fiber strand is elongated by an additional 4 %, randomly stretch-breaking the remaining filaments

The stretch-broken fiber sliver is sprayed with a water-soluble epoxy bath, dried and wound up on a spool.

CF (or AR) stretch-broken slivers are processed into unidirectional prepreg slivers or into textile semi-finished products, e.g. multiaxial non-crimp fabrics. During pricking with the needles, part of the torn fiber ends are pulled out in z-direction, during which the fibers are entangled with one another, creating a 3D reinforcement structure during fabric formation [49].

The stretch-broken fiber slivers can also be mixed with slivers made from thermoplastic fibers. These blends are processed into hybrid core-spun yarns (see Sect. 4.2.3) or prepregs, known by their trade name TPFL® [53].

Pharr Yarns LLC (Mills Inc. of McAdenville, North Carolina, USA): CF stretch-broken fiber sliver made from 24 to 80 K heavy tows, or AR (Du-Pont's Kevler) stretch-broken fiber sliver of 2.78–55.56 ktex (25,000–500,000 denier): The rovings or heavy tows are stretched by means of several roller pairs and stretch-broken [54]. The draft is low (usually 2.0). As heavy tows are preferred for processing, delivery speeds are usually high (0.51–2.54 m/s). The manufactured stretch-broken slivers contain staple fibers of lengths ranging from a few millimeters to 180 mm, with the average fiber length between 127.0 and 152.4 mm. The slivers are stored in sliver cans and serve as input material on the spinning machine. The company aims to produce finer rovings in the future (6.0 K, 3.0 K, 1.0 K, 0.8 K, and finer) [49].

Conventional ring spinning or friction spinning (see Sect. 4.2.3.3) are viable spinning methods.

Manufacture by Cutting (Cut Fiber Sliver)

Slice-broken fiber slivers are offered by Pepin Associates Inc. (Greenville, Maine) [55]. In the production of slivers from cut fibers, continuous-fiber strands are processed into staple fibers of defined length by cutting rollers, nip rollers or reciprocating cutting rolls by means of oblique trapezoidal cut. To ensure cohesion of the Sliver, the cut fiber slivers by Pepin are supplemented with additional thermoplastic filament yarns immediately after cutting, e.g. as a core between two cut rovings, and for wrapping of the cut fibers, relying on the same principle as wrap-spinning (Fig. 4.23). These yarns (trade name: DiscoTex™) made from CF, GF, or ceramic fibers are processed into textile semi-finished products [56].

Long-Staple Card Slivers

These are produced on *cards* from (preferably natural) fiber length above 65 mm (Fig. 4.13). In comparison to the flat card, the fiber treatment is gentler.

The carding machine, originally used in long-staple spinning, is also used for the manufacture of nonwoven fabrics (see Chap. 9). Before the card, the bales

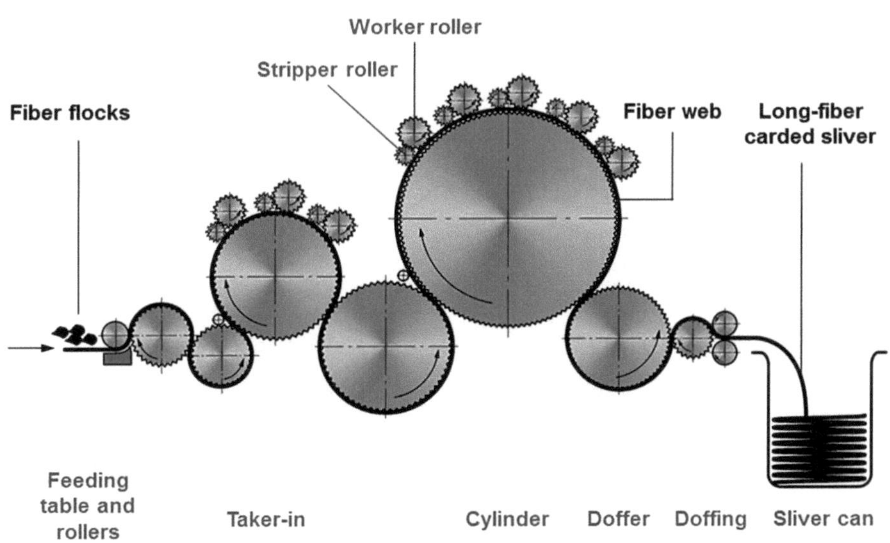

Fig. 4.13 Basic Principle of card sliver production

compacted into flock have to be opened thoroughly and cleaned. If necessary, they can be mixed with type-similar thermoplastic fibers (flock blend). The dissolution of the flocks into individual fibers is performed in the linear processing area between the rotating tambour, worker, and stripper rollers. Here, the fibers are also mixed in longitudinal direction, as they are only partially returned to the cylinder and therefore circulate multiple times. The resulting lap, which consists of largely parallel individual fibers, is transferred to the doffer and combined into to a sliver in the take-up mechanism, after which it is stored in a sliver can.

The card sliver is then processed on gills, where the blending with thermoplastic fibers (sliver blend) can alternatively be performed. The drafting between the drafting roller pairs is controlled by means of gills running along with the sliver and ensuring the guiding of the relatively long staple fibers (>65 mm).

Short-Staple Card Slivers

Short-staple card slivers are produced on flat cards from fibers shorter than 65 mm (Fig. 4.14). In comparison to the long-staple carding machine, the processing of the fibers on the flat card (which originates from short fiber spinning) is more intense. Before the flat card, the fiber bales of compressed flocks have to be opened and cleaned, similar to the long-staple carding machine. If necessary, they can also be mixed with type-similar thermoplastic fibers (flock blend) after cleaning. The dissolution of the flocks into single fibers is primarily performed in the planar processing area between moving flat and cylinder. Simultaneously, the fibers are mixed in longitudinal direction, as they are only transferred partially to the doffer. Afterwards, it is combined into a sliver and stored in a sliver can.

The *short-staple card sliver* is further processed on drawing frames, which offer an alternative option to perform the sliver blend.

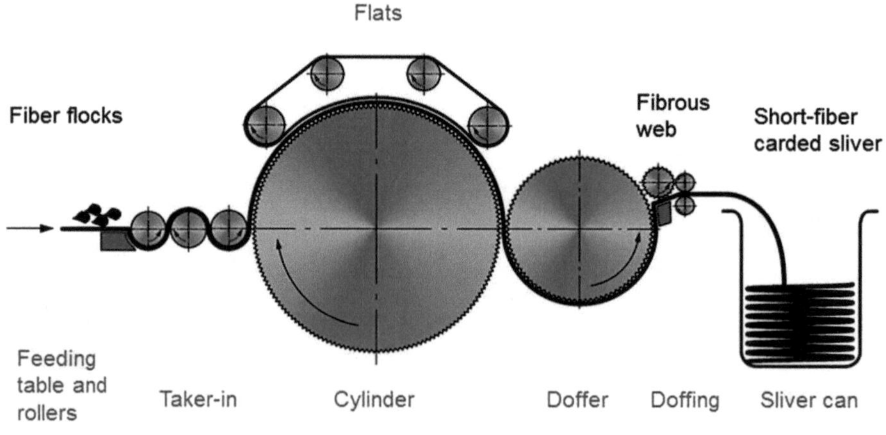

Fig. 4.14 Basic principle of short-staple card slivers

Pre-treatment of Natural Fibers for the Manufacture of Slivers

Steam Explosion Technology for the preparation of natural fibers before processing on short fiber spinning machines: To utilize natural fibers cost-effectively and economically, the fiber production chain in agriculture (retting, decortication) can be combined with its textile counterpart. The steam explosion technology can serve as the connecting process step in this combination. Here, the slivers are prepared for spinning after decortication by breaking them down into elementary fibers. For this, the fibers, e.g. hemp fibers, are treated in saturated water vapor for 10 min at 180 °C and a pressure between 1.0 and 1.2 MPa in a reaction vessel, and then blown out into a cyclone, which separates the sliver into elementary fibers of circa 50 mm in length. The prepared fibers can be processed (after mixing with similar type of fibers, if desired) on short fiber spinning machines (flat card) [57].

Slivers from Multiple Fiber Material Components

Sliver blend for the production of *hybrid staple fiber tapes*: The blend of reinforcement and thermoplastic fibers is usually performed in tape form on the drawing frames following the long-staple carding machine or short-staple carding machine. The blend ratio can be adjusted easily by varying the number of inserted tapes, with due regard to the tape fineness. The fiber tapes are combined, which is referred to as doubling. During Drafting, which then follows, the fiber tapes pass roller pairs of increasing peripheral speed rotating in production direction. There, the sliver is again refined, and the fibers are oriented in the direction of the sliver axis. As the blend across the sliver cross-section still retains considerable heterogeneity, the combining and drafting has to be repeated in additional drawing runs. Due to the relative movement of the fibers during drafting, the homogeneity of the fiber blend steadily increases. To blend and retain the homogeneity of the blend, the fibers to be blended should have similar properties regarding fiber length, finenesses, surfaces, and working capacity.

4.2.3.3 Spun Yarn

The slivers produced with the methods described above are the input product for the manufacture of *spun yarns*. The slivers are refined, drawn, and potentially blended with thermoplastic fibers by additional mechanical processes on short or long fiber spinning machinery.

The slivers can be introduced to the spinning machines directly (as in OE rotor and OE friction spinning machines). Alternatively, rovings have to be manufactured in an additional process step (flyer or finisher rubbing frame) (ring spinning machine). This additional process step is necessary because ring spinning is a top-down method, which complicates storage in the sliver cans.

As ring spinning remains the most commonly used method and is most suitable for the processing of reinforcement fibers such as aramid, it will be covered in detail below.

The spun yarns are generally stabilized by the insertion of twist. The *permanent twist* remain in the yarn after yarn production, the *false twist* is temporary and partially process-inherent. After a fashion, it is a supporting twist that grants the thin sliver made from parallel fibers stability during manufacture. Figure 4.15 [letters A and C designate the nips, letter B indicates the twisting element, which can simultaneously be a nip (Fig. 4.15a)] shows the schematics of creating permanent or false twists. When creating false twists, the fiber tape is given Z-twists above the twisting element or S-twists below the twisting element. This way, any Z-twist imparted before the twisting element is reversed by the S-twist created after the twisting element during vertical yarn take-up. By modifying the system (Fig. 4.15b), e.g. $l_1 \neq l_2$ or broadly fed fibers in l_1, which allow part of the fibers to avoid twist insertion, false twists can be used to stabilize.

The portion of fibers imparted with a twist, the number of twists, and their distribution across the cross-section of the yarn depend on the spinning method.

The more twists are imparted on the fiber tape, the higher will be the initial maximum load. The fibers are held together by the increasing static friction caused by the compaction of the fibers. With an increasing number of twists in the yarn, the fibers will be extended beyond the elastic elongation range until they finally break. Thus, yarn strength is reduced again. Therefore, the number of twists should be selected to not exceed the maximum set by yarn strength.

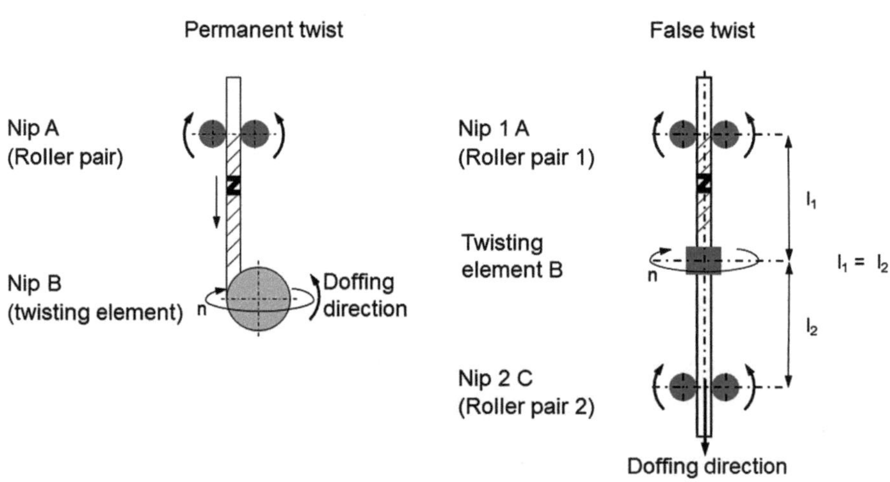

Fig. 4.15 Fundamental possibilities of twist insertion

Ring-Spun Yarn

Ring spinning is the oldest spinning method and remains the only process with universal versatility. Finest and coarsest yarns can be produced from short or long fibers. Furthermore, machine modifications have made the technology available for the manufacture of core ring-spun yarns (see Sect. 4.2.4) or for ring twisting (see Sect. 4.3.2).

The ring-spun yarn is produced in a multi-step process (Fig. 4.16). On the roving machine, the sliver is processed into the roving by reducing the fineness of the sliver by means of drawing and equipping it with a small number of twists to ensure the strength required for further processing.

Afterwards, the ring spinning machine (Fig. 4.17, ref. Fig. 4.15a) is used to draft the roving to its final fineness. The fiber tape is imparted with twists by means of a traveller (twisting element B) driven via the piece of yarn between bobbin and traveller by the rotating spindle. These twists travel through the fiber tape, which is exposed to high yarn tensions (creation of a yarn balloon), to the nip A of the last pair of drawing rollers. The number of yarn rotations T results from the quotient of rotational speed of the spindle n_{spi} and delivery speed v_L:

$$T = \frac{n_{s\,pi}}{v_L} \tag{4.8}$$

The winding-up of the yarn on the bobbin results from the rotational speed of the traveller being lower than that of the spindle. This is caused by the frictions between the spin ring and the traveller, the yarn tension, air resistance, et cetera. The winding-up is performed in conical layers from bottom to top, with the ring rail being switched up by one denier after every upward and downward motion.

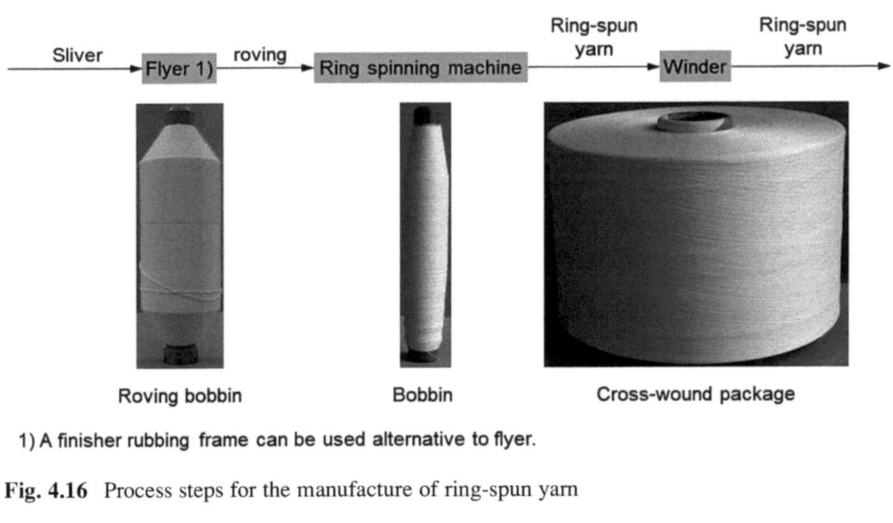

Fig. 4.16 Process steps for the manufacture of ring-spun yarn

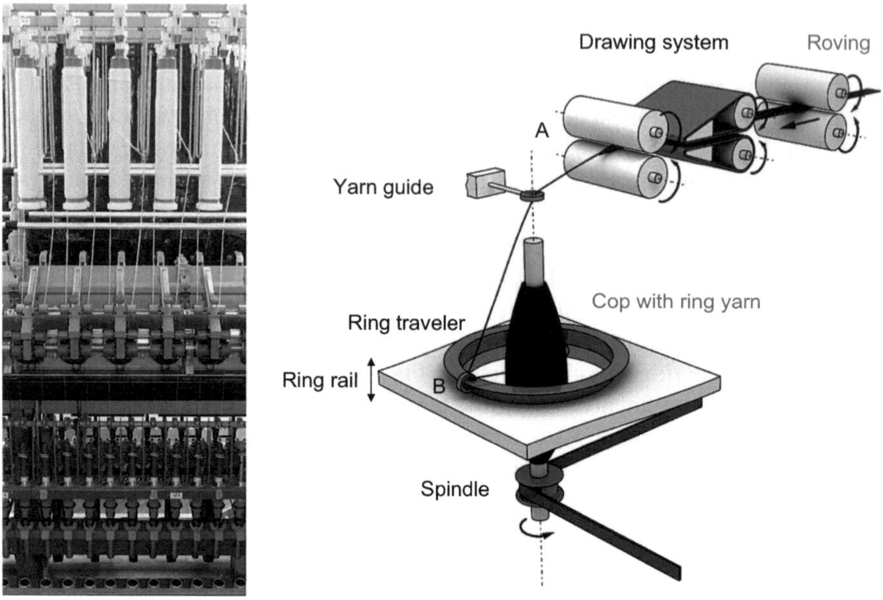

Fig. 4.17 Working elements of a ring spinning machine

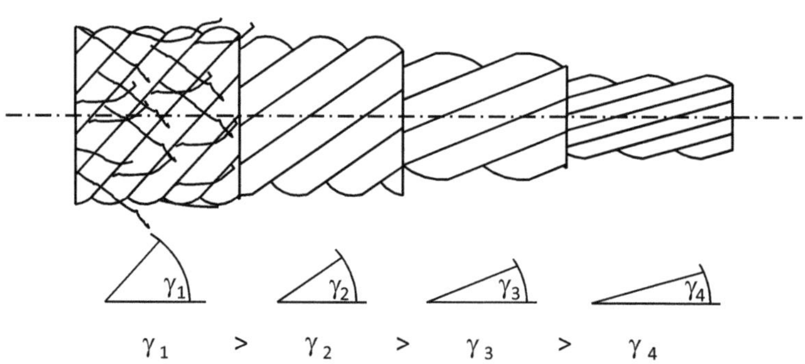

Fig. 4.18 Twist structure of the ring-spun yarn

During the processing of plantal natural fibers (especially long fibers from flax, ramie, jute), the roving is guided through a water bath before the drawing rollers in order to dissolve any pectins, making the fibers much easier to draft relative to one another.

The layered makeup of the ring-spun yarn twist structures (Fig. 4.18) results from the twist insertion, which works inward from outside. It gives the fibers of the exterior yarn structure the greatest gradient angle γ, while the gradient angle of the fiber layers decreases toward the yarn center. Especially in the outer layers of fibers wound around the yarn center, the fibers are elongated heavily. They seek to return to their original state, creating a high compacting pressure directed at the core. If fibers of the external yarn structure are loaded, e.g. by abrasion, the cohesion of the entire fiber compound is impaired. Due to their superior orientation in axis direction, the fibers in the yarn core primarily absorb tensile loads.

OE Friction-Spun Yarn Made from 100 % Staple Fibers

In contrast to ring-spun yarns, OE friction-spun Yarns are centrical core-sheath multi-component structures in which a variety of fibers can be used for core and sheath, while additional fibers (including short fibers >10 mm) can be mixed and specifically oriented in the yarn sheath. Usually, friction-spun yarns are produced as hybrid yarns.

The working area of the DREF 3000 is shown in Fig. 4.19 (ref. Fig. 4.15a, b). Optionally, a filament core can be added to the core (ref. Sect. 4.2.3). Several sheath slivers are opened into individual fibers by an opening roller, transported into the drum gap of the two perforated spinning drums rotating in the same direction by means of negative pressure, decelerated, and twisted into the open yarn end by a mechanical process in which the yarn rotates in the drum gap, where newly arriving fibers come into contact with it and are twisted in (friction). The yarn sheath is built

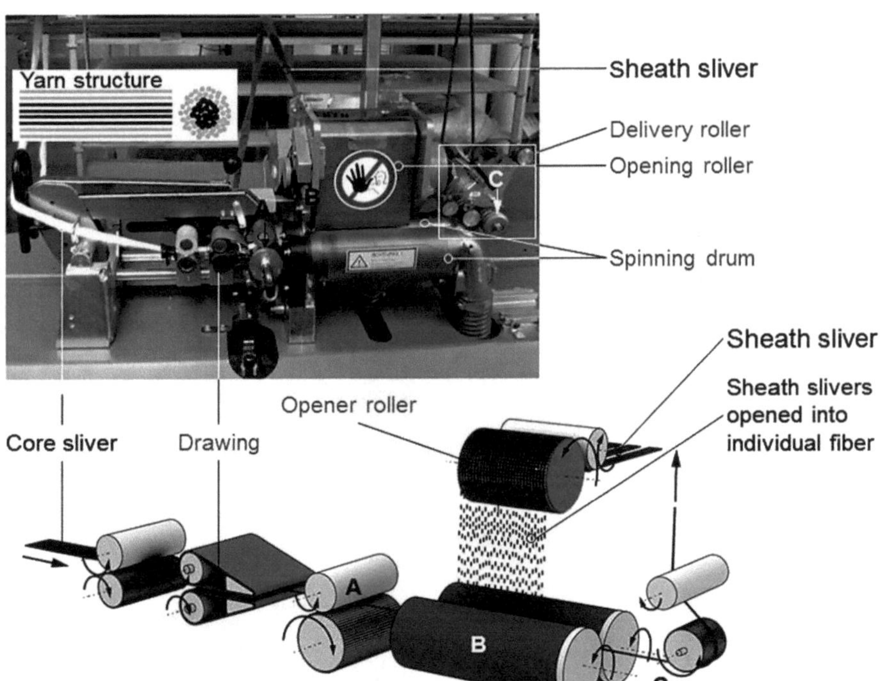

Fig. 4.19 Working area and functional principle of the OE friction spinning machine (DREF 3000) for the production of core-sheath structures made from 100 % staple fibers, and yarn structure (schematic, see inset in top picture)

up from the inside out in layer, i.e. by a specific introduction of various slivers a fiber blend in the sheath can be achieved. The lateral drawing rollers allow the drafting of a core sliver, preferably made from reinforcement fibers, to the desired fiber mass fraction in the yarn and its axial insertion into the spinning drums. The core fibers are given false twists by the spinning drums B in the area of the nip line of the final pair of drawing rollers A and the *twisting element*. This means that after passing the spinning drums, the fibers are again oriented largely parallel to the yarn axis. Therefore, the yarn strength results from the yarn twists of the sheath fibers, which compact the core fibers oriented in longitudinal direction. The imparting of twists to the sheath fibers is performed force-fit and depends on the pressure caused by the suction inside the spinning drums and the coefficient of friction between fibers and metal (spinning drum surface) or between the different fibers. The number of revolutions and the extraction of the spinning drums as well as the take-up speed have significant influence on the yarn structure in the yarn sheath.

In comparison to a ring-spun yarn, these yarns have a coarser and more voluminous surface, and a lower hairiness. If necessary, it can be spun into yarns of relatively high fineness, which will still be lower than that of a ring-spun yarn. An interruption of the fiber flow and the corresponding separation of twist insertion and winding, however, allow a significantly higher (up to tenfold) spinning speed than in ring spinning.

OE Rotor Yarn

Figure 4.20 (ref. Fig. 4.15 a, b) demonstrates the working area of the OE rotor spinning machine. In the rotor groove, a conical sliver is continuously formed from

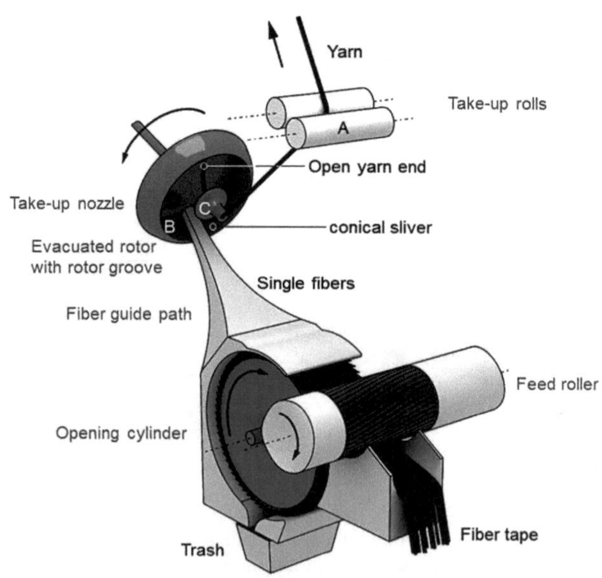

Fig. 4.20 Working area and functional principle of the OE rotor spinning machine

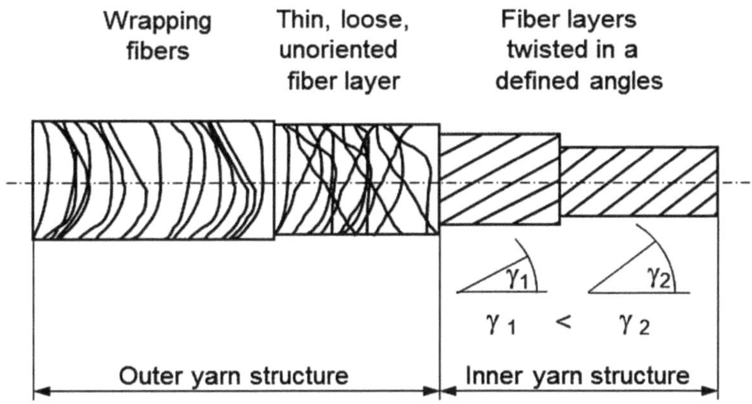

Fig. 4.21 Twist structure of an OE rotor yarn across the yarn cross-section

staple fibers opened in the opening roller, and oriented by means of air flow as well as centrifugal forces. The produced conical sliver displays a relatively constant fiber tension. At the thickest point of the sliver, fibers are twisted into the open yarn end rotating in the rotor B, with the twists spreading to the delivery rollers C.

The twist structure of the OE rotor yarn (Fig. 4.21) results from the imparting of the twists, which is performed from the inside out. The fibers located in the yarn core cannot evade the imparting of twists, which results in higher yarn twist and the corresponding higher packing density in the yarn core. Towards the outer yarn structure, the fibers can evade the imparting of twists increasingly well, making the yarn more voluminous. The most outer fibers, so-called wrapping fibers, are wound around the yarn core in the form of a band, with any orientation being possible. For process-inherent reasons, they are applied to the finished yarn rotating in the rotor. Underneath this is a thin fiber layer with few or partially opposing twists. The latter result from the superposition of permanent and false twists between the trumpet-shaped mouthpiece C (which often features a special notch design) for the imparting of false twists) and the rotating yarn take-up point in the rotor groove of the rotor B. If the fibers being applied to well-twisted yarn feature fewer permanent than false twists, the fibers are twisted in the opposite direction of the yarn twisting direction after the trumpet-shaped mouthpiece. As the outer fibers can therefore not contribute to yarn strength, the yarn has a lower maximum tensile load than a ring-spun yarn. OE rotor yarn is more abrasion-resistant because the inner, load-absorbing yarn structure is protected.

4.2.4 *Filament Spun Yarn*

Monofilament or multifilament yarns are combined with staple fibers in *filament spun yarns*. The basic classifications are as follows:

- Parallel staple fibers oriented in axis direction are wrapped with a monofilament or multifilament yarn: core-spun yarn,
- Parallel filaments oriented in axis direction are wound around with staple fibers—co-wrapped yarns: core ring-spun yarn, OE friction-spun Yarn with filament core, core OE rotor yarn, online hybrid co-wrapped yarn,
- Filament spun yarns without core-sheath structure as well as twist, and
- Intermingled filament spun yarns with or without elastomer.

Core ring-spun yarns in particular are often used as sewing yarns. In most yarns wrapped around with staple fibers, the resistance of the sheath fibers against abrasion between the filament core and the staple fiber sheath is small. This can cause problems in further processing, when the yarn is guided through yarn guides, needles and similar parts (which are also often mobile) at high speeds. The core of the yarn is partially uncovered, while the fibers are pushed together at one end of the exposed core, which can cause broken yarns during further processing [58, 59].

4.2.4.1 Core-Spun Yarn (Schappe Techniques, Charnoz, France)

Slivers or hybrid slivers made from parallel, homogeneously mixed reinforcement and thermoplastic fibers are stretched and wrapped around with a very fine filament yarn, e.g. made from the same thermoplastic polymer as the thermoplastic fibers, for example at 20–300 twists/m. The filament yarn makes up only a small percentage of the material composition of the yarn. It holds the fibers together by means of the applied outside pressure, thus ensuring yarn strength, which is commonly lower than that of ring-spun yarns. These yarns have a low packing density. Carpenter et al. (for natural fibers) [60] and Zhang et al. (for natural fiber/thermoplastic hybrid slivers) [61] have shown that, these yarns are more suitable for the use in composite materials than ring-spun yarns or similarly twisted structures.

The yarn finenesses are currently, for instance, 6 K, 3 K, or 1 K for CF. Figure 4.22 shows a typical core-spun yarn structure.

The production is more cost-effective that in Ring spinning and is usually performed on hollow spindle wrap spinning machines (Fig. 4.23, ref. Fig. 4.15a). After drafting by means of drawing rollers (nip A), the sliver is guided through a hollow spindle B, which is equipped with a spool holding the core-spun yarn. The spindle rotates and the filament yarn is wrapped around the sliver, which does not feature any twist after running through the delivery rollers (nip C).

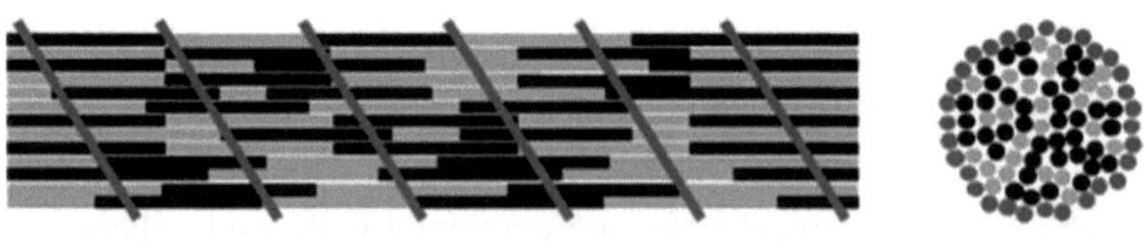

Fig. 4.22 Typical structure of a hybrid core-spun yarn (schematics)

Fig. 4.23 Working area of a hollow spindle wrap spinning machine

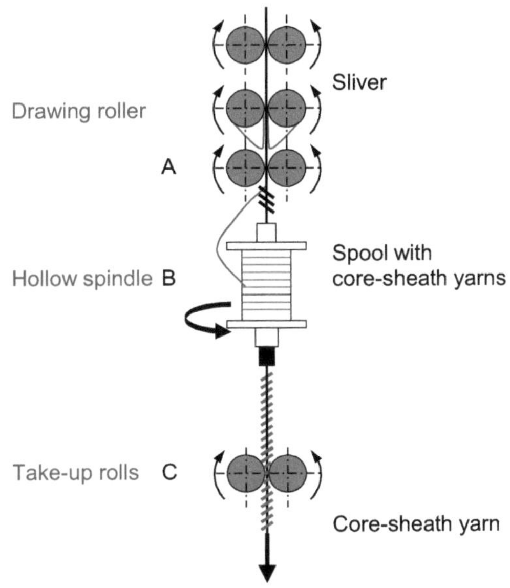

Thomanny et al. examine the influence of the yarn structure parameters (reinforcement fiber content ratio, fiber fineness, fiber length distribution, yarn fineness) on the characteristics of the fiber-reinforced plastic composite [62].

4.2.4.2 Core Ring-Spun Yarn

In the field of sewing yarn industry, a modified ring spinning process is common for the production of core yarns. At the final pair of the drawing rollers, the core yarn, e.g. an EL yarn or a CF filament yarn of defined stretching, which acts as the yarn core, is inserted separately and centrically [63].

To improve the resistance of the sheath fibers against abrasion it is recommended to set the pretension for filament yarn feeding relatively high and keep the twist in the filament yarn in the opposite direction of the rotational direction of the spindle [58].

4.2.4.3 Clustered Core Ring-Spun Yarn

The basic idea of this technology is based on the Solospun Technology developed in 1996 [64], which includes a modified filament feeding in which the delivery roller for filament insertion is furnished with fine grooves. The delivery roller separates the untwisted filament yarn into two to four filament groups called clusters. After the last pair of drawing rollers, these filament groups are merged again at different angles to the yarn axis, simultaneously forming several spinning

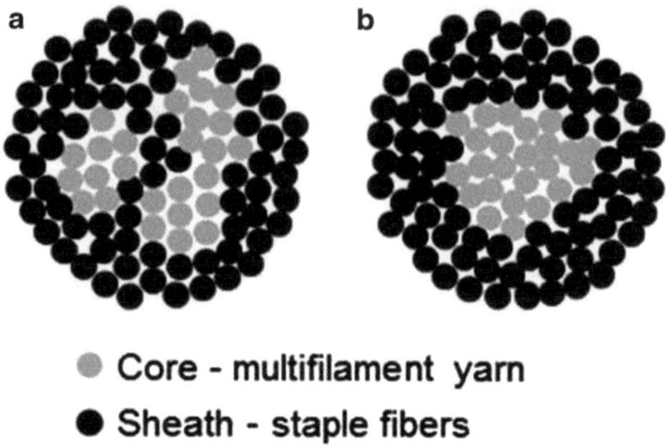

Fig. 4.24 Schematic yarn cross-section: (**a**) clustered core ring-spun yarn, (**b**) conventional core ring-spun yarn (based on [65])

triangles. The migration of staple fibers from the spinning triangle into a filament group or between the filament groups causes a good integration of the filaments into the yarn. The clustered core ring-spun yarn (Fig. 4.24) improves the mechanical yarn properties (maximum tensile load, maximum longitudinal elongation) [65].

4.2.4.4 OE Friction-Spun Yarn with Filament Core

The working area of the DREF 2000 is similar to that of the DREF 3000 (Fig. 4.19), but does not feature drawing rollers. The insertion of a core component is obligatory, while the DREF 3000 (see Sect. 4.2.3.3) offers the option of adding components such as monofilament, multifilament or spinning fiber yarns to the yarn core. These additional core components allow:

- Increased yarn tensile strength,
- A mechanical protection of the core component(s) by the staple fiber coating (usually from thermoplastic fibers), and
- The integration of functional components (see Sect. 4.6).

The components inserted into the core (Fig. 4.25) are imparted with false twists during their insertion between the yarn brake and the top of the perforated drums. These have to be set high enough to wrap the core component homogeneously and tightly, but have to be low enough to avoid filament breaks in the core component. Here, the pretension applied to the core yarn has significant influence on the false twist remaining in the core. It is highest in core yarns inserted without pretension [66]. Miao et al. recommend the following for higher resistance of the sheath fibers against abrasion:

- Application of slight pretension to the core yarn,
- Selection of identical twists direction for core yarn pre-twists and sheath fiber twists, and
- Inserting as many twists to the sheath fibers as possible.

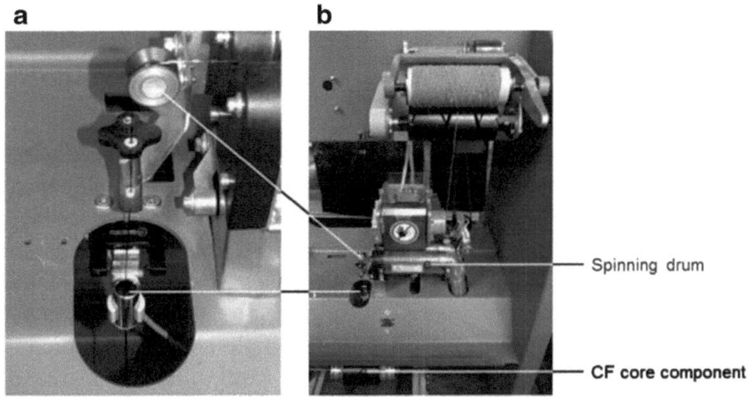

Fig. 4.25 Device for feeding a core component on the DREF OE friction spinning machine: (**a**) detail of the feeding device, (**b**) total view of spinning unit

Customary OE friction-spun yarns are offered by Schoeller Spinning Group Schoeller GmbH Co. KG and by Fischer Tech Garn GmbH.

4.2.4.5 Core OE Rotor Yarn

Core-sheath yarn in which a filament yarn, e.g. EL yarn is fed into the rotating rotor from behind through a hollow shaft during spinning, settling into the rotor groove in loop shapes. The staple fibers added during spinning are wrapped around this yarn core as usual. Fundamental researches are documented in [67].

In the comparison of core ring yarns and core OE rotor yarns, Yang et al. determine that the latter display a smoother surface, lower hairiness and high abrasion resistance, while the breaking load at lower coefficient of variation is smaller [68].

4.2.4.6 Online Hybrid Co-wrapped Yarns

Instead of the thermoplastic filament yarn as in online hybrid yarn spinning (see Sect. 4.2.2.3), the thermoplastic component is added in the form of staple fibers via a spinning vessel, and combined with the glass filament inside the spinning plant [69].

4.3 Plied Yarns for Semi-finished Yarn Products

4.3.1 Classification

Plied yarns are produced from two or more single yarns, with the yarns being twisted together or around each other. During this process, the fiber strand is geometrically modified. Cross-section, density, fiber orientation and surface roughness are changed, with the respective method resulting in different structures (Fig. 4.26). Thus, new yarn characteristics are created or properties of the single yarns are improved.

The fibers can be blended by combining single-component and/or multi-component yarns. The homogeneity of the blend across the yarn cross-section is lower than in single yarns, but linear functional components can be integrated easily, making unique multifunctional yarns in a great diversity of variants possible.

4.3.2 Flat Twisted Yarn

In comparison to single yarns, flat twisted yarns have the following advantages:

- Higher mass consistency/fewer yarn faults,
- Higher strength and elasticity,
- Higher volume and density,
- Rotational calming (removal of the snarling tendency), and
- Improved processability, e.g. lower sensitivity to abrasion, caused by good fiber incorporation, optimized spool format, no fabric distortion and no shifting of loops.

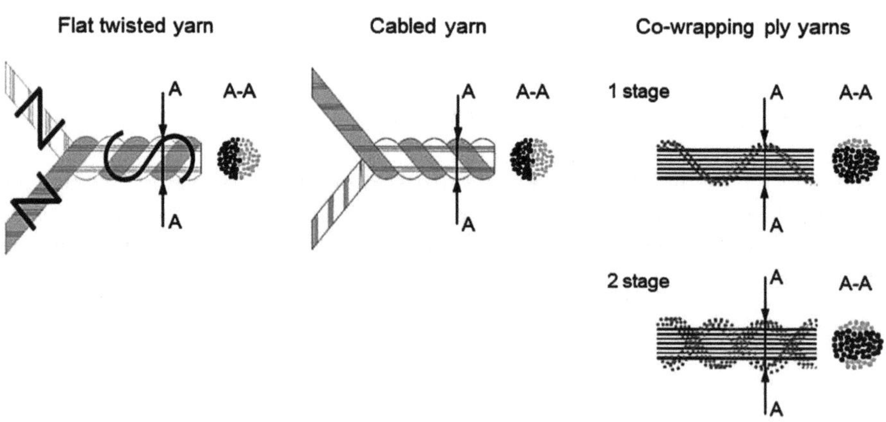

Fig. 4.26 Schematics of principal twisted yarn structures

Twisted yarns are often used as warp yarns in weaving, or as sewing threads, i.e. they are exposed to great strains during further textile processing and/or have to meet significant requirements in the composite material.

The surface of a twisted yarn is comparatively smooth. The yarns are twisted symmetrically with one another (Fig. 4.26a), usually using yarns of identical twist direction and fineness. A helical change of the yarns is performed on the respectively observed twisted yarn side. The frequency of the change depends on the construction of the twisted yarn. Construction variables include:

- Structure of the provided yarns (type, fiber material/blend, fineness),
- Folding (Number of merged yarns),
- Number and direction of twist of the yarns and of the twisted yarn, and
- Number of twisting processes (single-stage: from single yarns, multi-stage: from twisting, in combination with yarns, if applicable).

Table 4.3 contains examples for the classification of twisted yarns. The labeling of multi-stage twisted yarn will serve as a sample. The classification includes the first step of twisting three single yarns of 20 tex fineness and 1,055 twists/m in Z-direction at 420 twists per meter in S-direction. In the second step, two of these single twisted yarns are twisted in Z-direction at 280 twists per meter.

The most common twisted yarns are single-step two-ply and three-ply twisted yarns. The production of sewing thread is usually organized in two steps:

- 1st step: manufacture of the pre-twisted yarn, generally from single yarns made from identical or different fiber materials but identical yarn construction
- 2nd step: manufacture of the final twisted yarn from the initial twisted yarns, using single yarns where applicable

The twist imparted during twisting results in a change of length referred to as shortening by twisting e, arising from:

$$e = \frac{l_o - l}{l_o} \times 100\% \tag{4.9}$$

L (m) twisted yarn length

L_0 (m) Yarn length of the input yarns to be used (single yarn, pre-twisted yarn)

Information on the shortening caused by twisting is required for both twisting (to ensure a controlled yarn tension during winding) and planning the required twisted yarn length in further processing.

Table 4.3 Examples of typical twisted yarn labeling (tex system) [70]

Number of twisting processes	Abbreviated designation	Full designation
Single-stage	2 tex × 2	2 tex Z 600 × 2 S 400
Multi-stage, identical yarns		20 tex Z 1055 × 3 S 420 × 2 Z 280

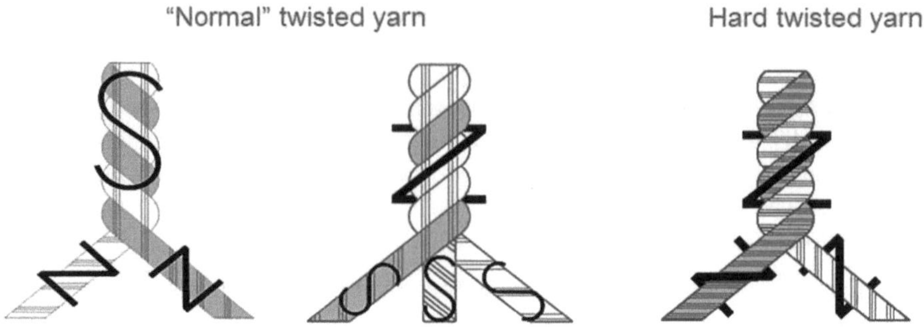

Fig. 4.27 Schematics of "normal" and hard-twisted yarns

In particular, the length reduction caused by twisting depends on the twist direction in the yarn and in the twisted yarn, with a distinction to be made between the "normal" twisted yarn and the hard-twisted yarn.

"Normal" twisted yarn (or counter-wire twisted yarn, as it is sometimes called) (Fig. 4.27a): The twist direction during twisting is opposed the yarn rotation direction. During the initial stages of the twisting process, the yarn (i.e. the pre-twisted yarn) is lengthened in the twisted yarn, and shortened again with increasing number of twists. When the twist torsion approximately matches the twist of the single yarn, the fibers are parallel to the twisted yarn axis after twisting. The reduced yarn twist of the single yarns in the twisted yarn results in:

- An increased volume of the twisted yarn,
- A softer handle in comparison to the single yarn,
- A rotational calming, and
- Prevention of product drawing and shifting of loops during fabric formation.

Stretch-broken slivers produced by Pharr (see Sect. 4.2.3.2) are processed into spun yarns (especially ring-spun yarns), which are then made into two-ply-"normal" twisted yarns. These are rotationally calmed and display up to 30 % increased in-plane shear properties (and thus improved impact strength), while tear strength and Young's modulus are circa 10–15 % lower than in filament yarns [49].

Hard-twisted yarn (Fig. 4.27b): The twists directions during yarn and twisted yarn production are identical. With increasing number of twists, the twisted yarn becomes shorter, and the yarn gradient angle increases, resulting in strong and hard twisted yarn. The fibers can be arranged almost perpendicular to the yarn axis after twisting. This creates a high tension of the fibers in the twisted yarn, with the ensuing counterforce causing a reverse torque upon unloading of the twisted yarn, which is referred to as snarling tendency. This snarling tendency increases with higher number of twists and impedes further processing (for example by causing a clinging between parallel-running yarns in the weaving process). Because of this snarling tendency, no such high twist as in "normal" twisted yarns is possible.

Apart from the twists number and twists direction, fineness of the used materials, the folding, yarn tension during twisting, the yarn type, and the fiber material characteristics (texturing, elongation, surface properties) all influence the shortening of the yarn by twisting.

Flat hybrid-twisted yarns: Twisted yarns produced by combining at least two yarns of different fiber material compositions, i.e.:

- Twisting of two yarns, where one yarn is preferably made from reinforcement filaments and the other of thermoplastic or staple fibers. In the twisted yarn cross-section, the reinforcement fibers and the thermoplastic fibers are alternately placed at the outside of the twisted yarn.
- Twisting of two hybrid single yarns.

Flat twisted yarns are produced by means of ring twisting, two-for-one or three-for-one twisting. With the exception of ring twisting, all of these are up-twisting processes, i.e. the twisted yarn is wound up on a cross-wound package. Usually, the yarns are folded in a separate process before twisting. For threefold and multifold twisted yarns, as well as in the processing of spun yarns, folding is always performed as a separate process. For instance, two to six yarns are wound up parallel on a spool as a hank and imparted with a small permanent twist (15–20 m^{-1}) or with false twist by means of an intermittent air flow. This has the following advantages:

- Ensuring identical yarn tension of the yarns to be twisted,
- Additional possibilities to eliminate yarn faults,
- Removal of twist faults caused by unwanted continued twisting after yarn break of a single yarn or after a feed spool runs empty, and
- Higher productivity of the twisting machine.

Ring twisting: A ring twisting machine is similar to a ring spinning machine in structure and working principle (Fig. 4.17). Instead of the flyer spools, cross-wound packages, carrying single yarns, folded yarns or twisted yarns are used. The drawing rollers are replaced by a delivery arrangement. The winding of the twisted yarn on the twine bobbin is performed either as on a ring spinning machine or simultaneously along the entire bobbin length. To achieve a smooth, radiant and closed surface, an integrated humidifier is used to apply a wetting agent to the yarns before the delivery arrangement.

Two-for-one twisting (Fig. 4.28, ref. Fig. 4.15a): The folded yarn is drawn off overhead from a stationary cross-wound package (up to six single yarns on one spool) held in place on the rotating spindle by means of magnets. Alternatively, single yarns placed on special spools can be used. The yarns run through a yarn brake and a hollow axle before reaching the protection pot through the opening in the storage disk and via the rotor. The yarn is deflected by 180° and rotates around the protection pot as a yarn balloon. In the twisted yarn, two twists are created in one spindle revolution: the first twist between yarn brake A1 and spindle rotor B, the second between spindle rotor and balloon yarn guide A2. The height of the yarn balloon is limited by the balloon yarn guide.

Three-for-one twisting (Tritec method, Fig. 4.29, ref. Fig. 4.15a): In contrast to two-for-one twisting, two pivoted spindle systems with cylindrical yarn guiding

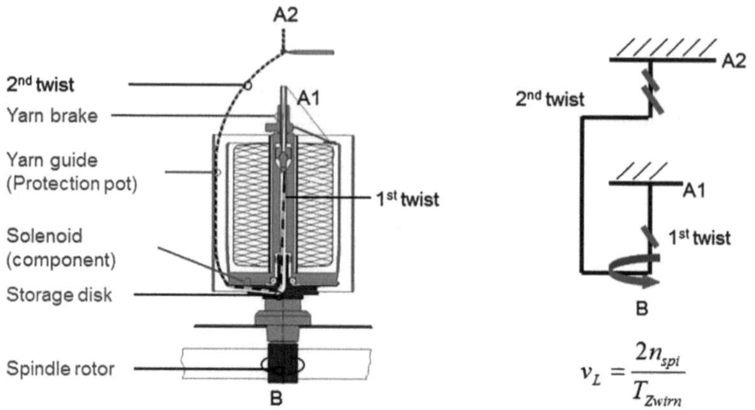

A1 Clamping point at the yarn brake
A2 Clamping point at the yarn balloon guide
B Clamping point and twisting element (spindle rotor)

Fig. 4.28 Working area and functional principle of a two-for-one twisting machine (*Source*: Oerlikon Saurer Allma und Volkmann product lines), (**b**) basic principle of twist insertion

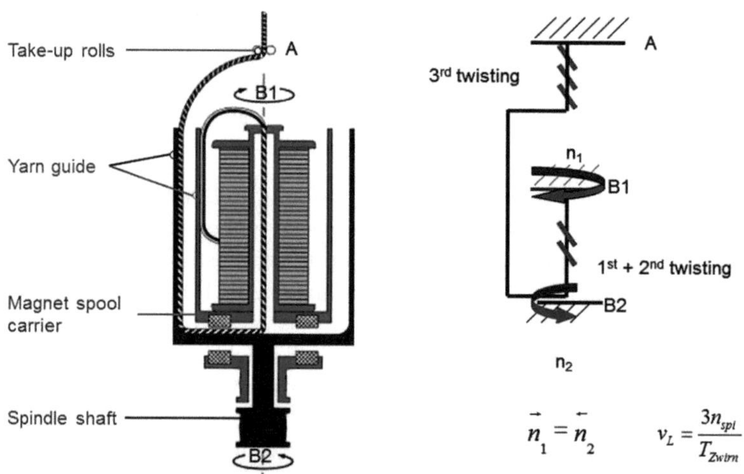

A Clamping point at the take-up rolls
B1 Clamping point and twister (spindle inner part)
B2 Clamping point and twister (spindle outer part)

Fig. 4.29 Working area and functional principle of a three-for-one twisting machine (*Source*: Oerlikon Saurer Allma und Volkmann product lines), (**b**) basic principle of twist insertion

elements are used. The outer and the inner system rotate in opposite directions have identical rotational speed. The folded yarn is thrown off the spool by the twist of the inner system and forms a self-adjusting storage on the inner yarn guiding element. It reaches the hollow shaft of the spindle and is imparted with two twists by the spindles B1 and B2 rotating in opposite directions. The outer yarn guiding element forms a second storage for the twisted yarn, which creates a third rotation of the twisted yarn between the outer spindle system B2 and the delivery rollers A.

Twisting methods for process shortening: *Sirospun* process (for staple fiber yarns): Spinning and twisting are performed in one process step. The rovings are separated and guided parallel through the drawing rollers on the ring spinning machine (ref. Fig. 4.17), and are thus drawn separately. The twists imparted by the ring spindle spread through both slivers to the nip of the last pair of rollers, while both slivers are simultaneously twisted with one another in yarn rotation direction, creating the twisted Siro yarn.

4.3.3 Cabled Yarn

Cabled yarn is often referred to as cordonnet or *tire cord*. The yarns, which are usually filament yarns of identical fineness, are twisted around one another, without a twist being imparted on the individual yarns (ref. Fig. 4.26b). This results in a twisted yarn of level tension, in which the fibers in the yarn are oriented in the axis direction of the twisted yarn. This results in the following advantages over conventional twisted yarns:

- Ensured parallel fiber orientation in yarn axis direction and resulting excellent utilization of substantial strength,
- Volume retention,
- Minimized risk of fiber damages during yarn processing, and
- More economic process management.

Direct cabling (Fig. 4.30, ref. Fig. 4.15): Cabling is a special wrapping process. The first yarn (outer yarn feed and nip A) is taken up from an external creel. By means of a drive, it rotates around the second yarn (inner yarn feed and nip C) above the spindle motor B. The second yarn is placed stationary on a spool and is taken up overhead. The outer yarn forms a yarn balloon after it leaves the deflection part of the motor spindle. In the clamping point, both yarns are cabled, i.e. wound around one another at identical yarn tension without changing the number of twists in the single yarn.

To ensure a symmetrical wrapping the yarn tensions of inner and outer yarn have to be identical. If this is not the case, cabling can also be used to produce co-wrapping ply yarns (ref. Sect. 4.3.4).

On the CC3 (Fig. 4.30b), cabling as well as the abovementioned processes of twisting around the central axis and two-for-one twisting can be performed without modifications of the machine.

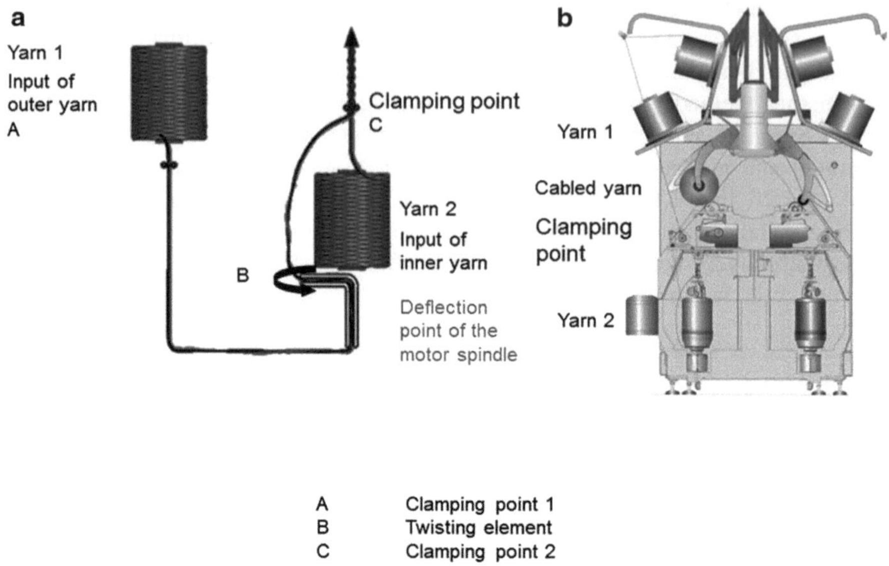

Fig. 4.30 (**a**) Basic principle of cabling, (**b**) machine cross-section of the Allma CC3 (*Source*: Oerlikon Saurer Allma und Volkmann product lines)

4.3.4 Co-wrapping Ply Yarns

The term *co-wrapping ply yarn* is consciously selected in deviation from DIN 60900-1 (July 1988) for reasons of the systematic classification of twisted yarns, as the standard classifies such yarns as core-spun yarns. The yarns are not symmetrically twisted with one another. A non-elastic or elastic yarn—the so-called filler thread—is located in stretched form in the center of the twisted yarn, helically twined around with one or several yarn or twisted yarn turns, which cover it more or less thoroughly. This results in yarns with heavily ribbed surface structure and larger surface (external yarn structure). An additional compression of the inner yarn structure can be achieved by the suitable selection of process parameters.

In a single-step co-wrapping ply yarn, the coating of the filler thread is performed with one winding layer in Z or S direction (helical rib structure), for a two-stage co-wrapping ply yarn, two stacked winding layers with opposite winding direction are used (ref. Fig. 4.26c).

The relevant constructive variables include:

- Yarn and twisted yarn construction of core and sheath yarn(s) (type, fiber material/blend, fineness),
- Number of sheath yarn/number of core yarns,
- Number of windings and twist direction of the sheath yarn(s), and
- Number of twisting processes (single-stage, two-stage or multi-stage).

Labeling: The number of twists is given by the letter W, the twists direction (S or Z), and the following number of twists per meter (nominal number of twists), e.g. W Z 400.

A *composite twisted yarn* has a filament yarn for a core (e.g. from 100 % reinforcement fibers), sheathed by a spun yarn made from 100 % natural fibers.

Co-wrapping ply yarns are usually produced on hollow spindle effect twisting machines (ref. Fig. 4.23), with a delivery arrangement replacing the drawing rollers. The filler yarn is guided through a hollow spindle and covered with the sheath yarn placed on a flanged bobbin at a defined number of twists per meter.

Alternatively, ring/two-for-one twisting and Tritec methods as well as cabling are possible options.

To achieve an asymmetric structure in ring/two-for-one twisting and Tritec methods, the following prerequisites have to be met:

1. Of the equal-length yarns to be used, one has an S-twist, the other a Z-twist. The yarn sharing a twist direction with the twisted yarn is shortened and forms the filler thread, while the other yarn und is wound helically around the filler thread.
2. The yarns are fed into the twist-creating device at different speeds. The slower yarn forms the filler thread and is thus covered by the faster yarn.
3. Of the equal-length yarns to be used, one yarn is of lower fineness. This yarn is wound helically around the finer yarn.

In cabling processes [71], the yarn tensions of both yarns are set to make the sheath yarn helically wrapped the yarn oriented stretched in the core. The gradient of the sheath yarn can be varied by means of the machine parameters of the rotational speed of the spindles and take-up speed.

Combination of all twisting methods in one machine (laboratory machine): On the DirecTwister 2B developed by AGTEKS [72], two independently operating working areas can be used to perform five different processes: Twisting around the central axis, ring twisting, two-for-one twisting, hollow spindle processes, and cabling. Up to eight yarns, including elastic material can be processed simultaneously.

4.4 Recommendations for the Processing into Semi-finished Yarn Products

Every yarn has a "biography" stored within it. In all processes, they are subjected to mechanical loads (e.g. by yarn tensions, friction, air resistance, flexural and compressive forces in the coil form), and the yarn structure is altered by the resulting fiber migration. This can irreversibly change the properties of the yarn (e.g. a change of molecular structure or macroscopic breaking of individual fibers by local overelongation of individual fibers and/or yarn sections). The consequences often only become apparent in later processing steps. If required (especially in case of existent snarling tendency), the yarns should be treated thermally to relax them. Rewinding improves the flow and economy of later process steps, as the yarn is:

- de-dusted,
- cleaned (removal of thick places, thin places and neps),
- paraffin-waxed (antistatic finishing, and decrease of friction, e.g. for a further processing in knitting), and
- given the required packaging.

The following steps are recommended for a gentle processing of the reinforcement fibers, to counter their low yarn elongation, the high coefficient of friction, high brittleness and stiffness in bending, and their sensitivity to transverse compression:

- To prevent demixing or the destruction of yarn morphology under mechanical loads, a high yarn cohesion in the yarn has to be ensured (e.g. by rotating or intermingling). Furthermore, yarn tension peaks must occur during yarn transport or yarn shaping.
- To avoid irreversible yarn damages, the following measures can be taken:
  - Minimization of the number of process steps,
  - Yarn tension reduction, e.g. by using positive storage feeders in knitting and weaving weft yarns, and stabilization of yarn tension in the process,
  - Yarn friction reduction, e.g. by modification of yarn guide elements (number, geometry, surface) and/or by finishing yarns with appropriate sizings (ref. Sect. 3.2.4),
  - Adjustment of machine speeds, and
  - Optimization of ambient conditions (temperature, humidity, lighting), including the thorough covering of UV-sensitive fibers (AR) on the machine, e.g. with black foil.

If electrically conductive fiber fragments (e.g. from CF and functional yarns) occur, the controls of the machine have to be encapsulated.

4.5 Sewing Thread for the Assembly of Textile Semi-finished Products

4.5.1 Introduction

A tendency to delaminate is a problem in all thermosetting and thermoplastic composites. One interesting option to minimize this tendency is the sewing of the multilayered textile reinforcement fabrics, providing reinforcement yarns in z direction, in addition to those reinforcement yarns present in the fabric. These additional reinforcement yarns counter the progress of the delamination.

Sewing (ref. Sect. 12.6) and embroidery (ref. Chap. 10) are proven textile joining methods for the positioning and assembly of textile reinforcement fabrics.

Sewing in order to insert reinforcement yarns in z direction is therefore a universal alternative to textile fabric formation methods producing spatial structures by means of braiding, multilayered weaving, 3D weaving or biaxial knitting [73].

Furthermore, sewing can be used to form any three-dimensional structures from textile fabrics in one or more layers, or to assemble them from several individual components.

Special *sewing threads* are used, as the sewing thread is exposed to extreme loads in the sewing machine. Stitch formation (ref. Sect. 12.6.2) is ensured by cyclically and actively moving a relatively long length of thread at only minimal thread consumption. Abrasion, thread flexions in the yarn guide elements with minimal transitional radii, and cyclical tensile loads in connection with reversals of direction affect the sewing thread before it assumes its connecting and bearing function in the seam. As processing conditions are critical in the formation of the double lock stitch used in sewing and embroidery techniques, the stability of the developed sewing threads for the double lock stitch is important. The double lock stitch is characterized by a great active thread length of the upper thread during stitch formation. An appropriately-sized needle thread loop has to be laid around the bottom thread spool and afterwards returned to the original position by means of the thread take-up lever. The strains during sewing can result in a significant deterioration of the sewing thread strength in comparison to the original strengths after thread production.

An urgent task is presented by the use of suitable sewing thread made from CF and GF with or without thermoplastic components (hybrid yarn). Only undisturbed and low-damage sewing and embroidery processes ensure a positive and reproducibly z-reinforcing effect of the reinforcement textiles on the delamination behavior of the composite component.

4.5.2 Sewing Thread Construction

4.5.2.1 Sewing Process

By means of the versatile parameter variation of stitch type, sewing needle geometry, sewing thread type, sewing thread fineness, stitch length, seam distance and seam direction can be set load-appropriate property characteristics of the textile seams during the sewing process.

Using reinforcement threads and fabrics, sewing process experiments render experiences and knowledge used for more complex material tests of the z reinforcements and pre-fabricated textile structures. These researched components can then be used in consolidated textile structures at optimized process parameters.

Here, packages of multilayered textile reinforcement semi-finished products with a total thickness of up to 20 mm are processed [74–77].

Stitch type selection: To sew textile reinforcement semi-finished products, the double lock stitch is the most suitable choice from a range of stitch types. The double lock stitch inherently features a yarn loop, which is usually located in the center of the work-piece and significantly reduces the strength of sewing threads made from brittle materials such as glass and carbon. The center of the composite is also the point of maximum interlaminary shear stresses in flexurally loaded fiber-reinforced plastic composites and of maximum principal stresses in z direction in joined textile-reinforced structural components (such as overlapping joints). By varying the tensions of upper and bottom thread, the loop can be displaced on the work piece in direction of top or bottom side, so that sewing threads are stretched in the critical section, thus avoiding this type of weakness.

Thread tension/thread brake settings: Apart from the position of the interlooping point, thread tension also influences the orientation of the seam in the consolidated fiber-reinforced plastic composite. The multilayered and sewn textile reinforcements are compressed to a fraction of the thickness during composite consolidation, creating a compact composite structure from an originally loose package of material layers. At low thread tension, however, the sewing thread in the needle hole has no chance to retain the stretched orientation created during sewing and will transition into a clinched random orientation after the reduction of composite thickness. This creates special requirements for the sewing process and selection of seam parameters.

Increasing the tension of upper and bottom thread and operating at a high sewing foot pressure can achieve a tendency of the multilayered reinforcement textiles to be compressed to the thickness resulting from consolidation.

Perforation effects on the textile reinforcement fabrics: The pricking of the layered stack of reinforcement textiles by the sewing needle is inevitably accompanied by a certain perforation effect. Apart from needle geometry, the crucial influence variable is stitch length or stitch density. Tensile tests of the reinforcement textile layers partially perforated transversely to the load direction by sewing should be used to minimize the degree of strength reduction caused by perforation.

Sewing thread strength reductions by strains of the sewing process: For the needle thread in particular, the sewing process is associated with abrasive and friction strains on active pair elements. Stronger (and thus, thicker) sewing threads can improve the safety of the sewing process, but they also require stronger needles, which increase perforation of the textile fabric, decreasing the strength of the textile composite. Assessing the loss of strength caused by processing strains is difficult, as the sewn thread has to be removed from the seam carefully. This is done, for instance, by cutting the bottom thread by means of severing every single stitch in order to be able to remove the needle thread from the seam without considerably further strains. A comparison of the maximum tensile load measurements (ref Sect. 14.4) of the sewed and unsewed thread with a tensile tester yields the desired statement, given a sufficient sample size.

If the seam is exposed to loads, the sewing thread in the stitch interlooping is strained in such a manner that makes an additional assessment of the interlooping strength of the sewing thread necessary.

4.5.2.2 Sewing Thread Construction

With regard to heat resistance, CF and GF filaments are suitable for sewing threads. Their suitability for a mechanical sewing process is limited by their brittleness. However, they have to function as reinforcement for seam joinings under corresponding thermal influence.

To counter mechanical strains during sewing and embroidering, the reinforcement threads are completed with PEEK thermoplastic matrix material.

With respect to the methods explained in Sects. 4.2 and 4.3, the following sewing thread constructions are suitable for the ready-made processing of textile semi-finished products:

Core ring thread: Core-sheath thread production uses the core ring yarn method. The aim is an optimized sheathing of the CF filament yarn with PEEK fibers (Fig. 4.31). By sheathing the filament thread with PEEK fibers, the yarn gains a certain flexibility which improves the sewing process.

OE friction-spun thread: OE friction spinning relies on the sheathing of filament threads with staple fibers (Fig. 4.32). The result is a hybrid thread with a core-sheath structure. The advantages of this sheathing method are primarily in the centric and twistless position of the filament core. In addition, the complete encasement of the core offers protection for the sensitive reinforcement filaments.

Co-wrapping ply thread: Co-wrapping ply threads (Fig. 4.33) are produced by inserting a ground thread (CF filament yarn) and an effect thread (PEEK filament

Fig. 4.31 Core ring thread, 152 tex fineness, CF filament yarn (T300, 67 tex) core, co-wrapped with PEEK fibers (85 tex)

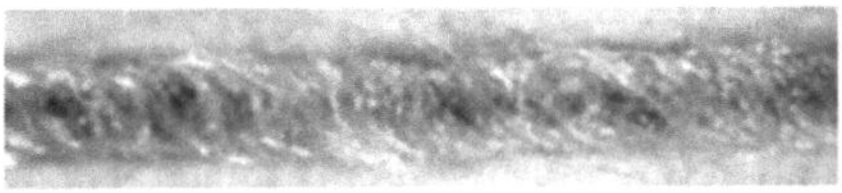

Fig. 4.32 OE friction-spun thread with filament core, 150 tex fineness, CF filament yarn (T300, 67 tex) core, wound about with PEEK fibers (83 tex)

Fig. 4.33 Co-wrapping ply thread, 116 tex fineness, CF filament thread (T300, 67 tex core), wound about with PEEK filament thread (49 tex)

Fig. 4.34 Multifold twisted thread, 232 tex fineness, initial thread made from CF filament thread (T300, 67 tex) and PEEK filament thread (49 tex), pre-twisted thread made from two initial twisted thread

yarn) into the hollow spindle of the twisting machine. To ensure a gentle enwinding of the CF filament yarn with the PEEK filament thread, the twisting element is removed. According to the selected machine parameters, the rotational speed of the spindles, and the delivery, the number of twists per meter can be varied.

Multifold twisted thread: The twisted thread is produced directly from the original spool, as the material is exposed to high mechanical strains during the folding process. During twisting, the CF and PEEK filament threads are processed into a two-step twisted thread (Fig. 4.34). The pre-twisted thread consists of different combinations of CF and PEEK filament thread. The gradient angle of $\alpha < 80°$ is deliberately kept small to prevent damages to the filaments. Despite gentle processing, damages of the CF filaments are occurred.

4.5.3 Sewing Thread Function

For the sewing of multiple textile layers for composite applications, the required seam functions are redefined:

Safety seam for cut edges: The cutting process involves a protection of the cut edges to prevent thread loss there. Overlock seams of stitch types 501–504 according to DIN 61400 are exceptionally suitable for this purpose, as well is the use of a double chain stitch seam of stitch type 401 (Safety Stitch). The overlock seams can be applied circumferentially to the cut contour on every individual cut part or directly on the stack arranged congruently with the component. After the plastic impregnation process, the yarn orientation is fixed by the matrix material, which makes the function of the cut edges safety seam a temporary requirement.

Positioning seam: To prevent displacements of the individual layers, they have to be positioned. The function of the positioning seam becomes obsolete after completion of the consolidation process.

Forming seam: Folding or other spatial deformations of the textile fabrics create geometrical reinforcement textile arrangements fixated by shaping seams. After the plastics process, the function of the forming seam becomes obsolete.

Reinforcing seam: The delamination tendency is countered by reinforcing seams, which can be distributed across the entire surface or limited to critical areas of the reinforcement textile layers. The reinforcing seam only serves its purpose in the consolidated composite, but also helps position the material during textile assembly. Forming seams also act as reinforcing seams in the consolidated composite.

Assembling seam: More complicated component geometries often require more than one process step. Here, one or more steps are necessary to manufacture a textile preform from several individual textile parts, which can contain other types of seams. The seams used in these processes are called assembly seams, and they fulfill their purpose primarily in the textile preform. Depending on component geometry, they can also act as reinforcement in the consolidated component. Assembly seams are not limited to use in the surface normal. Using special sewing technology, they can also be applied for shaping and reinforcement purposes at variably adjustable angles to the surface normal.

4.6 Functional Integration

Functional integration is achieved by fiber and/or yarn blending. Exemplarily, three instances will be considered in detail:

- The functional components, e.g. electrically conductive fibers, are inserted as monofilament or filament thread during air texturizing or as core components during friction, ring, or rotor spinning.
- Threads contain hollow glass filaments serving as reinforcement and containing agents for self-repair, for which the principal possibilities are given in [78].
- Special fibers made from shape-memory alloys are used for the vibration damping of FRPC components [79].

4.7 Thread Finishing

The after-treatment and surface preparation of filaments for improved processability by means of textile auxiliaries, such as finishing agents, are described in Sects. 3.2.4 and 13.5. Finishing processes can be integrated into yarn formation machines.

4.8 Significance of Threads for Fiber-Reinforced Plastic Composites

This section gives an overview of the Threads currently used primarily in lightweight construction, serving as basic elements for the production of textile semi-finished products and their assembly.

The Threads consist entirely of reinforcement fibers or a blend of reinforcement and matrix fibers (thermoplastic FRPCs) and can contain other functional fibers. They produced from filaments and/or staple fibers by means of various technologies, so that thread structure and yarn properties can be customized according to the

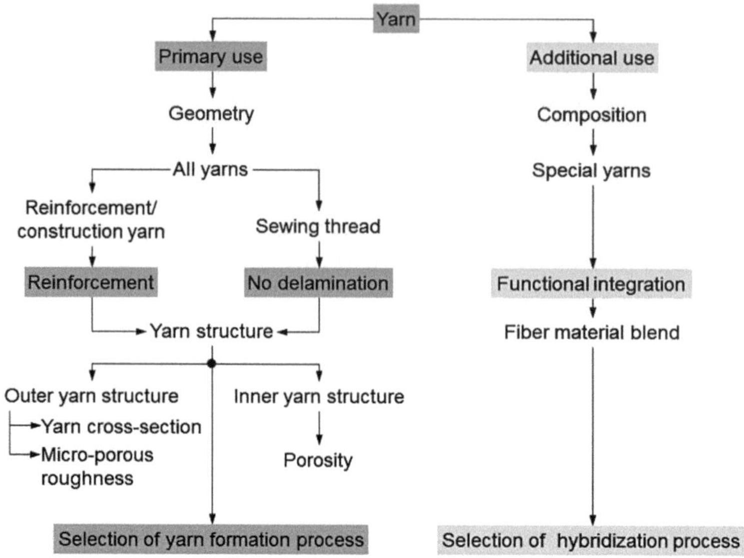

Fig. 4.35 Criteria for the selection of thread formation and hybridization methods, according to manner of modification (thread geometry, fiber material composition)

respective functional requirements. With due regard to the criteria detailed in Fig. 4.35, the yarn formation and hybridization methods can be preselected. Afterwards, the suitable process parameters have to be determined. Single threads can be processed into multiple yarns by further process steps of mechanical yarn processing in order to improve the mechanical characteristics.

During textile processing, composite manufacture and an in the final composite, the thread has to fulfill sometimes contrary requirements, which is determined by the thread structure. These are shown by the following examples:

- If the thread consists of filaments with twist and the resulting excellent thread cohesion, it is well-suited for textile processing. The impregnation with matrix material is complicated by the thread twist during composite production. In the composite, stiffness is limited, as the thread is characterized by a structural elongation. This improves drapeability, making single and even complex 3D geometries possible.
- If the thread consists of filaments without twist or chemical adhesion, they are much harder to process by textile means, but can be impregnated far more easily during composite manufacture. In the FRPC, the stretched orientation of the filaments results in excellent stiffness, but also signifies the lack of structural elongation for draping. Therefore, these structures are primarily suited for use in 2D geometries.

These two examples alone illustrate that textile engineers, FRPC manufacturers and users alike have to be involved in component development. All of them require basic knowledge of the thread construction parameters and the possibilities to adjust them by means of thread formation.

The ideal thread structure for the manufacture of semi-finished products is likely unachievable. The current thread formation technologies with their modification options offer numerous innovative possibilities to adapt yarns to the matrix-specific and functional requirements in the composite material, always with due regard to excellent textile processability. This enlarges the commercially available range of threads. Machine engineering will offer more flexible yarn formation solutions and machines, which will allow the utilization of various technologies. A tendency towards increasingly fine fibers is obvious, which will result in better properties in the composite at lower material consumption as well as a more homogeneous blending of different fiber materials in the hybrid threads.

References

1. LONG, A. C.;WILKS, C. E.; RUDD, C. D.: Experimental characterisation of the consolidation of a commingled glass and polypropylene composite. In: *Composites Science and Technology* 61 (2001), No. 11, PP. 1591–1603. DOI 10.1016/S0266–3538(01)00059–8166
2. FELEKOGLU, B.; TOSUN, K.; BARADAN, B.: Effects of fibre type and matrix structure on the mechanical performance of self-compacting micro-concrete composites. In: *Cement and Concrete Research* 39 (2009), No. 11, pp. 1023–1032. DOI 10.1016/j.cemconres.2009.07.007
3. SVENSSON, N.; SHISHOO, R.; GILCHRIST, M.: Manufacturing of thermoplastic composites from commingled yarns - a review. In: *Journal of Thermoplastic Composite Materials* 11 (1998), No. 1, pp. 22–56. DOI 10.1177/089270579801100102
4. BEYREUTHER, R.; BRÜNIG, H.; VOGEL, B.: Preferable filament diameter ratios of hybrid yarn components for optimized long fibre reinforced thermoplastic. In: *Proceedings. 17th Annual Meeting of the PPS*. Montreal, Canada, 2001
5. BRÜNIG, H.; BEYREUTHER, R.; VOGEL, R.; TÄNDLER, B.: Melt spinning of the fine and ultra fine PEEK-filaments. In: *Journal of Materials Science* 38 (2003), No. 10, pp. 2149–2153. DOI 10.1023/A:1023719912726
6. GOLZAR, M.; BRÜNIG, H.; MÄDER, E.: Commingled hybrid yarn diameter ratio in continuous fiber-reinforced thermoplastic composites. In: *Journal of Thermoplastic Composite Materials* 20 (2007), No. 1, pp. 17–26. DOI 10.1177/0892705707068069
7. YE, L.; FRIEDRICH, K.; KASTEL, J.; MAI, Y. M.: Consolidation of unidirectional CF/PEEK composites from commingled yarn prepreg. In: *Composites Science and Technology* 54 (1995), No. 4, pp. 349–358. DOI 10.1016/0266–3538(95)00061–5
8. ALAGIRUSAMY, R.; FANGUEIRO, R.; OGALE, V.; PADAKI, N.: Hybrid yarns and textile preforming for thermoplastic composites. In: *Textile Progress*. Cambridge : Woodhead Publishing Limited, 2006
9. CURBACH, M.; JESSE, F.: Eigenschaften und Anwendung von Textilbeton. In: *Beton- und Stahlbetonbau* 104 (2009), No. 1, pp. 9–16. DOI 10.1002/best.200800653
10. Protective right US 005182839 (2nd February 1993).
11. Protective right US 5200620 (6th April 1993).
12. JOU, G. T.; EAST, G. C.; LAWRENCE, C. A.; OXENHAM, W.: The physical properties of composite yarns production by an electrostatic filament-charging method. In: *Journal of the Textile Institute* 87 (1996), No. 1, pp. 78–96. DOI 10.1080/00405009608659058
13. Protective right US 4874563 (17th October 1989).
14. Protective right EP 0486884 A1 (7th November 1991).

15. Protective right EP 0364874 (25th April 1990).
16. BALASUBRAMANIAN, N.; BHATNAGAR, V. K.: The effect of spinning conditions on the tensile properties of core-spun yarns. In: *Journal of the Textile Institute* 61 (1970), No. 11, pp. 534–554. DOI 10.1080/00405007008630021
17. Protective right US 5094883 (10th March 1992).
18. KLETT, J. W.; EDIE, D. D.: Flexible towpreg for the fabrication of high thermal conductivity carbon/carbon composites. In: *Carbon* 33 (1995), No. 10, pp. 1485–1503. DOI 10.1016/0008–6223(95)00103–K
19. FITZER, E.; MANOCHA, L. M.: *Carbon Reinforcements and Carbon/Carbon Composites*. 1st Edition. Berlin, Heidelberg, New-York : Springer Verlag, 1998
20. Protective right US 005241731 (7th September 1993).
21. Protective right US 3704485A (5th December 1972).
22. Protective right US 4539249 (3rd September 1985).
23. WU, W. Y.; LEE, J. Y.: Effect of spread width on the structure, properties, and production of a composite yarn. In: *Textile Research Journal* 65 (1995), No. 4, pp. 225–229. DOI 10.1177/004051759506500406
24. Protective right DE 102009029437 A1 (25th March 2010).
25. CHERIF, Ch.: Entwicklung eines Verfahrens zur schädigungsarmen Durchmischung vonMultifilamentgarnen (IGF-No. 14686 BR) / TU Dresden, Institut für Textil- und Bekleidungstechnik. Dresden, 2008. – Final report
26. Cytec Industries Inc.: *Cytec Engineered Materials* http://www.cytec.com/engineeredmaterials/prepreg.htm (24-04-2010)
27. OCV Reinforcemcents: *Twintex*®. http://www.twintex.com/ (16-04-2010)
28. MÄDER, E.; ROTHE, Chr.; BRÜNIG, H.; LEOPOLD, Th.: Online spinning of commingled yarns - equipment and yarn modification by tailored fibre surfaces. In: *Key Engineering Materials* 334-335 (2007), pp. 229–232. DOI 10.4028/www.scientific.net/KEM.334–335.229
29. MÄDER, E.; RAUSCH, J.; SCHMID, N.: Commingled yarns - Processing aspects and tailored surfaces of polypropylene/glass composites. In: *Composites Part A: Applied Science and Manufacturing* 39 (2008), No. 4, pp. 612–623. DOI 10.1016/j.compositesa.2007.07.011
30. CHERIF, Ch.; RÖDEL, H.; HOFFMANNN, G.; DIESTEL, O.; HERZBERG, C.; PAUL, C.; SCHULZ, Ch.; GROSSMANN, K.; MÜHL, A.; MÄDER, E.; BRÜNIG, H.: *Textile Verarbeitungstechnologien für hybridgarnbasierte komplexe Preformstrukturen/Textile manufacturing technologies for hybrid based complex preform structures*. http://www.kunststoffe.de/directlink.asp?WAK090202 (20-04-2010)
31. ABOUNAIM, Md.; HOFFMANN, G.; DIESTEL, O.; CHERIF, Ch.: Development of flat knitted spacer fabrics for composites using hybrid yarns and investigation of two-dimensional mechanical properties. In: *Textile Research Journal* 79 (2009), No. 7, pp. 596–610. DOI 10.1177/0040517508101462
32. Standard DIN 60900 Teil 5 July 1988. *Garne: Texturierte Filamentgarne, Herstellungsverfahren und Begriffe*
33. ACAR, M.; TURTON, R.; WRAY, G. R.: Analysis of the air-jet texturing process. II. Experimental investigation of the air flow. In: *Journal of the Textile Institute* 77 (1986), No. 1, pp. 28–43. DOI 10.1080/00405008608658519
34. MIAO, M.; SOONG, M.-C. C.: Air interlaced yarn structure and properties. In: *Textile Research Journal* 65 (1995), No. 8, pp. 433–440. DOI 10.1177/004051759506500801
35. MÄDER, E.; ROTHE, Ch.; GAO, S.-L.: Commingled yarns of surface nanostructured glass and polypropylene filaments for effective composite properties. In: *Journal of Materials Science* 42 (2007), No. 19, pp. 8062–8070. DOI 10.1007/s10853–006–1481–x
36. Oerlikon Heberlein Temco Wattwil AG: *Taslan-Luftblastexturieren Heberlein® HemaJet-Düsenkerne patentiert*. 2010. – Company document
37. Protective right EP 0801159 A2 (15th October 1997).

38. CHOI, B. D.: *Entwicklung von Commingling-Hybridgarnen für langfaserverstärkte thermoplastische Verbundwerkstoffe*. Dresden, Technische Universität Dresden, Fakultät Maschinenwesen,Dissertation, 2005
39. ALAGIRUSAMY, R.; OGALE, V.: Development and characterization of GF/ PET, GF/Nylon, and GF/PP commigled yarns for thermoplastic composites. In: *Journal of Thermoplastic Composite Materials* 18 (2005), No. 3, pp. 269–285. DOI 10.1177/0892705705049557
40. OGALE, V.; ALAGIRUSAMY, R.: Properties of GF/PP Commingled Yarn Composites. In: *Journal of Thermoplastic Composite Materials* 21 (2008), No. 6, pp. 511–523. DOI 10.1177/0892705708091281
41. MANKODI, H.; PATEL, P.: *Study the effect of commingling parameters on glass/ polypropylene hybrid yarn properties*. http//www.autexrj.org/No3-2009/0316.pdf (28-04-2010). Version: 2009
42. COMFIL: *Yarns/Rovings*. http://www.comfil.biz/products/yarnsroving.php (22-04-2010)
43. Rieter Machine Works Ltd.: *Spinning documentation*. 2008
44. SEDLACIK, G.: *Beitrag zum Einsatz von unidirektional naturfaserverstärkten thermoplastischen Kunststoffen als Werkstoff für großflächige Strukturbauteile*. Chemnitz, Technische Universität Chemnitz,Dissertation, 2003
45. LEE, D. H.; CHOWDHARY, U.; SEO, M. H.; JEON, B. S.: Effect of the stretch breaking process on fiber length distribution. In: *Textile Research Journal* 79 (2009), No. 7, pp. 626–631. DOI 10.1177/0040517508099391
46. Protective right EP 1319740 A1 (18th June 2003).
47. NG, S.; MEILUNAS, R.; ABDALLAH, M.G.: Stretched broken carbon fiber (SBCF) for forming complex curved composite structures. In: *Proceedings. 51st International SAMPE Technical Conference*. Long Beach, USA, 2006
48. LEE, K.; LEE, S. W.; NG, S. J.: Micromechanical modelling of stretch broken carbon fiber materials. In: *Journal of Composite Materials* 42 (2008), No. 11, pp. 1063–1073. DOI 10.1177/0021998308090449
49. BLACK, S.: *Aligned discontinuous fibers come of age*. http://www.compositesworld.com/articles/aligned-discontinuous-fibers-come-of-age (20-04-2010)
50. http://www.google.de/search?hl=deei=8l7NS63lDIKAOJDd7KwPsa=Xoi=spellresnum=0ct=resultcd=1ved=0CBAQBSgAq=Hexcel+Stretched+Broken+fiberspell=1 (20-04-2010)
51. Protective right US 5910361 (8th June 1999).
52. ROSS, A.: *Will Stretch-broken Carbon Fiber Become The New Material Of Choice?* http://www.compositesworld.com/articles/will-stretch-broken-carbon-fiberbecomethe-newmaterial-of-choice (20-04-2010)
53. http://www.schappe.com/rubrique.php3?idrubrique=25lang=en (27-04-2010)
54. Protective right WO2006020404 A1 (23rd February 2006).
55. Protective right US 5487941 (30th January 1996).
56. Pepin Associates, Inc.: *Disco Tex*. http://www.pepinassociates.com/discotex.html (27-04-2010)
57. TOONEN, M.: Ökologischer Hanfanbau und Anwendungsmöglichkeiten im Textilbereich. In: *Proceedings. Tagungsband zur Fachtagung am 19. Juni 2007 in Kassel-Wilhelmshöhe / Schriftenreihe IBDF, Band 20*. Darmstadt : Verlag Lebendige Erde, 2007
58. MIAO, M.; HOW, Y.-L.; HO, S.-Y.: Influence of spinning parameters on core yarn sheath slippage and other properties. In: *Textile Research Journal* 66 (1996), No. 11, pp. 676–684. DOI 10.1177/004051759606601102
59. JOU, G. T.; EAST, G. C.; LAWRENCE, C. A.; OXENHAM, W.: The physical properties of composite yarns production by an electrostatic filament-charging method. In: *Journal of the Textile Institute* 87 (1996), No. 1, pp. 78–96. DOI 10.1080/00405009608659058
60. CARPENTER, J. E. P.; MIAO, M.; BRORENS, P.: Deformation behaviour of composites reinforced with four different linen flax yarn structures. In: *Advanced Materials Research* 29–30 (2007), pp. 263–266. DOI 10.4028/www.scientific.net/AMR.29–30.263

61. ZHANG, L.; MIAO, M.: Commingled natural fibre/polypropylene wrap spun yarns for structured thermoplastic composites. In: *Composites Science and Technology* 70 (2010), No. 1, pp.130–135. DOI 10.1016/j.compscitech.2009.09.016
62. THOMANNY, U. I.; ERMANNI, P.: The influence of yarn structure and processing conditions on the laminate quality of stampformed carbon and thermoplastic polymer fiber commingled yarns. In: *Journal of Thermoplastic Composite Materials* 17 (2004), No. 3, pp. 259–283. DOI 10.1177/0892705704041988
63. BABAARSLAN, O.: Method of producing a polyester/viscose core-spun yarn containing spandex using a modified ring spinning frame. In: *Textile Research Journal* 71 (2001), No. 4, pp. 367–371. DOI 10.1177/004051750107100415
64. NAJAR, S. S.; KHAN, Z. A.;WANG, X. G.: The new solo-siro spun process for worsted Yarns. In: *Journal of the Textile Institute* 97 (2006), No. 3, pp. 205–210. DOI 10.1533/joti.2005.0182
65. GHARAHAGHAJI, A. A.; ZARGAR, E. N.; GHANE, M.; HOSSAINI, A.: Cluster-spun yarn - a new concept in composite yarn spinning. In: *Textile Research Journal* 80 (2010), No. 1, pp. 19–24. DOI 10.1177/0040517508099916
66. MERATI, A.A.; KONDA, F.; OKAMURA, M.; MARUI, E.: False twist in core yarn friction spinning. In: *Textile Research Journal* 68 (1998), No. 6, pp. 441–448. DOI 10.1177/004051759806800609
67. MATSUMOTO, Y.-I.; FUSHIMI, S.; SAITO, H.; SAKAGUCHI, A.; TORIUMI, K.; NISHIMATSU, T.; SHIMIZU, Y.; SHIRAI, H.; MOROOKA, H.; GONG, H.: Twisting mechanisms of open-end rotor spun hybrid yarns. In: *Textile Research Journal* 72 (2002), No. 8, pp. 735–740. DOI 10.1177/004051750207200814
68. YANG, R.-H.; XUE, Y.;WANG, S.-Y.: Comparison and analysis of rotor-spun composite yarn and sirofil yarn. In: *FIBRES TEXTILES in Eastern Europe* 18 (2010), No. 1 (78), pp. 28–30
69. Protective right DE19915955 C2 (13th September 2001).
70. Standard DIN 60900 Teil 2 July 1988. *Garne: Beschreibung im Tex-System*
71. KOLKMANN, A.: *Methoden zur Verbesserung des inneren und äußeren Verbundes technischer Garne zur Bewehrung zementgebundener Matrices*. Aachen, RWTH Aachen, Fakultät Maschinenwesen,Dissertation, 2008
72. http://www.agteks.com (04-05-2010)
73. OFFERMANN, P.; DIESTEL, O.; CHOI, B.-D.: Commingled CF/PEEK hybrid yarns for use in textile reinforced high performance rotors. In: *Proceedings. 12th International Conference on Composite Materials*. Paris, France, 1999
74. HERZBERG, C.; RÖDEL, H.: *Beanspruchungsgerechte 3D-Verstärkungen durch funktionsgerechte Nähtechnik (Informationsblatt)*. 2000
75. ZHAO, N.; HERZBERG, C.; RÖDEL, H.: Assembly of textile preform for the composite material with sewing technology. In: *Proceedings. 5th Pacific Rim International Conference on Advanced Materials und Processing*. Beijing, China, 2004
76. HERZBERG, C.; RÖDEL, H.: New sewing threads for composite applications. In: *Proceedings. 4th International Conference IMCEP 2003*. Maribor, Slowenien, 2003, pp. 139–143
77. HERZBERG, C.; KRZYWINSKI, S.; RÖDEL, H.: Load-adapted 3D-reinforcement through function-adjusted stitching technique. In: *Proceedings. 13th International Conference on Composite Materials*. Beijing, China, 2001
78. BLEAY, S. M.; LOADER, C. B.; HAWYES, V. J.; HUMBERSTONE, L.; CURTIS, P. T.: A smart repair system for polymer matrix composites. In: *Composites Part A: Applied Science and Manufacturing* 32 (2001), No. 12, pp. 1767–1776. DOI 10.1016/S1359–835X(01)00020–3
79. PAUL, C.: *Funktionalisierung von duroplastischen Faserverbundwerkstoffen durch Hybridgarne*. Dresden, Technische Universität Dresden, Fakultät Maschinenwesen,Dissertation, 2010

Additional Literature

BAGHAEIA, B.; SKRIFVARS, M.; SALEHI M.; BASHIR T.; RISSANEN M.; NOUSIAINEN, P.: Novel aligned hemp fibre reinforcement for structural biocomposites: Porosity, water absorption, mechanical performances and viscoelastic behavior. In: *Composites: Part A*, 61 (2014), pp. 1-12

HASAN, MMB.; CHERIF, Ch.: Analysis of the influence of process parameters on the mechanical properties of carbon core friction spun hybrid yarns for composites. In: *FIBRES & TEXTILES in Eastern Europe* 19 (2011), no. 4 (87) pp. 59-64

HASAN, MMB.; DIESTEL, O.; CHERIF, Ch.: Electro-mechanical properties of friction spun conductive hybrid yarns made of carbon filaments for composites. In: *Textile Research Journal* 81 (2011), pp 1603-1616

HASAN, MMB.; NOCKE, A.; CHERIF, Ch.: High temperature resistant insulated hybrid yarns for carbon fiber reinforced thermoplastic composites. In: *Journal of Applied Polymer Science* 130 (2013), no. 2, pp. 1179-1184

HASAN, MMB.: *Development of new functional hybrid yarn constructions for structural health monitoring of high performance composites.* Dresden, Technische Universität Dresden, Fakultät Maschinenwesen,Dissertation, 2013

KRAVAEV, P.; STOLYAROV, O.; SEIDE, G.; GRIES, T.: A method for investigating blending quality of commingled yarns. In: *Textile Research Journal* (2012), pp. 1-8. DOI: 10.1177/0040517512456760

KRAVAEV, P.; STOLYAROV, O.; SEIDE, G.; GRIES, T.: Influence of process parameters on filament distribution and blending quality in commingled yarns used for thermoplastic composites. In: *Journal of Thermoplastic Composite Materials* (2014), pp. 1-14. DOI: 10.1177/0892705712446167

MAHMOOD, A. H.; GONG, R. H.; PORAT, I.: Development of air textured core-and-effect glass yarns for improving the bonding strength of laminated composites. In: *Fibers and Polymers* 14 (2013), no. 2, pp. 271-276

MILLER, W.; REN, Z.; SMITH, C.W.; EVANS, K.E.: A negative Poisson's ratio carbon fibre composite using a negative Poisson's ratio yarn reinforcement. In: *Composites Science and Technology* 72 (2012), pp. 761-766

MORIMOTO, T.; IIZUKA, H.: Tensile behaviors of pre-twisted composite strands. In: *Arch Appl Mech,* 83 (2013), pp. 193-205

RAJWIN, AJ.; GIRIDEV, VR.; RENUKADEVI, M.: Effect of yarn twist on mechanical properties of glass fibre reinforced composite rods. In: *Indian Journal of Fibre & Textile Research* 37 (2012), pp. 343-346

SHAH, D. U.; SCHUBEL, P. J.; LICENCE, P.; CLIFFORD, M. J.: Determining the minimum, critical and maximum fibre content for twisted yarn reinforced plant fibre composites. In: *Composites Science and Technology* 72 (2012), pp. 1909-1917

TORUN, A. R.; HOFFMANN, G.; MOUNTASIR, A.; CHERIF, Ch.: Effect of twisting on mechanical properties of GF/PP commingled hybrid yarns and UD-composites. In: *Journal of Applied Polymer Science* 123 (2012), no. 1, pp. 246-256

YAN, L.; CHOUW, N.; JAYARAMAN, K.: Flax fibre and its composites – a review. In: Composites: Part B 56(2014), pp. 296-317

Chapter 5
Woven Semi-finished Products and Weaving Techniques

Cornelia Sennewald, Gerald Hoffmann, and Roland Kleicke

This chapter introduces the structural description, weaving-technical manufacture and possibilities of woven fabric structure modifications for the development of practically suitable woven fabrics for application in lightweight construction. The basic structural set-up, methods for its description and the properties resulting from the structure of the woven fabrics will be clarified. An overview of basic methods of woven fabric manufacture and the corresponding weaving machines will demonstrate the diversity of technical solutions for a gentle and damage-free processing of special and high-performance fiber materials into a variety of woven structures. The focus of the chapter will be on an extensive introduction to woven structures and the respective structural modifications. This includes 2D structures as flat, grid, multiaxial, and polar woven fabrics, 3D structures as multilayered woven fabrics and spacer fabrics as well as shell-shaped 3D geometries. As early as during the production of the woven fabrics or the preforms, the inclusion of special fibers or external materials, such as electronic devices and inserts for mechanical connections, additional functions can be integrated.

C. Sennewald (✉) • G. Hoffmann
Institute of Textile Machinery and High Performance Material Technology, TU Dresden, Dresden, Germany
e-mail: cornelia.sennewald@tu-dresden.de; gerald.hoffmann@tu-dresden.de

R. Kleicke
KLEICKE ingenieure, Demitz-Thumitz, Germany

Formerly: Institute of Textile Machinery and High Performance Material Technology, TU Dresden, Dresden, Germany
e-mail: ingenieure@kleicke.de

C. Cherif (ed.), *Textile Materials for Lightweight Constructions*,
DOI 10.1007/978-3-662-46341-3_5

5.1 Introduction and Overview

Textile fabrics, which, beyond woven fabrics, include weft-knitted fabrics (see Chap. 6), warp-knitted fabrics (Chap. 7), braided fabrics (Chap. 8), and nonwovens (Chap. 9), are created by the interlacement (woven, braided fabrics) of several yarns, the formation of intertwined yarn loops (weft-/warp-knitted fabrics), or the mechanical, chemical or thermal connection of fibers (nonwovens). A fabric is only called a woven fabric if it consists of usually two interlacing yarns (warp and weft yarns). For the development of technical products with practically utilizable textile semi-finished products, a clear distinction has to be made between woven fabrics and other textile fabrics.

The manufacture of fabrics has been known to mankind for more than 7,000 years, making it the oldest textile fabric production technique. Particularly over the past 200 years with their new solutions in mechanical engineering, new materials and the development of control engineering and electronics, the production of woven fabrics has been evolving continuously from the handloom to the flexible high-performance weaving machines of today. In the past 50 years alone, performance was increased sevenfold. One weaver today controls up to 20 machines. Electronic control systems and mechatronic solutions significantly increased the flexibility of weaving machines and reduced machine set-up times. By relying on preparatory facilities, a machine operator can switch between products to be produced on a machine within 25 min. This allows manufacturers of woven fabrics to offer woven fabrics fit-for-use in small amounts (a few 100 m) or in large batch sizes at attractive prices. The standards working widths of the weaving machines range from a few millimeters (for so-called narrow woven fabrics) to 5.4 m (for so-called broad woven fabrics), or more than 30 m in special mechanical engineering.

Structure and properties of a woven fabric depend on the yarns used for their production and the manner of their interlacing. All filamentary materials with sufficient tensile strength and sufficiently low bending stiffness can be used as warp and weft yarns, e.g. finest silk or steel wire with a diameter of 2 mm. Colored warp and weft yarns allow the weaving of color-patterned fabrics. There are no weaving-technological boundaries to the size of the color patterns or the manner of the images on the product. The maximum structural density and planar character of the woven fabrics are very high in comparison to other textile fabrics. The densest plain-weave woven fabrics, e.g. for ink jet print heads, have a warp and weft yarn density of up to 1,200 yarns/cm. Weaving technology enables the production of light and strong fabrics with small air permeability, as well as the manufacture of open grid structures. Sieves from filamentary materials, with a defined mesh size, are preferably produced by weaving. By orienting additional yarns and technical modifications, woven fabrics with a three-dimensional structure and three-dimensional geometry are feasible. One well-known example of three-dimensional structures is looped woven fabrics for terry products (towels). By using several warp and weft yarn systems, compact woven fabrics are realizable. Spacer fabrics consisting of two woven fabric layers connected by yarns (pile yarns), can be manufactured with spacings of up to 10 cm. Spacings of several meters and more between the woven fabric layers are possible with auxiliary weft yarns. The required double rapier weaving machines are primarily used for the production of carpets by cutting the interior pile. Its flexibility regarding material selection, structural variety and product change make weaving the most commonly used fabric manufacturing process. Woven fabrics are extensively used for technical textiles and feature prominently in the fields of textiles in lightweight construction, semi-finished products for fiber material composites, and textile membranes.

5.2 Woven Fabric Structure

5.2.1 Definition

Conventional 2D woven fabrics consist of at least two yarn systems, which cross at a right angle. The yarn systems running in the direction of production are called warp yarns, the ones running transversely are the weft yarns. The interlacing of the warp yarns with the weft yarns interconnects the two yarn systems, resulting in the formation of the textile fabric.

The order of the interlacing of a fabric is referred to as the weave pattern. It significantly influences the appearance and properties of the woven fabric, such as its strength and drapeability.

5.2.2 Schematic Representation

The weave pattern of a woven fabric is usually not represented by drawn depictions of the appearance (Fig. 5.1a), but as simplified colored fields in a grid of squares

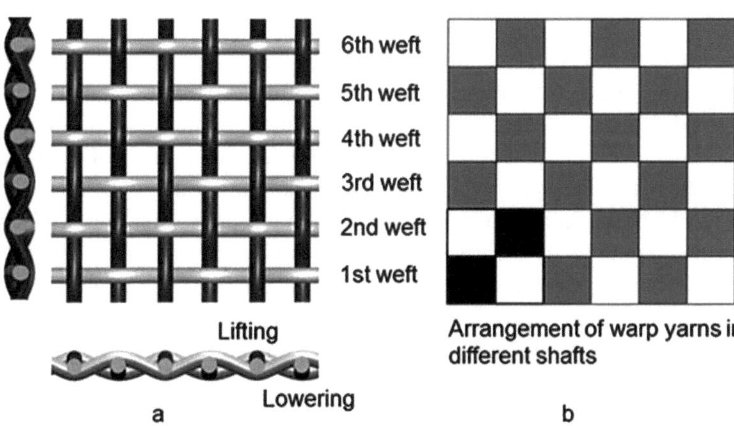

Fig. 5.1 Schematic representation of a woven fabric with (**a**) a cross-section in warp and weft direction and (**b**) weave diagram (9 repeats)

(Fig. 5.1b). The individual squares represent the interlacing points of the warp and weft yarns. A filled-in square denotes the warp yarn being positioned on top of the weft yarn at this position (referred to as warp lift). Empty squares designate a warp yarn going underneath the weft yarn in this place (referred to as warp lowering). Any representation of a weave pattern starts in the bottom left corner of the grid of squares, corresponding to the first weft and the first warp yarn. The warp yarns are therefore numbered from left to right, the weft yarn from bottom to top. A weave patterns is represented schematically by a weave diagram (Fig. 5.1b). As the majority of weave patterns are systematically repeated in warp and weft direction, the repeat unit (Fig. 5.1b, marked black) is sufficient for the unambiguous representation of a weave pattern. The repeat unit is the smallest unit of differently interlacing warp and weft yarns to the point of repetition of a weave diagram. A representation of the weave pattern can be supplemented by a cross-section view in warp and weft direction (Fig. 5.1a).

The schematic representation of the weave pattern is very clear and universally applicable. It can be complemented by representations, such as in matrix format or in views according with DIN ISO 9354, which are electronically processable [1]. Fundamentally, it has to be remembered that identically interlacing warp yarns can be lifted and lowered together in a warp yarn group, while differently interlacing warp yarns have to be lifted and lowered in different warp yarn groups. The maximum number of individually controllable warp yarn groups determines the maximum size of the repeat unit and depends on the equipment of the respective weaving machine. Special retractions of the warp yarns into the heald frames allow an extension of the repeat unit. By means of the Jacquard technology, each warp yarn can be lifted and lowered individually, which removes any limitations of repeat size.

5.2.3 *Basic Weave Patterns*

5.2.3.1 Introduction

The three basic weave patterns are plain weave, twill weave (warp and weft faced twill) and satin weave (warp and weft faced satin). All basic weave patterns are characterized by their repeat unit, which is constituted by the same number of yarns in warp and weft direction. The repeat unit indicates the number of warp and weft yarns necessary to repeat the weave pattern. Each warp yarn (end) interlaces with the weft (pick) in the same manner whereas the next corresponding adjacent warp yarn, starts interlacing by stepping up (or down) by one or more squares in the graph. These basic weave patterns are sufficient for most applications of woven fabrics in lightweight construction. Complex patterns can be derived from the basic weave patterns by means of expanding or derivation for special fields of application.

Plain weave is the simplest form of interlacing and the most common basic weave pattern (Fig. 5.2). It has a repeat unit of two warp and two weft yarns. The first warp yarn is raised over the first weft and then lowered under the second weft

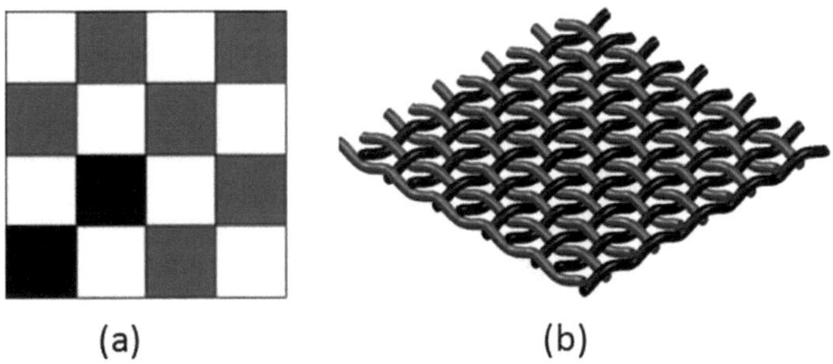

Fig. 5.2 (**a**) Plain weave diagram with 4 repeat units, (**b**) schematic diagram

whereas the second warp yarn is lowered under the first weft and raised over the second weft. The plain weave is the most closely interlaced weave pattern with the highest crimp. This results in a high shear resistance and good handling characteristics, but also causes a loss of strength and stiffness. The high crimp of plain weaves causes a significant difference between fiber orientation and angle of maximum stress and significantly reduces the load-carrying capacity.

5.2.3.2 Twill Weave

The twill weave (Figs. 5.3 and 5.4) is characterized by distinct diagonal lines. Its minimal repeat unit consists of three warp and three weft yarns. The simplest twill weave is the weft faced twill shown in Fig. 5.3a. One warp yarn is raised over one weft and then lowered under two wefts. The diagonal line runs from the bottom left to the top right side. Three additional basic twill weaves can be derived from the first one by negotiation and reflection:

- Weft faced twill with diagonal line going down,
- Warp faced twill with diagonal line going up,
- Warp faced twill with diagonal line going down

Twills with more weft yarns than warp yarns floating on the fabric surface are called weft faced (Fig. 5.3) and twills with more warp yarns than weft yarns floating on the fabric surface are called warp faced (Fig. 5.4), with the fabric back of a weft faced twill looking like a warp faced twill and vice versa.

5.2.3.3 Satin Weave

The third basic weave pattern is the satin weave (Fig. 5.5). Satin weaves are characterized by a smooth surface and a minimum amount of interlacing. Compared to plain weave and twill weave, the satin weaves enable higher warp and weft densities (see Sect. 5.3.1). The smallest repeat unit for satin weaves consists of five warp and five weft yarns. One warp yarn is raised over one weft yarn and lowered under four weft yarns. The interlacing points in satin weaves are scattered as far as

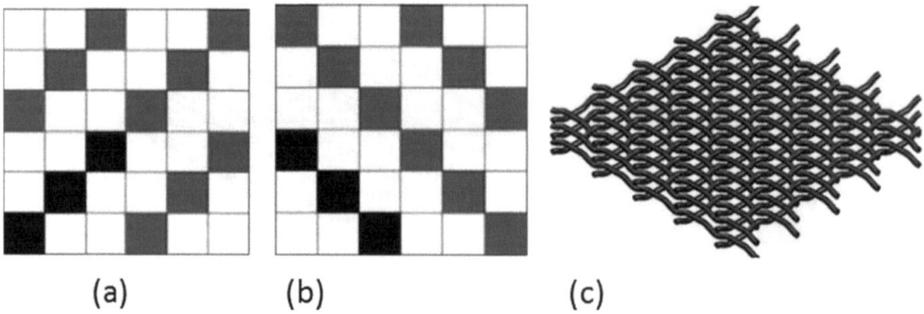

Fig. 5.3 (**a**) Weft faced twill with diagonal line going up, (**b**) weft faced twill with diagonal line going down (4 repeat units), and (**c**) schematic diagram

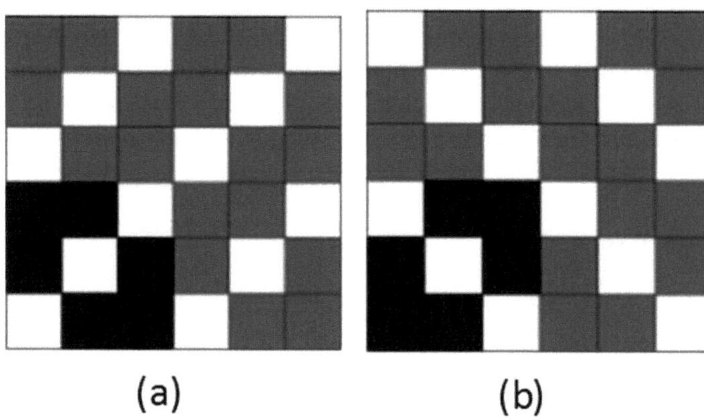

Fig. 5.4 (**a**) Warp faced twill with diagonal line going up and (**b**) warp faced with diagonal line going down (4 repeat units)

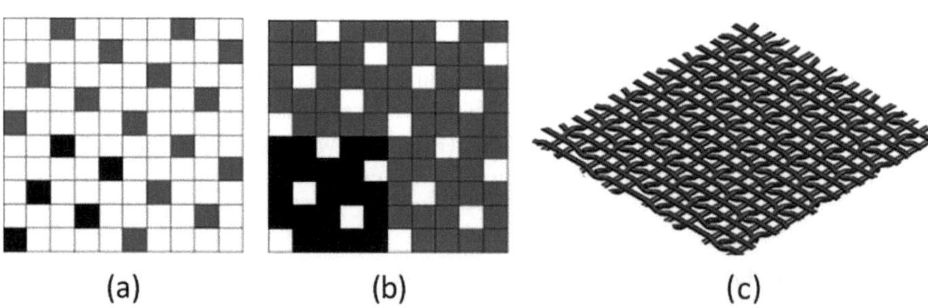

Fig. 5.5 (**a**) Weft faced satin, (**b**) warp faced satin (4 repeat units), and (**c**) schematic diagram

possible, e.g. the points of interlacing must touch neither side by side nor diagonal. Satins with more weft yarns than warp yarns floating on the fabric surface are called weft faced and satins with more warp yarns than weft yarns floating on the fabric surface are called warp faced, where the fabric back of a weft faced satin looks like a warp faced satin and vice versa. The relatively small number of interlacing points results in a lower crimp and causes a lower shear resistance, in comparison to plain weave (assuming same material and warp and weft density). This results in an outstanding draping behavior but also creates poor handling characteristics.

5.2.4 Extended and Derived Basic Weave Patterns

5.2.4.1 Extended Basic Weave Patterns

The three basic weave patterns can be extended into a multitude of other weave pattern as desired and required. *Extended basic weave patterns* are created by inserting or removing warp lifts in any direction. However, within each weave pattern, only one variant may be applied. For the plain weave, no extension exists. In contrast to the basic satin weave, an extended satin weave allows direct contact between the interlacing points. Examples for extended twill and satin weaves are given in Figs. 5.6 and 5.7.

5.2.4.2 Derived Basic Weave Patterns

Derived basic weave patterns are created by the enhancement of basic weave patterns or extended basic weave patterns. This can be achieved by orienting additional warp lifting and/or warp lowering, or by using offset numbers, which are not valid in basic weave patterns or extended basic weave patterns. Derived

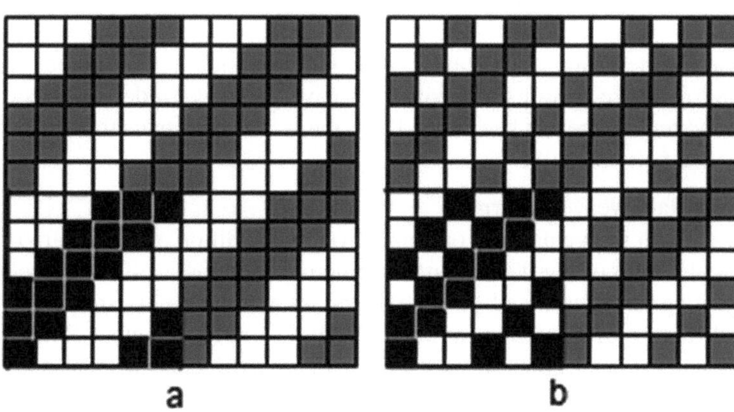

Fig. 5.6 Extended twill weaves—4/2 twill weave and equilateral stitched twill (4 repeats)

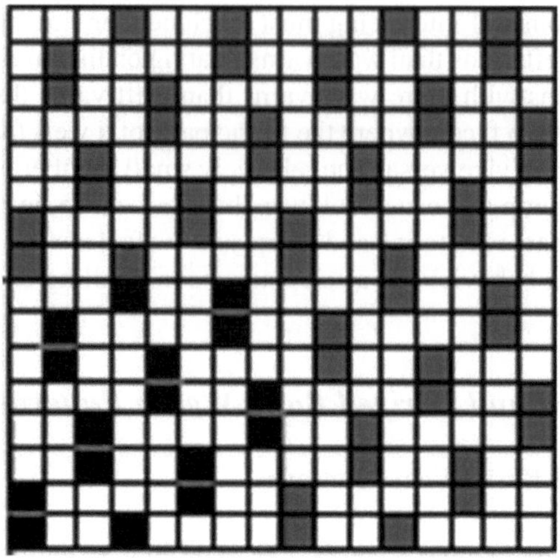

Fig. 5.7 Extended satin weave (4 repeats)

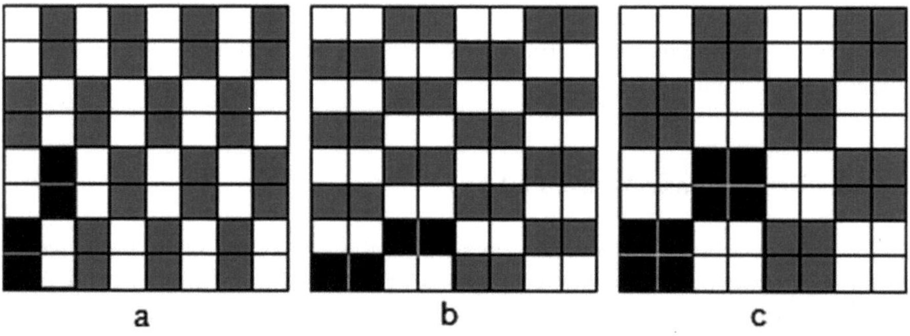

Fig. 5.8 Derivations of the plain weave—(**a**) warp rib weave pattern, (**b**) weft rib (8 repeats each), and (**c**) panama weave pattern (4 repeats)

basic weave patterns are also created by the joining, doubling, omission, mirroring or turning of weave patterns and parts of weave patterns. In derived basic weave patterns, the repeat size can be changed, and the weave pattern characteristics of the initial weave pattern can be lost.

- Derived plain weave patterns
  Derived plain weave patterns originate from the addition of extra warp lifts. The principal set-up of the weave pattern structure remains, while the repeat of the weave diagram is enlarged. If the existing interlacing points are extended by a warp lift in warp direction, a warp rib (Fig. 5.8a) is created. Extending the existing interlacing points by a warp lift in weft direction results in a weft rib (Fig. 5.8b). A panama weave pattern (Fig. 5.8c) is the result of adding a warp lift at existing interlacing points in both warp and weft direction.

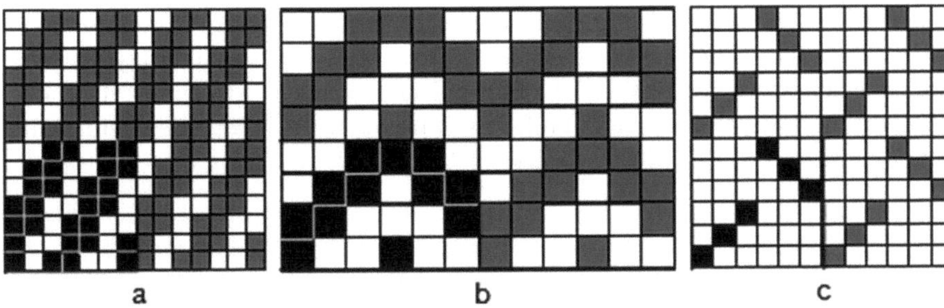

Fig. 5.9 Derivations of the twill weave—(**a**) steep twill, (**b**) transverse pointed twill, and (**c**) cross twill

- Derived twill weaves
  Derived twill weaves are realized by inserting additional warp lifts or changes in offset or ridge direction by turning or mirroring the weave diagram. Steep twill and reclining twill (Fig. 5.9a) are examples for changes in offset, in which the typical angle of inclination of the twill ridge is altered. Pointed and zigzag twill weaves (Fig. 5.9b) are the result of an alternating use of Z ridge and S ridge. Here, the twill ridges are converging at the reversal points. One example for a mirroring of the one repeat half is the cross twill (Fig. 5.9c).
- Derived satin weaves
  Derived satin weaves result from the application of a steadily changing offset. This aims to prevent false twill lines. The number of possible derivations of the satin weave is much smaller than for the twill weaves. Detailed descriptions can be found in [2].
- Special forms
  In addition to the derived basic weave patterns, weave patterns beyond the named classifications can be developed. These weave patterns, referred to as special forms, allow the realization of design or functional effects.

  By means of this large number of weave pattern possibilities, a practicable woven fabric with locally adjustable mechanical properties and drapability characteristics can be constructed. Since these weave patterns are usually created on dobbies (with a maximum of 28 heald frames), the pattern possibilities are limited to a weft repeat corresponding to the maximum number of heald frames.

5.2.5 *Jacquard Weave Patterns*

As opposed to the basic weave patterns, in which several warp yarns are lifted and lowered simultaneously in groups, jacquard-patterned woven fabrics allow the lifting and lowering of each individual warp yarn. The result is an unlimited variety of pattern possibilities with infinite repeat size. A repeat can therefore cross the entire width of the woven fabric. The use of various weave pattern variations on the same fabric and/or the patterned change from single-layer to multilayer woven fabrics enable the production of semi-finished stringer rib-reinforced shells, branched hollow profiles, and preforms with different properties within the woven fabric.

5.3 Woven Fabric Parameters and Properties

5.3.1 Woven Fabric Parameters

The following parameters have significant influence on the properties of the woven fabric:

5.3.1.1 Yarn Material

The properties of the yarn used for the woven fabric influence the properties of the woven fabric, according to the type of material, fineness, inner or outer yarn structure, and additional finishing. A more detailed explanation of the individual yarn parameters is given in Chap. 4.

5.3.1.2 Crimp

Crimp refers to the relation of the length of a representative strip of a woven fabric and the length of the yarn used to weave this strip. The crimp exists both in warp and weft direction, and they influence one another (Fig. 5.10). In general, crimp is a gage of how the yarn deviates from the straight orientation within the woven fabric. The smaller the crimp, the more stretched the yarns of the woven fabric are. By using the adjustment possibilities (warp and weft yarn tension) offered by the weaving machine, the crimp can be varied depending on the direction. Generally, the relation of warp yarn and weft yarn crimp is balanced (Fig. 5.10b).

5.3.1.3 Weave Pattern

The realized weave pattern of the woven fabric crucially influences crimp, making it a factor in the orientation of the individual yarns within the woven fabric. A higher interlacing point density generally means a higher crimp of the individual yarn systems. Woven fabrics with high crimp (e.g. plain-woven fabrics), display a much higher structural elongation than woven fabrics with low interlacing point density (e.g. satin-woven fabrics) (Fig. 5.11) [3]. Woven fabrics with a high

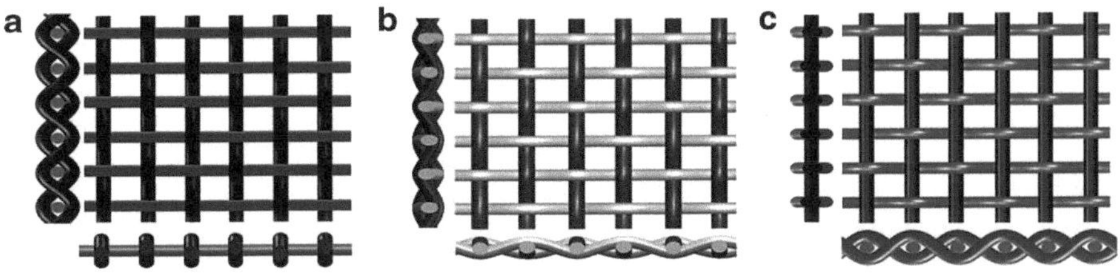

Fig. 5.10 Sectional views of plain-woven fabrics with (**a**) low crimp, (**b**) balanced crimp relation, and (**c**) high crimp

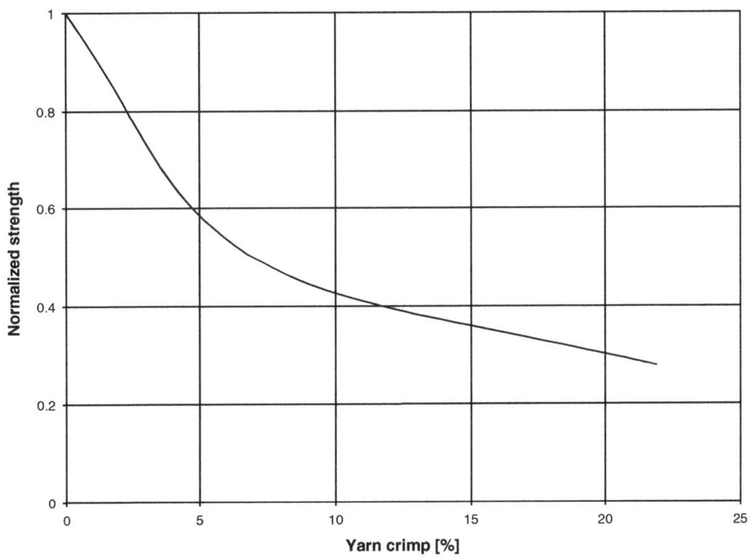

Fig. 5.11 Qualitative relation of achievable maximum breaking strength and crimp

interlacing point density (e.g. plain weaves) have a high slippage resistance, but the high crimp detrimentally affects mechanical parameters in the composite.

5.3.1.4 Woven Fabric Density

Generally, both warp and weft yarn density is given in yarns per centimeter. The higher these values are, the higher the slippage resistance of the woven fabric will be. The relative woven fabric density can be calculated with Eq. 5.1 by Walz and Luibrand [4, 5].

$$DG = \frac{1}{nkns}(ds + dk)\,2\,p \cdot 100\,\% \tag{5.1}$$

In which,

DG (%)	Woven fabric density
n (mm)	Yarn distance (k . . . warp, s . . . weft)
d (mm)	Diameter (k . . . warp, s . . . weft)
p (−)	Weave factor (weave pattern-dependent, tabled in [4])

The weave factor *p* denotes the number of warp and weft changes in relation to the plain weave. For the plain weave, the weave factor is $p = 1$. For basic twill weaves, it is 0.67, and 0.33 for satin weaves.

5.3.2 Woven Fabric Properties

Chapter 14 details the determination of the properties of textile materials. In the following, woven fabric properties relevant for composite materials and technical textiles will be considered in particular.

Table 5.1 compares the relationships of woven fabric weave patterns and properties at identical warp and weft density and identical yarn material. Depending on the realized weave pattern, the maximum achievable mass per unit area increases from plain to twill to satin weave patterns.

5.3.2.1 Characteristics of the Stress-Strain Behavior

The stress-strain behavior (Fig. 5.12) of woven fabrics is marked by two distinct regions: the region of structural deformation and the range of yarn deformation. Structural deformation in the initial part of the stress-strain behavior is distinguished by a high elongation at comparably small force. Here, yarns are oriented in load direction, without their material properties coming into effect. The elongation is effected exclusively by the structural elongation reserves of the woven fabric. The structural elongation depends on the weave pattern. For example, it is only ca. 2–3 % for plain-woven fabrics made from glass fibers, and only ca. 1 % for satin-woven fabrics made from identical material. Structural deformation seamlessly changes into yarn deformation. When the yarns are stretched the structural deformation ends and the material properties of the respective yarn can take effect.

Table 5.1 Overview of structural properties

	Plain	Twill	Satin
Structural deformation	++	+	0
Slippage resistance	++	+	0
Single yarn pull-out strength	++	+	0
Bending strength	++	+	0
Permeability	0	+	++
Drapability	0	+	++
Handling	++	+	0
Mechanical properties in the composite	0	+	++

Note: 0=low; +=average; ++=high

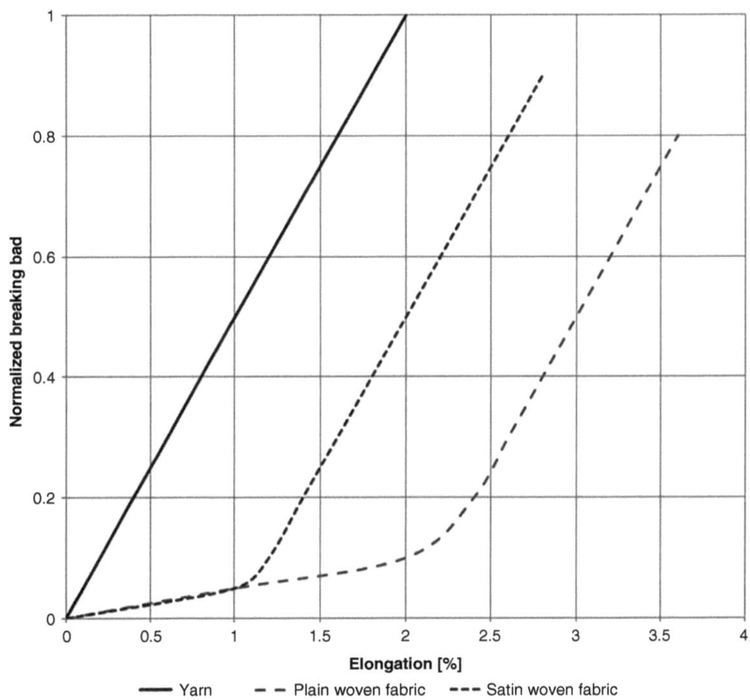

Fig. 5.12 Principal stress-strain behavior of yarns and woven fabrics of different weave patterns

5.3.2.2 Slippage Resistance/Single Yarn Pull-out Force

Besides shear (see Sect. 15.2.4) slippage resistance or single yarn pull-out force are suitable parameters to determine the drapability of woven fabrics. The force required to shift a yarn perpendicular to its longitudinal axis within the fabric plane is a measure for the slippage resistance (Fig. 5.13a) of a woven fabric. The single yarn pull-out force indicates how much force is required to draw a yarn from

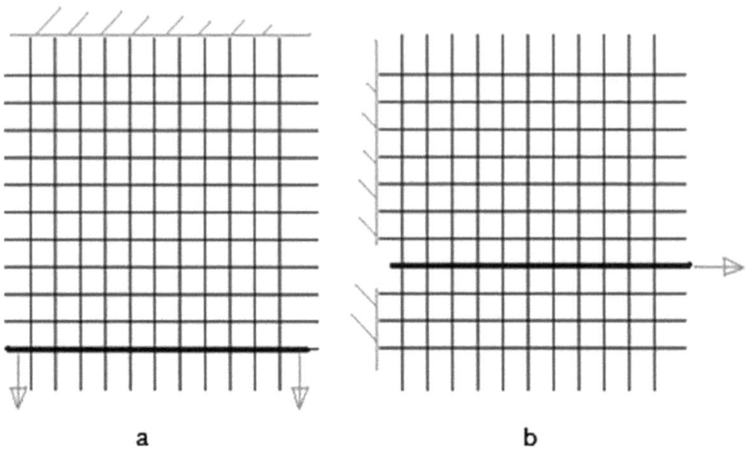

Fig. 5.13 Schematic principle for the determination of (**a**) slippage resistance and (**b**) single yarn pull-out force

the woven fabric parallel to its longitudinal axis (Fig. 5.13b). These two parameters are mutually dependent. However, a direct comparison is not readily possible as it requires further woven fabric parameters (e.g. crimp, weave pattern) to be regarded.

Based on the basic experiments for the characterization of the handling of textile reinforcement structures for composite materials by determining the slipping of individual yarns, a new test method was developed and patented [6, 7]. This patented test method is based on a defined rotation of a force-introducing area compared to a fixed-positioned textile structure (Fig. 5.14). This test method offers a possibility to determine a characteristic structural parameter for the handling of textile structures.

In general, the higher the slippage resistance and single yarn pull-out force are, the more alterations exist between lifting and lowering of the warp and weft yarn systems.

These two properties of the woven fabric are particular parameters for the handling. They are also important for drapability. The higher these two parameters are, the harder the woven fabric will be to drape.

5.3.2.3 Bending Stiffness

The bending stiffness of textile fabrics also offers indications of their drapability. Therefore, it can be used to assess the usability of the woven fabric for the manufacture of 3D geometries. The bending behavior of woven fabrics is primarily defined by the textile fiber material (yarn material, yarn construction, and fineness), by the construction (weave pattern and yarn density), and by the resulting adhesion and friction influences between the individual filaments or yarns. Generally, it is characterized as the relation of bending angle, bending length, and exerted force.

Fig. 5.14 Slipping test: before (*top*) and after the test (*bottom*)

There are two well-established test methods. The cantilever method (see Sect. 14.5.4), and the patent-protected method [7] differ with regard to the sample orientation (horizontal for the cantilever method, vertical for the patented method), the related external influences (such as gravity) and the testable textiles (flexible in the cantilever method, bend-proof in the patent-protected method). Principally, the following rule applies: the more bend-proof the yarn used, the denser the woven fabric, and the more alterations between lifting and lowering of warp and weft yarn systems there are, the higher the bending stiffness will be.

A high bending stiffness in a woven fabric translates to a larger required exertion of force for draping. The restoring forces occurring in draping negatively affect the manufacturing process, the reproducibility and the quality of the components. These forces cause a relaxation of the textile structure, making complex geometries extremely difficult to model permanently.

5.3.2.4 Permeability

The permeability is a gage for the penetrability of fabrics by fluids. This parameter is particularly important for infusion and barrier processes. A general rule applies: The lower the woven fabric density, the higher its permeability and the less difficult will be the impregnation of the woven fabric during composite formation.

5.3.2.5 Mass per Unit Area

The mass per unit area of any textile fabric is stated in grams per square meter (see Sect. 14.5.2). The mass per unit area of a woven fabric can be increased by using coarser yarns, selecting a weave pattern with a small weave factor, increasing the warp and weft densities, and producing multilayered fabrics. An increasing mass per unit area usually results in increased thickness of the woven fabric, as the yarns generally evade in thickness direction during fabric formation. This can be prevented by means of the weave pattern, but will cause a greater limitation of the thickness of the woven fabric.

5.3.2.6 Woven Fabric Thickness

The thickness refers to the vertical distance between the top and bottom sides of a fabric. Any fabric usually has a distinctive surface profile, which makes normative definitions necessary for an accurate determination of the thickness of a woven fabric (see Sect. 14.5.1). The thickness of the woven fabric is significantly influenced by the respective fiber material and the construction of the woven fabric, and it is closely connected with the areic mass. Special woven fabrics, such as pile and spacer fabrics, can be used to increase the thickness of woven structures while decreasing the density.

5.3.2.7 Resistance to Tear Propagation and Cut Resistance

The resistance to tear propagation and cut resistance quantifies the resilience and damage tolerance of woven fabrics to random and directed damages of membrane materials. In woven fabrics, initial tears continue to grow rapidly. Using specific materials, modified weave patterns and special coatings can significantly increase the resistance of woven fabrics to tear propagation. To increase the cut resistance, which denotes the resilience of a woven fabric to damages caused by sharp-edged objects, special high-strength fiber materials have to be integrated into the structure of the woven fabric in a suitable manner.

5.4 Manufacture of Woven Fabrics

5.4.1 Weaving Methods

To produce woven fabrics, rectangularly running warp and weft yarns have to be interlaced. The warp yarns, which are oriented parallel to the production direction, are separated into at least two groups. This separation is necessary to allow one group of warp yarns to be moved upwards and the other downwards, contributing to the weave pattern-related formation of the weave structure of the woven fabric. The space created by moving the separate groups of warp yarns is referred to as the shed. The shed provides the required space for the subsequent weft insertion and is to be kept as small as possible to reduce tensile forces on the warp yarns. After the weft yarn is inserted (weft insertion) at a right angle to the warp yarns, it is beaten up to the woven fabric edge, parallel to the previous weft yarn (weft beat-up). To manufacture highly dense woven fabrics, a high beat-up force is required, which can cause fiber damages when brittle high-performance fiber materials are used. In order to interlace the warp and weft yarns (see Sect. 5.2), the corresponding warp yarns change position in the weaving shed (changing of shed) and open the shed for the next weft insertion. Weft beat-up and shed change takes place almost simultaneously. The shed is usually already slightly open at the moment of the weft beat-up. This prevents recoiling of the yarn. By means of the take-up device, the finished piece of woven fabric is then taken up.

5.4.2 Basic Set-up of the Weaving Machine

Figure 5.15 shows the course of the warp yarn system from a warp beam over the back rail, the heald frames with the healds, through the reed and the fabric take-off to the fabric storage. The warp beam (1) stores the warp yarns and delivers them as needed. The warp yarns, especially if they are carbon or glass yarns, can also be guided to the weaving machine directly from a bobbin creel (Fig. 5.16). The back rail (2) and the fabric take-off (8) deflect the warp yarns to the weaving plane. The back rail (2) is usually mounted flexibly and compensates variations in warp yarn consumption by opening and closing the shed, which is particularly important when working with stiffer yarns or yarns with low tensile strength. The warp yarns are mostly guided to a heald (3) individually. The healds can be assigned to the heald frames (4) in groups, or individually to the harness cords of a Jacquard machine (see Sect. 5.4.4.4). The movements of the heald frames or the Jacquard healds cause the opening and changing of the shed. Different available weft insertion systems (5) transport the weft yarn through the shed. The reed (6), which is integrated into the sley, arranges the warp yarn and sets the theoretical warp yarn density of the woven fabric. By the movement of the reed (6) with the sley (7), the last weft yarn is beat-up to the fabric edge. The fabric take-off device (8) provides the take-off of the woven fabric at a preset speed and adjusts its weft density. In the fabric storage (9), the woven fabric is then stored temporarily before the next process steps. In the processing of high-performance fiber materials, the use of warp stop motions is usually omitted.

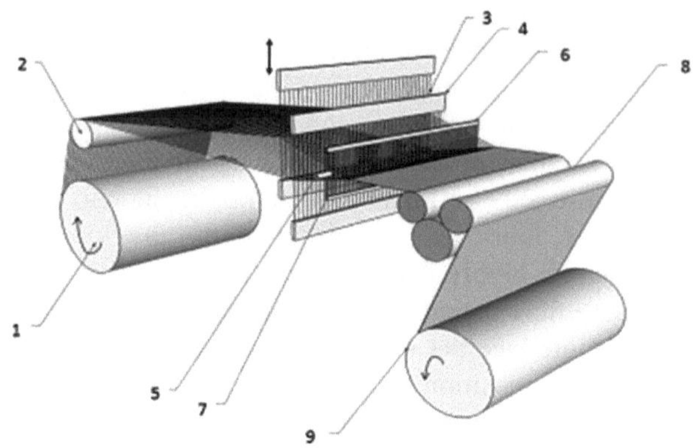

Fig. 5.15 Basic set-up of a weaving machine. (1) Warp beam, (2) back rail, (3) heald, (4) heald frames, (5) weft insertion system, (6) reed, (7) sley, (8) fabric take-off, (9) fabric storage

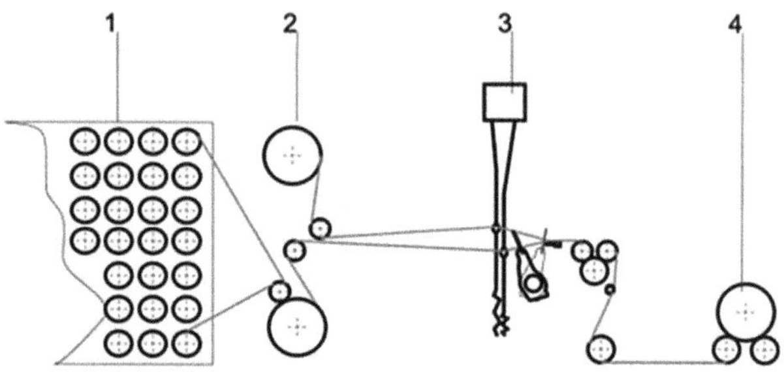

Fig. 5.16 Cross-section view of a weaving machine, including optional components. (1) Bobbin creel, (2) double warp beam, (3) Jacquard machine, (4) ascending batch winder

5.4.3 Weft Yarn Feeding

The warp yarn system is usually stored on a warp beam and can consist of a few hundred or even several thousand of parallel-running yarns. During warping of the warp yarn in a warp beam, special attention has to be paid to ensuring that all warp yarns are processed in the woven fabric at the same length, i.e. that the number of warp changes (weave structure-related) is, on average, the same for all warp yarns. Small differences in warp yarn consumption are only permissible for ductile yarns. If this requirement cannot be met, several warp beams (up to six) have to be used, or the yarns have to be fed from a bobbin creel.

Using warp beams reduces the bulk of the weaving plant. Producing warp beams requires a number of additional technological process steps, which will be omitted for the sake of brevity. The warp yarns are fed to the weaving machine at a constant tension by a controlled warp let-off drive. The tension measurement required for controlling the process can be measured indirectly using the position of a flexibly mounted back rail, by using pressure sensors in the back rail or the fabric take-off area, or with yarn tension measuring heads in the warp yarn group.

Using a bobbin creel is a sensible course for small numbers of warp yarns, sensitive yarns and for varying warp yarn consumption. Carbon filament yarns as well as carbon and glass rovings are categorically fed from the creel, as the additional process step of constructing a warp beam would lead to intolerable yarn damages of these shear force-sensitive yarns. The design and construction of the bobbin creel is adapted to the spatial conditions and the bobbin shape. The yarns are taken off in direction of the bobbin axis ("overend unwinding") if the bobbin is stationary, or tangentially if the bobbin is rotating. As an overend unwinding adds some twist to the yarn and would thus reduce the strength of the fiber-reinforced plastic composite, the production of woven semi-finished products for composite materials requires a twist-free processing of the yarns, using special creel with tangential take-off. From the creel, the yarns are fed to the weaving machine via yarn guiding systems. Instead of a warp beam, a delivery roller with friction lining can be used for the suitably tension-controlled warp yarn feeding from the creel. The monitoring of the yarns integrated into their course switches off the weaving machine in case of warp yarn breakage. On classic weaving machines and in the processing of spun fiber yarns, drop wires can be used. They can be installed between the back rail and the heald frames and are based on the short circuit principle. On weaving machines for technical textiles or filament yarns processing, optical systems are used to monitor the rear and front side of the shed. In case of sticking yarns or yarn breakage, the weaving machine is shut down.

5.4.4 Shed Formation Systems

5.4.4.1 Overview of Shed Formation Systems

The warp yarns are guided in the healds, which are arranged in the heald frames or coupled with the harness cords of the Jacquard machine. The healds are usually manufactured from flat-bar steel. To reduce strains on the warp yarns in the heald eyes (opening for guiding the warp yarn), round wire strands are available, which can be fitted with a ceramic heald eye. For ribbon-shaped yarns, especially for a twist-free processing of carbon rovings or splayed yarns, special healds with a rectangular yarn eye are used. The healds are lined up on the heald frames or connected to the harness cords of the Jacquard machine, corresponding to the warp density and the number of warp yarn groups. Since the maximum density of the healds in the heald frames (healds per cm) is limited, plain-woven fabrics requiring at least two heald frames are usually woven with four to six heald frames.

Cam machines, dobbies, and Jacquard machines are used for shed formation (Fig. 5.17). Depending on the respective shed formation system, differently-sized repeat units are feasible. Cam machines and dobbies are connected to the heald

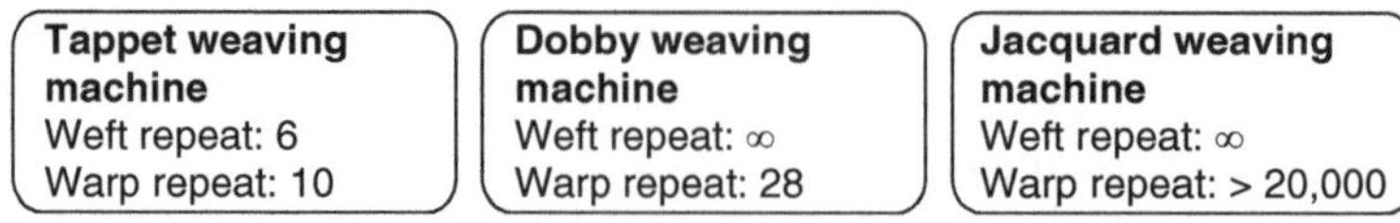

Fig. 5.17 Classification of shed formation systems

frames by a coupler. In Jacquard machines, the healds are individually coupled with the machine by harness cords. Dobbies are prevalent to produce woven fabrics for composite materials although Jacquard machines are seeing increasing use, especially in industrial applications.

Shed formation subjects the warp yarns to a strong alternating tensile stress, and to flex abrasion stress in the healds. To reduce yarn damage and stabilize the weaving process, warp yarns can be sized. The sizing agent increases the yarn compaction by gluing the fibers or filaments in the yarn together. It strengthens the yarn, enhances elasticity and makes it more resistant to abrasion. Special sizing agents have been developed for weaving yarns from glass or carbon filaments, meeting both the weaving technological requirements and ensuring adhesion between fiber and matrix in the composite. Due to the high brittleness and low transverse strength of glass and carbon yarns, the machine speed (number of revolutions) of the weaving machines is often reduced by up to 75 % when processing these materials.

5.4.4.2 Cam Machines

Cam machines with cam disks (Fig. 5.18a) present themselves as a cost-effective solution to drive the heald frames. The length of the weft repeat is limited to a maximum of six differently binding weft yarns through the number of heald liftings and lowerings on the cam disk. In a weft repeat of six yarns, the seventh weft yarn is crossed with the warp yarn in the same manner as the first weft yarn. Technologically, the selection of a maximum of eight heald frames is sensible due to the small weft repeat. This means that a maximum of eight warp yarns bind differently in the woven fabric and that the warp repeat at an even threading is a maximum of eight warp yarns. For a change of the weave pattern, the cam disks have to be replaced and the transmission between weaving machine and cam mechanism machine has to be altered for changes to the weft repeat length.

5.4.4.3 Dobbies

Dobbies (Fig. 5.18b), usually rotary dobbies, are used to control the heald frames electronically and lower or lift them according to the selected weave diagram. The drive of the each heald frame is either electromechanically coupled to the drivetrain of the weaving machine or equipped with its own motor. Through the individual

Fig. 5.18 Shed formation systems: (*left*) cam machine, (*right*) dobby (*Source*: Stäubli GmbH)

driving of the heald frames by means of a servomotor, the shed formation for each heald frame can be customized for each weft insertion. The variable shed geometry allows easier production of complex weave patterns or woven fabric structures, and allows a more efficient processing of yarns with low tension and elongation. The preparation and change of patterns are performed electronically. The maximum weft repeat depends on the size of the electronic pattern memory. With a maximum of 28 deployable heald frames, the warp repeat is limited to a maximum of 28 warp yarns.

5.4.4.4 Jacquard Machines

Unlimited patterning and structuring possibilities are offered by Jacquard machines, which are usually arrayed above the weaving machines (Fig. 5.19). The use of heald frames is thus avoided. Each individual heald is connected upward to the Jacquard machine by a harness cord and tensioned downward with a readjusting spring. By installing the Jacquard machine below the warp yarns in the weaving machine, harness cords and springs could be omitted. The mechanical complexity and high operating effort required for this type of set-up have hindered its establishment. The control boards of the Jacquard machine are driven either by electromechanical couplings from the main drivetrain of the weaving machine, an individual drive of the Jacquard machine itself, or an individual drive for each control board. The coupling with the main drive of the weaving machine is conventional, although an individual drive of the control boards allows the adjustment of the heald motion to the requirements of the weaving process and yarn material for each weft insertion and each warp yarn.

Through the modular set-up of the Jacquard machines, the number of individually selectable warp yarns can be increased nearly at will. Weaving machines equipped with Jacquard machines offer the greatest variety in terms of patterns, weave patterns, and structures. If a visual theme is repeated across the width of the woven fabric, the harness cords of identically binding warp yarns can be driven by the same control element of the Jacquard machine, reducing the overall dimensions

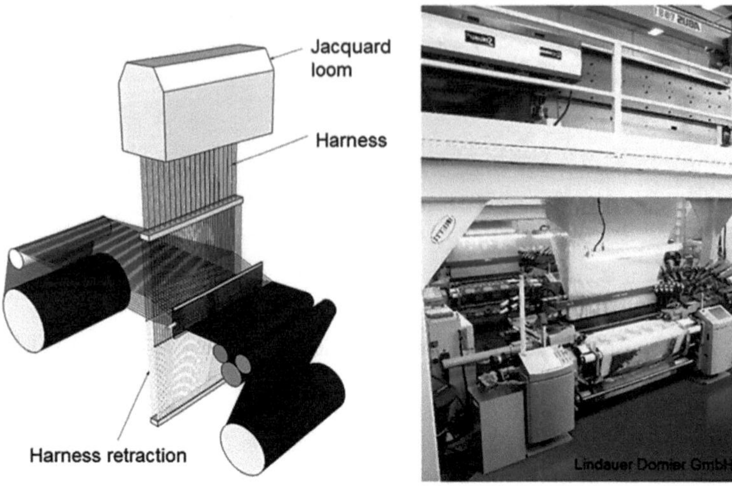

Fig. 5.19 Jacquard weaving machine, operating principle and view (*Source*: Lindauer Dornier GmbH)

of the machine. The ability to change the warp density without replacing the harness system or the Jacquard machine is restricted in weaving machines fitted with a Jacquard unit.

5.4.5 Weft Insertion Principles

5.4.5.1 Overview of Weft Insertion Principles

According to the respective principle of weft insertion, weaving machines are classified into the following types:

1. Shuttle weaving machines
2. Projectile weaving machines
3. Jet weaving machines
4. Rapier weaving machines
5. Ribbon needle weaving machines
6. Circular weaving machines
7. Wave-shed weaving machines, and
8. Shed course weaving machines

Circular weaving machines are sometimes used on a small scale to produce tubular fabrics for technical textiles. The production and development of wave-shed and shed-course weaving machines have been stopped due to insufficient flexibility. Therefore, these three machine types will not be included in the following pages.

The corresponding details for these weaving machines can be found in textbooks [8, 9]. Machine types 1–5 will be at the focus of the next sections.

5.4.5.2 Weft Insertion by Shuttles

Until the mid-1960s, shuttle weaving machines were used near-exclusively. The shuttle, also called weaver's shuttle, picks up the weft bobbin. By means of a beating device, the shuttle is shot through the open shed from one side of the machine to the other (Fig. 5.20). After a changing of the shed and reed beat-up, the shuttle is shot back to the original side. By reversing the uncut weft yarns at the border of the woven fabric, a firmly defined edge is formed on the fabric. Beyond that, shuttle weaving machines can be used to produce tubular woven fabrics by means of selected weave patterns and the weft reversal at the fabric edge, with the working width being one half or a quarter of the tube circumference. Shuttle weaving machines are well-suited to the weaving of fiber-reinforced plastic profiles (see Sect. 5.6.1). For these special applications, special machines capable of processing a maximum of 360 m of weft yarn per minute are built as shuttle weaving machines. Warp yarns are highly strained because of the necessary size of the shed, while the weft yarn strain is very small.

5.4.5.3 Weft Insertion by Projectile

In a projectile weaving machine, the weft yarn is located on a bobbin outside of the shed. The weft yarn is fixed to the clamp of a metal projectile, which is then shot through the shed by means of an acceleration system. The projectile is guided through the shed by guide blades and intercepted on the opposite side of the shed, where it releases the weft yarn (Fig. 5.21). It is then transported back to the machine side with the projectile thrower underneath the shed. Therefore, there are always

Fig. 5.20 Weft insertion by shuttle

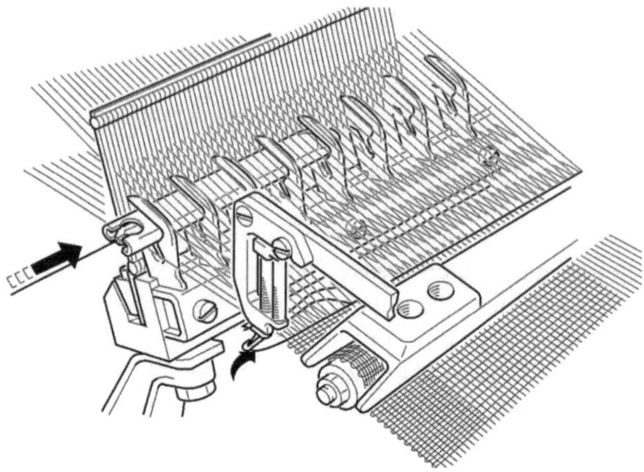

Fig. 5.21 Weft insertion by projectile (*Source*: ITEMA)

several projectiles in motion at the same time. The weft yarn is greatly accelerated during the launch of the projectile, exposing it to a tensile force. To ensure the laying of the next weft yarn, the current weft yarn has to be cut on the weft insertion side. Fabric edges are formed by means of a variety of methods for selvedge formation (see Sect. 5.4.6). Projectile weaving machines are especially effective in the production of woven fabrics wider than 6 m, e.g. for membrane materials. They can process up to 1,500 m of weft yarn per minute.

5.4.5.4 Weft Insertion by Jets

Jet weaving machines transport the weft yarn through the shed by air or water. The weft yarn is accelerated in one or several subsequent main nozzle tubes. To concentrate the air jet in the shed, the blades of the reed form a locating channel on three sides. Several relay nozzles are arrayed along the width of the weaving machine, projecting into the shed and controlled in such fashion as to ensure that the air jet is always at its highest speed at the yarn tip (Fig. 5.22).

Thus, the weft yarn is pulled through the shed. As soon as the preset weft insertion length is achieved, the weft yarn is stopped on the weft insertion side. This creates very high tension on weft yarn. In this method too, the weft yarn is cut after insertion. This makes additional processes for selvedge formation necessary. Monofilaments and highly opened weft yarns, such as rovings, cannot be processed. On the other hand, lightweight glass and carbon filaments are easily woven with this method. Despite their high energy consumption, jet weaving machines are most commonly used among the types of weaving machines because they are capable of processing up to 2,500 m of weft yarn per minute, even at a working width of more than 5 m.

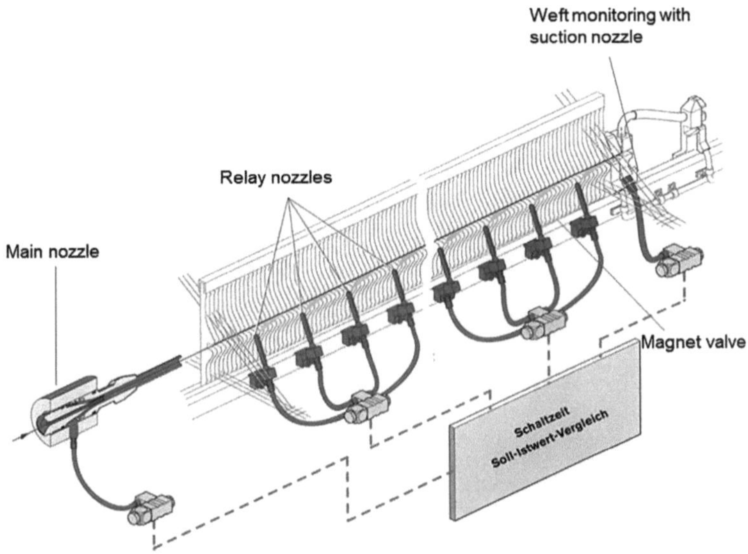

Fig. 5.22 Weft insertion by air jets (*Source*: Lindauer Dornier GmbH)

5.4.5.5 Weft Insertion by Rapier

The great advantage of weft insertion by rapier over other weft insertion methods is that the weft yarn is always connected non-positively to the weft insertion elements of the weaving machine and thus with the drive elements of the weaving machine. This way, the greatest process stability, flexibility regarding the processable weft materials, and a low strain on the weft yarn can be achieved. According to number and arrangement, various machine configurations are distinguished:

- Single-rapier weaving machine
- Double-rapier weaving machine
- Multi-rapier weaving machine

According to the number of rapier systems, the corresponding number of weft yarns can be inserted simultaneously. Single-rapier weaving machines are most common, especially in the double-rapier configuration (one carrying rapier and one taker rapier). The weft yarn is taken up by the carrying rapier on the weaving machine side and separated from the previously inserted weft yarn with scissors. The rapier transports the weft yarns to the center of the woven fabric. Simultaneously, the taker rapier is pushed into the shed from the other side. In the center, the taker rapier takes over the weft yarn. Both rapiers are removed from the shed, the clamp of the taker rapier is opened and releases the inserted weft yarn. Figure 5.23 shows this basic principle.

The use of rapier rods (Fig. 5.23) offers the advantage of giving the ability to safely process of very fine as well as very coarse weft materials, e.g. monofiles, rovings, heavy tows, and wires. As the rapier rods have to be removed from the shed

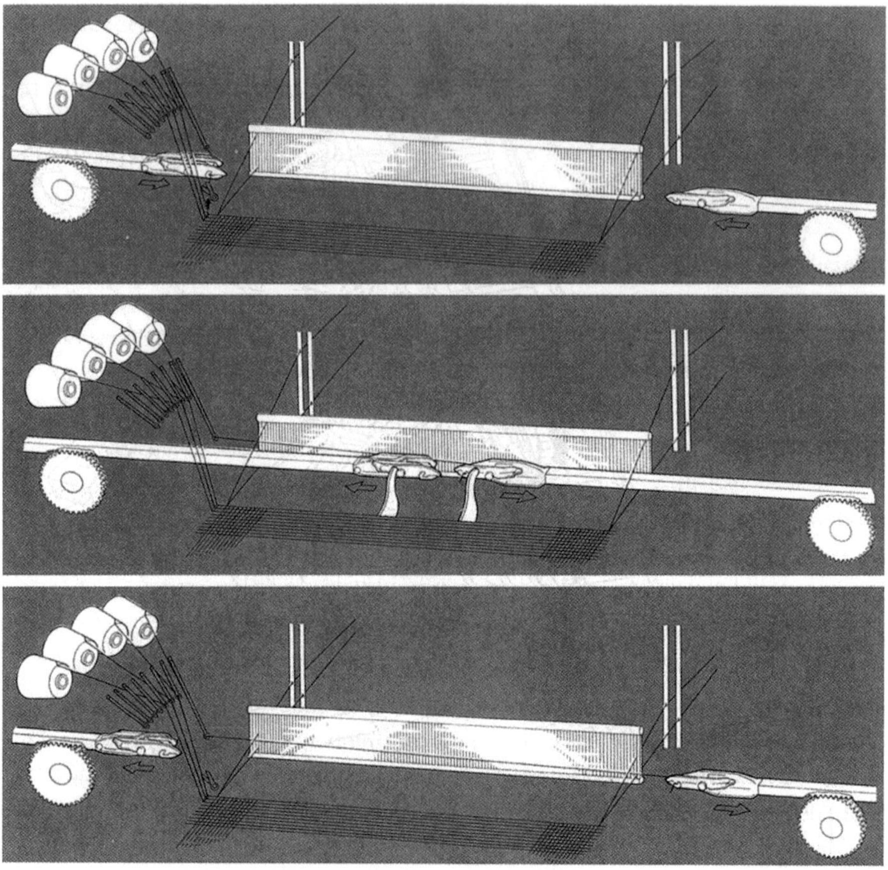

Fig. 5.23 Weft insertion with single-rapier system and active weft transfer (*Source*: Lindauer Dornier GmbH)

in a straight line, the width of the rapier rod weaving machines is twice that of the working width. In tape rapier weaving machines, the flexible rapier tape carrying the rapier head is deflected outside of the shed under the weaving plane.

This is why tape rapier weaving machines are far less wide than rapier rod weaving machines. The flexible rapier tape, however, requires additional tape guiding elements in the shed. While the clamps of the rapiers are always operated actively at the edge of the woven fabric, the central transfer can be performed with either active or passive clamps. Active clamps are operated by additional opening levers reaching into the shed. Rapiers with active transfer achieve a considerably higher stability in the weft yarn transfer and can be used for open weft yarns, untwisted filament yarns and especially glass or carbon rovings. They are also absolutely necessary for the processing of coarser high-performance yarn. With passive rapiers, the weft yarn is taken up from the carrying rapier by the hook-shaped yarn clamp of the taker rapier and pulled out of the yarn clamp of the carrying rapier. Rapiers with a passive transfer are smaller, resulting in a smaller shed, a gentler warp yarn processing, and a more efficient weaving performance. The processing of open yarns, particularly of rovings and heavy tows, is complicated by the passive central transfer. Rapier weaving machines process up to 1,400 m of weft yarn per minute.

5.4.5.6 Weft Insertion by Weft Needle

The weft yarn is inserted as a weft loop by means of a weft needle on ribbon needle looms for narrow woven fabrics, straps, and profiles (Fig. 5.24). Therefore, the insertion is always a double weft insertion. The weft yarn is not cut, which creates a solid edge on the insertion side. On the other side, the weft loops are transformed into a wale with a knitting needle, which solidifies the edge (see Sect. 6.2.2). When processing coarse weft yarns, an auxiliary yarn is transformed into stitches, and the weft yarn loop is incorporated into the wales. Ribbon needle looms can achieve up to 4,000 weft insertions per minute, depending on the width of the narrow woven fabrics and the yarn material.

5.4.5.7 Weft Change Device and Weft Feeder

All weft insertion systems can be combined with weft change devices, which allow the pattern-suited use of weft yarns of different colors, or of different types of yarn. On rapier weaving machines, up to 16 different weft yarns can be processed in accordance to the structure. By means of the weft change devices, membrane materials can be reinforced with grid-like assemblies of high-performance fiber materials, e.g. for an increased resistance to tear propagation. Furthermore, the weft change devices can be used to suitably integrate functional yarns (e.g. for sensor networks) into woven semi-finished products for fiber-reinforced plastic composites. Monitoring devices control the weft insertion. Weft yarn breakages are repaired manually and do no cause flaws in the woven fabric. Weft feeders are imperative for a secure and gentle weft insertion, which cache the weft yarn and reduce the yarn tension resulting from the high accelerations during weft insertion. For a non-damaging and twist-free processing of glass and carbon yarns, special feeders are necessary, which roll off the weft yarn actively and tangentially from the bobbin. To achieve an even yarn tension, a highly dynamic cache is installed between the unwinding and the weft insertion system. The unwinding system delivers the weft yarn to the weaving machine at a constant speed.

Fig. 5.24 Weft insertion by weft needle

5.4.6 Weft Beat-up

After the insertion of the weft yarn into the shed is completed by the weft insertion system, the weft yarn is beat-up to the fabric edge by means of the reed. The segmentation of the reed and the passing of the warp yarns into the reed determine the distance between the individual warp yarns or the warp yarn density within the woven fabric. When the reed beats the weft yarn up to the fabric edge, the heald frames (or the healds of the Jacquard machine) change their position according to the weave pattern. When the sley with the reed reaches the weft beat-up point, the heald frames already begin to open the next shed. This creates a warp yarn interlacing behind the weft yarn, which prevents the recoil of the weft yarn into the shed. The phase assignment of the heald frames on the weaving machine can be adjusted according to the yarn material and product requirements. For highly dense fabrics, the weft yarn has to be pressed into the woven fabric of the yarn interlacement with considerable force. For the high required weft beat-up force, a correspondingly high warp yarn tension is required as counterforce. For a gentle processing of glass and carbon yarns, high beat-up forces (shear stress) have to be avoided. Therefore, only woven fabrics with low to medium densities can be produced (see Sect. 5.3.2).

The weft yarn is always inserted into the shed in a stretched state. The stretched weft yarn is pressed into the weave pattern by weft beat-up. The relation between warp yarn and weft yarn tension determines the crimp of the weft yarn in the woven fabric (see Sect. 5.3.1). The yarn tension resulting from the difference in length between stretched feed and weft yarn crimp causes a contraction of the woven fabric over its width. For a secure weaving process, this contraction has to be prevented in the fabric formation zone by means of temple systems, which stretch the woven fabric at the edges or across its entire width in weft direction. The temple does not damage the woven fabric. The contraction of the woven fabric is thus delayed until after the fabric leaves the temple. The extent of the contraction depends on the yarn properties, the weave pattern, the density of the woven fabric, and the tension of the warp yarn.

In all modern weft insertion methods, the weft yarn is cut at the edge, creating an open fabric selvedge, which is likely to cause fraying of the edge warp yarns. For this reason, special selvedge devices are used to fix the edge warp yarns. In woven fabrics for garments and home textiles, the end of the final weft yarn (ca. 1–2 cm) is therefore inserted into the following shed and incorporated there (foundation lath). In this manner, a solid selvedge can be realized. The leno selvedge, in which two or more warp yarns changes their lateral position at the fabric edge at each weft insertion and thus clamp the weft yarn, is utilized for technical textiles. The principle of the leno weave pattern is detailed in Sect. 5.5.4. For the selvedge formation, a variety of leno designs and different technical solutions are used [9].

5.4.7 Woven Fabric Take-off and Storage

Woven fabrics are usually taken off by means of a three-roller-take-off (take-off roller with two deflection and clamping rollers). By controlling the take-off rollers, the weft density of the woven fabric is adjusted. The maximum theoretical weft density is determined by the used yarn, the warp yarn density, and the weave pattern (see Sect. 5.3.1). When the weft density, which is set at the take-off of the woven fabric, approaches the theoretical maximum, the strain on the warp yarns increases considerably. Highly dense woven fabrics with barrier effects make high demands on warp yarn materials and require optimum settings of all weaving machine parameters. In the production of glass and carbon woven fabrics, low to medium fabric densities achieve a gentle yarn processing.

The take-off rollers are equipped with different friction linings (silicone, sand-paper, metal surface irregularities, and needles) for a safe take-off of the woven fabric. The woven fabric is stored on a winder on the weaving machine for small lot sizes. For large lot sizes, separate ascending batch winders are used. These winders have either an internal or an external drive. The latter consists of two driven rolls on which the wound core or the wound-up woven fabric are situated. With special winders, which guarantee a constant rolling pressure by means of a controlled pressure roller, glass and carbon woven fabrics can be stored with little to no damage.

5.5 2D Woven Fabric Structures

5.5.1 Conventional 2D Woven Fabrics

Conventional 2D woven fabrics consist of two interlacing orthogonal yarn systems (see Sect. 5.2). Typical 2D woven fabrics for lightweight construction applications are:

(a) Plain weaves
(b) Twill weaves
(c) Satin weaves
(d) Woven fabrics in extended or derived basic weave patterns, and
(e) Jacquard-woven fabrics

Types of woven fabrics used for lightweight construction applications are described with regard to the woven fabric parameters (see Sect. 5.3.1). Particularly in aeronautics, the woven fabrics are standardized and defined with regard to material, appearance, weave pattern, and condition on delivery.

5.5.2 Two-Layer and Profiled Woven Fabrics

By using at least two warp and two weft yarn systems on one weaving machine, at least two woven fabrics can be produced on top of one another (Fig. 5.25). The systematic interlacing of selected warp and/or weft yarns of the one woven fabric with the warp and/or weft yarns of the other woven fabric (connection, Fig. 5.25a) allows the connection the two woven fabrics in parts or in their entirety. Substituting all warp and weft a yarn of the upper fabric for those of the bottom fabric (change of fabric, Fig. 5.25b) is another option to create a connection between the two fabrics.

The areas containing two fabric layers, connected fabric layers or a change of fabric can be selected freely from the entire area of the woven fabric by using a Jacquard technology. Subsequent process steps like cutting or molding enable the manufacture of woven semi-finished products with 3D geometries. Figure 5.26 shows the hollow profiles producible with this technique [10]. For instance, side airbags in vehicles are produced seamlessly with this technology [11]. The ready-made technological processing of two-layer woven fabrics makes 2.5D geometries for profiles feasible.

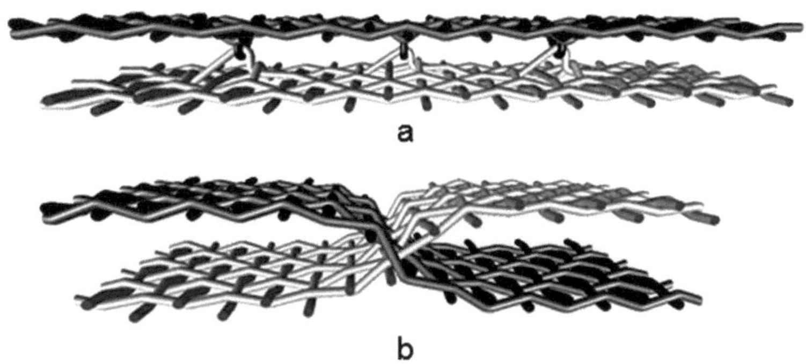

Fig. 5.25 Two woven fabric layers with connecting area: (**a**) connection, (**b**) change of fabric

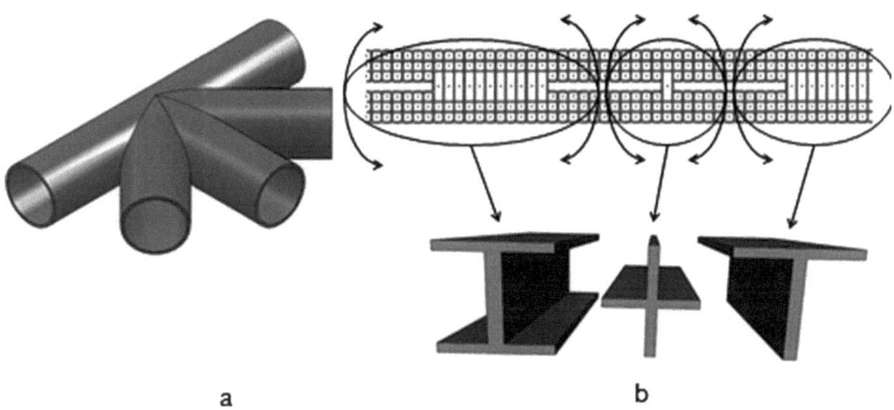

Fig. 5.26 Profiles producible with weaving technology: (**a**) tube joint profile, (**b**) carrier profiles

5.5.3 Two-Dimensional Multilayered Woven Fabric Structures

Multilayered woven fabrics consist of several stacked warp and weft yarn systems in the fabric thickness direction. Classification of the multilayered woven fabrics as 2D or 3D structures is handled variably in the literature. Planar multilayered woven fabrics and woven fabrics with 3D geometry without reinforcement yarns in Z direction are classified as 2D woven fabric structures in accordance with the definition in Sect. 2.3.1, while those 3D woven fabrics containing reinforcement yarns in Z direction are counted among the 3D woven fabric structures (see Sect. 5.6.1).

In textile technology, multilayered fabrics are classified according to the number of fabric layers. Each fabric layer consists of at least one warp and one weft layer. Thus, a two-layer woven fabric contains at least two warp and two weft layers. From the perspective of composite materials, a composite made from a two-layer

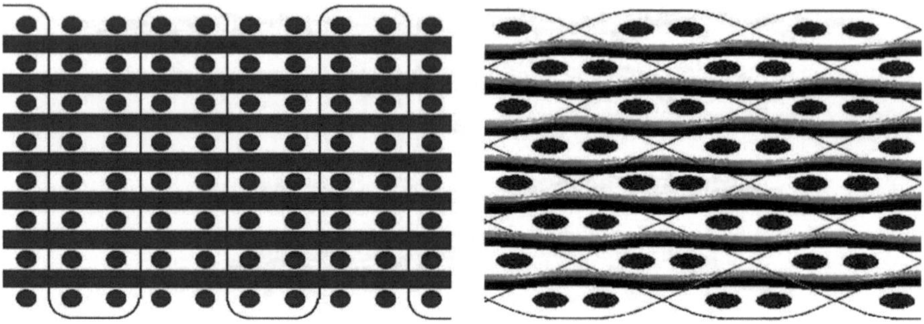

Fig. 5.27 Structural examples of multilayered woven fabrics (warp view)

woven fabric displays at least a four-layer structure (two reinforcement layers in X direction, and two reinforcement layers in Y direction).

The warp and weft systems of the multilayered woven fabrics are weave-technically interlaced to form a 2D woven fabric structure. While the warp and weft yarns in single-layer standard woven fabrics are always arranged side-by-side respectively, the warp and weft yarns of different layers of multilayered woven fabrics are also placed on top of each other. Multilayered woven fabrics are differentiated by the number of layers and the type of interlacement between them. Figure 5.27 shows two different typical structures.

The number of warp and weft yarns processable on weaving machines is limited, depending on the machine equipment and yarn properties. Multilayered woven fabrics from glass yarns with thicknesses of up to 18 mm can be produced by weaving technology [12]. For thicker woven fabrics, special technological measures have to be taken, which in turn will limit the width of the woven fabric. These measures are described in Sect. 5.6.4 (3D orthogonal woven fabric).

Multilayered woven fabrics offer the possibility to orient warp and weft yarns completely stretched in the structure. These structures are very similar to non-crimp fabrics. For this reason, they are also referred to as *Non-Crimp Weaves*. As the warp and weft yarns of the individual layers of these 2D woven fabric structures do not cross one another, an auxiliary yarn system (binding warp) (Fig. 5.27) is required. The stretched orientation of the warp and weft yarns gives the fiber-reinforced plastic composites produced from these woven fabrics higher stiffness and strength comparable to fiber-reinforced plastic composites manufactured from biaxial non-crimp fabrics. Using multilayered woven fabrics as reinforcement semi-finished products considerably reduces the cost-intensive handling of the layering as well as the expense of composite formation. However, an increased number of warp and weft layers will decrease the drapability of the 2D multilayered woven fabric structures.

5.5.4 Leno-Woven Fabric

In comparison to the conventional woven fabrics described in Sect. 5.2, *leno-woven fabrics* consist of at least two warp yarn systems and one weft yarn system. The two warp yarn systems (stationary yarns and binding yarn systems) are entwined after each weft insertion. For process-immanent reasons, these entwinements are realizable only in the warp yarn system. Leno-woven fabrics are defined differently depending on the manner of entwinement, the number of involved warp yarns, and the amount of incorporated weft yarns. The most common weave pattern in technical applications is the so-called single-weft two yarn standard leno (shortly, standard leno) [13]. It consists of two alternately entwined warp yarns with one incorporated weft yarn each (Fig. 5.28a) [8].

Leno-woven fabrics are usually described as healds or needle bar lenos. Structurally, the two variations do not differ. Heald twists can be realized on any kind of weaving machine, while needle bar lenos require special machines. Heald twists are not suitable for technical yarns and productive manufacturing technologies, since the yarns are subjected to high strains between the healds, which can damage the filaments and are likely to reduce productivity. Needle bar lenos are suitable for mass production. The yarns are not guided through healds, but through guide needles, which considerably minimizes the mechanical strain during the weaving process.

In contrast to conventional woven fabrics, the fabric density of leno-woven fabrics cannot be determined by means of the equations by WALZ and LUIBRANDT given in Sect. 5.3.1. It is necessary to calculate the fabric density as the relation of total area to overlapped area for a typical surface element (Eq. 5.2).

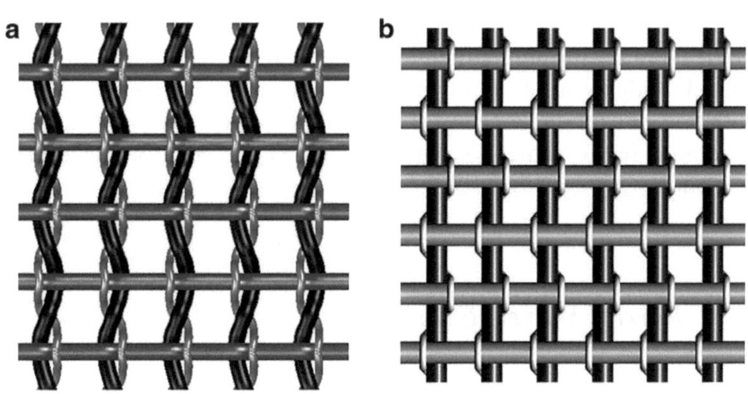

Fig. 5.28 (**a**) Conventional leno-woven fabric, (**b**) NCF leno-woven fabric

$$DG \approx n_k n_s \left(\frac{d_s}{n_k} + \frac{d_k}{n_s} + \frac{\pi d d^2}{2} - d_k d_s \right) p \cdot 100\% \tag{5.2}$$

In which,

DG (%)	Fabric density
n (mm)	Yarn distance (k: warp, s: weft)
d (mm)	Diameter (k: warp, s: weft, d: leno)

As the binding yarn system has to be guided through the layers of stationary yarns and weft yarn system, the maximum fabric density is lower than in standard woven fabrics. Conventional leno-woven fabrics display a much higher slippage resistance than conventional 2D woven fabrics due to the additional entwinement in warp direction. A different crimp of stationary yarn system and binding yarn system can be realized by using separate warp beams, resulting in a variation of the leno interlacing point position within the textile reinforcement structure (Fig. 5.28). This allows the production of woven fabrics with both very high and very low structural elongation in warp direction. Particularly woven fabrics with almost stretched oriented stationary yarns are well-suited for use as semi-finished products in fiber-reinforced composite materials, because their properties are similar to those of non-crimp fabrics (Figs. 5.28b and 5.29). Non-crimp fabrics are structures with stretched yarn positions, and are usually manufactured as UD tape (1D structure) or biaxial stitch-bonded fabrics (2D structure).

According to [14], woven fabrics in leno weave pattern are suitable for an even impregnation and deaeration in injection and infusion methods. Due to the asymmetric structure of single-layer laminates made from leno-woven fabrics with low structural elongation, preferred flow directions will occur along channels formed by the yarns. In contrast to conventional non-crimp fabrics, leno-woven fabrics do not experience broaching of the reinforcement yarns by the binding yarn system. This greatly reduces yarn damage. Beyond that, the drapability of the woven fabric and the yarn pull-out capacity in the composite are vastly improved. Today, leno-woven fabrics are primarily used in selected technical applications. For example, technical leno-woven fabrics with high structural elongation are used as reinforcing materials, in crack bridging of plastering mortar, or as textile membranes, while woven fabric with low structural elongation are utilized as coated (and thus stabilized) grid structures for the reinforcement of mineral matrices [15], and for injection molding reinforcement.

5.5.5 *Multiaxial Woven Fabric*

Multiaxial woven fabrics as 2D structures with at least three yarn systems arranged at non-right angles to one another were originally patented in 1974 (Fig. 5.30)

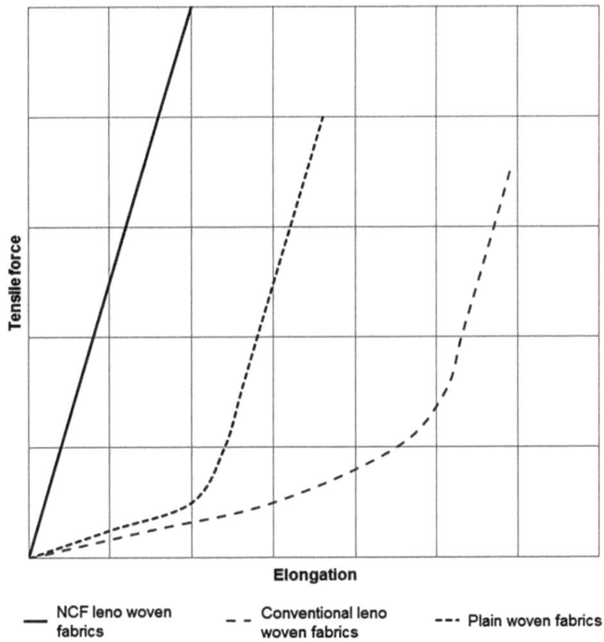

Fig. 5.29 Fundamental stress-strain behavior, depending on the structure

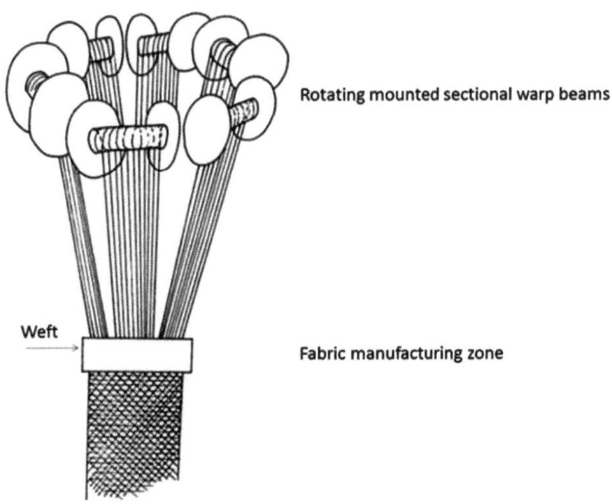

Fig. 5.30 Manufacturing process for the production of multiaxial woven fabrics, according to Norris

[16]. In contrast to orthogonal woven fabrics, they contain yarn systems in at least three directions within the plane. By increasing the number of axes in the plane, the slippage resistance is enhanced significantly, because each individual interlacing unit acts as a fully defined three-hinged arch, instead of a four-hinged arch. Triaxial woven fabrics with one weft yarn system and two warp yarn systems are typical. Each of the yarn systems encloses an angle of, for instance, 60° between two fiber

axes. This angle depends partly on the weft density and the rotational speed of the warp beams. Furthermore, multiaxial woven fabrics with more than three axes (tetraxial, pentaxial) can be produced. A higher number of axes results in increased isotropy of the multiaxial woven fabrics.

Multiaxial woven fabrics are produced on special weaving machines in which the warp yarns are fed by segmented, rotationally arranged warp beams or single spools. The reversal points of the diagonal warp yarns are situated at the edge of the fabric. The underlying machine technology is very complex, its operation requires a lot of effort, and its productivity is very limited. Shed formation of the diagonal yarns is not performed by healds, but by open yarn guides arrayed perpendicular to the warp yarn plane [17].

The weave patterns are classified analogously to the orthogonal woven fabrics. The plain weave pattern is defined as the basic weave pattern, i.e. the yarns of two axis directions alternately interlace below or above one another (Fig. 5.31a). Triaxial plain-woven fabrics are distinguished by their high dimensional stability and permeability. The created hexagonal opening are nearly twice the size of the effective diameter of the yarns used [18]. This is caused by the interlacing point between two systems preventing a telescoping of the yarns of the third system. The woven fabric densities of triaxial fabrics in plain weave are circa 67 % and decreases successively with increasing number of axes (tetraxial: ca. 55 %, pentaxial: ca. 20 %) [19]. More complex weave patterns, such as the triaxial woven fabric in twill weave can be manufactured analogously and may feature a closed surface. The fabric density of the depicted structure is 100 %. With an increased number of axes, the yarn density in axis direction will decrease, although dense structures remain feasible.

According to [20], triaxial woven fabrics can be draped much easier than orthogonal woven fabrics. Moreover, their use simplifies a layered structuring of multidirectionally reinforced composites. Especially for layers with an axis orientation other than 0° or 90°, the cutting waste is significantly reduced, which makes them particularly interesting for cost-intensive high-performance fiber materials.

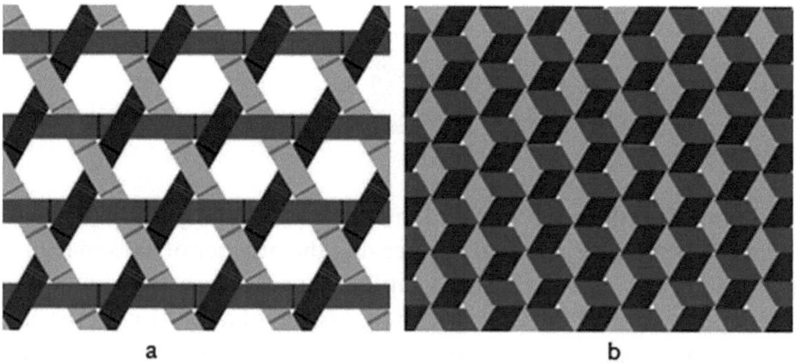

Fig. 5.31 (**a**) Triaxial plain-woven fabric, (**b**) triaxial twill-woven fabric

For example, ultra-lightweight, open-cell, foldable parabolic aerials are being made from coated triaxial woven fabrics in plain weave (0°, 60°, and 120°).

The manufacturing costs and production effort of multiaxial woven fabrics increase with the growing number of axes. Since triaxial woven fabrics are complicated and cost-intensive to produce, their applications are limited to niche and decorative uses.

5.5.6 *Open Reed Woven Fabrics*

To integrate additional warp yarn systems variably incorporated across the width into the weaving process, it is necessary to open the reed upwards. By means of special yarn guiding elements, this allows the insertion of additional yarns between the open reed and the heald frames, which are comparably more stable than standard reeds. The additional yarns are fed to the machine through a bypass (additional warp stop motions, deflection systems, yarn guiding elements) (Fig. 5.32). The yarn guiding elements are slidably mounted on a rail within a conventional heald frame, perpendicular to the direction of production. Thus, the heald frame drives of the weaving machine allow vertical travel, while additional linear drives provide lateral offset [21]. For dipping the additional yarns into the ground warp yarns, the following methods exist.

5.5.6.1 Open Reed Weaving

Due to positive locking effects the dimensions of the yarn guiding elements in open reed weaving create wider gaps between the warp yarns in the upper shed. Due to the warp yarn tension, corresponding gaps are opened in the reed, through which the additional yarns can change their position through the upper shed to the bottom shed without interlacing with the ground warp. The width of the yarn guiding elements is dimensioned to create a wide enough gap in the reed to allow the additional yarns to be positioned securely between the selvedge and the reed despite an angular yarn inlet. After dipping, weft insertion takes place, and the additional yarns are bound off. In the following weaving cycle, the yarn guiding elements are lifted back out of the shed and can be laterally displaced before the next lowering. This way, nearly unlimited varieties of patterning effects can be realized (Fig. 5.33).

The open reed weaving devices can be installed as a module in the weaving machine and does not limit the performance potential and application spectrum of the machine. Jacquard-like weave patterns can be realized (Fig. 5.33), obviating subsequent stitching processes of pattern-adapted yarns.

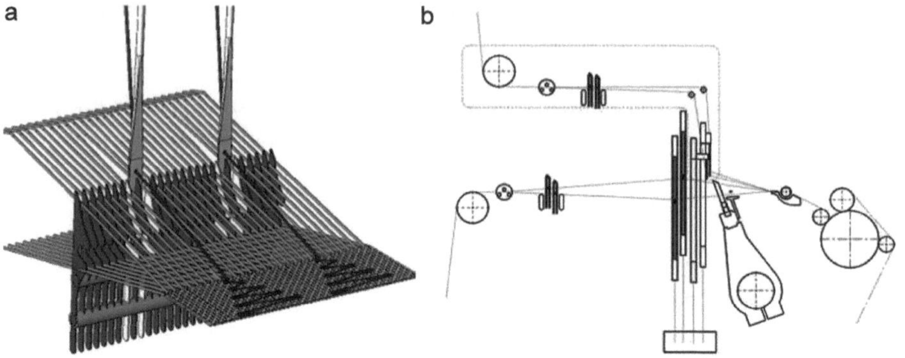

Fig. 5.32 Functional principle of open reed weaving (*Source*: Lindauer Dornier GmbH)

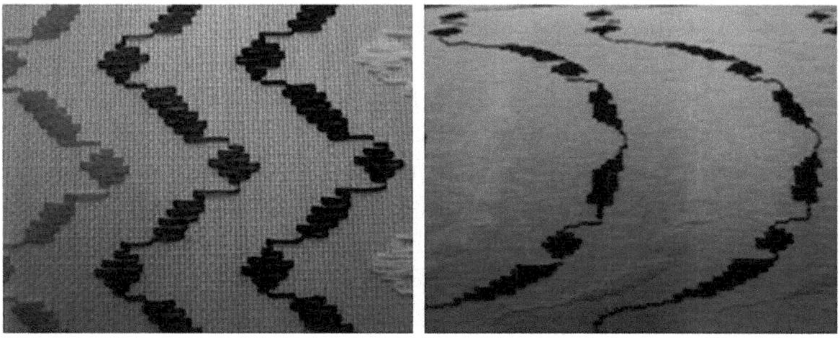

Fig. 5.33 Examples of open reed woven fabric for curtains (*Source*: Lindauer Dornier GmbH)

5.5.6.2 Open Reed Multiaxial Weaving

In open reed multiaxial weaving, the yarn guiding elements are dipped into the ground warps in a positively locking manner (Fig. 5.34). The remaining reed gaps are covered segment-by-segment by means of a specifically formed open reed with angled tips, and a funnel is formed to receive the additional yarns in the designated gap. Analogous to open reed weaving, the additional yarns are bound off by the weft yarns, and the yarn systems are separated and subsequently offset laterally by linear drives.

The additional systems are also integrated in the weaving machine as modules and do not affect the performance potential and application spectrum. The capability to laterally offset the additional yarns allows the manufacture of leno-woven fabrics in different types of weave pattern (standard leno, selvedge motion, and combined leno weave pattern). This method makes complex weave patterns feasible, significantly extending the weave pattern variety of wide woven fabrics. The economic manufacture of so-called selvedge motions is especially interesting, as it allows the realization of significantly more slippage resistant woven fabrics for agricultural and filtration applications or geotextiles.

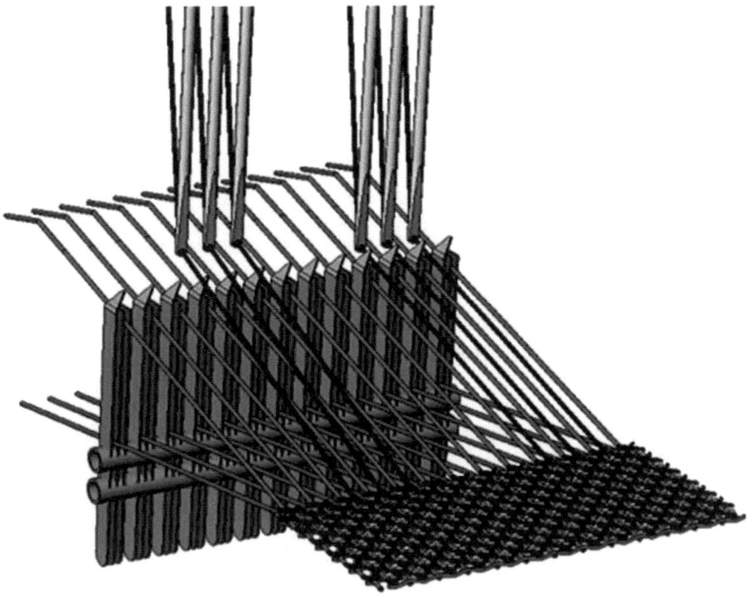

Fig. 5.34 Principle of open reed multiaxial weaving (*Source*: Lindauer Dornier GmbH)

On the other hand, multiaxial woven fabrics with two orthogonal yarn systems, and locally reinforced woven fabrics become feasible. These consist of an orthogonal basic woven fabric in any desired weave pattern (plain, twill, satin, leno; Fig. 5.35) and one or two diagonal warps oriented toward the former at an adjustable angle and fixed to the ground warp by binding-technical means. The incline of the diagonal yarns can be varied in a range of 10°–170° against the weft axis.

The open reed weaving technology offers the possibility to extend the patterning diversity for woven fabrics immensely, and subsequently reduces the most cost-intensive processes (e.g. embroidery, layer structure. The technology makes the development of leno-woven fabrics with complex weave pattern, triaxial and tetraxial woven fabrics, and locally reinforced woven fabrics feasible.

5.5.7 2D Polar Weave pattern

Polar woven fabrics consist of helically arranged warp yarns and radially running weft yarns (Fig. 5.36) [22]. These 2D structures will be referred to as 2D polar woven fabrics in the following. For reasons of their width, they are classified as narrow textiles. The individual weft yarns are inserted as needed, i.e. by single bobbins (creels). A specially modified fabric take-off with conical rollers creates the helix by taking off the woven fabric more from the outer than from the inner

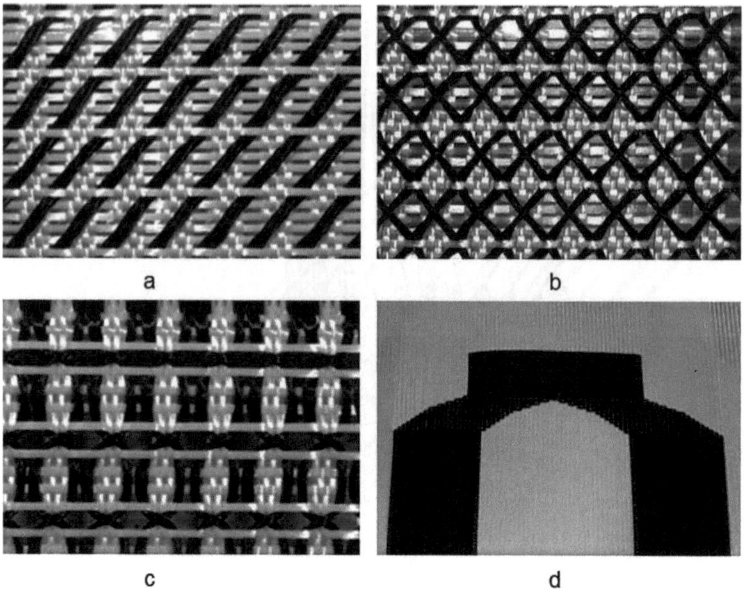

Fig. 5.35 Examples of the open-reed woven fabric: (**a**) triaxial, (**b**) tetraxial, (**c**) combined interlacing leno weave pattern, (**d**) local reinforcement (*Source*: Lindauer Dornier GmbH)

radius. For reasons of manufacturing, these woven fabrics have a homogenous structure in warp direction. The yarn density generally decreases gradually in weft direction from the central axis.

In principle, all basic weave patterns can be realized. 2D polar woven fabrics in plain weave have been marketed most successfully. The varying distribution of weft yarns influences the mechanical properties. According to [23], 2D polar-woven fabrics have a constant and usually higher specific modulus in warp than in weft direction. This is caused by the even warp yarn density and the varying weft yarn density across the width of the fabric caused by method-inherent reasons.

Since the warp yarns run on concentric circles, 2D polar woven fabrics are suitable for the load-adapted reinforcement of rotationally symmetric structures. In contrast to orthogonal woven fabrics, all fibers run along the main directions of loading (radial, tangential) and do not need to be shaped by costly and elaborate draping and cutting processes.

Usually, 2D polar woven fabrics are used for reinforcing the edges of disks or (partly) circular excavations. Woven cable harnesses are another example of application [24].

Fig. 5.36 Schematic illustration of a 2D polar woven fabrics

5.5.8 Woven 3D Geometries in Shell Shapes

The groundwork of near-net shape weaving of shell-shaped woven 3D geometries (with 2D structure) was laid by a development from 1993 [25]. With the help of a technical demonstrator unit, it could be shown that the manufacture of hemisphere shells without subsequent draping process is entirely possible. To produce shell-shaped, woven 3D geometries, it is necessary to guide the warp yarns to the shed individually. This requires a single yarn feeding and compensation. Three different methods are combined to create the three-dimensional shape:

- Variation of the relative woven fabric density: The weave diagram-related change of the woven fabric density factor influences the inner structure of the woven fabric insofar as increasing local structural tension due to varying crimp lengths, causing the unfixed woven fabric to bulge. This requires the individual selection of the warp yarns by means of a Jacquard machine.
- Variation of the warp yarn density: Using a reed with bars oriented in a fan shape (so-called V reed, Fig. 5.37) and adjustable height, the warp yarn density can be varied locally.

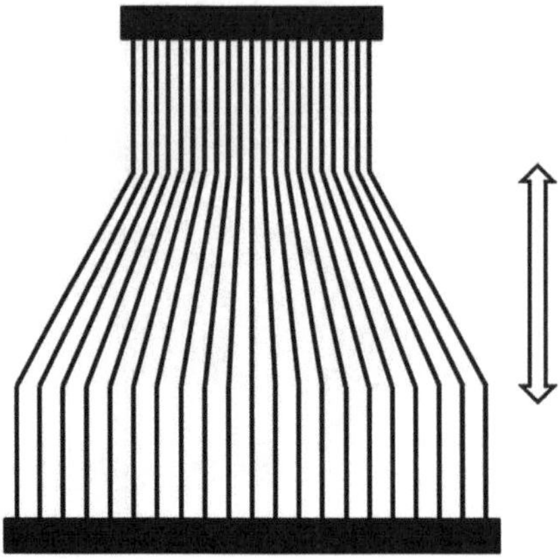

Fig. 5.37 V-shaped reed

- Variation of the take-up speed of the individual warp yarn: By varying the take-up speed of individual warp yarns or at least individual sections, the local weft density is changed. The fixation is then achieved by subsequent weft insertions, creating a specific three-dimensional formation of the woven fabric. Usually, auxiliary copies of the molded bodies are used for this purpose [25, 26]. For method-inherent reasons, neither closed geometries nor undercuts are producible this way, as the elements of the auxiliary copy of the molded body have to be removed again from the woven fabric.

Although this method requires extensive mechanical efforts (e.g. an adapted take-up system for each geometry), it enables the creation of near-net shape shell geometries for highly loaded component groups. The significant advantages of shell-shaped woven 3D geometries over draped 2D semi-finished products are in their own semi-finished products, which do not require cutting, are crease-free, and arranged reproducibly and load-oriented, giving them much better mechanical parameters.

Shell-shaped woven 3D geometries are primarily used in the manufacture of seamless shells (e.g. helmets, hemispheric shells or open cuboids) or strips with tapers, branching, or curvatures. By varying the weave diagram, the strips can also be manufactured as closed tubes, making pipes with variable diameter feasible.

5.6 3D Woven Fabric Structures

5.6.1 Three-Dimensional Multilayered Woven Fabrics with Integrally Woven Profiles

Multilayered woven fabrics as 3D structures consist of several stacked warp and weft yarn systems. Planar multilayered woven fabrics without reinforcement yarns

Fig. 5.38 Structure examples of 3D multilayer woven fabrics (warp cross-section view)

in Z direction are classified as 2D woven fabric structures (see Sect. 5.5.3), while those containing reinforcement yarns in Z direction are counted among the 3D structures. The geometries of 2D and 3D multilayered woven fabrics can be identical. The decisive factor in classification as a 3D woven fabric structure is the use of high-performance yarns in the binding yarn system (in Z direction). 3D multilayered structures also can be manufactured as so-called non-crimp weaves (Fig. 5.38).

Aramid, glass, and carbon yarns or high-strength yarns can be used as binding yarns in Z direction. This way, a simultaneous reinforcement of the structures in Z direction is achieved. The composite materials produced from these woven fabrics have better out-of-plane properties distinguished by higher delamination resistance and energy absorption under impact and crash loads. By adjusting the content ratio of high-performance yarns in the Z reinforcement and by changing the arrangement of the high-performance yarns in Z direction, these properties can be set to suit the specific requirements. Content ratio of high-performance yarns in Z-reinforcement can be controlled by binding technical means and/or by mixing with thermoplastic yarns [27].

Using the basic principle of 3D multilayer weaving technology, a variety of profiles with a range of geometries can be developed. With the two-layered woven fabric reinforcement structures introduced in Sect. 5.5.2, the complexity of the profiles can be extended. Implementing 3D multilayer weaving technology on needle ribbon looms, the subsequent cutting of the profiles required in broad weaving is omitted. Narrow woven profiles also display compact selvedges. As the maximum warp density on needle ribbon looms is higher than on broad weaving machines, semi-finished products for profiles with greater wall strengths can be produced. To manufacture profiles of varying widths along the production direction, a V reed (Fig. 5.37) is required. Further structural changes of the profile geometry are made possible by the reversal of individual weft yarns in the woven fabrics. Because a selected warp width at the selvedge does not form a shed at weft insertion due to the joint lift or reduction of all warp yarns, and because no shed change takes place after weft insertion, no yarn interlacing occurs in the area. The

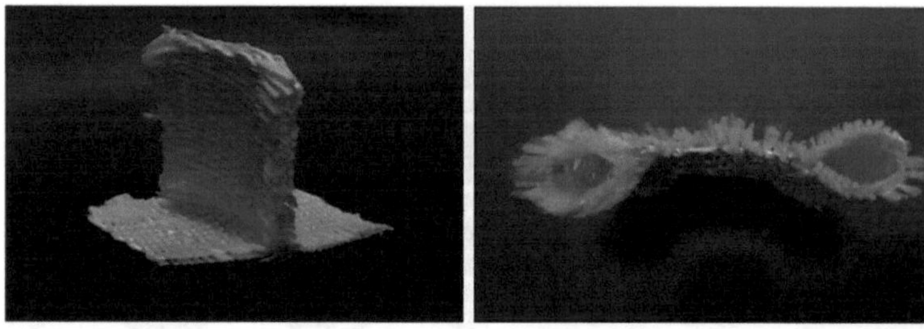

Fig. 5.39 Woven profiles

weft yarn reverses at the inner edge of this area. While this solution can only be used on the weft insertion side of needle ribbon looms, it is feasible on both selvedges on shuttle ribbon looms.

Using the above mentioned technological possibilities, shuttle ribbon looms can be used to produce tubes with variable and asymmetric diameters. Furthermore, a variety of profiles, such as L, T, and double-T profiles, can be manufactured (Fig. 5.39) [28]. From a weave technical perspective, it is possible to vary the profile shape along the tape. Additional weaving-technical modifications allow the production of helical tapes.

In the field of technical textiles, a number of tapes manufactured in this manner are used for belts, lifting equipment, load absorption and transfer elements, seatbelts, safety harnesses, and so on. In the composite material sector, tapes made from glass and carbon yarns are used for the partial reinforcement of composite components, and for specialized parts.

5.6.2 Woven Spacer Fabrics

Conventional woven *spacer fabrics* consist of two layers of woven fabrics connected by pile yarns (Fig. 5.40).

For the structure of the top and bottom fabrics, any weave patterns and materials used in flat-fabric production can be used. The pile yarns connecting both woven fabric layers are principally made from warp yarns. As the warp yarns cannot change their lateral position, the pile yarns are always oriented vertically in a sectional view in weft direction. In a sectional view in warp direction, they can also be oriented angularly (Fig. 5.41). On the contrary, the pile yarns are always arranged vertically in warp direction and randomly in transverse direction in warp-knitted spacer fabrics (see Sect. 7.5).

Woven spacer fabrics are preferably produced on weaving machine with double rapier systems (Fig. 5.42). Two sheds are arranged one above the other and two weft yarns are inserted simultaneously by two pairs of rapier rods. In the shed, the warp yarns of the ground warp switch from the top position to the center of the shed (top fabric), or from the center of the shed to the bottom position (bottom fabric). The

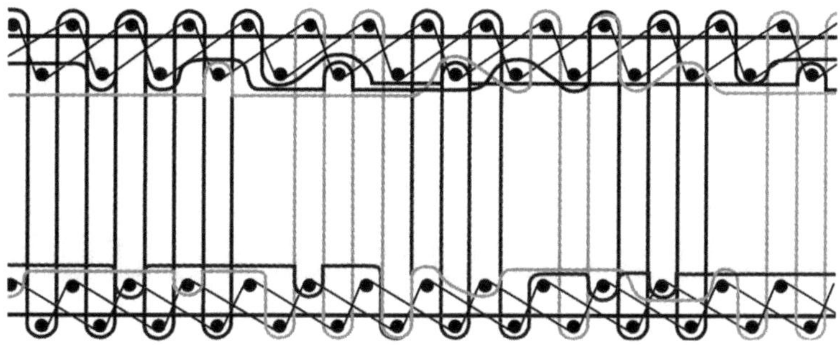

Fig. 5.40 Cross-section view of a woven spacer fabric (*Source*: Stäubli GmbH)

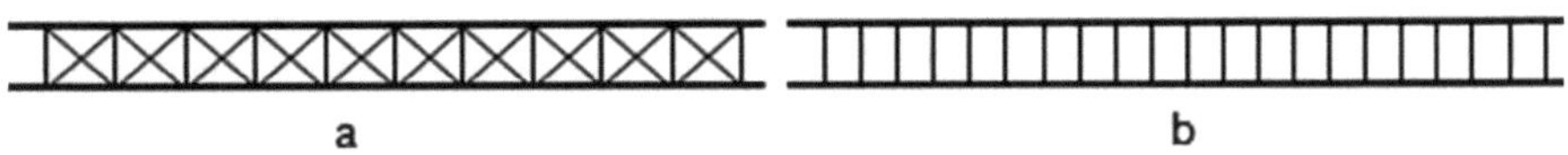

Fig. 5.41 Sectional view of a conventional woven spacer fabric: (**a**) warp direction, (**b**) weft direction

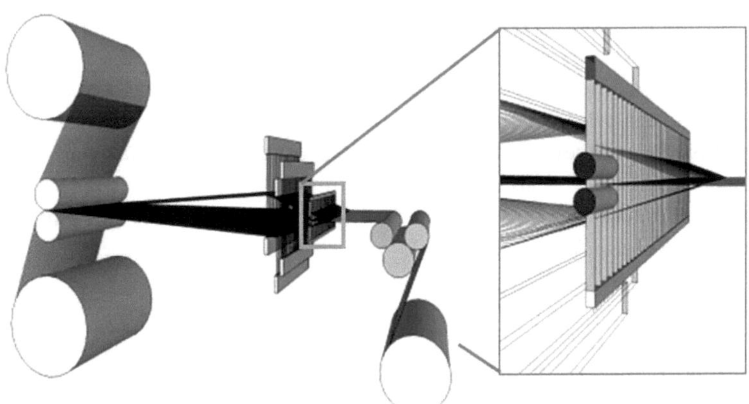

Fig. 5.42 Manufacture of woven spacer fabrics

pile warp yarns are selected by dobbies or Jacquard machines in manner that makes them freely arrangeable in the shed in the top, center, or bottom position.

The distance between both layers of woven fabric can be adjusted up to 100 mm. By using so-called loose weft yarns, distances of several meters can be realized in the product. As the pile yarn is integrated with one loose weft in top and bottom fabric, and the loose weft is subsequently removed from the woven fabric, the pile yarn will come loose from the woven fabric in these places. The free pile yarn length between the layers increases (for process-inherent reasons, it increases in uneven-numbered multiplications of the woven fabric distance), and the distance between the woven fabric layers increases correspondingly.

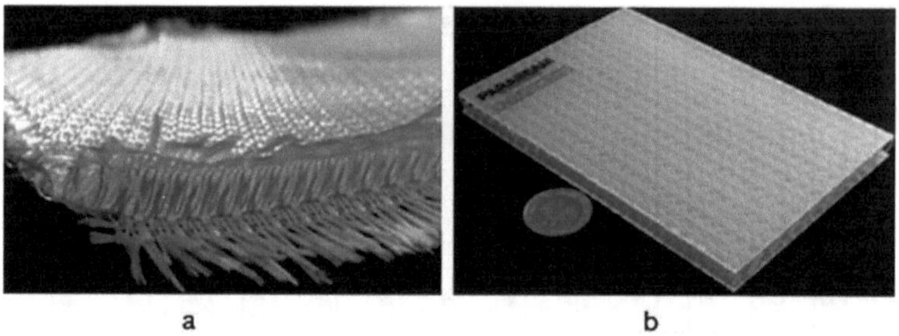

Fig. 5.43 (a) Woven spacer fabrics, (b) consolidated lightweight construction plate

The use of bend-proof yarns (e.g. monofiles) necessary for resilient surfaces in the pile requires more bend-proof weft yarns and very high warp yarn tensile forces to form the yarn interlacing and reversal of direction of the pile yarns in the surface layers.

Conventionally, double rapier system weaving machines are used for the manufacture of velours and carpets. For these products, the pile yarns are cut in the weaving machine, and two woven fabrics with velours surface are created. For the field of technical textiles, covers for paint rollers, woven spacer fabrics for air mattresses, air beds, or rubber dinghies are produced. Such spacer fabrics are classified as 3D geometries with a 2D reinforcement because of the use of thermoplastic yarns in the pile. For special applications, such as the production of double-walled tanks [29], spacer fabrics with high-performance fiber materials in all yarn systems are used (Fig. 5.43a). Such resin-laminated spacer fabrics are classified as 3D structures and are used, for instance, as lightweight construction plates (e.g. in boat building) (Fig. 5.43b).

5.6.3 Terry Weaving and Pleated Woven Fabrics

A special weaving technique called terry weaving can be used to produce 3D woven fabric reinforcement structures. In terry weaving, selected warp yarns or sections of the woven fabric are pushed out of the fabric plane. Figure 5.44 shows the basic principle used for the manufacture of terry woven fabric for towels and similar products. The formation of the pleated woven fabric is performed in three steps. In the first step, the distance between the two weft yarns is increased. With this distance, the height of the loop is adjusted. Afterwards, the ground warp yarns and the pile yarns forming the loop are usually connected by three wefts. In the third step, the gap created between the weft yarns in the first step is pushed together with the reed. The ground warp is kept tense during this time, while the pile warp is slackened. Depending on the weave pattern, the pile yarns jump upward or downward out of the woven fabric surface as pile loops. Larger weft distances and the subsequent pushing of the weft yarns can be realized technically by an offset of the

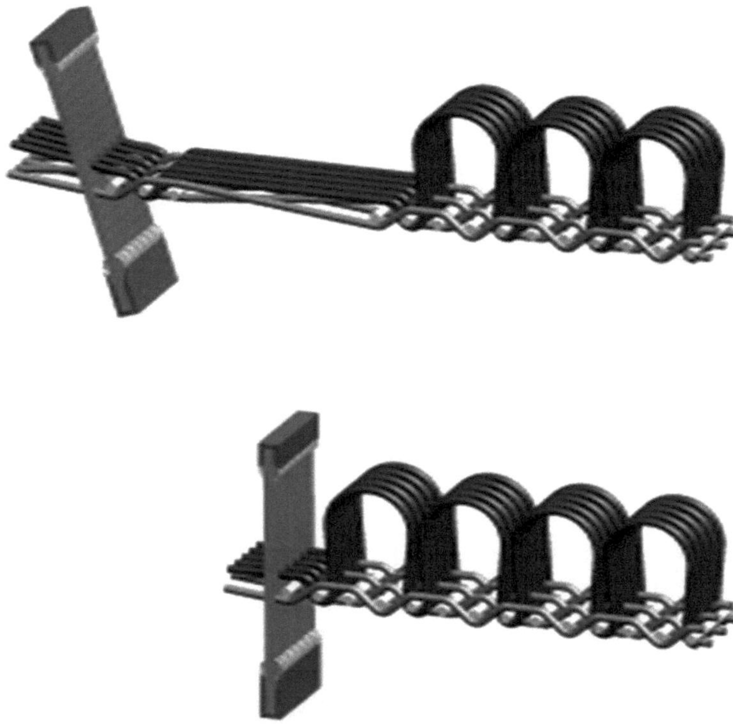

Fig. 5.44 Principle of terry weaving

entire woven fabric with the warp yarns in warp direction, or by a variable stroke of the reed.

To create fabric pleats, the warp yarn is excluded from weaving at the desired length of the pleat in the first step. After weaving the length of the pleat, the excluded floating warp yarns are inserted back into the woven fabric by means of one to three weft yarns. In a last step, a terry weaving device is used to relocate the entire woven fabric against the processing direction according to the length of the pleats to be formed. While the floating yarns are kept tense, the reed is used to push the weft yarns together in a way that eliminates floating. The woven fabric formed parallel to the floating then jumps out of the woven fabric surface in the shape of a pleat. This basic technology can also be implemented on double rapier system weaving machines with two woven fabric layers with an inside pleat connecting both woven fabric layers. Figure 5.45 shows the principle [30].

In a first step, the two surfaces are woven. Afterwards, the rib is woven, with a portion of the warp yarns arranged floating on top and bottom. In the center of the rib, the rib fabrics change position and are connected at the end to the floating yarns by one to three wefts after reaching a predefined rib length. Afterwards, the pleats are formed, with the fabric pleat being set upright as a rib between the two woven fabric layers. Figure 5.46 shows a corresponding rib fabric after consolidation, fit

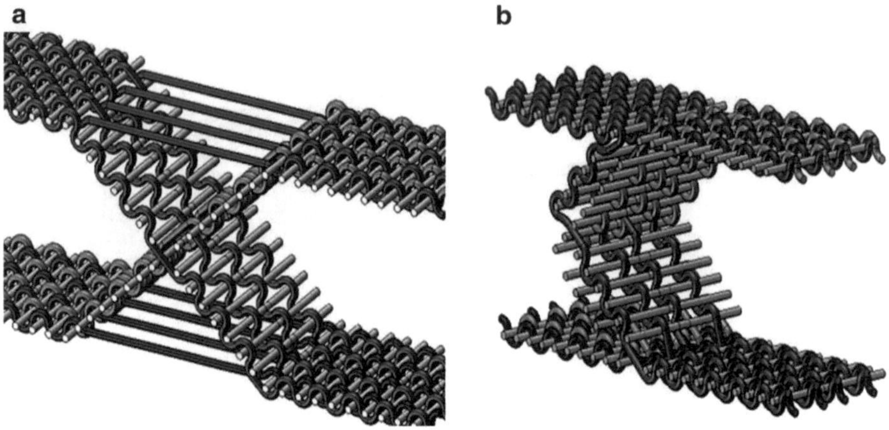

Fig. 5.45 Principle of terry weaving for spacer structures with (**a**) floating warp yarns, and (**b**) closed pleat

Fig. 5.46 Spacer structure as (**a**) rib fabric and (**b**) consolidated composite plate

for use as a lightweight construction plate. The take-up of such spacer fabrics requires the use of special systems protecting the 3D geometry of the woven fabrics by a system of support rods, and linearly taking up the spacer fabric, with integrated support rods, by roller pairs. The take-up direction can be supplemented by an automated cutting and magazine system for gathering and storing the spacer fabric preforms. With this technique a wide range of different geometries (u-shaped, v-shaped, and curved) can be realized [31, 32].

5.6.4 3D Orthogonal Woven Fabrics

3D orthogonal woven fabrics consist of yarn systems standing on another in the three spatial directions. They are also called "through-the-thickness fabrics", as stretched reinforcement yarn layers are realizable in any spatial direction. They are manufactured by feeding warp yarns in production direction by means of a slotted mask. The yarn feeding can be performed in nearly any desired cross-section

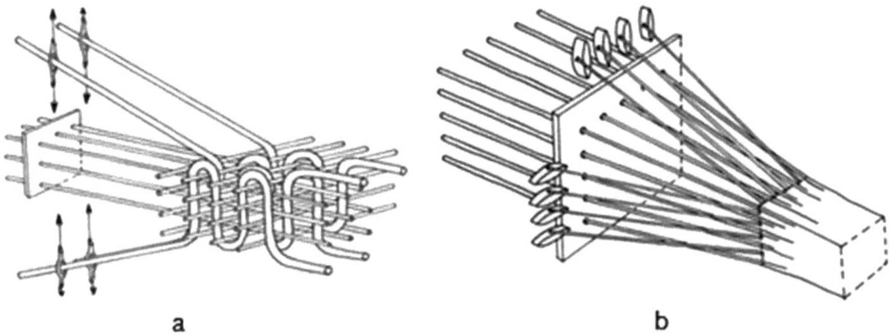

Fig. 5.47 3D orthogonal weaving (**a**) production method according to 3Tex, (**b**) production method according to Biteam—Tape Weaving Sweden AB

geometry and is mainly orientated towards the cross-section of the planned component. In general, two production methods are distinguished:

(a) Two warp yarn systems and one weft yarn system
In the market, the 3Weave™ method is well-established. It weave-technically interlaces a non-undulating weft and warp yarn system each with an additional binding warp (Fig. 5.47a) [33]. The healds undergo a significant deflection in this process, which greatly limits the processing of shear-sensitive fiber materials (e.g. carbon) and the variety of viable weave patterns.

(b) One warp yarn system and two weft yarn systems
Several weft yarns are inserted perpendicular to the fed warp yarns, first in Y, then in Z direction, and then bound off by the warp yarns (Fig. 5.47b). For this reason, the shed formation has to be performed in Z or Y direction, which requires elaborate weaving machine technology. For this reason, the production of 3D orthogonal woven fabrics is limited in width, height, and productivity. With the patented method described in [34] directly interwoven semi-finished products of up to 60 × 60 warp yarns can be realized.

3D orthogonal woven fabrics have a regular reinforcement in three spatial directions. Additionally, stretched yarns can be integrated, which are irrelevant for the formation of the woven fabric, but increase volume and stiffness. Due to the spatial arrangement of the individual yarns and the related displacement of orthogonal yarn systems, 3D orthogonal woven fabrics cannot be produced at any desired density. Using a multifilament yarn, and at identical yarn content ratio and identical type of yarn in all three spatial directions, a maximum fiber volume count of 68 % (compared to 80 % UD non-crimp fabrics) can be achieved. In contrast, composites made from 3D orthogonal woven fabrics exhibit very good delamination behavior across the entire cross-section. Due to the limited dimension, this method is primarily suitable for the manufacture of complex-geometry profiles with limited cross-sections for quasi-isotropic composite components. A special challenge is posed by the complete impregnation during composite formation, as the flow of the matrix is complicated by both the transverse yarn systems and the high degree of compacting of the voluminous structures.

5.6.5 3D Polar Woven Fabrics

3D polar woven fabrics consist of three yarn systems running in axial, circumference, and radial direction of a cylinder.

During the 1990s, 3D polar woven fabrics were introduced at conferences and in textbooks [33, 35, 36]. However, the method has not been established in the market and is rarely used. Only a few research institutions in Japan are still performing basic research on 3D polar woven fabrics. Recent publications regarding this process are not available.

Typical properties of 3D polar woven fabrics are given in tables in [35]. They are suitable for the manufacture of cylindrical hollow structures (Fig. 5.48). Similar to 2D polar woven fabrics (see Sect. 5.5.7), the yarn density decreases in radial direction at increasing distance from the central axis. By adjusting the fiber materials to be used, the yarn distances and yarn numbers, it is principally possible to produce load-adapted textile semi-finished products. An automated manufacture requires extensive effort and does not represent the state of the art.

5.7 Functional Integration

Functional integration refers to the creation of an additional functionality beyond the original function of the woven fabric. This additional functionality is achieved by introducing additional materials with special properties into the woven fabric. In general, any material that is not used as the basic material of the woven fabric and

Fig. 5.48 3D polar woven fabrics

has specific properties which can generate an additional functionality in the woven fabric can be used as functional material. Additional materials can take the form of yarns or component parts. The basic prerequisite for the integration of filamentary materials is their textile processability. Examples of such weaving-technically processable yarns include thermoplastic yarns, electrically conductive yarns on a metal basis, or electrically conductive carbon rovings. They can be inserted locally as warp or weft material, realizing sensor networks. Apart from inserting filamentary materials to generate an additional function, components can also be integrated in the woven fabric. This is based on the progressing miniaturization of both electrical and mechanical construction elements. Examples of such functional elements include RFID tags, sensor knots, or inserts for mechanical connections. One possibility to integrate functional elements is the formation of pockets in the woven fabric structure. This can be performed by weave-technical means by alternating single-layer and multilayer woven fabric, where only the section of the functional element features multiple layers.

References

1. Standard DIN ISO 9354 Oktober 1993. *Gewebe - Bindungskurzzeichen und Beispiele*
2. KIENBAUM, M.: *Bindungstechnik der Gewebe - Konstruktion und Gestaltung mit warenkundlichen Beispielen (Bd. 1-3)*. Berlin : Schiele Schön GmbH, 1990-1999
3. BÖHM, R.; GUDE, M.; HUFENBACH, W.: A phenomenologically based damage model for textile composites with crimped reinforcement. In: *Composites Science and Technology* 70 (2009), No. 1, pp. 81–87. DOI 10.1016/j.compscitech.2009.09.008
4. WALZ, F.: *Die Gewebedichte."Textilpraxis 2"*. Stuttgart : Robert Kohlhammer-Verlag, 1947
5. AIBIBU, D.: *Charakterisierung, Modellierung und Optimierung der Barriereeigenschaften von OP-Textilien*, Technische Universität Dresden, Fakultät Maschinenwesen, Dissertation, 2005
6. SCHIERZ, M.; FRANZKE, G.; WALDMANN, M.; OFFERMANN, P.; HES, L.: Charakterisierung der Handhabbarkeit textiler Bewehrungsstrukturen. In: *Technische Textilien* 46 (2003), No. 2, pp. 141–144
7. Protective right CZ 3383U1 (12th July 1995).
8. TEAM OF AUTHORS: *Gewebetechnik*. Leipzig : VEB Fachbuchverlag Leipzig, 1978
9. ANONYM: *Webereitechnik- Herstellen von Geweben*. Eschborn : Arbeitgeberkreis Gesamttextil, 1997
10. TAYLOR, L.; CHEN, X.: 3D Woven Nodal Hollow Truss Structures. In: *Proceedings. 13. Internationales Techtextil-Symposium 2005*. Frankfurt/M., Germany, 2005
11. Protective right US7770607B2 (10th August 2010).
12. HÖRSCH, F.: Dreidimensionale Verstärkungsmaterialien für Faserverbundwerkstoffe. In: *Kunststoffe* 80 (1990), No. 9, pp. 1003–1007
13. KLEICKE, R.; LOTZMANN, H.; METZKES, K.; HOFFMANN, G.; CHERIF, Ch.: Dreherweben-Anknüpfen an alte Wurzeln mit neuer Technologie. In: *Melliand Textilberichte* (2007), No. 6, pp. 423–425
14. BECHTHOLD, G.; YE, L.: Influence of fibre distribution on the transverse flow permeability in fibre bundles. In: *Composites Science and Technology* 63 (2003), pp. 2069–2079
15. OFFERMANN, P.; KÖCKRITZ, U.; ABKADER, A.; ENGLER, T.; WALDMANN, M.: Anforderungsgerechte Bewehrungsstrukturen für den Einsatz im Betonbau. In: CURBACH, M. (Hrsg.): *Proceedings. Textile Reinforced Structures. Proceedings of the 2nd Colloqium on*

Textile Reinforced Structures (CTRS2). Dresden, Germany : Technische Universität Dresden Sonderforschungsbereich 528), 2003, pp. 15–28
16. Protective right DE2548129A1 (20th May 1976).
17. Protective right US3799209 (26th March 1974).
18. SCHWARTZ, P.; FORNES, R. E.; MOHAMED, M. H.: Tensile Properties of Tri-axiallyWoven Fabrics Under Biaxial Loading. In: *Journal of Engineering for Industry* 102 (1980), pp. 327–332
19. FRONTCZAK-WASIAK, M.: Characteristics of Multi-axial Woven Structure. In: *Fibers and Textiles in Eastern Europe* 13 (2005), No. 4, pp. 27 ff.
20. FLEMMING, M. et al.: *Faserverbundbauweisen, Halbzeuge und Bauweisen*. Berlin : Springer Verlag, 1996
21. WAHHOUD, A.: Innovative production of textile fabrics through process integration in weaving operation. In: *Melliand International* (2001), No. 1, pp. 28–29
22. RHYNE, M.: Trends in dryer fabric design. In: *Proceedings. Hi-Tech Textiles Exhibition Conference*. Greenville, USA, 1993, pp. 163–174
23. NEITZEL, M.; MITSCHANG, P.: *Handbuch Verbundwerkstoffe*. München, Wien : Carl Hanser Verlag, 2004
24. PD Group: *Verbindungen schaffen mit Gewebebandleitungen*. http://www.pdlappsystems.de/ (06-04-2011)
25. Protective right DE3915085A1 (November 1990).
26. BÜSGEN, W.-A.: *Neue Verfahren zur Herstellung von dreidimensionalen Textilien für den Einsatz in Faserverbundwerkstoffen*. Aachen, Rheinisch-Westfälische Technische Hochschule Aachen, Fakultät für Maschinenwesen, Dissertation, 1993
27. CHEN, X..; POTIYARJ, P.; MATHER, R. R.; MCKENNA, D. F.; MNOX, R. T.: Modeling and simulation for 3D complicated woven structures. In: *Proceedings. World Textile Congress, Industrial, Technical High Performance Textiles*. Huddersfield, Great Britain, 1998, pp. 227–235
28. ISLAM, M. A.: 3-D woven near net shape and multi axial structures. In: *Proceedings. 80th World Conference of The Textile Institute*. Manchester, Great Britain, 2000, pp. 1–12
29. SWINKELS, K.: Die kostengünstige Produktion von doppelwandigen, faserverstärkten Kunststofftanks im Wickelverfahren mit einem Abstandsgewebe. In: *Proceedings. 35. Internationale Chemiefasertagung Dornbirn*. Dornbirn, Österreich, 1996
30. TORUN, A. R.; HOFFMANN, G.; ÜNAL, A.; KLUG, P.; CHERIF, Ch.; BADAWI, S.: Technologische Lösungen zur Entwicklung von spacer fabrics mit Flächenstrukturen als Abstandshalter/ Technological solutions to the development of spacer fabrics including fabric structures as the spacer. In: *Proceedings. 8. Dresdner Textiltagung*. Dresden, Germany, 2006
31. MOUNTASIR, A.; HOFFMANN, G.; CHERIF, Ch.: Development of Weaving Technology for Manufacturing three-dimensional Spacer Fabrics with High-Performance Yarns for Thermoplastic Composite Applications: An Analysis of two-dimensional Mechanical Properties. In: *Textile Research Journal* (2011). DOI 10.1177/0040517511402125
32. GROSSMANN, K.; MÜHL, A.; LÖSER, M.; CHERIF, Ch.; HOFFMANN, G.; TORUN, A. R.: New solutions for the manufacturing of spacer preforms for thermoplastic textile-reinforced lightweight structures. In: *Production Engineering Research and Development* 4 (2010), No. 6, pp. 589–597. http://dx.doi.org/10.1007/s11740-010-0267-9. – DOI 10.1007/s11740–010–0267–9
33. Protective right DE69122967 (7th May 1997).
34. Protective right US6889720B2 (10th May 2005).
35. LEE, S. M.: *Handbook of Composite Reinforcement*. New York : Villey-VCH, 1992
36. Protective right US4052913A (11th October 1977).

Additional Literature

BOGDANOVICH, A.E.; MOHAMED, M.H.: Three-Dimensional Reinforcements for Composites. 3TEX, Inc., Cary, NC. SAMPE Nov/Dec 2009.

MCHUGH, Ch.: Carbon fibre textiles and textile structures for light-weight fibre reinforced composites. *1. Denkendorfer Kolloquium Garn- und Flächenerzeugung*, Denkendorf, May 7-8, 2014

3-D woven reinforcements update: Composites World. http://www.compositesworld.com/articles/3-d-woven-reinforcements-update (04-04-2014)

KHOKAR, N.: Merging Fabric-forming Technologies: From 2D Sheets to 3D Objects *(Textile Institute's Centenary Conference)*, Manchester, 2010

MISHRA, R.; BAHETI, V.; BEHERA, B.K.; MILITKY, J.: Novelties of 3-D woven composites and nanocomposites. In: *The Journal of the Textile Institute* 105 (2014), no. 1, pp. 84-92

ZENG, X.; BROWN, L. P.; ENDRUWEIT, A.; MATVEEV, M.; LONG, A. C.: Geometrical modelling of 3D woven reinforcements for polymer composites: Prediction of fabric permeability and composite mechanical properties. In: *Composites: Part A* 56 (2014), pp. 150-160

WEISE, D.; DÖBRICH, O.; HOFFMANN, G.; CHERIF, Ch.: Non-crimped leno woven fabrics for high-performance fiber-reinforced composites. *AUTEX, AUTEX World Textile Conference*, 14 / 2014

MOUNTASIR A.; HOFFMANN G.; CHERIF Ch.: Development of multilayered woven panels with integrated stiffeners in transverse and longitudinal directions for thermoplastic lightweight applications. In: *Textile Research Journal* 83(2013). DOI 10.1177/0040517512474367

Chapter 6
Semi-finished Weft Knitted Fabrics and Weft Knitting Techniques

Wolfgang Trümper

This chapter details the most important development steps, binding elements, and basic bindings of weft knitting as well as the basic properties of weft knitted fabrics. This overview will include the various ways of influencing the properties of a mechanical engineering-based production of knitted fabrics. One focus of the chapter will be to provide an introduction to the wide range of possibilities in the realization of suitable, near-net shape, semi-finished, weft knitted fabrics, particularly for use in fiber composite components. One crucial prerequisite for that is the load-adapted integration of stretched reinforcing yarns into the knitted structure. In connection with the extensive technological shaping possibilities during manufacture, and with a subsequent drapability adjustable by means of stitch length, ideal conditions are created for a wrinkle-free production of weft knitted fabrics for complex component geometries. Due to the structure of weft knitted fabrics, such components exhibit outstanding properties, in particular with regard to impact loads.

6.1 Introduction and Overview

Knitting boasts a long tradition in the production of garments and presumably originated in the Middle East. Knitted stockings unearthed in the region have been dated to the time of the second to third century AD. In Europe knitting needles and the hand-knitted products manufactured with them have been known since the fourth century AD. For the production of hand-made knitted fabrics, two-needle and four-needle techniques are known [1].

W. Trümper (✉)
Institute of Textile Machinery and High Performance Material Technology, TU Dresden, Dresden, Germany
e-mail: wolfgang.truemper@tu-dresden.de

C. Cherif (ed.), *Textile Materials for Lightweight Constructions*,
DOI 10.1007/978-3-662-46341-3_6

William Lee's invention of the stocking hand frame in 1589 was the first step in the mechanization of stitch formation. The stocking frame invented by Lee contained one needle bed and was able to produce roughly 600 stitches per minute, equaling the sixfold productivity of an experienced manual knitter. An improved stocking hand frame produced by Lee in 1609 more than doubled that number to 1,500 stitches per minute. In 1758, Jedediah Strutt patented a stocking frame with a second needle bed arranged at a right angle to the first one [2].

Stitch formation became much easier after Matthew Townsend patented the latch needle in 1847. Up to this point, an especially designed needle tip had to be pressed against the needle shaft to enable stitch formation by closing the needle hook. The invention of the latch needle also served as the stepping stone to the creation of the first flat knitting machine by William Lamb in 1863 [3].

The automation of weft knitting began in the 1950s with the manufacture of stocking products as complete garments. At roughly the same time, the first semi-automatic weft knitting machines for gloves were constructed [4]. The recent weft knitting machine generations, through the possibility of single needle selection and the use of special yarn guides, allow a wide variety of realizable knitting patterns [1]. Apart from complex structural and color patterns, the machines can also be used for the manufacture of complete garments such as pullovers or swimwear in a single production step. Beyond these, weft knitting technology offers ideal conditions for the production of near-net shape technical textiles, for instance for use as reinforcement structures in lightweight construction [5, 6].

To facilitate understanding, a general classification of warp and weft knitting machines will be included, which can be done according to different criteria. Figure 6.1 shows a classification of the machines by principle-inherent needle movement (individual or joint), and by yarn presentation at the needles (transverse and longitudinal). Other criteria for classification of weft and warp knitting are the technology of the stitch formation and the required process cycles (Fig. 6.2).

Generally, flat and circular geometries of the needle bed are known for all machine types except warp knitting machines. Suitable machines are available for the basic bindings given in brackets in Fig. 6.1, but are currently not being built anymore or do not play any major role in production.

The various weft knitting techniques for conventional weft knitted structures like pullovers are the foundation of any new machine concept developments for the realization of semi-finished products for sophisticated technical applications, such as fiber composite components for lightweight construction.

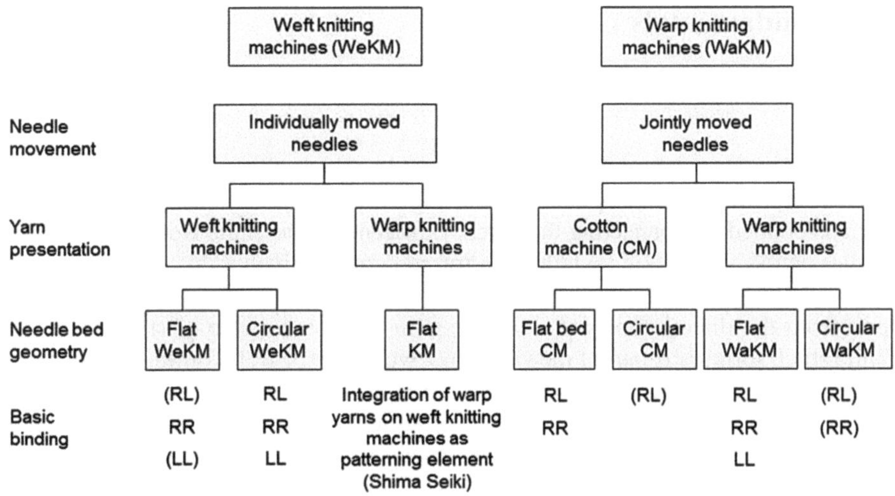

Fig. 6.1 Systematization of the classification of warp and weft knitting machines according to needle motion (abbreviations of the standard structures, *RL* single jersey, *RR* double jersey, *LL* purl stitch, see also Sect. 6.2.3)

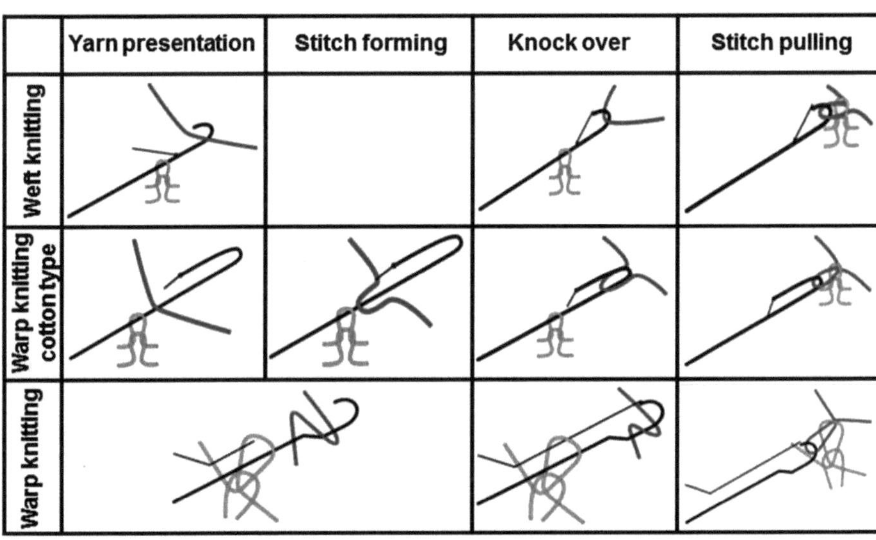

Fig. 6.2 Work steps of the stitch formation in knitting, cotton-type knitting and warp knitting (see [7])

6.2 Fundamentals

6.2.1 General Remarks

Weft knitting is a fabric formation method in which the yarn material is fed transversely on the needles and formed into stitches by the needles. The textile fabric is created by connecting the stitches. In general, the same work cycle on the needles across the working width is performed consecutively.

The difference between weft *knitting, cotton-type knitting,* and *warp knitting,* apart from the direction of yarn feeding, is in the process step of stitch *forming.* While stitch forming is part of the *knockover* work step of weft knitting, the stitch is formed in a separate step in Cotton-type knitting. In warp knitting, the steps of stitch forming and *yarn guiding* are performed simultaneously (Fig. 6.2). In addition, the same work step is performed on all needles across the working width at the same time in warp knitting (compare Chap. 7).

In the following sections of the chapter, the binding elements of weft knitting will be introduced, basic bindings will be shown, and the main parameters to influence and describe the properties of weft knitted fabrics during manufacturing will be derived from the previous information. Emphasis will be placed on those properties crucial for the manufacture of textile reinforcement semi-finished products by weft knitting.

The terminology for the binding elements is based on the specifications laid out in standards EN ISO 4921 and EN ISO 8388.

6.2.2 Binding Elements

6.2.2.1 Stitch

The basic binding element of weft knitting is the *stitch* (Fig. 6.3). It consists of the head, two legs and two feet. *Stitches* always contain head and foot intermeshing points. The geometry of the stitch can be described with regards to the width (b_M) and height (h_M). The length of the yarn material contained in a stitch is referred to as stitch length (l_M). Deformations of the weft knitted fabrics, e.g. by application of force, result in changes of stitch width and stitch height, but do not in any way influence the stitch length.

Depending on the manufacturer and type of machine, latch needles or compound needles are used in weft knitting machines (Fig. 6.4), with latch needles being more popular at the time of publication.

For the stitch formation, a yarn presentation in the opened needle hook is required. In latch needles, the opening and closing of the needle hook is performed by a motion of the needle tongue, usually initiated by the head of a previous stitch carried on the needle. Brushes at the carriage in the area of the stitch-forming

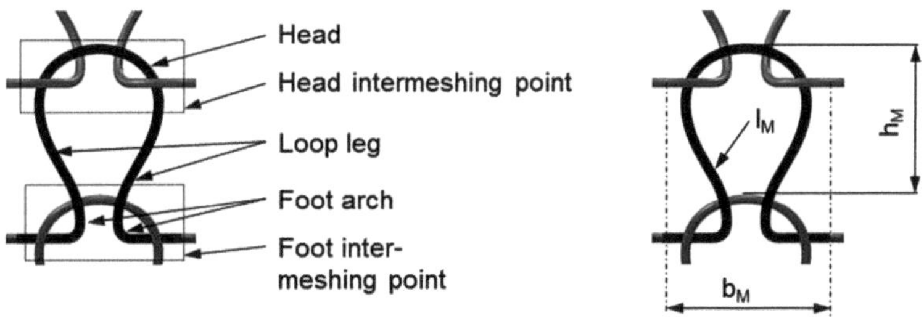

Fig. 6.3 Components of a stitch and stitch geometry

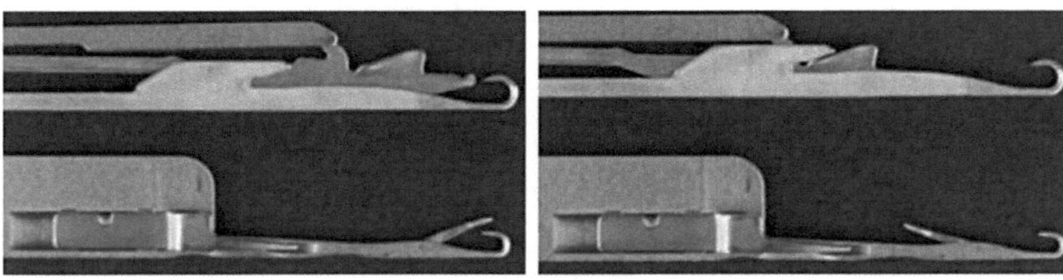

Fig. 6.4 Compound and latch needles in the basic position (*left*), and with opened needle hook (*right*)

elements support the opening and closing of the needle hook especially during the initial insertion of the yarn into empty needles or in critical knitting situation. For the stitch formation, only the motion of the needle needs to be controlled.

Compound needles are equipped with an independently controllable slider to open and close the needle hook. For stitch formation, both the needle and the slider have to be moved. In comparison to latch needles, this manner of stitch formation requires a much smaller maximum clearing path. The resulting smaller relative motion between needle and yarn material can positively affect the friction stresses exerted on the yarn material. Therefore, compound needles are tools with a high potential for the processing of more sensitive yarn materials or for increasing machine speed during knitting.

Figure 6.5 shows the schematics of a cam box driving latch needles in a flat knitting machine with one needle bed. The numbers signify the individual needle positions (compare Fig. 6.6).

The needle positions traversed by the needles during stitch formation are given with their designations in Fig. 6.6.

At the beginning of the stitch formation, the needle is in its basic position (1). A yarn stitch with foot intermeshing point, but without head intermeshing point (half stitch) is positioned in the needle hook. In a first step, the needle is cleared into the tuck position (2). The half stitch slips across the needle shaft onto the needle tongue, which is opened by the motion. After that, the needle is moved into the clearing position, the highest position during stitch formation.

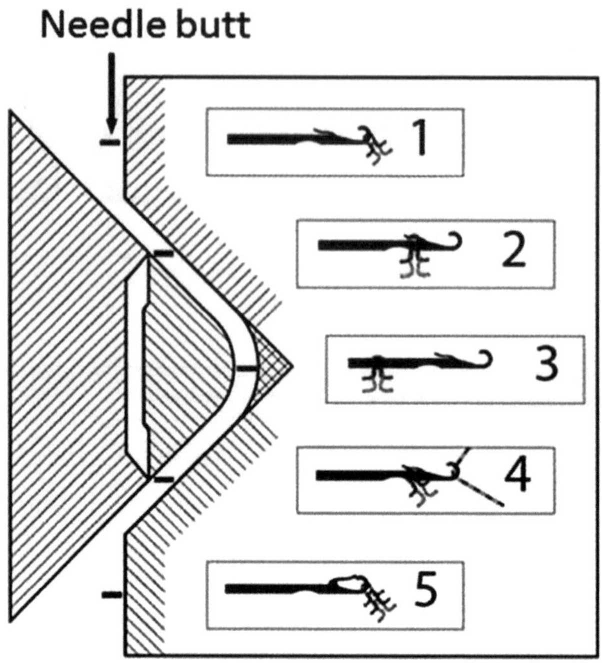

Fig. 6.5 Schematics of a cam box with needle running sequence

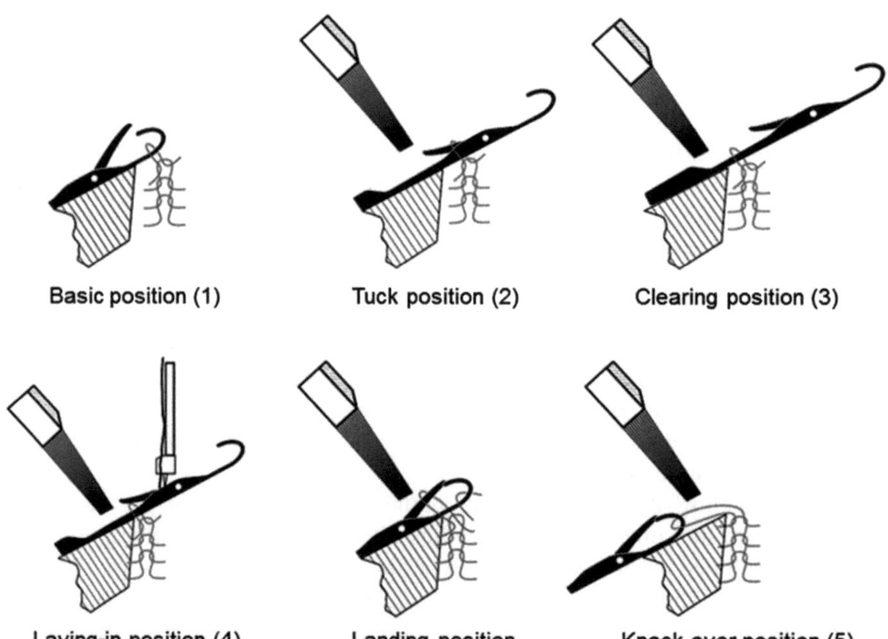

Fig. 6.6 Stitch formation and needle positions on a flat knitting machine equipped with latch needles

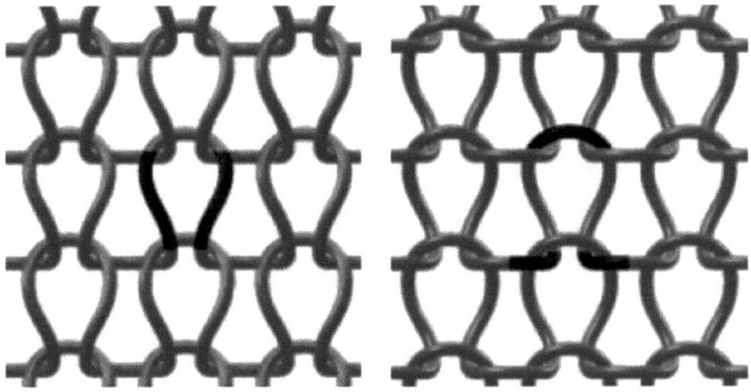

Fig. 6.7 *Right* and *left* stitch side of a right-left knitted fabric

In this position, the half stitch is located on the needle shaft behind the needle tongue. After this, the needle is retracted back into the yarn insertion position (4). Here a piece of the yarn, which is fed to the needles by a yarn feeder, is inserted into the needle hook. The half stitch is located directly behind the needle tongue. Due to the subsequent backward motion of the needle, the half stitch closes the needle tongue. The half stitch is then landed on the needle tongue (landing position). In a final step, the stitch formation is completed by the knockover of the half stitch off the needle. For this, the yarn piece in the needle hook is pulled through the half stitch and a new half stitch is created by moving the needle to the knockover position (5). Adjustment of the stitch cam enables variable stitch length.

For method-inherent reasons, the stitch has two differing sides. The *right stitch side* shows the legs, while the *left stitch side* shows the *heads* and *feet*. In Fig. 6.7, the respective parts of the stitch are shown in a darker shade.

Within one knitted fabric, horizontally arranged stitches are referred to as *course* (CD), while vertically arranged stitches are called *wale* (WD) (Fig. 6.8).

6.2.2.2 Tuck Stitch

A tuck or *tuck stitch* contains only a head intermeshing point, a foot intermeshing point does not exist (Fig. 6.9, left). Technically, a tuck is created by not knocking over a yarn loop or half stitch held by the needle hook, but by insertion of additional yarn loops into the needle hook while the needles are in tuck position. The needles do not achieve the clearing position and are moved into the knockover position immediately after yarn insertion (Fig. 6.6). Depending on needle geometry and the yarn material used, the number of tucks that can be formed consequently on one needle is limited.

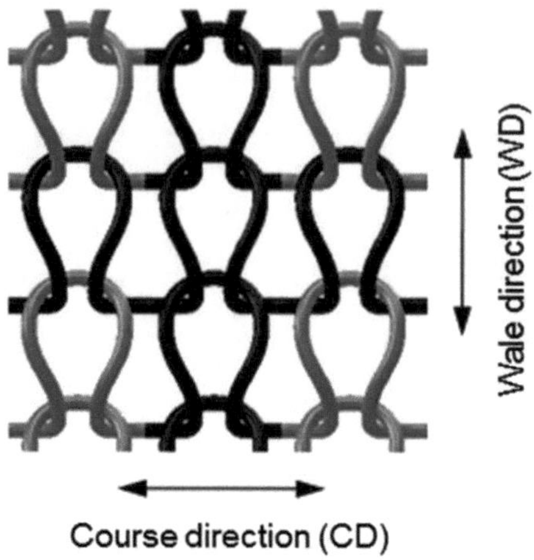

Fig. 6.8 Terminology of knitted fabric directions: *horizontal*—course direction (CD), *vertical*—wale direction (WD)

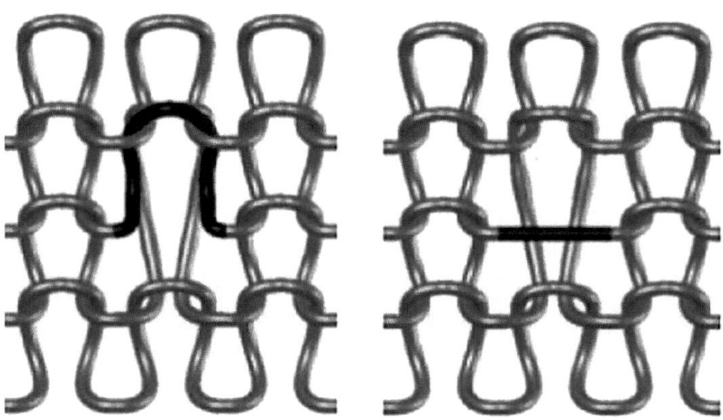

Fig. 6.9 Binding elements: Tuck (*left*) and miss stitch (*right*)

6.2.2.3 Miss Stitch

A *miss stitch* is a piece of yarn stretched across at least one needle and having neither head nor foot intermeshing point (Fig. 6.9, right). A miss stitch is bordered on both ends by a tuck or stitch.

6.2.2.4 Weft Inlay

A *weft inlay* is a piece of yarn inserted stretched in course direction, but not bordered by any other binding element like tuck or stitch. A weft inlay can be inserted across the entire width of the knitted fabric or across sections only (*partial weft inlay*) (Fig. 6.10).

Fig. 6.10 Binding elements weft inlay (*left*), and partial weft inlay (*center*), and combined with warp yarns (*right*)

6.2.2.5 Warp Yarn

A *warp yarn* is a piece of yarn usually inserted stretched in wale direction, and not bordered by binding elements like tuck or stitch (Fig. 6.10, right). Warp yarns are usually inserted between two needles. By means of a warp yarn displacement, a diagonal arrangement of the warp yarns deviating from the wale direction can be achieved in the knitted fabrics.

The formation of a textile fabric in knitting as such is always based on the combination of stitches. The other binding elements serve the purposes of the patterning and local alteration of fabric properties.

6.2.3 Basic Bindings

6.2.3.1 General Remarks

The term *binding* refers to the manner of yarn interlacing or the arrangement of binding elements in a knitted fabric. Only the binding element *stitch* is used in basic bindings. Depending on the arrangement of right and left stitch sides of the knitted fabric, there are four distinguishable basic bindings. In addition to the basic bindings, the literature gives a wide range of bindings which are derived from theses, e.g. by using the binding elements tuck and/or miss stitch. Beyond those, a free arrangement of binding elements can be used to develop so-called "fancy bindings". When realizing bindings, a logical, knitting-technically feasible sequence of binding elements on each needle has to be kept.

6.2.3.2 Single Jersey

Single jersey knitted fabrics are single-layer knitted fabrics created on one needle bed by a generally identical knockover direction of all stitches off the needles. They show the legs on the right side of the knitted fabric, and the heads and feet of the stitches on the left side of the knitted fabric (Fig. 6.11). Knitting machines with at least one needle bed are required for the production of single jersey fabrics.

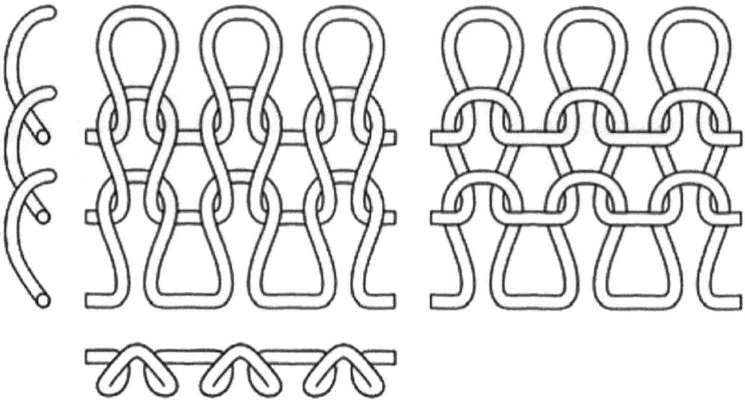

Fig. 6.11 Single jersey binding: *right* and *left* side of the knitted fabric

Due to the binding-related arrangement of the stitches, single jersey fabrics display a high tendency to roll from the left to the right fabric edge and from the bottom to the top edge. This is caused by interior states of stress resulting from the deformation of the yarn material into space curves, and by the attempts of the fabric to attain a state of lowest possible stresses.

6.2.3.3 Double Jersey

Double jersey knitted fabrics are two-layered knitted fabrics created by the wale-wise alternation of the knockover direction of the stitches off the needles. Two needle beds on the knitting machines are arranged in a manner that places a needle of the second needle bed centrally between two needles of the first needle bed. In a no-stress state, both sides of the knitted fabric show the right stitch side (Fig. 6.12). In a transversely elongated state, not only the legs, but also the heads of the respective other side are visible.

Due to the opposing arrangement of two knitted fabric layers in a knitted fabric, the forces introduced by yarn deformation neutralize one another at an exact geometry of the two layers. The knitted fabrics will therefore display a less pronounced tendency to curl.

6.2.3.4 Interlock

The *Interlock* binding can be understood as a combination of two double jersey bindings overlaying in a way that, in a wale direction elongated state, no heads or feet of the stitches are discernible. On both outer sides, the legs of the right side of the knitted fabric are visible (Fig. 6.13). The needles are directly opposite, arranged in two needle beds.

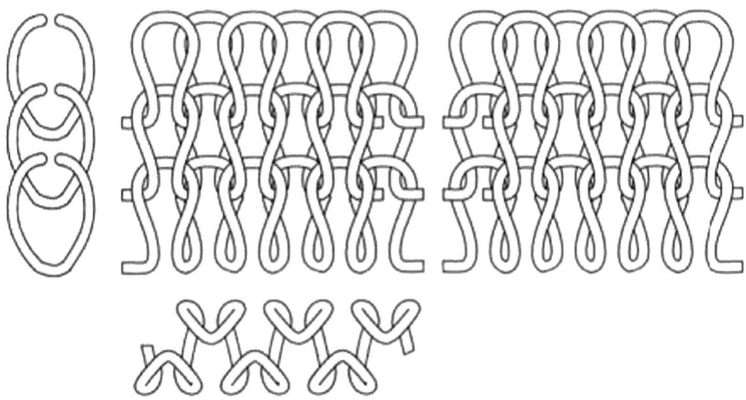

Fig. 6.12 Double jersey binding: *right* and *left* side of the knitted fabric

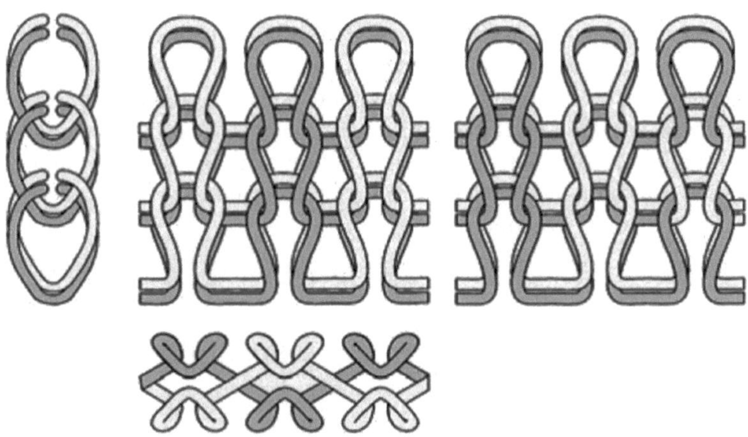

Fig. 6.13 Interlock binding: *right* and *left* side of the knitted fabric

6.2.3.5 Purl Stitch

Knitted fabrics with a basic binding in purl stitch show course-wise alternation of the stitch heads on the two outer sides of the fabric (Fig. 6.14). This is achieved by changing the knockover direction of the stitches off the needles in each course.

6.2.3.6 Multi-layered Weft Knitted Fabrics Bindings

With regard to the use of weft knitted fabrics as semi-finished products in fiber-reinforced composites, a derived binding for the production of multi-layered weft knitted fabrics [MLG—abbreviation taken from the German word **M**ehr**l**agen**g**estrick within the literature also referred to as biaxial weft knitted fabrics

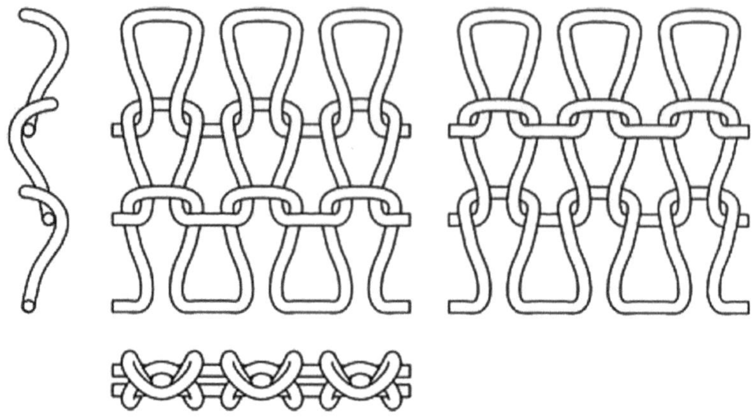

Fig. 6.14 Purl stitch basic binding: *right* and *left* side of the knitted fabric

(BWK), multiaxial weft knitted fabrics (MWK), Multi-layered knitted fabrics (MLKF) or non crimp knit (NCK)]—the single jersey MLG, or interlock-MLG binding, can properly be introduced in this context. These bindings are based on the single jersey or interlock basic binding, in each case extended with stretched weft and warp yarns integrated into the stitch structure (Fig. 6.10). Depending on stitch configuration, different numbers of warp yarn layers and/or weft yarn layers can be integrated in the stitch structure [8].

6.2.4 Knitted Fabric Properties and Knitting Parameters

The binding describes the arrangement of the binding elements for the formation of a knitted textile surface. The selection of the binding, therefore, fundamentally defines the structure of the weft knitted fabric. Apart from this, the yarn material and chosen knitting parameters, such as machine gauge, stitch cam adjustment, take-down settings or yarn tension, exert considerable influence on the structure and characteristics of the knitted fabric.

The m*achine gauge* describes the number of needles per reference length, which is usually given in inches. The higher the machine gauge, the higher the *stitch density* of the knitted fabrics. The take-down settings have to be selected in a manner that ensures the knockover of the stitches off the needles as well as the removal of the weft knitted fabric from the stitch forming area. A suitable yarn tension during yarn feeding to the knitting area, which is also constant on all needles, is the most important prerequisite for the realization of an equal *stitch length* in all sections of the knitted fabric.

The characterization and description of weft knitted fabrics also rely on textile-physical properties like stitch density in course and wale direction, stitch length, fabric thickness, or areic mass. According to the projected application of the knitted fabrics, the statement of other properties of the knitted fabric can be required. Particularly for the use as reinforcement semi-finished products, the following properties of the knitted fabrics are important:

- the curling behavior,
- the stress-strain behavior, and
- the shear behavior

Knowledge of the stress-strain behavior and the shear behavior allows a broad description of the draping behavior of a weft knitted fabric. The *drapability* refers to the spherical deformability without wrinkling of textile structures. Knowing the draping behavior is necessary for a first assessment of the effort required to adapt the knitted fabrics to, for instance, defined three-dimensional semi-finished product geometry.

The curling behavior depends primarily on the selected binding, and influences how easy the knitted fabrics are to handle during further processing (see also basic bindings in Sect. 6.2.3).

The stress-strain behavior of a textile structure describes its deformation by yarn stretching mechanisms, yarn elongation, yarn slippage, and shear. The yarn stretching and yarn displacement are mainly influenceable by the selection of the binding and knitting parameters [9]. The resulting elongation is referred to as *structural elongation* ($\varepsilon_{Structure}$). The yarn elongation depends on the yarn material used and is referred to as *material elongation* ($\varepsilon_{Material}$). The material elongation results from the difference between elongation at break and structural elongation. Figure 6.15 shows the schematics of the stress-strain behavior of a knitted fabric loaded to failure, including a labeling of the various calculated elongations.

Generally, the stress-strain behavior differs greatly in both directions of the knitted fabric due to the stitch structure. With the exception of the purl stitch basic binding, at the same forces all unreinforced knitted fabrics display higher elongations in course direction and lower elongations in wale direction. Comparing the basic bindings, knitted fabrics in single jersey basic binding display the lowest

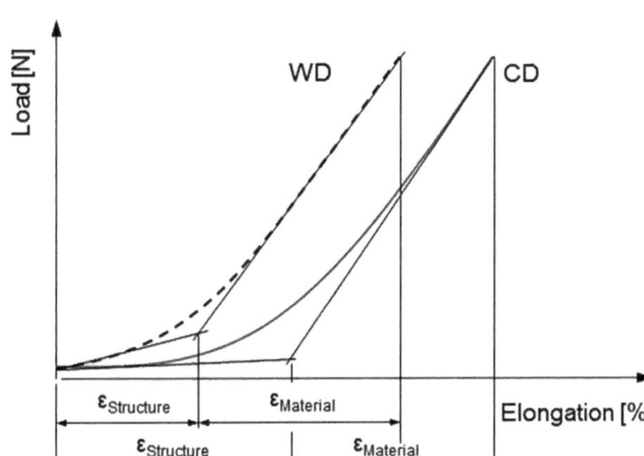

Fig. 6.15 Schematics of the stress-strain behavior of unreinforced knitted fabrics

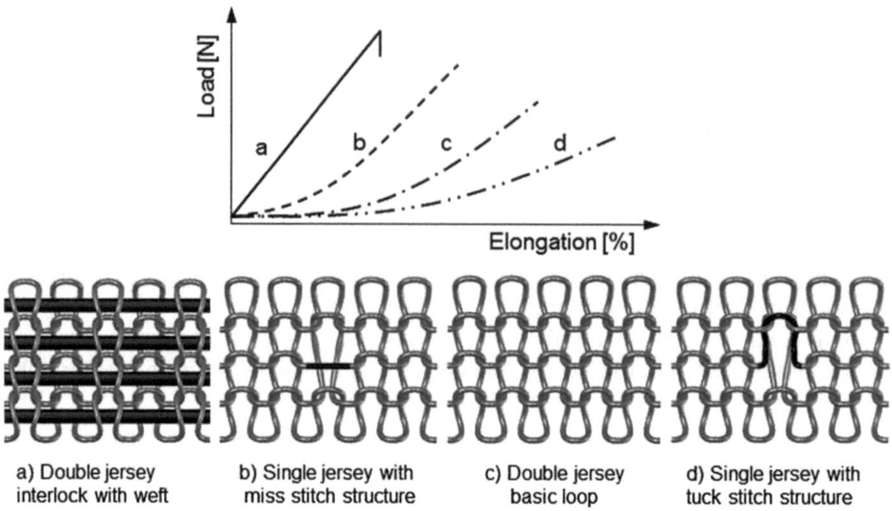

Fig. 6.16 Stress-strain behavior of knitted fabrics in course direction, depending on the binding elements used. (**a**) Double jersey interlock with weft. (**b**) Single jersey with miss stitch structure. (**c**) Double jersey basic loop. (**d**) Single jersey with tuck stitch structure

elongations in course direction and wale direction. Knitted fabrics with a double jersey basic binding in course direction, and a purl stitch basic binding in wale direction, display high elongations. By combining two double jersey bindings into an interlock basic binding, knitted fabrics with low elongations in both directions of the knitted fabric can be realized.

Building on a basic binding, the properties of the knitted fabric can specifically be adjusted by an aimed arrangement of additional binding elements such as tuck, miss stitch or weft yarn [10]. The principal influence of the arrangement of various binding elements in the knitted fabrics on the stress-strain behavior is shown in Fig. 6.16, exemplified by a single jersey basic binding. Symbolically, only one corresponding binding element is given.

The different curves of the diagram show that the initial tensile strength of the basic structure is reduced by the inclusion of tuck stitches, and increased by the inclusion of miss stitches. The integration of a stretched weft inlay in the weft knitted fabrics ideally leads to a stress-strain behavior matching that of the respective weft inlay.

For unreinforced weft knitted fabrics, a statement concerning the drapability can often be derived from the stress-strain behavior. Here, the general assumption is that a high elongation at low forces or a high structural elastic limit, are associated with good drapability. By determining the correlation between shear angles and the required force, the draping behavior can be described more precisely.

Describing the draping properties of MLGs requires a determination of shear characteristics due to the weft yarns and/or warp yarns integrated in the stitch

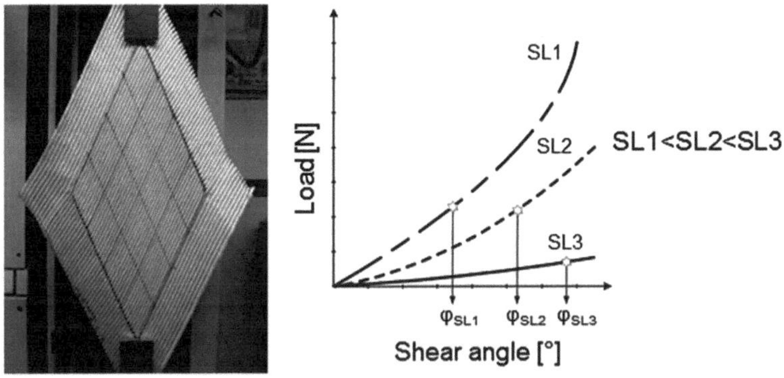

Fig. 6.17 Shear behavior of an MLG in dependence on loop length

structure and the resulting limited deformation possibilities of the stitches. In basic researches, the stitch length was identified as the crucial parameter for an adjustment of the shear behavior of MLGs [11]. This is clarified in Fig. 6.17, where three MLGs are compared. They have identical arrangements of weft yarns and warp yarns integrated in the stitch structure, but their stitch lengths (SL) increase from SL1 to SL3. From the shear angle (φ_{SLx}) given in the diagram, wrinkling has to be expected during draping of the respective MLG variant.

Knitting parameters, such as stitch cam adjustment, take-down settings, or yarn tension, can be used to adjust the characteristics of the knitted fabric in a wide range. Primarily, this concerns the stitch length and, by extension, the dimensions, areic mass or stitch densities. In conjunction with suitable machine and material selection, knitted fabrics with an extensive variety of characteristics can be generated, including different drapabilities or specifiable mechanical properties or energy absorption capacities.

6.2.5 Knitting Processes

Various criteria can be used for the classification of knitting methods. This chapter will separate them by needle bed geometry, and by the number and arrangement of needle beds required for the manufacturing of knitted fabrics.

6.2.5.1 Needle Bed Geometry

Referring to the geometry of the needle bed, knitting methods can be classified into circular and flat knitting processes.

Circular Knitting Processes

Circular knitting processes are distinguished by the formation of a first needle bed in the shape of a cylinder, and by the arrangement of the needles at the circumference of the cylinder. Additionally, a second needle bed, either cylindrical or in the shape of a dial can be arranged above the first cylinder. Knitted fabrics produced in this manner are usually tubular fabrics.

Flat Knitting Processes

Flat knitting processes are characterized by the side-by-side needle arrangement in plane needle bed. Depending on the machine design and manufacturer, up to four needle beds are arranged horizontally or at angles. The knitted fabrics produced in the basic bindings are finished as rolled goods.

6.2.5.2 Number and Arrangement of Needle Beds

If machines are classified based on the number and arrangement of the needle beds, the machines are referred to by the basic bindings producible on them. In the following schematics concerning the respective knitting methods, the geometry and arrangement of the needle bed is shown for flat and round knitting methods.

Single Jersey Technology

A needle bed is required for the stitch formation. The used needles have a needle hook at one end (Fig. 6.18). The single jersey basic binding and its derivations can be produced.

Purl Stitch Technology

Implementing the purl basic binding and its derivations requires knitting machines with at least two needle beds. To produce knitted fabrics, special knitting needles with a needle hook at each end are needed. The needle bed arrangement has to allow a transfer of the needles from one to the other needle bed (Fig. 6.19). Knitting machines operating on the purl stitch technology are theoretically suitable for the realization of knitted fabrics in single jersey and some kinds of double jersey basic bindings. Especially purl flat knitting machines, however, are rarely used anymore in the industry.

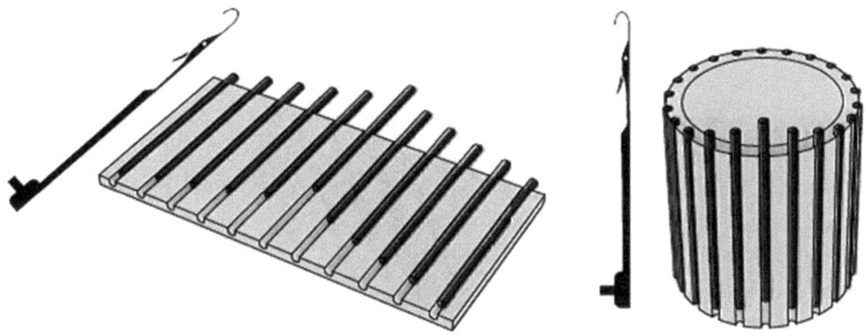

Fig. 6.18 Arrangement of the needle bed on plain flat (*left*) and circular (*right*) knitting machines

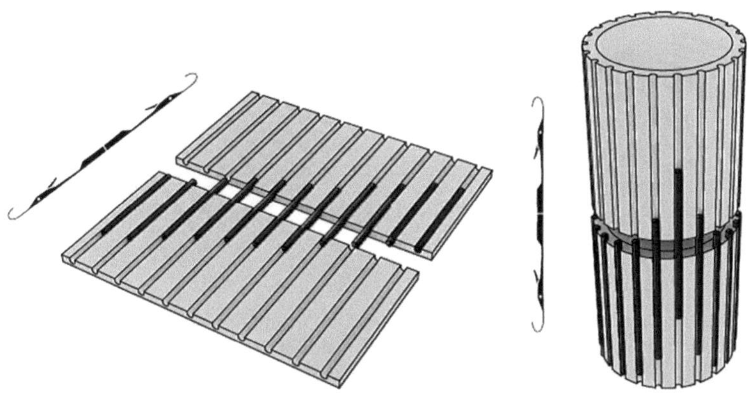

Fig. 6.19 Arrangement of the needle bed on purl flat (*left*) and circular (*right*) knitting machines

Double Jersey Technology

To produce knitted fabrics with the double jersey technology, at least two needle beds are required (Fig. 6.20). The needles remain in their respective needle beds during the knitting process. Due to the high flexibility during the realization of the various knitted fabric structures, modern knitting machines usually operate using double jersey knitting. These machines can produce knitted fabrics in the basic bindings single jersey, double jersey, and interlock. Machines with the possibility of stitch transfer between the needle beds can also be used for the production of the purl stitched knitted fabrics.

Furthermore, machines working on a double jersey method with more than two needle beds are available, particularly in flat knitting (Fig. 6.21). The additional needle beds allow a significantly extended variation of the realizable knitted fabric structures. Their use includes the temporary parking of the half stitches during stitch transfer in the fully-fashioned production of "classical" knitted fabrics like pullovers as complete garments.

Fig. 6.20 Double jersey knitting process: flat (*left*) and circular (*right*) knitting

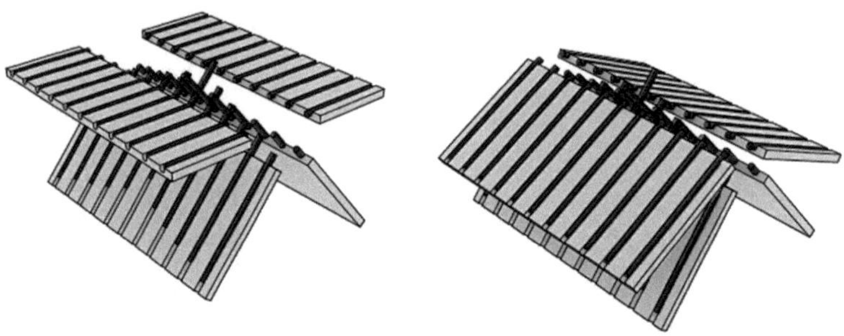

Fig. 6.21 Arrangement of needle beds on flat knitting machines for the production of complete garments, basic principles by companies Stoll (*left*) and Shima Seika (*right*)

6.3 Flat and Circular Knitted Fabrics

6.3.1 Shaping Possibilities

The shaping of knitted fabrics according to requirements can be completed either in a separate ready-made process or during the knitting process (fully-fashion).

Shaping in a separate assembly process is usually achieved by cutting the individual components from rolled goods and subsequent sewing into a final product (see Sect. 12.6). While flat knitting machines usually produce rolled goods during manufacture, circular knitting machines sometimes need additional devices for the cutting the tubular knitted fabrics and winding them up as rolled goods. The advantage of shaping in a separate assembly process is an often higher flexibility compared to a fully-fashion process. However, the assembly processes are additional process steps in the manufacture of knitted products. The cutting of individual components results in a greater cutting loss than in fully-fashion manufacture.

According to the degree of shaping during knitting, various terms are commonly used, which are listed below:

- *Garment length*:
  The knitted fabrics are manufactured in wale direction at the desired length. Cutting to length is therefore not necessary. If multiple knitted fabric parts are manufactured subsequently, a separating yarn is often inserted between the parts. This separating yarn is easily removable sideways to separate the knitted fabrics. As required, the complete shaping of the knitted fabric contour can be performed by an additional cutting process. All knitted fabric parts have a fixed beginning, but may have an open end due to process-inherent reasons. To prevent destruction of the knitted fabric, additional process steps have to be taken to secure the open edges.
- *Fully-fashioned*:
  This is a broad term and can refer to the shaping of the outer contour of a planar knitted fabric, but also to the three-dimensional shaping of a shell-shaped or tube-shaped knitted structure during knitting. Fully-fashioned knitted fabrics do not necessarily have fixed edges.
- *Regular*:
  This term is defined more narrowly. Here, the knitted fabric knitted according to a sectional drawing. The knitted fabrics have fixed edges, due to the use of half-stitch transfer. A cut to length or width is not necessary.

The following approaches can be used for shaping during the fabric formation process:

- Structural variation and
- Variation of the number of stitches in course and/or wale direction.

The structural variation includes the shaping by adjusting the stitch length, the use of different basic bindings in neighboring sections of the knitted fabric, and the variation of the binding elements in the fabric [12]. Figure 6.22 shows knitted fabric structures realizable by the use of these methods.

The following methods are known for varying the number of stitches in course and/or wale direction:

- widening,
- narrowing,
- binding-off, and
- flechage knitting

In *widening,* additional needles are activated at the edge of the knitted fabric and included in the knitting process. In *narrowing*, the width of the knitted fabric is reduced. For this, half stitches on the needles are transferred to neighboring needles at the fabric edge, and the empty needles are deactivated (Fig. 6.23).

Binding-off allows the production of a fixed straight edge of the knitted fabric in course direction. As in narrowing, half stitches are transferred to the neighboring needles, beginning at the right or left edge of the knitted fabric. The empty needles are deactivated. In the following, stitch formation is performed on these needles, and additionally on few neighboring needles. Thus, a destruction of the knitted fabric structure after removal from the knitting machine can be avoided.

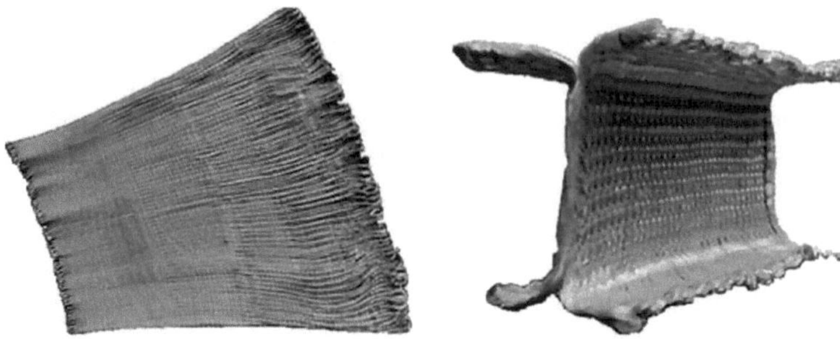

Figure 6.22 Shaping by means of structural variation by adjustment of the stitch length (*left*) and of the binding (*right*)

Fig. 6.23 Illustrations of widening, narrowing, and flechage knitting

These shaping possibilities allow the realization of knitted fabrics according to a predefined outer contour, while simultaneously creating fixed edges on all borders of the fabric as well as minimizing cutting waste. This technique of near-net shape manufacture of knitted fabrics is also referred to as the "fully fashioned" method [13].

Flechage knitting is used to change the number of stitches in wale direction. Here, no stitches are formed on a number of adjacent needles of a section, and the half stitches are parked on the needles until their re-introduction into the knitting process. Stitch formation continues on the remaining needles. Thus, more needles are formed locally in wale direction on the knitting needles, resulting in a spatial contour of the knitted fabrics.

6.3.2 Production on Flat Knitting Machines

Currently available flat knitting machines occupy the same basic construction scheme as shown in Fig. 6.24.

The yarn material is removed from the yarn storage (1) and, by way of corresponding elements imparting the required yarn tension (2), fed to the stitch forming area (3). Usually, the yarn is meanwhile monitored for yarn breakages,

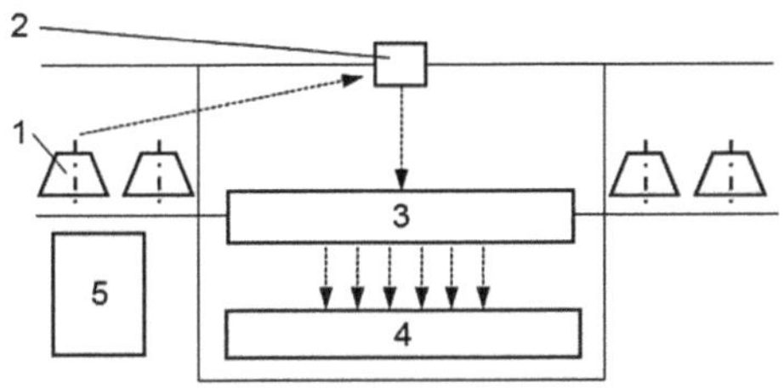

Fig. 6.24 Basic construction principle of a flat knitting machine (according to [7])

knots or thick spots, in order to prevent damaging or destruction of both finished product and knitting machine.

Yarn guides feed the yarn material to the needles arranged in planar needle beds in the stitch forming area. Depending on machine type, the yarn guides are driven by individual motors or by yarn carrier selection devices arranged along the carriage.

The design of the yarn feeder can differ according to their individual function. For a single-needle insertion of the various yarn materials on neighboring needles, so-called *intarsia yarn feeders* are used. They can be activated by suitable mechanisms, such as individual motor drives or swinging devices. A multitude of designs are also know for the covering of a first yarn by a second (plating of yarn materials), for instance by the use of consecutively arranged yarn guides (Fig. 6.25).

The selection and driving of the needles used in the knitting process are performed by the cam box, which is attached to the carriage and consists of sliding and curved mechanisms. In flat knitting machines, the carriage oscillates during knitting. Depending on the type of the machine, it can have several cam boxes in one carriage or several carriages, generally resulting in greatly increased productivity.

The selection of the needles arranged in the needle bed can be performed jointly, in groups, or individually. A grouped selection of needles can, for example, be effected by needle butts of varying heights, or be adjusting the arrangement of the needle butts in different tracks. For this, the needles are often allocated manually to the different groups during the fitting of the needle beds.

Individual needle selection grants the greatest flexibility regarding achievable structural parameters and patterning possibilities, as the knitting operation of each needle is freely programmable. Each machine manufacturer uses an individual system. Figure 6.26 shows a carriage and a cam box with single needle selection as a combination of permanent magnets and electromagnets (H. Stoll GmbH & Co. KG, Reutlingen).

The take-down of the knitting machines creates the necessary fabric tension for stitch formation and ensures the knock-over of the stitches off the needles. The specific design of the take-down system can differ depending on the manufacturer of the machine. Conventional take-down systems, for instance, can consist of two opposing shafts on which rolls or straps are arranged. The knitted fabric is clamped between the rolls or straps. At least one of the rolls is driven by a motor, and a

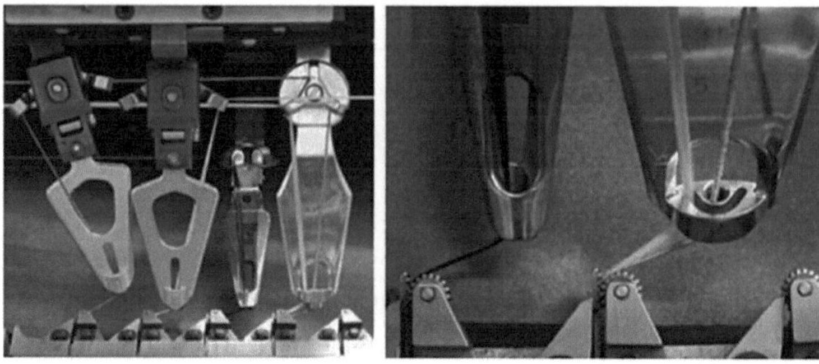

(a) (b)

Fig. 6.25 (**a**) Intarsia yarn guides (*left*: parking position, 2nd from *left*: knitting position), standard yarn guide (3rd from *left*) and plating yarn guide of a flat knitting machine; (**b**) detail of a yarn feeder edge in standard yarn guide and plating yarn guide

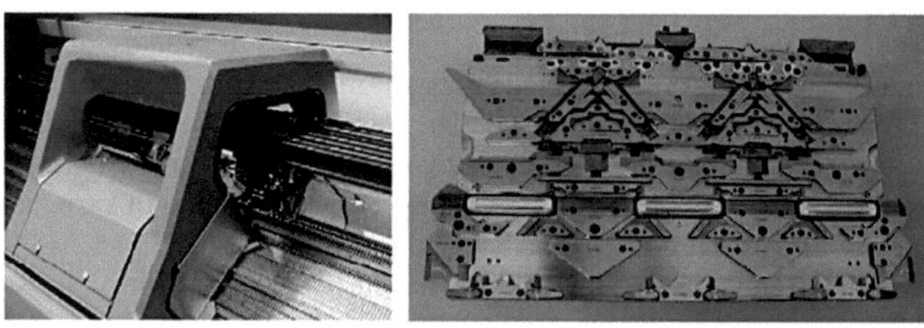

Fig. 6.26 Carriage and cam box with electromagnetic single needle selection

suitable control unit allows adjustments to the rotation angle and the force affecting the knitted fabric structure. Often, the take-downs also offer the possibility to locally adjust the clamping pressure and thus influence the take-down efficiency across the width of the knitted fabric.

The use of *sinkers* is a significant extension of the conventional take-down systems. They are arranged in the needle beds in the area of the needle hook. Their purpose is to fixate the knitted fabric in the vicinity of the knock-over edge, and support the knock-over of the stitch during the clearing of the needle. With a suitable design and controlled motion of the sinkers, they can also be used to achieve a specific local setting of the stitch length. Due to the spatial proximity of the sinkers to the stitch forming area, the stitch length can very effectively be influenced in early stages of the fabric manufacture, which is particularly advantageous for a knitting-technical shaping. In some realizations of sheer stitch structures the use of controlled sinkers makes additional take-down elements (e.g. rollers) obsolete.

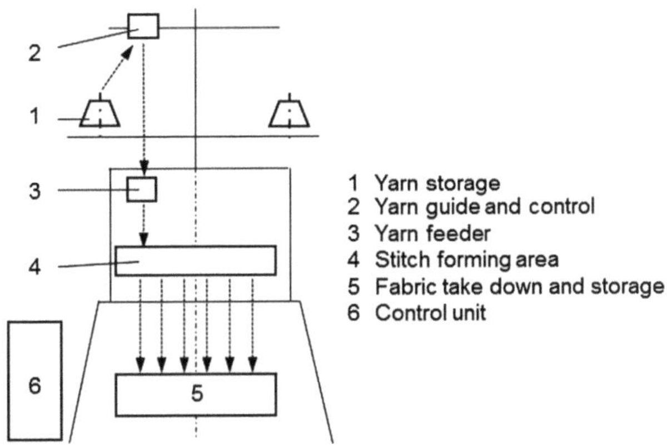

Fig. 6.27 Basic structure of a circular knitting machine (according to [7])

6.3.3 Production on Circular Knitting Machines

6.3.3.1 General Remarks

Generally, circular knitting machines are distinguished by their high productivity resulting from the large number of knitting systems that can be arranged in the circumference of the cylinder-shaped needle bed. In each revolution of the machine, one course can be created per knitting system. Two important parameters for productivity are the cylinder circumferential speed and the knitting system density, which is the relationship of the number of knitting systems on the cylinder circumference and the nominal diameter in inches [3]. Figure 6.27 represents the basic structure of such a large-scale circular knitting machine.

As in flat knitting machines, the majority of circular knitting machines are equipped with latch needles. Machines with compound needles are also available. One development especially for single-jersey circular knitting machines is the so-called relative movement technology for the reduction of yarn deflection points in the path of the yarn from spool to stitch forming area (Fig. 6.28). Furthermore, the use of the relative movement technology can reduce the strains on the yarn material and on the parts responsible for stitch formation at the cam box. During loop forming, the stitch comb sinkers and knock-over sinkers complete a motion opposed to the needles. The Cotton-type weft knitting motion for the realization of the stitch length results from the superposition of the needle and sinker motions. The relative movement technology is therefore used to increase production speeds and allows the processing of sensitive yarn materials.

6.3.3.2 Small Circular Knitting Machines

Small circular knitting machines have a cylinder diameter in the range of 1/12″–7″ (between 2 and 177.8 mm). Usually, these machines work according to the single jersey or purl knitting techniques, while double jersey knitting is rare due to

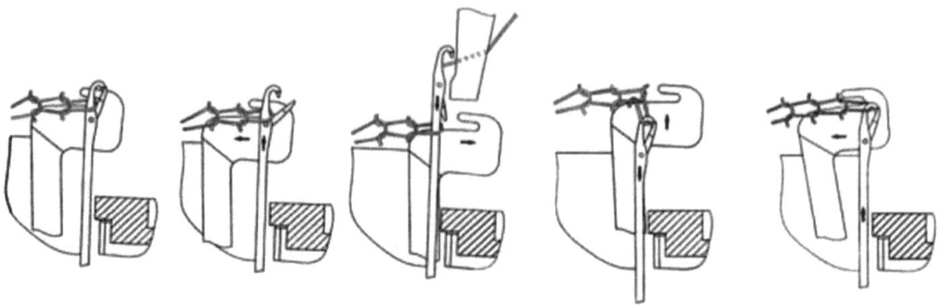

Fig. 6.28 Relative movement technology on single jersey circular knitting machines [14]

limitations in construction space. These machines are commonly designed according to product specifications and fitted with the corresponding control and patterning possibilities. Normally the cylinder is moved for stitch formation, and the cam boxes are fixed to the circumference of the cylinder.

The main area of application of small circular knitting machines is the production of socks and tights for garment and medical industries. Beyond that, the machines can be adapted for the processing of high-performance fiber materials or metal wires. This way, circular knitted fabrics for use in technical tubing, air filters, catalyzers or attenuators can be produced. Small circular knitting machines with very small cylinder diameters are also used to manufacture artificial blood vessels such as stents [15].

Specifically for the production of socks, various patterning and shaping possibilities are available. Depending on machinery equipment, these possibilities may also allow the manufacture of socks and tights in a single process step. By using additional devices and special knitting procedures means, the heel of the sock can be formed, and the tip of the sock can be closed.

The sock heel can be produced in knitting as a reciprocated heel or as a heel pocket. In a *reciprocated heel*, the additionally required partial courses in the area of the heel are formed by a repeated change of the rotational direction of the knitting cylinder during knitting, which requires symmetrically designed cam boxes. The reciprocated heel is predominantly used in the production of high-quality products. Greater productivity can be achieved by completing the heel as a so-called *heel pocket*. For this, the required additional partial courses in the heel section are formed by activating additional knitting systems. Machines working according to this principle can be conceptualized as machines running without changes to the rotational direction. Changing the number of stitches at the circumference of circular knitted fabrics is usually impossible during production. The advanced adjustments to the fit of the socks are therefore made by a specific setting of the stitch cam adjustment.

The shaping on small circular knitting machines is also supported by the additional use of combined stitch comb sinkers and knock-over sinkers arranged in the area of the needle hook. The use of such sinkers allows the productions of the beginning of a knitted fabric on empty needles without take-down. The patterning variety in the production of structural patterns is extended by the use of special transfer mechanisms for the realization of pattern-adapted holes, or by the use of plush mechanisms in the manufacture of sports socks.

On small circular knitting machines, take-offs are usually pneumatic. Apart from creating the tension required for knitting, the transport of the finished knitted goods to the storage unit is another task of the take-down devices.

The technical and technological versatility of circular knitting, therefore, offers a great potential for the manufacture of form-fit, near-net shape knitted fabrics in a single process step.

6.3.3.3 Large Circular Knitting Machines

The cylinder diameter of large circular knitting machines ranges from 7″ to 55″ (177.8–1,398 mm). Usually the machines with diameters from 10″ to 20″ are used for the production of T-Shirts or bathing suits in body widths, producing the tubular knitted fabrics in the corresponding product circumference. On machines with cylinder diameters greater than 20″, knitted fabric tubes are produced as web material for later processing. In further ready-made processes, this web material is then cut into individual parts, which are later joined to form the final product.

As in small circular knitting machines, their larger counterparts are designed specific to the product or binding. The needle selection mechanism is geared to that. While Jacquard machines have an individual needle selection and can therefore be used for a wide range of bindings, so-called *rib machines* or *Piqué machines* with grouped needle selection are available for the realization of special knitting-technical bindings such as the rib binding for undergarments. The grouped needle selection generally allows higher knitting speeds, resulting in productivity advantages.

A more recent development in circular knitting technology is a type of machine called *seamless machines*, which is suitable for the knitting-technical production of complete, primarily elastic items for bathing suits and undergarments. The machines offer the possibility of single needle selection and stitch transfer between the needle beds. On circular knitting machines with single needle selection, knitted fabrics of adjusted lengths (for example pullovers) can be produced. The patterning possibilities are largely identical to those flat knitting techniques, although the realization of displacement patterns is limited in scope.

The majority of circular knitting machines is equipped with a revolving needle cylinder and fixed knitting systems. By contrast, circular knitting machines on which shaping can be undertaken, such as seamless machines, are often equipped with fixed needle cylinders and mobile knitting systems, analogous to the movement relations on flat knitting machines.

On circular knitting machines for the manufacture of seamless garments, take-downs systems are usually pneumatic and simultaneously responsible for the transport of the items to the storage unit. Drum pull-offs are commonly used on machines producing web material. The product is wound up by the corresponding winding units.

Innovations in circular knitting allow the realization of a greater variety of knitting patterns, even without single needle selection. By using special yarn guides, weft yarns can be inserted into the knitted fabric. The covering (*plating*) of a ground yarn with a second yarn can be achieved by using a plating yarn carrier. Plush-knitted fabrics for a number of applications can be manufactured with special stitch comb sinkers and knock-over sinkers. The use of so-called carding units allows the presentation of fibers individualized from fiber bands to the needles and their integration into the stitch structure, which creates the possibility to produce fur knitwear.

The variety of knitting systems also allows the introduction of dyed yarn material for colored stripes in wale direction. The number of courses in a certain color or between two colors is determined by the yarn laying on the knitting systems. Using additional devices such as striper units, a color switch in wale direction can be freely programmed. For this purpose, the knitting systems are equipped with several yarn guides. The color switch is created by integrating the yarn of the new color and simultaneously clamping and cutting the previously used yarn. Using the wrapping finger system, longitudinal stripes can be introduced into circular knitting fabrics. Depending on the machine gauge and the pattern, the stripe can extend over one or more wales.

6.3.4 Development Tendencies in Knitting Technology

In knitting technology, new development approaches aiming at a simplification of the stitch forming process are frequent. Many of these approaches aim to move the stitch-forming elements independently without the use of driving components like carriages. This is to result in greater productivity and an extended variety of patterning. As an example, this section will look at the examples of stitch forming elements for circular knitting (Fig. 6.29, left), and single needle drive for the field of flat knitting technology (Fig. 6.29, right).

Such innovative solutions are particularly interesting for a use-specific local setting of the knitted fabric properties by means of a single-needle adjustability of the stitch length. However, they are far from providing the means for a reproducible knitted fabric manufacture. One remaining challenge is the ascertainment and quantification of the changing friction conditions between the elements in the stitch forming area. To implement a single needle drive, knowledge of these friction conditions is a crucial control parameter for the realization of an even stitch structure with a consistent length of all stitches. To ensure a high acceptance of such solutions in industrial use, the required controlling effort is to be limited to a minimum.

Furthermore, the machine developments aim to largely automate the production of knitted fabrics, focusing on the extension of patterning possibilities in the

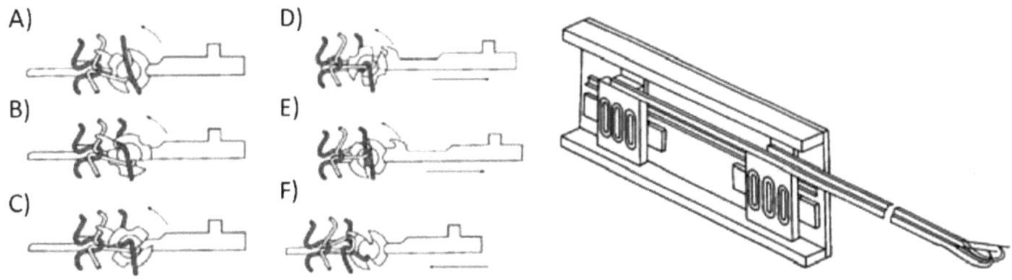

Fig. 6.29 Stitch formation using rotating stitch-forming elements [16], single-needle drive by linear motors [17]

production of seamless garments or on the possibility to sectionally adjust the take-down conditions to a variety of different manufacturing conditions across the fabric width. Apart from the direct knitting to shape of semi-finished products according to component geometry, the suitable insertion of reinforcement yarns in freely selectable knitting directions is another future research focus for the production of textile semi-finished products in lightweight constructions.

6.3.5 2D and 3D Knitted Fabrics with Integrated Reinforcement Yarns

6.3.5.1 General Remarks

In general, knitted fabrics with reinforcement yarns integrated into the stitch structure are referred to as *Multi-layered knitted fabrics* (MLG). MLG combine the advantages of knitted fabrics, such as drapability which is adjustable by stitch length, with those of non-crimp technology, for example the stretched orientation of reinforcement yarns. This allows the wrinkle-free forming of the knitted fabrics even for very complex component geometries with knitted semi-finished reinforcement products. The mechanical characteristics in the respective knitted fabric directions, for example in case of a tensile load, are usually dominated by the yarn materials used. Due to the stitch structure and the resulting properties, such as reinforcement yarn content ratio in thickness direction and knitted fabric-typical structural elongations, MLG semi-finished products show great potential for use in lightweight construction components exposed to crash and impact loads.

6.3.5.2 Monaxially Reinforced Multi-layered Knitted Fabrics (MLG)

Monaxially reinforced MLG feature reinforcement yarns integrated in the knitted fabrics in one direction. Apart from the reinforcement in course direction (weft yarns) or in wale direction (warp yarns), reinforcement yarns arranged at an angle to these can also be integrated.

The integration of weft yarns into knitted fabrics has been known for a long time and is described as state-of-the-art in patent documents from as early as 1934 [18]. The yarn material utilized can be used to specifically adjust, for instance, the strength of the compression effect in knitted bandages or ortheses. In knitted fabrics for seating furniture, weft yarns serve to limit the maximum elongation of the knitted fabric.

The integration of warp yarns as reinforcement yarns in knitted fabrics has not been adopted in industrial practice. However, patent specification DE700291 (1937) shows possibilities for the integration of stretched warp yarns running between two wales to reduce the elongation of the knitted fabric in course direction [19]. Patent specification DE31117362A1 (1981) extends the introduction of warp yarns by yarn displacement. The yarn displacement also serves the purpose of tighter integration of the warp yarn into a single jersey stitch structure [20].

Knitted reinforcement structures for lightweight construction can be realized using yarns from high-performance fiber materials such as glass, carbon or aramid. The forces usually straining a component from various directions can be dispersed by the targeted layering of numerous MLG with different reinforcement fiber orientations.

The integration of weft yarns requires a knitting machine with at least two needle beds, and the possibility of a half-stitch transfer according to the basic binding used. A weft yarn can be integrated into the stitch structure using the single jersey, double jersey or interlock basic binding (Fig. 6.30). The process of the weft yarn integration for the various basic bindings is shown in the following figure.

Independent of the selected basic binding, the weft yarn (light grey) is inserted into a net row. While the net row is created as standard when using the double jersey (Fig. 6.30b-1) and interlock basic bindings (Fig. 6.30c-1, c-2), it has to be provided in the single jersey basic binding by transferring (Fig. 6.30a-1, a-2), for instance, every second half stitch to the opposite needle bed. To fixate the weft yarns after its insertion, the half stitches are transferred back onto the first needle bed (single jersey), or another pattern-conforming stitch formation is completed (double jersey, interlock).

The use of the double jersey and interlock basic binding is usually more productive due to the transfer process not being required, but for technological reasons the produced knitted fabrics have a higher loop yarn proportion. When using the single jersey basic binding, repeated knock-over of the half stitches off the needles during the transfer process causes high strain on the yarns. This is compensated by the use of more robust yarn materials and the reduction of manufacturing speed.

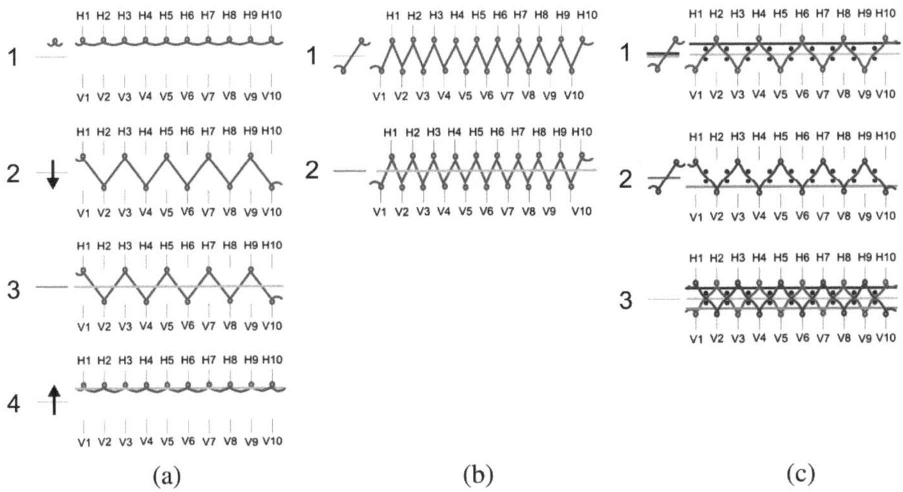

Fig. 6.30 Process of weft yarn (*light grey*) integration into the stitch structure, using (**a**) single jersey, (**b**) double jersey, (**c**) interlock basic bindings

6.3.5.3 Biaxially Reinforced Multi-layered Knitted Fabrics

Biaxially reinforced multi-layered knitted fabrics contain reinforcement yarn systems in two spatial directions, which are usually the direction of manufacture (0°) and the transverse direction (90°). First development steps in mechanical engineering towards the production of biaxially reinforced MLG based on circular knitting technology are documented in the patents US3859824 and DE2933851 from 1973 and 1984, respectively. In the United States, biaxially reinforced MLG with two reinforcement layers are available commercially. The development of MLG structures with different numbers of reinforcement layers, based on flat knitting technology and using the corresponding shaping possibilities, is researched at the Institute of Textile Machinery and High Performance Material Technology (ITM) of TU Dresden. Currently, up to 11 reinforcement layers can be integrated into the MLG (Fig. 6.31).

The knitting-technical fixation of the reinforcement yarns is secured using either the MLG single jersey or the MLG interlock binding, depending on the number of reinforcement layers. The following illustration shows biaxially reinforced, basic MLG structures with various numbers of reinforcement layers.

For a secure and fixed connection of all reinforcement yarns with the stitch structure, related to at least one of the active needle beds, the following technology-inherent arrangement of the stitch-forming elements or yarn systems has to be adhered to: needles of an active needle bed as well as weft, warp, and loop yarn systems (see Fig. 6.31). Exemplarily, the technological process of the manufacture of biaxially reinforced MLG structures for various basic bindings and numbers of layers is shown in Fig. 6.32.

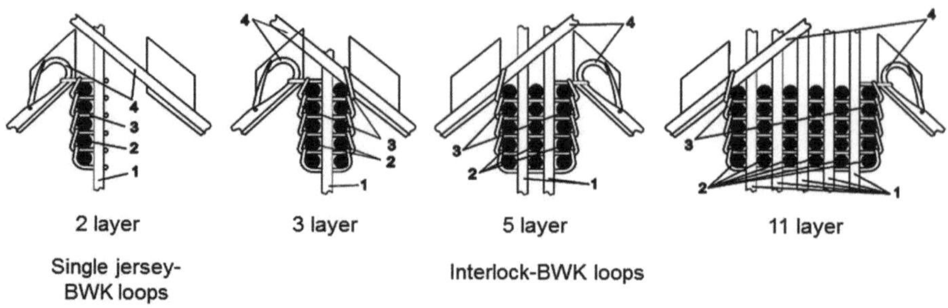

Fig. 6.31 Basic biaxial MLG structures with various numbers of reinforcement layers (1—warp yarns, 2—weft yarns, 3—loop yarns, 4—needles)

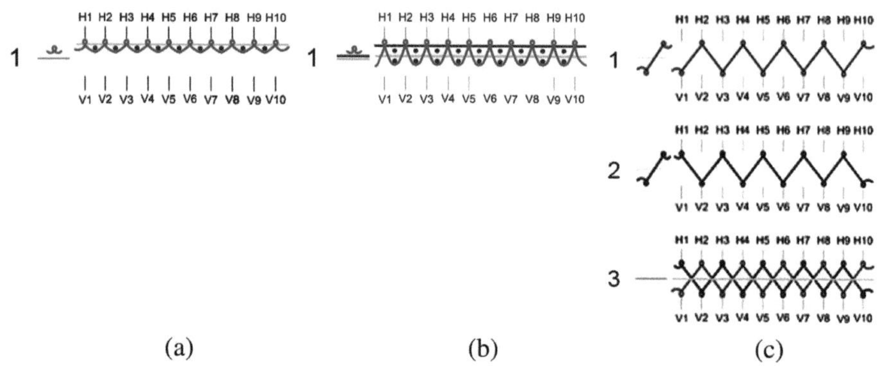

Fig. 6.32 Process of the integration of reinforcement yarns in biaxially reinforced MLG, using (**a**) single jersey-MLG-binding (2-layered), (**b**) single jersey-MLG-binding (4-layered), and (**c**) interlock-MLG-binding (5-layered)

The realization of symmetrical MLG structures requires the use of two active needle beds and the utilization of the double jersey-MLG or interlock-MLG bindings. The respective outer reinforcement layers of symmetrical MLG are weft yarn layers.

Regardless of the selected binding and the utilized knitting method, the warp yarns are fed to the stitch forming area from above, between the needle beds of a knitting machine. This requires knitting machines with a divided carriage. A correspondingly modified flat knitting machine and the respective stitch forming area are shown in detail in Fig. 6.33.

6.3.5.4 Multiaxially Reinforced MLG

The consequent enhancement of the possibilities of reinforcement yarn insertion in more than two spatial directions allows the realization of multiaxially reinforced knitted fabrics. The implementation of a production principle was performed on a

Fig. 6.33 Flat knitting machine modified for the integration of warp yarns and detailed view of the stitch forming area

laboratory flat knitting machine at the ITM (Fig. 6.34). A displacement mechanism allows a single-needle positioning and setting of the angle at which the diagonal warp yarns are inserted into the knitted fabrics. Apart from warp yarns in 0° direction, weft yarns in 90° direction can also be inserted. The fixation of the reinforcement layers against each other is ensured by the loop yarns.

Figure 6.34 illustrates the fixation of the reinforcement yarn arrangement by the interlock-MLG binding. The introduction of a weft yarn layer as reinforcement layer on each of the outsides enables the realization of symmetrically structured MLG.

6.3.5.5 Near-Net Shape Knitted Fabrics

Near-net shape MLG can be classified into plane knitted fabrics with a two-dimensional contour, and spatial knitted fabrics with a three-dimensional shape or geometry. The aim of the manufacture of near-net shape MLG for use as reinforcement semi-finished products is the reduction of production effort of composite components, while simultaneously suitably arranging the reinforcement yarns in these components. Beyond that, the formation of the net shape of the MLG can help achieve a much improved material efficiency.

For a formation of the two-dimensional near-net shape, the shaping methods of narrowing and widening can be used (see Sect. 6.3.1). Examples for the realization of near-net shape knitted fabrics are shown in Fig. 6.35.

By means of a suitable yarn guide control, the weft yarns can be fed single-needle into the stitch structure according to the net shape. The adaption of the stitch structure to the required net shape is performed by narrowing and widening. This way, the edges of the stitch structure can at the same time be secured against

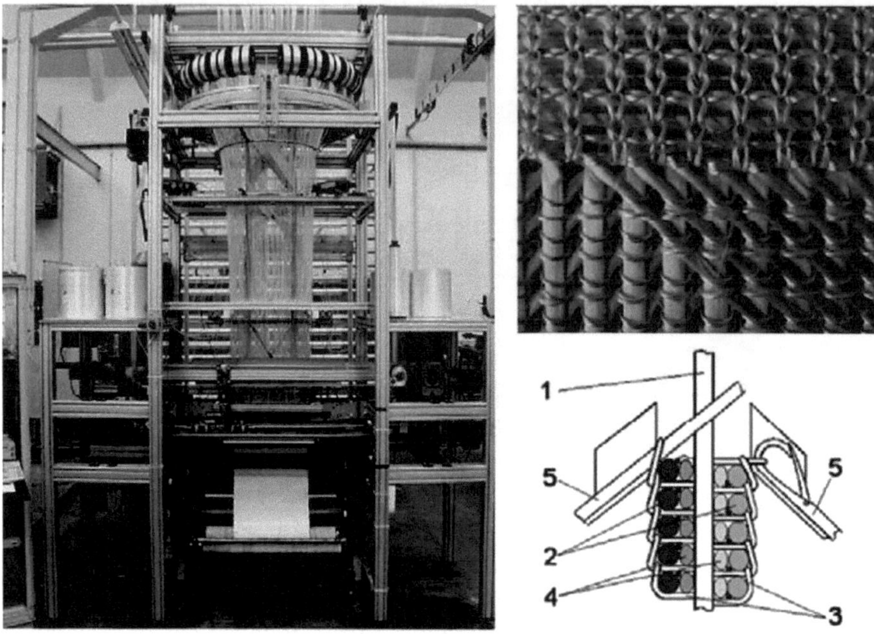

Fig. 6.34 Multiaxially reinforced MLG, and laboratory flat knitting machine (1: warp yarns, 2: weft yarns, 3: diagonal warp yarns, 4: loop yarn, 5: needles)

	Contour target	Semi-finished product after knitting	Semi-finished product after cutting	Close-up view
Outer contour without warp yarn displacement	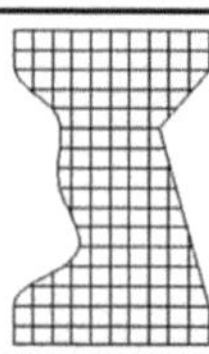	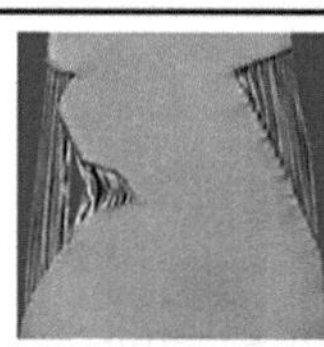	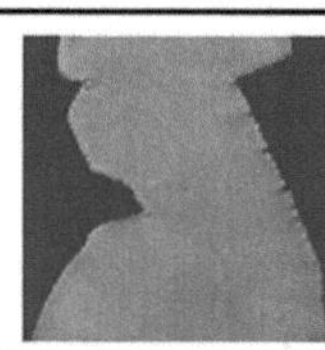	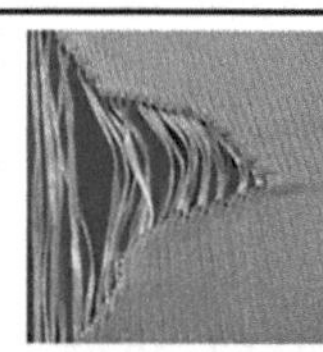
Inner contour without warp yarn displacement	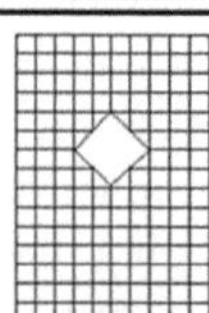			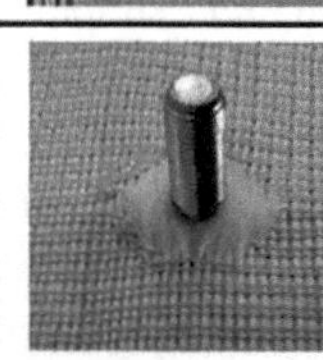
Outer contour with warp yarn displacement		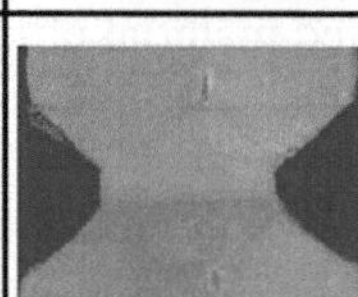		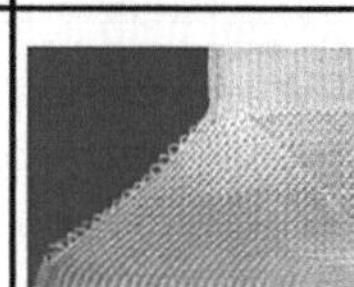

Fig. 6.35 Knitting-technical realization of near-net shape MLG

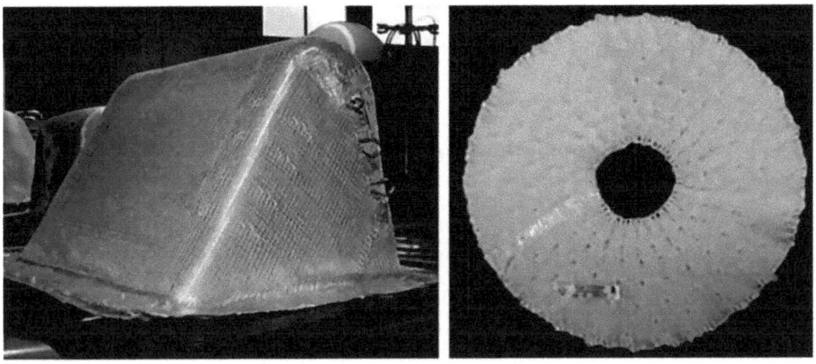

Fig. 6.36 Examples of near-net shape, biaxially reinforced MLG: bulk material container (*left*), and circular knitted fabric for juncture reinforcement (*right*)

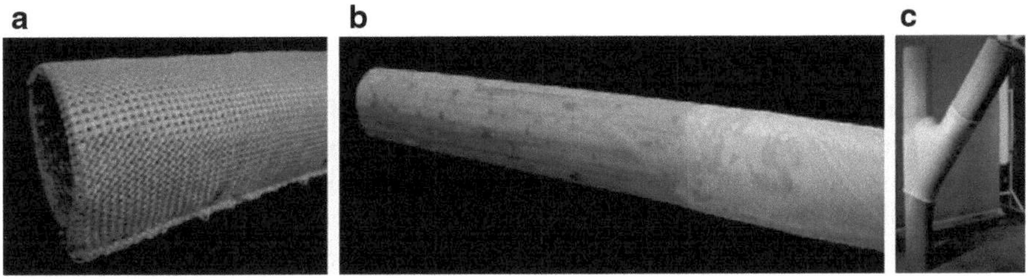

Fig. 6.37 Biaxially reinforced MLG as tubular structures—(**a**) textile semi-finished product, (**b**) during application on the surface of a molded wood profile, (**c**) Y structure as crotch reinforcement

unravelling. Weft yarns and/or warp yarns not inserted into the stitch structure at some places for reasons of the shape of the MLG have to be removed manually after production [12].

The formation of a three-dimensional geometry of the MLG is executed by needle parking. As defined by the pattern, a weft yarn is inserted only locally into parts of the MLG (partial weft inlay). Half stitches in neighboring parts are parked on needles until they are reincorporated into the active knitting process. Figure 6.36 shows a few examples of realized MLG semi-finished products in near-net shapes.

By combining the partial weft insertion and the warp yarn displacement, the required cutting of the semi-finished products can be avoided almost entirely. Furthermore, particularly the sections of the MLG in which an integration of inserts (e.g. as fixtures) is projected, can be suitably reinforced with an arrangement of warp yarns aligned with force directions.

Apart from the formation of the net shape of textile semi-finished products by needle parking as a shaping technique, knitting technology also allows the net shape-conforming production of tubular semi-finished products with reinforcement yarns integrated in various directions (Fig. 6.37). Generally, such knitted fabrics can be realized by circular as well flat knitting technology. While the use of circular knitting technology usually achieves higher productivities, flat knitting allows a change of the semi-finished product diameter during manufacture [21, 22].

6.3.5.6 Weft Knitted Spacer Fabrics

Conventional weft knitted spacer fabrics are 3D constructions consisting of two surface layers with a pile yarn layer between them (Fig. 6.38). The majority of such knitted fabrics are used as padding material in protectors, garments and mattress toppers. The knitted fabric properties, such as pressure stability or air permeability, can be adjusted in a wide range by means of controlling the knitting-technical binding, the pile yarn density or the yarn material of surfaces and pile yarn layer. The pile yarn is usually integrated into the surfaces by means of tuck stitches.

The variety of the knitting-technical possibilities results in a multitude of patterns regarding contour and thickness in the realization of suitable spacer fabrics. However, the manner of pile yarn integration limits the maximum possible distance between the two surfaces. Furthermore, an increased distance between the surfaces goes hand in hand with a decreased pressure stability of the spacer fabrics.

The development of weft knitted spacer fabrics, i.e. spacer fabrics in which the surfaces are connected by knitted connection yarns, helps overcome the mentioned disadvantages for lightweight construction applications [23]. The use of flat knitting technology for the realization of the structures allows the production of a variety of cross-sections and contours of the chambers created between the connection yarns (Fig. 6.39).

Spacer fabrics are made from high-performance fiber materials such as glass, carbon or aramid and are most frequently used in lightweight construction structures. The integration of stretched reinforcement yarns into the connection yarns and surfaces by means of the warp and weft yarn binding elements improves the achievable mechanical characteristics, while maintaining a small weight of the component [24].

Such spacer fabrics can be consolidated using both thermoplastic and thermosetting matrices. The chambers of the spacer fabrics can be used for the laying of

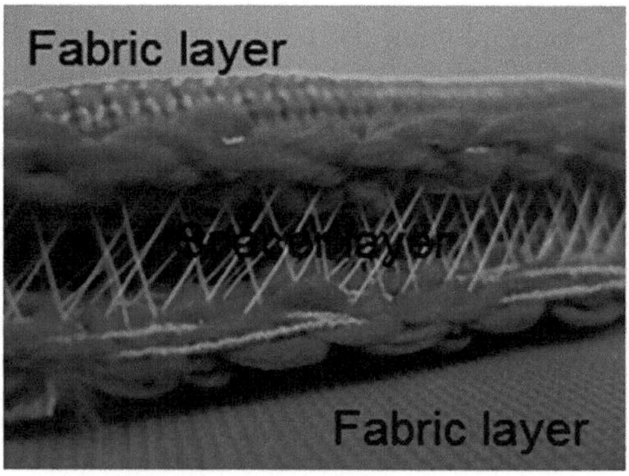

Fig. 6.38 Structure of a conventional weft knitted spacer fabric

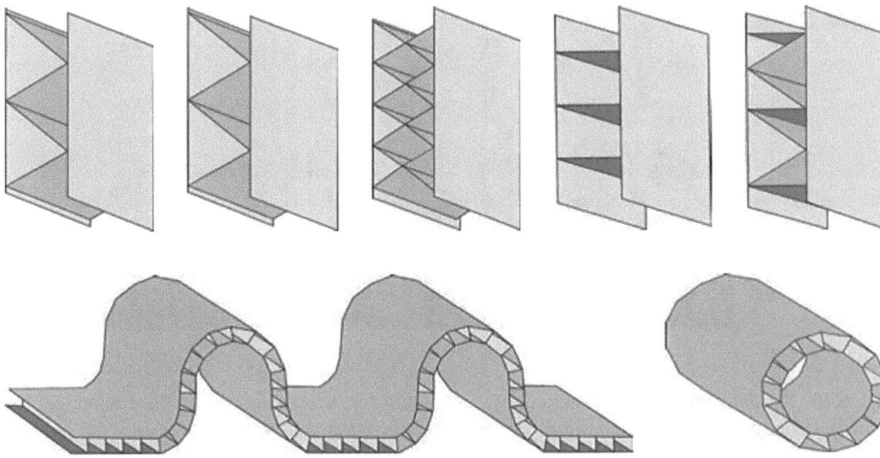

Fig. 6.39 Examples for possible cross-sections and geometries of spacer fabrics with knitting-technically achievable connecting yarns between the surfaces [24]

cables or the adjustment of thermal or acoustic properties of the components by being filled with suitable insulants. As all knitted fabric-based reinforcement structures, spacer fabrics display outstanding adjustable impact properties.

6.4 Functional Integration

The production of knitted fabrics is conducted course-wise by means of deforming yarn material fed transversely to the direction of production into yarn loops and then connecting these yarn loops. For process-inherent reasons, the inserted yarn material can be varied in each course. If the individual knitting machine is suitable for the production of Intarsia patterns, the yarn material can also be introduced single-needle within the course. Knitting technology is therefore ideal for any functional integration during manufacture of the knitted fabrics. In the following, this will be shown for a number of examples.

According to the respective product and application, the implementation of a variety of functions can be required. For example, the medical effectiveness of bandages or ortheses depends on the achievable compression effect, which can be adjusted by inserting elastic weft yarns of defined pre-tension into the knitted fabrics (Fig. 6.40a). The realization of knitted fabrics with, for instance, an integrated heating functionality requires the local insertion of and connection with thermally conductive yarn materials (Fig. 6.40b). The integration of electrically conductive yarn material allows the realization of conductive structures and even sensor networks. Beyond that, the locally varied use of yarn materials enables the production of load-adapted knitted, textile semi-finished products for composite materials with sectionally varied reinforcement effects (Fig. 6.40c).

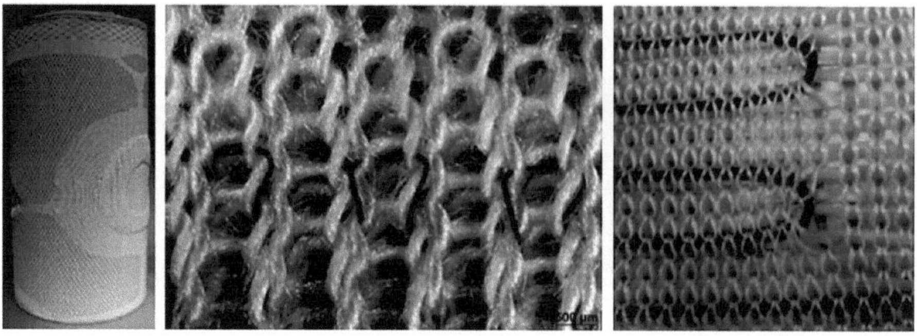

(a) (b) (c)

Fig. 6.40 Knitted fabrics with functionality integrated by yarn material: (**a**) knee bandage, (**b**) electrical conductor, (**c**) adapted reinforcement effect or integrated sensors

Against the backdrop of the usually brittle failure behavior of composite materials, possibilities to monitor the load cases of such components are currently being researched. The aim is an integration of the sensor material into the textile semi-finished products during textile manufacture. One approach is the knitting-technical processing of yarn materials displaying a characteristic change of parameters, for example their electrical resistance, in specific load cases such as tensile strain. Following a corresponding connection of the sensors, the monitoring functions can be realized.

References

1. *Zur Geschichte des handgestrickten Strumpfes.* http://www.deutsches-strumpfmuseum.de/technik/01handgestrickt/handstrick.htm (03-12-2010)
2. SPENCER, D.: *Kitting technology a comprohensive handbook and practical guide.* 3rd edition. Cambridge : Woodhead Publishing Limited, 2001
3. IYER, C.; MAMMEL, B.; SCHÄCH, W.: *Rundstricken: Theorie und Praxis der Maschentechnik.* 2nd edition. Bamberg : Meisenbach, 2000
4. SHIMA SEIKI: *Company History.* http://www.shimaseiki.com/company/history/ (12-01-2011)
5. BENDER, W.: Stoll CMS-Flachstrickmaschinen stricken technische Textilien. In: *Mittex* 107 (2000), No. 2, pp. 24–26
6. *Abstandsgestricke.* http://www.faiss.de/downloads/FaiFlyer3DFlex03.pdf (03-12-2010)
7. OFFERMANN, P.; TAUSCH-MARTON, H.: *Grundlagen der Maschenwarentechnologie.* Leipzig: Fachbuchverlag Leipzig, 1977
8. Protective right DE 4419985C2 (08th August 1994).
9. ERMANNI, P.: *Composite Technologien Version 4.0, Vorlesungsunterlagen.* Zürich, 2007
10. ÜNAL, A.; OFFERMANN, P.: Einfluss der Bindung auf die Verformungsverhältnisse von Verstärkungsgestricken. In: *Melliand Textilberichte* 86 (2005), No. 4, pp. 258–260
11. ORAWATTANASRIKUL, S.: *Experimentelle Analyse der Scherdeformationen biaxial verstärkter Mehrlagengestricke*, Technische Universität Dresden, Fakultät Maschinenwesen, Dissertation, 2006

12. CEBULLA, H.: *Formgerechte zwei- und dreidimensionale Mehrlagengestricke mit biaxialer Verstärkung - Entwicklung von Maschine, Technologie und Produkten*, Technische Universität Dresden, Fakultät Maschinenwesen, Dissertation, 2004
13. RAZ, S.: *Flat Knitting Technology*. Westhausen : C. F. Rees, 1993
14. Groz-Beckert: *Nadeltechnik Maschenbildung*. Albstadt, 1993
15. ABIZAID, A.: Innovative Stenting Approach for the Treatment of Thrombus-containing Lesions in Acute Myocardial Infarction, Saphenous Vein Grafts and Acute Coronary Syndromes. In: *Interventional Cardiology* 3 (2008), No. 1
16. Protective right EP2192219A1 (26th November 2008).
17. Protective right EP717136A1 (14th December 1994).
18. Protective right DE614839 (6th. March 1934).
19. Protective right DE700291 (15th March 1937).
20. Protective right DE3117362A1 (2nd May 1981).
21. TRÜMPER, W.; DIESTEL, O.; CHERIF, Ch.: Stricktechnische Lösungen für die Herstellung endkonturnaher Preformen / Solutions for flat knitting of near net shape textile preforms. In: *Melliand Textilberichte* 88 (2007), No. 7/8, pp. 537–538, E103–E104
22. HALLER, P. et al.: *Hochleistungsholztragwerke - HHT - Entwicklung von hochbelastbaren Verbundbauweisen im Holzbau mit faserverstärkten Kunststoffen, technischen Textilien und Formpressholz (BMBF 0330722A-C)* / TU Dresden. Dresden, 2011. – Research report
23. ABOUNAIM, M.; DIESTEL, O.; HOFFMANN, G.; CHERIF, Ch.: Thermoplastic composites from curvilinear 3D mulit-layer spacer fabrics. In: *Journal of Reinforced Plastics and Composites* 29 (2010), No. 24, pp. 3554–3565. DOI 10.1177/0731684410378541
24. ABOUNAIM, M.: *Process development for the manufacturing of flat knitted innovative 3D spacer fabrics for high performance composite applications*, Technische Universität Dresden, Fakultät Maschinenwesen, Dissertation, 2011

Additional Literature

WEBER, M.O.; WEBER, K.-P.: Wirkerei und Strickerei, 6 Auflage. Deutscher Fachverlag, Frankfurt am Main, 2014 - ISBN 3866412991.

JIANG, G.; GAO, Z.; MA, P.; MIAO, X.; ZHU, Y.: Comparative study on the mechanical behavior of carbon weft-knitted biaxial fabrics stitched by polyester fibers and preoxidized polyacrylonitrile fibers. In: Journal of Industrial Textiles 44/1 (2014), pp. 5–21.

XUE, D.; HU, H.: Mechanical properties of biaxial weft-knitted flax composites. In: Materials & Design 46 (2013), pp. 264–269.

SCHULZ, C.: Entwicklung einer Technologie für die Herstellung textiler Halbzeuge für thermoplastische Verbundwerkstoffe auf Basis der Mehrlagenflachstricktechnik. Dresden: Technische Universität Dresden, Fakultät Maschinenwesen, Institut für Textilmaschinen und Textile Hochleistungswerkstofftechnik, Dissertation, 2012.

DEMIRCAN, O.; KOSUI, T.; ASHIBE, S.; HAMADA, Y.; NAKAI, A.: Effect of stitch and biaxial yarn types on the impact properties of biaxial weft knitted textile composites. In: Science and Engineering of Composite Materials (2013), pp. 1–10.

TAESLER, J.; TRÜMPER, W., et al.: Development of a technology enabling highly productive manufacturing of multilayer weft knitted reinforcing structures for lightweight applications, pp. 115–116.

QI, Y.; LI, J.; LIU, L.: Tensile properties of multilayer-connected biaxial weft knitted fabric reinforced composites for carbon fibers. In: Materials & Design 54 (2014), pp. 678–685.

ZHU, L.; SUN, B.; HU, H.; GU, B.: Ballistic impact damage of biaxial multilayer knitted composite. In: Journal of Composite Materials 46/5 (2012), pp. 527–547.

FUJIMURA, T.; ASHIBE, S. T.; Demircan, O.; Nakai, A.: Development of three-dimensional shape forming in biaxial weft knitted composites. In: Proceedings of the Aachen-Dresden International Textile Conference (2012), S. Paper-No. CO-18.

LIN, H.; KRZYWINSKI, S.: 3D multilayer knitted preforms for composite applications with load-adapted thread orientation. In: AUTEX World Textile Conference 2014 (2014), pp. 1–11.

CHERIF, C.; KRZYWINSKI, S.; DIESTEL, O.; SCHULZ, C.; LIN, H.; KLUG, P.; TRÜMPER, W.: Development of a process chain for the realization of multilayer weft knitted fabrics showing complex 2D/3D geometries for composite applications. In: Textile Research Journal 82/12 (2012), pp. 1195–1210.

RISICATO, J.-V.; TRÜMPER, W.; KELLY, F.; SOULAT, D.; LEGRAND, X.; COCHRANE, C.; KONCAR, V.: A Complex Shaped Reinforced Thermoplastic Composite Part Made of Commingled Yarns With Integrated Sensor. In: Applied Composite Materials (2014)

Chapter 7
Warp-Knitted Semi-finished Products and Warp-Knitting Technologies

Jan Hausding and Jan Märtin

The production of warp-knitted semi-finished products is based on classical warp knitting processes, in which warp yarns of one or several groups are simultaneously and parallelly transformed into loops. This basic principle is used to connect yarn layers and/or other fabrics, such as nonwovens or pre-impregnated fiber layers by means of these loops. The main advantages of warp-knitted semi-finished products are their highly productive manufacture, the adjustability of the angles at which the individual yarn layers can be arranged respective to one another, and the versatile combinations of layer structuring and layer arrangement. Typical products are made from glass or carbon filament yarns and find applications in rotors in wind energy plants, in boat and vehicle construction, sports equipment, and construction.

7.1 Introduction and Overview

Warp-knitted semi-finished products for lightweight construction applications are an important part of the warp-knitted textile product range, which covers many fields from conventional garment and home textiles to numerous technical applications. Warp-knitted textiles are classified as knitted fabrics (see Sect. 2.2.2.5), although in lightweight construction contexts the loops are primarily used to connect yarn layers to each other or to other textile fabrics. In contrast to weft knitting, warp-knitting forms all loops simultaneously from yarns running parallel and lengthwise to the needles (warp yarn fabrics). On this foundation, the possibilities for designing warp-knitted semi-finished products are more diverse than in any other fabric manufacturing process. The possibilities to arrange different yarn

J. Hausding • J. Märtin (✉)
Formerly: Institute of Textile Machinery and High Performance Material Technology, TU Dresden, Dresden, Germany
e-mail: i.textilmaschinen@tu-dresden.de; i.textilmaschinen@tu-dresden.de

C. Cherif (ed.), *Textile Materials for Lightweight Constructions*,
DOI 10.1007/978-3-662-46341-3_7

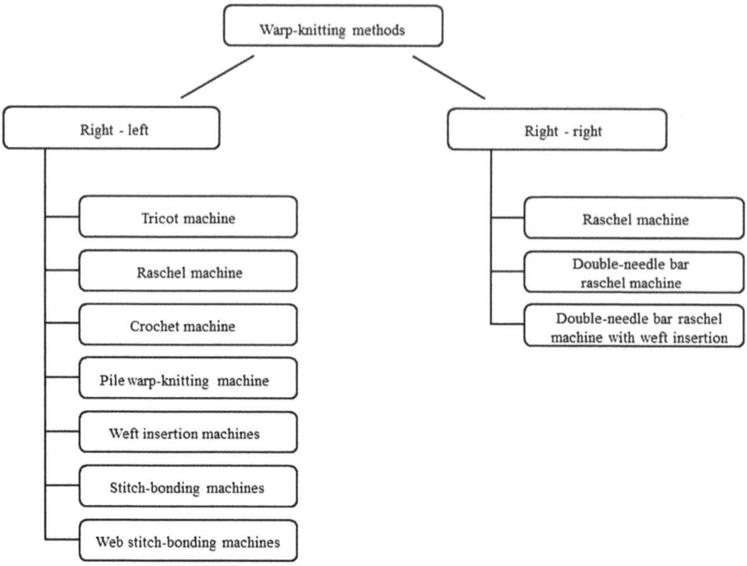

Fig. 7.1 Classification of warp-knitting methods

materials in multiaxial orientation, to adapt the arrangement of the yarns to predetermined loads, and the option to integrate, for instance, nonwovens as outer layer of the fabric, deserve special attention in this regard. The warp-knitting technique is among the most productive textile manufacturing processes. Figure 7.1 gives an overview of the currently common types of warp-knitting machines. The usual classification of these machines is based on the number of needle bars. Machines equipped with one needle bar or two needle bars are referred to as right-left (RL) and right-right machines (RR), respectively. Not all machine types listed in Fig. 7.1 are primarily used in the production of technical textiles for lightweight construction applications. For this reason, in the following, the focus will be placed on RL weft insertion and RL stitch-bonding machines as well as RR double-needle bar raschel machines. Stitch-bonding of fiber webs is treated in Chap. 9.

The first focus of this chapter treats the warp-knitted semi-finished products themselves, after which the machine technology will be considered. Basically, warp-knitted semi-finished products consist of yarn layers as well as other fabrics, and a loop yarn system is used to connect them. The manner, form, and arrangement of the loops are referred to as the pattern (see DIN 62050). Warp-knitting offers immense pattern variation possibilities. As adjustments to the pattern offer the possibility to alter the characteristics of the semi-finished and the final product, basic knowledge of the available designing possibilities is crucial. These possibilities will be introduced next, followed by descriptions of typical warp-knitted products for lightweight construction and machine engineering applications.

7.2 Pattern Construction in Warp-Knitting

7.2.1 *Representation of Patterns*

The representation of warp-knitted fabrics has to describe the respective pattern correctly and unambiguously in order to enable the construction of the pattern and ensure the storage and transfer of the information. This information includes details of the design, machine fineness, the material used as well as type and shape of the pattern design. The following types of representation are commonly used:

- Looping diagram
- Design pattern diagram
- Lapping pattern diagram
- Lapping diagram
- Chain notation, and
- Threading

These are based on uniform and mandatory principles, sometimes amended by company-specific representations and notations. The looping diagram is an enlarged, two-dimensional linear representation of the yarn course in the pattern (Fig. 7.2) and graphically represents the warp-knitted fabric, especially for simpler patterns. However, the creation of a looping diagram is an elaborate process, and presentiveness deteriorates with a growing number of yarn systems to be depicted.

Design and lapping pattern diagrams are abstracted visualizations of the pattern design of warp-knitted fabric on quadrille or honeycombed paper, and are used primarily for Jacquard patterns and curtain or tulle fabrics. DIN 62050 contains more detailed information.

Lapping diagrams are used for simpler pattern designs on warp-knitting machines with one or more guide bars (see Sect. 7.3.3). This is based on the representation of the yarn guide movement around the needles. The yarn guides of a yarn system are mounted to the same guide bar and therefore complete identical and simultaneous movement. Thus, the movement of a yarn guide and the corresponding yarn in it are representative of the movement of all yarn guides of the corresponding guide bar. The lapping diagram is based on rows of points

Fig. 7.2 Examples of looping diagram (RL pillar, RL tricot, RL cord)

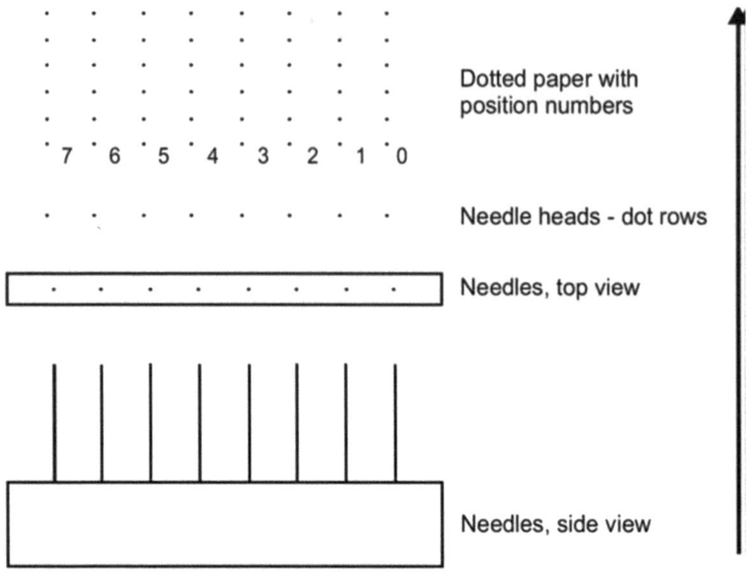

Fig. 7.3 Representation of needles on dotted paper

representing the needle heads in top view (Fig. 7.3). A horizontal line of the dotted paper grid represents the needle bar, one swing cycle of the guide bar around these needles and the resulting row of loops. Points above each other symbolize the same needle and give the subsequent rows of loops. The lapping diagram is always drawn and read bottom to top. With due regard to the machine fineness (horizontal dot distance) and the number of courses per unit of length (vertical dot distance) a true-to-scale lapping diagram with a realistic representation of the warp-knitted fabric is created.

The spaces between needle lanes are numbered consecutively from zero (0, 1, 2...). The numbering starts from the right of the rightmost row of dots, as the patterning device of the machine is usually located on the right-hand side. If it is located on the left-hand side, numbering will begin on that side. The numbers serve to control the lateral shogging movement of the guide bar. The completed motions are laid out in Fig. 7.4. These motions are entered into the lapping diagram. Straight lines are used exclusively for the representation of the guide needle movement, while the progress of movement is depicted with rounded lines (Fig. 7.5).

The yarn guide movement is depicted until repetition occurs. This so-called *repeat unit* is the basic unit of the pattern and is repeated continuously in height (height repeat unit) and width (width repeat unit) direction. When using several guide bars for a pattern, the bars are drawn in different colors on top of each other.

Due to the numbering of the spaces between needle lanes (position numbers), a drawn representation for the patterning on the machine is not necessary. A numerical depiction of the guide bar movement, the so-called chain notation, which is

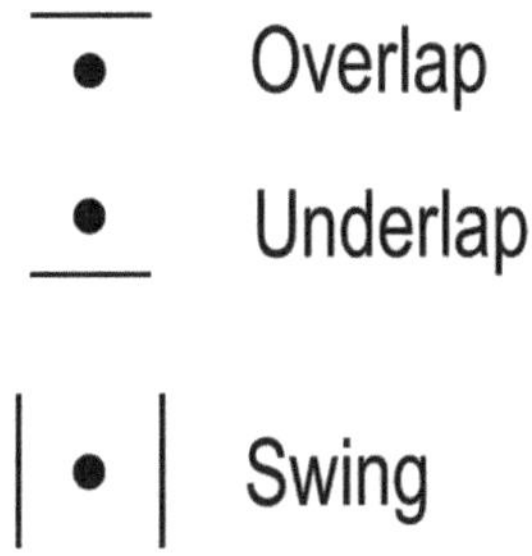

Fig. 7.4 Guide needle movement

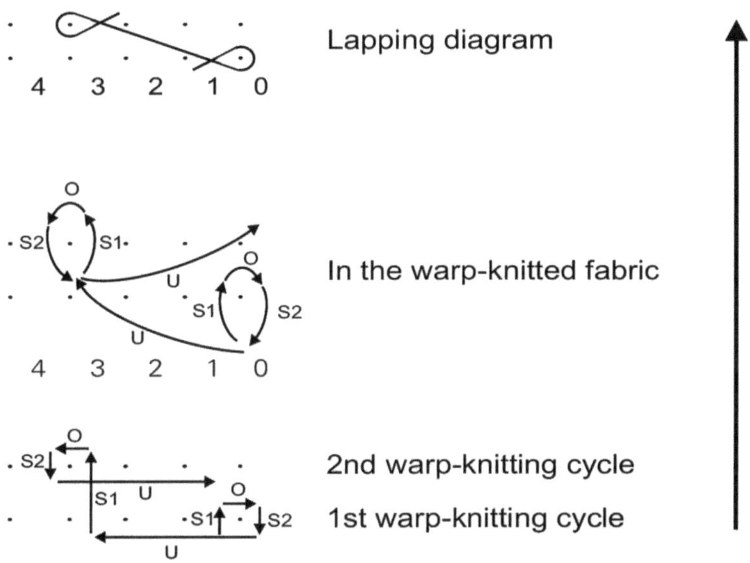

Fig. 7.5 Guide needle movement and lapping diagram (*O* Overlap, *U* Underlap, *S* Swing, *S1* Swing-in, *S2* Swing-out)

derived from the lapping diagram, is sufficient for this purpose. It is the basis for the design of the chains or cams, and for the programming of the servo or linear drives controlling the guide bar shogging movement. Two position numbers indicate the direction and width of the overlap (e.g. 1–0: overlap from needle lane 1 to needle lane 0). The numbers are written below or next to each other. A dash in the vertical notation, or an oblique in the horizontal notation designate an underlap. If the overlaps are depicted correctly, the underlaps follow from them automatically, as an underlap always leads from the end of the first overlap to the beginning of the second one. The end of the repeat unit is designated by a double vertical line. The chain notation is read from top to bottom or left to right respectively. When stated in several side-by-side columns, these are also read from left to right (Fig. 7.6). For each guide bar, a chain notation has to be created and is numbered with the number

Fig. 7.6 Chain notation for the RL double tricot pattern, in opposition

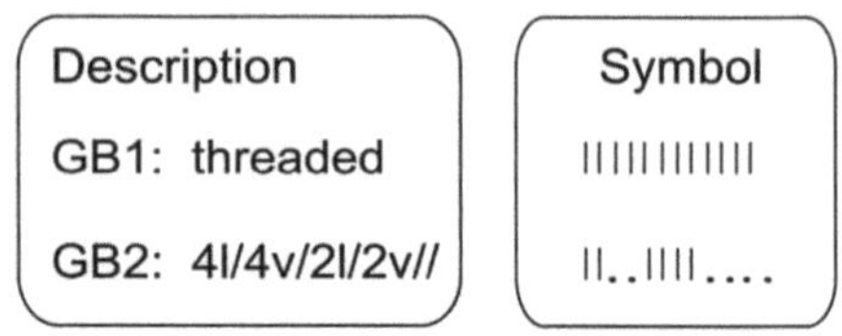

Fig. 7.7 Threading representation

of the guide bar (GB1, GB2, etc.). This numbering starts with one, on the operating side.

The threading is a representation of the occupation of the yarn guides in yarn systems (DIN 62050). A yarn guide is either threaded (v) and designated by a vertical line, or empty (l), and denoted by a dot (Fig. 7.7). The draft repeat is the repeat in the width of the design pattern and has to be stated individually for each guide bar. In complicated design patterns with different yarns (color, fineness, etc.), letters are used for designation.

Often, patterns are also stated by name, on their own or supplementing numerical and graphical representations. The commonly used terminology is explained in the next chapter. Fundamentally, the pattern group (RL, RR) is stated before the pattern designation. In combined patterns, the sequence of the guide bars is taken into account when designating the pattern (e.g. RL cord-tricot). The designation can be specified by additional indications such as the relative direction of the underlap (in-unison or in-opposition arrangement) or the type of loops (open, closed).

7.2.2 Pattern Elements

Patterns in warp-knitted fabrics are composed of individual pattern elements, the most important of which is the loop. Depending on whether the legs of the loop (i.e. the incoming and outgoing yarn) are crossed or not, it is called a closed or open loop (Fig. 7.8). Head and legs of the loop can be subsumed under the term needle loop, while the connection of two loops is referred to as a sinker loop. Loops arranged side by side form a course, and loops arranged one above the other constitute a wale.

Other pattern elements are tuck stitches, miss stitches, weft inlays (across the knitting direction), and miss-lap (in knitting direction). Wefts can be inlaid across a few needle gauges (part-width weft) or across the entire width of the textile (full-width weft). Except for the binding element of tuck stitches, which requires special

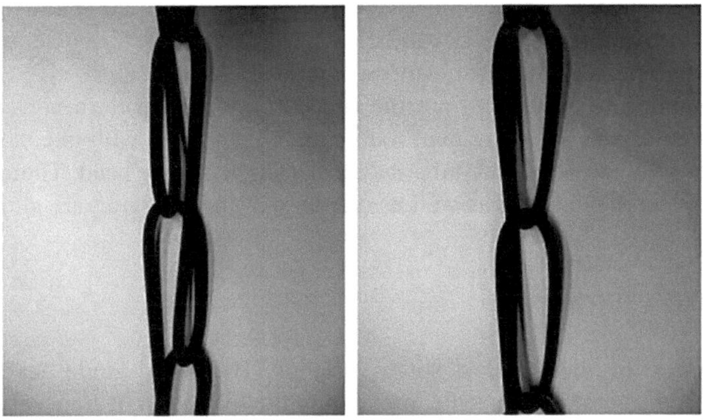

(a) (b)

Fig. 7.8 Loops: (**a**) open, (**b**) closed

additional devices, all pattern elements are formed by the motion of the yarn guide in relation to the needles.

7.2.3 Patterns for the Production of Technical Textiles

7.2.3.1 Classification

Basically, the following parameters can be varied to develop a warp-knitted pattern:

- Lapping (the sequence of overlap and underlap), in the following variations:
  - Overlap followed by underlap in the opposite direction (closed loop)
  - Overlap followed by underlap in the same direction (open loop),
  - Only overlaps (open loops)
  - Only underlaps (as part-width weft or transverse weft)
  - Neither overlaps nor underlaps (miss-lap)
- Shogging distance (number of needle lanes by which the yarn guide is laterally offset), changeable in full incremental steps beginning at 0
- Change of direction (the number of rows after which the shogging direction is changed), starting at 0, changeable in full incremental steps,
- Needle lapping (lapping over one or two needle at the same time), a lapping over two needles at once (twill) is rare,
- Threading

These parameters can be combined across all categories. Furthermore, the characteristics within a category can be changed in every new course, resulting in an enormous versatility of combinations and resulting patterns.

The so-called basic stitches are the foundation of the pattern construction of warp-knitted fabrics. They are pure loop patterns produced with one yarn system. The yarn is fully threaded and only placed in a single needle head. Therefore, only the lapping and the distance as well as direction of the shogging are altered.

7.2.3.2 Basic Stitch: Pillar

Pillar stitches are the simplest stitch variation. Here, one guide needle always serves the same compound needle, preventing the formation of transverse connection between the individual wales and thus necessitating auxiliary yarns in transverse direction to form fabrics. The pillar stitch is therefore used with other pattern elements coming from a second guide bar or to connect yarn layers. If only overlaps are used for lapping, a closed pillar stitch is created. If an overlap is followed by an underlap in opposite direction, the pillar stitch is closed (see Table 7.1).

7.2.3.3 Basic Stiches: Tricot, Cord, Satin, and Velvet

In the second group of basic stitches, the shogging direction changes regularly in each row, with the length of the underlaps ranging from one to four needle lanes. Greater shogging widths of the guide bars are possible, but unusual. Neighboring loops are connected with each other and fabrics are formed. Depending on the length of the underlap, the stitches are referred to as *tricot* (by one), *cord* (by two), *satin* (by three), and *velvet* (by four). Here, too, the loops can either be opened or closed. The corresponding representations are given in Table 7.2.

7.2.3.4 Combined Stitches

Combined stitches are created by the use of two or more guide bars. A different stitch can be created on each guide bar. When producing warp-knitted semi-finished products for lightweight construction, combined stitches with two yarn systems on two guide bars are common. In other areas of application, warp-knitting machines with several dozen guide bars are in use. The guide bars can be shogged in the same direction (in unison) or in different direction to one another (in opposition). Here, only the combination of several basic stitches with one another and with the weft binding element will be considered. Table 7.3 contains examples for the possible combinations of basic stitches.

In order to include transverse wefts and miss-laps in a fabric, overlaps to a second yarn system are necessary. A transverse weft is a stretched yarn section laid across the knitting direction of the machine, while a miss-lap (also referred to as a

Table 7.1 Pillar stitch formation

Designation and chain notation	Lapping diagram, open	Lapping diagram, closed
RL pillar stitch, open GB1 1–0/0–1// RL pillar stitch, closed GB1 0–1//	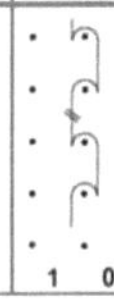	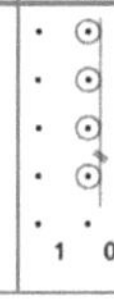

Table 7.2 Basic bindings with regular change of direction of the guide bar

Designation and chain notation	Lapping diagram, open	Lapping diagram, closed
RL tricot, open GB1 0–1/2–1// RL tricot, closed GB1 1–0/1–2//	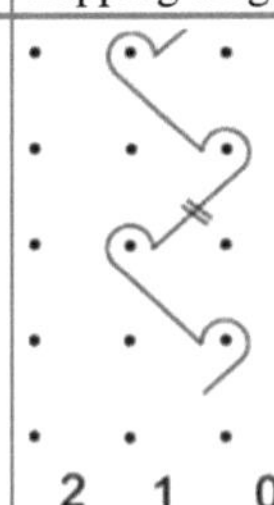	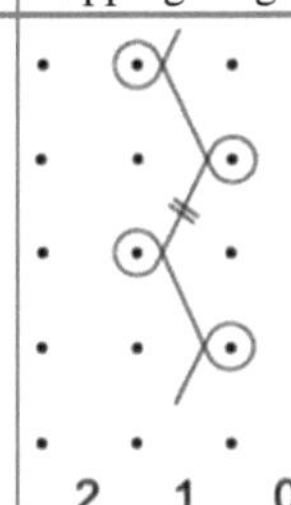
RL cord, open GB1 0–1/3–2// RL cord, closed GB1 1–0/2–3//	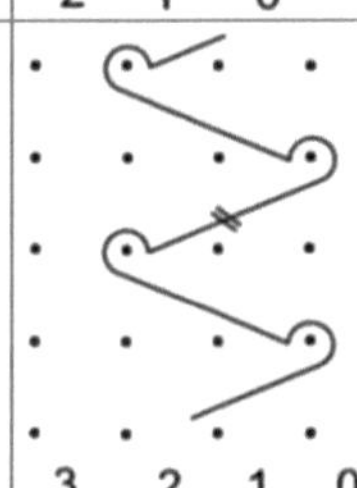	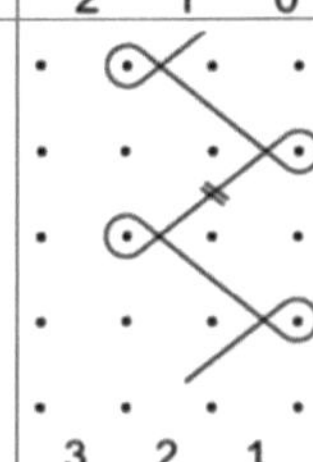
RL satin, open GB1 0–1/4–3// RL satin, closed GB1 1–0/3–4//	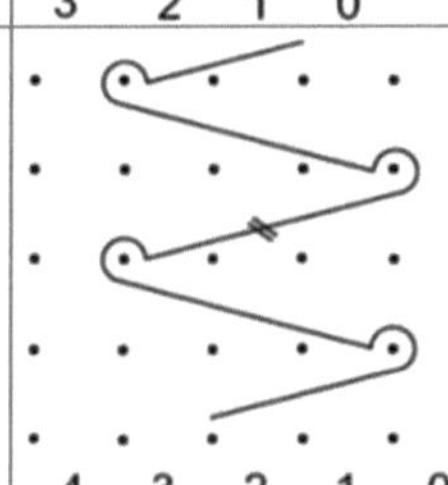	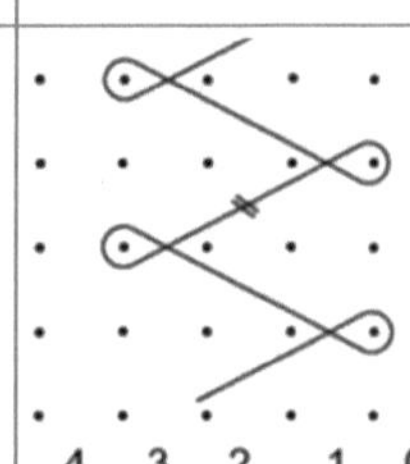
RL velvet, open GB1 0–1/5–4// RL velvet, closed GB1 1–0/4–5//	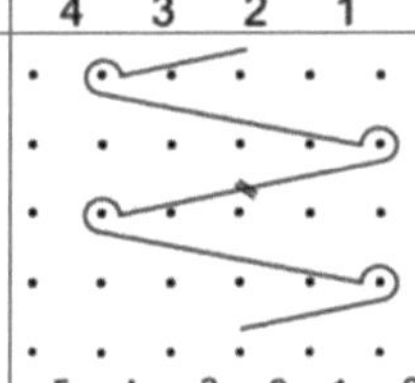	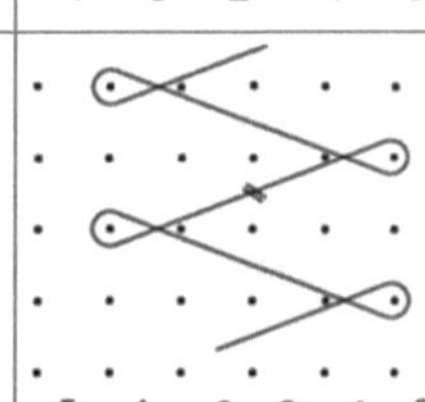

longitudinal weft) runs in knitting direction. The loop-forming basic stitch is always created by guide bar 1, the wefts are laid using guide bar 2. The manner of weft insertion can be varied regarding the shogging width and alternation in the repeat

Table 7.3 Examples of basic stitch combinations

Designation and chain notation	Lapping diagram
Double tricot (RL), closed, in opposition GB1 1–2/1–0// GB2 1–0/1–2//	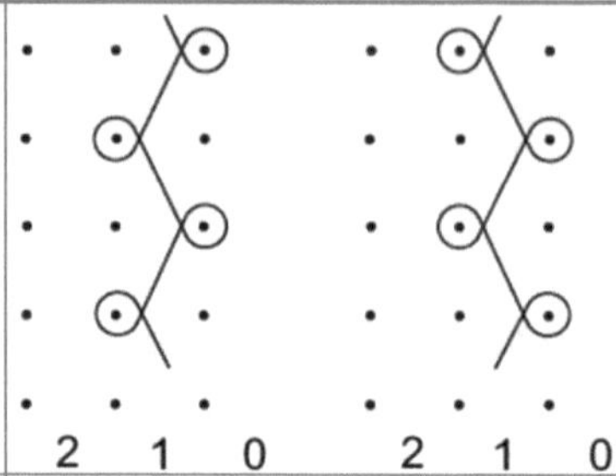
Double tricot (RL), closed, in unison GB1 1–0/1–2// GB2 1–0/1–2//	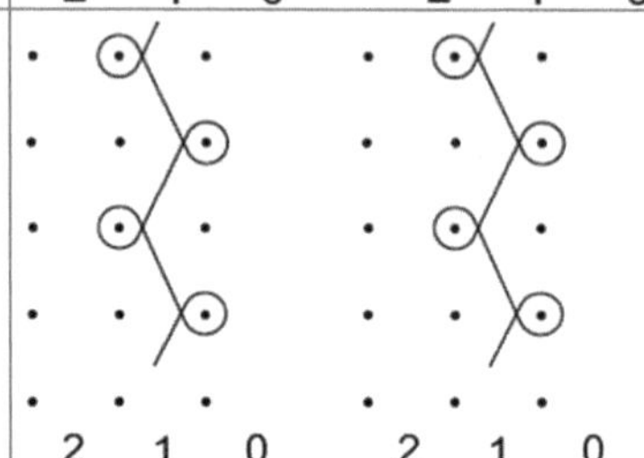
Cord tricot (RL), closed, in opposition (Charmeuse) GB1 1–0/2–3// GB2 1–2/1–0//	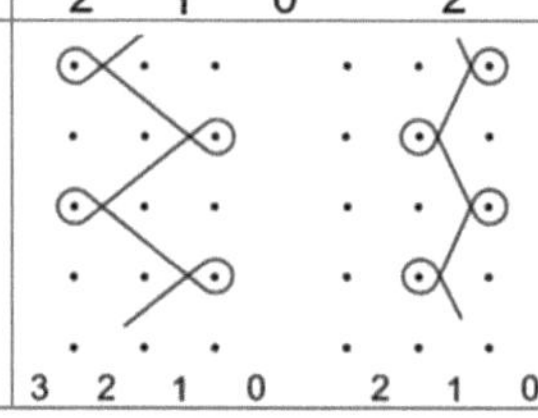

unit. Depending on the number of needle gauges by which the guide needle is offset during weft insertion, the stitches are referred to as weft under 3, weft under 4, and so on. The weft stitching element is often used in combination with the RL pillar stitch and RL tricot stitch (Table 7.4) and provides the textile semi-finished product with improved crosswise stability, due to the higher number of underlaps.

7.3 Manufacturing Basics of Warp-Knitted Fabrics

7.3.1 Introduction

Warp-knitting machines is the umbrella term for textile machines on which the production of textile fabrics is performed according to the principle of loops forming from one or more groups of yarns running in the knitting direction of the machine. The functional structure of warp-knitting machines in general comprises: storage of the base materials (usually yarns), feeding as well as transport of the base materials and their bonding into a fabric, if necessary the separation of the fabric

Table 7.4 Shogging widths and changes at weft inlays (examples)

Designation	Lapping diagram
Part-width weft under 3	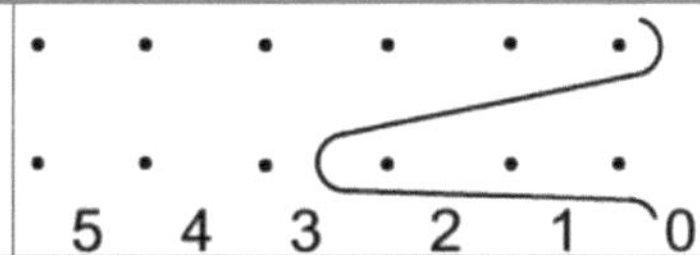
Part-width weft under 5	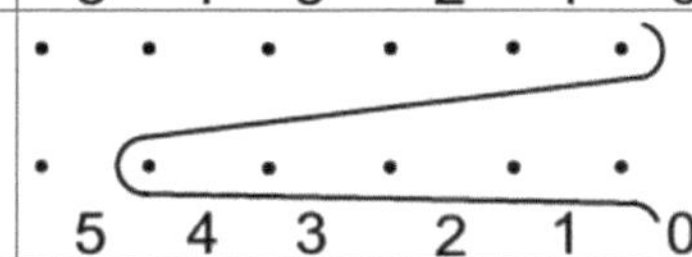
Part-width weft under 5, change after 3 courses	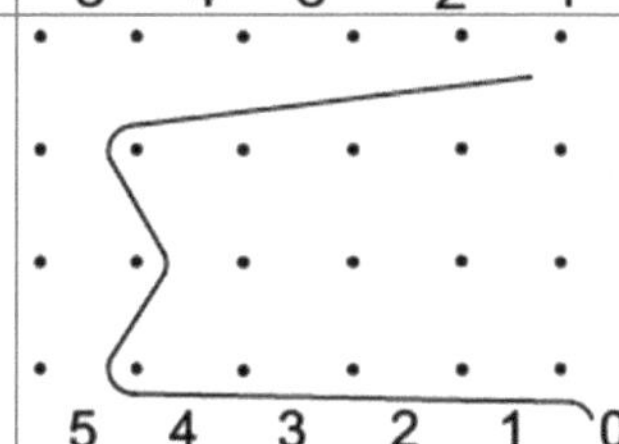
RL pillar stitch, open, weft under 3 (weft change after each course) GB1 1–0/0–1// GB2 0–0/3–3//	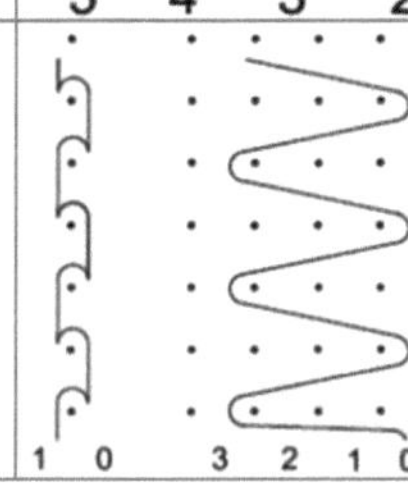

from the machine frame, and finally the take-up, winding as well as storage of the fabric. These functions will be explained in the following.

Warp-knitted fabrics fundamentally consist of at least one knitting yarn system, commonly with a high number of individual yarns as its basic structure. For the application in lightweight construction, the knitting yarn primarily serves the purpose of joining yarn layers, chopped glass fiber mats or planar materials such as nonwovens or films. Knitting yarn is the term for yarns which are treated by the needles of the warp-knitting machine, i.e. formed into loops. The warp-knitted fabrics can also contain a so-called base layer of stretched and straight yarn groups, referred to as either warp yarns (if arranged in production direction) or weft yarns (if across or oblique to the production direction) (Fig. 7.9).

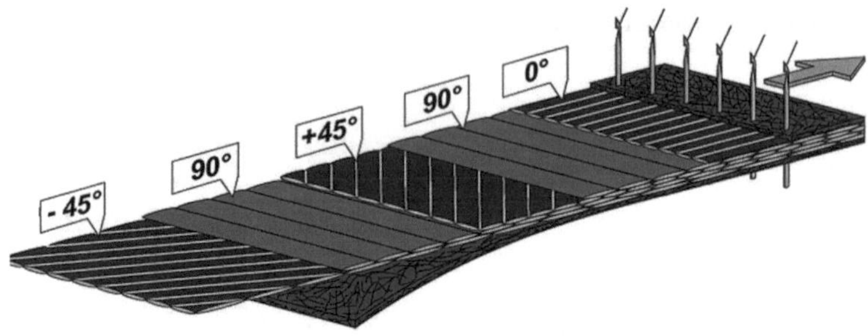

Fig. 7.9 Schematic representation of warp-knitted semi-finished products made from yarn layers and a loop system (*Source*: LIBA Maschinenfabrik GmbH)

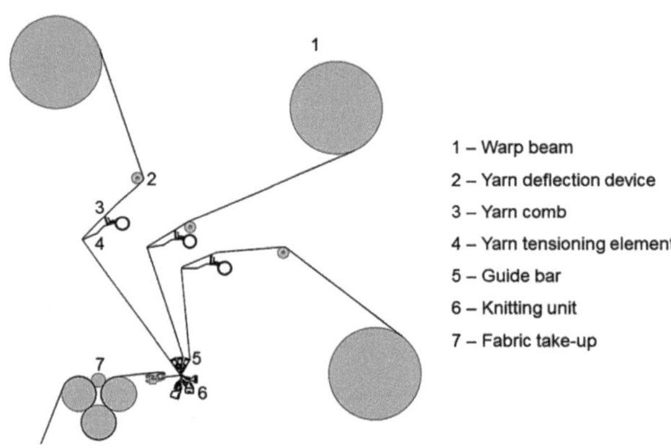

Fig. 7.10 Feeding of the yarns to the knitting unit

7.3.2 Storage, Feeding, and Transport of Base Materials

7.3.2.1 Knitting Yarn

By means of yarn feeding units, the knitting yarns for the formation of loops are transported from warp beams (1) to the knitting unit (6) via yarn guiding devices (2), the yarn comb (3), and yarn tensioning elements (4), to the respective guide bar (5). They are finally removed by the take-up unit (7). All of this is shown in Fig. 7.10.

The task of the yarn feeding units is the adjusted delivery of the amount of yarn required for the loop formation process. The yarn feeding units can be classified into active and passive systems. In the passive system, the needles draw the necessary length of yarn from packages (usually warp beams). The movement of the packages is commonly triggered by the increase of yarn tension until they surmount retarding and inertial forces. Rope brakes and v-belt brakes support the controlling of yarn tension. However, due to fluctuations in yarn tension during the process and the inertia of the warp beams, these systems are only rarely used.

In the active systems, the required amount of yarn is fed to the knitting unit by means of an active drive. The warp beams are moved by their own drives. The continuous yarn feeding is adjusted to the respective required yarn length for the loop formation process by control means. The realization of the set value is performed either mechanically (e.g. by a progressive transmission gear) or electronically (e.g. by means of a servo motor). According to the manner of feeding, electronically controlled yarn feeding systems are classified into three groups [1, 2]:

- Constant feeding (Singlespeed type),
- 2-step feeding (Simplified multispeed type), and
- Sequential feeding (Multispeed type)

In the *singlespeed* type, a set value is constantly compared to the actual value and correspondingly re-adjusted. The control process of the *simplified multispeed* type is analogous to that of the singlespeed type, with the exception of two different available yarn feeding values, which are controlled by means of a two-step drive. In the *multispeed* type, any desired amount of yarn to be fed can be set for each loop row by means of a servo motor.

Currently, the knitting yarn is mostly fed to the knitting unit of the warp-knitting machines at a constant feeding speed (singlespeed type).

The yarn guiding device (Fig. 7.11) deflects and guides the knitting yarns from the warp beam to the knitting unit. To keep process yarn tensions small, the number and arrangement of the yarn guiding elements is adapted to the respective machine type, and all surfaces are modified to reduce friction. The yarn comb used on warp-knitting machines separates the warp yarns and is placed either between the yarn tensioning element and the guide bar or above the yarn tensioning element.

Setting an optimal knitting yarn tension is a crucial factor of ensuring quality and smooth operation. For a correct execution of the warp-knitting process, all knitting yarns have to be fed from the warp yarn to the knitting unit at a consistent yarn

Fig. 7.11 Yarn guiding device (*top*) and yarn comb (*bottom*)

tension. Excessive yarn tensions cause yarn breaks, while insufficient tensions lead to snarling and faulty lapping. Since the amount of yarn fed from the feeding device at constant speed is processed discontinuously during loop formation, the fluctuating lengths of yarn between warp beam and knitting unit have to be compensated by a yarn tensioning element. This spring-based yarn tensioning element temporarily stores the excess yarn lengths. The following factors decisively influence the effectiveness of the yarn tensioning element:

- Spring (trapezoid or spiral spring)
  - Spring stiffness,
  - Number of springs,
  - Characteristic frequency,
  - Angular position
- Yarn feeding
- Strain behavior of the yarn
- Fabric take-up
- Movement of the knitting unit (see "Joining of the base materials")
  - Needle movement,
  - Guide needle swing movement,
  - Guide needle shogging movement,
  - Sinker movement

The yarn tensioning elements usually used on warp-knitting machines consist of a hollow-profile tensioning bar (tensioning shaft) fixed to trapezoid or spiral springs (Fig. 7.12) of defined number, thickness, and length. The absorption and transfer of forces depends on the material properties (spring stiffness, mass) and the geometrical shape of the spring. Trapezoid and spiral springs are able to store mechanical work as potential energy and then release it. The selection of the spring type as well as the number and arrangement of springs depend on the textile-physical characteristics of the yarn material to be processed.

7.3.2.2 Warp and Weft Yarns

Warp and weft yarns are usually stored on bobbins in bobbin creels. These creels are equipped with yarn brakes to ensure uniform yarn tension during unwinding and they often also feature yarn monitors to automatically shut down the machine in case of yarn breaks. The yarns can be unwound from the bobbins either over the circumference (external; tangential take-up) or over-head (external or internal unwinding). Over-head unwinding does not require additional installations but inserts rotations into the yarns. These rotations can differ along the run length of the yarn and can persist even in the finished textile structure. For an unwinding over the circumference of cylindrical yarn bobbins unwinding devices are necessary, which in turn require a larger set-up area. In quality-sensitive applications, such as the processing of carbon yarns, rotations in the yarn can thus be avoided.

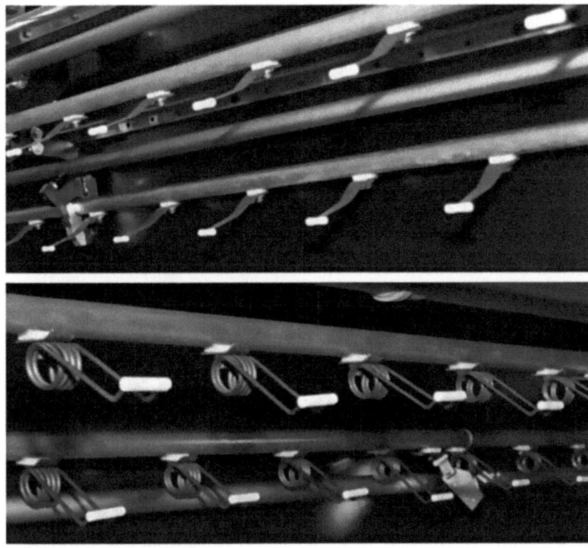

Fig. 7.12 Yarn tensioning elements (*top*: trapezoid springs, *bottom*: spiral springs; *Source*: LIBA Maschinenfabrik GmbH)

Unwinding over the circumference is also a prerequisite for the use of so-called spreading devices used to spread the yarns during feeding. This way, yarns with a high number of filaments can be processed into very thin and light fiber layers.

Creel placement always has to be adapted to the circumstances in the production plant. In general, long feeding distances with frequent deflections are to be avoided, as the involved friction can damage the often quite sensitive yarn materials. Therefore, weft yarn creels are usually placed next to the machine, directly at the weft yarn laying device. When processing warp yarns, the creels can be set up behind, in front of or next to the machine. A set-up of the creels on a pedestal or in an elevated level above the machine is sensible, especially for larger machines with more than one weft yarn laying device. The warp yarns are directly fed to the knitting unit via suitable guide elements, or they can be guided over special feeding devices, i.e. rollers arranged above the knitting unit and driven by electric motors. This manner of feeding homogenizes yarn tensions across the entire group of warp yarns, resulting in an improved quality of the textile fabric.

The use of the so-called magazine weft insertion is a typical feature of weft yarn feeding in the manufacture of warp-knitted semi-finished products for lightweight construction purposes. Here, the weft yarns are not directly fed to the knitting unit after unwinding from the bobbin creel, but first placed in a transport system and delivered from there to the knitting unit (Fig. 7.13). The transport systems fundamentally consist of two holding devices running around both sides of the machine frame and fixing the weft yarn in position under tension. These holding devices can be designed as hooks, pegs, needles or clamps (Fig. 7.14). The weft laying device usually places the weft yarns continuously in the transport system. This means that the transporting device moves steadily from one side of the transport system to the other, each time placing several weft yarns in the holding devices without them

Fig. 7.13 Magazine weft insertion of the Karl Mayer Malitronic® Multiaxial stitch-bonding machine

Fig. 7.14 Example of a holding device in the weft transport system of the Karl Mayer Malitronic® Multiaxial stitch-bonding machine

being cut. For the processing of spread carbon filament yarns, solutions are available in which the weft yarns are placed individually, exactly measured, cut, and placed in the holding system by means of an insertion device.

In magazine weft insertion the weft yarns can be laid either parallel to one another (parallel weft) or slightly crossed and overlapped (cross weft). A parallel insertion is the result of the weft laying device moving from one machine side to the other while compensating the advance motion of the transport system. Without the compensation, a cross weft is created. Another differentiation can be made with regard to the pricking of the yarns into the base layer. If the yarns are positioned between the knitting elements in such a fashion that prevents pricking during loop formation, the yarn insertion is in line with the loops and vice-versa. Weft insertion with transporting devices achieves only a very slow yarn unwinding per unit of time in the reversal point of the transporting device. During the movement between both sides of the transport system, the unwinding is sped up considerably. This results in an uneven yarn unwinding with clear yarn tension peaks. The use of compensation devices (Fig. 7.15), which build up yarn reserves during the reversal, can homogenize the weft yarn unwinding.

Fig. 7.15 Compensation device for weft yarns (picture shows a yarn tensioning device of the Karl Mayer Malitronic® Multiaxial stitch-bonding machine)

The weft insertion can be performed perpendicular or diagonal to the knitting direction of the machine. By arranging several weft insertion systems consecutively, the production of fabrics with several yarn layers of different orientations can be produced. The weft insertion systems can be fixed in their position and orientation, or they can be adjusted by manual or electronic controls.

7.3.2.3 Additional Fabrics

When using suitable knitting elements (see Sect. 7.3.3), warp-knitting machines can be used to connect nonwovens, films or woven fabrics to the warp and weft yarn layers. These fabrics are fed off fabric rolls, with the feeding devices arranged in a way that allows fixation both above and beneath the warp and weft yarn layers. Chopped glass fiber mats can also be produced on warp-knitting machines. These mats consist of chopped glass fiber filaments, and the chopping can be performed by chopping rollers installed in the machine. If the production of the chopped fiber mat constitutes the first step of production, the chopped fiber pieces are laid out as a homogenous mat on a transport belt stretched in the frame of the transport system. They can also be spread onto the unsolidified weft yarn layer if the chopping rollers are positioned directly before the knitting unit. This way, the glass mat is integrated on either the top or bottom side of the fabric. The transport belt also supports the weft yarn layers, particularly with greater working widths. For the latter purpose, parallel straps are an alternative.

7.3.3 *Joining of the Base Materials: Loop Formation Process*

7.3.3.1 Overview of the Knitting Elements

The joining of the base materials in the warp-knitting process, i.e. the connection of individual yarn layers to each other or to other fabrics, is performed during the loop formation process. The loop formation process consists of coordinated movements of the individual knitting elements. The influence of the knitting yarn with its mechanical characteristics, friction behavior in relation to the knitting unit, and other machine parts also plays a significant role. During the loop formation process, the knitting yarns are transformed into loops by the interaction of all knitting elements. The type of knitting elements depends on the respective machine type, and the following knitting elements are commonly used in warp-knitting machines (Fig. 7.16):

- Needles (arranged on the needle bar),
- Sliding tongues (arranged on the tongue bar),
- Guide needles (arranged on the guide bar),
- Sinkers depending on the machine type
  - combined holding-down and knocking-over sinkers (compound sinkers) (tricot machine),
  - trick plate and holding-down sinker (raschel machine),
  - knocking-over sinkers and supporting element (multiaxial warp-knitting machine),

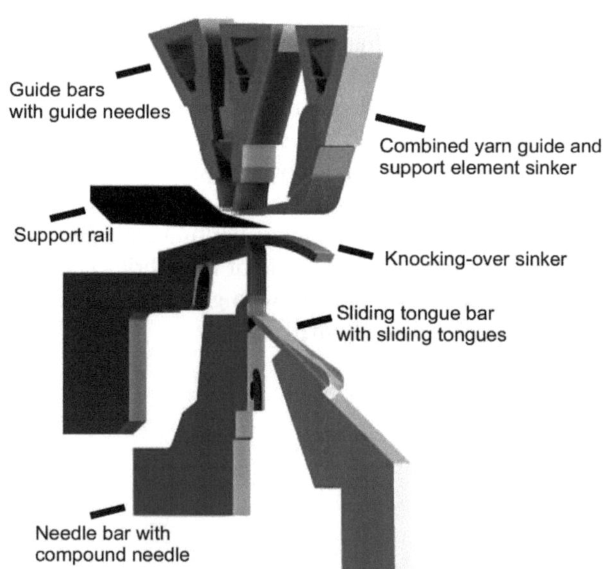

Fig. 7.16 Knitting elements exemplified by a multiaxial warp-knitting machine

- Filler yarn bar for the warp yarns (partly combined with retainers),
- Weft extension assists

7.3.3.2 Knitting Element: Guide Bar

Guide bars are rails made from metal or fiber-reinforced plastic, on which yarn guide elements are fixed. On warp-knitting machines, these are referred to as *guide needles*. They move on spatially elliptical paths consisting of swinging and shogging motions. The separate movement stages are called underlap, swing-in, overlap, and swing-out (see Sect. 7.2). The exact process of the swinging movements depends on the machine type and the number of guide bars as well as the stitch and the machine gauge. The swinging movement of the guide bars is realized by gears connected to the main shaft, depending on the rotational angle of the machine. The following systems are common for driving the guide bar shogging movement:

- Pattern disks,
- Pattern chains,
- Linear motors, or
- Servo motors with additional gears

A *pattern disk* is a type of cam read off by a slider. The advantage of the pattern disk is in the high speeds which can be achieved, and in the precision of operation, while the low flexibility is rather disadvantageous. For each binding and machine fineness, a different pattern disk has to be used, and the repeat is limited to the circumference of the pattern disk. In this regard, pattern chains are much more flexible. The individual chains can be assembled according to the pattern and the repeat. This allows pattern designs with very long repeats. The low speed of pattern chains brought about by clearance fits of the individual chain links and the decrease of precision at increasing speeds is a clear disadvantage. Linear and servo motors with additional drives offer greater flexibility and efficiency (wide variety of bindings and infinite repeat lengths) can be used to replace chain links. However, they are only being used at low machine speeds due to their restricted dynamics. In addition, they require higher investment costs.

7.3.3.3 Knitting Elements: Needles and Sinkers

The task of the needles arranged on the needle bar is to form and shape the loops. Depending on the type of warp-knitting machine and the intended use, compound needles or piercing compound needles (Fig. 7.17) as well as latch needles (see Sect. 6.2.2) are common choices in the field of warp-knitting. The sliding tongue used to close the needle hook, or depending on the needle design, the tongue or tip of the needle safely carry the yarn material until the half loop is knocked over. The compound needle-sliding tongue system is most common on warp-knitting machines with a lightweight construction product range. This solution allows

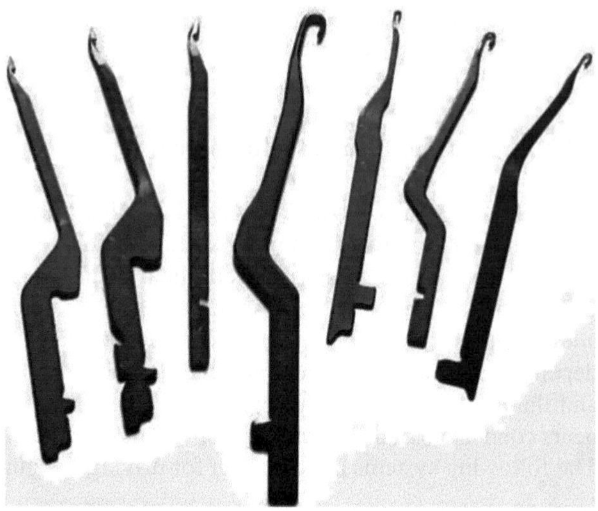

Fig. 7.17 (Piercing) compound needle types (*Source*: LIBA Maschinenfabrik GmbH)

high speeds, smooth movement sequences, and yarn-protecting processing. A piercing compound needle design with a sharp tip can penetrate as well as connect several yarn layers and fabrics.

According to the machine type (see Sect. 7.4), different sinkers can be used as knitting elements on warp-knitting machines. They support the formation of loops by needles and yarn guides during the warp-knitting process, retain those loops already formed and prevent the displacement of yarn layers by the movement of the needle during warp-knitting processes. The loops and the manner of their arrangement in the warp-knitted fabric, are created by the interaction of the mentioned knitting elements on the basis of pre-set paths through the drive of the mainshaft. At each revolution of the mainshaft, a new loop is formed. The process of loop formation will here be explained using the example of an RL tricot machine with compound needle/sliding tongue system (Fig. 7.18) [1]:

- Position 1 (0°—Knocking-over position): The compound needles and the sliding tongues are stationary in the bottom knocking-over position. The guide bars perform the underlap, and the sinker swings forward into the holding-down position.
- Position 2 (60°–120°—Holding-down of the fabric): The compound needles move upward, the fabric is held in place by the sinker, while the sliding tongues remain in the bottom position. The guide bars finish the underlap.
- Position 3 (120°–195°—Guide bar swinging): The compound needles are now in top position. The sliding tongues start to rise in the groove of the compound needles. The guide bars have finished the underlap and commence the swing movement.
- Position 4 (195°–255°—Overlap of the knitting yarns): The compound needles and the sliding tongues remain in the top position. The sliding tongues emerge

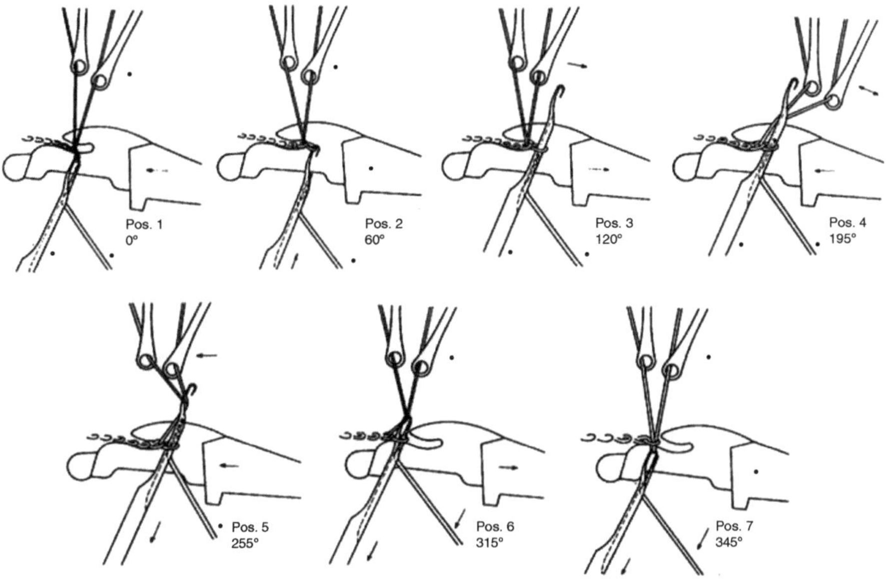

Fig. 7.18 Loop formation process, exemplified by an RL tricot machine (*Source*: LIBA Maschinenfabrik GmbH)

from the sliding grooves. The guide bars swing to the rear reversal position, and the overlaps are completed.

- Position 5 (255°–315°—Catching the knitting yarns): The compound needles and the sliding tongues are still in top position. The guide bars have swung forward and have placed the knitting yarns in the needle hook.
- Position 6 (315°–345°—Closing the needles): The compound needles have moved halfway down, the sliding tongues have moved from the sliding groove of the compound needles, beginning to close the needles. The half loops slide off the needle shaft, across the needle breast and onto the sliding tongue, while the sinker bar swings into the back position.
- Position 7 (345°–360°—Knocking-over of the half-loop): The compound needles and the sliding tongues dip into the sinker bar together. The half-loops are knocked over by the closed needles heads via the sinkers. The half loops are turned into loops, and the inlaid yarns are formed into half-loops. The guide bars commence the underlaps.

7.3.4 Separation, Take-up, and Winding of the Fabric Web

In the production of warp-knitted semi-finished products manufactured with magazine weft insertion (see Sect. 7.3.2.2), it is necessary to separate them from the transport system after the knitting unit. This is done by cutting the selvedge in the holding system either immediately after the knitting unit or before it reaches the fabric take-up.

The warp-knitted fabric is transported out of the knitting unit by the take-up unit, which can also provide the yarn tension required for processing warp and weft yarns as well as the fabric. Forces are introduced by the friction between the fabric and a group of two or three rollers. Looping around the rollers ensures a secure and non-slip fabric take-up. Structure and coating of the roller surfaces can be varied, and adjusted to the desired product. The take-up rollers can be driven by means of:

- the mainshaft (the desired take-up speed is set mechanically by the change of gear wheels), or
- a separate servo motor [the desired take-up speed is set electronically by computer input (EAC system)].

In addition to selvedge cutting, a longitudinal separation of the warp-knitted fabric can be performed, which allows the simultaneous manufacture of several parallel fabric webs. In a final process step, the web advanced by the take-up is wound onto a tube or a roller by means of a winding unit.

7.3.5 Process-Integrated Production

To economize later process steps and to achieve improved textile properties, additional functions can be integrated in the warp-knitting process. These usually serve the purpose of textile finishing, such as coating, (partial) impregnation, drying, or heat treatments. Fundamentally, a large number of diverse additional functions can be combined with warp-knitting machines, if they are arranged between the warp-knitting machine and a spatially separated winding unit (Fig. 7.19). Here, the fabric is guided through the additional machine module.

If there are particularly high requirements concerning the accuracy of the fabric geometry of warp-knitted fabrics with weft yarns, it is recommended to arrange the finishing steps before the separation of the fabric web from the weft yarn transport system. This way, a fixation of the weft and warp yarn layer geometry in the required shape can be ensured by coating or heat treatment. Only afterwards the fabric web is separated from the transport system and wound up. This requires an extension of the machine frame between knitting and fabric take-up unit. A modular machine structure provides the user with great freedom of design in this context.

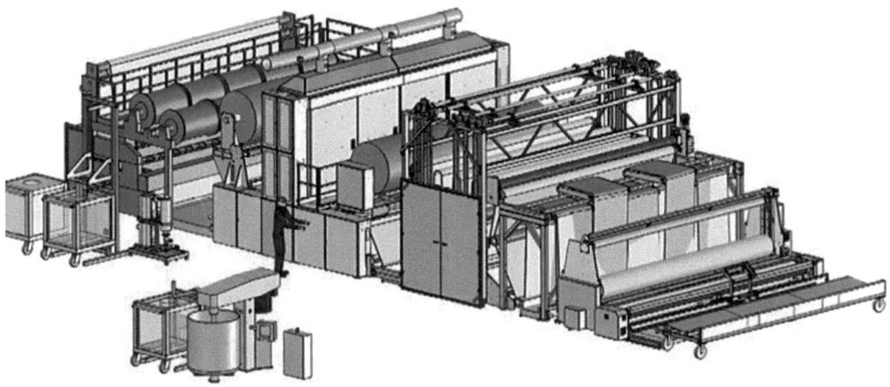

Fig. 7.19 Example of the integration of coating and drying in the warp-knitting process, exemplified by the manufacture of technical textiles for road construction and geotextiles (*Source*: LIBA Maschinenfabrik GmbH)

7.4 Machine Technology for Two-Dimensional Warp-Knitted Semi-finished Products

7.4.1 Conventional RL Warp-Knitting Machines for the Manufacture of Warp-Knitted Semi-finished Products

Tricot machines and *raschel machines* are the common types of RL warp-knitting machines used for the manufacture of technical textiles. In contrast to the multiaxial warp-knitting machines described below, these machines can produce fabrics consisting purely of loops, although they also offer the possibility of integrated warp and weft yarn layers. Raschel machines also offer the opportunity to connect other types of fabric such as nonwovens with yarn layers. The essential differences are:

- the knitting elements
  - in tricot machines: with combined holding-down and knocking-over sinker
  - in raschel machines: with trace comb bar and trick plate
- the number of guide bars
  - in tricot machines: up to five
  - in raschel machines: up to 64
- the angle between warp yarn and fabric
  - in tricot machines: ca. 45°–90°
  - in raschel machines: ca. 170°

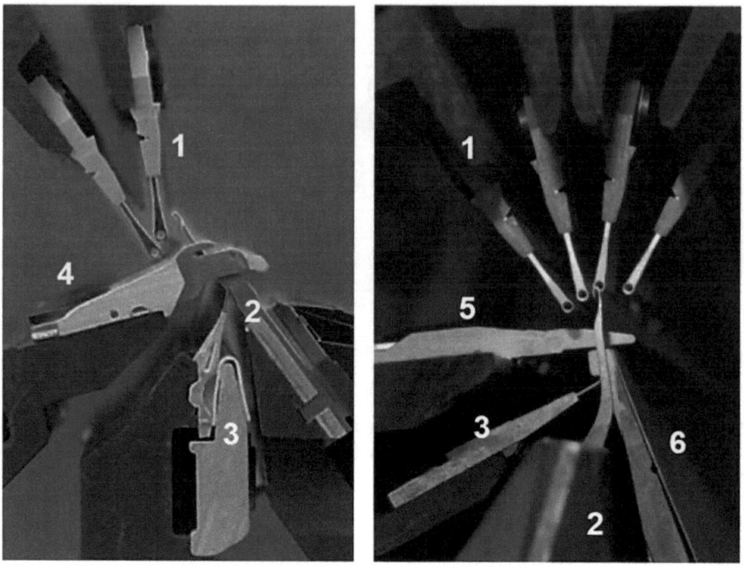

Fig. 7.20 Knitting elements of a tricot machine (*left*) and at raschel machine (*right*) (*Source*: LIBA Maschinenfabrik GmbH). 1: Guide bar with guide needles, 2: Compound needle bar, 3: Tongue bar, 4: Combined holding-down and knocking-over sinker bar, 5: Stitch comb bar, 6: Trick plate

Figure 7.20 shows the knitting units of a tricot machine and a raschel machine. In tricot machines, the combined holding-down and knocking-over sinker supports the brushing and knocking-over (see Sect. 7.3.3). The dimensions of the combined holding-down and knocking-over sinker in relation to the needle stroke define the knocking-over depth and essentially the loop size. On a raschel machine, trace comb and trick plate assume the tasks of retaining the previously formed half-loops and defining the loop size. Apart from the standard knitting elements, additional devices, such as magazine weft insertion systems, Jacquard systems, pile sinkers, and knock-over plates can be assembled according to the desired product. The design types resulting from additional modules can be summarized as follows [2]:

Tricot machines

- Standard machines with up to five guide bars
- Tricot machines with weft insertion
- Tricot machines for pile fabrics
- Tricot machines with cut-press technology
- Standard machines with two knitting units

Raschel machines

- Standard raschel machines
- Raschel machines with Jacquard unit
- Raschel machines with multibar and Jacquard unit

- Double-bar raschel machines
- Raschel machines with weft insertion

For technological and construction reasons tricot machines are characterized by a higher machine speed in comparison to raschel machines, due to the smaller number of guide bars and the shorter paths run by the loop formation elements. The pattern design possibilities are limited by the smaller needle stroke and the small number of guide bars. In these aspects, raschel machines offer a much wider range of design possibilities. On both machine types, an additional weft insertion in longitudinal or crosswise direction allows the realization of monaxially or biaxially reinforced warp-knitted fabrics similar to those of multiaxial warp-knitting machines. This way, textile properties such as strength, elongation, or flexural stiffness can be influenced specifically. For an efficient insertion of a high number of weft yarns into narrow fabrics, the *crochet* technology was developed. In contrast to conventional warp-knitting machines, the movements of the weft guide bars (which are pivoted by 90°) on crochet machines are separated from the movements of the loop-forming guide bars (of which usually just one or two are installed). During the removal of the knitting needle, the needle passes between the yarn guide elements of the weft laying device. Immediately afterwards, the weft guide bars swivel outwards and begin the shogging movement, which has to be finished before the renewed removal of the needle. Before the knitting needle removal, the weft laying devices are pivoted back into the area of the knitting needle, and the obliquely placed weft yarn is arranged underneath the needle, where it is integrated into the loops during the following loop formation process. The wefts are laid under electronic controls across partial widths or across the entire fabric width. In contrast to conventional warp-knitting, narrower fabric widths of up to 600 mm are common. The crucial advantage of crochet machines is their ability to integrate even very coarse yarns, such as glass and carbon rovings as cross, diagonal or longitudinal weft, in-line with the loops (i.e. without pricking of the yarns). Crochet machines are therefore suitable primarily for the manufacture of special-purpose, narrow preform structures akin to non-crimp fabrics.

7.4.2 Multiaxial Warp-Knitting Machines for the Manufacture of Warp-Knitted Reinforcement Structures

Multiaxial warp-knitting machines (alternatively referred to as stitch-bonding machines) are the typical platform for the manufacture of multiaxial structure for lightweight construction, as the stretched yarns running in various directions result in a high stiffness of the textile fabric and the final composite component. Multiaxial warp-knitting machines usually have one needle bar (right-left) and two guide bars to connect the fed yarn groups and/or fabrics with knitting yarns by piercing them with the needles and joining them by means of loops. This

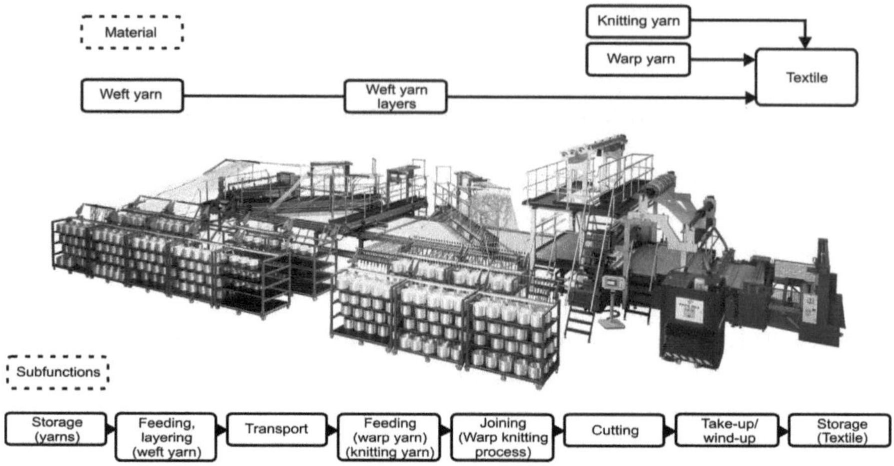

Fig. 7.21 Functional structure of a multiaxial warp-knitting machine (*Source*: Karl Mayer MALIMO Textilmaschinenfabrik GmbH)

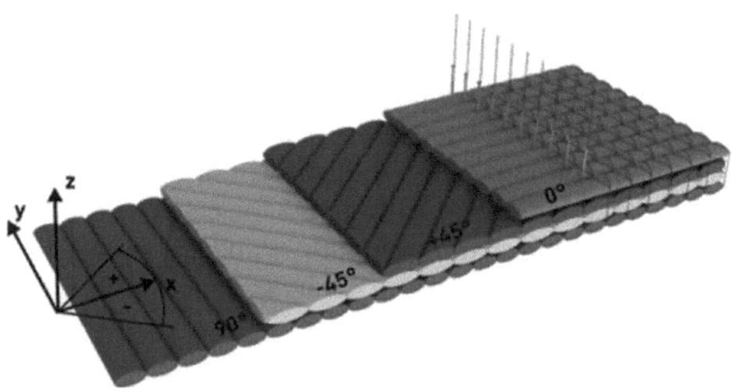

Fig. 7.22 Yarn layer orientation in stitch-bonded fabrics

principle of operation increases the productivity of the fabric formation process. The functional structure of a multiaxial warp-knitting machine is represented in Fig. 7.21. One typical feature of the structure of a multiaxial warp-knitting machine is the weft transport system, feeding the yarn layers to the knitting unit. The machines are equipped with several weft laying devices (see Sect. 7.3.2.2), of which at least one, but usually two can insert diagonal wefts. This means that the yarn layers can be placed on top of each other as required and then be joined by the knitting unit (Fig. 7.22). Machines with only one fixed weft laying device for weft insertion across the working direction are often called biaxial warp-knitting machines as yarns in the fabric run in only two directions. With these technical circumstances, the prerequisites are met to manufacture tailor-made textile semi-finished products with suitable fiber orientation for composite components exposed to multiaxial load cases with normal, shear, flexural, and torsional stresses. After the weft yarns are inserted in the transport chain, the warp yarns are introduced as the uppermost layer, and the layer stack is then connected by loops in the knitting unit. The connection of yarn layers through the loops allows versatile possibilities for specific designs of the textile semi-finished products with regard to their adaption to the final component form and, when using suitable knitting yarns, can increase their resistance to impact loads and the corresponding danger of delamination. The working width of the machines is commonly between 1.25 and 2.5 m, with most commercially available machines not offering a change of the working width. Machines for greater working widths are also available.

The fabrics produced on multiaxial warp-knitting machines are referred to as *stitch-bonded fabrics* in accordance with DIN 61211. In practice, a number of other terms are used, such as multiaxial non-crimp fabric or warp-knitted non-crimp fabric—all of them indicate that stitch-bonded fabrics are a sub-group of non-crimp fabrics. The umbrella term non-crimp fabric is used for a fabric fixed by chemical or mechanical (frictional and/or form fit) bonding, and comprising one or more layers of parallel, stretched fibers and/or other fabrics. Each yarn layer can be oriented differently and feature different yarn densities. Therefore, a *stitch-bonded fabric* is a non-crimp fabric bound by warp-knitted loops. The orientation of the yarn layers in the multiaxial non-crimp fabric does not necessarily have to be straight. Using devices for the manipulation of warp and weft yarns the yarn course can be adjusted in a wide range [3, 4].

With regard to the textile product, multiaxial non-crimp fabrics can be classified into yarn layer multiaxial non-crimp fabrics and combined multiaxial non-crimp fabrics, depending on the fabrics to be joined, in accordance with DIN 61211. Yarn layer multiaxial non-crimp fabrics consist exclusively of warp and weft yarn layers. Combined multiaxial non-crimp fabrics can contain yarn layers as well as fabrics. Apart from the straight and stretched yarn course of multiaxial non-crimp fabrics and the nearly freely adjustable angular orientation, this freedom of combination is one of the essential advantages of multiaxial non-crimp fabrics when used in composite materials. Such a combination of characteristics cannot be achieved by any other textile fabric formation process.

According to the respective yarn layer orientation, the designation of multiaxial non-crimp fabrics can be made more precise, namely:

- monaxial non-crimp fabric with one or more yarn layers in one direction (1D structure),
- biaxial non-crimp fabric with two or more yarn layers in two directions but in the same plane (2D structure),
- multiaxial non-crimp fabrics with more than two yarn layers in more than two directions but on the same plane (2D structure), and
- multiaxial non-crimp fabrics with more than two yarn layers in more than two directions, with knitting yarns as reinforcement yarns (3D structure)

The orientation of the individual layers is usually stated in the form of ($\alpha/\beta/\gamma$), with α, β, and γ corresponding to the insertion angle of the yarn layers. The orientations of the individual yarn layers are stated in relation to the production direction. The production direction is defined as X direction or 0° direction, the transverse direction as Y direction or 90° direction, and the stacking sequence as Z direction. Yarn orientations of yarn layers running from the origin of ordinates in the direction of the positive x and y axes are labeled "+", yarn layer orientation for layers running in direction of the positive x axis and the negative y axis are labeled "−" in relation to the 0° direction (Fig. 7.22). A four-layered multiaxial non-crimp fabric with the individual layers oriented at 0°, +45°, −45°, and 90° is therefore designated as (0/+45/−45/90).

The additional use of reinforcement yarns as knitting yarns will reduce interlaminary delamination (i.e. the failure of the composite between the yarn layers), enable damage-tolerant components, and improve the crash and impact behavior of the composites in the final composite component.

The yarn layer orientation is set via the weft yarn laying device, with three different available systems:

- multiaxial warp-knitting machines with fixed weft laying device arrangement [e.g. (+45/90/−45)],
- multiaxial warp-knitting machines with manually adjustable arrangement of the weft-laying device, i.e. the orientation is set before production by shifting the laying device relative to the transport frame; the laying devices are fixed during production, and
- multiaxial warp-knitting machines with electronically controlled weft laying devices, i.e. the insertion angle is controlled via the machine controls, the laying devices are moved in working direction of the machine, and the orientation of the weft yarns results from the adjusted travel (Fig. 7.23)

The realizable number of yarn layers and their orientations corresponds to the number of weft laying devices as well as warp yarn feeding devices on the machine. Common multiaxial warp-knitting machines feature three weft laying devices, but configurations with five to seven devices plus the option to feed one warp yarn layer are offered by some manufacturers. Thus, the strength characteristics of the textile semi-finished products can be adjusted specifically from anisotropic to orthotropic and quasi-isotropic. This is represented schematically as a polar diagram in Fig. 7.24, showing the qualitative distribution of the mechanical composite characteristics depending on the strain angle for a monaxial, a biaxial, and a multiaxial non-crimp fabric.

Basically, the knitting elements of a multiaxial warp-knitting machine are similar to those of all other warp-knitting machine. Deviations arise from the principle of connecting the fed yarn layers and/or fabrics. Since these have to be pierced by the needles, the compound needles are equipped with a tip and are therefore referred to as piercing compound needles. Another typical feature is the knocking-over sinker fitted on the needle side and retaining the yarn layers during the backward movement of the needles. For the same purpose, the guide bar side is equipped with supporting pins to retain the yarn layers during the piercing of the needles. The knitting unit design varies between manufacturers. Two examples are shown in Fig. 7.25. All recent multiaxial and biaxial warp-knitting machines use a horizontal yarn layer feeding, although older machines might feed the yarn layer

Fig. 7.23 Electronically controlled adjustment of the weft laying device orientation on multiaxial warp-knitting machines (*Source*: LIBA Maschinenfabrik GmbH)

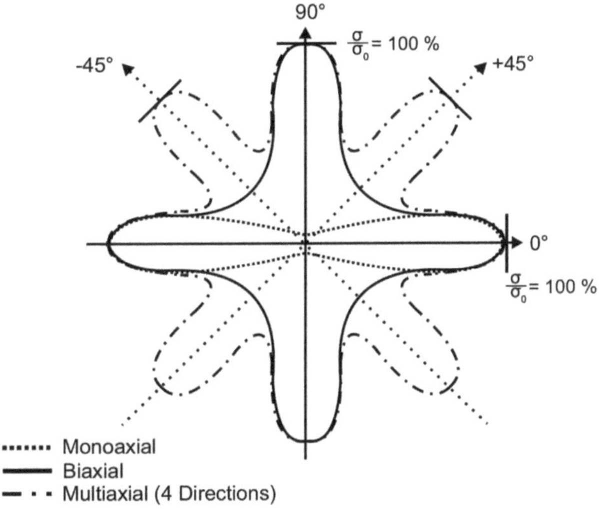

Fig. 7.24 Correlation of yarn layer orientation and structural composite strengths of multiaxial non-crimp fabrics, related to their respective strength in reinforcement direction

vertically. In biaxial warp-knitting machines, the weft yarns in transverse direction can also be introduced in a manner that prevents their piercing by the needles and allows the formation of the loops precisely around the yarns (weft yarn insertion in-line with the loops). Comparable systems for multiaxial warp-knitting machines with diagonal weft yarn laying devices are currently not available, but a few older machines with weft yarn insertion of all layers in-line with the loops are still in industrial use.

The warp yarn is fed directly to the knitting unit by means of special warp yarn guides. The design of these guides is matched with the yarn material to be

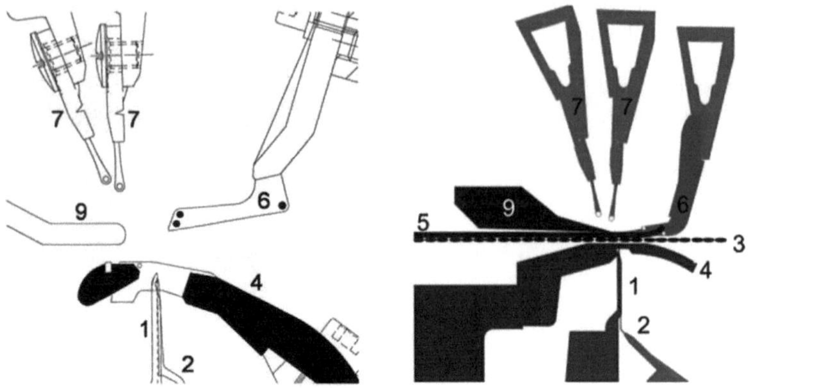

Fig. 7.25 Schematic representation of the knitting elements on current multiaxial warp-knitting machines (*left*: LIBA Copcentra MAX 3, *right*: Karl Mayer Malitronic® Multiaxial). 1: Piercing compound needle, 2: Sliding tongue, 3: Weft yarn, 4: Knocking-over sinker, 5: Warp yarn, 6: Warp yarn guide with supporting needles, 7: Guide bar (knitting yarn guide), 8: Knitting yarn, 9: Support rail

processed, and these guides can be designed to be able to complete a shogging motion, which allows tying off warp yarn layers with pillar stitch. A special warp yarn offset system is used for the production of multiaxial non-crimp fabrics with force flow-oriented warp yarns. This way, individual warp yarns or warp yarn groups can be arranged on the top side of the multiaxial non-crimp fabric in way that matches the pre-calculated force flow in the component [3, 4].

The purposeful adjustment of the piercing compound needle movement can also enable the user to influence the characteristics of the multiaxial non-crimp fabric. One option is to move the needle along the working direction of the machine during the piercing of the yarn layers. This reduces the size of the openings created by the piercing motion and minimizes damages to the warp and weft yarns. The targeted influencing of the yarn layer arrangement in multiaxial non-crimp fabrics can also be attained by laterally offsetting the compound needle bar during the extended warp-knitting process. This allows the user to influence the binding of the knitting yarn on the compound needle side as extensively as on the guide bar side. This makes the realization of, for instance, symmetrical multiaxial non-crimp fabrics with warp yarn layers on the outsides feasible, which would not be an option of conventional multiaxial warp-knitting machines [5].

Recent developments in the field of glass fiber-processing multiaxial warp-knitting machines, which are the dominant type available on the market, are focused on increasing productivity, which is currently at a production speed of 6 m/min, and on improving quality of the warp-knitted semi-finished products, primarily by ensuring a low-damage processing of the yarn material. Simultaneously, specialized machines are being developed for the processing of carbon fibers, focused on the fail-safe and high-quality spreading of the filament, and on the secure, non-twisted and stretched insertion of the weft yarns into the transport mechanism. Here, filament yarns with very high numbers of filaments (so-called heavy tows) are seeing increased use. The integration of additional process steps into the warp-knitting process, for example textile finishing, has caused an increased modularization and flexibilization of the manufacturing process. All of these developments aim at enabling textile producers to react flexibly to new product requirements while ensuring increased profits. The use of special yarn laying systems allows the insertion of stiffening elements and additionally reinforced force introduction areas. Simultaneously, efforts are underway to separate the warp-knitting process from the production of multiaxial non-crimp fabrics in order to produce textile semi-finished products with properties similar to those of chemically solidified non-crimp fabrics. The greatest challenge in this context is the setting of drapeability comparable to that of warp-knitted non-crimp fabrics.

Apart from online textile finishing in the warp-knitting process, the integration of additional functional elements like sensor networks can shorten the entire production process in many lightweight construction applications. Another tendency in this context is the manufacture of 3D semi-finished products close to the final component geometry, aimed at reducing the number of consequent process steps.

Quality assurance becomes more and more important with the development of application fields with very high quality requirements, such as the manufacture of automobile parts. Here, practical solutions are currently being developed for online control methods for damage-free material testing across all production steps. It can therefore be assumed that the use of warp-knitted semi-finished products for lightweight-construction applications will increase in the future.

7.5 Machine Technology for Three-Dimensionally Warp-Knitted Semi-finished Products

The manufacture of three-dimensional warp-knitted fabrics is becoming increasingly important for the area of technical textiles. This is due to the possibility to combine different, sometimes even contrary characteristics such as closed or open, smooth or strongly structured, dimensional stability on one fabric side and high elasticity on the opposite side, in the same three-dimension semi-finished product. Due to the range of adjustable parameters, the properties of the warp-knitted semi-finished products can be influenced in a wide variety of ways and allow for requirement-adapted three-dimensional textile structures for a number of applications.

Three-dimensional warp-knitted semi-finished products are manufactured on double-bar raschel machines (RR double-needle bar raschel machines). Based on the RR warp-knitting technology, two independent textile fabrics are connected into a spatial textile construction by means of a 3D yarn system (pillar yarn system).

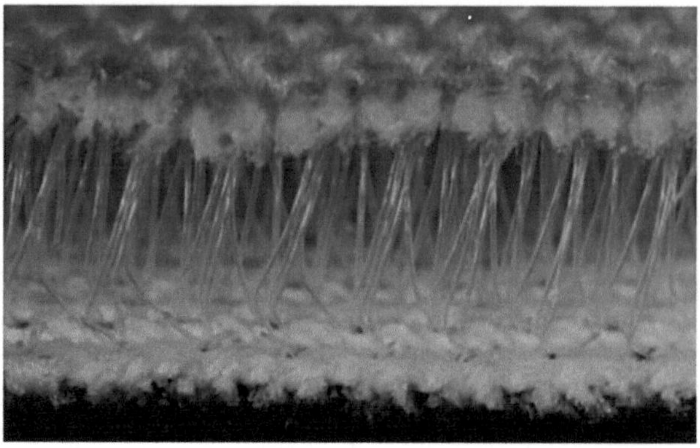

Fig. 7.26 Cross-section of a three-dimensional warp-knitted semi-finished product

The 3D yarn system usually consists of monofilaments to realize a pressure-resilient connection between the two warp-knitted fabric surfaces (Fig. 7.26).

The knitting elements of a double-bar raschel machine and their construction design are analogous to those of RL raschel machines. Double-bar machines feature an additional, mirror-imaged knitting unit. The distance between the needle bars can be adjusted up to 65 mm, depending on the machine type and the manufacturer. By adjusting the needle bar distance during production, three-dimensional warp-knitted semi-finished products with a cross-section variable along the production direction are created.

During loop formation, both needle bars are in engagement and asynchronously offset by 180° of main shaft rotation angle. The loop formation is performed as on RL warp-knitting machines described previously in this chapter, with the difference being that yarns run between the two needle bars. Typical double-bar raschel machines are equipped with five to seven guide bars. At least the first and last guide bars are used to manufacture the respective surface of the spacer fabric. The remaining guide bars form the pile yarn system and connect the surface into a three-dimensional warp-knitted fabric. The knitting unit of a double-bar raschel machine with tongue needles is given as an example in Fig. 7.27.

The additional insertion of warp and weft yarns on one or both machine sides (depending on the machine type), can be used to combine the advantages of multiaxial warp-knitted fabrics with the possibilities of three-dimensional warp-knitted semi-finished products. These axially reinforced 3D warp-knitted fabrics are perfectly suited for applications requiring not only a defined strength and stiff in longitudinal and transverse direction, but also a defined pressure-resilient spacer layer. One current application for such semi-finished products is in textile reinforcements for concrete (see Sect. 16.4).

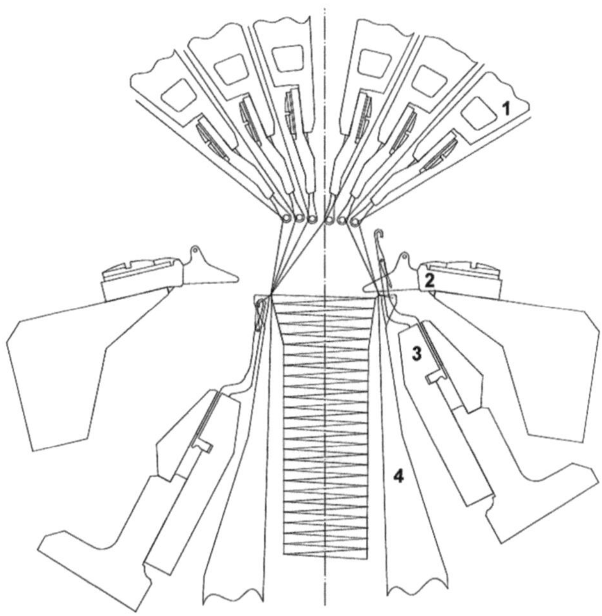

Fig. 7.27 Knitting elements of a double-bar raschel machine (*Source*: LIBA Maschinenfabrik GmbH). 1: Guide bar with guide needles, 2: Stitch-comb bar, 3: Needle bar with tongue needles, 4: Trick plate

7.6 Parameters and Characteristics of Warp-Knitted Semi-finished Products

The warp-knitting process for the manufacture of two- and three-dimensional semi-finished products offers a variety of parameters, which can be varied to influence the characteristics of the desired final products extensively and adapted to the requirements. For instance, in multiaxial warp-knitting the following factors can be adjusted for the manufacture of warp-knitted semi-finished products:

Yarn Material

- Design
  - fiber length in the yarn (staple fiber yarns, filament yarns)
  - yarn cross-section (round, oval, tape-shaped)
  - special designs (smooth yarns, texturized yarns, effect yarns, hybrid yarns, wires)
- Material
  - natural fibers
  - classic synthetic fibers (PAN, PES, PP)
  - high-performance fiber materials (carbon, glass, aramid)

Feeding of Base Materials

- Storage medium of the yarns (warp beam, bobbin, type of bobbin)
- Removal from storage medium (across the circumference or overhead, internal or external, active or passive)
- Knitting yarn delivery amount (stated in mm/Rack, 1 Rack = 480 min^{-1}, related to the revolutions of the main shaft)
- Knitting yarn tension (depends, among other factors, on the knitting yarn delivery amount, the yarn brakes, the existing places of friction, and the pattern)
- Weft yarn take-up from the creel (continuous, intermittent)
- Weft yarn insertion into the transport system (infinite, sequential, spread)
- Warp and weft yarn tension
- Number of warp and weft yarns per yarn guide (one or more)

Arrangement of Base Materials

- Orientation of the knitting yarn (loop density)
  - the number of wales across the production direction is adjusted by guide needle and needle gauges, and by the threading. For example, the guide needle distance for a machine gauge E7 (7 per inch) is ca. 3.6 mm. With a threading of 1 full and 1 empty, this results in a theoretical wale number of ca. 3.5 per inch at a distance of ca. 7.2 mm.
  - the number of courses in production direction, for instance on multiaxial warp-knitting machines, is adjusted by stitch length (as the relation of machine rotational speed to take-up speed). These adjustments can be performed in very small steps (e.g. 2 mm).
- Yarn layer arrangement
  - number (e.g. two)
  - orientation of the individual yarn layers [e.g. (0/90)]
  - yarn density within the individual yarn layers (e.g. 5 per inch)
- Arrangement of additional materials
  - chopped glass fibers
  - nonwoven fabrics
  - films
  - UD prepregs
  - multiaxial non-crimp fabric (for the realization of high layer numbers or symmetrical lay-ups)

Pattern

- Pattern (pillar, tricot, etc.)
- Manner of stitching (in yarn layer feeding)
  - in-line with the loops (rare), i.e. no piercing of the warp and weft yarns by the piercing compound needles occurs, instead the needles penetrate the yarn

layers in between the yarns, incorporating the warp and weft yarns into the loops
  - not in-line with the loops (common), i.e. the warp and weft yarns are principally pierced by the piercing compound needles

Take-up, Post-treatment, Storage

- fabric take-up speed
- process-integrated finishing steps (heat treatment, coating)
- winder type (surface or central winders)

The influence of the mentioned parameters on the properties of the final products of multiaxial warp-knitting, which usually are composite components on a thermoset, thermoplastic or mineral basis, cannot be phrased universally at this time. The range of requirements imposed by the different applications, the numerous influencing factors and their interplay have entailed a diversity of insights regarding individual cases and certain parameter combinations, not all of which are suitable for generalization. Therefore, this section will give a short overview of how important process parameters in multiaxial warp-knitting influence product properties. Decisive characteristics for fiber-reinforced plastics composites are high specific values for tensile strength, Young's modulus, flexural strength, resistance to delamination and impact as well as easy impregnation with typical matrix materials. Reference values are often taken from the corresponding specific values of UD prepregs (see Chap. 11), which exhibit very high tensile strength, high Young's modulus, but a low resistance to delamination.

Experiments with multiaxial non-crimp fabrics show that the use of high-strength knitting yarns (e.g. glass or aramid filament yarn) results in a significant improvement of delamination resistance compared to UD prepregs. Several other specific values are positively affected by such a loop-shaped connection, due to the high-strength yarns arranged in thickness direction. Thus, suitable knitting yarns improve damage tolerance, fracture toughness as well as compression and tensile strength after impact loads [6–9].

The connection of layers by knitting yarns impairs relevant mechanical parameters of the corresponding composite materials in fiber direction (in-plane characteristics). It is well known that the knitting yarns cause fiber constrictions on the outsides of the multiaxial warp-knitted fabric and undulations of the reinforcement fibers at the points where they penetrate the textile layers [9, 10]. This creates openings in the textile surface, which usually cannot be closed during further processing. These fault spots can cause resin accumulations and air pockets in fiber-reinforced plastic composites, bearing the risk of stress concentrations. In general, this increases the danger of defects in the composite material. As a result, multilayer composites often exhibit impaired characteristics in fiber direction in comparison to UD prepregs, particularly with regard to stiffness, tensile strength, compressive, shear, and flexural strengths [9–14]. New researches also show that a targeted selection of the machine parameters can significantly reduce unwanted influences of the knitting yarn on the characteristics in fiber direction [15]. One possibility to realize this is the use of needles that can be moved along with the process, compensating the forward movement of the yarn layers and thus reducing the size of the pricking point to the dimensions of the needle.

Little attention has so far been paid to the influence of the pattern. It has been established that the RL tricot stitch causes greater angular deviation of the fibers in the yarn layers than the RL pillar stitch binding. The influences of the stitch type on the tensile strength and interlaminary shear strength of multilayer composites made from multiaxial non-crimp fabrics could not be verified yet. With regards to the impregnation properties of the multiaxial non-crimp fabric, the pattern influences permeability. Irregular yarn sections such as in a RL pillar -tricot pattern cause fluctuations in thickness and can close channels to the flow of the resin. This results in randomly distributed permeability properties in the multiaxial non-crimp fabric. When using a pattern with a regular yarn course (e.g. RL tricot), no displacements can be detected, and the resulting impregnation behavior is significantly improved [16]. The pattern also influences shear capacity. For instance, a multiaxial non-crimp fabric in RL tricot stitch displays a better deformation capacity and better drapeability than a fabric in RL pillar stitch or combined RL pillar -tricot stitch [17].

The length of the loops in a multiaxial non-crimp fabric influences the degree of fiber deviations in the warp and weft yarn layers. The induced deviations are smaller in case of shorter stitch lengths. Tensile strength, Young's modulus, and shear strength in longitudinal and transverse directions increase with decreasing stitch length [12]. The delamination behavior is also positively influenced by shorter stitch lengths [8, 18].

References

1. ANONYMOUS: *Ausbildungsunterlagen Firma LIBA Maschinenfabrik GmbH (30-11-2010)*
2. ÜNAL, A.: *Analyse und Simulation des Fadenlängenausgleichs an Kettenwirkmaschinen für die optimale Konstruktion von Fadenspanneinrichtungen.* Dresden, Technische Universität Dresden, Fakultät Maschinenwesen, Dissertation, 2003
3. HEINRICH, H.; VETTERMANN, F.: Gestaltungsmöglichkeiten für bionische Verstärkungsstrukturen durch variable Filamentablage auf Multiaxialgelegen. In: *Kettenwirk-Praxis* 43 (2009), No. 4, pp. 27–29
4. HUFNAGL, E.; BÖHM, R.; KUPFER, R.; ENGLER, T.; CHERIF, Ch.; HUFENBACH, W.: Mehraxiale Gitterstrukturen als Funktionselemente für Kunststoffbauteile. In: *Kunststoffe* (2011), No. 4, pp. 85–88
5. HAUSDING, J.: *Multiaxiale Gelege auf Basis der Kettenwirktechnik - Technologie für Mehrschichtverbunde mit variabler Lagenanordnung.* Dresden, Technische Universität Dresden, Fakultät Maschinenwesen, Dissertation, 2010. http://nbn-resolving.de/urn:nbn:de:bsz:14-qucosa-27716
6. AYMERICH, F.; PANI, C.; PRIOLO, P.: Effect of stitching on the low-velocity impact response of [03/903]s graphite/epoxy laminates. In: *Composites: Part A* 68 (2007), pp. 1174–1182
7. CHEN, L.; SANKAR, B.; IFJU, P.: A new mode I fracture test for composites with translaminar reinforcements. In: *Composite Science and Technology* 62 (2002), pp. 1407–1414

8. DRANSFIELD, K.; JAIN, L.; MAI, Y.: On the effects of stitching in CFRPs - I. Mode I delamination toughness. In: *Composites Science and Technology* 58 (1998), pp. 815–827
9. MOURITZ, A.; LEONG, K.; HERSZBERG, I.: A review of stitching on the in-plane mechanical properties of fibre-reinforced polymer composites. In: *Composites: Part A* 28 (1997), pp. 979–991
10. LEONG, K. et al.: The potential of knitting for engineering composites - a review. In: *Composites: Part A* 31 (2000), pp.197–220
11. LOENDERSLOOT, R. et al.: Carbon composites based on multiaxial multiply stitched preforms Part 5: Geometry of sheared biaxial fabrics. In: *Composites: Part A* 37 (2006), pp. 103–113
12. CHUN, H.; KIM, H.; BYUN, J.: Effects of through-the-thickness stitches on the elastic behaviour of multi-axial warp knit fabric composites. In: *Composite Structures* 74 (2006), pp. 484–494
13. LOMOV, S. et al.: Carbon composites based on multiaxial multiply stitched preforms - Part 1: Geometry of the preform. In: *Composites: Part A* 33 (2000), pp. 1171–1183
14. TRUONG, T. et al.: Carbon composites based on multi-axial multi-ply stitched preforms. Part 4: Mechanical properties of composites and damage observation. In: *Composites: Part A* 36 (2005), pp. 1207–1221
15. MEYER, O.; GESSLER, A.; WEGNER, A.; VETTERMANN, F.: Influence of sewing of multiaxial textile structures in regard of mechanical properties. In: *Proceedings. SEICO 08 - Sampe Europe International Conference and Forum.* Paris, France, 2008
16. LOMOV, S.; VERPOEST, I.; PEETERS, T.; ROOSE, D.; ZAKO, M.: Nesting in textile laminates: geometrical modelling of the laminate. In: *Composite Science and Technology* 63 (2003), pp. 993–1007
17. KONG, H.; MOURITZ, A.; PATON, R.: Tensile extension properties and deformation mechanisms of multiaxial non-crimp fabrics. In: *Composite Structures* 66 (2004), pp. 249–259
18. JAIN, L.; MAI, Y.: Determination of mode II delamination toughness of stitched laminated composites. In: *Composites Science and Technology* 55 (1995), pp. 241–253

Additional Literature

BEL, S.; BOISSE, P.; DUMONT, F.: Analyses of the Deformation Mechanisms of Non-Crimp Fabric Composite Reinforcements during Preforming. In: Applied Composite Materials 19 (2012), No. 3-4, pp. 513–528

HAUSDING, J.; CHERIF, C.: Application of stitch-bonded multi-plies made by using the extended warp-knitting process: influence of pattern and knitting yarn tension on the geometry of the textile fabric. In: Journal of the Textile Institute 103(2012), No. 2, pp. 179–192

LI, L.; ZHAO, Y.; Vuong, H.-g.-n.; CHEN, Y.; YANG, J.; DUAN, Y.: In-plane shear investigation of biaxial carbon Non-Crimp Fabrics with experimental tests and finite element modeling. In: Materials & Design 63(2014), pp. 757–765

MÄRTIN, J.: Analyse, Simulation und Weiterentwicklung des Marterialtransportes in Hochleistungskettenwirkautomaten zur Steigerung der technologischen Leistungsparameter. Dresden, Technische Universität Dresden, Fakultät Maschinenwesen, Dissertation, 2012

POHLEN, V.; SCHNABEL, A.; NEUMANN, F.; GRIES, T.: Optimisation of the warp yarn tension on a warp knitting machine. In: Autex Research Journal 12 (2012), No. 2, pp. 29–33

VALLONS, K.; ADOLPHS, G.; LUCAS, P.; LOMOV, S. V.; VERPOEST, I.: The influence of the stitching pattern on the internal geometry, quasi-static and fatigue mechanical properties of glass fibre non-crimp fabric composites. In: Composites Part A: Applied Science and Manufacturing 56 (2014), pp. 272–279

Chapter 8
Braided Semi-finished Products and Braiding Techniques

Ezzedine Laourine

Traditionally, braiding is regarded as a manufacturing method for narrow fabrics like cords and ropes. New braiding methods allow the manufacture of structures with complex geometries which are used in lightweight construction solutions, for instance in vehicle engineering. Braiding plays a special role in the production of reinforcement structures due to the adjustment of the angular orientation in the braiding and to intertwine threads in nearly all three spatial directions at continual fiber orientation. 3D-braiding methods facilitate the manipulation of fiber orientation and thus ensure high strengths and stiffnesses at reduced mass. This chapter describes the various braiding technologies for the manufacture of 2D-and 3D-structures. The principle and functionality, as well as the most important braiding machine components are explained in detail. Potential application areas of this process based on different examples are shown in this chapter, also, possibilities for functional integration are discussed.

8.1 Introduction and Overview

Braids are produced by the regular intertwining of intertwining at least three yarns running diagonal to the production direction. In DIN 60000, they are defined as products from braiding threads with even thread density and closed fabric appearance [1]. Braids can be supplemented by additional reinforcement threads in axial direction (so-called 0° or pillar threads) and can be constructed as planar or volume-

E. Laourine (✉)
Albstadt-Sigmaringen University, Albstadt, Germany

Formerly: Institute of Textile Machinery and High Performance Material Technology, TU Dresden, Dresden, Germany
e-mail: laourine@hs-albsig.de

C. Cherif (ed.), *Textile Materials for Lightweight Constructions*,
DOI 10.1007/978-3-662-46341-3_8

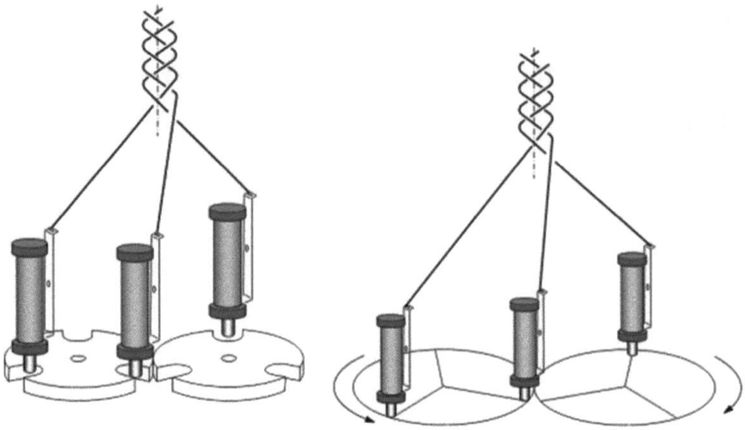

Fig. 8.1 Principle of the braiding machine

forming structures [2]. In contrast to other textile processes, braiding relies on open yarn ends [3].

The distinguishing property of braids is the intertwining of their threads diagonal to the structural main axis, which imparts high flexibility and resulting formability.

At least three rovings alternately intertwined in regularly repeated steps are required to create a braid. Figure 8.1 shows the steps required to form a simple braid.

Repeat denotes a repeated pattern created by the above process after conclusion of a certain number of steps. In the method shown in Fig. 8.1, a repeat is created after only two steps. Industrial braiding commonly uses more than three threads (and up to several hundreds of threads), which are often called *braiding threads* or *bobbin threads*. For complex, three-dimensionally braided structures with longitudinally variable cross-sections, a *repeat* can require many steps. It is created after all carriers return to their original position on the braiding plate. Depending on component size and complexity, some braiding methods allow nearly unlimited pattern repeats.

8.2 Classification of Braiding Methods

Braided structures can be classified into two- and three-dimensional constructs. The dimension of the textile structure is not defined by its exterior design, but by the thread delivery. If the threads are delivered in just one plane, the structures are classified as 2D-braids.

If the threads interlace in all three spatial directions, they form a 3D-braided structure.

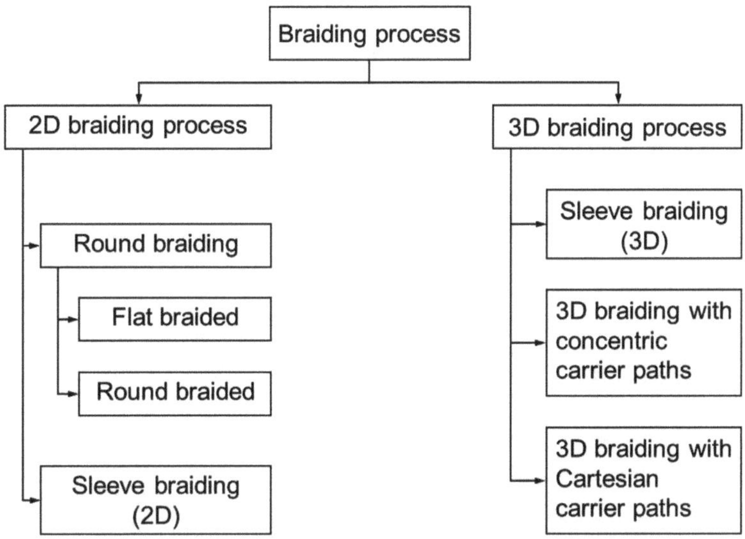

Fig. 8.2 Classification of the braiding process

Sleeve braiding and round braiding are two 2D-braiding methods. They are primarily used for purely textile applications, such as the production of cords, ropes, or simple cord braids. They can also be used for reinforcement structures in special application. It has to be noted that sleeve braiding also allows the manufacture of three-dimensional geometries.

More complex structures become feasible with 3D-braiding methods. These methods are mainly used for the production of three-dimensional reinforcement structures for lightweight construction. Figure 8.2 gives an overview of current braiding methods.

8.3 Functional Principle of Braiding Machines

The production sequences of a braiding machine are fundamentally inspired by the principle of the manual labor. The first machine of this kind was patented as early as the eighteenth century [4]. Original machines were made of wood. Even today machines built from metal are still based on the same principles. Each braiding thread is moved along a pre-defined path on a braiding plate. The thread is stored on a spool moved by a carrier. To attain homogeneous fabric appearance, the carrier ensures constant thread tension during the movement across the braiding plate.

In round braiding, the threads are intertwined by transporting one half of them in one direction, while the other half is moved in the opposite direction of the impeller wheels. The carriers complete a sinusoidal motion and thus enable an intertwining of the threads.

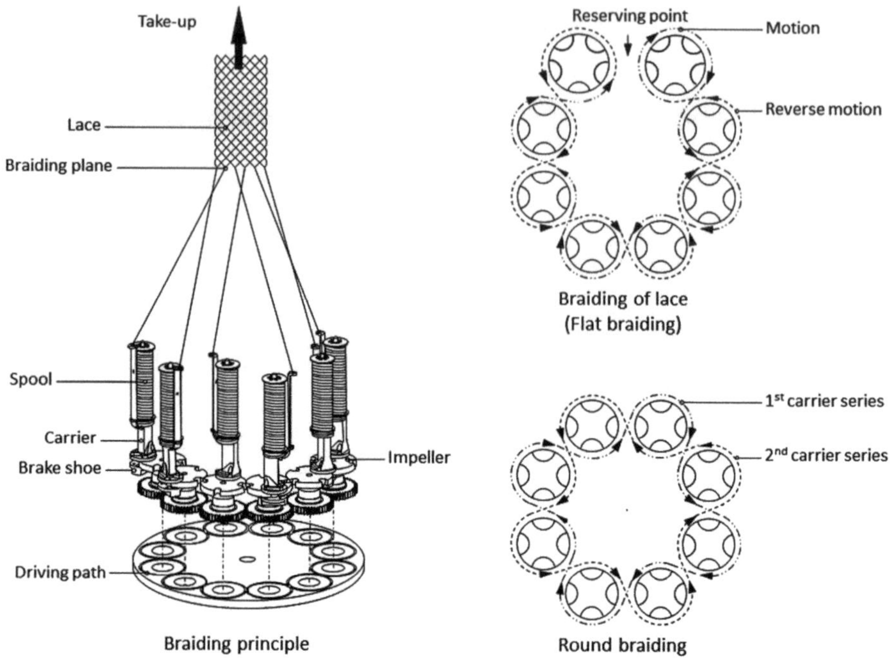

Fig. 8.3 Braiding principle

Figure 8.3 shows the schematic construction of a round braiding machine as well as the arrangement of the *impeller wheels* for the manufacture of a *strand* (flat braided fabric) and a round braided fabric. While in round braiding the respective half of the carriers turns in a continuous path in one direction, the production of a *strand* requires the integration of a reversal point in the braiding plate, causing a reversal of the direction at both ends. In practice, two larger impeller wheels are used on the common machines in order to create more space for the reversal carrier.

The braiding is produced in the so-called braiding plane (or braiding point), from where it is subsequently taken off. Depending on the take-up direction, braiding machines are categorized as horizontal or vertical machines. The relation of take-up speed and carrier circulation speed on the braiding plate determines the intertwining angles of the braiding threads.

There is an abundance of different carrier constructions available, all of them customized to suit the respective machine types and the braids to be produced. The principle of a carrier is described in detail in Sect. 8.6.

8.4 2D-Braiding Methods

Round and *sleeve braiding* are two 2D-braiding methods. Usually, the technique is used for tubes and simple cord braids. Sleeve braiding requires a core, around which the threads are then braided. The outer geometry of the braiding is determined by the shape of the core.

8.4.1 Round Braiding

In *round braiding*, two basic braiding structures have to be distinguished as possible products: flat braids (cord braids) and round braids.

Flat braiding consists of a single thread system. All threads interlace diagonally and reverse at the terminal points of the path. When subjected to tensile and compressive loads in longitudinal or transverse directions, flat braids change their width or cross-section and length. To prevent changes of the spatial dimensions by tensile loading, horizontal (*pillar*) threads can be included in the braiding.

In *round braids*, the diameter of the structures changes under the influence of tension. Diameter reduction continues until the available threads either support one another or come into contact with the interior structure (e.g. the core yarns). Figure 8.4 shows a schematic comparison of round, flat, and 3D-braindings.

8.4.2 Sleeve Braiding

In *sleeve braiding*, the threads are placed around a mandrel. The mandrel geometry determines the final shape of the braiding. This method is suitable for the production of both rotationally symmetrical profiles (with constant and variable cross-sections) and arcuate profiles. For lightweight construction purposes, mandrels are commonly made from foam materials and can remain in the finished component after the braided structure is impregnated. If the geometry features indentations or

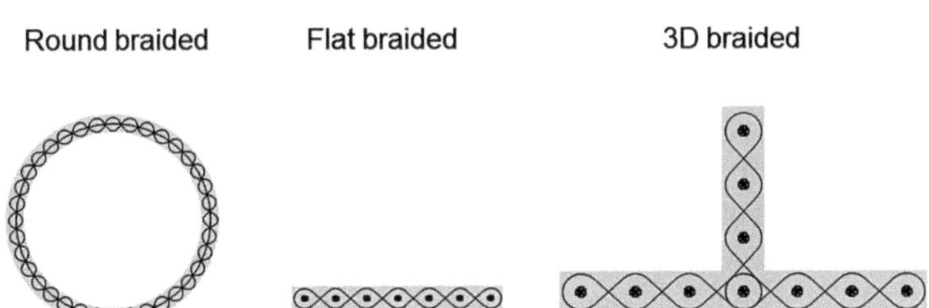

Fig. 8.4 Round, flat, and 3D braids

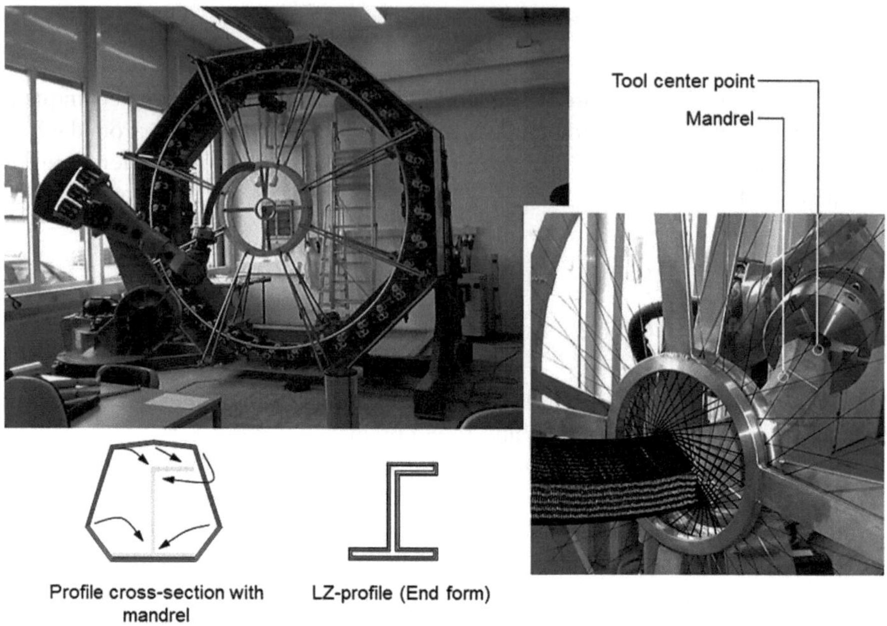

Fig. 8.5 Sleeve braiding machine with mandrel feed (*Photos*: August Herzog Maschinenfabrik GmbH & Co. KG)

bulges, and if the mandrel cannot remain in the component, they are destroyed and removed.

The diameter of the braided semi-finished products partly depends on the number of braiding threads and the braiding angle. There are circular braiding machines capable of controlling several 100 carriers at once. A sleeve braiding machine with a robot-assisted mandrel is shown in Fig. 8.5. This machine is able to handle a maximum of 144 carriers. The robot arm replaces the take-off unit [5]. During braiding, the robot arm moves a mandrel, which is fixed to the *tool center point* through the braiding plane, so that the threads are placed on the mandrel. Carbon fiber rovings are used as braiding yarns in the given example. Repeated sleeve braiding results in the desired component wall thickness. The mandrel in this component is shaped so that after repeated sleeve braiding and removal of the mandrel, a curved LZ profile can be produced by folding the profile edges of the 2D-braiding inward (see Fig. 8.5). These profiles were developed as stringer reinforcements for airplane frames. The curvature of the mandrel matches that of the frame to be produced.

The advantage of sleeve braiding of the LZ profiles in this manner is in the stretched placement of the rovings, which ensures their orientation in load direction despite the braid's curvature. Thus, the forces acting on the profile are largely absorbed by the stretched fibers. Although the component has a three-dimensional geometry, it is a curved 2D-braided structure with stacked reinforcement layers according to the classification in Sect. 2.3.1.1, as it lacks a reinforcing effect in thickness direction.

8.5 3D-Braiding Methods

As defined by Sect. 2.3.1.1, only structures in which threads are intertwined in all three spatial directions are classified as 3D-braided structures. A 3D-geometry has a voluminous structure developed without transforming and does not depend on the number of thread systems and the structure produced with them. All compact profiles with or without variable cross-sections are counted among 3D-braided structures.

8.5.1 Sleeve Braiding of 3D-Structures

Multilayered round braids are classified as 3D-braided structures if the individual layers are connected in thickness direction. To clarify this, Fig. 8.6 schematically illustrates the difference between 2D-and 3D-round braided structures.

Based on the principle of round braiding, new methods for the manufacturing of 3D-braids are being developed. They aim to produce thick-walled structures in which the individual layers are connected with one another, and make use of enhanced round braiders with several concentric rings, circumferential guiding grooves as well as radial grooves for carrier transport between the individual layers.

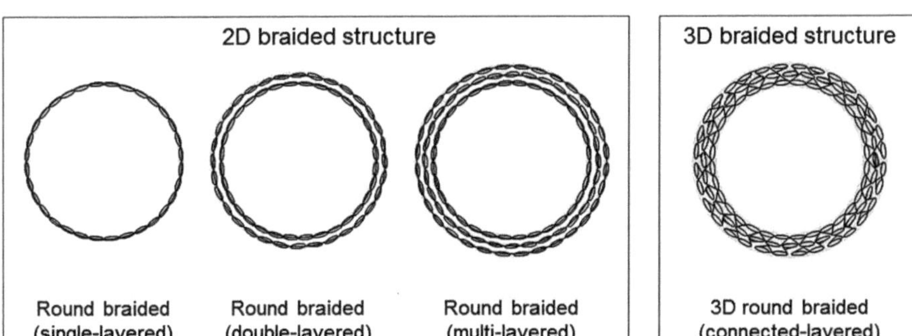

Fig. 8.6 Schematic depiction of the difference between two-dimensional and three-dimensional sleeve braided structures

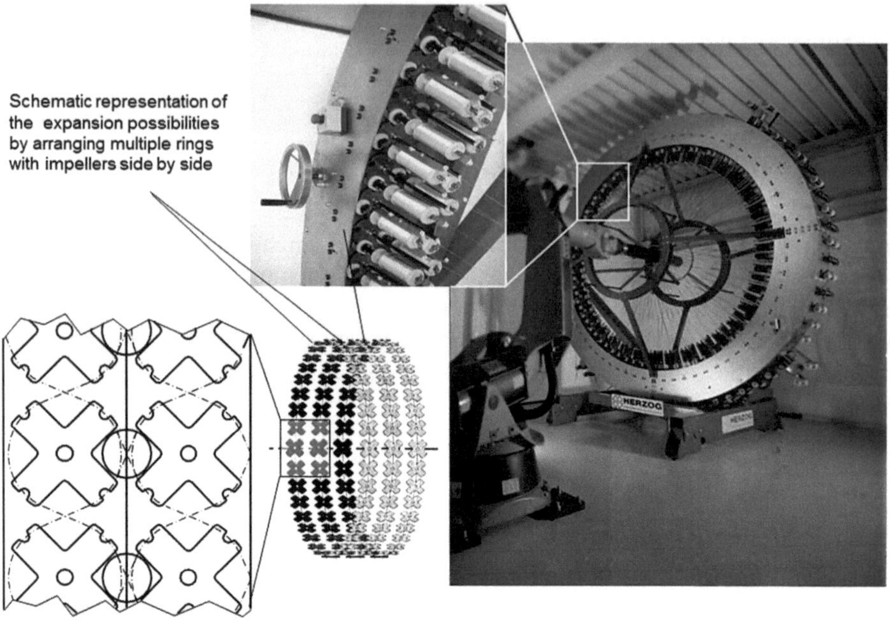

Fig. 8.7 Radial sleeve braiding (*Photo*: August Herzog Maschinenfabrik GmbH Co. KG)

8.5.2 3D-Braiding with Concentric Carrier Paths

3D-braiding with concentric carrier paths is matched with *radial sleeve braiding*, which is an enhancement on basic sleeve braiding. In a radial sleeve braiding machine, the carriers are attached to the inner circumference of a circular machine frame. In comparison, the carriers of a conventional sleeve braiding machine are fixed to the frontal plane of the machine. Figure 8.7 shows the positions of the carriers on the circular casing of the radial sleeve braider for 2D-braids.

One advantage of this arrangement is a possible increase in the number of rings. Several rings with impellers can be assembled into an extended multilayer interlock braider with concentric carrier paths. Crossings are in place for carrier movement between the individual rings. This allows the placement of connecting threads between the layers. Additional spools can be placed on the outside of the multilayer interlock braider on the circumference. These threads are guided through the axes of the individual impellers and integrated into the structure as pillar threads through the braiding threads.

8.5.3 3D-Braiding with Cartesian Carrier Paths

8.5.3.1 Packing Braiders

In a packing braider, the impellers are affixed to a defined surface. The arrangement of the carriers corresponds with the cross-section of the desired component. To transfer the carriers, the neighboring impellers counter-rotate. All involved threads are processed into a braid. Profiles can be realized with the help of a *packing braider*. A 3D-braid is formed with constant, usually square or round cross-section along the entire profile length.

8.5.3.2 2-Step Braiders

To manufacture more complex, profile-like structures with constant cross-sections in component axis direction (e.g. for I-girders), the *two-step* method is developed. It is automatable and can realize profile-like structures much quicker than any packing braider [6].

This method requires pillar threads for the intertwining with the braiding threads, since, as opposed to classical braiding, the threads are not braided by the sinusoidal motion of the carriers. In two-step braiding, one half of the braiding threads move diagonally through the braiding plate in a single step (first direction). In the next step, the other half of the braiding threads are moved across the entire braiding plane transversely to the first carrier direction. After this second step, the carriers are back at their original arrangement on the braiding plate. The braiding is formed by the repetition of these two steps [7]. Figure 8.8 gives a visual representation of the principle. For the realization of a double-T profile, the positioning of the pillar threads and the motions of the carrier threads are shown schematically. For a close connection of braiding and pillar threads, high thread tensile forces are required. However, handling becomes difficult in complex cross-sections. The profiles (their cross-sections) will therefore not be very pronounced [8].

8.5.4 3D-Braiding with Modular Carrier Drives

The first-ever patented idea for a *3D-rotary braider* [9] is based on the principle of a packing braider with added flexibility regarding the geometry of the braid and the yarn angles variable in component axis direction within it. This is realized by individually driving the impellers and controlling the carrier transfer with switches [10].

In a *3D-rotary braiding machine*, the braiding plane is formed by the arrangement of several impellers. The impellers can be operated separately.

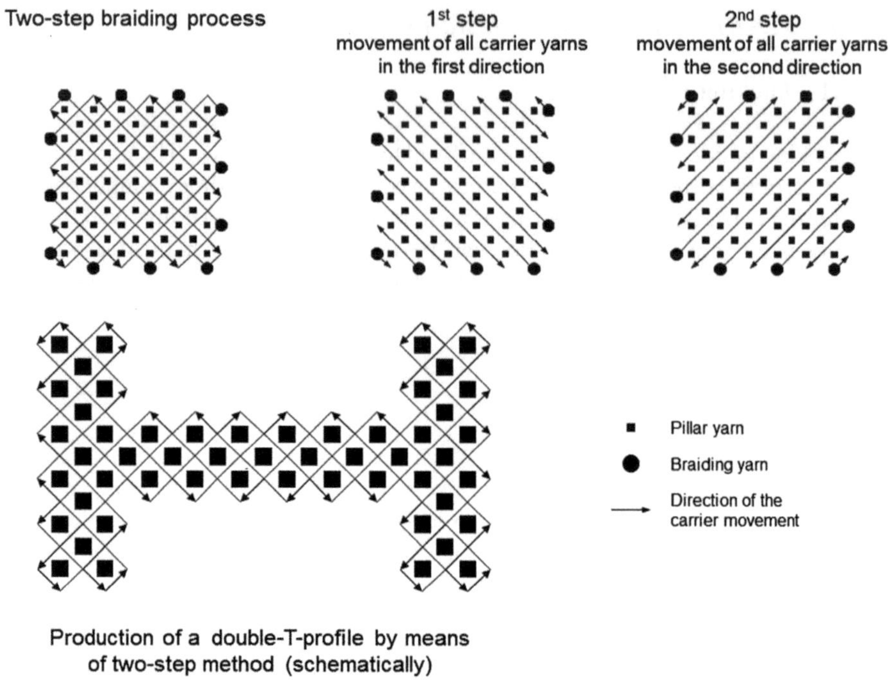

Fig. 8.8 Principle of two-step braiding

The transfer of a carrier to the next stationary impeller is controlled by the so-called switches [11, 12]. The impellers can also be put into a parking position. The axes of the impellers are fitted with yarn guides for pillar threads. In the corners between the impellers, additional guide sleeves for pillar threads are attached in grids. The complex, profile-shaped product is formed in the braiding plane, and then taken up.

Figure 8.9 shows a horizontal and a vertical 3D-braiding machine for the manufacturing of complex reinforcement structure from glass and carbon filament yarn. On the left side, the prototype for 3D-rotary braiding technique with vertical take-up is shown. This machine consists of four modules arranged in a square, with 5×5 impellers each. The individual modules can be arranged as required. This allows the positioning of the individual segments for the manufacturing of large L or T profiles [12]. On the right, another machine variation is shown. The machine is arranged vertically and the take-up is performed horizontally. This braiding machine is also constructed modularly, and consists of a group of nine modules arranged next to and above one another, each with 4×4 impellers. This method is distinguished by its high flexibility for the manufacturing of 3D-braids with defined thread angles and variable cross-sections. Complex geometries can be realized by using the corresponding machine controls.

Fig. 8.9 3D-rotation braiding (*Photo*: August Herzog Maschinenfabrik GmbH Co. KG)

8.6 Carrier Constructions

A braid is produced in the braiding plane by the movements of the braiding threads and the take-up of the braid. Therefore, the essential components of a braiding machine, are the carriers, the impellers, the drive and the take-up device, as well as the winder for the final product (see Fig. 8.3). These are the main functional elements necessary to reproduce a braid.

The constructions of these parts are versatile, and can include other elements required to ensure special effects and realize special products.

In the *carrier*, the thread material for the formation of the braid is stored on spools. The spring system of the carrier ensures a defined tension of the braiding threads during braiding. The movements of the individual carriers causing the threads' intertwining, in combination with the take-up, make the formation of a closed reproducible fabric appearance possible.

The carrier is made up of two parts, the carrier bottom and the carrier top. The carrier bottom is in direct contact with the impellers, constituting the interface of spools and carrier-guiding elements. The carrier top is responsible for thread material storage and the regulation of thread tension.

To ensure an uninterrupted braiding process, modern carrier constructions allow a breakdown of the braiding process in case of thread breakage or depletion of the thread material stored on the spool. A clamping slide, which is moved along with the carrier, drops to a lower position in case of thread breakage, makes contact with the electrical cutout switch during carrier rotation and triggers the termination of the braiding process.

A decisive criterion for the manufacturing of braids is the tension of the threads, which influences the formation and appearance of the braid by determining the intensity of the thread compression in the braiding plane. Here, the tension of the threads results from the force component of the thread compensation mechanism and the acting frictions. The frictions act contrary to the take-up direction and are caused by the direct contact of the thread material with guiding elements and deflection points. The maximum thread tension is determined by the stiffness of the exchangeable thread tensioning spring in the compensation mechanism. Depending on the maximum tensile force of the thread material to be processed, various thread tensioning springs are available.

There are several carrier designs in existence. The lever carrier and slide carrier designs are among the most popular [8]. Other special braiding methods, such as 3D-rotary braiding, require a high thread compensation length and thus special carrier designs. Figure 8.10 shows the lever carrier, the slide carrier and the TDF carrier, which are used in 3D-braiding.

For the braiding process, the thread material is wound on parallel bobbins or cylindrical cross-wound bobbins. The sinusoidal motion of the carriers in the braiding plate changes the distance between carrier and braiding plane, as illustrated for the lever carrier in Fig. 8.10. For slide carrier and lever carrier, thread length is compensated by a pulley mechanism. From the bobbin, the thread is led to the central fixed thread guide, is deflected by a mobile thread guide and finally passes the upper fixed thread guide. The mobile thread guide is relocated mechanically and tensioned by a spring. Its motion keeps the braiding thread taut. When the mobile thread guide reaches its upper position, a mechanism is triggered and the bobbin locking is deactivated so that the thread material can be released from the bobbin (Fig. 8.11).

8.7 Impeller

The function of the impeller in the machine is to guide the carriers across the braiding plate. The impellers are arranged on the plate according to two criteria: the outer shape of the braid and the desired braid appearance. For this, the incisions on neighboring impellers have to face each other after every interval. The course of the carrier is determined by the groove cut into the braiding plate. By means of the impellers, kinetic energy is transferred to the carriers. The drive movement is initiated by the gearwheel incorporated in the lower part of the impeller unit, which is in mesh with the neighboring wheel. This creates a counter-rotational operation from one wheel to the next, which is necessary for the intertwining of the threads.

Since the arrangement of the impellers on the braiding plate is decisive for the final shape of the structure, there is an abundance of arrangement variations. It has to be kept in mind that the arrangement of the impeller wings and the distribution of the impellers themselves on the braiding plate ensure an unimpeded transfer of the

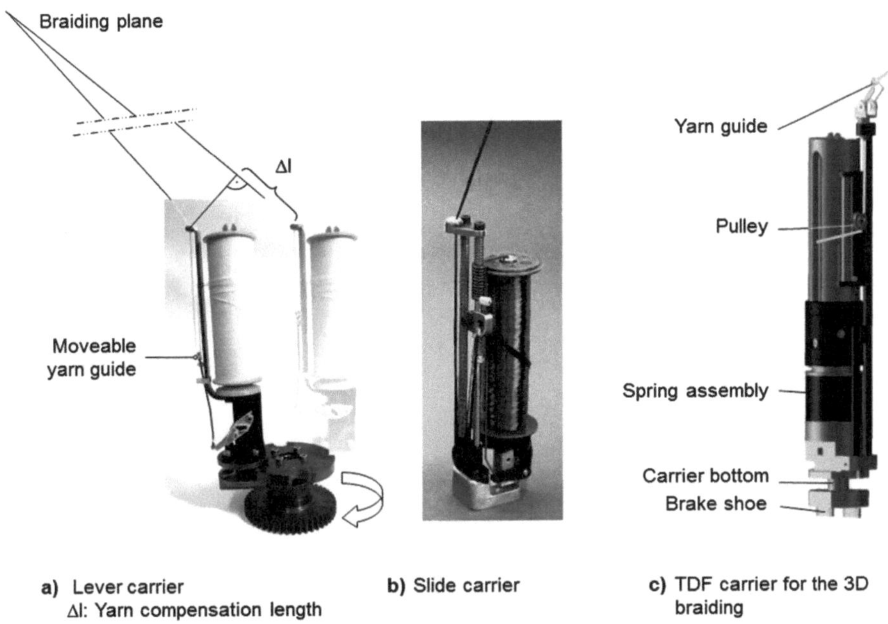

Fig. 8.10 Carrier constructions (photo of the TDF carrier courtesy of August Herzog Maschinenfabrik GmbH & Co. KG). (**a**) Lever carrier Δl: Yarn compensation length. (**b**) Slide carrier. (**c**) TDF carrier for the 3D braiding

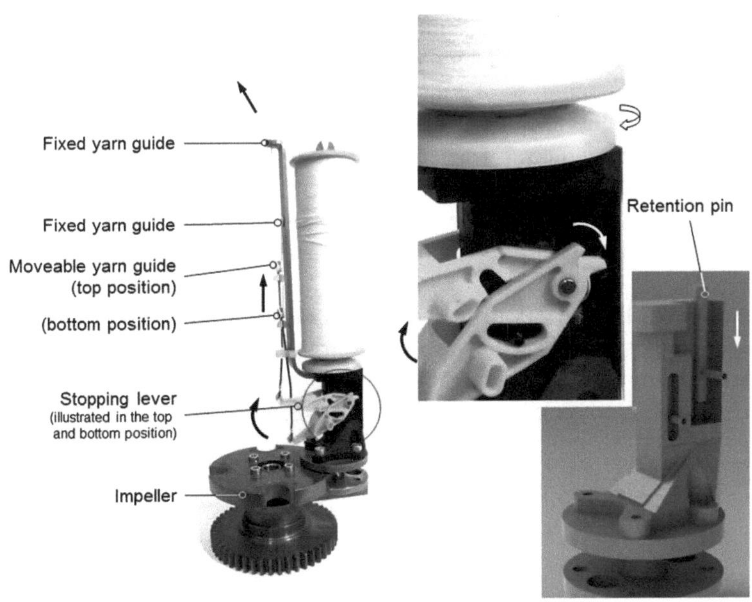

Fig. 8.11 Functioning of the yarn compensation and deactivation of the spool retention

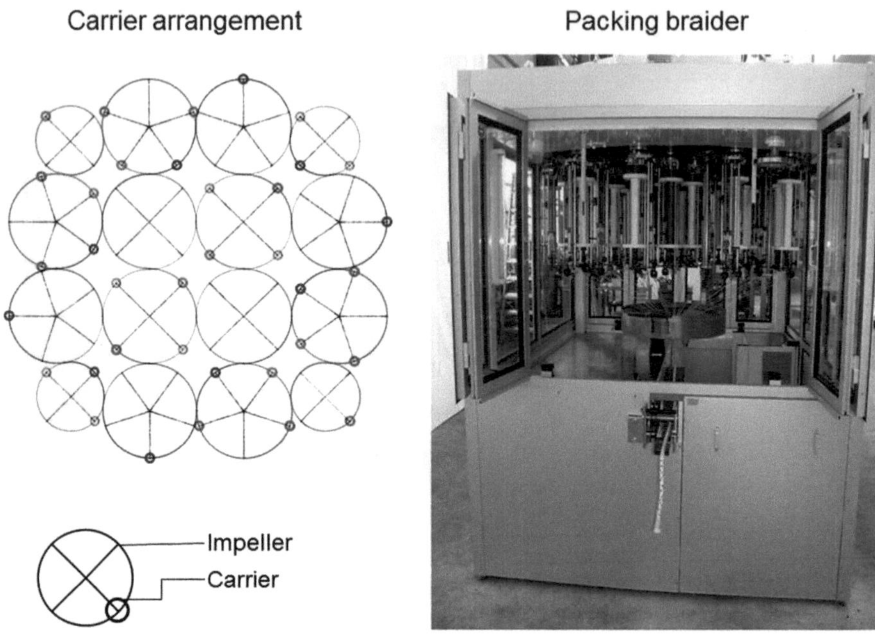

Fig. 8.12 Impeller shapes and carrier occupation for packing braiders (*Photo*: August Herzog Maschinenfabrik GmbH & Co. KG)

carriers. It can therefore be sensible and necessary to combine impellers with different numbers of wings on a single machine, as in the case of packing braiders (Fig. 8.12). Braiding machine manufacturers are increasingly aspiring to give their machines a modular setup, so that the combination of multiple modules allows the realization of different machine sizes, as is the case with the 3D-Cartesian braider.

Beyond that, the choice of carrier arrangement determines the braid construction. As an example, Fig. 8.13 shows the full, half, and tandem carrier occupations and their individual effects on the appearance of the round braiding. More basic braiding constructions and the principles of braiding representations are shown in detail in [13].

To introduce the carriers to their path, an insertion groove is required in flat braiding machines, as all carriers can move in the same direction. Round braiding machines contain two insertion grooves, which assign two opposing directions to the carriers. The grooves can be blocked by a wedge.

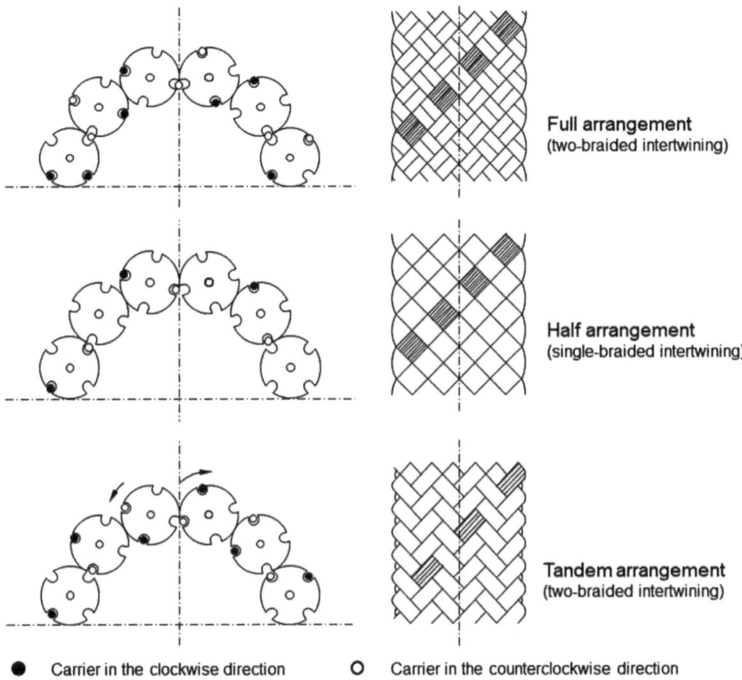

Fig. 8.13 Carrier arrangement and intertwining of braided fabrics

8.8 Drive

Historically, braiding machines were driven manually, using a large hand wheel to transfer human strength to the central drive wheel. Traditional products include slim braids for outerwear as well as shoe laces.

With the advent of the machine age, steam machines and electric motors were introduced to drive braiding machines. Their improved performance allowed an increase in the number of bobbins, which in turn resulted in a wider choice of patterning techniques. Today, modern electric motors ensure a controlled drive for braiding machines. As a rule, one motor is used per machine, although smaller round braiding machines with two heads also require just one motor. One part of the driving power is mechanically channeled off for the fabric take-up.

In medium-sized and heavy machine variants, take-up and impellers are individually driven by separate motors, both of which are synchronized by control mechanisms. The driving motion is transferred to the neighboring impellers by means of the pinion gear drives connected to the impellers.

There are also braiding machines with several drives evenly distributed on the braiding plate. Not only does the use of several synchronized drives ensure the required power, but it also allows a more precise impeller positioning. This reduces a possible accumulation of the necessary functional clearance between the gear wheels and prevents a collision during the transfer of the carrier from one impeller to the next.

8.9 Applications of Braided Structures

Braiding technology is a time-tested method, historically used for the production of tubes and ropes. Braids are also applied, for instance, as aerials (radio communication), electric wiring (telephone and submarine cables), cords (garments, interior design), medical products (net bandages, artificial limbs, sutures, catheters), nets (automobile, sports), reinforcement tubes (mechanical engineering, automobile), and ropes (shipping, mountain climbing, guy ropes, fishing lines, candlewick, shoelaces). The use of electrically conductive braiding threads makes other products, such as Smart Textiles, feasible [8, 11].

Over the past years, intense research has been conducted in the field of 3D-braiding. One of the central research and development subjects was the processing of carbon and glass filament yarns into braided structures for lightweight construction components [11, 14].

Braiding is advantageous because the braiding angle and the surface density are adjustable and pillar threads can be added. This results in textile fabrics and three-dimensional structures capable of absorbing forces in two or three directions. Therefore, braids are predestined for structural components. The special structural characteristic of braids is the diagonally crossing threads suitable for the transfer of torsional moments [15]. Round braiding can transfer the acting forces diagonally within the structure along the thread directions. By means of the flexible 3D-rotary technology, individual, load-adapted 3D-structures can be realized.

Crash elements with a high energy absorption capacity can also be realized excellently by braiding. The braiding pultrusion method allows the braiding of preforms as continuous-fiber-reinforced profiles in an online process, and their impregnation with thermoplastic and thermosetting matrix materials [16, 17]. The method offers a cost-effective production method for continuous-fiber-reinforced plastic profiles with constant cross-section. For thermoplastic matrix systems, pultrusion can be applied for the production of more complex component cross-sections, e.g. curved profiles with remolded, locally variable cross-section [18].

8.10 Functional Integration

Functional integration denotes the addition of a secondary function to the structure, beyond the primary function. An additional function can be integrated by, for instance, adjusting the arrangement of bolting elements, metallic force transmission elements, or inserts between the revolving, uninterrupted, continuous fibers of the braided structure. These fastening elements serve as assembly interfaces when used in composite components made from braided products.

Further possibilities for the attainment of secondary functions result from the integration of conductive threads being inserted in the braiding process. For example, they can act as force sensors, supervising stresses in the braiding online. The aim is to improve the functional safety of braided ropes and lines made from synthetic fibers, and to enable an objective assessment of strain and wear conditions [19].

References

1. Standard DIN 60000 January 1969. *Textilien. Grundbegriffe*
2. MIRAVETE, A.: *3-D textile reinforcements in composite materials.* Cambridge : Woodhead Publishing, 1999
3. HUFENBACH, W. (Hrsg.): *Textile Verbundbauweisen und Fertigungstechnologien für Leichtbaustrukturen des Maschinen- und Fahrzeugbaus.* Dresden : SDV - Die Medien AG, 2007
4. BRUNNSCHWEILER, D.: Braiding Technology. In: *Skinner's Silk and Rayon Record.* Cambridge: Journal of the Textile Institute, 1954, pp. 666–686
5. LAOURINE, E.; SCHNEIDER, M.; WULFHORST, B.; PICKETT, A. K.: Computerunterstützte Berechnung und Herstellung von 3D-Geflechten. In: *Band- und Flechttechnologie* (2001)
6. Protective right US4719837 (19th January 1988).
7. BUCKLEY, J. D. (Ed.): *Fiber-Tex 1991 - The Fifth Conference on Advanced Engineering Fibers and Textile Structures for Composites : proceedings of a conference sponsored by the National Aeronautics and Space Administration, Office of Management, Scientific and Technical Information Program.* Hampton, USA : Langley Research Center, 1992
8. ENGELS, H.: *Handbuch der Schmaltextilien. Die Flechttechnologie. Teil 2: Maschinen und Verfahren zur Erzeugung von Flechtprodukten mit speziellen physikalischen und chemischen Anforderungen.* Mönchengladbach : Institut für Textil- und Bekleidungswesen Mönchengladbach, 1994
9. Protective right DE4201413A1 (22nd July 1993).
10. BÜSGEN, W. A.: *Neue Verfahren zur Herstellung von dreidimensionalen Textilien für den Einsatz in Faserverbundwerkstoffen.* Aachen, RWTH Aachen, Dissertation, 1993
11. SCHNEIDER, M.: *Konstruktion von dreidimensional geflochtenen Verstärkungstextilien für Faserverbundwerkstoffe.* Aachen, RWTH Aachen, Fakultät Maschinenwesen, Dissertation, 2000
12. LAOURINE, E.; SCHNEIDER, M.; WULFHORST, B.: Production and Analysis of 3D Braided Textile Preforms for Composites. In: *Proceedings. Texcomp 5.* Leuven, Belgium, 2000
13. ENGELS, H.: *Handbuch der Schmaltextilien. Die Flechttechnologie. Teil 1: Maschinen und Verfahren zur Erzeugung konventioneller Geflechte.* Mönchengladbach : Institut für Textil- und Bekleidungswesen Mönchengladbach, 1994
14. LAOURINE, E.; SCHNEIDER, M.; PICKETT, A. K.; WULFHORST, B.: Numerische Auslegung und Herstellung von 3D-Geflechten für Faserverbundbauteile. In: *DWI Reports.* Aachen: DWI, 2001
15. AYRANCI, C.; CAREY, J.: 2D braided composites: a review for stiffness critical applications. In: *Composite Structures* 85 (2008), Nr. 1, S. 43–58

16. MILWICH, M.; LINTI, C.; PLANCK, H.: Thermoplast-Braid-Pultrusion of thin walled hollow profiles. In: *Proceedings. SAMPE EUROPE CONFERENCE EXHIBITION*. Paris, France, 2005
17. MILWICH, M.; LINTI, C.; PLANCK, H.; SPECK, T.; SPATZ, H.-C.; SPECK, O.: Thin Walled Hollow profiles. In: *Proceedings. 8th World Pultrusion Conference*. Budapest, Ungarn, 2006
18. ILLING-GÜNTHER, H.; HELBIG, R.; ARNOLD, R.; ERTH, H.; MILWICH, M.; FINCK, H.; PLANCK, H.: Pultrusion Processing of New Textile Structures Getting Functionally Graded Materials. In: *Proceedings. Aachen-Dresden International Textile Conference*. Aachen, Germany, 2007
19. STÜVE, J.; GRIES, T.: Threadlike Sensors-or How to Realize an Intelligent Rope. In: *MST NEWS* 1 (2007), pp. 42

Additional Literature

BILISIK, K.: Three-dimensional braiding for composites: A review. In: Textile Research Journal 83 (2013), No. 13, pp. 1414-1436

BRANSCOMB, D.; BEALE, D.; BROUGHTON, R.: New Directions in Braiding. In: Journal of Engineering Fibers and Fabrics 8 (2013), No. 2, pp. 11-24

GAO, Y.; LI, J.: Effects of braiding angle on modal experimental analysis of three-dimensional and five-directional braided composites. In: Composites, Part B 43 (2012), pp. 2423-2428

HELLER, L.; VOKOUN, D.; SITTNER, P.; FINCKH,H.: 3D flexible NiTi-braided elastomer composites for smart structure applications. In: Smart Materials and Structures, 21 (2012), No. 4

LI, J.-C.; CHEN, L.; ZHANG, Y.-F.; PAN, N.: Microstructure and finite element analysis of 3D five-directional braided composites. In: Journal of Reinforced Plastics and Composites, 31 (2012), No. 2, pp. 107-115

SOHEL, R.; EMILIJA Z.; CRISTIANA, P.; RAUL, F.; GOMES CORREIA, A.: Development of Hybrid Braided Composite Rods for Reinforcement and Health Monitoring of Structures. In: The Scientific World Journal 2014 (2014), Article ID 170187, http://dx.doi.org/10.1155/2014/170187

XU, L.; KIM, S.; ONG, C.-H.; HA, S.: Prediction of material properties of biaxial and triaxial braided textile composites. In: Journal of Composite Materials 46 (2012), No. 18, pp. 2255-2270

Chapter 9
Nonwoven Semi-finished Products and Nonwoven Production Technology

Kathrin Pietsch and Hilmar Fuchs

The properties of nonwoven-based semi-finished products are influenced much more significantly by their production methods than any yarn-based semi-finished products. Due to the multitude of available production methods, nonwoven semi-finished products offer a comparably specific and diversified property profile. In order to fully utilize this property potential of nonwoven-based semi-finished products for the characteristics of the composite material, basic knowledge of the combination between the structure and the characteristics of the nonwoven fabrics, in relation to various manufacturing methods is absolutely necessary. This chapter, beginning with the basic technological principles, reviews the combination of construction and structural as well as processing properties of the nonwoven semi-finished products in interaction with the manufacturing process. Conclusively, selected nonwoven-based lightweight construction solutions will be presented.

9.1 Introduction and Overview

9.1.1 Terminology

Nonwovens and *mats* in the widest sense are flat semi-finished fabrics made from fibers or filaments, whose cohesion relies on form-fit, friction-fit, or inter-fiber bonding. They are distinguished from other flat textile fabrics, such as woven,

---

K. Pietsch
Institute of Textile Machinery and High Performance Material Technology, TU Dresden, Dresden, Germany
e-mail: kathrin.pietsch@tu-dresden.de

H. Fuchs (✉)
Sächsisches Textilforschungsinstitut e.V., Chemnitz, Germany
e-mail: hfuchs@stfi.de

C. Cherif (ed.), *Textile Materials for Lightweight Constructions*,
DOI 10.1007/978-3-662-46341-3_9

braided, or knitted fabrics by not necessarily containing any type of yarn. Virtually any type of fiber of any desired length, and made from any raw material basis can be processed.

The use of the terms "nonwovens" and "mats" in practice is informed by specialist use. In textile technology, "nonwoven" refers to a manufactured sheet, web or batt of directionally or randomly orientated fibers, bonded by friction, and/or cohesion and/or adhesion", according to ISO 9092 [1]. A defined distinction from paper is made. In contrast to papers, wet nonwovens do not feature hydrogen bonds (wet-laid nonwoven process) between the web fibers. In plastics technology, a "mat" is a fabric used to reinforce thermoset and thermoplastic material, although the term is by definition limited to glass fibers [2]. In practice, it is often used for other fiber materials, especially those made from plants.

From a textile-technological perspective, semi-finished products referred to as "nonwovens" and "mats" are structured equivalently. Therefore, only the term "nonwoven" will be used below.

9.1.2 Overview

The manufacture of nonwoven semi-finished products includes the following sub-processes:

- processing of raw materials,
- nonwoven formation,
- nonwoven bonding, and
- nonwoven finishing

In contrast to textile fabrics made from yarns, nonwovens are produced in one continuous process using the dry, wet, or extrusion technique and omitting the process step of yarn production (Fig. 9.1).

The finishing of the nonwoven fabric web stored in roll form is usually performed discontinuously. The production methods differ primarily with regard to the presentation of the fiber raw material, and the manner of nonwoven fabric formation. Both fibers (fiber nonwovens) and filament (spun-bonded nonwoven) can be processed into nonwoven semi-finished products. The fibers can be formed into a plane nonwoven either in a dry or in a wet manner by means of fiber suspension. In the extrusion method, spinning the fibers and filaments is integrated into the nonwoven formation process. The bonding of the web fibers into a nonwoven fabric can be ensured by mechanical, chemical, or thermal methods.

The multitude of available production methods, fiber materials, and composite technologies allows the realization of a wider range of nonwoven material properties than for fabrics made from yarns (Fig. 9.2). In particular, *special structural properties*, such as porosity, high volume, and deformability can be realized, which is important for both semi-finished product characteristics and within the process chain.

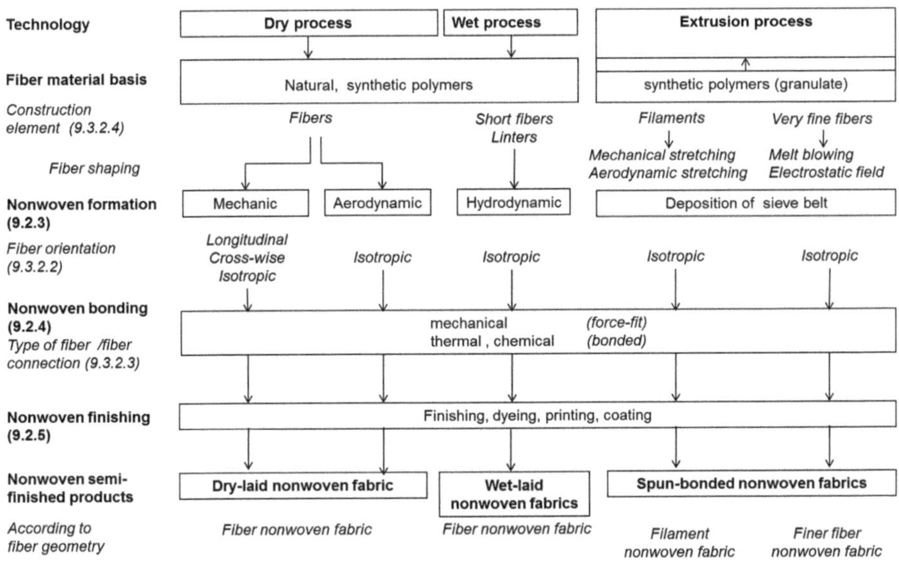

Fig. 9.1 Overview of the manufacturing processes and terminology of nonwoven semi-finished products

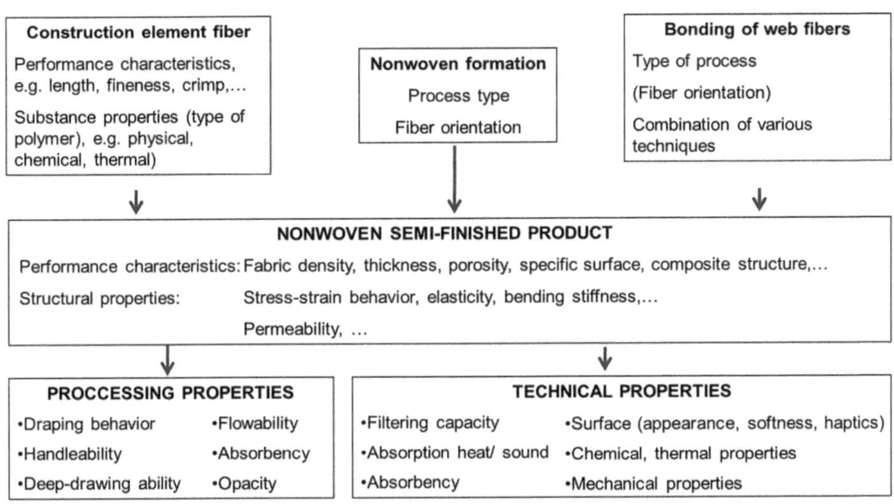

Fig. 9.2 Influencing variables on the construction and properties of nonwoven semi-finished products

Due to the relatively high production speeds during nonwoven production, which is several times faster than the manufacture of yarn-based products, nonwovens require product niches with large purchase quantities (mass markets).

Table 9.1 Fields of application for nonwoven fabrics in Europe (Consumption 2009: 1,615 million ton or 48.6 million m^2) (numerical data from [31])

	Market share related to		Other areas	Market share related to	
Lightweight construction-relevant fields	Mass	Area		Mass	Area
Construction	16.8	5.3	Hygiene/medicine	37.2	61.6
Filtration	6.4	6.4	Cleaning wipes	15.9	11.5
Upholstery/household products	4.7	3.0	Flooring	1.9	0.5
			Coating carriers	2.3	1.0
Automobile	3.5	1.3	Garments/shoes	3.7	2.1
			Agriculture	1.9	3.1
			Others	5.6	4.2
Total	31.4	16.0	Total	68.6	84.0

The *applications for nonwovens* are varied (Table 9.1) Important markets include hygiene, medicine, cleaning wipes, construction (construction engineering, automobile manufacture), and filtration. Traditionally, the fields of medicine and hygiene are the main areas of application. Here, primarily very light nonwoven materials are used as disposables, such as diapers, incontinence products, bandages, and surgical textiles. Nonwoven products with a longer time of use (durables) are used in filtering technology, automobile engineering (interior trims), and civil engineering (e.g. as insulating materials). The main raw fiber material is man-made fibers with diverse specifications on crude oil basis, primarily cost-effective polyolefin fibers such as polyethylene, polypropylene and viscose fibers (hygiene, medicine) and polyester and polyamide fibers (technical applications) [3].

In comparison to the abovementioned classical fields of application, the market share of lightweight semi-finished products in relation to the total nonwoven application amount is relatively small, both related to mass (ca. 30 %) and area (16 %) (see Table 9.1). Due to their special structural properties (Fig. 9.2), their primary importance is not as single materials, but in combination with other textile and non-textile semi-finished fabrics for lightweight construction. Often, the resulting realizable synergy effects allow functional integration in the component.

Nonwoven fabrics are therefore particularly important for the automobile industry. Nonwoven fabric composites are used in various automobile systems (engine, passenger, and luggage compartment), assuming such diverse functions as filtration or heat and sound insulation. Furthermore, special nonwoven fabric composites are an ecologically beneficial substitute for PU foam materials and can improve emission values in the interior of the vehicle.

In the context of limited availability of resources and the resulting increased requirements of product sustainability, lightweight construction solutions with materials based on renewable resources are becoming more and more important. For instance, natural-fiber-reinforced plastics (NFRP) are of interest to the automobile industry. Especially nonwoven-based semi-finished products with their

Table 9.2 Application fields for nonwoven semi-finished products in lightweight construction

Application fields	Function
Composites in automobile and machine engineering	Air filtration Insulation (sound, heat) Loading capacity (moldings) Substitution of PU foam material
Surface nonwovens (high performance-fiber-reinforced plastics)	Flow medium for plastic matrix Surface sealant (component)
Thermoset prepregs (SMC) Thermoplastic prepregs (GMT)	Reinforcement component for standard requirements
Adhesive nonwovens	Fixation of load-adapted yarn structures Joining of several fabrics into a laminate (composite material)
Natural-fiber-reinforced plastics (NFRK)	Automobile interior (sound/heat insulation, preforms)
Insulating materials	Structural engineering (sound insulation)

favorable emission and energy absorption behavior entail great innovation potential [4].

The application of nonwoven fabrics for the realization of textile-based lightweight constructions mainly comprises of five main fields, where the nonwoven fabrics assuming different function in both the semi-finished product composite and the processing (Table 9.2). Selected application examples are given in Sect. 9.4.

9.2 Production Methods for Nonwovens

9.2.1 Overview

The classification of production methods shown in Fig. 9.1 is the most common and is based on the basic technological principles of nonwoven formation with regard to the characteristic construction features (see Sect. 9.3).

Apart from these established processes, a number of special methods have been developed due to the increasing demand for lightweight construction materials, especially in automobile engineering. The airlaid blow molding method blows fibers onto a molded body. This resource-efficient method avoids material waste and allows a gradient construction with varying density and thickness of the nonwoven material [5]. With these technologies, e.g. NET forming [6] or fiber injection molding [7], fibers (especially natural and recycling fibers) are directly processed into molded bodies without the process step of nonwoven formation, which is required for conventional nonwoven production methods.

This Sect. 9.2 will give a comprehensive overview of the classical methods (Fig. 9.1), with which nonwovens are produced for the subsequent manufacture of semi-finished products.

9.2.2 Fiber Preparation

The processing of short-length fibers (compare Fig. 9.1; Dry method) into nonwoven-based semi-finished products first requires a defined preparation of the fibers. The purpose of this process step is to transfer the spun or recycled fibers in a suitable condition for the subsequent nonwoven formation. To realize the necessary sub-processes, several machines are available, which are connected modularly into continuously working production lines. Their configuration is determined by the requirements of the projected semi-finished product and the qualities of the fibers to be processed. The individual technological sub-processes can be performed separately, parallelly, or repeatedly. The sub-processes as well as the required machines are known from the production of spinning yarns (see Chap. 4). Technological and installation engineering details are given in [8].

9.2.3 Basic Principles of Nonwoven Formation

9.2.3.1 Preliminary Remark

The process step of *nonwoven formation* aims to form a homogeneous, even, and plane nonwoven fabric with a cohesion based at first entirely on the effects of friction between the short length fibers or, respectively, continuous nonwoven fibers. The technological subtasks of the nonwoven formation process are:

- Fiber dissolution of the fiber tufts into individual fibers (nonwoven fabric made from staple fibers)
- Formation of a nonwoven fabric web with:
  - defined fiber orientation,
  - defined areic weight, and
  - defined dimensions (width, thickness)

The quality of the nonwoven semi-finished products and the projected lightweight construction component is primarily influenced by the homogeneity of the fiber distribution across the width and along the length of the nonwoven fabric web. The structural properties of the nonwoven semi-finished products are influenced by the principle of nonwoven formation and by the process parameters. The formation of the fibers into a nonwoven fabric web can be performed according to various basic principles.

9.2.3.2 Mechanical Nonwoven Formation

The *mechanical nonwoven formation* consists of two sub-processes: card web formation and doubling.

Card web formation takes place on a carding unit, using the carding principle, in which the pre-dissolved fiber tufts are further dissolved into single fibers by the friction effects between rotating rollers with saw tooth surfaces (carding elements). This separation takes place between the main rollers, the working roller and the stripper roller (Fig. 9.3a). The carding process consists of several steps and is performed by several pairs of working and stripper rollers arranged at the upper circumference of the main roller. After the carding process, the now parallel, individualized fibers are located in machine direction (longitudinal orientation) on the entire surface of the working roller, from where they are removed as a cohesive fibrous web by the delivery roller.

The sub-processes of carding and removal of the fibrous web are achieved by different circumferential speeds of the rollers, and by different positions of the set of saw teeth. By placing a random card roller between main roller and the subsequent delivery roller (Fig. 9.3b), the position of the parallelized and production-direction-oriented fibers can be influenced. Different circumferential speeds and reverse rotational directions of the rollers allow a random placement of the fibers on the delivery roller (i.e. without any preferential direction). The physical fundamentals of web formation are included in the carding theory [8].

To create a fibrous web, usually modularly constructed carding units are used, whose working elements are adjusted to the fiber material to be processed. Fundamentally, all fibers of a length between 20 and 150 mm can be processed into a fibrous web by means of the carding process. The suitability of brittle fibers, such as mineral fibers, is limited. The same goes for the processing of carbon fibers, whose

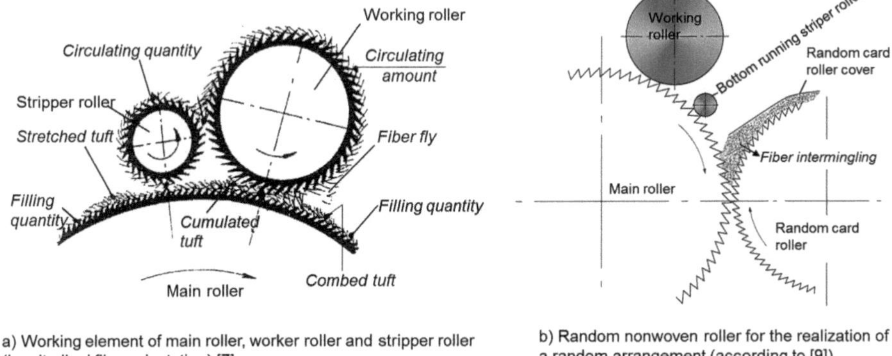

Fig. 9.3 Working elements and sub-processes in the carding process (according to [8, 9]). (**a**) Working element of main roller, worker roller and stripper roller (longitudinal fiber orientation) [7]. (**b**) Random nonwoven roller for the realization of a random arrangement (according to [9])

electrical conductivity makes special machine engineering safety precautions necessary.

To ensure cohesion of the fibers, a minimum number of fibers in the fibrous web is required. This number is determined by the fineness and defines the minimum realizable fibrous web weight. Nowadays, cards are also often used for the formation of fibrous webs.

The carding process is followed by *doubling*, which turns the fibrous web into a nonwoven fabric with higher fabric density, greater nonwoven width, and a defined fiber orientation adjusted to the requirements of the processing of semi-finished product. The nonwoven fabric construction is done by stacking individual layers of fibrous web, which also helps homogenize the fabric density across the nonwoven fabric web. To realize the process, which is also referred to as web forming, machines with a crosslapping design are most commonly used nowadays.

In *crosslapping*, the fibrous web is laid in a zig-zag pattern on a removal belt by oscillating feeding belts located above (Fig. 9.4). Fabric density, fiber orientation and width of the nonwoven fabric can be set by adjusting the speed ratios of the transport belts for the fed fibrous web and the formed nonwoven fabric as well as by varying the stroke of the oscillating nonwoven belt [8].

Crosslapped nonwovens often are characterized by a transverse fiber orientation. The fiber orientation can also be influenced by a downstream nonwoven drafting system according to the desired isotropic or anisotropic semi-finished product characteristics.

In practice, the carding unit, crosslapping, and optionally the drafting system required for mechanical nonwoven fabric formation are assembled into continuously working lines. The web speed of a formed nonwoven fabric with a width of ca. 6 m can be up to 100 m/min.

The crosslapping principle (Fig. 9.4) creates a two-dimensional (2D) fiber orientation. By using special laying methods, such as rotary laying, the nonwoven

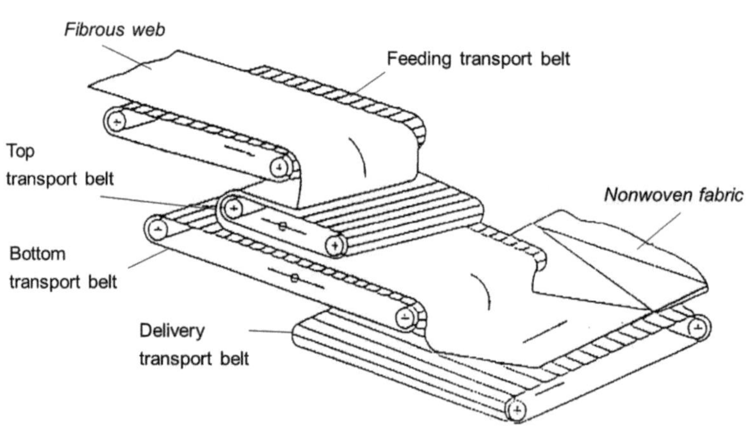

Fig. 9.4 Crosslapping principle [8]

fabric fibers can be oriented in Z direction [8], making three-dimensional (3D) nonwoven fabric structures possible.

9.2.3.3 Aerodynamic Nonwoven Fabric Formation

The *basic aerodynamic principle* is characterized by realizing the sub-processes disentangling, separation, and laying of the individual fibers by means of air flows (Fig. 9.5). Here, fibers (e.g. from upstream carding units) are removed from the surface of an opening roller by centrifugal forces and a tangentially directed air flow, and transported to the sieve belt surface (place of nonwoven formation). By means of a negative pressure, the transporting air is separated, compressing the fibers the moment they come into contact with the sieve belt surface and fixing them in position. In contrast to the carding method, this results in a more intensive isotropic fiber orientation. By means of a suction drum located above the fiber laying point, the fibers can additionally be oriented in Z direction. As the laid nonwoven already has the fabric density needed for the semi-finished product, no doubling is necessary. Nonwovens formed in this aerodynamic manner therefore have a compact 3D fiber structure and do not display the distinctive delamination tendency exhibited by mechanically produced nonwovens. This is important for the processing into a semi-finished product as well as for the properties of the projected composite.

The quality of the finished nonwoven semi-finished product is crucially influenced by the degree of opening of the pre-dissolved fiber tufts, and by the

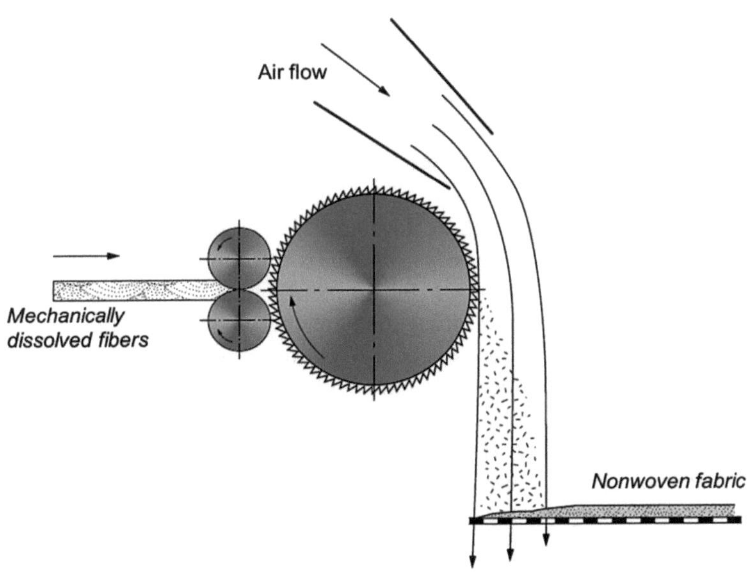

Fig. 9.5 Aerodynamic nonwoven formation principle (according to [9])

homogeneity of the air-fiber volume flow applied to the sieve belt surface. Particularly the compact 3D surface has a beneficial influence on the properties of the finished semi-finished product, as it results in isotropic strength characteristics combined with volume and compression elasticity.

The commercially available methods differ with regard to the manner of pre-disentanglement of the fiber tufts, and in the guiding of the air flows [8]. Due to the abovementioned characteristic nonwoven fabric properties and technological aspects, the aerodynamic method has economic advantages over other processes. Fundamentally, short as well as long fibers on natural or chemical basis, or recycling fibers can be processed into nonwoven semi-finished products cost-efficiently and at high productivity.

9.2.3.4 Hydroentangled Nonwoven Formation

The nonwoven formation in a wet manner is similar to the process of paper production. It includes the sub-processes of dispersion and laying of the fibers on a sieve belt as well as drying.

The nonwoven fibers are provided in fiber suspensions. To ensure homogeneity of the nonwoven semi-finished products, a homogeneous distribution of the fibers or a good dispersion capacity of the fibers in an aqueous medium is necessary. These are significantly influenced by the following fiber parameters:

- fineness ratio (fiber length, fiber fineness),
- wet stiffness,
- crimp
- wettability of the surface, and
- cutting quality

The nonwoven fibers (nonwoven formation) are placed on an inclined sieve belt. There, the water underneath the sieve belt is separated by means of negative pressure (dewatering boxes, Fig. 9.6a). The *hydroentangled method* is particularly suitable for the processing of short fibers (ca. 5 to ca. 25 mm) on natural or man-made basis as well as high-performance fibers such as GF, AR, and CF, often from production waste or recycling fibers into nonwovens. To connect the nonwoven fibers, binding agents in the form of fibers or plastic as dispersions (acrylates, butadiene styrene acrylonitrile) are mixed with the fiber suspension. Pulps, cellulose fibers as well as specially designed binding fibers are used as binding fibers; nonwoven fibers are connected by various binding mechanisms (hydrogen bonds, thermoplasticity, and solubility). By arranging additional headboxes over the sieve belt (Fig. 9.6b), multilayered web structures from different fiber materials can be manufactured in a single process.

The water remaining the formed nonwoven is removed by a subsequent drying process. The heat can be applied to the nonwoven fibers by means of various methods (radiation, convection, contact), simultaneously activating the binding agents. For this, a variety of dryer systems is available [8].

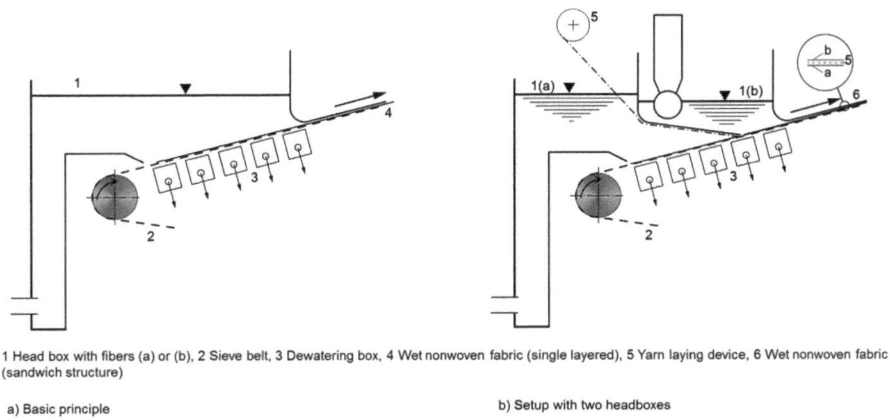

Fig. 9.6 Hydroentangled nonwoven formation principle (according to [9, 10]). (**a**) Basic principle. (**b**) Setup with two headboxes

Wet nonwovens are characterized by sufficient strengths to usually make a subsequent nonwoven consolidation obsolete (see Sect. 9.2.4). The homogeneous concentration of the fiber-water suspension is essential for the properties of the wet-laid nonwoven semi-finished products. For technological reasons, the majority of the fibers is orientated in machine direction in the resulting 2D fiber orientation. The fiber orientation can be adjusted by means of incrementing the negative pressure in the dewatering boxes. Further special characteristics of the wet nonwoven fabrics are a highly constant density at low mass level (20–400 g/m^2), and the adjustable high porosities.

The wet method also offers potential for high-performance fibers such as AR and CF. Corresponding to the growing demand, the amount of production waste of these high-priced fibers has increased. The hydroentangled method is an opportunity to recycle high-performance fiber waste for high-quality applications, e.g. in the form of semi-finished products used in the LFT-D process. Feasibility studies with recycled glass fibers show that the property parameters of composite materials produced in such a manner are comparable to those of primary fibers [11]. Wet nonwoven semi-finished products made from recycled CF can be processed using injection molding, the LFT-D process, laminating, or RTM methods [12]. Furthermore, wet nonwovens with highly fibrillated AR pulps (0.2–1.6 mm) can be functionalized in various ways, for instance to achieve special filtration properties or electromagnetic shielding effects [13].

Machine lines for the production of wet nonwovens, which also referred to as inclined sieve conveyors or hydroformers, consist of individual zones for material preparation, nonwoven formation, and drying/impregnating. Depending on the respective product line, the production speed for glass fiber nonwovens can be between 100 and 400 m/min.

9.2.3.5 Extrusion Process

In the extrusion process, extruded polymer melt flows are formed immediately into continuous-fiber-based fabrics. In contrast to the dry and wet nonwoven formation processes, in which length-limited fibers are processed into a nonwoven web, the extrusion process starts with granulated polymer as base materials (see Fig. 9.1). As the process step of filament formation is integrated into the nonwoven formation process, this method is often referred to "direct technology".

The whole process can be separated into the sub-steps of spinning, stretching, and nonwoven formation, all of which are performed simultaneously (Fig. 9.7). Fundamentally, all polymer materials on synthetic or mineral basis (glass, ceramics, rocks), from whose melt a yarn can be drawn, are processable into a spun-bonded nonwovens, by means of the extrusion method.

The filament formation is performed using the melt spinning and dissolution spinning procedures known from use in the field of chemical fibers: (see Chap. 3). The melt spinning method is most common due to the simple process design. PP, PES, and PA fibers are spun using this method, with PP being the most important base material in the industry for reasons of its low price and advantageous properties (see Sect. 9.1.2).

The granulated base material is plasticized using an extruder, and then extruded through a multitude of spinnerets. After the exit of the molten yarn from the spinneret, it is non-positively stretched by air. Thereby the molecule chains are directed, which is of fundamental importance for the mechanical properties of the later semi-finished product. The stretched and almost solidified filaments are

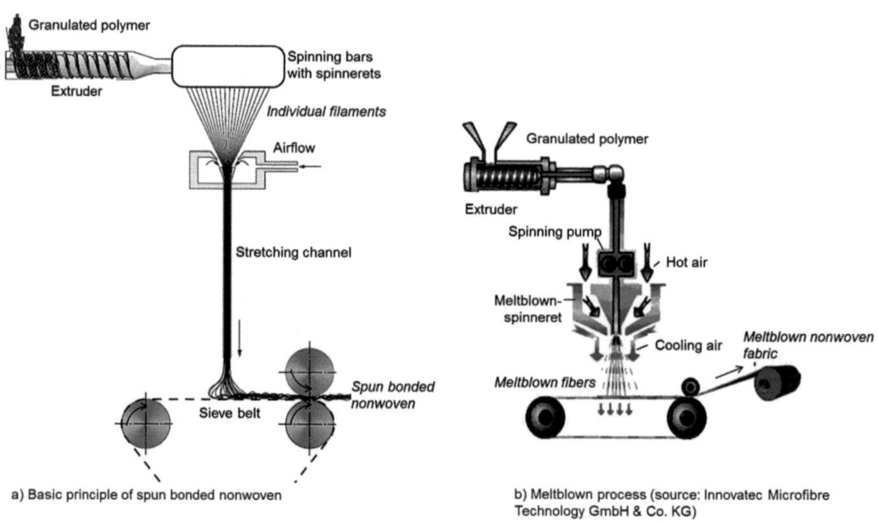

Fig. 9.7 Process schematics for the production of filament nonwovens and very fine fiber nonwovens. (**a**) Basic principle of spun bounded nonwoven. (**b**) Meltblown process (*Source*: Innovatec Microfibre Technology GmbH & Co. KG)

tubularly placed on a perforated band, where they are fixed by a suction stream of air (nonwoven formation).

Figure 9.7 is to be understood as the basic principle of the entire process. The commercially applied methods differ with regard to the process-technological design of the individual process steps and the construction of the separate modules, which are detailed in [8]. The configuration of the individual modules and the design of the sub-processes have to be adjusted to the respective spun-bonded semi-finished product.

Filament nonwoven materials are characterized by a 2D continuous-fiber orientation, which can be controlled by means of altering the speed ratios (yarns, nonwoven web) at the time of filament placement. Different process modifications also allow the spinning of very fine fibers. By means of a special spinneret and hot air flows, the polymer melt can be blown into microfine fibers (meltblown process, Fig. 9.7b). The microfibers are placed on a sieve surface. By means of the electrospinning method, nano-scaled fiber layers can be created [8]. Often, these processes are integrated into other nonwoven formation processes such as carding or the spunbonded nonwoven process. This way, nonwoven material composites with gradient design, for example regarding their porosity, can be realized. Important technical properties, such as the separation behavior of the acoustic insulation effect can therefore be adjusted in a manner that suits the respective requirements of the application.

9.2.4 Basic Principles of Nonwoven Bonding

9.2.4.1 Overview

The cohesion of the fibrous web depends entirely on the active friction forces between the fibers or filaments. The resulting mechanical properties do not meet the requirements of the projected nonwoven semi-finished product. To improve the mechanical properties, it is important to increase the strength of the fiber bonds. For this, several physical and chemical technologies are available to strengthen the nonwovens into nonwoven fabrics. The characteristics of the nonwoven semi-finished product are significantly influenced by the bonding technology. Figure 9.8 illustrates that even only mechanical bonding technologies (compare a–c), with which the nonwoven fibers are bonded in friction-fit and form-fit manner, can create a number of quite different nonwoven material structures. This gives reason to expect correspondingly differing semi-finished product properties.

The combination of different bonding methods allows the definition of specific nonwoven fabric properties. Furthermore, the bonding methods are used to connect different types of fibrous non-wovens to each other or with different types of fabrics (see Sect. 9.4.2). By utilizing synergy effects, this allows the manufacture of requirement-adapted, nonwoven-based semi-finished products.

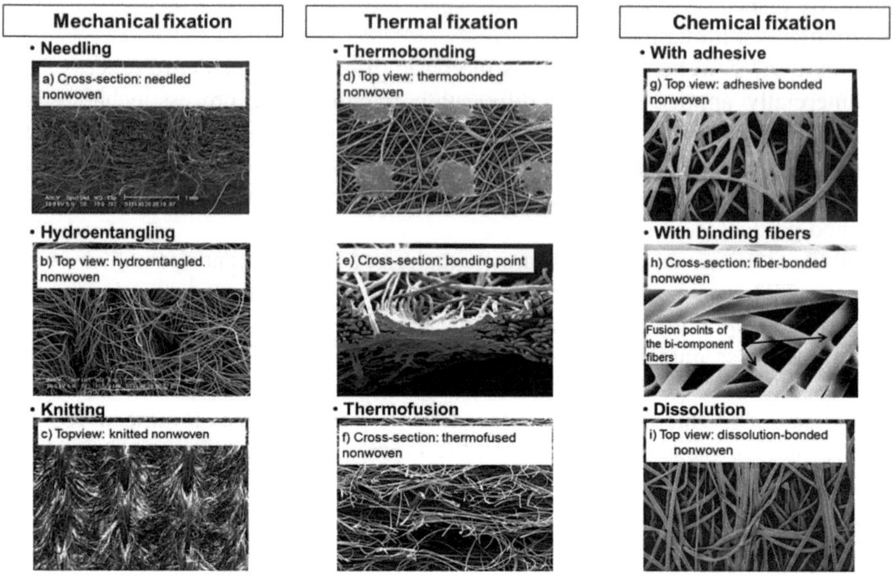

Fig. 9.8 Overview of the bonding methods and the resulting nonwoven fabric structures (*Photos*: Sächsisches Textilforschungsinstitut e.V.). (**a**). Cross-section: needled nonwoven. (**b**) Top view: hydroentangled nonwoven. (**c**) Top view: knitted nonwoven. (**d**) Top view: thermobonded nonwoven. (**e**) Cross-section: bonding point. (**f**) Cross-section: thermofused nonwoven. (**g**) Top view: adhesive bonded nonwoven. (**h**) Cross-section: fiber-bonded nonwoven. (**i**) Top view: dissolution-bonded nonwoven

9.2.4.2 Mechanical Bonding

By comparison, mechanical bonding methods are the most important process in use. Mechanical bonding is realized by the basic principles of needling, knitting, and entangling.

Needling

In needling, the web fibers are entangled by means of special-geometry needles (Fig. 9.9). The barbed needles pierce the fibrous web and condense it. The fibers carried by the barbs are oriented in the needle hole channel (orthogonal to the web plane) during the downward movement of the needle. A special barb geometry ensures the preservation of the orthogonal orientation of a part of the web fibers in the needle hole channel during the upward movement of the needle. Thus, a 3D fiber orientation is achieved (see Fig. 9.8a).

For lightweight construction applications, fibrous webs made from glass fibers and filaments are often needled. Such "mats" (see Sect. 9.1.1) are used for the production of thermoplastic prepregs, so-called "glass mat-reinforced thermoplastics" (GMT, Chap. 11).

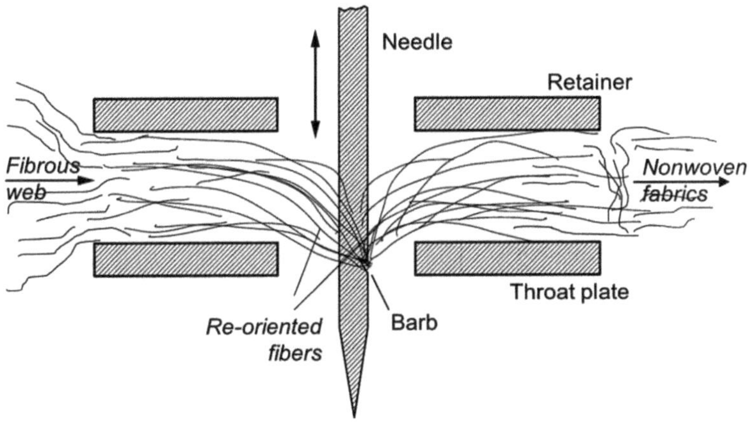

Fig. 9.9 Basic needling principle

The structure and properties of needled nonwovens can be adjusted to requirements by various needling parameters, such as the geometry, needle hole depth, and density of the needles, or the speed of the fibrous web. Furthermore, a variety of modifications to the working elements allow the manufacture of nonwoven fabrics with specific structural effects [8].

Stitch-Bonding

By stitch-bonding and warp-knitting, fibrous webs can be bonded or processed for composite production.

The warp-knitting-based MALIMO stitch-bonding process has become most important for the bonding of fibrous webs. Several process modifications have been developed for this method, which are given in an overview by Fig. 9.10.

Loops are the strengthening structural element, with a distinction to be drawn between loops made from knitting yarns and loops made from web fibers (Fig. 9.10). Despite the different process modifications, the loop formation cycle basically consists of the sub-phases known from knitting processes: holding-down, yarn laying, loop forming, and knocking-over. The working principles of the most important methods for nonwoven bonding are briefly introduced below. A comprehensive overview of nonwoven-processing knitting methods and process-technological details are given in [8].

By means of *nonwoven stitch-bonding*, the preferably transversely oriented web fibers are integrated into a meshed network of knitting yarns, without being part of the loop formation process. Figure 9.11a shows the working elements of the MALIMO stitch-bonding process of the *Maliwatt* type, with which the knitting yarns can be integrated into a pillar or tricot lapping. The nonwoven fabric bonded in this manner is a composite of knitting yarns and fibrous web ("Nonwoven composite", Fig. 9.11b). By using additional knitting yarn systems, other basic

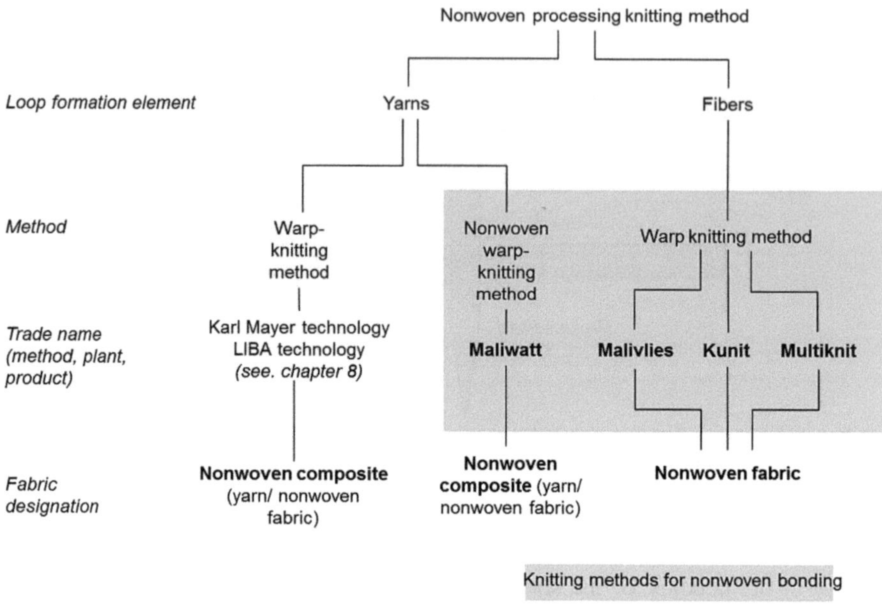

Fig. 9.10 Overview of the MALIMO stitch-bonding process of nonwoven bonding [8]

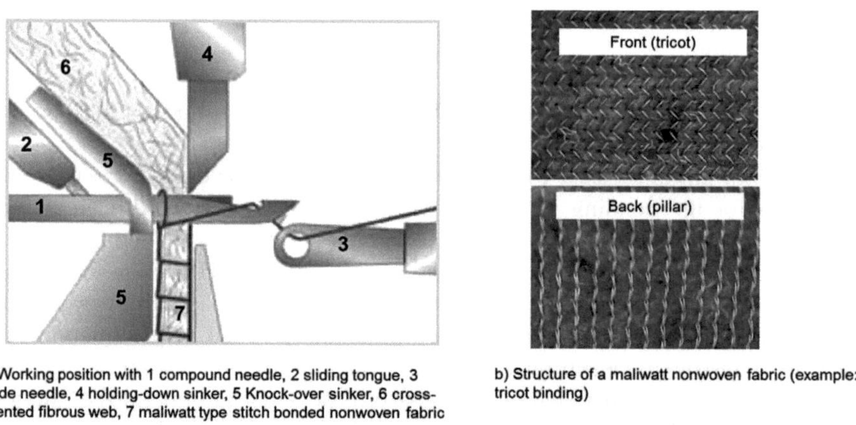

Fig. 9.11 Nonwoven stitch-bonding technique, Maliwatt type (*Source*: Karl Mayer MALIMO Textilmaschinenfabrik GmbH). (**a**) Working position with 1 compound needle, 2 sliding tongue, 3 guide needle, 4 holding-down sinker, 5 knock-over sinker, 6 cross-oriented fibrous web, 7 maliwatt type stitch bonded nonwoven fabric. (**b**) Structure of a maliwatt nonwoven fabric (example: tricot binding)

stitches like plain, velvet, or satin stitch are possible, as are parallel weft insertion and the resulting requirement-adapted adjustment of the stress-strain behavior of the nonwoven semi-finished products. Various adjusted installation-engineering

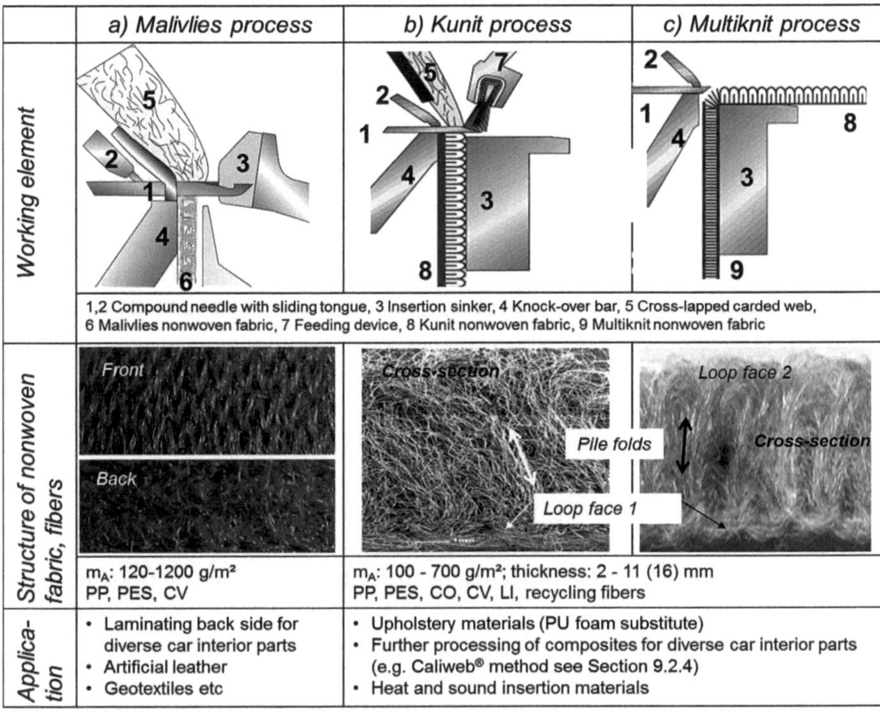

Fig. 9.12 Basic nonwoven knitting principles and structures of fiber nonwoven knitting fabrics (*Source*: Karl Mayer MALIMO Textilmaschinenfabrik GmbH)

solutions allow the bonding/processing of short-staple primary fibers (production waste), secondary fibers, and glass fibers ("glass mats").

In contrast, *nonwoven stitch-bonding* forms the web fibers themselves into loops (structural elements). This can be achieved by another process modification known by their trade names Malivlies, Kunit, and Multiknit (see Fig. 9.10), which differ with regard to the construction of the working elements and the realized loop structures.

In the *Malivlies technique* (Fig. 9.12a), the crosswise-fed fibrous web is pierced by compound needles. During the loop formation cycle, a part of the web fibers is retained at the side facing the insertion sinkers and formed into loops, while the other part is only integrated into the fiber loops. Malivlies nonwoven fabrics are characterized by the visible loop legs on their front.

The working elements of the *Kunit technique* (Fig. 9.12b) feature an additional oscillating stuffer feeding device with which the web fibers are laid into pile folds before loop formation, orienting them in Z direction. The length of the pile folds, and thus the thickness of the Kunit nonwoven fabric is variably adjustable.

Corresponding to the length of the pile folds, 3D nonwoven fabric geometries can be realized. The *Multiknit* technique (Fig. 9.12c) allows the further processing of fabrics with pile fiber structures, in which the pile fibers are formed into loops. This way, single- or multi-layer nonwoven fabric structured can be realized, even in combination with other types of fabric. The orientation of the pile fibers in Z direction, and the loop structure result in a voluminous, compact 3D fiber structure. The corresponding compression elasticity is utilized in alternative upholstery materials. Especially in automobile engineering, they are an ecologically viable substitute for PU foam material. A process solution based on the Multiknit method and referred to as Caliweb (see Sect. 9.4.2) has been developed for the economic production of seat components and textile car interior components.

Hydroentanglement

The nonwoven fibers are bonded by means of water jets. The high-pressure jets make contact with the web surface at a right angle, seizing the fibers, re-orienting and entangling them with other fibers elements (Fig. 9.13). Water-jet-bonded nonwovens (compare Fig. 9.8b) are characterized by their special softness, specific strength, and haptics. Therefore, they are primarily used in the fields of medicine, hygiene, and household products.

Similar to the needling technology, the hydroentangling technology has been undergoing an especially dynamic development over the past 20 years. Hydroentangling technology can be used to connect textile fabrics of different kinds for the purpose of composite production (Sect. 9.4.2). The method is suitable for surface structuring for the sake of functionalization, for instance to improve insulation effects or to realize spacers.

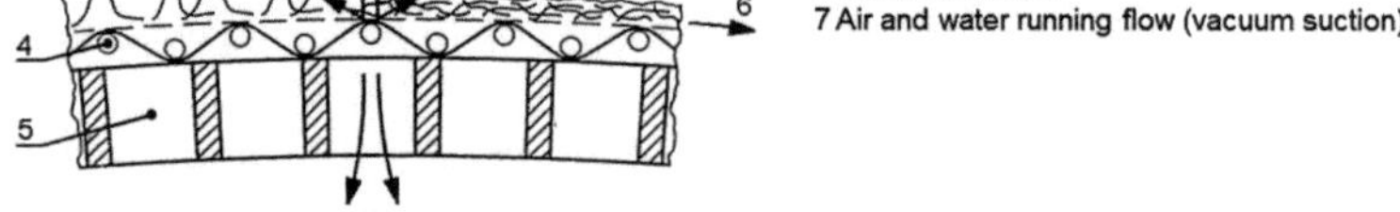

Fig. 9.13 Basic principle of hydroentanglement [8]

9.2.4.3 Chemical Bonding

By means of binding agents, web fibers can be connected adhesively by chemical means. The binding agent can be a liquid (plastic dispersion) or a solid (binding fiber, or powder) on a thermoplastic basis.

Various methods are available for applying the binding agent. *Padding* (Fig. 9.14) is a classical, well-established method to apply liquid binding agents, in which the fibrous web is thoroughly impregnated in a trough filled with a binding agent bath. After removal from the bath, the excess is mechanically removed by a pair of squeeze rollers. The binding agent content can be adjusted by the nonwoven speed and roller spacing, but with due regard to the wetting properties of the web (see Sect. 9.3.3.4).

Binding agents of non-liquid consistency are applied by means of other methods, including blade coating (pastes), spraying, slop padding, impregnation (bath, foam), printing, or spreading (powder). The application of binding agents is followed by a heat treatment phase to active the liquid (coagulation, cross-linking) or solid binding agents (melting). In the case of liquid binding agents, this also removes the remaining humidity. Usually, dryer systems familiar from textile finishing are used for this (sieve cylinder or conveyor setup).

Fibrous webs from cut or continuous glass fibers ("mats", see Sect. 9.4.3) are often bonded with liquid binding agents. If the nonwoven fabric is created in a wet process (see Sect. 9.2.3.4), the binding agent is added to the fiber suspension. The strength properties of the bonded nonwovens are determined by the cohesion and adhesion characteristics of the binding agent, and by the content ratio of the binding agent. The arrangement of the binding agent at the fiber crossing points can be influenced to range from "holohedral" or "sail-like" (compare Fig. 9.8g) to "point-like" (compare Fig. 9.8h) by altering the wetting behavior, which allows a definition of the mechanical properties, e.g. flexural properties, of the nonwoven fabric semi-finished products.

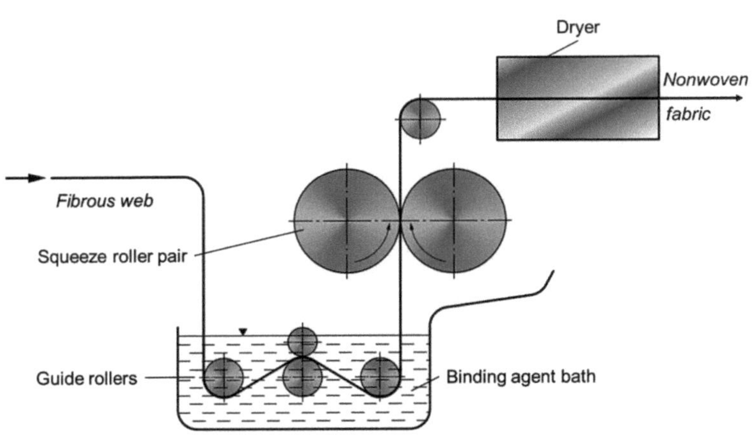

Fig. 9.14 Basic principle of applying liquid binding agents (according to [8])

9.2.4.4 Thermal Bonding

Heat treatment is another possibility to bond nonwovens in a physical manner. This bonding of web fibers relies on thermoplasticity, which requires the use of synthetic fibers with thermoplastic properties. Non-thermoplastic fibers can be bonded thermally by using so-called binding fibers. In contrast to conventional thermoplastic fiber polymers, their softening and melting characteristics for achieving the optimal thermal binding behavior are specially adjusted. Depending on the type of web fibers and the desired properties of the projected nonwoven semi-finished product, various types of thermoplastic binding fibers are available (Table 9.3). Bi-component fibers with a core-sheath structure (see Chap. 3) result in an especially voluminous, soft, and spring-elastic nonwoven fabric character, as the nonwoven and core fibers are only connected at points by the low-melting sheath polymer (see Fig. 9.8h).

The heat transfer into the nonwoven to be bonded is performed according to thermal conduction, convection, or radiation principles. Calendar and thermofusion technologies are the most important methods in practice. In *calendar bonding* (Fig. 9.15a), the web fibers are plasticized in the nip of the roller under pressure and temperature effects, resulting the bonding. The use of a steel roller with surface gravure allows the manufacture of highly flexible nonwoven fabrics. Here, the fibers are only compressed underneath the embossed gravure points. Between the points, the web fibers retain their fiber structure and their textile properties (Fig. 9.8d, e). Using the geometry of the gravure surface, the flexural rigidity can be set specifically. The method is primarily suitable for the bonding of light PP webs.

Thermofusion is usually realized by means of driers, often convection driers. The fibrous web to be bonded is perfused orthogonally to the surface by hot air, which condenses it only slightly (Fig. 9.15b). Using bico fibers (see Table 9.3), particularly voluminous and compression-elastic nonwoven fabric structures are achieved.

Due to their simplicity and eco-friendliness, thermal methods are used more and more frequently in comparison to bonding with binding agents.

Table 9.3 Thermoplastic binding fibers

Binding fiber type	Processing temperature (°C)
Homopolymer fibers (PP)	>120
Copolymer fibers (Co-PA, Co-PES)	150 ... 210
Unstretched, amorphous PES fibers	>80
Bi-component (Bico) fibers, e.g. with core/sheath polymer PP/PE PES/Co-PES PA6/PA6.6	80 ... 115 150 ... 210 220 ... 250

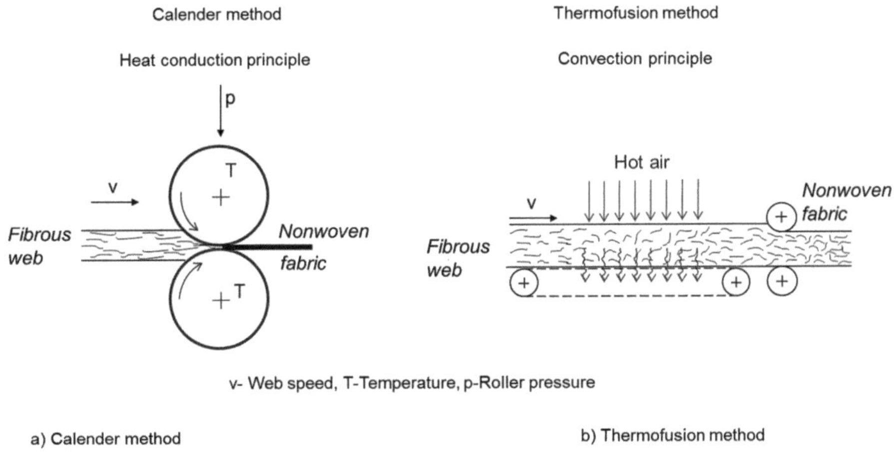

Fig. 9.15 Thermal bonding methods. (**a**) Calender method. (**b**) Thermofusion method

9.2.5 Finishing

Nonwoven fabrics can be finished with all known mechanical, thermal, and chemical methods. The technological basic principles for finishing, printing, dyeing and coating are given in Chap. 13. This way, specific functionalities, such as hydrophobicity and hydrophility, can be applied to the surface of the nonwoven fabric.

9.3 Structure and Characteristics

9.3.1 Overview

The construction and *structural characteristics* of nonwoven fabrics are far more significantly influenced by production than is the case for yarn-based materials. Due to the immediate construction from fibers on the one hand and the lack of a "yarn" sub-structure on the other hand, comparably very special structural properties, e.g. defined porosity, are realizable. Even though the mechanical property level of yarn-based materials cannot be achieved, the utilization of nonwoven fabrics allows the implementation of specific technical property parameters, such as absorbency into the semi-finished product, which is important for further processing and for the characteristics of the semi-finished products/finished components. Figure 9.16 depicts the basic construction parameters, characteristic structural properties, and technical features.

Simulation methods and software tools enabling predictions and visualizations of structure-property relations on the basis of virtual nonwoven fabric models have been developed. Therefore, technical features, e.g. inflow and perfusion behavior,

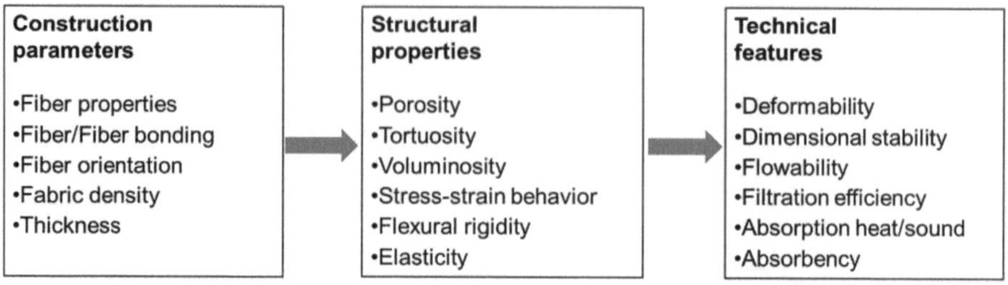

Fig. 9.16 Construction parameters and characteristic properties of nonwoven fabrics and nonwoven semi-finished products

particle transport, or isolation capacity (heat, sound), all in relation to the macro- and microstructural parameters of the nonwoven fabric structure, can be pre-calculated by means of specialized software modules [14].

Despite the availability of modern simulation technology, basic textile-technological and material-technological knowledge of the correlations between the structure and the properties of the nonwoven fabric constructions (structure-property relations) is an important tool in life cycle-oriented product development They are essential to the thorough utilization of the property potential of nonwoven fabrics efficiently for the characteristics of semi-finished products and finished components.

9.3.2 Construction Parameters

9.3.2.1 Fabric Mass and Thickness

The macrostructure of a nonwoven fabric is generally characterized by the parameters of fabric mass [15] and thickness [16], which are to be determined according to the ISO9073 standard series. Depending on their fabric mass, nonwoven fabrics are classified as "light" (<60 g/m^2), "medium heavy" (up to 150 g/m^2), and "heavy" (>150 g/m^2).

9.3.2.2 Fiber Orientation

The orientation of the web fibers in relation to the machine direction (MD) is another basic structural parameter. The orientation has significant influence on the mechanical structure properties of a nonwoven semi-finished product such as strength, stress-strain behavior, elasticity and flexural stiffness. The fibers can be undirected or definitely oriented longitudinally or transversely (Fig. 9.17). The mechanical properties of the nonwoven semi-finished products are therefore adjustable in a wide range between isotropic and anisotropic.

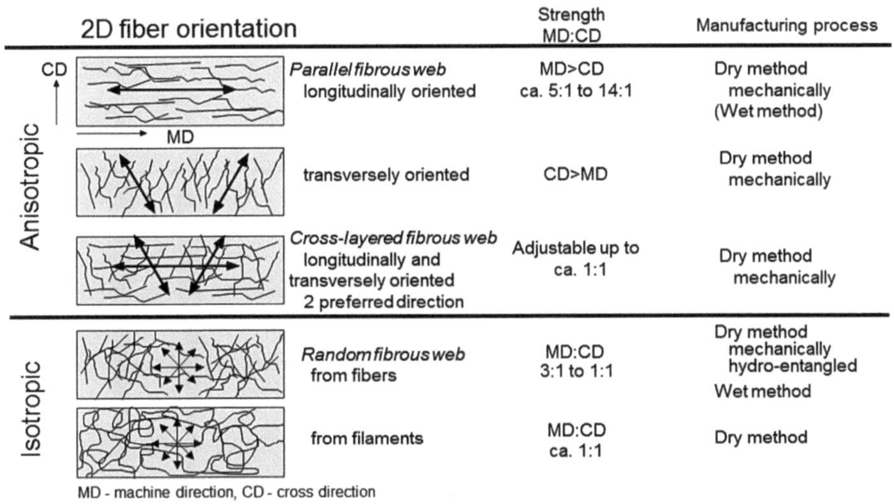

Fig. 9.17 2D fiber orientation in nonwoven fabrics

As a sophisticated image analysis is required to directly determine the fiber orientation, the relation of longitudinal and transverse strength (which is determined according to ISO 9073-3 [17]) is used as the parameter for describing the 2D fiber orientation in nonwoven fabrics. The homogeneity of the distribution and orientation of the fibers in the nonwoven semi-finished product are important quality parameters for nonwoven fabrics.

Special methods such as aerodynamic method (see Fig. 9.5), special laying techniques (see Fig. 9.4), or knitting methods (Kunit, Multiknit see Fig. 9.12) also allow an orientation of the fibers in Z direction, or a 3D-isotropic arrangement of the web fibers. This helps achieve compression-elastic properties in the nonwoven fabric plane, which makes the fabrics a viable option for replacing PU foam materials.

9.3.2.3 Type of Fiber/Fiber Bonding

The bonding structures realizable with the variety of bonding structures between the web fibers are based on the effective principles of frictional, form-fit, and adhesive connection. Figure 9.18 contains an overview of the morphologies on bonding points and their characteristic properties. The flexibility of the bonding points significantly defines the structure-mechanical nonwoven fabric properties, which are especially important for the semi-finished product processing.

Nonwoven fabrics with friction-based and form-fit connections are characterized by low strength and elasticity. The flexing binding points allow great drapeability and deformability as well as sufficient flowability of the web fibers during molding. Fixed binding points, as those of webs bonded with binding fibers and binding

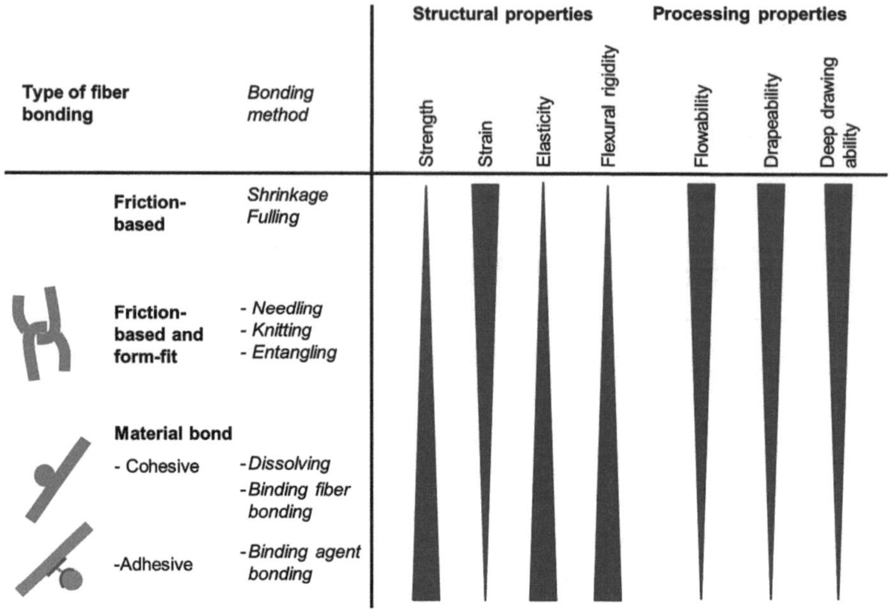

Fig. 9.18 Type of bonding between web fibers, and their qualitative influence on the properties of the nonwoven semi-finished products

agents, result in qualitatively higher specific mechanical property values. Such nonwoven semi-finished products are used for instance as temporary fixation of reinforcement structures, due to their comparably higher dimensional accuracy (see Sect. 9.4.3.3).

9.3.2.4 Construction Element Fiber

The parameters of the web fibers, which are the actual design element, influence the nonwoven fabric-specific structural characteristics (see Sect. 9.3.3) like voluminosity and porosity, in a variety of ways. By selecting the corresponding fiber parameters, differentiated absorption and permeability characteristics, such as sound insulation and absorbency, can therefore be implemented in the nonwoven semi-finished products. Figure 9.19 gives an overview of the fiber parameters relevant for structural and nonwoven fabric properties.

The *fiber length* primarily influences the cohesion effect between the web fibers. A length increase of the fibers improves tensile strength, elasticity, and other mechanical characteristics. The flowability in the mold is impeded by longer fibers.

Among others, *fiber fineness* is a relevant factor in the specific fiber surface and porosity, among others. With finer fibers, comparably large active specific fiber surfaces and smaller pore sized in the nonwoven fabric can be realized. Nonwoven fabric properties, such as the filtration efficiency and absorbency can be improved.

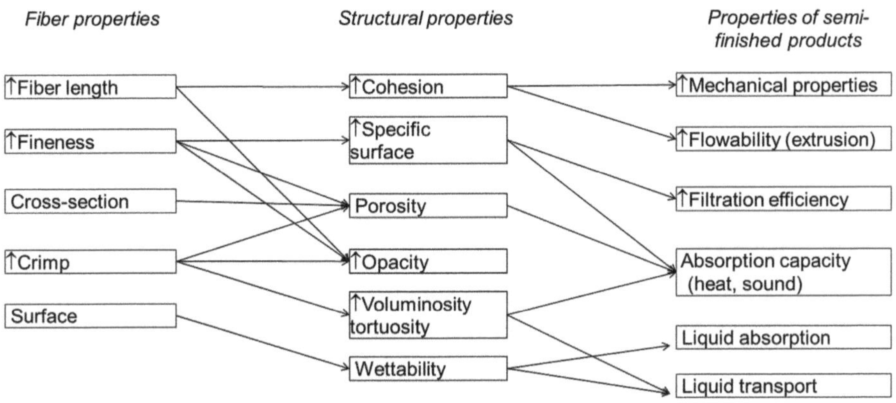

Fig. 9.19 Qualitative relations between fiber characteristics, structure, and properties of the nonwoven fabrics

Especially in surface nonwoven fabrics, finer fibers allow a defined absorption and higher opacity. Furthermore, finer fibers make smaller bending radii in the mold and very high filtration efficiency possible.

Fiber crimp, characterized by the number and size of the crimp curves, is important to the cohesion between the web fibers themselves. Increases in crimp lead to enhanced strength, elasticity, and voluminosity. In synthetic fibers, defined crimp effects can be created by means of texturizing processes. Plant fibers show hardly any crimp, and fiber cohesion is facilitated by partial separations of fiber bundles and elementary filaments.

With the *fiber cross-sections*, especially with profile cross-sections, the specific fiber surface and the opacity can be increased. Highly profiled fibers and hollow fibers improve heat insulation capacity. The ability of the hollow fibers for capillary humidity transport gives the nonwoven fabric construction a good sorption behavior.

The type of fiber polymer is decisive for the substance-associated nonwoven fabric properties, e.g. the resistance to chemicals and temperatures. The texture of the fiber surface has considerable influence on friction and sliding processes between the nonwoven fibers (flowability) and with other materials, and on the adhesion and wettability properties.

9.3.3 Selected Structural Properties

9.3.3.1 Voluminosity

Voluminosity is crucial for various technical characteristics, especially for the absorption behavior for heat and sound. In general, nonwoven fabrics containing more air than fibers can be referred to as voluminous [18]. The voluminosity of a nonwoven fabric with the geometric dimensions of width, length, and thickness can be quantified by means of various parameters (Eqs. 9.1–9.4):

Loft factor (according to Jordan) [18]

$$L_F = \frac{V_V}{V_F} = \frac{d}{m_A} \cdot \rho_F \tag{9.1}$$

Raw density [18]

$$\rho_V = \frac{m_A}{d} \tag{9.2}$$

Air volume [18]

$$V_L = \left(1 - \frac{\rho_V}{\rho_F}\right) \cdot 100\,\% \tag{9.3}$$

Proportion of enclosed air [18, 19]

$$\begin{aligned} v_L &\approx \left[1 - \frac{1}{L_F}\right] \cdot 100\,\% \\ v_L &= P \end{aligned} \tag{9.4}$$

Where,

D (m)	Nonwoven fabric thickness
V_V (m^3)	Total volume of nonwoven fabric
V_F (m^3)	Fiber volume
V_L (m^3)	Enclosed air volume
L_F (−)	Loft factor according to Jordan
ρ_F (kg/m^3)	Fiber material density
m_A (kg/m^2)	Fabric density nonwoven fabric
ρ_V (kg/m^3)	Raw density Nonwoven fabric
V_L (%)	Proportion of enclosed air
P (%)	Porosity

In general, nonwoven fabric constructions are referred to as voluminous or "*highloft*", if their solid material proportions is maximum 10 % of their volume, or if they have a Loft factor >40 at thickness above 3 mm. Highloft nonwoven fabrics are characterized by great heat insulation capacity, which can be adjusted diversely to meet the requirements by the production method (e.g. aerodynamic nonwoven formation, vertical placement) and specific fiber selection.

9.3.3.2 Porosity

Porosity is another important structural property and crucial for the various transport and separation processes of solid particles, liquids, and gasses. The porosity of nonwoven fabrics can be determined by measuring the air permeability (or the air flow resistance) [20]. By means of porometric measuring methods, the pore structure of the nonwoven fabrics can be quantified on the basis of pore size and pore size distribution [21, 22]. Pore size is an important parameter, especially for filtration processes. Nonwoven fabrics are used to separate gasses, liquids, and solid in surface and deep filtration. Furthermore, pore geometry is important for wetting processes (see Sect. 9.3.3.4).

Apart from *pore size*, the interconnection of the pores (interconnectivity) is important for the barrier effect of the nonwoven fabric structure. In contrast to yarn-based materials, nonwoven fabrics are characterized by a highly complex system of winding pore channels with changing cross-sections (pore labyrinth). An approximated value for the complexity of a pore labyrinth is given by the *tortuosity factor*. The tortuosity factor τ described for nonwoven fabrics by Batchu [23], is the reciprocal value of porosity P (Eq. 9.5) or the deviation from the ideal cylindrical-shape channel:

$$\tau = \frac{1}{P} \tag{9.5}$$

τ (−) Tortuosity factor
P (%) Porosity

The length of a pore channel is the product of the nonwoven fabric thickness and the tortuosity factor [19, 23]:

$$l_P = d \cdot \tau \tag{9.6}$$

lp (μm) Length of a pore channel
τ (−) Tortuosity factor
d (μm) Nonwoven fabric thickness

Nonwoven fabric constructions with a 3D-isotropic fiber arrangement and high loft characteristics, such as aerodynamically produced nonwoven fabrics, are characterized by a particularly pronounced tortuosity. The labyrinth structure is important for a variety of technical processes, such as perfusion with resin or air, and sound absorption (compare Sect. 9.3.3.3).

9.3.3.3 Sound Insulation

Noise and sound insulation are important from the perspective of health protection issues in industrial applications, and in practically all other spheres of life such as construction or traffic. In vehicle engineering, noise comfort is an essential consideration. Nonwoven fabrics make an important contribution in these fields, which has given sound protection a central place in the creation of standards for technical nonwoven fabrics [8].

The insulation effect for sound and noise is based on processes of sound absorption and sound dampening. Sound is absorbed by converting sound energy into heat, with the energy being absorbed by the absorbing material by means of friction effects. Porous materials with open pores are particularly suitable for this purpose. As the sound energy cannot always be absorbed completely on the source of the sound, dampening measures are necessary.

Sound-reflecting barriers or sound-dampening measures aim to impede the diffusion and transmittance of sound waves, e.g. between two rooms [8, 24].

The principle of footstep sound insulation relies on the elastic decoupling the source of the sound disturbance at the bearing or linking by means of a spring-mass system. Especially insulating materials with low dynamic stiffnesses are suitable for footstep sound insulation. Nonwoven fabrics made from finest fibers in a meltblown process (Fig. 9.7b) or an electro-spinning process [25], provide great absorption effects. Such highly-efficient absorber systems are used as "mats" or moldings. In a composite with heavy layers, very effective spring-mass systems with significantly reduced semi-finished product weights are feasible. The combination of several nonwoven fabric layers allows the realization of different absorption degrees in one composite structure, which can then absorb sound in wide range of frequencies. High values of thickness and fabric density facilitate sound insulation. Micro-structure parameters, e.g. fiber orientation, tortuosity, pore structure, influence the sound absorption efficiency [24].

For the acoustic optimization of viscoelastic insulation structures, a special simulation module for nonwoven fabrics was developed. It is based on the micro-structure simulation software "GeoDict" [14] and allows the pre-calculation of acoustic material properties depending on the micro-structure parameters of the nonwoven fabric [26]. Multi-layered compression-molded components are used for sound dampening in car interiors. Using the "AcoustoDict" simulation module and other software tools, the nonwoven-fabric-based composite structure from pure PES could be generated according to automobile industry requirements and realized commercially as car headliners in various renowned vehicle types. This composite solution displays similar sound-absorbing, mechanical, and optical properties, but is much more recycling-friendly than conventional absorber systems [27]. This approach, which also allows time-saving and more economical development and design of highly efficient absorber systems without prototype production and extensive measuring series, can also be used for other areas of use, e.g. in construction or machine engineering.

9.3.3.4 Absorbency

The capacity to absorb and transport liquids is another characteristic property of nonwoven fabrics determined by the chemical properties of the fiber surface and by the nonwoven fabric structure.

A wetting of the surface is a prerequisite for liquid absorption. The process of wetting a porous system, such as a nonwoven fabric, with a liquid can be described using the Laplace equation (9.7):

$$p_B = \frac{4 \cdot \sigma_1 \cdot \cos \Theta}{d_{p,\max}} \tag{9.7}$$

P_B (Pa)	Wetting pressure
K (−)	Correcting factor without dimension depending on the capillary form in the fabric; ($k \approx 1$)
σ_1 (N/m)	Surface tension of the wetting liquid
Θ (°)	Contact angle
$d_{p,max}$ (m)	Maximum pore diameter

The *wetting pressure* designates the resistance of the nonwoven fabric surface to wetting with a given liquid, e.g. resin. It morphologically depends on the maximum pore diameter $d_{p,max}$ as well as the physical/chemical parameters of contact angle Θ and the surface tension δ_l of the wetting liquid. Thus, the wetting pressure p_B increases for a given liquid with high contact angles and smaller maximum pore diameters.

The wetting behavior on the nonwoven fabric surface can be characterized using static and dynamic methods, utilizing the wetting angle [28, 29].

The penetration of a liquid through the nonwoven fabric construction can be approximately described using the regularities of laminary capillary flow in porous solid bodies according to HAGEN-POISEULLE and WASHBURN. For the impregnation process, the kinetics of wetting the inner nonwoven fabric pores with liquid resin matrix is particularly relevant. For a first approximation of this process, the modified equation according to WASHBURN can be taken as a basis (Eq. 9.8) [30]:

$$v = \frac{r \cdot \sigma_l \cdot \cos \Theta}{4 \cdot l \cdot \delta} + \frac{p \cdot r^2}{8 \cdot l \cdot \eta} \tag{9.8}$$

v (m/s)	Speed of the capillary flow
r (m)	Pore diameter
Θ (°)	Contact angle

P (Pa)	Hydrostatic pressure
l (m)	Capillary length
η (Pa s)	Dynamic viscosity of the liquid
σl (N/m)	Surface tension of the wetting liquid

It becomes clear that the dispersal speed of the capillary flow through the nonwoven fabric cross-section is determined by the diameter and length of the pore channels in the nonwoven fabric. The contact angle and the hydrostatic pressure act as driving forces.

To realize homogeneous fiber volume content ratios in the component, a homogeneous resin distribution (constant fiber volume content ratio) is essential. Special sizings and finishing agents (see Chap. 13) are used to facilitate the wettability of the fiber surface by the liquid resins.

Apart from the pore structure, the flow resistance of the nonwoven fibers to perfusing media (e.g. resin) determines the liquid distribution. The flow resistance partly depends on the orientation of the nonwoven fibers and increases with growing degrees of anisotropy. Homogeneously distributed and isotropically oriented web fibers (homogeneous pore structure) can support resin distribution to all sides in the nonwoven fabric structure. On the other hand, the degree of anisotropy can be used to control the flow resistance for the purpose of directed resin spread, if desired [14].

9.4 Selected Application Examples for Nonwoven Semi-finished Products

9.4.1 Overview

The multitude of nonwoven-based constructions can mainly be classified into the following application fields:

- fiber-reinforced composites,
- special functional layers for filtration and energy technology,
- various insulation materials for sound, heat, and cold (automobile engineering, construction),
- substitution for PU foam materials

Special composite technologies on the basis of bonding technologies allow functional integration and a reduction of the number of individual semi-finished product components.

Due to the great variety of applications, the following will only give an overview based on exemplary applications. This is in no way a complete list, but only intended to show the range of applications and inspire innovative, nonwoven-based composite solutions, which is based on previous chapter contents.

9.4.2 Technologies for the Production of Nonwoven Fabric Composites

In order to meet application-specific components requirements, the production of textile semi-finished products is increasingly based on *composite technologies*. Nonwovens as composite parts in particular allow the implementation of special additional functions in the semi-finished product, such as sound absorption, defined filtration and suction properties, particle retention and storage in combination with fiber substance-specific characteristics like temperature resistance, electrical conductivity, or chemical resistance. Composites with yarn-based materials, non-crimp fabrics and other nonwoven components allow a high degree of functional integration while simultaneously reducing component weight. Composite materials therefore hold great innovative potential, especially in automobile engineering.

In principle, all bonding methods are suitable to connect nonwoven fabrics with one another or with textile and non-textile materials. In light of the multitude of bonding and connection technologies [8] and their possible combinations, the realizable composite effects seem hardly assessable.

Needling, knitting, and entangling are the primary methods for mechanical connections. Here, yarn-based materials can be integrated as reinforcement component. Figure 9.20 shows the cross-sections of nonwoven fabric materials, in which the interior woven fabric structure and the exterior nonwoven fabric layers are connected into a compact composite by needling (Fig. 9.20a) or hydroentanglement (Fig. 9.20b).

Furthermore, binding agents (powders, pastes, and foils) can be used to connect textile and non-textile fabric components under exposure to heat and pressure. Special adhesive nonwoven materials, which transition into an adhesive state at

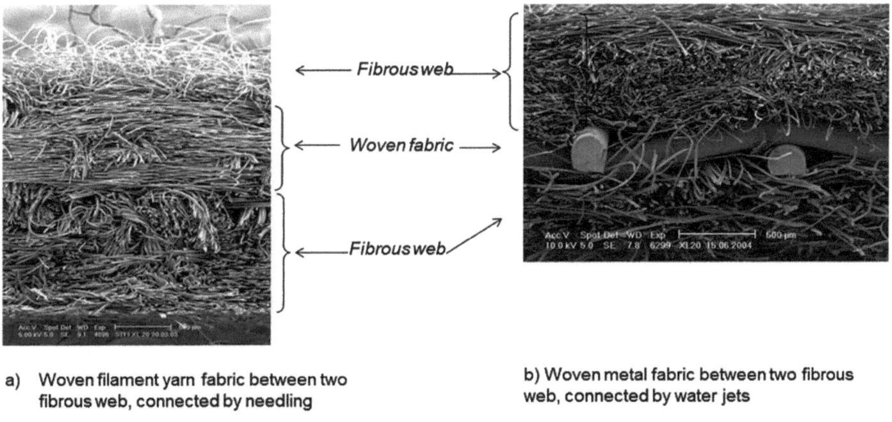

Fig. 9.20 Cross-sections of nonwoven fabric composites for surface filtration (*Source*: Sächsisches Textilforschungsinstitut e.V.). (**a**) Woven filament yarn fabric between two fibrous web, connected by needling. (**b**) Woven metal fabric between two fibrous web, connected by water jets

specific temperatures, can be used to create a fabric composite. Often, the laminating process can be integrated into the shaping process. For some composite solutions, complete machine line concepts have already been developed and partly realized to ensure an economical production of nonwoven-based laminated composites for car interiors. The use of nonwoven-based laminated composites, sometimes combined with back-injection molding in car interiors, can therefore be expected to increase in the future.

The substitution of PU foam materials in automobile design is an important development to meet the requirements of recyclability and emission reduction in the interior of the car. A nonwoven-based material composite made from PES (single-compound system) for upholstery materials in car seats has been developed. It is based on a multiknit PES nonwoven fabric (see Sect. 9.2.4.2). On a flatbed laminating machine using the patented Caliweb method [31], this nonwoven fabric is connected with a PES decorative fabric by means of an adhesive system. The thermal treatment during the laminating process serves to simultaneously set the defined thickness and beneficial compression-elastic properties of the composite structure [32]. Interior cladding parts (headliners, doors) can be manufactured requirement-adaptedly with this single-step method, which can also connect other nonwoven materials bonded by the nonwoven knitting method, such as Malivlies nonwovens [33] and Optiknit [34] as decorative optical component faces, with a Multiknit nonwoven fabric as sound-absorbing component.

9.4.3 Nonwoven Semi-finished Products for Fiber-Reinforced Composites

The use of nonwoven semi-finished products in the field of fiber-reinforced composite materials is primarily useful for the following application areas:

- components for the optimization of composite performance,
- reinforcement components, and
- auxiliary materials for temporary fixation of the reinforcement structure

9.4.3.1 Components for the Optimization of Composite Performance

So-called "surfacing nonwoven fabrics" and "core nonwoven fabrics" are used as composite component, usually in glass fiber reinforced plastics (GFRP), with their primary function being the support of the composite effect.

Surfacing nonwoven fabrics (also referred to as *surfacing mats, surfacing nonwovens or surfacing veils*) have various functions. As interfaces, they serve as corrosion protection by shielding the reinforcing yarn structures from a variety of mechanical (e.g. rockfall) and chemical environmental influences (weather, chemicals, radiation). Furthermore, nonwoven fiber covers can be used to create smooth and paintable component surfaces meeting the requirements of surface quality in the finished lightweight component. A third task is the absorption of liquid resin matrix and the homogeneous transfer thereof into the reinforcement structure during impregnation process (see Sect. 9.3.3.4). This is of crucial importance with regard to ensuring homogeneous fiber volume content ratios in the component and their reliability. To ensure corrosion-protective effects, hydrophobic textile fibers (C-, E-, and ECR-GF) and man-made fibers from PAN and PES are used for this purpose. To achieve good impregnation capability, the fibers have to be finished with special sizing and finishing agents (see Chap. 13). This can improve wettability of the outer and inner fiber surfaces of the nonwoven fabric by the liquid resins.

Apart from surface applications, nonwoven fabrics are also used as core materials in sandwich structures. Core nonwovens consist of glass fibers or man-made fibers with homogeneously distributed micro-spheres and are used to ensure a stabilizing infilling of sandwich cores. Thus, the required amounts of resin and the weight of the semi-finished products can be realized without impairing the properties of the material.

9.4.3.2 Reinforcement Components

The starting point for the development of composite materials was resins reinforced with glass fiber nonwoven fabrics. Even though nowadays yarn structures made from high-strength fibers are used for the production of heavy-duty lightweight construction components, nonwoven fabrics still are important reinforcement components for construction parts with standard requirements.

In GFRP applications, so-called “glass mats” are used as reinforcement components, classified according to fiber length and bonding method. Correspondingly, the terms “*cutting mat*”, or “*endless mat*” in needled form (“needled mats”, i.e. without binding agent) or with binding agents are commonly used. The processing of glass fibers into mats can be performed in several manners (Fig. 9.21). “Endless mats” are produced analogously, using direct technology (see Sect. 9.2.3.5). The mixture of raw materials (quartz powder, lime etc.) is melted. The molten glass is drawn into fine glass filaments through nozzles, and the yarns are combined. The continuous yarns are placed randomly on a transport belt, creating a “mat”. They are then needled, or otherwise bonded by means of a liquid or powdered binding agent. Therefore, production lines include needling machines or unit applying and activating the powdered or liquid binding agents. Rovings, which are produced in a separate process, are the initial material for the production of “cutting mats” (see Fig. 9.21). The continuous fibers or rovings are cut to a defined fiber length (25–50 mm) by means of a special cutting unit, and randomly placed across the width of a transport belt, where mechanical or chemical bonding takes place.

The nonwoven fabrics (“mats”), as described above (Fig. 9.21), are impregnated with thermoplastic (GMT) or thermoset (SMC) matrices, using a variety of

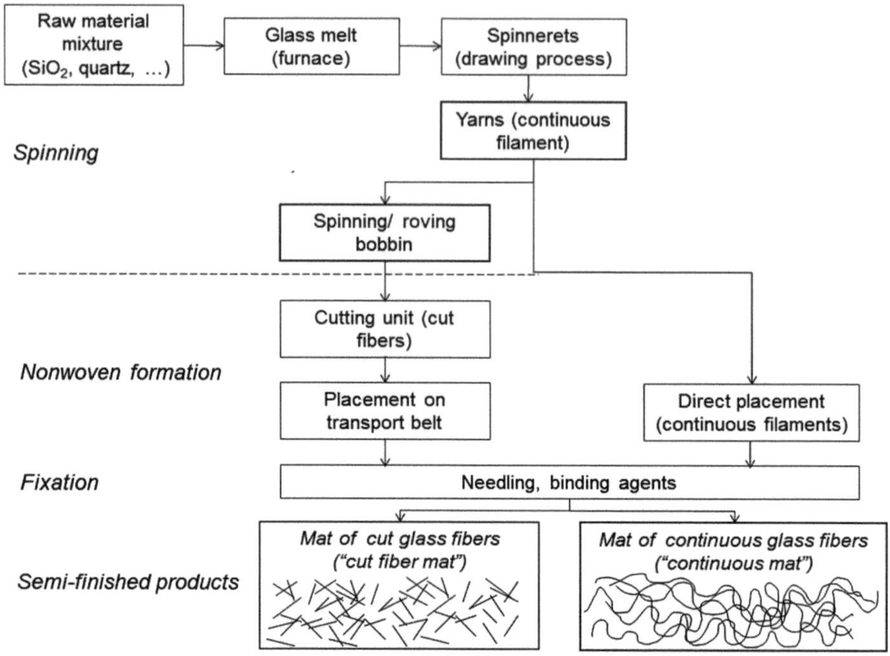

Fig. 9.21 Overview of production possibilities for mats and nonwoven materials made from glass fibers

methods. They are then processed into prepregs by extrusion or compression molding (see Chap. 11).

As the glass filaments are not packaged in the mats, their impregnation with the liquid matrix is neither complete nor homogeneous. This causes inhomogeneous areas and high deviations from the standard mechanical component properties. This is also problematic in GMT panels, in which the restoring forces of needled glass mats create air bubbles, distributed unevenly in the matrix [35].

New possibilities for the manufacture of nonwoven-based thermoplastic semi-finished products are offered by the so-called *mixed nonwovens* made from reinforcement fibers (web fibers) and matrix fibers. This approach, which is analogous to yarn material-based hybridization technology, aims to achieve an improved micro-structure, which helps optimize the flow paths for the thermoplastic matrix and evenly distribute air pores for draft-free preforms with reproducible properties. Methods have been developed for an efficient manufacture of textile-based thermoplastic semi-finished products [35, 36]. The production of nonwoven semi-finished products uses carding, airlaid, and hydroentangling processes, with the mixing of the reinforcement fibers (e.g. GF, AR, CF) and matrix fibers (e.g. PP, PEI) being realized in the nonwoven formation process. The hybrid webs are bonded by needling or with a binding agent [35, 36]. These hybridization technologies allow the manufacture of easily recyclable material composites, in which reinforcement fibers and matrix fibers are made from the same type of polymer, e.g. PP ("single-compound system"). For a cost-effective production of PP/PP composites, concepts for the manufacture of semi-finished products based on hybrid nonwovens were developed [37].

9.4.3.3 Structural Fixation

The requirement-adapted arrangement of the reinforcement yarns achieved in the fabric formation process can be disturbed by handling and transport processes during further processing into 3D preforms and fiber composite components. This can impair the reproducibility of the semi-finished product properties. To temporarily fixate the yarn arrangement and to stabilize the yarn material-based semi-finished product, adhesive nonwovens ("webs", "adhesive webs") are used. These are produced from Co-PA, Co-PES, PU, or other materials, in an extrusion process (see Sect. 9.2.3.5). Both the polymer substance and the spun-bonded nonwoven process are adjusted to achieve an optimized bonding capacity. Usually, an open-pore structure is realized, which allows a selective point-by-point fixation of the reinforcement yarns. The adhesive effect is activated by temperature and/or humidity.

9.4.4 Nonwoven-Based Functional Layers for Filtration and Energy Engineering

Due to intensified measures to minimize emissions, filter media are becoming more and more important. About 90 % of all filtration products are nonwoven-based constructions, which are used as deep and surface filtration sheets in dry and wet filtration [8]. The range of applied fibers (e.g. PA, PES, PEI, PPS, AR, GF, CF, MTF), production technologies, and realizable nonwoven fabric constructions is extremely broad (Table 9.4). Composite constructions are increasingly prevalent to adjust to the specific filtration task. By integrating nonwovens from nano-scale or micro-scale fibers, defined structural gradients can be generated, which allow staged separation efficiencies over the filter medium.

Fuel cells play crucial role in the realization of alternative concepts for future energy supplies and mobility. They can transform hydrogen into electricity, efficiently and without emissions. The degree of efficiency of a fuel cells is determined by a gas diffusion layer ensuring special transport functions between the reaction layer and the bipolar plate (electric conductor) inside the cell. The gas diffusion layer consists of a carbon fiber nonwoven fabric. For the production of this fabric, an innovative technology [38] was developed, in which precursor fibers (e.g. oxidized PAN fibers) are carded into a web and bonded with water jets. The web is then subjected to a carbonization process, in which the PAN fibers are

Table 9.4 Overview of nonwoven-based filter media

Dry filtration		Liquid filtration	
Surface filter	Deep filter	Surface filter	Deep filter
Needled nonwoven fabric	Thermo-bonded nonwoven fabric Stitch-bonded nonwoven fabric Filament and fine-fiber spun-bonded nonwoven fabric	Needled nonwoven fabric Wetlaid nonwoven fabric	Needled nonwoven fabric
Both types with carries fabric; also with micro- and nano-fiber coating (meltblown, electro-spinning method)	Usually in composite structures	Both types with finest fiber coating (meltblown method)	

transformed into carbon fibers. By means of various coating technologies, special functional layers, e.g. binding agents or a wet-laid fibrous web [39] can be applied. This special method creates a nonwoven fabric structure meeting the requirements of water management and gas diffusion within the fuel cell. A special software module was developed for the structural optimization of the gas diffusion layer [40].

9.5 Development Tendencies

The growing use of nonwoven fabrics can be traced back to the increasing requirement values for insulation and emission behavior. Increasingly life-cycle-oriented value chains will contribute to further nonwoven-based innovations. Nonwoven fabric composites have the potential to meet the demand for eco-friendliness and recyclability in automobile engineering and construction applications.

In the automobile industry, three-dimensional nonwoven fabrics (spacer nonwovens) are steadily becoming more important as semi-finished products for active and passive personal protection inside and outside of the car.

References

1. Standard ISO 9092 May 1988. *Textiles. Nonwovens. Definition*
2. Standard ISO 472 November 1999. *Plastics. Vocabulary*
3. European Disposal Association Nonwoven: *Edana 2009 nonwoven production statistics released.* http://www.edana.org/content/default.asp?PageID=75DocID=4221 (01-02-2011)
4. ANONYMOUS: Naturfasern in Vliesstoffen und technischen Textilien. In: *avr - Allgemeiner Vliesstoff-Report Nonwovens Technical Textiles* (2009), No. 3, pp. 58–60
5. ANONYMOUS: Spezialverfahren für Vliesstoffe. In: *avr - Allgemeiner Vliesstoff-Report Nonwovens Technical Textiles* (2009), No. 3, pp. 67

6. ANONYMOUS: News Views. In: *avr - Allgemeiner Vliesstoff-Report Nonwovens Technical Textiles* (2009), No. 6, pp. 7–8
7. ANONYMOUS: Faser-Einblastechnik: Von der Faser zum Akustikdämmteil für die Fahrzeugindustrie/ Fibre blowing technology: from fibres to acoustic insulation parts for the vehicle industry. In: *avr - Allgemeiner Vliesstoff-Report Nonwovens Technical Textiles* (2007), No. 5, pp. 15–18
8. ALBRECHT, W.; FUCHS, H.; KITTELMANN, W.: *Vliesstoffe*. Weinheim : WILEY-VCH, 2000
9. LÜNENSCHLOSS, J.; ALBRECHT, W.: *Vliesstoffe*. Stuttgart : Georg Thieme, 1982
10. BÖTTCHER, P.; SCHRÖDER, G.;MÖSCHLER,W.: *Vliesstoffe*. Leipzig : VEB Fachbuchverlag Leipzig, 1976
11. RÖSKE, M.; REUSSMANN, T.; LÜTZKENDORF, R.: Nassvliese aus Hochleistungsfasern und ihr Potenzial für Verbundwerkstoffe. In: *Technische Textilien* 52 (2009), No. 5, pp. 249–250
12. KNOBELSDORF, C.; LÜTZKENDORF, R.: Chancen und Möglichkeiten für neue Produkte nach dem Nassvliesverfahren. Teil 2: Carbonfaservliese. In: Technische Textilien 52 (2009), No. 2, pp. 72–73
13. KNOBELSDORF, C.; LÜTZKENDORF, R.; SCHMITT, M.: Chancen und Möglichkeiten für neue Produkte nach dem Nassvliesverfahren. Teil 3: Pulpe in Nassvliesanwendungen. In: *Technische Textilien* 52 (2009), No. 3, pp. 152–153
14. Fraunhofer-Institut für Techno- und Wirtschaftsmathematik ITWM: *GeoDict. Geometric Material Models and Computational PreDictions of Material Properties*. http://www.geodict.com (24-03-2011)
15. Standard ISO 9073-1st July 1989. *Textiles. Test methods for nonwovens - Part 1: Determination of mass per unit area*
16. Standard ISO 9073-2nd March 1995. *Textiles. Test methods for nonwovens - Part 2: Determination of thickness*
17. Standard ISO 9073-3rd July 1989. *Textiles. Test methods for nonwovens - Part 3: Determination of tensile strength and elongation*
18. WATZL, A.: *Thermofusion, Thermobonding und Thermofixierung für Nonwovens-Theoretische Grundlagen, Praktische Erfahrungen, Marktentwicklung (Firmenschrift Fa. Fleissner GmbH & Co., Egelsbach)*. 1995
19. EPPS, H.; LEONAS, K.: Pore Size and Air Permeability Of Four Nonwoven Fabrics. In: *International Nonwovens Journal* 9 (2000), No. 2, pp. 55–62
20. Standard ISO 9073-15 July 2007. *Textiles. Test methods for nonwovens - Part 15: Determination of air permeability*
21. GROSSE, S.; RUDOLPH, A.; PETERS, C.: Messung von Porengrößenverteilungen als Mittel zur Produktcharakterisierung, Produktoptimierung und Qualitätssicherung. In: *Filtern und Separieren* 21 (2007), No. 3, pp. 152–155
22. GROSSE, S.; RUDOLPH, A.: Bubble Point and Pore Size Distribution Measurements of Filter Papers, Wovens and Nonwovens using a Pore Size Meter PSM. In: *Proceedings. 10th World Filtration Congress*. Leipzig, Deutschland, 2008
23. BATCHU, H.: Characterization of Non-Wovens for Pore Size Distributions Using Automated Liquid Porosimeter. In: *Proceedings. Nonwovens Conference*. Atlanta, USA, 1990, pp. 367–381
24. KUMAR, R. S.; SUNDARESAN, S.: *Acoustic Textiles – sound absorption*. http://textination.de/de/document/1130003803282734/1.0/Acoustic20/Textiles20/sound20/absorption.pdf (16-03-2011)
25. KALINOVA, K.: Influence of nanofibrous membrane configuration on sound absorption coefficient and resonant frequency. In: *Proceedings. 6th AUTEX Conference*. Raleigh (NC), USA,2006

26. SCHLADITZ, K.; PETERS, S.; REINEL-BITZER, D.; WIEGMANN, A.; OHSER, J.: Design of acoustic trim based on geometric modeling and flow simulation for non-woven. In: *Computational Materials Science* 38 (2006), pp. 56–66
27. ANONYMOUS: *Innovationsreport: Forum für Wissenschaft, Industrie und Wirtschaft*. http://www.innovations-report.de/html/berichte/informationstechnologie/bericht-17496.html (24-03-2010). Version: 2003
28. ZILLES, J.: *Charakterisierung der Benetzbarkeit und Oberflächeneigenschaften von Textilien und Fasern (Firmenschrift Fa. Krüss GmbH, Hamburg)*. 2005
29. CASSIE, A.; BAXTER, S.: Wettability of Porous Surfaces. In: *Transactions of the Faraday Society* 40 (1944), pp. 546–551
30. WASHBURN, E.: The Dynamics of Capillary Flow. In: *The Physical Review* 17 (1921), No. 3, pp. 273–283
31. SCHMIDT, H.; RIEDEL, B.; SCHMIDT, G.: Requirements for and evaluation of CALIWEB non-woven fabric padding in cars. In: *Proceedings. 40th Dornbirn Man-Made Fibers Congress*. Dornbirn, Austria, 2001
32. FUCHS, H.: 3-D automotive textiles - a comparative evaluation. In: *Proceedings. 42nd Dornbirn Man-Made Fibers Congress*. Dornbirn, Austria, 2003
33. A.: Strukturierte Kaschierverbunde für Automobilinnenraumauskleidungen. In: *Proceedings. 24. Hofer Vliesstofftage*. Hof, Germany, 2009
34. ERTH, H.; SCHMIDT, G.F.: OptiKnit - Maschenvliesstoff mit neuen Möglichkeiten. In: *Melliand Textilberichte* 87 (2006), No. 5, pp. 333–335
35. Property right EP1373375 (16th February 2004).
36. Property right EP1719611 (8th November 2006).
37. BECKMANN, E.; REUSSMANN, T.; LÜTZKENDORF, R.: PP needle-punched nonwovens - new application fields for reinforced materials. In: *Technical Textiles* 48 (2005), No. 4, pp. E193–E196
38. WAGENER, S.; QUICK, C.; BUTSCH, H.: Fuel cells - a challenge for technical nonwovens. In: *Technical Textiles* 53 (2010), No. 6, pp. E193–E195
39. Property right DE 102005022484 A1 (16th November 2006).
40. ZAMEL, N.; XIANGUO, L.; SHEN, J.; BECKER, J.; WIEGMANN, A.: Estimating effective thermal conductivity in carbon paper diffusion media. In: *Chemical Engineering Science* 65 (2010), pp. 3994–4006

Additional Literature

ALBRECHT, W.; FUCHS, H. (Eds): *Vliesstoffe / Rohstoffe, Herstellung, Anwendung, Eigenschaften, Prüfung*. 2. vollständig überarbeitete Auflage. Weinheim : WILEY-VCH, 2012

ALBRECHT, W.; FUCHS, H.; KITTELMANN, W. (Eds): *Nonwoven Fabrics. Raw Materials, Manufacture, Applications, Characteristics, Testing Processes*. Weinheim : WILEY-VCH, 2002

CHEN, S.; YE, W.; HOU, H.: Chapter 4: *Electrospun Fibrous Membranes as Separators of Lithium-Ion Batteries*. In: Ding, Bin, Yu, Jianyong (Eds.): Electrospun Nanofibers for Energy and Environmental Applications. Series: Nanostructure Science and Technology 2014, pp. 91-110. Springer, 2014.

DAS D.; ISHTIAQUE, S.M.; AJAB RAO, S.V.; POURDEYHIMI, B.: *Modelling and Experimental Studies of Air permeability of Nonuniform Nonwoven Fibrous Porous media*. In: *Fibers and Polymers* 2013, vol. 14, No. 3, pp. 494-499.

GULICH, B.: Waste to Airlaid. In: *avr - Allgemeiner Vliesstoff-Report*, 05/2013, p. 28

ILLING-GÜNTHER, H.; GULICH, B.; HOFFMANN, M.: Nonwovens made of recycled carbon fibres as basic material for composites. In: *Nonwoven Innovation Academy*, Tourcoing (F), 27th - 28th November 2013.

SCHILDE, W.; GULICH, B.; NAUMANN, R.: Latest Developments in Nonwovens Technologies. In: *IFAI Advanced Textiles Europe 2011*, Barcelona (ES), 26 – 27th September 2011

SCHILDE, W.; SPRANGER, J.: Antimicrobial functionalization of polypropylene spun-bonded nonwovens. In: *avr – Allgemeiner Vliesstoff-Report*, 3/2011, pp. 67–70.

SCHMALZ, E.; HEIDENREICH, R.; RUDOLPH, H.: New filter media for reducing fine dust during biomass incineration. In: *avr – Allgemeiner Vliesstoff-Report*, 1/2011, pp. 21-25

SCHOLZ, R.; SIGMUND, I.: Steering of dispersibility of spunlaced nonwovens with fibre properties and spunlacing parameters. In: *Lenzinger Berichte*, 90/2012.

SIGMUND, I.: Inherently flame resistant melt blown nonwovens made from melamine resins: manufacture and selected applications. In: *Nonwovens Research Academy*, Aachen (DE), 16th – 17th November 2010.

TAUBNER, R.: Application potentials of PLA spunbond bi-components and film composites. In: *NatureWorks Conference 2012*, Orlando (USA), 20th – 22nd February 2012.

TAUBNER, R.: Process development to produce spunbond nonwovens with crimping ability. In: *Man-Made Fiber Year Book* 2010, p. 84.

TAUBNER, R.: Spunbond-film-composites made from renewable resources. In: *avr – Allgemeiner Vliesstoff-Report*, 2/2011, pp. 68-69.

Chapter 10
Embroidered Semi-finished Products and Embroidery Techniques

Mirko Schade

Since antiquity, embroidery has been known as a familiar textile method for applying yarns to textile surfaces, mostly as decoration. There are many variations as to the manner and direction in which the yarn material is placed and how much of it is used. Owing to modern drive and calculation technology, embroidery machines offer nearly unlimited pattern variety at high productivity and reproducibility. One enhancement of embroidery technology is constituted by Tailored Fiber Placement (TFP). This technically fully developed method enables a targeted local reinforcement or functionalization of textile semi-finished products and the production of textile preforms with required reinforcement yarn arrangement for the manufacture of fiber composite components. This chapter provides an insight into the embroidery technology of technical textiles for fiber composite applications. It will deal with process-relevant parameters regarding mechanical properties and conveys an overview of two- and three-dimensionally embroidered semi-finished products. The potential of semi-finished products and preforms functionalized with embroidery technology will be illustrated by examples.

10.1 Introduction

One special characteristic of fiber-reinforced plastic composites (FRPC) is their anisotropic material behavior. The mechanical properties of the reinforcement yarns can only be fully utilized in their longitudinal direction. If the load direction and the fibers' longitudinal direction differ, the degree of utilization of fiber properties is reduced. The textile and plastic industries are therefore striving to

M. Schade (✉)
Formerly: Institute of Textile Machinery and High Performance Material Technology, TU Dresden, Dresden, Germany
e-mail: i.textilmaschinen@tu-dresden.de

C. Cherif (ed.), *Textile Materials for Lightweight Constructions*,
DOI 10.1007/978-3-662-46341-3_10

position the reinforcement yarns in a manner which maximizes load carrying capacity within the fiber composite component.

Depending on application and due to different load cases and/or multiple force-introducing elements, complex load directions arise for fiber-reinforced parts. Analytical and numerical calculation methods enable a qualitative and quantitative determination of the stresses within the component [1, 2]. The derived ideal fiber directions and locally differing fiber amounts which are required in textile structures are insufficiently reproduced with more conventional textile semi-finished products such as woven, weft- and warp-knitted, or braided fabrics [3].

The requirements for load-adapted reinforcement of FRPCs motivated the development of the *Tailored Fiber Placement* (*TFP*) technology [2]. This enhancement of embroidery technique facilitates the textile-technological manufacture of complete preforms for fiber composite parts, as well as an aimed local reinforcement and functionalization of textile semi-finished products with pre-defined arrangement and alignment of functional material.

10.2 Basic Principles of Embroidery

Regarding methodology, classical embroidery differs only slightly from sewing (ref. Chap. 12). While sewing is usually employed to join multiple textile layers with commensurate, unidirectional stitches, embroidery is used to apply decorations to a textile surface (*embroidery ground* or *substrate*) at a freely variable size and direction of the stitches. The main difference is constituted by the clamping of the substrate into a frame. Using the frame, the substrate can be moved in x and y direction relative to the stitch formation device and according to the embroidery pattern. This controlled movement also defines the manner of stitch formation. The embroidery pattern is created (*punched*) with company-specific CAD/CAM programs before the embroidery process. All established vector and image formats can be used as a basis for the creation of an embroidery pattern. Virtually all company-specific programs have become inter-configurable in the meantime [4].

Most embroidery machines work with a two-yarn system [upper and lower (bobbin) yarn] and form a seam based on the *double lockstitch* (DLS). As in sewing, a wide range of materials can be used as upper and lower yarns in embroidery. Light gauzes, non-woven fabrics, films, foams, textile reinforcement semi-finished products, artificial leathers or automotive carpets of thicknesses up to 10 mm can be used as substrates [4, 5]. The needle has to be chosen depending on the upper yarn (e.g. material, appearance, yarn count) and the material to be embroidered. Embroidery machine needles are nowadays offered in a number of special variants concerning needle eye and needle point design as well as material [4].

10.3 Tailored Fiber Placement (TFP)

10.3.1 Principle

For an optimum reinforcement of highly stressed fiber composite parts, the reinforcement yarns are inserted into the component as follows [1, 2]:

- as stretched as possible (without undulation and twist in a filament yarn)
- aligned in the load direction, and
- with uniform stress (component cross-section and forces are equivalent)

The abovementioned requirements of the reinforcement yarn are only partially fulfilled by conventional textile semi-finished products, as the principle prevents the yarn arrangement from being wholly flexible. The basic principle of embroidery is used in TFP to fix an additional functional material to the substrate [6, 7]. This permits the effective realization of qualitatively and quantitatively determined stresses in the component into a textile structure, and enables a targeted application-specific functionalization of textile semi-finished products. These semi-finished products manufactured by TFP technology can display locally variable fiber orientations and amounts. The principle of this technology is illustrated in Fig. 10.1.

The following advantages of embroidery by means of TFP can be used for the manufacture and functionalization of textile semi-finished products for fiber composite production [4, 8]:

- geometrical application of the upper yarn on the substrate is virtually unlimited,
- freely reproducible control of the embroidery frame,

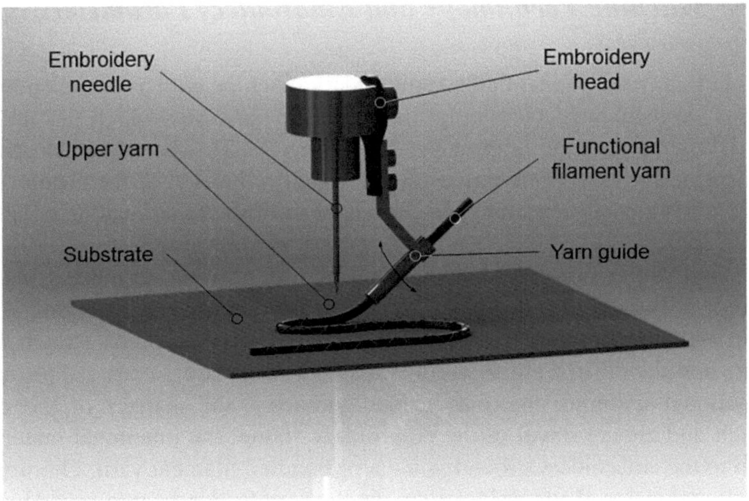

Fig. 10.1 Principle of the tailored fiber placement technology

- customized application of yarn material,
- creation of coherent forms and surfaces,
- simple integration of non-textile elements,
- relative independence from the substrate, as well as
- very small development efforts for the realization of yarn course in a company-specific CAD/CAM program.

This results in the following advantages of the TFP technology over other textile manufacturing methods [3, 9]:

- deposition of functional material independent of angle,
- high positioning precision (functional material deposition in modern CNC embroidery machines at ±0.3 mm),
- possible manufacture of two- and three-dimensional textile semi-finished products with load-adapted and locally variable reinforcement yarn arrangements in the x-, y-, z-directions,
- prevention of fiber material and matrix aggregation in the finished component by a component-adapted reinforcement yarn deposition,
- near-net shape production for economic material use and waste reduction, and
- unproblematic processing of natural, glass, aramid, carbon and ceramic fibers, as well as non-textile elements (e.g. optical fiber, metal wire) is possible.

Due to low productivity in comparison to other textile production methods, TFP technology is currently used mainly for selective local reinforcement, for example of perforated structures, and in the production of complex preforms for small structural parts in small quantities [10, 11].

10.3.2 Machine Technology and Embroidery Parameters

For the TFP, a conventional flatbed embroidery machine with a special embroidery head was supplemented by adding an extra yarn guide by which the functional material is guided to the embroidery head (Fig. 10.2). The adjustable pendular movement of the yarn guide moves the material to the side of the needle, creates constant sewing conditions, and prevents damage of the functional material by the needle.

To ensure a function- and load-adapted fiber deposition on the x-y plane, the yarn guide is hinged in front of the stationary needle. By implementing a zigzag function for the embroidery frame, the functional material is applied stretched to the substrate and the desired stitch width is realized. Depending on the appearance of the functional material, the yarn feeder geometry, the manner of the double-lockstitch and zigzag travel of the embroidery frame, the functional material can be fixed to the embroidery ground as a flat or bundled filament yarn. The storage of the functional material and its insertion into the yarn feeder are performed by means of a yarn-guiding tube, from an external, conventional bobbin or from a creel. The

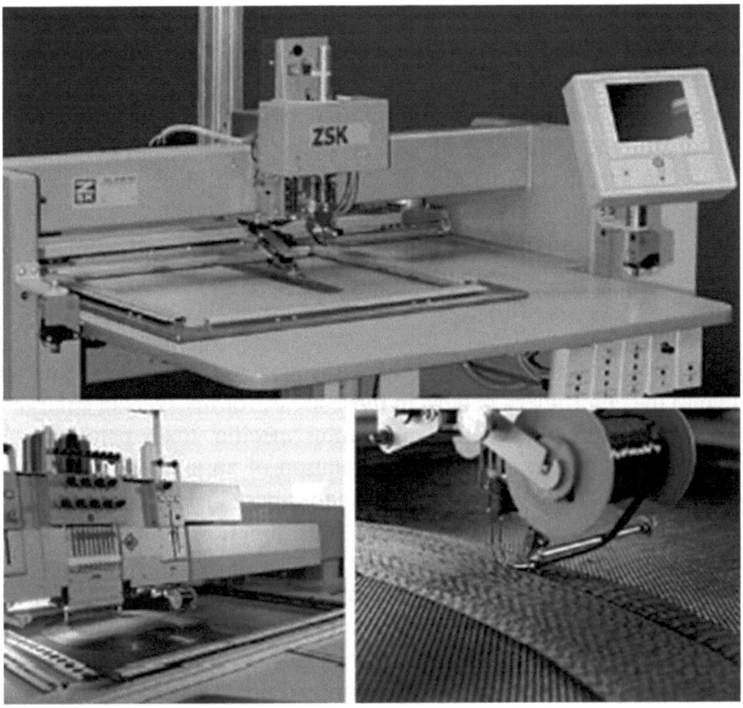

Fig. 10.2 Embroidery machine (*Source*: ZSK Stickmaschinen GmbH)

guidance along the bobbin integrated into the embroidery head and the available space cause a special bobbin geometry, which requires an additional winding process and limits the continuously processable yarn amount. The advantage over tube-guiding is in the free rotatability of the yarn feeder system around the embroidery needle, which makes special attention to system rotations during embroidery pattern creation obsolete. In tube guiding, it has to be kept in mind that the tube could accidentally be wound around the embroidery head, for example during the creation of embroidered spirals [7, 12].

The current maximum speed of these special embroidery machines is up to 1,000 stitches per minute. To increase productivity, the technology was transferred onto multi-headed embroidery machines, so that analogous textile semi-finished products can be produced by multiple embroidery heads arranged side-by-side at constant distances. Currently, embroidery machines with up to 30 embroidery heads for technical textiles are available from ZSK Stickmachinen GmbH and Tajima GmbH [13, 14]. The embroidery field size is limited by the stitch frame. However, it possible to produce one preform with several embroidery heads by overlapping. An automatic conveyor device for the substrate with a special hydraulic clamping system is optionally available for the embroidery of rolled goods. Here, the gentle and damage-free clamping of already embroidered semi-finished products is noteworthy. Depending on the substrate's material, weight and thickness, a variety of clampings can be used, e.g. border or magnetic clamping fixtures. A clamping tailored to the substrate is necessary to prevent any pattern-distorting shifting of the substrate during embroidery.

Considering the functional material, particularly its count, tensile strength and flexural strength, the radii of embroidery pattern reversion points have to be observed, as the functional material is embroidered in place quasi-continuously. In case of wide filament yarns due to the rigidity of materials, the curve radius may tilt up, which in turn might constrict, resulting in a deviation from the desired, application-specific embroidery pattern structure [11]. This can be compensated by reduced machine speed at critical reversion points or by adjustments to the embroidery pattern. Modern CNC embroidery machines automatically adjust their machine speed in critical reversion areas. However, it is crucial to remember that any change to the embroidery pattern in these areas, for instance an enlargement of the radius, may shift the ideal, application-specific fiber pathway. Another machine-technical solution is the implementation of an automatic separation and positioning unit. This device, built into the embroidery head, permits the automatic storing, separating and repositioning of the functional material at suitable points [15].

Top and bobbin yarn materials (type, count, sizing) and especially their tensions are also great influences on the appearance of the embroidery pattern. Excessive tensions cause constriction and undesired displacement of the functional material [11, 16], which negatively affect preform permeability and the mechanical properties of the composite. A more homogeneous functional material deposition in the fiber composite part to be produced can be attained by using dissolvable embroidery yarns [11, 16]. The typical embroidery yarn ratio in FRPCs is between one and four volume percent [9].

10.4 Embroidered Semi-finished Products

10.4.1 Two-Dimensional Embroidered Semi-finished Products

Using the TFP technology, complete preforms can be manufactured for fiber composite parts with high mass/strength ratios. The mechanical properties of the TFP preforms at a free, load-independent fiber deposition are comparable with conventional multi-axial non-crimp fabrics and minimally lower than those of prepreg materials [17, 18]. A finite element (FE) method analysis of the expectable load cases of the fiber composite part to be constructed allows the calculation of global and local stress states and optimal courses of fibers. Various approaches are used for this purpose such as generic algorithms, topology optimization, and the principal stress method. Subsequently, TFP permits an exact realization of the course of the yarn derived from the locally calculated principal stresses, leading

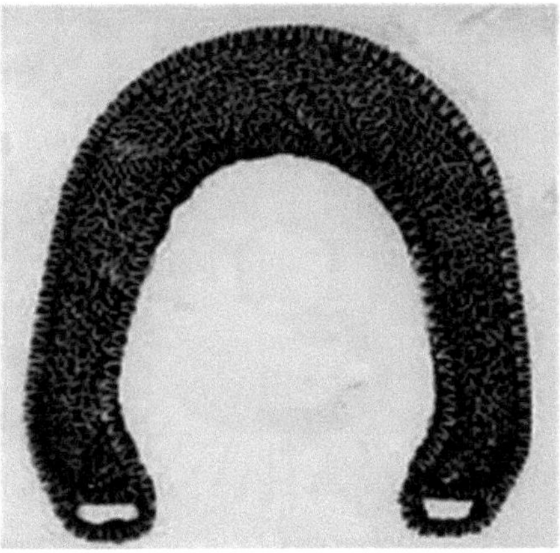

Fig. 10.3 Embroidered brake booster (*Source*: HightexVerstärkungsstrukturen GmbH)

to the production of the preform (exemplary structure in Fig. 10.3) [19, 20]. Recent research works serve to integrate embroidery-technological basic conditions, such as the disposition behavior of the functional material in tight radii, and specific fiber composite properties, for example the local fiber volume count, into the calculation programs [21]. Currently, this is often performed by means of a manual feedback of optical and mechanical results of embroidered semi-finished products and components into the calculation program followed by an iterative simulation. The attainable performance of the TFP technology concerning occurring load cases becomes apparent during preform production for fiber composite parts with specific design of force transmission areas and linkage points. By determining the states of stress using the FE method, and by applying the reinforcement yarn along the principal load direction, local stress peaks within the component can be reduced and specific strengths can be improved significantly [6, 9, 20]. When using load-adapted, two-dimensional, embroidered semi-finished products, for instance, the properties of the connecting support of the A340 horizontal tail in tension and tension/pressure load cases can be improved by more than 60 % over conventionally produced components [20].

In other research activities, the possibility of producing continuous-fiber-reinforced thermoplastic components by means of TFP is being explored. The distinctive feature in this case is the expansion of the TFP technology from two-dimensionally embroidered semi-finished products into the hybridization by a type of *Film Stacking* method, or into hybrid yarn formation. This requires reinforcement and matrix yarns to be guided parallel into the filament yarn guide before being used for embroidery simultaneously, resulting in a *side-by-side* hybrid structure [22].

Another suitable field for TFP technology is the selective local reinforcement or the functionalization of textile semi-finished products. By combining the freedom of design of TFP with the high productivity of textile fabric manufacturing

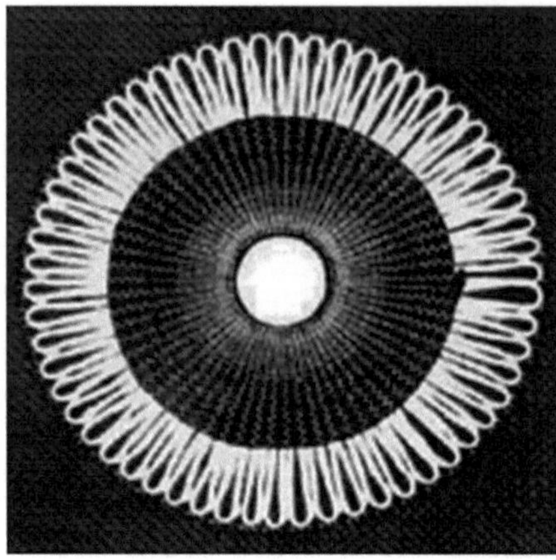

Fig. 10.4 Embroidered hole reinforcement (*Source*: HightexVerstärkungsstrukturen GmbH)

(e.g. weaving, warp-knitting, weft-knitting) for the creation of embroidery grounds, the retroactive application of functional material onto textile semi-finished products has great potential [6, 9, 11, 19, 23]. By aligning the fiber courses in the main stress directions, even subsequent local applications of reinforcement yarns can result in significant improvements of the mechanical properties (exemplary structure Fig. 10.4). This is shown in [9] for an embroidered hole reinforcement in a tension plate. The tensile strength can be increased by 45 %, and the bearing stress can be improved by 68 % in comparison to non-reinforced multi-axial non-crimp fabrics. The two-dimensional reinforcement of textile semi-finished products by TFP improves the fatigue behavior of the structure in the x and y planes, regarding radial and tangential stresses by centrifugal forces and moments. TFP preforms have to be additionally reinforced in the z direction for forces applied orthogonally or diagonally to this plane, as well as for the improvement of damage tolerance regarding shear and peel forces, and thermal tensions [3].

Carbon fibers (CF) are preferred for FRPCs due to their high specific stiffness and strengths. In a small share of applications, the electrical conductivity of CF in fiber composite components is used for structural health monitoring and heating structures [24–29]. For this, resistance fields are established by proper selection of the CF yarn, correct spacing of the filament yarns and the length of the embedded conductor. The CF yarns are applied to the textile semi-finished products, e.g. by TFP (Figs. 10.5 and 10.6) [27–29].

By stacked embroidery of several CF yarns, a parallel circuit can be realized to reduce the electrical resistance at this point of the resistance field. This possibility of resistance adjustment is made specific to the application, for instance in realizing different heating outputs for fiber composite tools for fiber composite production, or for forming devices in preform manufacture. This allows the tool to be heated homogeneously and thus reduces internal stresses in the finished FRPC part. Furthermore, the use of such tools reduces energy consumptions and costs compared to classical metal tools. The concepts realized so far could attain a maximum

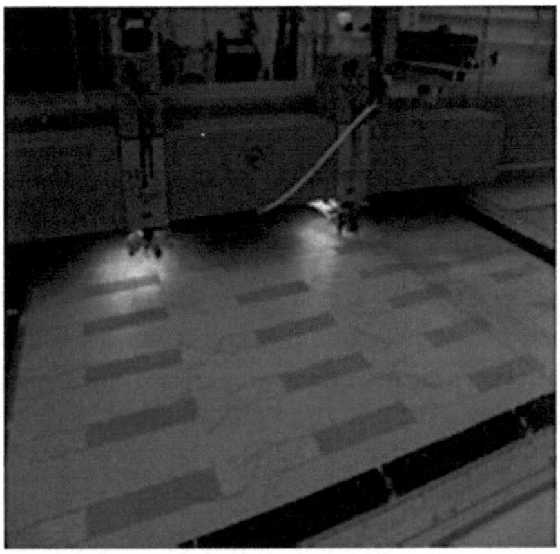

Fig. 10.5 Embroidered sensor fields for structural health monitoring of FRPCs

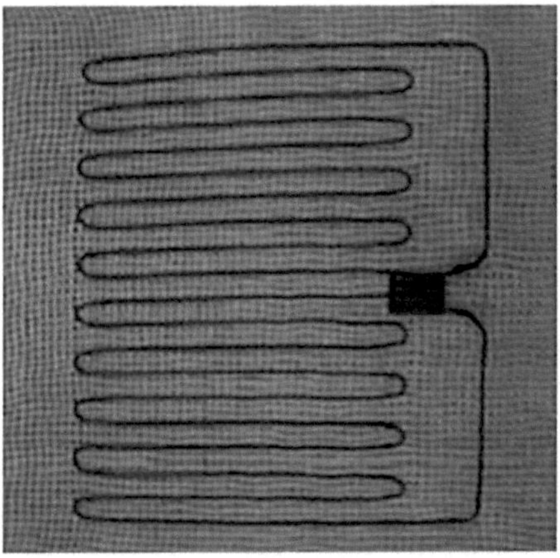

Fig. 10.6 Meandering-pattern sensor field for structural health monitoring

panel heating output of 22 kW/m^2 and temperatures of 230 °C [25, 26]. For preform production, energy consumption and costs could be reduced by using more thinly laminated, heated machine linings. These advantages are already exploited in the series production of the connection fittings for A330/340 wing spoilers and in the related necessary preform production [26]. Other applications for embroidered heating structures are found primarily in aeronautics and in wind energy plant construction, for de-icing or anti-icing purposes, for example on wing slats.

For functional integration into fiber composite components by means of embroidery, not only the material deposited by the yarn guide, but also the embroidery yarn itself is used. For this, depending on the bending sensitivity of the material to be processed, special conductive yarns with a diameter of 0.03 mm are applied to the textile semi-finished product as upper and bobbin yarn. This enables the one-step production of sensor structures, for example for the detection of shape changes, temperature or humidity, as well as of functional structures like antennas or switches [30].

10.4.2 Three-Dimensionally Embroidered Semi-finished Products

In three-dimensionally embroidered semi-finished products, a distinction has to be made between semi-finished products with three-dimensional geometry and those with a three-dimensional structure (ref. Sect. 2.3.1.1).

Within the framework of one research project, a method was developed for the embroidering of a monaxially curved shell geometry [31]. This method allows the direct three-dimensional manufacture of preforms without relying on the conversion of two-dimensional embroidered semi-finished products. This has the advantage of providing an exact and stretched yarn orientation for geometrically complex parts. For the realization, a cap embroidery mechanism was modified so that the embroidery ground could be stretched into the shape of a cylindrical shell. The potential of this method was proven on selected demonstrators. Compared to converted two-dimensional preforms, three-dimensional embroidery leads to an improvement of component quality due to the minimization of matrix-rich zones and application-specific yarn orientation. As a result an increase of force absorption potential during drop impact testing is achieved. The greatest restrictions are currently being imposed by the flexibility of the producible preform geometries.

Another way of producing a three-dimensional geometry by embroidering is used in the reconstruction of pipes by *sliplining*. In this case, a directed structuring of the inner lining of the pipes is performed in order to provoke the fluid's turbulence, ensuring improved removal of solid materials and thereby preventing sedimentation. The three-dimensional geometry is created by an embroidery-technical production of structural bodies on a GF-reinforced textile semi-finished product using GF embroidery yarn and special foams. The transport of media depends on the shape and the orientation of the structural bodies, which can be purposefully varied [8].

The demand for textile semi-finished products that can absorb multi-axial tensions within the FRPC in order to compensate the low mechanical properties of conventional, two-dimensionally reinforced FRPCs perpendicular to the reinforcement plane can be met by producing three-dimensional structures, among others by TFP [3]. One approach includes the insertion of reinforcement yarns in the z direction into the two-dimensional semi-finished product. The type and extent of *out-of-plane* property improvements depend largely on the material and the ratio of reinforcement yarns, as well as the angle to the x-y plane at which they are introduced [32]. For multilayered TFP semi-finished products, reinforcement in the z direction is possible by inserting reinforcement yarns as upper and bobbin yarn. Due to the altered fiber ratio and the damage of the previously applied reinforcement yarns by needle, the in-plane properties are reduced. Therefore it is prudent to include the z-directional reinforcement yarns only locally and adapted to the expected loads. Consequently, the mechanical properties in the x-y plane of the component are only partially influenced [11, 18, 32].

The orientation of the embroidery yarn in the z-direction can also be used for the production of functionalized semi-finished products. In [5], a three-dimensional sensor structure made from a precision resistance wire was used for a direction-dependent deformation measurement (shear, compression, elongation measurement) in the component.

10.4.3 Embroidered Semi-finished Products with Yarn Reserves

Embroidery technology allows the production of deep-draw-capable preforms with load-adapted yarn deposition for complex-shape components [32, 33]. For this, thread reserves for critical formation points are integrated in the embroidery pattern or, by means of non-embroidered yarn loops in the embroidered fabric. The 3D component geometry is mathematically processed two-dimensionally; the embroidery program is created with regard to the material-specific restrictions (e.g. functional material properties, shrinkage of the substrate) and executed two-dimensionally. The distinctive feature here is the application of the embroidery pattern on a substrate soluble in cold water. After the substrate's dissolution, the 3D structure is formed.

The potential of embroidered semi-finished products with yarn reserves is apparent in the reinforcement of hollow-cast parts, of so-called metal-matrix composites. Compared to massive components, a mass reduction of up to 30 % in hollow-cast components has been attained without impairing mechanical properties. In a research project concerned with metal-matrix composites, geometrical 3D embroidered fabrics such as hemispheres or cones from glass, basalt or carbon fibers are produced and later embedded in the molten bath as reinforcement structures (Fig. 10.7) [33].

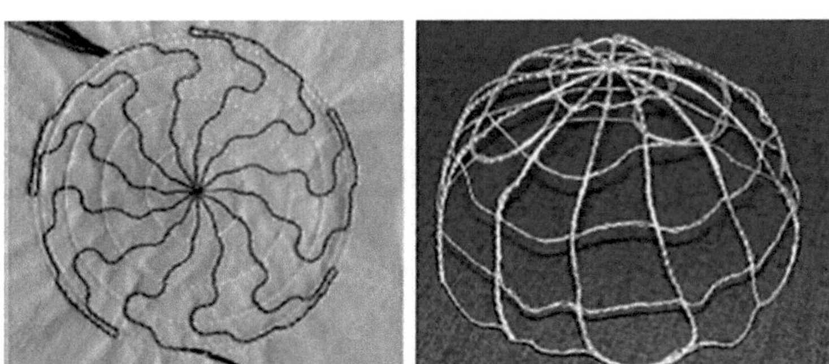

Fig. 10.7 Embroidered hemisphere and formed three-dimensional hemisphere (*Source*: Gerber Spitzen und Strickereien GmbH)

References

1. EHRENSTEIN, G. W.: *Faserverbund-Kunststoffe: Werkstoffe, Verarbeitung, Eigenschaften.* München, Wien : Carl Hanser Verlag, 2006
2. GLIESCHE, K.; ROTHE, H.; FELTIN, D.: Technische Gesticke als kraftflußgerechte Textilkonstruktionen für Faserverbund-Bauteile. In: *Konstruktion* 48 (1996), pp. 114–118
3. WITT, G.: *Taschenbuch der Fertigungstechnik.* München, Wien : Carl Hanser Verlag, 2006
4. BÜHRING, L.; GRAWITTER, N.: *Fachlexikon Stickerei und Spitze, CD, Deutsch/Englisch.* Greiz : Textilforschungsinstitut Thüringen-Vogtland e.V., 2007
5. ELSNER, H.; REINHARDT, A.: Sensorintegration in Leichtbauverbundstrukturen mittels Sticktechnologie. In: *Proceedings. Symposium Technische Textilien"Innovative Fügetechniken".* Reichenbach, Germany, 2008
6. CROTHERS, P. J.; DRECHSLER, K.; FELTIN, D.; HERSZBERG, I.; KRUCKENBERG, T.: Tailored fibre placement to minimise stress concentrations. In: *Composites Part A: Applied Science and Manufacturing* 28 (1997), No. 7, pp. 619–625. DOI 10.1016/S1359–835X(97)00022–5
7. FELTIN, D.: *Entwicklung von textilen Halbzeugen für Faserverbunde unter Verwendung von Stickautomaten.* Dresden, Technische Universität Dresden, Fakultät Maschinenwesen, Dissertation, 1998
8. GRAWITTER, N.; MÖHRING, U.; LABAHN, J.; MÜ LLER, M.; BERGER, W.: Ein neuer, strukturierter Rohrliner sorgt für Turbulenzen. In: *Kettenwirk-Praxis* 43 (2009), No. 2, pp. 20–21
9. MATTHEIJ, P.; GLIESCHE, K.; FELTIN, D.: Tailored Fibre Placement-Mechanical Properties and Applications. In: *Journal of Reinforced Plastics and Composites* 17 (1998), No. 9, pp. 774–786. DOI 10.1177/073168449801700901
10. HERSZBERG, I.; BANNISTER, M. K.; LEONG, K. H.; FALZON, P. J.: Research in Textile Composites at the Cooperative Research Centre for Advanced Composite Structures. In: *The Journal of The Textile Institute* 88 (1997), No. 3, pp. 52–73. DOI 10.1080/00405009708658587
11. MEYER, O.: *Kurzfaser-Preform-Technologie zur kraftflussgerechten Herstellung von Faserverbundbauteilen.* Stuttgart, Universität Stuttgart, Fakultät Luft- und Raumfahrttechnik und Geodäsie, Dissertation, 2008
12. FELTIN, D.; GLIESCHE, K.: Preforms for Composite Parts made by Tailored Fibre Placements. In: *Proceedings. 11. International Conference on Composite Material.* Gold Coast, Australia, 1997
13. http://www.zsk.de (06-08-2010)
14. http://www.tajima.de (06-08-2010)
15. Property right DE 102006021425 A1 (8th November 2007).
16. HAZRA, K.; POTTER, K.: Design of Carbon Fibre Composites Aircraft Parts Using Tow Steering Technique. In: *Proceedings. SAMPE Europe 29th International Conference and Forum SEICO 08.* Paris, France, 2008
17. UHLIG, K.; SPICKENHEUER, A.; GLIESCHE, K.; KARB, I.: Strength of CFRP open hole laminates made from NCF, TFP and braided preforms under cyclic tensile loading. In: *Plastics, Rubber and Composites* 39 (2010), No. 6, pp. 247–255. DOI 10.1179/174328910X12647080902772
18. MATTHEIJ, P.; GLIESCHE, K.; FELTIN, D.: 3D reinforced stitched carbon/epoxy laminates made by tailored fibre placement. In: *Composites Part A: Applied Science and Manufacturing* 31 (2000), No. 6, pp. 571–581. DOI 10.1016/S1359–835X(99)00096–2
19. GLIESCHE, K.; ORAWETZ, H.; HÜ BNER, T.: Application of the tailored fibre placements process for a local reinforcement on an open-hole tension plate from carbon/epoxy laminates. In: *Composite Science and Technology* 63 (2003), No. 1, pp. 81–88. DOI 10.1016/S0266–3538(02)00178–1

20. ASCHENBERGER, L.; TEMMEN, H.; DEGENHARDT, R.: Tailored Fibre Placements Technology - Optimisation and computation of CFRP structures. In: *Proceedings. Advances in Design and Analysis of Composite Structures - ESAComp Users' Meeting*. Braunschweig, Germany, 2007
21. KRÄGER, L.; KLING, A.: Feedback Method transferring manufacturing data of TFP structures to as-build FE models. In: *Proceedings. IV European Congress on Computational Mechanics*. Paris, France, 2010
22. SCHIEBEL, P.: Ressourceneffiziente Fertigung hochbeanspruchter CFK Bauteile für den Fahrzeugbau. In: *Proceedings. 3. Internationale Messe und Symposium für Technische Textilien im Fahrzeugbau*. Chemnitz, Germany, 2010
23. GLIESCHE, K.; FELTIN, D.: Die Leichtigkeit des Seins. In: *MM MaschinenMarkt, Das Industriemagazin* (2002), No. 9, pp. 48–51
24. PRASSE, T.;MICHEL, F.;MOOK, G.; SCHULTE, K.; BAUHOFER,W.: A comparative investigation of electrical resistance and acoustic emission during cyclic loading of CFRP laminates. In: *Composites Science and Technology* 61 (2001), No. 6, pp. 831–835. DOI 10.1016/S0266–3538(00)00179–2
25. GLIESCHE, K.; ORAWETZ, H.; KUPKE, M.; WENTZEL, H. P.: Entwicklung eines Werkzeuges aus Faserverbunden mit gleichmäßiger Temperaturverteilung auf der Werkzeugoberfläche. In: *GAK Gummi Fasern Kunststoffe* 58 (2005), No. 3, pp. 167–173
26. KOHSER, C.: Innovative Heatable Composites Based on Carbon Fibre Structures. In: *Railway Technology International* (2010), No. 1, pp. 67–68
27. KUNADT, A.; STARKE, E.; PFEIFER, G.; CHERIF, Ch.: Messtechnische Eigenschaften von Dehnungssensoren aus Kohlenstoff-Filamentgarn in einem Verbundwerkstoff. In: *tm-Technisches Messen* 77 (2010), No. 6, pp. 113–120. DOI 10.1524/teme.2010.0014
28. KUNADT, A.; HEINIG, A.; STARKE, E.; PFEIFER, G.; CHERIF, Ch.; FISCHER, W. J.: Design and Properties of a Sensor Network embedded in thin fibre-reinforced Composites. In: *Proceedings. IEEE Sensors 2010 Conference*. Waikoloa, USA, 2010
29. CHERIF, Ch.; SCHADE, M.; HOFMANN, G.; FISCHER, W. J.; KUNADT, A.: Sustainability of the European Textile Industry Through Textile Based Lightweight Construction in Multimaterial Design with Function Integration: Visions and Chances. In: *Proceedings. 9th World Textile Conference AUTEX 2010*. Vilnius, Lithuania, 2010
30. ELSNER, H.: Sticktechnologie - Zukunftsbilder zu Möglichkeiten und Grenzen für technische Anwendungen in Kunststoffen. In: *Proceedings. Fachtagung Technisches Sticken - zwischen Forschung und Markt*. Plauen, Germany, 2010
31. GRIES, Th.: Grundlegende textilphysikalische Untersuchungen zur Konstruktion von dreidimensionalen gestickten Verbundwerkstoffen - 3D-Sticken (AiF-No. 15475 N) / Institut für Textiltechnik der RWTH Aachen. Aachen, 2010. – Research report
32. ORAWETZ, H.: Grundlagen und Anwendungen zur Tailored Fibre Placement Technologie. In: *Luftfahrttechnisches Handbuch*. LTH Koordinierungsstelle, 2007
33. HESSBERG, S.: Sticken mit Hochleistungsfaserwerkstoffen: Herstellung von 3-D-Gesticken randverstärkter Metallhohlkörper. In: *Proceedings. Symposium Technische Textilien"Innovative Fügetechniken"*. Reichenbach, Germany, 2008

Additional Literature

HERRMANN, A.S.; SCHIEBEL, P.: Analyse des Verformungsverhaltens von Kohlenstoff-Rovings im textilen Fertigungsprozess für belastungsgerechte CFK-Strukturen. In: *Proceedings 13. Chemnitzer Textiltechnik Tagung*, Chemnitz (2012)

HEß, H.; HIMMEL, N.: Structurally stitched NCF CFRP laminates. Part 2: Finite element unit cell based prediction of in-plane strength. In: *Composites Science and Technology* 71 (2011), pp. 569–585 http://www.laystitch.com (29.08.2014)

KARAMUK, E.; MAYER, J.; WINTERMANTEL, E.: *Sticktechnologie für medizinische Textilien und Tissue Engineering. Medizintechnik – Life science engineering*, 5th Edition, Springer Verlag Berlin Heidelberg, 2009. – ISBN 978-3-540-93935-1

KÄRGER, L.; KLING, A.: As-built FE-simulation of advanced fibre placement structures based on manufacturing data. In: *Composite Structures* 100 (2013), pp. 104 –112.

SELVARAJ, K.; SPICKENHEUER, A.; BITTRICH, L.; HEINRICH, G.: Bolted joints made of carbon fiber reinforced plastics using the tailored fiber placement technology: numerical and experimental failured analysis. In: *Proceedings 7. Aachen-Dresden International Textile Conference*, Dresden (2012)

SPICKENHEUER, A.: *Zur fertigungsgerechten Auslegung von Faser-Kunststoff-Verbundbauteilen für den extremen Leichtbau auf Basis des variabelaxialien Fadenablage-verfahrens Tailored Fiber Placement*. Dresden, Leibnitz-Institut für Polymerforschung Dresden e. V., Dissertation, 2014

UHLIG, K.; SPICKENHEUER, A.; BITTRICH, L.; HEINRICH, G.: Development of a highly stressed bladed rotor made of CFRP using the tailored fiber placement technology. In: *Mechanics of Composite Materials* 49 (2013), No. 2

UHLIG, K.: Wege zum inflitrierten TFP-Bauteil. In: *Proceddings. 1. TFP - Workshop 2012 - Tailored Fiber Placement meets Industrial Partners*, Dresden, 2012

Chapter 11
Pre-impregnated Textile Semi-finished Products (Prepregs)

Olaf Diestel and Jan Hausding

Pre-impregnated semi-finished textile products (prepregs), are an important base material for the manufacture of thermoset and thermoplastic composite materials. They consist of a usually flat and planar textile reinforcement structure and are combined with thermoset or thermoplastic matrices required for the production of the final component. Both short fibers or continuous filaments and textile fabrics, such as woven fabrics or multiaxial warp-knitted fabrics, can be used as base material. The basic principle of using this special textile semi-finished product is to separate the step of impregnating the reinforcement structure with the matrix during composite material production from the final step of producing the component form.

11.1 Introduction

Pre-impregnated textile semi-finished products are required for the further processing into thermoset or thermoplastic fiber-reinforced plastic composite components. They are often referred to as prepregs, which is the abbreviation for *pre-impregnated fibers* or *pre-impregnated materials*. Prepregs are pre-fabricated, mostly planar semi-finished products with a reinforcement structure produced from short or continuous fibers and are combined with thermoset or thermoplastic

O. Diestel
Institute of Textile Machinery and High Performance Material Technology, TU Dresden, Dresden, Germany
e-mail: olaf.diestel@tu-dresen.de

J. Hausding (✉)
Formerly: Institute of Textile Machinery and High Performance Material Technology, TU Dresden, Dresden, Germany
e-mail: i.textilmaschinen@tu-dresden.de

C. Cherif (ed.), *Textile Materials for Lightweight Constructions*,
DOI 10.1007/978-3-662-46341-3_11

matrix required for the production of the final component. Further processing into components is performed under heat and pressure, usually by extrusion, compression molding, or in an autoclave method. For thermoset prepregs, the textile structures are impregnated with thermoset resin systems, whose crosslinking reaction proceeds at a considerably slow rate at low temperatures. When properly stored, they remain suitable for component manufacture even after several months to up to 1 year or longer. They are also used as semi-dry prepregs for high-performance applications, where the strength and stiffness of the reinforcement fibers are to be utilized to their full extent.

The basic principle of using this special form of textile semi-finished product is to separate the impregnation of the reinforcement structure with matrix during composite material production from the actual production of the component form. Textile semi-finished products, such as woven fabrics or multiaxial non-crimp fabrics, as well as unidirectional prepregs (*UD prepregs*) can be processed into pre-impregnated reinforcement materials [1, 2]. The latter are distinguished by the highest strengths and stiffness because of their fully stretched yarn orientation, and serve as the yardstick for all textile semi-finished products. The impregnation of the prepregs, which is performed in special prepreg plants, can achieve high precision and reproducibility due to the very small thickness of the textile structures. The matrix of the finished composite component acts as a chemical binder, whose task is to bind and protect the fibers in the composite material. It transfers loads to the fibers and retains them in their predefined position and orientation. The matrix also determines the resistance of the component to environmental conditions during its lifetime, in particular exposure to extreme temperatures.

A wide range of resin systems is available for different application areas, such as automobile construction, aeronautics, sports gear as well as ship building. Apart from the previously mentioned high strengths and stiffnesses of the composites resulting from the parallel fiber orientation and the consequently attainable high fiber volume fraction (usually 60 %), another advantage of thermoset UD prepregs is their very small tolerance in stacking into multilayered composites or laminates. The latter is crucial for the utilization of the lightweight construction potential of fiber-based semi-finished products, as a precise selection of available layer thicknesses is necessary to realize the pre-calculated component dimensions. Customary thicknesses for UD prepregs are 0.125 and 0.250 mm [2, 3].

For *thermoplastic prepregs*, the textile reinforcement structure is combined with the thermoplastic matrix so that both components are intermingled as thoroughly as possible in the semi-finished product. It is crucial to achieve short flow paths for the usually highly viscous thermoplastic melt and to ensure an optimal impregnation and wetting of the reinforcement structure during component manufacturing. This is a necessary prerequisite for the full utilization of the performance potential of the high-performance fibers. Thermoplastic prepregs are either textile, more or less flexible reinforcement structures based on hybrid yarn (Chap. 4) or powder-impregnated textile semi-finished products (in which the matrix component is present in fibrous or powder form), or plate-shaped semi-finished products impregnated and consolidated by various methods, which are often referred to as *organic sheets*.

Thermoset as well as thermoplastic prepregs are classified into free-flowing and non-free-flowing pre-impregnated textile semi-finished products. Free-flowing prepregs are based on a reinforcement structure made from short fibers. This ensures that the fibers can slide off one another during component production, allowing the fiber/matrix/filler mixture to fill the mold cavity during manufacture and adjust to it. Attention has to be paid to the fact that the fibers can orient themselves in flow direction in critical zones, which is usually not desirable. In general, the flowability decreases with increasing fiber length.

Non-free-flowing prepregs contain a reinforcement structure made from oriented, practically continuous reinforcement fibers, reinforcement yarns or planar textile semi-finished reinforcement products. They do not allow significant flow processes during molding. For this reason, they require an exact allocation of parts in the molding tool to accurately match the cavity of the molding form. Non-free-flowing thermoplastic prepregs can either be fully impregnated and consolidated, plate-type semi-finished products (organic sheets), or not fully impregnated or fully consolidated textile semi-finished products.

Figure 11.1 provides a schematic overview of the numerous paths from base material through thermoset or thermoplastic prepregs to the finished fiber-reinforced component.

In the following, the most important semi-finished products and their further processing by common pressing processes are introduced. This is supplemented by free-flowing molding or pressing compounds based on short fibers. Special methods, such as back foaming and back injection of foils will not be treated in this chapter.

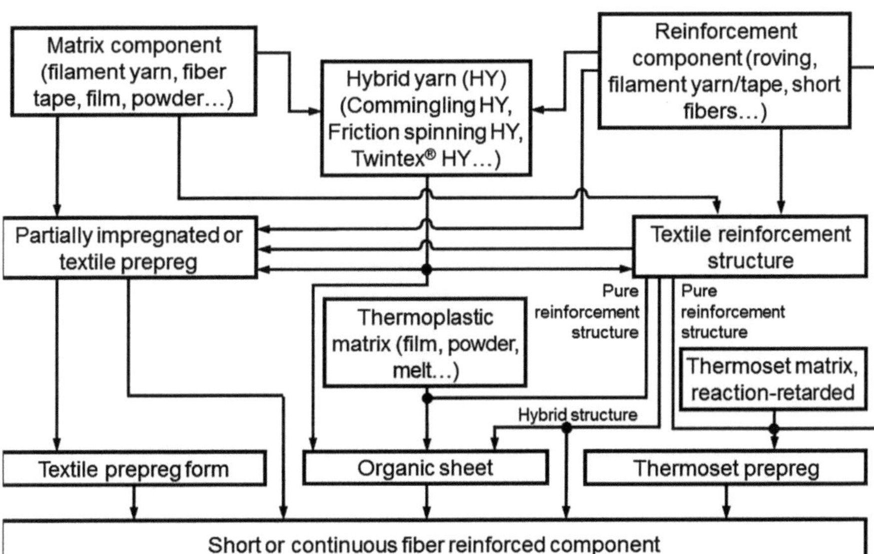

Fig. 11.1 Overview of base materials, thermoset and thermoplastic semi-finished products, and fiber-reinforced FRP component

11.2 Thermoset Prepregs

11.2.1 Free-Flowing Thermoset Prepregs and Molding Compounds (SMC/BMC)

The *SMC method* (*Sheet Molding Compound method*) is practiced widely in industries for the production and processing of free-flowing prepregs. The majority of SMC production is needed for automotive construction and electronics applications [4]. SMC prepregs are processed into planar semi-finished products made from glass fiber nonwoven fabrics or glass fiber mats, which are impregnated with a cross-linkable thermosetting resin system and fillers. Depending on the application, a variety of recipes can be used, which will not be detailed here.

Unsaturated polyester resin systems are most commonly used as matrix material. They are intermixed with mineral fillers and other ingredients depending on the requirement of further processing and application. The glass fibers are cut to lengths between 25 and 50 mm. Their mass fraction can amount to a maximum of 30 %. So-called directional SMC semi-finished products (SMC-D) with reduced flowability are produced by using longer glass fibers of up to 200 mm.

The production of the prepreg is carried out on continuously operating SMC impregnation plants (Fig. 11.2). The resin/curing agent/filler material is knife-coated onto the release film. Glass fiber rovings are chopped into short fibers by a wide chopping machine. These short fibers are dispersed across the entire production width and covered by an upper release film. Afterwards, this structure passes through the impregnation line, where the components are intermixed and the prepreg is compacted. After winding up and a maturing time of 2–4 weeks under

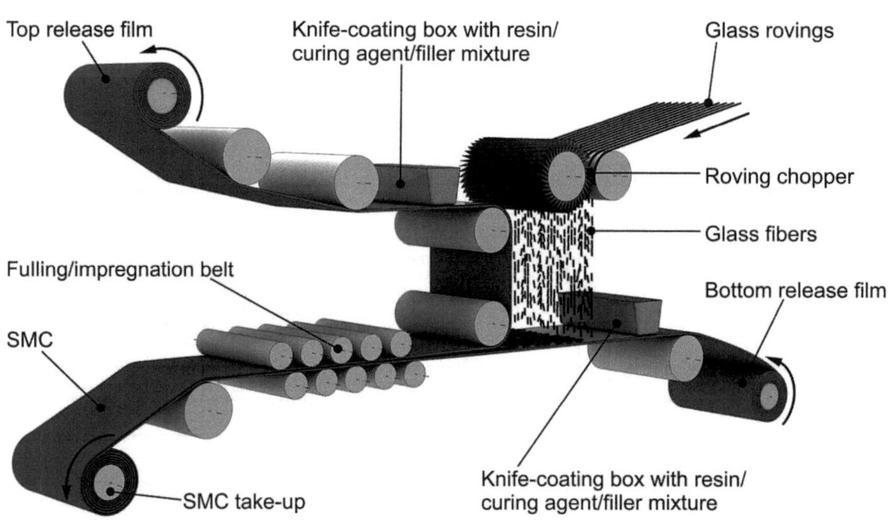

Fig. 11.2 Principle of SMC production

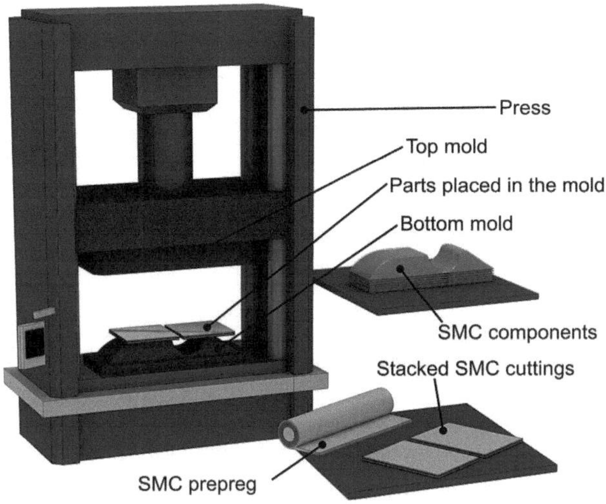

Fig. 11.3 Process chain for the manufacture of SMC components (*left*) by compression molding, according to [9]

defined storage conditions to ensure chemical thickening, the semi-finished products can be processed further. Depending on the individual mixture and storage conditions, the storage period can range from only a few weeks to several months [5, 6].

The additional integration of continuous reinforcement fibers or rovings in the direction of production of the SMC semi-finished products results in C-SMC semi-finished products, which are suitable for components to meet extreme mechanical requirements. However, these additional yarns significantly reduce the flowability [3, 7]. The production and processing of SMC semi-finished products with a reinforcement consisting entirely of carbon fibers (CF-SMC) is also an option and offers improved mechanical properties in comparison to SMC on a glass fiber basis but at higher cost [8].

The so-called BMC (*Bulk Molding Compound*) is produced by simply striping off the mixture of fibers, resin, curing agent, and fillers at the end of the SMC plant. These molding compounds are also further processed using compression molding after plasticization. Here, the maximum length of the glass fibers is 12 mm, which results in significantly reduced mechanical composite properties compared to SMC [7]. Beyond these methods, short fiber-reinforced thermoset components are produced by DMC (Dough Molding Compounds) or GMC (Granulated Molding Compound) using either injection-moldable compounds without chemical thickening or dry, granulated compounds [3].

The further processing of both SMC semi-finished products and BMC molding compound into components is performed by compression molding. For this purpose, pressures of 25–250 bar are applied on parallel-controlled servo-hydraulic presses with two-part steel molds, which are heated to processing temperatures of about 140–160 °C (Fig. 11.3). In the processing of SMC, mostly rectangular or

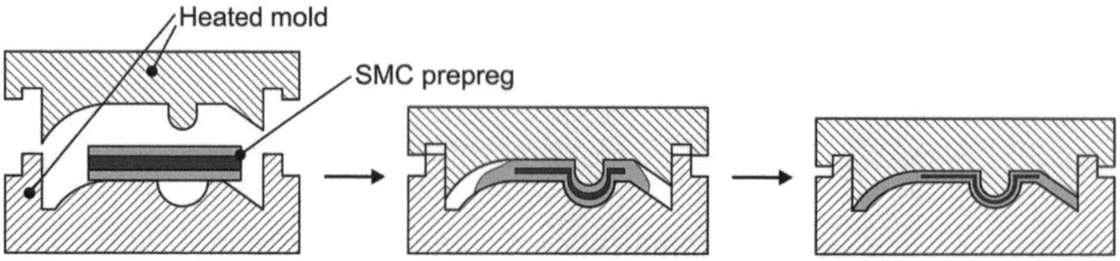

Fig. 11.4 Flow behavior of the fiber/resin/curing agent/filler mixture in the processing of SMC semi-finished products by compression molding, according to [3]

trapezoidal SMC cuttings are stacked in sufficient numbers and inserted into the pressing mold at defined positions [1].

During compression molding, the viscosity of the matrix system is reduced due to the influence of pressure and temperature and the semi-finished products start flowing from the top and bottom sides of the pressing mold. Afterwards, the closing movement of the pressing mold causes the cavity to be completely filled with the fiber/resin/curing agent/filler mixture, and the matrix is then cured by chemical cross-linking under holding pressure (Fig. 11.4).

To ensure that the mixture reaches all sections of the mold cavity, the cuttings have to cover at least 70 % of the mold surface [1, 10].

When using SMC prepregs with long or continuous fiber reinforcements, the entire mold surface is covered with prepreg cuttings. The processing is done by compression molding without significant flowing.

After cooling and demolding of the components, post processing is necessary such as deburring, drilling, tilling or varnishing. Depending on the amount of material and the size of the component, a suitable automation can achieve cycle times of usually just a few minutes or as short as 1 min [1, 6].

SMC components are used in automobile construction (bonnets, doors, trunk lids, cylinder head covers, noise mufflers, or spoilers) and for railed vehicles (interior and exterior cladding), in electronics (switchboards, emergency call boxes), and in the sanitary industry. Engine parts, headlight reverberators, terminal boxes, or insulators are components based on BMC.

11.2.2 Non-free-flowing Thermoset Prepregs from Yarns and Flat, Planar Semi-finished Products

For the production of non-free-flowing thermoset prepregs, the same fiber materials are used as in woven or knitted fabrics for lightweight construction applications: mostly glass, carbon or aramid fibers. These are processed either directly into UD prepregs or in a two-step process—firstly, woven or multiaxially warp-knitted into textile semi-finished products and then processed into prepregs with multiaxial reinforcement (see Fig. 11.1). The continuous fibers in the form of rovings are taken

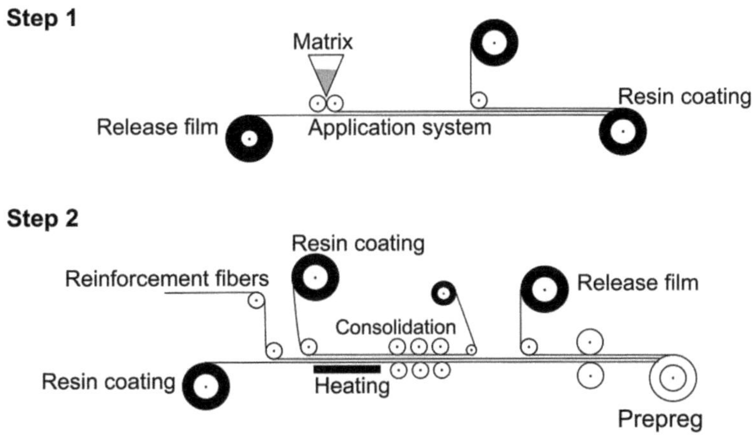

Fig. 11.5 Production of UD prepregs by means of a release film (step 1: coating of the release film, step 2: production of the prepreg)

off from bobbins placed on bobbin creels. Two methods are available for the application of thermosetting matrix materials: by means of release films (film transfer process) or by means of the solvent impregnation [11]. In the resin film method, the first step consists of applying the matrix as a resin coating to the siliconized release film (Fig. 11.5, step 1). The filament yarns drawn from bobbin creels are combined into a tape, then parallelized by combs, and finally brought into contact with the release film. Under pressure and heat (ca. 60–90 °C), the filament tape is thoroughly impregnated with the resin on the release film (Fig. 11.5, step 2). Then, a release paper is applied, and the UD prepreg is wound up continuously on a roll. It is important to cool the preimpregnated semi-finished product immediately, to stop the further course of the curing reaction of the resin. This method has become widespread industry standard [11].

In *solvent impregnation,* the rovings, which are spread out into filament tapes next to each other, are guided through a matrix dip (vertical process arrangement), or fibers are placed on a release film coated with matrix, emulating the manufacturing process with release films (horizontal process arrangement). In the matrix dip, the resin/curing agent mixture is blended with solvent to achieve the required viscosity for impregnation. In both methods, a heating device is used for the removal of the solvent and a preliminary cross-linking of the resin, before the UD prepregs are fitted with a paper protector and rolled up (Fig. 11.6). These prepregs display a highly viscous impregnation at room temperature.

UD prepregs can be produced in a variety of forms: as preimpregnated individual yarn (*single tow*), *tape*, or *stripe* (Fig. 11.7). While single tows are not classified as textile structures, tapes and stripes are classified as semi-finished textile products. Single tows of widths above 1 mm are commonly processed by means of the winding process. Tapes of very small areal masses of about 100 g/m^2 are usually used in applications of very high quality such as load bearing structures in airplanes,

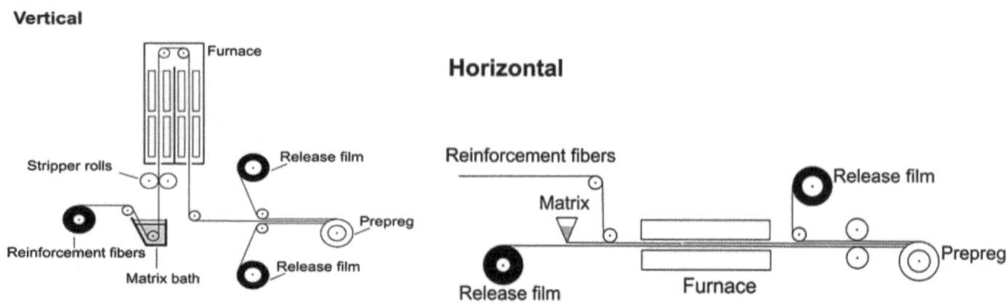

Fig. 11.6 Production of UD prepregs in a solvent impregnation process

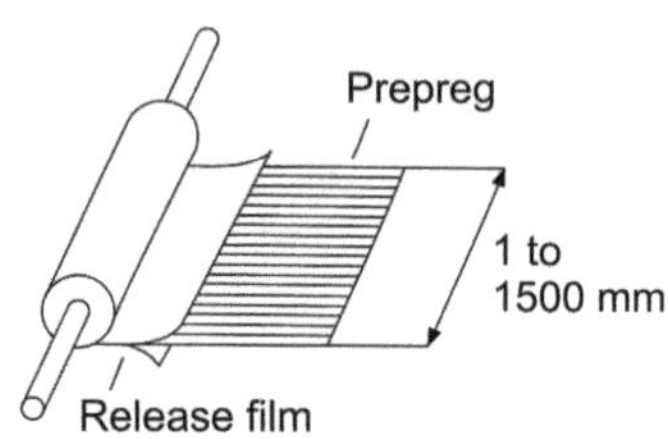

Fig. 11.7 Structure of UD prepregs

whereas stripes, at areal masses ranging from 500 to 1,500 g/m^2, are applied primarily in the manufacture of boats and rotors for wind energy plants. The areal mass also determines the layer thickness which can be achieved in the cured multilayer composite. For instance, the layer thickness of a UD prepreg with a fiber volume content of 60 % and an areal mass of 200 g/m^2 is about 0.18 mm after curing [11].

One specific aspect of composite manufacture based on prepregs with a reactive matrix is the necessity of permanent cooling during prepreg storage, which is required to slow the curing of the resin. In particular cases, air-conditioned rooms are required during processing. Commonly used resin systems allow storage periods of up to 12 months at refrigeration at −18 °C. Before further processing, the prepregs have to defreeze. This should be done in the film packaging under exclusion of air in order to prevent the condensation of atmospheric humidity on the prepreg surface, which can take up to 48 h to complete [5]. Afterwards, the prepregs can be assembled according to the specific component. Due to the pre-impregnation, the prepreg surface features a defined stickiness (tack). This allows their secure positioning in complex molds, even under the influence of gravity.

The limited flowability of the matrix of UD prepregs and the resulting limited mobility of the fibers to each other limit the producible component geometries, hindering or complicating the use of these materials for complex-shape components. However, the use of UD prepregs achieves the highest stiffnesses and strengths [3], which are entirely unattainable with non-pre-impregnated semi-

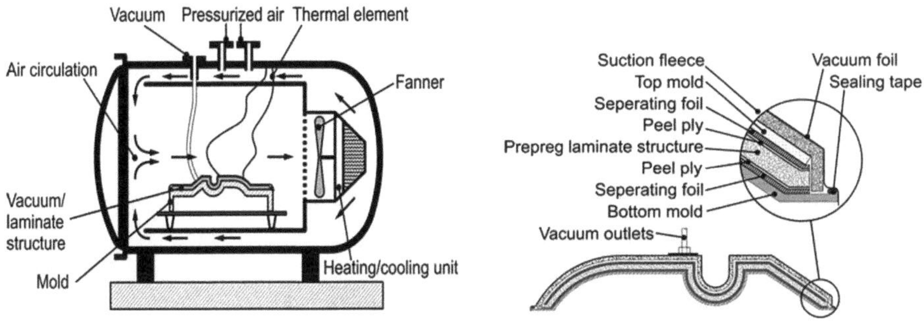

Fig. 11.8 Basic structure of an autoclave, and exemplary layer structure for the component manufacture by autoclave method

finished textile products (dry preforms). Here, the selection of semi-finished products has to achieve a balance between the requirements of drapability and strength. There are various approaches to enhance the limited deformation capabilities, such as laser-micro perforation, although losses of strength occur under tensile strain [12].

Another limitation of the use of UD prepregs can be their lack of resistance to delamination, which is caused by the missing reinforcement effect in thickness direction (see Chap. 7). Therefore, further developments concern the insertion of so-called Z reinforcements into the unconsolidated prepregs. This can be realized by sewing, pinning with metal tacks, tufting, or the so-called Z-pinning with special steel, titanium, glass, or carbon elements [13, 14]. The carbon elements are inserted into the previously produced layer structures by ultrasound. The resulting losses of strength in longitudinal fiber direction are greater than in multiaxial non-crimp fabrics produced by warp-knitting methods, as the insertion process causes significant damages [15, 16].

The further processing of the non-free-flowing thermoset prepregs made from yarns or flat, planar textile reinforcements into (usually) simple-geometry, shell-shaped, and mostly heavy-duty components for aeronautics, prototype construction or sports and leisure applications is nowadays commonly performed by means of the *low-pressure autoclave method* [3].

The low pressure autoclave method requires, an inherently stiff and forming half-mold, on which the cut and suitably oriented thermoset prepregs are placed according to the component requirements in a very complex process. It has to be considered that a multiaxial draping of these prepregs is not possible, and there is no material flow during the consolidation process. The structure, fitted with a release film and possibly with a top mold part or a textile suction structure, is packaged in a vacuum foil, sealed, and compacted by vacuum (Fig. 11.8, right). For a reproducible and swift placement in the mold for the production of larger structural components in aeronautics, specially developed robot-controlled tape laying machines are used these days. Despite such technological possibilities, this method remains a time-consuming one.

The actual component production is done in the *autoclave plant* (Fig. 11.8, left), whose pressure container can apply pressures of up to 70 bar and temperature from 80 to 180 °C for the processing of thermoset resin systems. The method is time consuming due to the complex placing of material on the mold and the usual cycle times of 3–12 h. However, it guarantees the best component properties at reproducible quality [3].

Further developments of the manufacturing process concern the decrease of temperatures and pressures required for component production, which would simplify the production process and allow the use of UD prepregs for larger and more complex components, for instance in boat building [2, 17].

Apart from the previously mentioned prepregs based on yarns, woven fabrics, or multiaxial warp-knitted fabrics, there are also semi-finished products with adhesive matrix. In the manufacture of these textile structures, warp and weft yarns are placed on top of each other as in multiaxial warp-knitting (see Chap. 7). Afterwards, they are connected not mechanically, but with adhesives, such as acrylate, ethylene-vinyl acetate, polyurethane, polyvinyl alcohol, polyvinyl acetate, polyvinyl chloride, or styrene-butadiene. In this manner, biaxial or multiaxial non-crimp fabrics as well as grid structures from all fiber materials, such as glass carbon, or aramid fibers, can be manufactured. The manner and sequence of layer arrangement is freely selectable, making symmetrical layer arrangements easily realizable. These semi-finished products are used for the reinforcement of roof sheeting, sails and flooring, more often than in typical lightweight construction applications.

11.3 Thermoplastic Prepregs

11.3.1 Free-Flowing Thermoplastic Prepregs and Pressing Compounds (GMT/LFT)

11.3.1.1 Prepregs Based on Non-woven Fabrics (GMT)

In industrial applications, various forms of fiber-reinforced thermoplastic semi-finished products are used for processing by compression molding. Due to their increasing use in the automobile industry, they have experienced above-average growth of demand over the past years [4].

Glass mat-reinforced thermoplastics (GMT) are plate-shaped thermoplastic semi-finished products with a non-woven or mat-shaped glass reinforcement prebonded by needling. Near-exclusively, polypropylene is nowadays used as thermoplastic. The production of these semi-finished products is performed by melt impregnation or by a multi-step wet processing similar to the one used in paper production.

Semi-finished products are produced in two sub-processes. Step 1 is a non-woven fabric production process (see Sect. 9.4.3.2). In this step, glass rovings are taken-off from bobbins and deposited on a conveyor belt across the entire

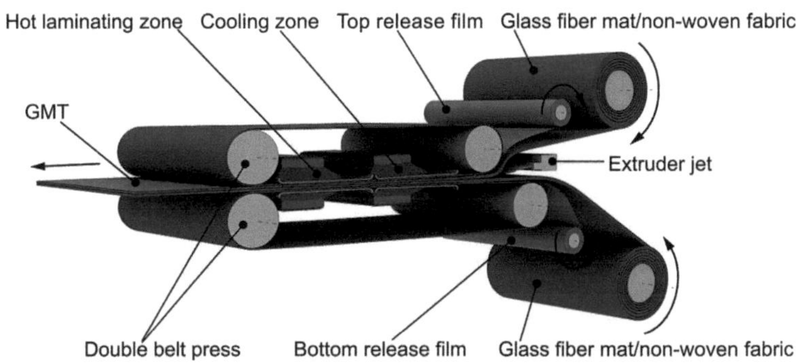

Fig. 11.9 Manufacturing process of GMT prepregs on a double belt press

working width. The subsequent fixation is performed by needling not only to ensure fixation, but also to dissipate fiber bundles into individual fibers. The needling also shortens overly long or continuous fibers in order to achieve sufficient matrix flow, or to create a three-dimensional fiber architecture. This process has a significant influence on the processing, material, and component properties. The resulting fiber lengths usually range from 25 to 50 mm. In step 2 (Fig. 11.9), two glass fiber non-woven fabric webs are impregnated with extruder-melted polypropylene using flat dies. Afterwards, using a double belt press, the semi-finished products are consolidated under temperature and pressure. The cutting of the GMT plates is the final step.

As in SMC manufacture, continuous fibers oriented in production direction can be integrated unidirectionally into the GMT in order to increase their suitability for higher mechanical requirements. These will, however, significantly reduce the flowability [6]. The local integration of one or multiple layers of unconsolidated hybrid yarn woven fabrics for an enhanced structural integration is also possible. These products are referred to as GMTex [18].

With classical GMT, fiber content ratios of up to 40 mass percent are achievable. Using unidirectionally oriented continuous rovings, this number reaches up to 52 mass percent [6].

The further processing of the consolidated GMT semi-finished products into components is usually performed in five sub-processes on servo-hydraulic presses with two-part steel molds [3]:

- Creation of suitably sized cuttings,
- Heating of the cuttings to 230–235 °C in a convection or infrared furnace,
- Transfer of the molten semi-finished product into the pressing implement, heated to 25–80 °C (depending on matrix),
- Forming and consolidating by extrusion or compression molding, and simultaneously cooling down,
- Demolding of the component, and post processing.

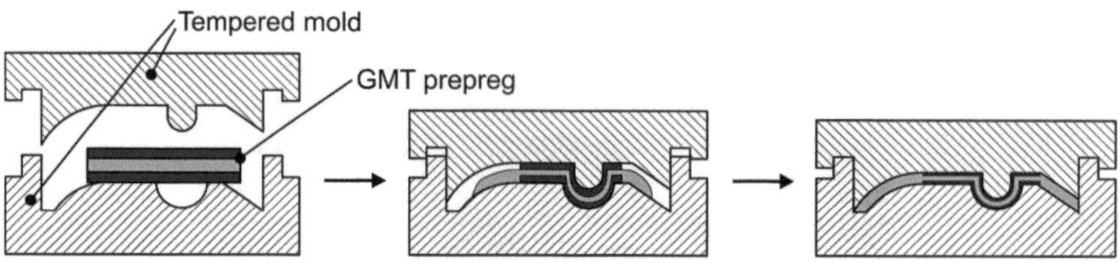

Fig. 11.10 Flow behavior of the fiber/thermoplastic mixture during the processing of GMT prepregs by extrusion, according to [3]

The flowing process during pressing of the heated plastificate (Fig. 11.10) differs significantly from the flowing process of the SMC. Immediately after placement in the mold cavity, the melt begins to solidify from the mold surfaces inward due to the difference in temperature between the semi-finished product and the mold. The force exerted by the press squeezes the still-liquid matrix (holding the reinforcement fibers) into the remaining cavity, where it solidifies. This requires quickly closing presses [3].

11.3.1.2 Inline-Compounding (Long Fiber Thermoplastic Direct Method)

Long fiber-reinforced thermoplastic granulates (LFT-G) are generally produced by a pultrusion process. In this process, a continuous fiber roving (glass, carbon, or aramid fibers) is impregnated with polymer melt (usually polypropylene or polyamide) in a pultrusion mold and then processed into a 10–25 mm-long rod granulate by chopping. Fiber mass percent ratios of up to 80 % are feasible. Alternatively, the application of the matrix can be performed by impregnation or coating. Before the further processing of the LFT-G by compression molding (Fig. 11.10), the granule material is plasticized in extruders with special screw geometry, aiming to minimize fiber damage. The dosed plastificate is then inserted into the pressing mold. As in LFT processing, the final steps consist of compression molding, demolding, and finishing [3, 6].

For the manufacturing of *long fiber-reinforced thermoplastics* (LFT-D) in a direct process, the usual granulate production is omitted. The process is directly coupled to component production. The realization of the heated, fiber-filled pressing compound is completed by combined feeding and processing of thermoplastic granulated material and glass rovings in a plasticization extruder. In the extruder, the glass filaments are separated into short fibers and the mixture is heated to near-melting temperature. In contrast to the LFT-G process, the filling of the heated pressing mold with the dosed plastificate, compression molding (to give the shape) as well as consolidation are performed immediately, and are followed by demolding and finishing of the component [3, 6].

One new development of the LFT-D method is the LFT-D/ILC method for the back compression molding of coextruded or varnished foils with LFT [19]. Here, a modified sub-process for the provision of the plastificate with a separate fiber chopping device is used to achieve a thorough separation of fibers in the plastificate. The back compression molding process is performed by extrusion, where the foil is first placed in the mold on its visible surface. This method is very promising for the achievement of Class-A surfaces [18].

Another enhancement is the so-called *Tailored LFT*, which is distinguished by the insertion of local reinforcements from continuous fiber-reinforced thermoplastics (e.g. rovings or reinforcement textiles based on hybrid yarn, pre-consolidated profiles or wound reinforcement elements) in the cavity before the pressing process [18].

The material properties and free choice of design allow a versatile use of extrusion-manufactured GMT and LFT components, particularly in automobile construction. In comparison to SMC or BMC components, they are lighter and tougher. The cost of production is significantly lower for larger components. In comparison to injection-molded components, the longer fibers of GMT and LFT provide them with a better static and dynamic long-term behavior at high endurance and alternating stresses [6].

Exemplary application of CMT and LFT components are found predominantly in automobile construction as mounting and dashboard support, crash absorption structures, underbody trims, or usually hidden covers.

11.3.2 Non-free-flowing Thermoplastic Prepregs with Continuous Fiber-Reinforcements from Yarns or Flat, Planar Semi-finished Products

11.3.2.1 General Remarks

Continuous fiber-reinforced thermoplastic semi-finished products are distinguished by the type and arrangement of their reinforcement fibers, by the matrix material, and by the degree of impregnation and consolidation. Glass, carbon, aramid, or other fibrous materials based on continuous filaments are used as reinforcement material. Usually, these are processed as textile fabrics such as woven fabrics, non-crimp fabrics, or warp-knitted fabrics with biaxial or multiaxial reinforcement yarn arrangement. The usability of biaxially reinforced multilayered weft-knitted fabrics has already been proven experimentally (see Chap. 6).

For the matrix component, the entire range from standard (PP or PA6) to high-temperature thermoplastics (e.g. PEEK or PPS) can be used [20–23]. Fundamentally, the classification differentiates between fully impregnated and consolidated plate-shaped semi-finished products (organic sheets) as well as not fully impregnated and consolidated prepregs (partially impregnated or textile prepregs) (Fig. 11.1). Additionally, specific properties of the organic sheets can be developed by integrating not only continuous fiber-based reinforcement layers, but also others based on nonwoven fabrics such as hybrid nonwovens (see Sect. 9.4.3.2).

11.3.2.2 Organic Sheets

The aspects of manufacturing and processing fully impregnated and consolidated plate-shaped semi-finished products are exemplarily detailed in [3]. The semi-finished products are manufactured in thermopressing processes, using discontinuous (e.g. static presses), semi-continuous (e.g. interval hot presses or transfer presses), or continuous (e.g. double belt presses, Fig. 11.9) systems. Specific process parameters for the production of the consolidated organic sheets depend on the matrix material used, the reinforcement structure, the degree of pre-impregnation, the thickness of the composite, and the specific pressing method. At small lot size, either partially impregnated/textile prepregs are processed or the film-stacking method is used. In this case, the press is fed with stacked cuttings, alternating between textile reinforcement structures and matrix films. For high material performance, the textile reinforcement component and the matrix are brought together directly at the entry to the press [3].

The further processing of the organic sheets into continuous fiber-reinforced thermoplastic components is usually performed by discontinuous or continuous thermoforming (heating, forming and consolidation, demolding). Here, it has to be considered that the demolding process depends decisively on the draping behavior of the textile reinforcement structure while the matrix is softened, which in turn depends on the effective forming mechanisms of the textile reinforcement structure: fiber elongation, fiber stretching, fiber slippage, and shear [24, 25].

For the discontinuous forming, the semi-finished product is heated above melting temperature of the matrix material and transferred into the pressing mold with a quick-acting press. The molds are heated at a temperature of 60 °C below the melting temperature, where the semi-finished product is formed and cooled immediately (Fig. 11.11). Special fixation, holding-down, or tensioning elements are used to avoid the formation of creases during forming.

Finally, the component enters into post processing. To successfully separate the time-consuming heat-up process from forming, component-dependent cycle times of 15 s should be achieved [3]. For large quantities and small forming radii two-part metallic molds can be used as forming tools to achieve short cycle and extended service times. To avoid creases or incompletely consolidated parts, the mold cavity height has to be accurately adapted to the thickness of the respective semi-finished product and to the draping behavior of the reinforcement structure. Forming with an elastomer or silicon punch onto a metallic bottom mold is also suitable for short cycle times and allows the manufacture of differently structured organic sheets with a single mold. But smaller forming radii cannot be achieved, and a high surface quality is possible on one side only. Other forming methods for organic sheets are diaphragm forming, hydroforming, and pressure-assisted thermoforming [26, 27].

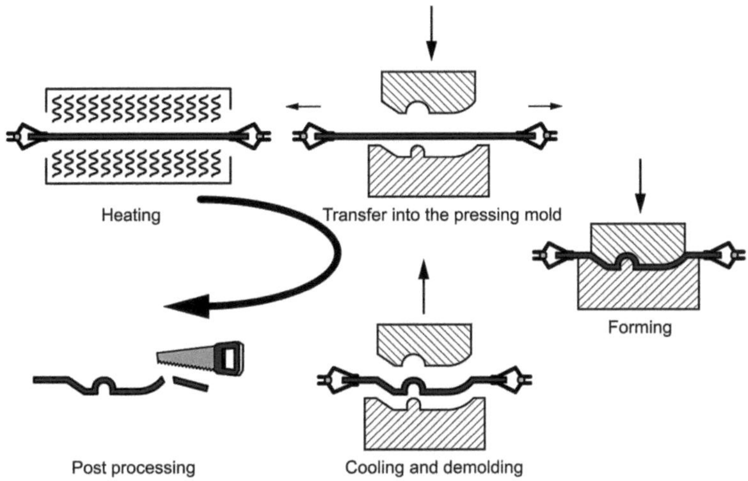

Fig. 11.11 Typical processing of thermoforming organic sheets

Recent industrial efforts have led to the development of an innovative method for the processing of organic sheets into heavy-duty thermoplastic-based hollow body structures. According to the FIT hybrid method, the forming of the organic sheets by forming molds is combined with an injection-molding process. A subsequent blowing or inflation process enables the realization of the final component shape in a single process step, making complex components with integrated hollow profiles feasible [28].

The continuous thermoforming of profiles from organic sheets is realized, for instance, by the roll forming method commonly used in metal processing. For this, the semi-finished product has to be heated prior to forming, for example by means of infrared radiation fields [3].

Components manufactured from organic sheets have a wide range of applications, among them structural and energy-absorbing automobile parts (bumper supports, crash and structural elements), in aeronautics (airplane seat components, interior components), in electronics (casings, speaker components), or in sports applications (safety helmets, shoes, bicycle components). Beyond those, other practical examples are safety and orthopedic applications.

11.3.2.3 Partially Impregnated or Textile Prepregs

Partially impregnated thermoplastic prepregs can be manufactured by means of a variety of (predominantly continuous) methods. The textile reinforcement fabrics or UD continuous fiber tapes are partially impregnated with the thermoplastic matrix by:

- Coating with polymer powders,
- Solvent impregnation,

- Dipping or coating with molten matrix,
- Coating with films or nonwoven fabrics,
- Insertion of thermoplastic yarns, or
- Mixing of reinforcement fibers and thermoplastic fibers (hybrid yarn partially impregnated with thermoplastic matrix component)

Here, the reduced formability of the partially impregnated prepregs (depending on the degree of impregnation) in comparison to the textile semi-finished products at room temperature and improved handling have to be taken into account for the further processing into three-dimensional preforms.

For further processing into components, the same methods are used as in the processing of organic sheets. In contrast, the complete composite consolidation has first to be achieved during pressing. Thermoplastic UD tape prepregs (tow prepregs), are usually processed by winding or tape laying processes.

Powder coating of textile semi-finished products is performed by scattering of the powder, in an impregnation bath, or by electrostatic binding. By heating the coated textile structure, the powder is melted, or the solvent is removed. The intensity of the partial penetration of the matrix into the textile semi-finished products depends on the additionally applied pressure during pressing. UD tape prepregs manufactured on a polymer powder basis can be fully or partially impregnated. Powder impregnation is of special importance for the incorporation of thermoplastic powders in a reinforcement yarn. Several methods are available for the opening of the yarn and the processing of the powder-air mixtures [29, 30]. The durable fixation of the powder to the reinforcement filaments is achieved by an additional coating of thermoplastic or by melting the powder [31].

For the *solvent impregnation*, a solvent suited to the respective thermoplastic material is required. As the viscosity of the polymer solution is very low, impregnation quality is very good. The solvent has to be vaporized after impregnation, dictating strict adherence to the corresponding industrial and fire safety regulations.

By impregnating (e.g. padding) or coating with molten matrix, the reinforcement textile is cross-linked with the molten thermoplastic, which later cools down and solidifies. In case of surface coating, thermoplastic films or nonwovens are heated up and laminated or pressed onto the semi-finished textile products in a separate process. The application of the thermoplastic matrix is commonly performed on one side in both methods.

Textile or hybrid yarn prepregs are distinguished by the continuous reinforcement filaments and the matrix component in fiber or filament form. Depending on the textile structure, they exhibit good drapability during further processing, which makes them ideally suited for the manufacture of complex-shape components with or without intermediary preform production.

Compared to the simple possibility of manufacturing textile prepregs by simultaneous or parallel processing of reinforcement and thermoplastic yarns during textile fabric formation, the application of hybrid yarns (see Sect. 4.1.3) offers a more homogeneous mixture of reinforcement and matrix component. Particularly the use of hybrid yarns manufactured by integrating the polymer filaments in glass rovings (Twintex®) during glass production or by a commingling process relying on modified air texturing technology, allows a largely homogenous fiber/matrix allocation and a bending stiffness ideal for further processing [21, 32, 33]. This enables the production of specifically designed textile semi-finished products and preforms with low-damage, short flow paths for the often highly viscous molten matrix during composite consolidation, and high formability. Thus, the crucial requirements for a high composite quality are met.

Non-free-flowing partially impregnated or textile thermoplastic prepregs with continuous fiber-reinforcement can either be processed directly or by the detour of a consolidated organic sheet (see Fig. 11.11). For this, the same pressing processes under temperature and pressure as in the processing of organic sheets are used. No flow processes occur. Distortions of the reinforcement structure resulting from the forming have to be considered and can be quite significant, particularly for complex-shape components.

For the further processing of partially impregnated or textile prepregs directly into components, the forming process, the complete impregnation and the consolidation have to be linked by a suitable process flow. For the impregnation, the temperature of the mold has to be significantly higher than the melting temperature, and significantly lower for demolding. The component manufacture is often performed by servo-hydraulic presses with two-part steel molds. Besides classic pressing technology, diaphragm or autoclave technology are also used. Afterwards, the components in each case have to be post-processed [26].

To realize very short cycle times, several methods are known. For one method, the partially impregnated or textile thermoplastic prepregs are heated in a separate station (for instance by infrared radiation), and then inserted into a cold, two-part steel or silicone mold, where they are formed and consolidated [34]. The Quicktemp concept is based on a permanently heated outer mold and a thin, fast-cooling inner mold. The direct impregnation method also uses a thin, but unheated inner mold. This is heated up together with the prepreg material by a forming/impregnation mold heated beyond melting temperature and then transferred into the significantly cooler consolidation mold, where the actual pressing takes place [3]. To manufacture continuous fiber-reinforced thermoplastic spacer components with one ground and one surface layer connected by links made from suitably designed textile hybrid yarn preforms and molding boxes with special kinematics are used. These allow the reproducible impregnation as well as quick cooling and consolidation of the structure after separate heating [35].

References

1. FLEMMING, M.; ZIEGMANN, G.; ROTH, S.: *Faserverbundbauweisen - Fertigungsverfahren mit duroplastischer Matrix*. Berlin, Heidelberg : Springer Verlag, 1999
2. PARTINGTON, N.: Prepregs for race boats: Adapting aerospace materials to the marine world. In: *JEC Composites Magazine* (2009), No. 51, pp. 30–33

3. NEITZEL, M.; MITSCHANG, P.: *Handbuch Verbundwerkstoffe*. München,Wien : Carl Hanser Verlag, 2004
4. BÜLTJER, U.: Produktion von GFK und duroplastischen Formmassen in Europa. In: *Proceedings.7. Internationale AVK-TV Tagung*. Baden-Baden, Germany, 2004
5. EHRENSTEIN, G.: *Faserverbund-Kunststoffe*. München, Wien : Carl Hanser Verlag, 2006
6. KANNEBLEY, G. et al.: *AVK-TV-Handbuch*. Frankfurt, 2004
7. LIEBOLD, R.: Harzmatten in engen Toleranzen herstellen. In: *Kunststoffe* 81 (1991), No. 10, pp. 923–928
8. REUTHER, E.: Kohlefaser SMC für Strukturbauteile. In: *Proceedings. 7. Internationale AVKTV Tagung*. Baden-Baden, Germany, 2004
9. DAVIS, B. A. et al.: *Compression Moulding*. München, Wien : Carl Hanser Verlag, 2003
10. HELLRICH, W.; HARSCH, G.; HAENLE, S.: *Werkstoff-Führer Kunststoffe - Eigenschaften, Prüfungen, Kennwerte*. München, Wien : Carl Hanser Verlag, 2004
11. ANONYM: *Prepreg Technology, Publication No. FGU 017b (Company document Hexcel Corporation)*. 2005
12. FORD, R.; GRIFFITHS, B.: Formable aligned-fibre composite materials. In: *JEC Composites Magazine* (2009), No. 50, pp. 52–54
13. PARTRIDGE, I. K.; CARTÍE, D. D. R.: Delamination resistant laminates by Z-Fiber® pinning: Part I manufacture and fracture performance. In: *Composites: Part A* 36 (2005), No. 1, pp. 55–64. DOI 10.1016/j.compositesa.2004.06.029
14. MOURITZ, A. P.: Review of z-pinned composite laminates. In: *Composites: Part A* 38 (2007), No. 12, pp. 2383–2397. DOI 10.1016/j.compositesa.2007.08.016
15. BYUN, J.-H.; SONG, S.-W.; LEE, C.-H.; UM, M.-K.; HWANG, B.-S.: Impact properties of laminated composites stitched with z-fibers. In: *Proceedings. 15th International Conference on Composite Materials*. Durban, South Africa, 2005
16. GRASSI, M.; ZHANG, X.; MEO, M.: Prediction of stiffness and stresses in z-fibre reinforced composite laminates. In: *Composites: Part A* 33A (2002), No. 12, pp. 1644–1653. DOI 10.1016/S1359–835X(02)00137–9
17. MILNER, S.: Innovative prepreg systems for the marine industry. In: *JEC Composites Magazine* (2007), No. 35, pp. 37–39
18. ERNST, H.; HENNING, F.; GEIGER, O.: Neueste LFT-D Direkttechnologie für hochbelastbare Bauteile und Komponenten mit hoher Oberflächengüte. In: *Proceedings. 10. Dresdner Leichtbausymposium*. Dresden, Germany, 2006, pp. 1–62
19. GEIGER, O.: Hinterformen von Folien mit langfaserverstärkten Thermoplasten. In: *Proceedings. 7. Internationale AVK-TV Tagung*. Baden-Baden, Germany, 2004
20. PAUL, Ch.; CHERIF, Ch.; HANUSCH, J.: Dreikomponenten-Hybridgarn und –Hybridgarngestricke für komplexe Leichtbauanwendungen. In: *Proceedings. 8. Dresdner Textiltagung*. Dresden, Germany, 2006
21. CHOI, B. D.: *Entwicklung von Commingling-Hybridgarnen für faserverstärkte thermoplastische Verbundwerkstoffe*. Dresden, Technische Universität Dresden, Fakultät Maschinenwesen, Dissertation, 2005
22. KALDENHOFF, R.: *Friktionsspinn-Hybridgarne als neuartige textile Halbzeuge zur Herstellung von Faserverbundkunststoffen*. Aachen, RWTH Aachen, Fakultät für Maschinenwesen, Dissertation, 1996
23. PAPPADÀ, S.; RAMETTA, R.; SUPPRESSA, G.; PASSARO, A.; MAFFEZZOLI, A.: PPScarbon reinforced panals with improved damage tolerance. In: *Proceedings. 32nd International SAMPE Conference - New Material Characteristics to cover New Application Needs*. Paris, France, 2011
24. ORAWATTANASRIKUL, S.: *Experimentelle Analyse der Scherdeformation biaxial verstärkter Mehrlagengestricke*. Dresden, Technische Universität Dresden, Fakultät Maschinenwesen, Dissertation, 2006

25. HÖRSTING, K.: *Rationalisierung der Fertigung langfaserverstärkter Verbundwerkstoffe durch den Einsatz multiaxialer Gelege*. Aachen, RWTH Aachen, Fakultät für Maschinenwesen, Dissertation, 1994
26. FLEMMING, M.; ZIEGMANN, G.; ROTH, S.: *Faserverbundbauweise - Halbzeuge und Bauweisen*. Berlin, Heidelberg : Springer Verlag, 1996
27. HENNINGER, F. H.: *Beitrag zur Entwicklung neuartiger Fertigungsverfahren zur Herstellung von kontinuierlich faserverstärkten Thermoplasten*. Kaiserslautern, Technische Universität Kaiserslautern, Fachbereich Maschinenbau und Verfahrenstechnik, Dissertation, 2005
28. HOFFMANN, L.; RENN, M.; DRUMMER, D.; MÜLLER, T.: FIT-Hybrid - Hochbelastbare Faserverbundbauteile großserientauglich hergestellt. In: *Leightweightdesign* (2011), No. 2, pp.38–43
29. RAMANI, K.; HOYLE, C. H.: Processing of thermoplastic composites using a powder slurry technique, I. Impregnation and preheating. In: *Materials and Manufacturing Processes* 10 (1995), No. 6, pp. 1169–1182. DOI 10.1080/10426919508935100
30. SALA, G.: Heated chamber winding of thermoplastic powder-impregnated composites, Part 1, Technology and basic thermomechanical aspects. In: *Composites. Part A* 27 (1996), No. 5, pp. 387–392. DOI 10.1016/1359–835X(95)00036–2
31. OSTGATHE, M.; MAYER, Ch.; NEITZEL, M.: Organobleche aus Thermoplastpulver. In: *Kunststoffe* 86 (1996), No. 12, pp. 1838–1840
32. MÄDER, E.; ROTHE, C.: Maßschneidern von Hybridgarnen für effektive Verbundeigenschaften. In: *Proceedings. 8. Dresdner Textiltagung*. Dresden, Germany, 2006
33. CHERIF, Ch.; RÖDEL, H.; HOFFMANNN, G.; DIESTEL, O.; HERZBERG, C.; PAUL, Ch.; SCHULZ, Ch.; GROSSMANN, K.; MÜHL, A.; MÄDER, E.; BRÜNIG, H.: Textile Verarbeitungstechnologien für hybridgarnbasierte komplexe Preformstrukturen /Textile manufacturing technologies for hybrid based complex preform structures. In: *Journal of Plastics Technology* (2009), No. 2, pp. 103–129
34. ADAM, F.; HUFENBACH, W.; GROSSMANN, K.; MODLER, K.-H.; HANKE, U.; LIN, S.; MODLER, N.; KRAHL, M.: Processing of Novel 3D Hybrid Yarn Textile Thermoplastic Composites Using Process-adapted Consolidation Kinematics. In: *Proceedings. 8. Dresdner Textiltagung*. Dresden, Germany, 2006
35. HUFENBACH, W. A.; MODLER, N.; KRAHL, M.; HORNIG, A.; FERKEL, H.; KURZ, H.; EHLEBEN, M.: Leichtbausitzschalen im Serientakt. Integrales Bauweisenkonzept. In: *Kunststoffe* 100 (2010), pp. 56–59

Additional Literature

ADVANI, S. G.; HSIAO, K.T. (Editors): *Manufacturing techniques for polymer matrix composites (PMCs)*. Woodhead publishing, 2012, ISBN 9780857090676

BALASUBRAMANIAN, M.: *Composite materials and processing*. CRC Press, 2014, ISBN 9781439879351

LUKASZEWICZ, DIRK H.-J. A.; WARD, C.; POTTER, K. D.: The engineering aspects of automated prepreg layup: History, present and future. *Composites: Part B* 43 (2012) pp. 997–1009

SEFERIS, J. C.; VELISARIS, C. N.; DRAKONAKIS, V. M.: *Prepreg manufacturing*. Wiley Encyclopedia of Composites, 2011

Chapter 12
Ready-Made Technologies for Fiber-Reinforced Plastic Composites

Hartmut Rödel

Ready-made technology processes are used to cut the semi-finished products from the textile fabric manufacturing processes, give them the shape of the near net-shape dry preform, assemble, and prepare them for composite material production process. This includes pattern design of the individual preform parts, nesting, and spreading as cutting preparation, as well as cutting and textile assembly of the preform by sewing, welding, and bonding. To ensure mechanical functionality of the composite component, the semi-finished products have to be selected carefully and integrated into the preform structure in the directions of forces. This has to be managed without fold formation and with only a defined change of yarn orientation during draping. In cutting and assembly, CNC-controlled machines or robot-guided handling and joining technologies, among them one-side sewing and industrial handling are required for reasons of reproducibility. Component properties are influenced by the assembly processes. Z-reinforcements are advantageous, while the pricking and piercing reduce in-plane characteristics due to perforation.

12.1 Introduction

Inspired by ingenious natural construction, mankind has designed, constructed, and manufactured fiber-reinforced materials in numerous areas of technology.

Fiber-reinforced plastic composites (FRPC) can be manufactured in two primary manners, differentiated by the respective matrix polymer. FRPCs consist of matrix polymer and a reinforcement textile, and are thus transformed into textile-reinforced plastics or composites. Duromers, i.e. curing, once-shapeable resins, or

---

H. Rödel (✉)
Institute of Textile Machinery and High Performance Material Technology, TU Dresden, Dresden, Germany
e-mail: hartmut.roedel@tu-dresden.de

C. Cherif (ed.), *Textile Materials for Lightweight Constructions*,
DOI 10.1007/978-3-662-46341-3_12

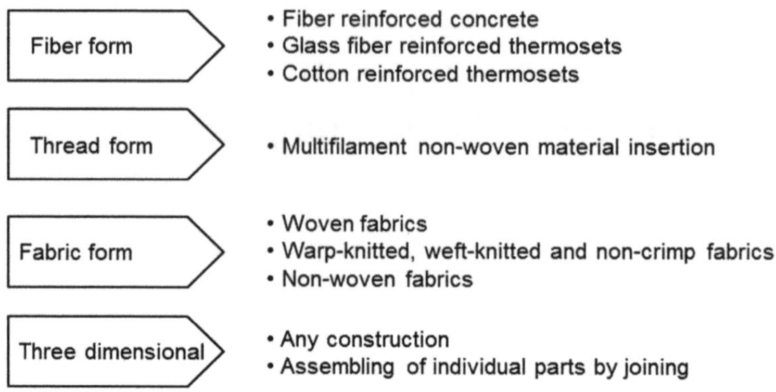

Fig. 12.1 Packaging variations of reinforcement textile, and application examples

thermoplastics in the form of granules, foils, or textiles can be used as matrix polymers. The repeated formability of the thermoplastic matrix is an advantage in these processes.

For about 20 years, analogous processes have been in development in the field of textile-reinforced concrete [1]. Textile-reinforced metals such as magnesium, ceramics, or biological tissues for medical applications are the subject of intense research. Details on processing and application aspects of FRPCs and textile-reinforced concrete have been discussed in Sects. 16.3 and 16.5. To realize the reinforcement function, textile research offers a packaging range for textile fiber materials and textile structures, which are given in Fig. 12.1. It is crucial to arrange the load-absorbing fibers or filaments in a geometry that allows them to resist the active forces in the component.

Textile and ready-made technologies allow the production of a diversity of textile materials and semi-finished products as reinforcement textiles for fiber-reinforced composites (FRC), which can differ significantly in their possible reinforcement effects and geometric complexity (Fig. 12.1).

The manifold restrictions, and especially the manual handling of pre-impregnated textile semi-finished products (Prepregs, see Chap. 11) make clear that innovations are necessary in the field of FRPCs. Preform technology offers one such sustainable innovation.

Complex *preforms* can often be produced by joining individual reinforcement textiles, which match the projected composite component net shape, using ready-made technologies. Furthermore, the stacks of joined individual parts can be protected against *delamination* in the component by reinforcements in Z direction. Sewing seams at the right points and in correct orientation improve interlaminar shear strength as a characteristic material parameter and impede crack propagation in the composite structure.

After textile technical manufacturing of the preform, it is impregnated, and consolidated in a plastics process. Resins can be infiltrated into the preform by a variety of means, and then cured into a thermoset matrix. Thermoplastic matrix materials are unsuitable to fully impregnate a dry preform, due to their limited flowability in comparison to thermoset resin systems. Here, the thermoplastic component has to be integrated in the dry preform in the form of granules, foils (also referred to as film stacking), or textiles in order to assume the matrix material function with minimized flow paths of the thermoplastics after heating to melting temperature.

For better understanding, a few definitions are necessary:

Textile ready-made technologies include the processing of textile fabrics into ready-for-use final products in the form of garments, home and room textiles, and technical textiles. The German term used for these processes, *Konfektion*, is derived from Latin "confire" (to complete, to finish) documents the conclusion of textile production by the manufacture of ready-for-use final products. These can either be used immediately or become part of a more complex technical system consisting of a variety of materials, for instance as airbags or preforms.

Textile ready-made technologies are a constructively and technologically coordinated sequence of processes for the development and construction of final products, the cutting of (usually plane) textile semi-finished products suited to the requirements, and the connection of several individual parts made from textile semi-finished fabrics, supplemented by non-textile components. Non-textile components can include inserts, safely and durably connecting the FRCs components to other parts of a more complex superordinate technical system in a manner that allows their dismantling for maintenance purposes.

The engineering specialty of ready-made technology therefore includes the product development for products made from flexible textile semi-finished products, the technological methods, as well as the construction and use of machines for the processing of these flexible textile semi-finished products in an industrial, serial, and economic manner. Product organization and quality assurance of these processes, as well as the relevant test engineering are also part of the field of ready-made technology.

The working process of ready-made technologies is usually broken up into the following steps (Fig. 12.2):

12.2 Product Development

The use of industry-specific CAD technology for the product development of flexible textile semi-finished products is described in detail in Sect. 15.3. For this reason, this chapter will only mention it in passing.

For the product development of textile preforms for FRCs, the following initial information is required:

- final form and dimensions of the composite component,
- selected reinforcement textiles, characterized by fiber material, yarn construction, and fabric construction,

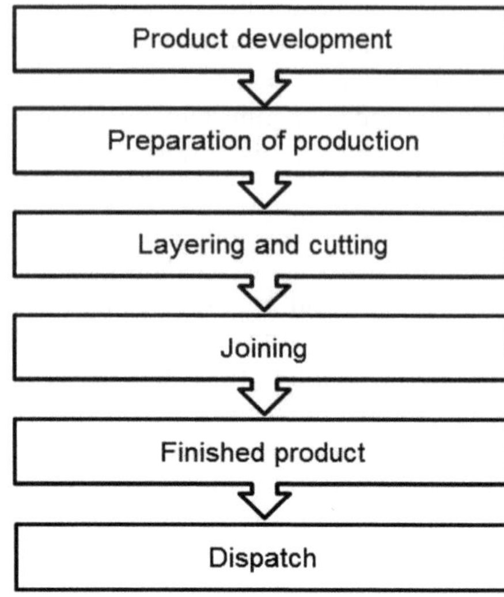

Fig. 12.2 Process steps of ready-made technology [2]

- spatial orientation of the reinforcement yarns in the preform,
- variable stacking of several fabric (constructions), and
- designing of connecting positions for the connection of composites and other elements, as well as integration of inserts and other force-introducing elements

For FRC construction, the following fiber materials are of interest and often used in ready-made technology:

- multifilaments
  - carbon fiber
  - glass fiber
  - aramid fiber
- fibers from natural cellulose like jute or flax

The palette of textile semi-finished products, which differ with regard to technology, structure, and geometry, is very versatile in the special properties:

- non-woven, also referred to in the industry as mats
- woven fabrics (2D = plane)
- (weft/warp-) knitted fabrics
- non-crimp fabrics
- braided fabrics
- braiding
- tapes (narrow fabrics)
- Tailored Fiber Placement (TFP) structures
- biaxial and monaxial structures
- biaxial and monaxial reinforced multi-layered knitted fabrics
- knitted spacer fabrics

- woven spacer Fabrics
- 3D woven fabrics

For the determination of the contours of the individual cutting parts by means of *pattern design*, the dimensions, and the number of layers of the preform are required. Furthermore, attention has to be paid to the yarn orientation in the individual layer (e.g. 0°, 45°, 90°), and to the following semi-finished product properties quantifiable by suitable measuring equipment:

- Shear stiffness,
- Flexural rigidity, and
- Tensile stress-strain behavior

Unlike usually, these specific values are not concerned with failure parameters, but with characterizing the material behavior at small loads. Ultimately, they deal with the quantification of drapeability, which describes the fold-free covering of a free geometric area with the textile semi-finished product.

To ensure an improved precision of the composite parts, it can be necessary to locally change the draping behavior of the preform during the processing. Various methods are used for this purpose. For example, the draping behavior of multiaxial non-crimp fabrics can be improved locally by a defined severing of loop yarns [3]. Alternatively, local limitations in shearing ability can be created by the targeted application of adhesives [4, 5].

The result of pattern design is a complete set of cutting patterns for all cutting parts in relation to yarn orientation. The cutting patterns are used in cutting layout. Pattern design, also called nesting in other technical fields, is an optimization task of determining the minimally required semi-finished product area, with the marginal conditions to be considered being:

- yarn course,
- width of the reinforcement textile semi-finished product web,
- quantities of the individual cutting parts,
- cutting medium (technological minimum spacing),
- determination of the optimized cutting path,
- cutting edge quality, and
- process time minimization by means of minimal connecting paths between patterns

A cutting path software module transforms the optimum cutting path into control information for a CNC cutter. Additionally, the product development has to be documented in a product data management (PDM) system, including materials, process step sequence, process parameters, quality criteria, seam lengths, used equipment, et cetera.

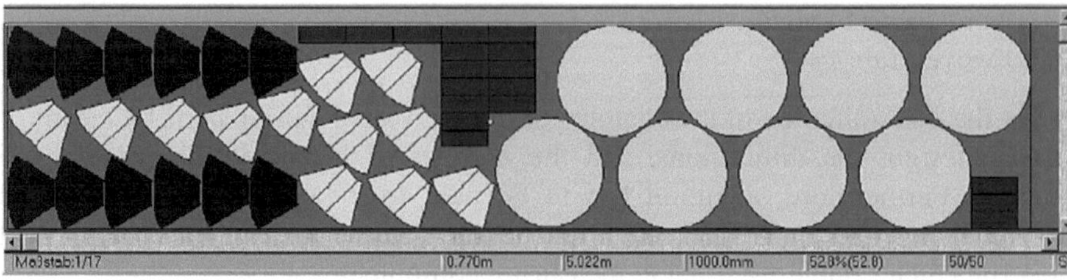

Fig. 12.3 Example of a cutting layout for cutting parts of a CF/PEEK rotor, consisting of circular base and top surfaces, and rectangular blade parts with seam allowances in yarn orientations of 0° and 45° [6]

12.3 Cutting Layout and Material Utilization

Some software-unrelated information regarding cutting layout (see Fig. 12.3) and material utilization are necessary, as the cutting layout has significant influence on the material economy and the disposal cost of material waste. Particularly in large-series production, it is economically recommendable to use semi-finished product widths adapted to the dimensions of the cutting parts.

The mathematical basics of cutting layout are given comprehensively in [7].

In-depth research [8] has shown that product-specific web widths and orientation variations of the cutting part patterns on the web contribute to optimized material utilization. Considering the high cost of carbon fiber materials and the use of their leftovers, optimization is an aspect that needs to be addressed more prominently, especially with the ever-increasing use of this particular material in mind.

12.4 Spreading

12.4.1 Purpose of Spreading

Plane textile semi-finished products are usually wound up during production and transported between the manufacturing and processing companies in this wound-up form.

Before realizing the construction cutting layout in production, it is necessary to unwind the textile semi-finished product and place in on a level surface without interior deformations and without changing the yarn orientation. This placement has to match the length of lay, including small safety additions on both ends. It is also possible to stack several layers of the semi-finished product, if doing so is justified by the demand for cutting parts and acceptable with regard to cutting edge quality. The available cutting medium has to be able to handle the height of the material stack during the cutting process.

The fabric stack is to be arranged "three edges straight", i.e. aligned on one long side and both orthogonal cutting edges, or on the halfway line of the semi-finished fabric web as well as both orthogonal cutting edges. Due to width fluctuations, a fabric stack aligned on all edges cannot be realized.

Spreading is an auxiliary process for the preparation of the cutting process and can be connected with the identification of fabric faults, at least for the top surface of the semi-finished product. Here, it is an advantage if faults occurring during fabric formation are already marked clearly.

12.4.2 Spreading Methods

Four spreading methods are known and in use for the processing of textile fabrics [2]:

- zigzag spreading
- left/right spreading
- right/right spreading
- stepwise spreading (to account for quantitative demand for specific parts)

Especially back-to-face spreading is suitable for the processing of reinforcement textiles. When using a cutting layout divided into zones, the cutting patterns can be arranged according to the different numbers of cutting parts required. This allows the stepped stacking of layers of different lengths, resulting in the creation of different number of cut parts. In this respect, step spreading is a special form of left/right spreading.

12.4.3 Spreading Process Variations

Generally, a distinction has to be made between pull-up and unrolling (spreading). Due to the strains on the fabric, unrolling should always be the preferred method. Both types of spreading can be realized manually or by means of the technology described below (Figs. 12.4 and 12.5).

Unrolling is ideally performed with a spreading machine equipped with additional devices. Manual spreading machines can reduce investment costs, but do not include such additional features as length measurement, edge regulation, fold straightening modules, or orthogonal cutting devices.

The roll of semi-finished fabric can be stored on bars or in a hollow for narrow fabrics within the spreading device. For reasons of the circumferential drive of the semi-finished product roll and the gentle spreading, the hollow for narrow fabrics is to be preferred. If the semi-finished product is stored on bars (preferably oriented with conical centering elements), the required length of semi-finished product

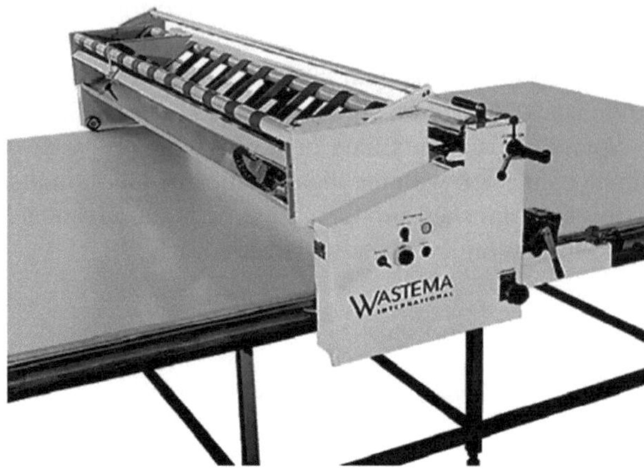

Fig. 12.4 Manual spreading machine with hollow storage of the fabric coil [9] (*Source*: Wastema International Steinhauser Spezialmaschinen GmbH)

Fig. 12.5 Spreading machine with additional modules: travel platform for operating staff, edge regulation, fold removal device, orthogonal cutting edge [9] (*Source*: Wastema International Steinhauser Spezialmaschinen GmbH)

should be provided by a center drive adjusted to the current diameter of the semi-finished product roll.

The fabric roll storage should be placed in close proximity to the spreading unit. As textile semi-finished products for composite cannot be discerned by pattern or color (as would be possible with garment materials), a defined storage of the individual semi-finished product rolls with precise labeling is of utmost importance to prevent mistakes in processing. Comparing RFID tags or label bar codes to the data of product data management/production planning software and ensuring clearance is recommended for quality assurance purpose.

At the same time, storage technology has to guarantee the gentle handle and storage of the semi-finished fabric rolls in order to prevent involuntary damages of the outer layers of the semi-finished product by improper contact or abrasive friction. The mass of such rolls also has to be taken into account. Therefore, mass-adapted and gentle storage, handling and spreading have to be part of project planning for spreading and cutting. Fabric roll paternosters combine the orderly, gentle storage with suitable handling technology and, if applicable, withdrawal of incompletely processed semi-finished product rolls into the storage system.

The width of the spreading table has to be designed in a manner that ensures a proper handling of the semi-finished fabric rolls and spread of semi-finished products. To transfer the spread layer or layer stack to the cutting machine, it is sensible to provide an auxiliary layer of packing paper for transportation and to decrease friction between table surface and bottom layer of semi-finished fabric. If the cutting machine fixates the layer or stack by means of a vacuum to absorb the cutting forces, the packing paper has to be suitably perforated. Alternatively, the table surface can be equipped with air nozzles to form an air cushion between table surface and layer stack, which allows for a slightly levitating mobility of the semi-finished product layer or layer stack.

Tables with a belt surface, also known as conveyor constructions, can also be used to spread and transport the semi-finished product layer. Some manufacturers offer conveyor constructions with several belts located atop each other, making a variation of process sequences between spreading and cutting processes possible, giving the production process increased flexibility.

12.4.4 Treatment of Faults in the Reinforcement Structure

Faults in textile semi-finished products can never be eliminated entirely. Their inevitable existence requires a reliable marking by the manufacturers of the semi-finished products as well as method to avoid these fault locations during the spreading and cutting processes. Comprehensive investigations of the subject were performed [8] in the 1980s. In this context, the term fault rate was defined as number of faults/unit of length (1/m). However, this exclusive states the number of faults, but neither their geometrical shape nor their extent.

Faults in the textile semi-finished fabric are assessed according to their cause, geometry, and significance in the textile product (in this case, the reinforcement textile preform). The marking of the faults should be made with suitable means, in the simplest case with a contrast-colored, sufficiently thick yarn at the edge of the semi-finished product. Metallic or otherwise machine-detectable markings can be detected during spreading by a sensor integrated into the spreading machine.

Single-color reinforcement semi-finished products are not ideal for fault identification during spreading. (Un-)Rolling the reinforcement semi-finished products for a separate fabric inspection process is not sensible due to the additional strain and damage risk. Therefore, quality control on the textile machine is imperative.

So far, no automated technology (for instance based on image processing) is available for fabric inspection. The human eye and mind are used to assess the respective semi-finished product under suitable lighting and in continual feeding. The width and length of the semi-finished product roll are measured; the coordinates of detected faults can be entered into a computer protocol. Incident and transmitted light-based image processing methods for a widespread use in fabric inspection, which were a distinct possibility few years ago, have not been realized in practice yet.

Regarding fault geometry, the following classification can be made:

- point-shaped faults
- line-shaped faults (warp faults, weft faults, runs)
- area-shaped faults (contaminations, abrasion spots caused by transport)

Five methods have been established [8] for fault treatment:

- multilayer method—cutting of additional parts in the manufacture of large series ("suspicion stock"),
- operative fault treatment—displacement of the individual layer during spreading, in order to put the faulty areas into the cutting waste sections; can be done only once per layer,
- patch method—covering the faulty area with a fabric piece of sufficient size, to cut an additional, fault-free part,
- orthogonal cutting of the faulty section and re-arrangement that ensures simultaneous cutting of faulty and fault-free parts
- compensation part cutting—faulty cutting parts serves as a template for a manually cut replacement part.

Often, cutting without remnants is named as a fault treatment variation. The comprehensive optimization calculation necessary for cutting without remnants requires a complete data base with roll lengths, fault positions, and other data. The often varying conditions of the manufacturing process do not allow the realization of such a broad data base. Furthermore, it is not clear whether the data gathering efforts and possible economic benefits are in acceptable proportions.

An improved computerization of the production processes, particularly in single-layer cutting, allows the identification of faults and fault coordinates in the early phase of inserting the semi-finished product into the cutting area of the CNC cutting machine. This enables an operative cutting layout change to exclude the faulty section from further processing while still providing all required cutting parts.

12.5 Cutting Technology

12.5.1 General Remarks

To cut the plane semi-finished fabric into the calculated cutting part layouts, hand-operated cutting technology and CNC cutting machines are available.

Hand-operated cutting can be performed with:

- shears,
- electro-powered shears for single-layer cutting,
- circular blade machines

- vertical knife machines, also referred to as straight knife machines
- vertical band knife machines

Hand-operated cutting technology can also be fitted with devices to reduce the effort. Servo-Cutters, for example, integrate the vertical knife machine in a swivel-arm bearing suspended above the cutting table.

For the CNC cutting machines established in the industry, various cutting media with individual advantages and disadvantages are available:

- stitch knives and (driven) circular blades—mechanical cut by means of wedge action
- ultrasound sonotrodes—thermal cut by mechanical oscillations at ultrasound frequencies (20–30 kHz)
- Laser beam—thermal cut
- Plasma jet—thermal cut
- Compressed water jet—mechanical cut by abrasion

Long-term form-stable cutting part layouts, e.g. for the necessary reinforcement components in every product, die-cutting technology can be used, which is categorized into step and continuous die-cutting.

12.5.2 Cutting of Reinforcement Textiles

The composite-relevant fiber materials, carbon and glass fibers, require shears with a material-adjusted blade geometry. The long-term suitability of the shears relies on their exclusive use for the respective fiber material. Otherwise, the special cutting ability is quickly lost. To cut the reinforcement textile semi-finished products, all hand-controlled knife/blade machines can be used. For handling reasons, a circular blade machine with small blade radius and minimal machine mass is to be preferred in pattern production or cutting of sub-parts in single-layer cutting. CNC cutting machines have been in use for computer-assisted cutting for decades. Detached stitch knives (without bar guides, as in vertical blades) are well-suited to the special fiber materials of the reinforcement textiles. Additional tools in the cutting head, e.g. for creating bores and holes of defined sizes, are currently becoming more prevalent.

By means of ultrasound, laser beams and plasma jets, thermal cuts can be performed. This is especially interesting if the reinforcement textiles contain some percentage of thermoplastic fiber materials, which melt under heat and create a solid and secure cutting edge. The blade geometry angle of the wedge-shaped sonotrode in ultrasound cutting influences the effective width of this tool from the cutting edge towards the fabric plane of the cutting parts. Small angles concentrate the energy into cutting, while angles above 90° amplify the effects of the ultrasound even a few millimeters away from the cutting edge. Especially the application of laser and plasma beams is accompanied by exhaust gasses which have to be

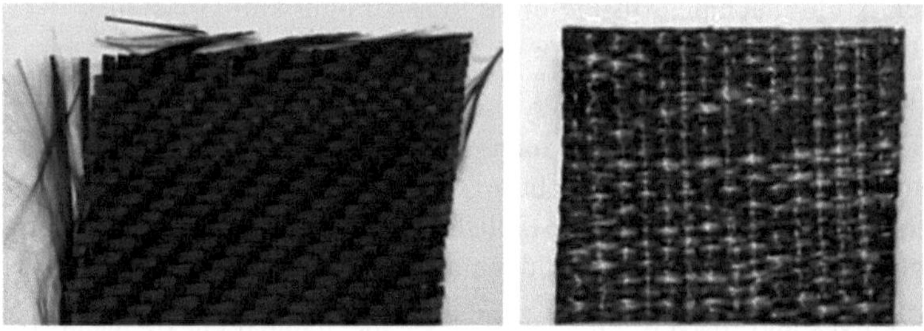

Fig. 12.6 Cutting edge of a CF/PEEK woven fabric after cutting by knife and laser beam [6], (**a**) after cutting by knife, (**b**) after cutting by laser

extracted immediately at the working elements and from manufacturing facilities in general.

A compressed air jet for cutting is a method well-known from manufacturing engineering, where it is used as a cutting medium for CNC cutting machines. The cutting of relevant textile reinforcement semi-finished products is possible, but does not create a thermally secured cutting edge. There is no risk of a critical moisture penetration of the textile reinforcement structure, as the rate of flow of the water is very low. The compressed water jet is well-suited to the processing of consolidated FRC components.

Figures 12.6 and 12.7 show the cutting edge quality achievable with the various cutting media, exemplified by reinforcement textile with a thermoplastic fiber material component.

The rotating demonstrator shown in Fig. 12.8 is mounted manually: the problem of unsecured cutting edges is clearly visible.

The cutting table has to be adjusted to the respective cutting medium:

- The stitch knife (Fig. 12.9) requires a table surface with a field of bristles, as the stitch knife dips into this field, ensuring the cutting of the bottom layer. The positional fixation is ensured by a vacuum, making a covering of the cutting area with foils a sensible option for reasons of energy efficiency. Due to the stitch knife, each cutting part consists of a closed cutting contour. Connecting cuts are not required, but can be included in the programming to shape the otherwise reticular cutting waste.
- Ultrasound is the term used for waves of frequencies above 18 kHz, which are inaudible to the human ear. The knife oscillates at ultrasound frequency and induces mechanical vibrations in the material. A stable, plane surface should be used for support (Fig. 12.10). The exclusion of glueing together, especially in cutting resin-preimpregnated textile fabrics, is the most valuable feature of ultrasound cutting. Conventional cutting wedges have a tendency to glue together because of the resin. The ultrasound cutting machines are being used by the composite manufacturers because of the following advantages [10]:

Fig. 12.7 Cutting edge of a CF/PEEK woven fabric after water jet cutting; cutting edges are not secured, risk of threads dropping out, especially at the edges in yarn system direction [6]

Fig. 12.8 Rotating demonstrator I preform without secured cutting edges [6]

Fig. 12.9 CNC cutter with stitch knife [9] (*Source*: Wastema International Steinhauser Spezialmaschinen GmbH)

- high cutting performance without disadvantageous cutting edge influence,
- reduced cutting edge friction,
- low cutting forces,
- high cutting edge quality,
- small cutting part spacing in the cutting pattern (nesting), and
- no glueing together by thermoplastic or other substances

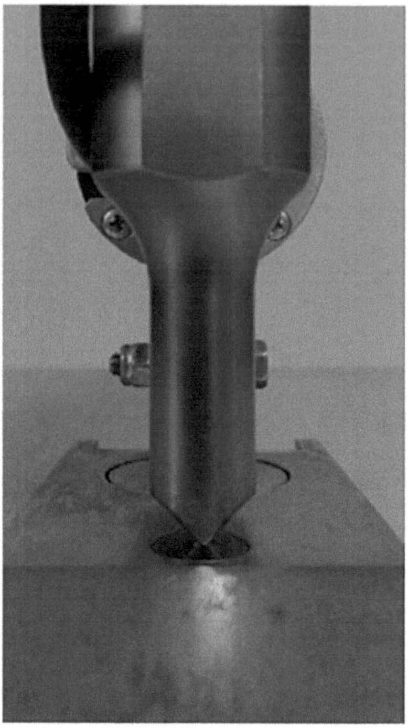

Fig. 12.10 Hand-controlled ultrasound knife by Rinco, Switzerland

- Laser beams have to be absorbed by the material to be cut, and thus transformed into thermal energy. As appropriate, the quality of the cut might require the removal of cutting remnants from the cutting gap by means of an air jet.
- Plasma beams are accompanied by a gas beam, which requires the table surface to allow unobstructed permeation.
- Water jet cutting requires a table design allowing unobstructed drainage of the water, while withstanding the cutting effect of the water jet over extended periods of time. High-grade steel honeycomb structures have been proved to meet these requirements.

One unsolved problem is the mechanized or even automated removal of the cutting parts of stacks (in multilayer cutting) from the cutting table, and the connected separation of cutting waste, foil remnants or other auxiliary materials.

One research approach [11] is based on the idea of removing the coherent cutting waste upward from a table with a conveyor surface, while special elements prevent the simultaneous removal of the cutting parts.

12.6 Textile Assembly by Sewing

12.6.1 Terminology

Sewing is the typical joining method for flexible textile cutting parts and the achieved seam properties are almost optimal for the useable final product manufactured from the cutting parts. Sewing has been done manually for millennia. In 1790, the first sewing machine patent was granted. Since then, sewing technology has gone through enormous developments. The requirements of assembling textile preforms have triggered a rapid surge of innovations in sewing technology during the 1990s.

In contrast to cutting technology, the technical evolution of sewing technology has led to a comprehensive standardization with the aim of organizing and precisely defining the interactions of sewing machine, sewing thread, and sewn material required in sewing.

The relevant standard works contain:

Terms of sewing technology	DIN 5300 Part 1 Sewing
Terms of sewing technology	DIN 5300 Part 2 Machine
Terms of sewing technology	DIN 5300 Part 3 Sewing material, sewing thread
Stitch types	DIN 61400/ISO 4915
Seam types	DIN ISO 4916
Sewing machines	DIN 5307
Needles	DIN 5330

According to the norm, *sewing* is a process in which one or more sewing threads are guided through the sewing material in accordance to certain regularities (DIN 61400). Sewing is used to create a connection, and the *sewn connection* between two or more sewing material parts or layers is generally created by one or several seams of equal distances to each other. This sewn connection is usually executed linearly. Locally limited seams allow a virtually pointwise connection. Alternatively, several parallel or fan-shaped seams can be used for planar assembly applications.

To realize the sewn connection, *stitches* have to be formed. This is generally achieved by piercing the sewing material with a needle and simultaneously passing a sewing thread through the material and creating a loop with the thread, either through the material, with the thread itself, or with other sewing threads. Depending on the stitch type, several needles and more than one sewing thread can be included in stitch formation.

Common knowledge of sewing and household sewing machines creates the image of the sewing needle penetrating the sewing material at a right angle, creating entry and escape points of the needle on either side of the sewing material, where the needle and the thread are inserted and removed from the material. However, preform manufacture also relies on technical variations, in which the entry and escape points are located on the same side of the material or the stacked reinforcement textile layers.

In a sequence of stitch formation processes, the *sewing seam* is created as a succession of sewing stitches or stitch types in one or more material layers. According to ISO 4916, the seams are classified into eight categories, which mainly focused on garment-related sewing tasks.

According to DIN 61400/ISO 4915, the stitch types are grouped into six stitch type classes and designated by a three-digit number system. The possible classifications distinguish the individual sub-classes. For example, the number 301 defines the double lock stitch used in household machines, while 304 designates the zig-zag double lock stick. The synonymous "Single" indicates stitch creation from a single thread, i.e. the needle thread. The term "Double" refers to the processing of needle thread and an additional lower thread:

- Class 100 Single chain stitches
- Class 200 Single lock stitches
- Class 300 Double lock stitches
- Class 400 Double chain stitches
- Class 500 Overedge stitches
- Class 600 Covering stitches

Before explaining in detail the individual stitch type classes and their properties, it has to be stated that only two principal variations of thread connection exist for sewing.

In all types of chain stitches, *chain connection* is used (Fig. 12.11) and modified according to the specific class of stitch type. Chain stitches are characterized by the supply of the threads from theoretically infinite-size bobbins to avoid technology-inherent interruptions of the sewing process. This makes chain stitches particularly interesting for automated sewing.

The double lock stitch on the other hand is based on *loop connections* between needle thread and lower thread (Fig. 12.12), with the thread storage of the lower thread being stored on a lower thread bobbin and limited available thread length, making system-inherent bobbin replacement inevitable. In some garment industry sewing machines, this flaw is compensated by an automatically changing magazine for lower thread bobbins. Despite this device, each change of bobbins results in an interrupted seam formation and a fault point in the seam.

Stitch type class 100 single chain stitches (Fig. 12.11) are relevant for preform assembly. The 101 single step stitch is also suitable for the explanation of stitch formation.

The demand of "sewing in the composite tool" during preform assembly makes blind stitches important, which are performed analogously to the 103 single chain stitch with a curved needle, in a manner that places entry and escape points of the needle on the same side of the sewn material. By definition, the piercing direction of a 103 single chain stitch is at a right angle to the seam formation direction. The two-needle blind chain stitch used in preform assembly is also performed in this geometrical constellation, while the blind stitch with a curved needle on the robot

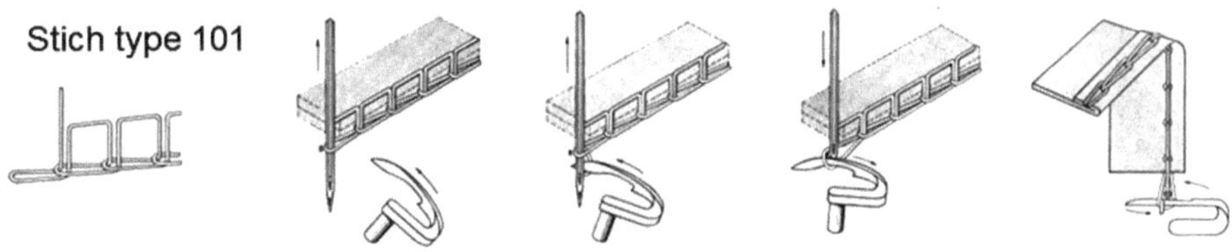

Fig. 12.11 Single chain stitch formation principle [12] (*Source*: Pfaff Industriesysteme und Maschinen AG)

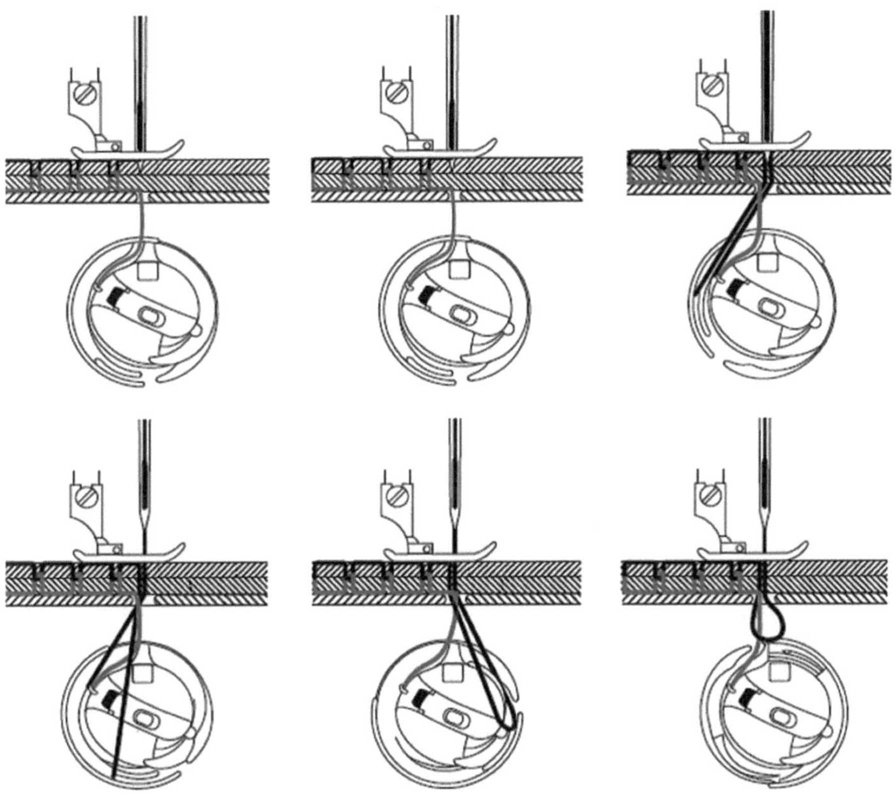

Fig. 12.12 Thread arrangement in the double lock stitch—*top*: upper thread, *bottom*: lower thread of the bobbin stored in the gripper system—and stitch formation [12] (*Source*: Pfaff Industriesysteme und Maschinen AG)

sewing head of the KSL Group moves the needle in seam formation direction for preform assembly.

For the sake of completeness, it has to be pointed out that the blind stitch can also be performed as double lock stitch. In this case, the special hook gripper of the blind single chain stitch has to be replaced by a gripper system with thread storage on a bobbin for double step stitches.

The 200 single lock stitch has not found extensive industrial use, since it is almost exclusively performed manually as an exception in garment manufacturing. The mechanical performance of the single lock stitch is hindered by the need to move the entire thread storage through the sewing material for each stitch, which is alternately performed from both sides of the sewing material.

The 301 double lock stitch (Fig. 12.12) uses inline-sewing to create strong seams, which are characterized by a stretched orientation of both threads in the stitch layout, minimizing their stretchability. The double lock stitch is very commonly used in preform production.

In some double lock stitch sewing machines, the needle only makes an upward/downward motion derived from the crank assembly of the mainshaft. To complete a 304 zig-zag double lock stitch, the needle bar has to complete an additional offset motion at a right angle to the seam formation direction, creating the eponymous zig-zag motion. A technical alternative is a lateral offsetting of the sewing material, as it is used on x/y cross table sewing machines.

To create the 401 double chain stitch (Fig. 12.13), an altered gripper system is required, as the chain connection of the double chain stitch is created with a thread-guiding hook gripper. The double chain stitch has a larger thread length reserve due to the threefold thread length at the bottom side of the stitch layout. Therefore, the double chain stitch seams are characterized by greater longitudinal seam elasticity than double lock stitches. The double chain stitch can also be performed as zig-zag stitch.

The overedge stitch is usually used to sew the cutting edge in order to prevent the threads of the textile structure from dropping out. Simultaneously, the insertion of several cutting parts of equal edge lengths allows the creation of a connecting seam. In garment manufacture, so-called safety sewing machines are being used, which parallelly produce one double chain stitch seam and one overedge chain stitch each, relying on two synchronously working needles and suitable hook grippers. This results in improved seam safety/durability, as expressed by the name of the corresponding machines.

Overedge stitches can be performed with one, two, or three threads. Stitch type 501 is standardized form of the single-thread overedge stitch, numbers 502 and 503 account for the two-thread overedge stitches, while number 504 and 505 designate the three-thread variations. The two-thread and three-thread variations differ only in the spatial position of the connection loops. The connection loop positions are set by adjusting the thread tensioners. While needle and thread-guiding hook

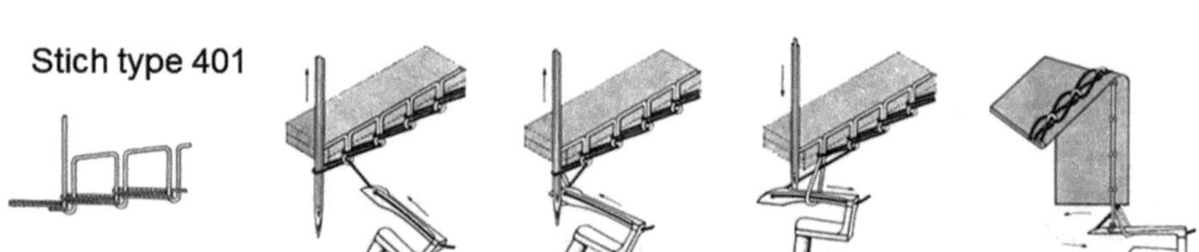

Fig. 12.13 Double chain stitch [12] (*Source*: Pfaff Industriesysteme und Maschinen AG)

gripper are sufficient for the two-thread overedge stitch, a three-thread overedge stitch requires a needle as well as thread-guiding top and bottom hook grippers to work in co-ordination. To apply the overedge stitch in preform manufacture, research [13] has shown that stacked reinforcement textiles with individual seams create matrix-rich zones around these seams, which impair the properties of the component.

Class 600 covering stitches consume the most thread per seam length, making them the most stretchable stitch type. Covering stitches are performed with several needles creating parallel arrangements of the needle threads. Hook threads from the hook gripper, and a cover thread at the top surface of the sewing material create a spatial thread arrangement able to connect even laterally joined cutting parts without overlap. So far, no application in perform production is known.

The position and shaping of the cutting parts to be joined as well as the position of the individual seams can be ensured with drawings and designation of the *sewing seam types* as described for garment-relevant examples in DIN ISO 4916. The representations show whether the sewing can be done simultaneously on a multi-needle machine or only in sequential sewing processes.

In preform production, information about the thread orientation of the individual reinforcement textile layers is required in order to complete the sewing assembly of the preform adjusted to the loads.

The exterior design of the sewing machine casing helps fulfill the requirements of optimum handling. For example, free-arm constructions can complete seams at the circumference of tubular products. The diameter of the free arm has to be smaller than the diameter of the tube for this. Post-bed and feed-off-the-arm sewing machines enable the assembly of fabric into a tube with longitudinal seams.

As a rule, sewing machines are designed in the characteristic C-form, which imposes handling limitations due to the limited clearance. Long-arm sewing machines used in the manufacture of technical textiles such as tents can reduce, but not remove, these limitations. In Japan, the specialized, patent-protected, z-construction of the casing [14] is used for the assembly of larger textiles without limitations, given by a free space for optimum handling. This z-sewing machine is offered as a double lock stick and double chain stitch variation with up to six simultaneously working needles. An application in preform assembly is not yet known.

Another alternative for the removal of handling deficits is offered by the dissolution of the conventional sewing machine casing by using multi-motor drives instead of the central motor drive with extensive mechanics.

Sewing machines used in the general ready-made industry are without exception driven electrically and the advancements in control engineering have boosted the use of electronically controlled drives. Furthermore, the development trends are focusing on CNC-motor drives for the individual working elements instead of mechanically coupling solutions regarding their number of revolutions and rotating angle. However, their introduction into service is slowed by their cost and the lack of experience of technical staff regarding the drive and control technologies. The required precision and reproducibility of the assembly works during preform manufacture will soon make such drive and control technologies with corresponding feed systems indispensable in sewing technology.

For the product-specific seam formation, a relative movement between the sewing needle and the sewing material is required.

To ensure this movement, the following principles have proven suitable over the past decades:

- Intermittent feeding of the sewing material to the fixed-positioned sewing machine is the standard manner of feeding in industrial and household applications. Usually, the feeding device acts upon the sewing material from below. Additional feeding effects can be achieved by upper and needle transport and their combinations together with manual sewing material guiding. Furthermore, additional feeding rollers or wheels are an option working behind the sewing zone, which work synchronously to the sewing process. Defined equalizing of length differences is technically reproducible and incrementally controllable by means of so-called differential feeding.
- If only small offset distances between the sequential piercing points are to be realized during sewing, the movement of the sewing material in the fixed-location sewing machine by a continuously working feeding system is a technically acceptable method proven for button holes in garments. The related deviation of the needle from the vertical direction is negligible.
- Usually, the sewing needle is fixed to the needle bar, which is moved exclusively in vertical direction by a straight crank assembly. A co-ordinate-axis-related segmentation of the relative movement between sewing material and machine can be performed in a manner that realizes the zig-zag stitch from an offset movement of the needle bar at a right angle to the seam direction and to the sewing material feeding in seam direction. At larger dimensions, this principle is used in sleeping blankets sewing machines by moving the sewing machine in x-direction, i.e. the axial direction of the main shaft, while the sewing material is intermittently offset in longitudinal product direction. The drives are subject to great dynamic requirements. The moved masses have to be at rest when the needle pierces the sewing material and while it is in contact with the sewing material. The offset movement of a few millimeters is only possible if the sewing needle is outside of the sewing material. About a third of the time of the main shaft revolution is available for this.
- The intermittent movement of the sewing material in both co-ordinate directions is another possibility to produce the seam contour. Machines for short and shaped seams are an example from garment production, overlaying the movement in both co-ordinate directions with two mechanical controlled cam disks. Other dimensions are produced with CNC x, y-cross table automated sewing machines, which are used successfully in airbag production.
- With sewing machines moved along the spread-out fabric webs, awnings and other large-format products can be produced with parallel or central seams. The co-ordination between the offset movement of the mobile sewing machine table and the cyclical working method of the sewing machine is crucial for this

process. The needle transport of the machine allows a uniform relative movement, which is very advantageous for this method.
- Alternatively, moving fabric webs are fed to fixed sewing machines positioned on one or both sides, and the edges are sewn, for example in bordering sleeping blankets with a narrow fabric. For a continuous edge processing, fluctuations in width can be compensated by equipping the sewing machines with edge guide systems. For the simultaneous processing of both sides, two sewing machines (in left-side and right-side positions) of identical construction should be used.
- Sewing robots, i.e. articulated-arm robots with sewing heads, were introduced in the mid-1980s. With these robots, it was possible to produce seams in three dimensions, not just in the plane, as offered by the previously mentioned methods. The prototype brought forth by a Japanese research project was conceptualized to insert sleeves into sports jackets, whose industrial applications are not yet known. Further development steps by German researchers were dedicated to industrial series production for car seat covers and headrests as well as simple-cut women's skirts. Introduction into industrial practice was equally unsuccessful due to reliability, flexibility and cost issues. This technology gained importance for the assembly of textile preforms shortly before the turn of the millennium, even though it still lacks the broad application and high-quantity capacity. The basic idea of using robots for sewing is to realize seams along cutting parts held in near-product or near-preform geometry by molded bodies. The sewing technology is used for forming the stitches, while the robot assumes the feeding function. While the sewing needle is in contact with the sewing material, the sewing mechanism stays immobile, while the robot head is kept in continuous motion to avoid unnecessary vibrations. The sewing mechanism, along with the exterior casing, has to move after and catch up with the robot head after completing the contact. For this purpose, the sewing mechanism is mounted with glide bearings.

The controls of sewing machines are traditionally mechanical, in the form of cam disks or eccentric elements. Pneumatic controls can be used both in the form of pneumatic cylinders for greater displacements, and in the form of pneumatic logic controls. For this, it is crucial that air blows the textile fiber fragments and fiber dust away, instead of sucking them in. For reasons of contamination, hydraulics is not used in sewing and textile processing. A few years ago, G. M. PFAFF AG developed an air-bearing needle bar to prevent the formation of an oil droplet at the needle points after extended machine downtimes, which would have resulted in product contamination.

Electrical controls are nowadays the standard equipment on automated pieces of sewing technology. This CNC technology allows geometry- and dimension-flexible work and can be connected online and offline to the CAD systems used in product development and construction, in order to utilize seam lengths and contours for controlling the machine. For a reproducible machine setting, multiphase and linear motors are becoming increasingly common. For example, the SRP system by G. M. PFAFF AG sets the preload spring of the pressure foot in dependence of the main shaft number of revolutions to ensure optimal feeding effects. It has to be kept in mind that the electrical conductivity of the carbon filaments and their unavoidable fragments require the highest possible degree of electrical protection to avoid short circuits and damage to circuit boards. The controls are either filled with air from a compressor, preventing any inflow of air and thus keeping the fiber filaments outside, or the hermetically closed switchboards have a very large surface area to dissipate the thermal energy resulting from energy loss. In the latter case, the switchboard is to be cleaned thoroughly before each necessary opening of the doors. In this context, the mandatory use of vacuum cleaners protected against explosive reactions has to be pointed out.

12.6.2 Stitch Formation

12.6.2.1 Sewing Machine Needles

The geometry of sewing machine needles is a decisive factor, which determines the degree of unavoidable perforation of the sewing material, especially in the case of the carbon- or glass-filament reinforcement textiles, caused by the piercing of the needle. The so-called in-plane properties of the reinforcement textile are reduced, as filaments are either damaged in the point of piercing or bent out of the stretched orientation by the needle and the inserted sewing thread.

The prerequisite of mechanical sewing was, of course, the invention of the sewing needle with the needle eye located close to the point of the needle. Sewing needles are defined in the standard DIN 5330, manufacturers offer broad ranges of needles and often support customers in needle selection. Sewing needles are produced from a wire semi-finished product using cutting and forming production methods (Fig. 12.14a, b). The needle eye is usually deburred chemically. High-quality needles are fitted with a mechanically polished ear to ensure optimum contacting points between needles and thread (Fig. 12.14c–g). Currently, the traditionally rotationally symmetric needles are given shapes deviating from the axial direction, and cross-sections beyond the usual circular form.

The sewing machine needle has a cylindrical shape. The needle piston is used to fixate the needle in the needle bar of the needle drive, the shaft connects piston and needle point (Fig. 12.15). The position of needle eyes is close to the point. The point ensures the piercing of the sewing material. The needle features two needle grooves in the needle shaft, which guide the needle thread to and from the eye. In a long needle groove (Fig. 12.15b), the needle thread is guided to the eye from the thread bobbin via the thread tensioner and thread take-up. From the eye of the needle, the needle thread is led through the short needle groove (Fig. 12.15a, c), through the top surface of the sewing material, and then positioned in the seam as specified by the stitch type. To ensure completion of the sewing process, the cylindrical piston has to be inserted in such a manner that the short needle groove faces the gripper device [15]. Thus, only one of the 360 possible rotational angles around the needle axis is the correct one. Needle and sewing machine manufacturers expect professional

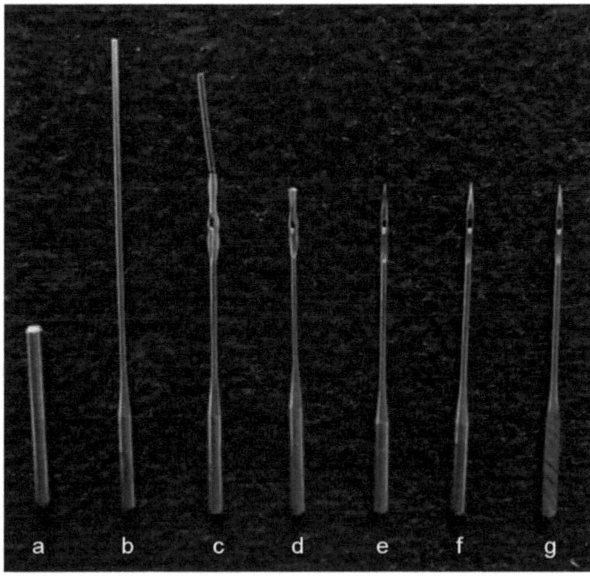

Fig. 12.14 Sewing needle manufacture sequence by (**a**, **b**) forming technology, (**c**) die-cutting, (**d**, **e**) cutting, and (**f**, **g**) electroplating

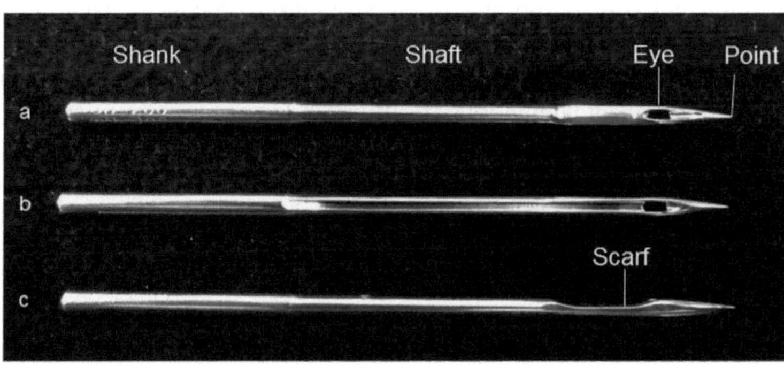

Fig. 12.15 View of a sewing machine needle from three sides: (**a**) back side with scarf left from need eye, (**b**) needle with long groove to the guide the need thread to eye, (**c**) side view with scarf left from needle eye

users to know about this requirement, while household sewing needles are whetted in a defined spot to ensure the correct position by form-fit fixation (see Fig. 12.14g).

The correct use of sewing machine needles relies on several specifications:

- Needle system—The needle system takes into account the geometrical conditions of the respective sewing machine construction, i.e. the distance to be bridged between the needle bar and the gripper system of the sewing machine. The needle system to be used is determined by the sewing machine manufacturer.

- Needle fineness—The needle fineness is stated in Nm (Number metric). The added number corresponds to the 100-fold diameter of the needle shaft, e.g. Nm 90 designates a needle shaft diameter of 0.9 mm.
- Needle point design—The term needle point evokes the mathematically exact end of a cone. Practically in sewing, however, a distinction is made between displacement points and cutting points. Displacement points, as their name suggests, are supposed to displace the yarns of the reinforcement textiles instead of damaging or severing them. The displacement creates sufficient room for the sewing machine needle and the inserted sewing threads. For this displacement function, conical needle points are rounded at corresponding transition radii, which are classified as curved point (CP), blunt curved point (BCP) or acute curved point (ACP) according to fuzzy logic. Cutting points feature a sharpened edge preparing defined openings in the sewing material, in which the sewing thread is placed in a defined manner. Cutting points are used in leather processing and the optical effects of the seams in leather products are well-known to every user. For the processing of reinforcement textiles made from carbon or glass fibers, displacement tips should be preferred due to the minimized in-plane strength losses.
- Needle surface finishing—The surface finishing of sewing machine needles can extend their life time, especially of the needle tip, and influence the frictional behavior between sewing needle and sewing material. Steel needles are usually coated with nickel or chromium in electroplating processes. Titanium nitride coatings, as used for the past 20 years and commercialized as Gebedur® by Groz-Beckert, offer longer life time for the needles as well as less material damage, smaller needle deviations (resulting in fewer faulty or missed stitches), fewer thread breaks and thread rips, and higher productivity due to reduced downtimes [16]. Another way to minimize frictional behavior is offered by manufacturer Schmetz with their so-called Blukold coating [17] preventing the adhesion of thermoplastic melt residue from the sewing material on the sewing needle. A Blukold needle is a sewing needle with a roughened, phosphated surface and a Teflon coating. With the NIT coating offered by Schmetz, it is possible to achieve a particularly smooth, easily slipping, and corrosion-resistant needle surface with homogeneous coating thickness. This special surface result in very good wear resistance, excellent slipping properties in the needle eye, and easy piercing even of hard materials. The NIT coating is also very resistant to aggressive chemical finishing of the sewing material.

Needle penetration force can be measured as a quantitative variable of the sewing process and may help support the selection of the sewing needle. For this, the needle plate of a sewing machine is equipped with strain gauges to record the effect of the needle penetration force causing a bending of the needle plate in relation to the main shaft rotation angle and at a common number of revolutions of the main shaft [18]. Bending and needle penetrating force are assigned by a calibration with defined forces which are to be introduced in the area of the needle hole. If necessary, the cross-section of the needle hole has to ground down to achieve sufficient and analyzable reactions caused by the needle penetration force in the system. The area of the determined function of the needle penetration force in relation to time corresponds to the needle penetration work. The maximum needle penetration force is achieved in the phases of initial penetration of the sewing material during penetration with the needle point and the distending of the piercing channel. The technical cycle of the sewing machine is crucially related to this, as only few degrees of the main shaft revolution are available for piercing and penetration, depending on sewing material thickness. Thus, the needle penetration force increases with the number of revolutions of the main shaft. The structure of the sewing material, which determines the mobility of the filaments and fibers of the fabric, and the number of layers to be assembled also influence the needle penetration force.

12.6.2.2 Looping Stroke in Sewing Machines

The sewing machine moves the needle by means of a straight crank shaft driven directly by the main shaft. About half-way from the top reversal point (TP) to the bottom reversal point (BP), the needle pierces the sewing material. During the piercing of the sewing material by the sewing needle, the sewing thread touches both sides of the sewing needle, forming a U shape. The piercing is performed at maximum speed, resulting in the needle penetration force and the related needle warming caused by the friction between sewing needle and sewing material. After passing the bottom reversal point, a needle thread loop is created on the needle side with the short needle groove, as the needle thread is fixed on that side by the sewing material. After crossing a certain length of the needle path to the BP, the so-called looping stroke, the stitch-type-specific gripper device grips the needle thread loop for a chain-connection or loop-connection of the needle thread with itself or other threads (Fig. 12.16). In garment sewing machines, the looping stroke is roughly 2 mm [2]. The looping stroke has to be adjusted to the sewing thread to be used. On the currently available sewing technological solutions, looping stroke can only be changed manually. Technically elegant solutions with a multimotor drive for needle and gripper device are not economically viable [15].

12.6.2.3 Thread Take-up

During the piercing motion of the thread-carrying needle, needle thread is required. This length of thread is provided by the motion of the thread take-up. In one rotation of the main shaft, thread is supplied during 2/3 of the rotation and then this thread element is taken back quickly during the remaining 1/3 of the rotation. The thread take-up movement (Fig. 12.17) is performed in co-ordination with the needle movement. The thread length required for the stitch formation itself, which is short in comparison to the length of the actively moved thread element, slips from the thread tensioner if the thread tension exceeds the braking force of the thread tensioner during the stitch formation.

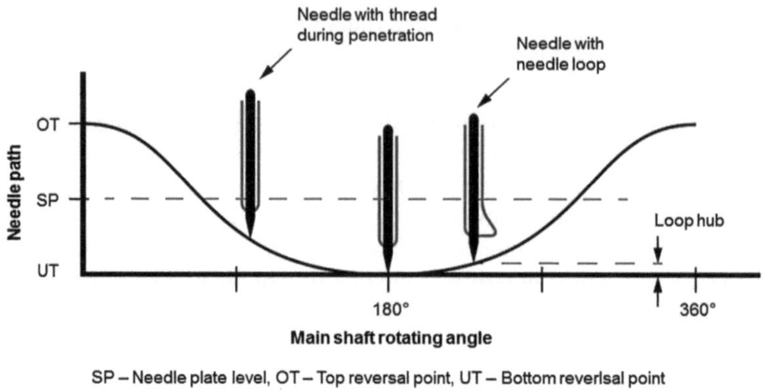

Fig. 12.16 Formation of a needle thread loop during the escape of the sewing machine needle [15]

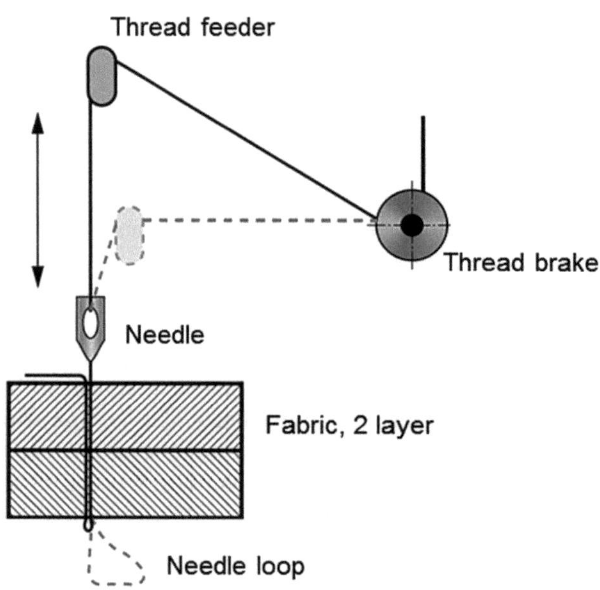

Fig. 12.17 Thread take-up principle [15]

Thread passage number is the quotient of length of the active thread element and the length of needle thread consumed in one stitch. The thread passage number is a measure for the strain on the needle thread by accelerations during changes of direction and friction on the thread guiding elements of the sewing machine during the sewing process.

12.6.2.4 Thread Tensioner

In sewing technology, disk-type thread tensioners with manual pretension of the spiral spring are common. To remove the processed sewing material, the thread tensioner is released parallel to the lifting of the pressure foot, allowing the sewing thread to follow smoothly before cutting.

In some cases, barrel tensioners are used as thread tensioners, which rely on belt friction on several dowel pins. Controlled thread tensioners, whose braking force is adjusted according to other process parameters for example by multiphase motors, never went past the testing phase due to high costs.

12.6.2.5 Sewing Process Parameters

The possible maximum main shaft number of revolutions, also given as the number of stitches (=number of sewing stitches/unit of time, stitches/min), is an expression of the sewing machine productivity. However, it has to be considered that the handling of the sewing material during sewing as well as the processing characteristics of the sewing thread can necessitate a reduction of the main shaft number of revolutions.

The term stitch length (DIN 5300, part 1) defines the distance between two consecutive piercings of the sewing material.

Sewing speed (DIN 5300, part 1) can also be regarded as seam formation speed, resulting from the product of stitch length and number of stitches (m/min or mm/s).

As the stitch length cannot be measured precisely, stitch density (number of stitches per unit of length, e.g. stitches/10 cm) is often chosen as a more practical variation.

Sewing thread tension is another process parameter depending on the number of revolutions of the main shaft. The maximum tension is reached during stitch formation, i.e. during the creation of the loop or chain connection. The maximum has to be well below the sewing thread strength, and sewing thread tension can be adjusted by changing the settings of the thread tensioners of the sewing machine. The main criterion of the tensioner settings is the optically evaluated stitch-type-specific location of the loop or chain connection points of the sewing threads. Measuring the sewing thread tension provides information regarding the process forces during sewing. Registering the thread tension maximum during loop or chain connection can be regarded as a sure sign of successful stitch formation and stored for later analytical confirmation of the product. The preferable position of the thread tension measuring head is between thread feeder and thread guide of the needle thread to the sewing needle.

12.6.3 General Remarks Regarding Sewing Technology in Preform Manufacture

The use of sewing technology in preform manufacture is initiated by the demand for protection of the textile-reinforced composites from *delamination*. The first idea is sewing in the consolidated form immediately after the spread of the reinforcement textile fabrics in different fiber orientations in the tool. Here, the reinforcement textile manufacturers assume that the reinforcement textiles, which are spread with a defined thread orientation, do not move any more in the tool. This "sewing in the tool" resulted in the notion of fixating the sewing threads in the stacked layer package and other only locally present cutting parts by means of one-side sewing starting from the open side of the tool. During consolidation as a part of the composite component production process the sewing threads would be fixed in position by being embedded in the matrix.

However, related drawings from publications and patent documents (Fig. 12.18) do not show any sewn connection. Due to ignorance regarding sewing terminology and standards, the depicted technique was referred to as a sewing technique. However, the sewing thread is neither loop- nor chain-connected with itself. The correct term would be *tufting technique*, as only the thread loops are inserted into the layer stack. Due to the missing loop or chain connection of the sewing thread, the position of the inserted thread loops is only defined permanently when the matrix is cured (thermoset) or solidified (thermoplastic).

Realizing "sewing in the tool" as tufting does not guarantee the users the ability to move or even transport the textile stacks again. This tufting preform is not assembled with sufficient stability for these process steps. Height differences in the stack can be easily compensated by tufting, if the penetration depth is varied locally.

First remarks regarding the use of sewing in the composite area were made by NASA—Preform Sewing-NASA ACT Program 1989. The CNC sewing plant shown in the context implies large-scale components [20].

The tendency of moving from prepreg (wet) technology to preform (dry) technology allows the utilization of all steps of textile production, including textile assembly, in order to manufacture load-adapted near-net shape complex preforms.

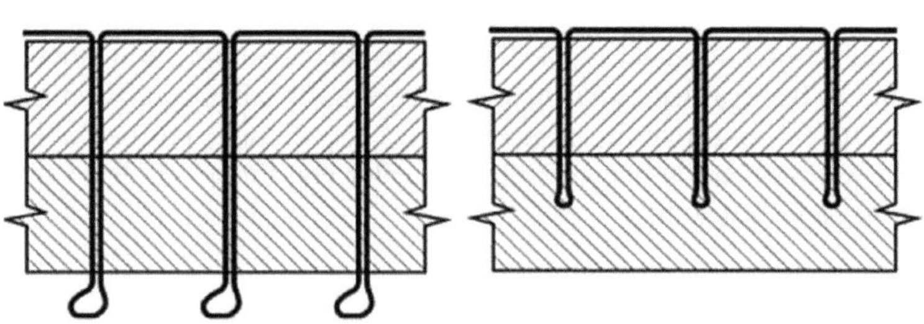

Fig. 12.18 Arrangement variations of the tufting thread in the stacked reinforcement textile structure [19]

These complex preforms have to be manageable during the production steps without as few additional auxiliary devices as possible. Sewing threads secure each other or themselves by looping or chain connection in the seam. This preference of sewing over tufting has to be brought into effect.

12.6.4 Seam Functions

Seams produced by sewing can be used in preform production as shown in Sect. 4.5.2.1 from a different perspective. Possible functions include:

- Edge seams—In textile processing, it is common to prevent the dropping of fiber sections by seams running parallel to the cutting contour. Usually, overedge stitches are used, but they are unsuitable for preform applications due to the development of matrix-enriched zones at the margins after edge sewing, stacking, and consolidation. These zones can cause cracks and faults in the composite structure [13].
- Forming seam—Seams can support forming or even create the desired form. One example is the formation of a tubular textile product with a sewn longitudinal seam made from a planar textile band of a width matching the circumference of the tube plus seam allowance.
- Z-reinforcement seam—This seam is used for the original purpose of sewing in composite manufacture, which is to prevent the propagation of tears. Tests [21] have shown that the propagation of tears caused by impact loads matches the grid of rectangularly arranged seams. Without these seams, tear propagation is most pronounced.
- Assembly seam—Assembly seams are indispensable for the assembly of complex preforms. For example, a multi-layered T-profile can be fixed to multi-layered base textile by means of perpendicularly arranged assembly seams. Under component loading, the forces are distributed between the assembled elements by the sewn seams, which also serve as tear propagation protection.
- Diagonal seam or Z-reinforcing assembly seam—The consideration that threads can only absorb tensile forces resulted in the thought of arranging the sewing threads diagonally in the layered structure of the composite in order to achieve a more advantageous orientation in a certain direction of loading. A first diagonal sewing machine was developed from a standard sewing machine [22], a CNC-controlled diagonal machine named PSN 3020 was manufactured by the Cetex-Institut für Textil und Verarbeitungsmaschinen gemeinnützige GmbH (Cetex Institute nonprofit for textile and processing machines) (see Fig. 12.19).

Fig. 12.19 Programmable diagonal sewing machine PSN 3020 (source: Cetex-Institut für Textil- und Verarbeitungsmaschinen gemeinnützige GmbH)

12.6.5 Sewing Technology for Composite Assembly

The following reasons inevitably lead to a demand for suitable CNC and robot technology:

- High production precision requirements,
- High reproducibility requirements,
- Higher masses per area than in classical ready-made technologies,
- Multiple stacking of fabric with higher mass per area,
- Obligation to keep the thread orientation in the individual layers according to the load profile of the component,
- 3D sewing tasks, and
- large preform dimensions

12.6.5.1 CNC and Robot Sewing Technology

CNC and robot sewing technology can be classified by several aspects:

- By guide machine type—Robot-guided sewing technology or CNC-x, y-cross table automated sewing machines with multi-directional or tangential seam formation are used as guide machines. The CNC-x, y-cross table automated sewing machines create seams in the plane, while robot technology allows 3D seams.
- By sewing material geometry—A distinction has to be made between free geometric areas spread in the plane, and those deformed three-dimensionally.
- By the angle of the needle to the textile stack—Usually, the needle pierces the surface of the textile stack at a right or almost right angle. The aforementioned

Sect. 12.6.4 clarifies the advantages for the properties of the composite component offered by a diagonal piercing and diagonal positioning of the sewing threads.
- By the number of sewing elements—Two-side sewing technology is the common design of sewing machines, while one-side sewing technology has so far only been used for special seams, such as blind stitch seams. In preform manufacture, several variations of one-side sewing technology have been developed in the past few years.

Multi-directional sewing by means of CNC-x, y-cross table automated sewing machines is characterized by the sewing head retaining a certain orientation in the co-ordinate system. The relative movement needed to realize the projected seam contour is created by overlaying offset of the product area in both co-ordinate directions. This requires clearance in the fixed-positioned sewing head along the width of the product including the tenter. The required area of material, therefore, is four times the product area. The needle and gripper of the two-side sewing head are driven by long shafts and, for example, drivebelt transmission at the required transmission ratio.

In technical development of drive technology with controlled motors, needle and gripper can be driven by two separate motors with electronically regulated transmission ratio and rotation angle regulation of the two driveshafts. This results in the possibility to carry out an offset of the sewing head instead of offsetting the product in one or both co-ordinate directions. The area required for these automated sewing machines is thus reduced to twice the product area, or even further to only the product area.

One example of this development are the SNA CNC-x, y-cross table automated sewing machines by Parker Hannifin GmbH Offenburg [23], which have gone through the described development steps. Originally, these machines were conceptualized for the production of quilts. But the technology is also used within the framework of the sponsored project "Process development and integrated lightweight construction concept for a waste-free, continuous preform RTM production (PRO-Preform-RTM)" at the IVW GmbH Kaiserslautern, concerned with the sewing-technical processing of layered reinforcement textiles. The reinforcement textiles are fed directly off the rollers. The cutting edges are secured with parallelly created seams (double lock stitch) before being cut with a CNC cutter (Fig. 12.20) [24].

In [25], on the other hand, the CNC-x, y-cross table automatic sewing machine KL 121 by Keilmann Group is used, which is characterized by the realization of tangential sewing. Tangential sewing means that the entire sewing head can be rotated around the needle axis without stops. As the sewing needle and gripper system always synchronously orient themselves towards the tangent of the sewing contour during the rotation, the sewing direction is always forward. In particular for the sewing thread, this ensures an optimized thread passage from the needle into the sewing material continually. This also creates ideal processing conditions for the generally very brittle sewing threads made from carbon or glass.

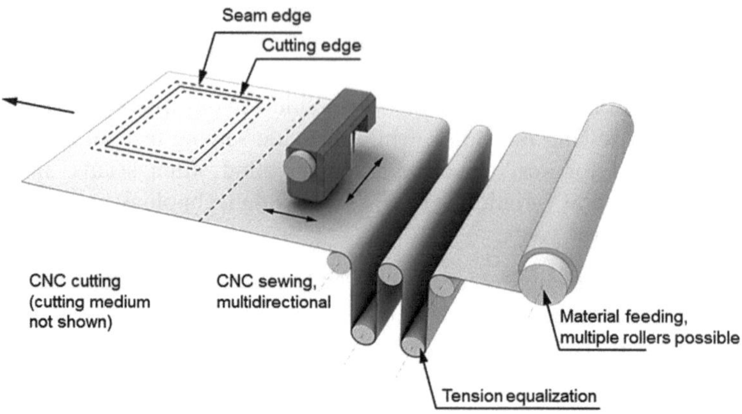

Fig. 12.20 Concept of sewing-technical assembly before cutting

Another example for the creation of seam contours in tangential sewing is the programmable circular sewing machine PRN 500, conceptualized, constructed, built, and successfully used by the DFG Research Group 278 [6]. The machine is suitable for the assembly and Z-reinforcement of circular preforms of up 500 mm diameters with a central opening, and can process stack heights of up to 20 mm (Fig. 12.21).

For CNC controls, a task-adapted program ensures that radii, stitch lengths, starting/ending angles, or other relevant seam contour parameters can be chosen freely within the construction-inherent limits. In the specific case of this circular sewing machine, a linear axis offsets the central drive in the direction of the sewing needle in such a manner as to allow the manufacture of varying radii. The central drive realizes the circumferential movement, with the respective seam radius having different angles of rotation for a constant stitch length. Especially for smaller radii, these angles of rotation are very wide, which might require a reduction of the number of revolutions of the machine for these intermittent offset movements at smaller radii.

The optimal working conditions of forward sewing could be used in other applications, e.g. for the sewing-technological assessment of composite sewing threads produced by various yarn formation methods [26].

12.6.5.2 Reasons for the Development and Application of One-Side Sewing

During the mid-1990s, the increasing interest of the composite industry gave sewing technology a development boost towards one-side sewing. The following reasons were crucial for this trend:

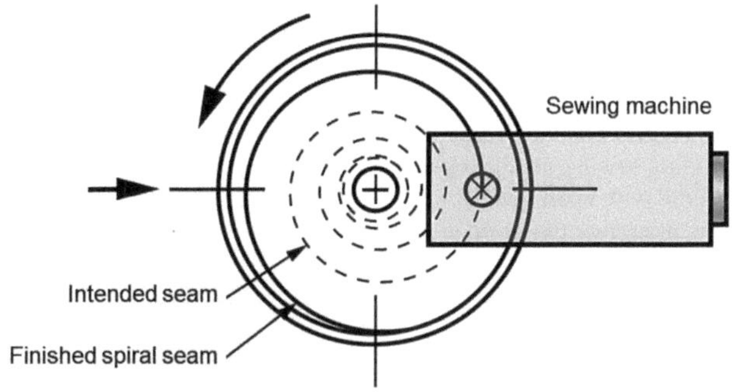

Fig. 12.21 Working principle of the programmable circular sewing machine PRN 500

- complex, near-net shape geometries of the preforms with cavities and undercuts,
- low flexibility of stacked reinforcement textiles, resulting in a handling problem with conventional sewing technologies,
- high production precision by use of CNC technology, and
- desire to sew in the tool

The latter demand is only met by some of the newly developed reinforcement and sewing technologies, which were released to the public in the following order:

At ALTIN Nähtechnik GmbH Altenburg, a one-side sewing technology with two co-operating needles was developed and initially released to the international research community market (Fig. 12.22). The thread-guiding needle with the eye (a) moves toward the sewing material at an angle of 48° and realized the loop stroke at the reverse side of the sewing material, forming a needle thread loop. A hook needle, i.e. a sewing needle with an eye open on one side (b) serves as gripper for a single chain stitch, piercing the sewing material at a right angle while being turned around this vertical axis. For this technology, the trademark OSS® One Side Sewing® was registered. In adapted form, the developments of ALTIN are now part of the product range of the Keilmann Group.

The so-called ITA sewing head [26] is equipped with another constellation of two needles. Here, both needles carry a needle thread, and both threads are alternately chain-connected with one another. Both needles assume the chain stitch gripper function during the alternating penetration, while the needle thread loop is formed during the escape of the needle from the respective needle thread taken up by the other needle. Due to this doubled function, the needles have to keep certain minimum distances between each other, which prevents any significant change of stitch length. Both needles are arranged in a V-shape to each other, putting the sewing threads in a V-orientation in the seam cross-section. Further developments were made [27]. Within the scope of a comprehensive technical revision work, two phase-shifted identical drives were introduced. Due to this change, the angular position of the needles toward one another is now variable within limits. There is no information available regarding the use of the ITA sewing head by other users.

Both technologies are characterized by the requirement of clearance underneath the sewing area, and can therefore not be used for sewing in the tool.

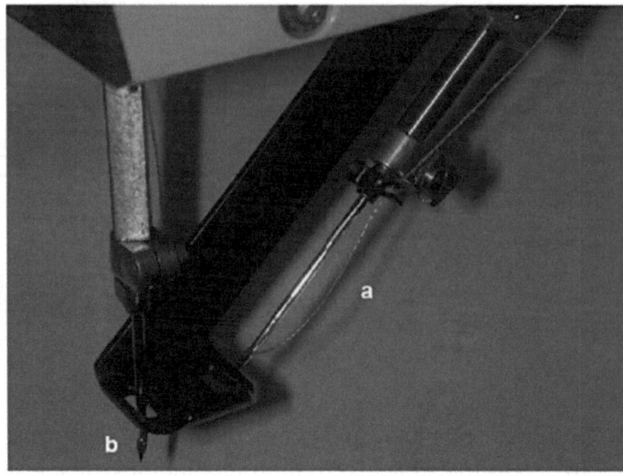

Fig. 12.22 Details of the sewing head used by ALTIN Nähtechnik GmbH with two needles, (**a**) diagonally piercing, thread-guiding sewing needle, (**b**) needle with open needle eye, assuming the gripper function

The Keilmann Group (see also KSL Keilmann Sondermaschinenbau Lorsch GmbH) introduced a blindstitch sewing head to the research market for composite technology. A needle curved by almost 180° (Fig. 12.23) and with a much smaller radius than that required by conventional blindstitch sewing machine needles for garment production is used. While the stitch and seam directions in classic blind stitching are at right angles, this innovative method is distinguished by identical stitch and seam directions. Concerns regarding the risk of damaging the previously inserted needle thread with the needle during piercing are removed by a slightly oblique position (ca. 5° in relation to the seam direction) of the needle path.

12.6.6 Sewing Material Fixations and Composite Production Tools

Regardless of the type of sewing technology, sewing material fixations and production tools suited to the special preform geometry have to be conceptualized, constructed, and provided for an application in sewing-technological preform assembly. On the x, y-cross table automated sewing machines, the sewing material fixations should preferably be designed with the possibility in mind of creating a high-quality multi-layered preform construction during the filling with cutting parts of defined thread orientation (Fig. 12.24). The seam formation, including the piercing of the layer stack, requires suitable openings of linear or areal design in the individual metal parts.

For three-dimensional seams on preforms, more complex production tools are necessary, which allow a multi-layered insertion and/or insertion of partially pre-assembled elements of the preform. Figure 12.25 shows the tool used to pre-assemble the rotor cover, without applied cutting parts. Figure 12.26 shows the production tool for the pre-assembly of the double-curved rotor demonstrator

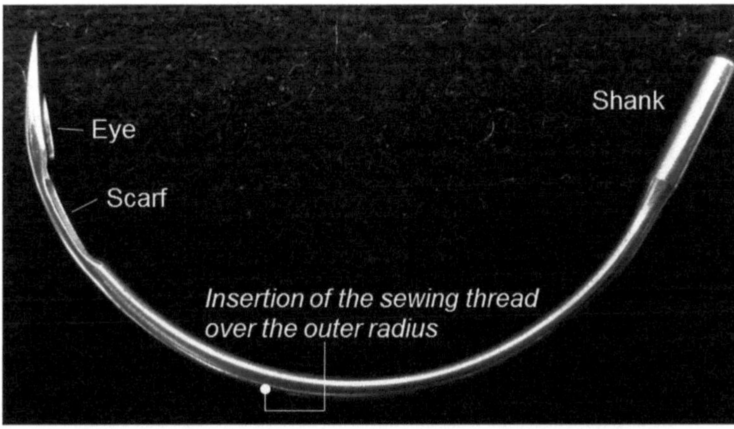

Fig. 12.23 Sewing needle of the blindstitch sewing head developed by the Keilmann Group

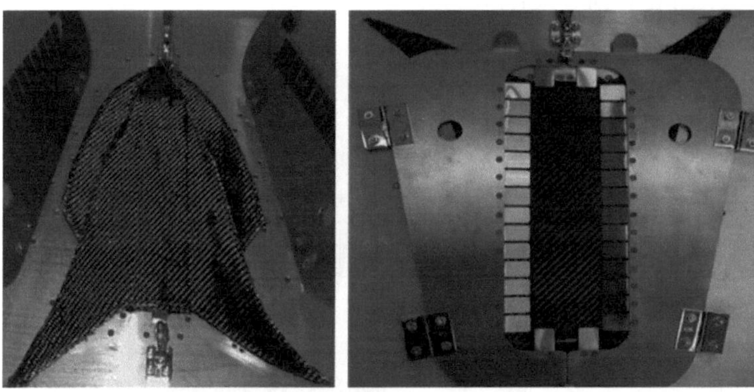

(a) open sewing fixation with cutting parts (b) closed sewing material fixation with the c for seam creation

Fig. 12.24 Sewing material fixations for cutting parts, example: prefabrication of the blades of the rotor demonstrator II made by DFG-FOR 278 [6], (**a**) open sewing fixation with cutting parts, (**b**) closed sewing material fixation with the rectangular field left open for seam creation

two cover, with glass woven fabric (a), and the tool for the assembly of the rotor cover to the upper edge of the rotor blade (b). The sewing robot is positioned to assume either sewing task as required.

To allow the use of OSS® one-side sewing, the tools are required to feature a groove with a 10 mm × 10 mm cross-section, as seen on the tool for rotor cover pre-assembly. This groove ensures smooth thread loop transfer during sewing.

All sewing material fixations and production tools shown in Figs. 12.24, 12.25 and 12.26 are fitted with snap tensioners or screwable clamping devices. These fixation variations are only suitable for testing purposes. In a serial preform assembly, new, quickly and automatically working clamping designs, such as

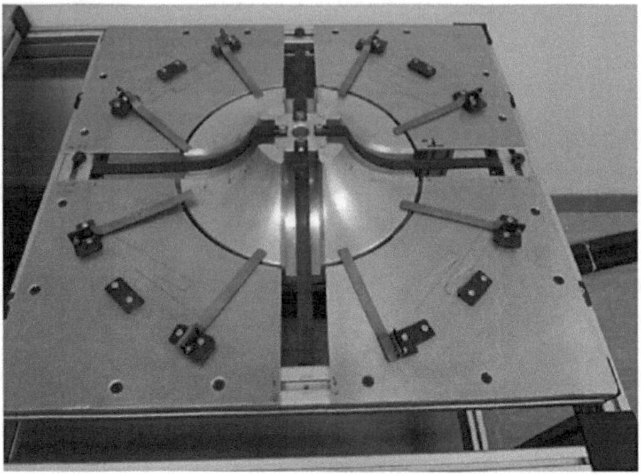

Fig. 12.25 Production tool with grooves for the proper execution of stitch formation, example: rotor demonstrator II by DFG-FOR 278 [6]

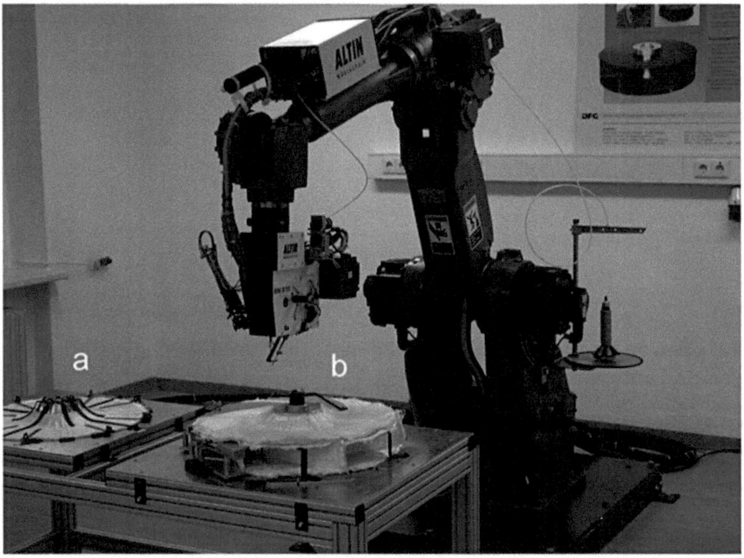

Fig. 12.26 Sewing robot with production tools for (a) the pre-assembly of the rotor cover, and (b) assembly of the upper blade edge and rotor cover of the rotor demonstrator II by DFG-FOR 278 [6]

pneumatic devices, are required. These are often combined with automated handling of the textile cutting parts of pre-assembled preform elements.

12.6.7 Sewing-Technical Processing Centers

The production of complex preforms for textile-reinforced plastics or other matrices requires product-oriented concepts for sewing production technology. In [25] it is shown what an assembling-technological department for preform manufacture in an aircraft factory can look like regarding technology, equipment, logistics, and quality assurance.

The sewing-technical processing of a rotor preform by means of a robot-guided sewing head shown in Fig. 12.26 is an example of creating one seam or a small number of defined seams. The re-clamping of the preform for further processing reduces production precision.

International, industry-owned research institutions, and the automotive industry, to an increasing extent, are occupying themselves with the production of dry, complex textile preforms. For this purpose, suitable, preform-adapted processing centers (analogous to metalworking) are being developed. Here, robot positioning in addition to preform-specific sewing material fixation is only a valid solution if the robot position allows the performance of all processing tasks.

Extensions of the working area of the robot-guided tufting or sewing technology are realizable by offsetting the robot on a rail system or, at even greater increase in mobility, by a hanging, re-locatable positioning of the robot above the preform-specific sewing material fixations. Internet presentations such as [19] show first concepts of complex processing centers in product-specific geometric design. Related to this, of course, there is a technical option of swift sewing head replacement to realize different stitch type classes and seam variations in a load-adapted manner.

In this context, the accessibility of the respective preform spots by the sewing heads has to be considered. Figure 12.27 shows that, depending on the dimensions, accessibility for tufting and sewing heads is often limited from the inside, but usually unproblematic from the outside.

Both [25] and [6] feature sewing material fixations with manually operated tensioning fasteners. This is a sign for further development requirements regarding the sewing material fixations, the defined, thread-orientation-adapted placement of the textile cutting parts (see Sect. 15.3), and of the automated handling of the textile cutting parts.

12.6.8 Automated Handling of Textile Cutting Parts

The automated handling of textile cutting parts for an automated assembly has been a subject of relevant research for decades. In the 1970s and, to a higher degree, in the 1980s, matters of sewing automation were at the center of international ready-made technological research attention, which was expedited with equal fervor in three important economic zones. No absolute breakthrough was achieved, which is

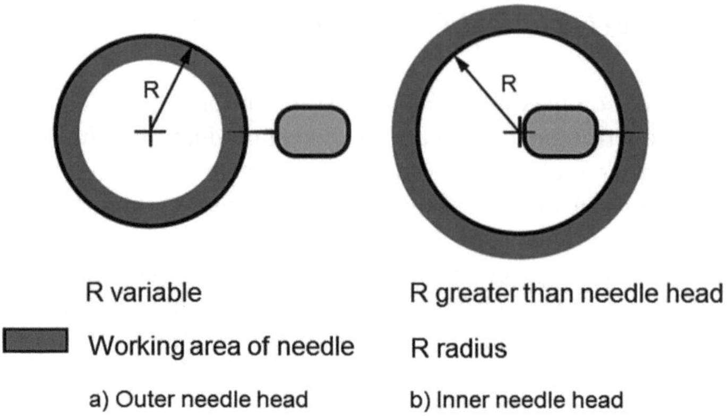

Fig. 12.27 Sewing head accessibility, (**a**) from outside, and (**b**) from inside

the reason for the garment industry still using cost-efficient manual labor instead of relying on automated processes in the manufacture of garments and similar products. In preform assembly, handling tasks are analogous, but characterized by significant differences in the textile structures to be handled, their mass per area, their shear, bending, and torsional load properties, their fiber materials, and their dimensions.

The tasks of handling technology in ready-made processes are parts of several technological process steps and can differ partially:

- handling of material bales during spreading,
- handling of cutting parts and stacks after cutting (removal from the cutting table),
- insertion into joining devices, including the position in the production tool and in the sewing material fixations,
- removal of the joined preform parts and distribution to the next assembly process, and
- transfer of the finished preform to the subsequent composite component process (if in close proximity) or to the transport containers

In the following, the transfer of the cutting parts to the production tools will be at the center of attention.

The following physical principles have been applied in textile-relevant handling technology for the fastening of textile cutting parts by mechanical means [28]:

- Needle principle—At least two reversely diagonal needles penetrates the fabric structure. A light movement of the needles away from each other tensions the textile structure and fastens it to the needles by friction. The offset motion can now be realized by means of a robot. The precise penetration depth is problematic, as only a single textile cutting part should be fastened securely. Various mechanisms for needle movement are known.

- Multihook principle—The multihook principle relies on the multiplied effect of a large number of hook-shaped needles arranged extensively and in reverse orientation on the surface of at least two gripping plates. Both gripping plates are pressed down and moved in opposite directions to tension the textile structure. The offset is then realized by means of robot technology.
- Clamping principle—Two opened clamps working in opposite directions are pressed down on the textile structure. Between the clamps, the textile structure bulges upward, especially when using a compressible padding, such as foam material, underneath it. The clamps can be closed when an applied sensor registers the fixation of the material. If necessary, i.e. in case of a missed clamping of the material, the processes can be programmed to repeat the clamping attempt. The bulging of the structure during the downward movement of the clamps must not reach a point where a second layer is clamped [29].
- Vacuum principle—The vacuum principle uses the effect of negative pressure in pneumatic systems. Suction air is known to be energy-intensive, making such vacuum grippers with sealing lips for the prevention of unwanted air currents perfectly suitable for coated materials.
- Aerodynamic paradox principle—The application of the aerodynamic paradox was experimentally tested in [30] and modeled for stacked porous textile structures.
- Adhesive tape principle—Adhesive grippers use the effect of adhesive tape, which is positioned by a suitable roll system in a manner that creates contact between the adhesive tape of the gripper and the textile structure when the gripper is pressed onto the textile structure. The connection can be disengaged with a fork-shaped holder. With regard to the quality of the composite, no adhesive residues should remain on the fiber or filament surface. Depending on the state of consumption, the adhesive tape is displaced intermittently. A new role has to be inserted as required.
- Cryogrippers—Accidents reports have often mentioned wet limbs or tongues freezing to metal structures below freezing point. This effect is utilized technically by spraying a small amount of water onto the textile structure. The gripper surface is cooled below freezing point, and the resulting ice created a connection between the gripper and the textile surface. In cryogrippers, Peltier elements are used, which can be cooled or heated up with changing electrical energies. Especially textile structures with very fine fibers and capillaries complicate the use of this method, water can move out of reach of the cooled gripper surface by capillary action. Despite this circumstance, cryogrippers are currently used extensively in the composite industry.

A problem is posed by the reliability of the fastening process, as until now the maximum reliability is achieved with a special fastening system/textile material combination of 99.8 % on one gripper. Usually, the large dimensions of the cutting parts require the use of several grippers. The existence of other error factors further reduces the reliability of the whole system. Because of this, the automated handling of textile cutting parts has not yet made a breakthrough in practical application.

Often, systems working on such concepts were disassembled and coupled with manual work.

Additional technical devices can increase the reliability of automated systems, as shown in paper processing, printing houses, and office technology, where the automated handling of paper sheets is successfully and reliable accomplished. Therefore, inspiration can and should be drawn from other technological fields:

- sensors to register the gripping of the material,
- control algorithms to trigger a repeat of the gripping process if necessary
- measuring and utilizing process parameters of the gripping process (e.g. clamping force, clamp opening distance),
- Removal of adhesion at the cutting edges due to multi-layered cutting of thermoplastic structures,
- reduction of adhesion between individual layers of the cutting part stack due to surface effects by means of suitable intermediate layers or other technical measures,
- optimization of mechanical gripper elements to avoid filament destruction by the gripper effects,
- suitable adhesive selection to avoid interface problems between the matrix and adhesive-contaminated fiber/filament structure caused by adhesive residues from adhesive grippers
- use of suitable CNC technology for high spreading precision to ensure compliance with designated thread orientation
- quality assurance by optical or other means to identify overturned corners, and to ensure the sequence and thread orientation of the layers

In summary, it has to be stated that technically reliable, acceptably error-free, and automated handling of textile cutting parts is still in the development phase, despite individual successful solution such as the handling of thick glass fiber mats by needle grippers.

The sewing-technical assembly of textile preforms requires a controlled compromise of reinforcement textile structure damage by perforation (during penetration) and assembly of the complex preform. Constructors have to design this compromise—the reduction of in-plane properties at simultaneous improvement of out-of-plane properties—in a manner that ensures, for instance, the required interlaminar shear strength or respective impact behavior of the individual application without critical changes to other characteristics.

12.7 Textile Assembly by Means of Welding

Textile welding is the bonded connection of two or more parts, or just one part with itself (tube production) from usually identical materials using heat and pressure. The welding seam is created without unrelated additives. Man-made fiber materials with thermoplastic properties, or coating materials of the textile fabrics with similar

Table 12.1 Welding methods and operating principle [31]

Welding method	Operating principle
Hot-air welding	Continuous
Hot-wedge welding	Continuous
Ultrasonic welding	Intermittent or continuous
High-frequency welding	Intermittent
Laser welding	Intermittent or continuous

properties, are required for the use of textile welding. The aim is to create a textile-adapted, flexible, and sufficiently strong seam.

The welding processes can be classified into continuous and intermittent welding methods (Table 12.1).

The development of welding methods is closely tied to the market relevance of thermoplastic materials. Hot-air welding and hot-wedge welding were already known in the 1940s. During the late 1960s, dimensionally consistent seam contours, as in button holes for nylon shirts, were created using high-frequency welding, which is still the established method for manufacturing polyester-fabric tarpaulins with PVC coating. The high-frequency welding method is only partially suitable for some fiber materials and polymers, as the interactions between the high-frequency electrical field and the polymer material requires polar macromolecules as well as sufficient internal friction between the molecules [32]. Ultrasonic welding is another time-tested welding method. However, it is to be preferred for welding fiber material blends, provided that material contains at least two thirds thermoplastics.

The suitability of laser beams for defined heat introduction is well-known and was first researched comprehensively during the 1970s [33]. Since the year 2000, textile laser welding research was repeatedly publicly funded.

In continuous welding, the seam is formed without interruption, as known from the sewing process. Intermittent welding, on the other hand, requires a closed tool to create the necessary welding pressure. The energy needed to achieve melting or welding temperature in the textile material is provided either by mechanical or electrical vibrations or laser irradiation.

For simple welding tasks like repairs, manual welding machines are commercially available, operating with hot-air and ultrasonic methods. Heating up the thermoplastic material in the welding line to melting temperature is the prerequisite to join together the welding components under method-specific pressures before cooling them down again. A secure and load-transferring connection is established after lowering the welding line temperature under glass temperature (Fig. 12.28).

Due to the partial melting of the fiber materials, the fiber structure is transformed into a hardened polymer melt in the area of the welding line. This melt does not exhibit the textile-typical characteristics. Therefore, it can be useful to set local welding points, ensuring partial textile properties along the seam, which can include also seam elasticity along the seam. Fluid-impermeable seams, however, have to be uninterrupted. The geometrical arrangement of the seam additions of the

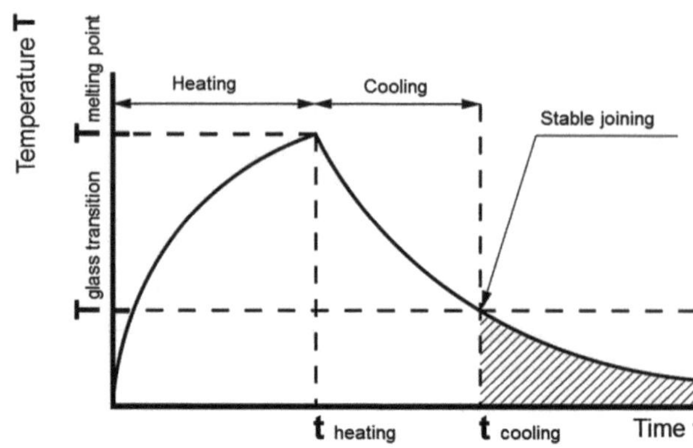

Fig. 12.28 Welding procedure [31]

cutting parts should be overlapped, as peeling seams can only transfer limited loads [34].

In preform production, welding technology can only be used to process hybrid yarn textile structures such as glass fiber/polypropylene. The thermoplastic fiber material content is transformed from a textile to a thermoplastic matrix state during processing via the preform to the composite component. A completion of this transformation in advance by creating a local welding line can support the handling and assembly process of the textile preform. However, no welded seam can assume Z-reinforcement or crack propagation prevention functions in the composite component, as they are offered by sewn seams.

After suitable positioning in a preform-specific tool, basting assembly processes can be performed, using thermal energy. By means of ultrasonic or hot-air welding equipment, thermal energy is inserted locally, until the thermoplastic component of the hybrid yarn starts to flow. After pressing and cooling, a connection based entirely on the matrix material between the respective cutting parts is created (Fig. 12.29). It has to be noted that this connection can support handling during preform assembly and preparation of the composite component production. The properties of the FRPC component, for instance to improve delamination behavior, are not feasible.

12.8 Textile Assembly in Preform Manufacture by Means of Adhesive or Binder Technology

Adhesive technology is successfully being used for two tasks in the field of textile-reinforced plastics. The assembly of textile preforms from dry reinforcement textile can be performed with adhesive technology, as an alternative to sewing. Furthermore, adhesives are used to fixate the structures of reinforcement textiles, since the carbon and glass filaments in the fabrics can cause undesirable shifts in the thread arrangements due to low friction between them.

Fig. 12.29 Local fixation of layered reinforcement structures made from hybrid yarns, after ultrasonic treatment

Adhesive bonding (glueing) is defined as the creation of a connection between joining parts by means of an additional material, the adhesive. Adhesives are defined as "non-metallic substance capable of joining materials by surface bonding (adhesion) in such a way that the join has adequate internal strength (cohesion)".

The adhesive seam between the joined materials is characterized by adhesion and cohesion. Adhesion refers to the connection between the reinforcement textile surface and the adhesive, while cohesion refers to the bonding in the adhesive itself. The interface of the reinforcement textiles is of primary importance for the adhesion between the reinforcement textile surface and the adhesive.

Another important aspect for composite manufacture is the compatibility of the adhesive with the future matrix material, as the adhesive must not become a disturbing factor in the composite, causing a gradual destruction of the composite component. In addition, the adhesive must not impair the infiltration of the liquid matrix material by locally lowering the permeability of the porous reinforcement textiles.

With regard to the durability of the adhesive connection, a distinction has to be made between permanent adhesive bonding (even in the composite component), and a temporary function of the adhesive, also referred to as a "basting" connection. The basting function to support the processing can be made ineffective, for instance by process temperatures during matrix consolidation, which remove the adhesive and evacuate its gaseous components by fume extraction during the last process steps. On the other hand, reactive connections between the adhesive and the matrix material, or the absorption of adhesive by dissolution during infiltration in the still-liquid matrix, are variations of integrating the adhesive into a composite component.

In the composite industry, adhesive are often called binders. Binders are applied to the reinforcement textiles in the form of powders or dispersions, and can usually be activated thermally due to thermoplastic properties. The application technique has to be matched to the aggregate state of the binder. Powders are spread on, while dispersions can be sprayed onto the textile.

The binders can be applied globally to the textile webs for the purpose of a general structural reinforcement, or locally on pre-calculated spots of the reinforcement textile cuttings, supporting the draping of the textile into the 3D shape of the preform without undesirable structural deformations by shear forces [4]. The transfer of this method into industrial practice of the composite industry is in progress.

The binder application can result macroscopically in superficial contacts between reinforcement textile and binder, or microscopically in partial penetration into the depths of the reinforcement textile structure.

The thermoplastic binders for structural fixation of the fabric webs can be temporarily activated by suitable heated tools locally during the draping of the cutting parts into the preform geometry. This allows pressing into the desired position and secures this preferred position after hardening of binders until the consolidation of the preform into the plastic component.

Chapter 13 offers detailed insights regarding the interactions of binder and matrix systems.

References

1. Deutsche Forschungsgemeinschaft (DFG): *Liste der laufenden Förderung der Deutschen Forschungsgemeinschaft (see link Sonderforschungsbereiche SFB 528 and SFB 532).* http://www.dfg.de/foerderung/programme/listen/index.jsp (25-02-2011)
2. RÖDEL, H.: *Analyse des Standes der Konfektionstechnik in Praxis und Forschung sowie Beiträge zur Prozeßmodellierung.* Aachen : Shaker Verlag, Habilitation, 1996
3. Cetex-Institut für Verarbeitungs- und Textilmaschinenbau gemeinnützige GmbH: *Cetex Informationen.* http://www.cetex.de/html/institut/deu/download/doc/cetexinfo/cetexinfo2006.pdf (25-02-2011)
4. Property right DE102007032904 (14th July2007).
5. GIRDAUSKAITE, L.: *Lokale Strukturfixierung im Preformherstellungsprozess für komplex gekrümmte Faserkunststoffverbundbauteile.* Dresden, Technische Universität Dresden, Fakultät Maschinenwesen, Dissertation, 2011
6. Technische Universität Dresden: *DFG-Forschergruppe FOR 278 Textile Verstärkungen für Hochleistungsrotoren in komplexen Anwendungen.* http://www.tu-dresden.de/mw/ilk/fg (25-02-2011)
7. TERNO, J.; LINDEMANN, R.; SCHEITHAUER, G.: *Zuschnittprobleme und ihre praktische Lösung.* Leipzig, Frankfurt/Main : Fachbuchverlag Leipzig und Verlag Harry Deutsch Thun, 1987
8. SCHLEGEL, W.: *Beitrag zur technologischen Optimierung des Zuschnittvorbereitungsprozesses in der textilen Konfektion.* Dresden, Technische Universität Dresden, Fakultät für Maschinenwesen, Habilitation, 1985
9. Fa.WASTEMA International Steinhauser Spezialmaschinen GmbH: *Foto-CD.* Veringenstadt. - Company documents
10. http://www.gfm.at/gfm/de/index.html (25-02-2011)
11. BLANK, S.; WÜSTENBERG, D.: Automatische Zuschnittischentsorgung. In: *Bekleidungstechnische Schriftenreihe* (1992), No. 89
12. G. M. PFAFF AG: *Stichbildung.* Kaiserslautern. – Company document

13. MITSCHANG, P.: Synergy Effects in the Field of Sewing Technology for the Manufacturing of Tailored Reinforcements. In: *Proceedings. 27. Aachener Textiltagung*. Aachen, Germany, 2000
14. http://www.queenlight.co.jp/hpeng/htmeng/headpage.htm (19-02-2011)
15. RÖDEL, H.: Stichbildung an Nähmaschinen. In: GRIES, Th. (Hrsg.); KLOPP, K. (Hrsg.): *Füge- und Oberflächentechnologien für Textilien - Verfahren und Anwendungen*. Berlin, Heidelberg: Springer Verlag, 2007, pp. 10–30
16. Groz-Beckert: *GEBEDUR-Nadeln. Revolution in Gold.* http://www.groz-beckert.com/website/media/de/media master 374 low.pdf (06-02-2011)
17. Schmetz: *The world of sewing*. http://www.schmetz.com/ (06-02-2011)
18. ARNOLD, J.: *Beitrag zum Bestimmen des Nadeltemperaturverhaltens bei Industrie-Doppelsteppstich-Schnellnähmaschinen unter Anwendung der Infrarotmesstechnik.* Dresden, Technische Universität Dresden, Fakultät für Maschinenwesen, Dissertation, 1985
19. WITTIG, J.: Recent Development in the Robotic Stitching Technology for Textile Structural Composites. In: *Journal of Textile and Apparel, Technology and Management* 2 (2001), No. 1
20. NASA: *The Advanced Stitching Machine: Making Composite Wing Structures Of The Future.* http://www.nasa.gov/centers/langley/news/factsheets/ASM.html (19-02-2011)
21. WEILAND, A.: *Nähtechnische Herstellung von dreidimensional räumlich verstärkten Preforms mittels Einseitennähtechniken.* Dresden, Technische Universität Dresden, Fakultät Maschinenwesen, Dissertation, 2003
22. RÖDEL, H.: Einsatz der Nähtechnik zur Herstellung von 3D-verstärkten Mehrschicht-Verbundstrukturen mit linearen Verstärkungen in z-Richtung unter variablem Winkel (1303/4-1) / TU Dresden, ITB. Dresden, 1999. – DFG application
23. Paker Hannifin Corporation: *Textile Machines.* http://www.parker.com/portal/site/PARKER/menuitem.7100150cebe5bbc2d6806710237ad1ca/?vgnextoid=f5c9b5bbec622110VgnVCM10000032a71dacRCRDvgnextfmt=defaultvgnextdiv=A87859vgnextcatid=3188941vgnextcat-=TEXTILE+MACHINESWtky= (06-02-2011)
24. MITSCHANG, P. et al.: Prozessentwicklung und ganzheitliches Leichtbaukonzept zur abfallfreien, durchgängigen Preform-RTM-Fertigung – PRO-Preform-RTM (02PP2475) / Institut für Verbundwerkstoffe GmbH. Kaiserslautern, 2004. – Final report
25. KÖRWIEN, Th.: *Konfektionstechnisches Verfahren zur Herstellung von endkonturnahen textilen Vorformlingen zur Versteifung von Schalensegmenten.* Bremen, Universität Bremen, Dissertation,2002
26. MOLL, K.-U.: *Nähverfahren zur Herstellung von belastungsgerechten Fügezonen in Faserverbundwerkstoffen.* Aachen, RWTH Aachen, Dissertation, 1999
27. LAOURINE, E.: *Einseitige Nähtechnik für die Herstellung von dreidimensionalen Faserverbundbauteilen.* Aachen, RWTH Aachen, Dissertation, 2005
28. PAKULAT, D.: *Beitrag zur Untersuchung des Vereinzelungsprozesses von textilen Flachformgütern.* Dresden, Technische Universität Dresden, Fakultät Maschinenwesen, Dissertation, 1983
29. RÖDEL, H.: *Grundlagenuntersuchungen zum Vereinzeln von Flachformgütern.* Dresden, Technische Universität Dresden, Fakultät für Maschinenwesen, Diploma thesis, 1980
30. HYKEL, K.: *Vereinzelung gestapelten porösen Flachformgutes unter Anwendung des Aerodynamischen Paradoxons mit gekoppelter Filterströmung, aufgezeigt am Beispiel textiler Zuschnitteile.* Dresden, Technische Universität Dresden, Fakultät für Maschinenwesen, Dissertation, 1974
31. RÖDEL, H.: *Konfektionierung technischer Textilien.* 2011. – Course script
32. AUTORENKOLLEKTIV: *Konfektion.* Leipzig : VEB Fachbuchverlag Leipzig, 1979
33. POLLACK, D.; WIEDEMANN, G.: *Beitrag zum Lasereinsatz, insbesondere des CO2-Lasers, in der Textil-, Bekleidungs- und Chemiefaserindustrie*, TU Dresden, Fakultät für Maschinenwesen, Dissertation, 1978
34. RÖDEL, H.: Konfektionierung Technischer Textilien. In: KNECHT, P. (Ed.): Technische Textilien. Frankfurt/Main : Deutscher Fachverlag, 2006, pp. 119–137

Additional Literature

RÖDEL, H.; HERZBERG, C.: Function-adjusted stitching technique for reinforced composite materials. In: *TEXtalks* (2012)7/8, pp. 48-52

TAIT, N.: Ultrasonics - an alternative method for seam construction. In: *Technical Textiles International* 23 (2014), no. 1, pp. 23-26, 28-29

TAIT, N.: Latest innovations for processing textiles and flexible materials. In: *Technical Textiles International* 22 (2013), no. 4, pp. 57-63

COSSON, B.; DELEGLISE, M.; KNAPP, W.: Numerical analysis of thermoplastic composites laser welding using ray tracing method. In: *Composites, Part B: Engineering* 68 (2015), pp. 85-91

BILISIK, K.; YOLACAN, G.: Experimental characterization of multistitched two-dimensional (2D) woven E-glass/polyester composites under low-velocity impact load. In: *Journal of Composite Materials* 48 (2014), no. 17, pp. 2145-2162

KOUSTOUMPARDIS, P.-N.; ASPRAGATHOS, N.-A.: Intelligent hierarchical robot control for sewing fabrics. In: *Robotics and Computer-Integrated Manufacturing* 30 (2014), no. 1, pp. 34-46

YANG, L.-J.; HOU, C.-J.; CHEN, W.-Q.; LI, L.; WANG, Y.: Study on Laser Cutting Carbon Fibre Reinforced Composites. High Speed Machining VI, ICHSM, International Conference on High Speed Machining, 6. In: *Materials Science Forum* 800-801 (2014), pp. 832-837

Chapter 13
Textile Finishing and Finishing Technologies

Heike Hund and Rolf-Dieter Hund

Principally, textile-reinforced composites consist of a form-giving matrix (polymers or inorganic) and the reinforcement structures embedded in it. The third and less clearly distinguishable component, the interface between the aforementioned elements, is decisive for the quality and properties of the manufactured composite part. This layer is formed by the surfaces and interfaces (phase boundaries) of reinforcement fibers and matrixes as well as in the space between them. This is influenced by the interactions between fibers and the neighboring molding compound. The distance between the interfaces can be on the molecular level so that direct interactions are possible. The insertion of further mediating substance layers is also possible. This chapter gives an overview starting from the consideration of the involved materials at the molecular level, via the pre-treatment of textile surfaces, up to the application of functional finishing.

13.1 Introduction and Overview

The finishing of textile materials for the application in lightweight construction and membrane manufacturing means the processing of outer material layers for the purposes of activation, functionalization, and modification. The effects to be achieved by finishing range from a simple adhesion improvement to highly complex interface design. The methods and processes applied for this purpose include three main areas. They can be used individually or in combination with:

- wet chemical processing for pre-treatment, finishing, and coating of textile reinforcement structures,

H. Hund (✉) • R.-D. Hund
Institute of Textile Machinery and High Performance Material Technology, TU Dresden, Dresden, Germany
e-mail: heike.hund@chemie.tu-dresden.de; rolf-dieter.hund@chemie.tu-dresden.de

C. Cherif (ed.), *Textile Materials for Lightweight Constructions*,
DOI 10.1007/978-3-662-46341-3_13

- physical/chemical methods based on corona/plasma technology for the processing of textile surfaces, and
- chemical surface activation and functionalization by reactive gas treatment

Apart from the selection of suitable methods, it is necessary to develop target-oriented processing able to guarantee the required result for the achievement of special and precisely defined goals by finishing. For example, the wet chemical surface modification can be carried out continuously or discontinuously, in which the contact between textile and treatment solution (liquor) can be ensured by dipping, spraying, or padding. Furthermore, the variable process parameters such as concentration of chemical substances, temperature, and treatment time are very important for the design process [1]. For the use of corona/plasma plants to treat textile substrates, it is necessary to select the energy input, treatment time as well as the possible feeding of gases and also the soluble substances in a way that allows the required surface functionalization without substantial material damage [2]. For the application of reactive gases for surface activation and functionalization, continuous and discontinuous methods are available in which gas composition, pressure, fumigation, and temperature are the important influencing factors [3].

13.2 Chemical-Physical Fundamentals

For the manufacturing of textile materials, the primary raw materials belong to the class of organic polymers. Furthermore, there are applications of inorganic and metal materials, especially of glass, carbon, ceramics, basalt, and steel. In most cases, the textile products manufactured from them have a surface sizing whose constituents include a wide range of substances [4]. For the efficient finishing of these materials, it is necessary to differentiate the given material characteristics from each other to ensure optimized processing results. The focus of the considerations lies primarily on the physical and chemical properties in the areas of textile surfaces. The surface energy, the chemical structure at the molecular level, and the topography are regarded as the important factors and give an overall statement concerning the adhesion properties (see Fig. 13.1) in terms of the ability to combine with a given matrix system.

Topography designates the geometrical structuring, particularly at a micro level. Furthermore, these variables provide information regarding the finishing strategies of the textile materials by means of processes and methods of textile finishing. These variables will be regarded more specifically in the following discussion.

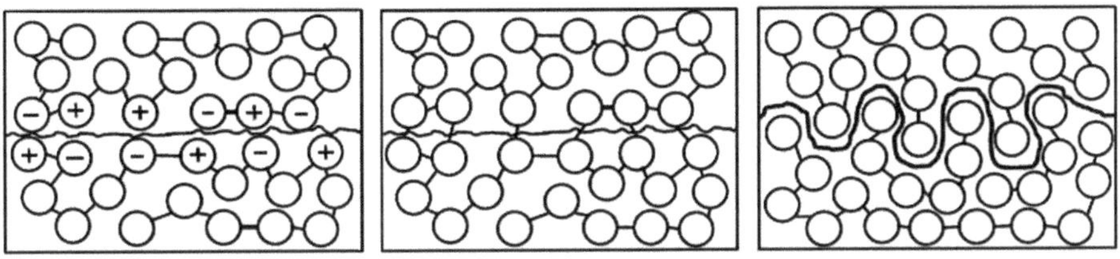

Fig. 13.1 Interpretation of adhesion, *left*: dipolar forces, *center*: covalent bonds, *right*: mechanical anchoring (according to [5])

13.2.1 Surface Energy, Surface Tension

Atoms and molecules on the surface of a condensed phase (solid or liquid) do not have similar neighboring atoms or molecules in the same spatial direction which would be suitable for chemical or physical interactions to reduce their energy potential. This means that a surface atom or molecule is characterized by considerably higher energy compared to its counterparts in the inner phase, resulting in a specific quantity for the interaction capacity with other adjacent phases (solid, liquid, or gas). This quantity is described as surface energy σ_s (solid) or surface tension σ_l (liquid) and possesses the dimension of an energy/area (Nm/m^2), which yields the unit N/m, the fraction of which is often given as mN/m (obsolete: dyn/cm). Since the energy properties of surfaces arise from different interactive forces, the surface energy of solids or the surface tension of liquids can be differentiated into disperse and polar parts. The disperse part of the surface energy/tension σ D depends on the electronic interactions of non-polar molecular structures, which induces weak dipoles and as a result causes relatively weak attraction forces at the phase surface. Its designation as London force was inspired by the works of physicist FRITZ LONDON [7]. Phase boundary atom groups with permanent dipoles are responsible for the polar part of the surface energy/tension σ P. Permanent dipoles arise from of the covalent bonds of atoms with different electro-negativity [8] or in the extreme from charged atom groupings. Another significant contribution to surface energy is provided by acid-base interactions according to the LEWIS model [9], and by hydrogen bonds [10]. The quantities of surface energy of solids as well as surface tension of liquids are easily accessible by indirect measurements and subsequent calculation, which will be discussed in Sect. 13.4.1.

13.2.2 Surface Energy

A surface energy σ_{12} resulting from two immiscible condensed phases (liquid/liquid, solid/liquid, solid/solid) is determined by the characteristics of the individual surface energy or surface tension, as well as their disperse parts and polar parts. OWENS and WENDT [11] have formulated this relationship as follows:

$$\sigma_{12} = \sigma_1 + \sigma_2 - 2\left(\sqrt{\sigma_1^D \times \sigma_2^D} + \sqrt{\sigma_1^P \times \sigma_2^P}\right) \quad (13.1)$$

σ_{12} = Surface energy between Phase 1 und 2
σ_1 = Surface energy of phase 1
σ_2 = Surface energy of phase 2
σ_1^D = Disperse part of surface energy phase 1
σ_2^D = Disperse part of surface energy phase 2
σ_1^P = Polar part of surface energy phase 1
σ_2^P = Polar part of surface energy phase 2

The surface energy plays an important role for the wettability of liquid phases such as thermoplastic melts, reactive resins, coating pastes or finishing liquids, for the wettability of solid phases such as textile structures and for the adhesion phenomena of two phases with each other. The exact relationship will be treated in Sect. 13.3.1.

13.2.3 Chemical Properties of Textile Surfaces

The surface energy of condensed phases discussed in the previous section is closely related to the chemical composition of materials from which the textile structures, matrix system as well as coating materials are comprised. The chemical structures of interfaces are also interesting for the assessment of their chemical reaction potential for direct utilization or to develop and execute functionalization strategies. Primarily, three material layers should be considered, each of which represents in turn the fiber surface, sizing, or matrix.

13.2.4 Chemical Properties of Fiber Materials

13.2.4.1 Fibers from Polyethylene (PE) and Polypropylene (PP)

PE and PP are polymers belonging to the so-called polyolefins and consisting entirely of hydrogen (H) and carbon (C) atoms, which are bonded to each other with single bonds, forming aliphatic structures (Fig. 13.2).

Polyolefins are characterized generally by their high chemical resistance (inertness). Therefore, no chemical reactions with composite partners are to be expected. Moreover, PE and PP exhibit only weak disperse interactions and thus only allow the wetting with non-polar liquids. However, the C–H bond is suitable for the

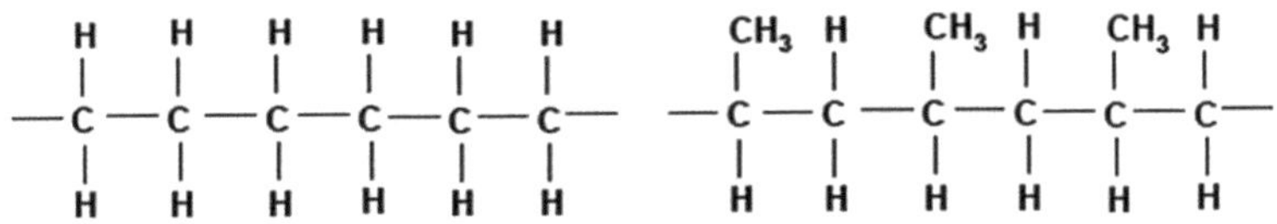

Fig. 13.2 Structural elements of PE (*left*) und PP (*right*)

oxidative functionalization by means of various processes (see Sect. 13.5.2) in order to achieve considerable improvement of wetting and adhesion properties. Furthermore, it is possible to incorporate chemically reactive molecule groups through physical anchoring or methods of chemical/physical treatment at the surface of fibers made from PE and PP [12, 13].

13.2.4.2 Fibers from Aliphatic Polyamides (PA)

These fiber materials are formed from the molecule chains of aliphatic hydrocarbon portions connected over carbonamide units (see Fig. 13.3 and Sect. 3.2.3.1). This type of bonds is also called amide, peptide or protein bond and results in hydrocarbon chains, which are regularly interrupted by nitrogen atoms (N). The number of C-atoms between the N-atoms of monomer units (sequences) gives the nomenclature of the polyamides built from them [14].

Polar (carbonamide) and unipolar (hydrocarbon) structures are connected with each other in aliphatic polyamides. The carbonamide units are capable of creating hydrogen bonds as donor (N–H) as well as acceptor (C=O). Furthermore, polar amino (–NH_2) and carboxyl (–COOH) groups are available at the terminals. The number of these groups on the surface of a PA fiber can be increased by hydrolysis in order to develop reactive centers. The hydrocarbon portion (segments) in PA can be functionalized oxidatively in respective structural elements of polyolefins (see Sect. 13.5) (Fig. 13.4).

13.2.4.3 Fibers from Aromatic Polyamides (Aramid)

The combination of monomers in aromatic polyamides occurs via the carbonamide bond, similarly to aliphatic polyamides. The intermediate links consist of either meta- or para substituted aromatic phenyl units (Fig. 13.5). The resulting polymers possess a strong tendency to form hydrogen bonds, which strongly influence their internal strength [15].

Due to their alternative polar/non polar structures, *aromatic polyamides* contain a pronounced polar part in their surface energy, which causes a good wettability of desized fiber materials [16]. Aromatic diamine constituents and amide groups are especially accessible for the chemical modification of aramid fibers.

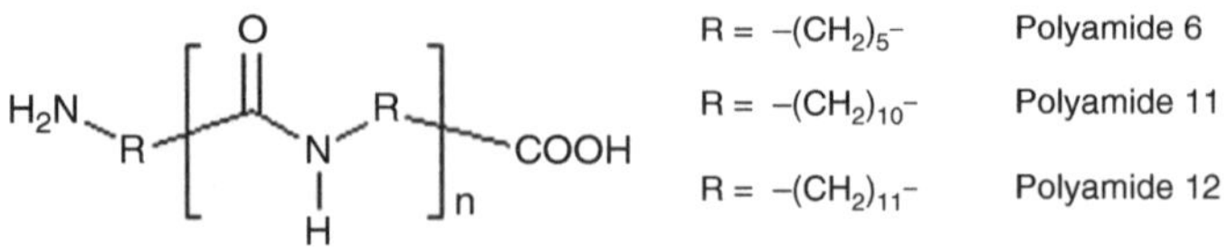

Fig. 13.3 Aliphatic polyamide from one monomer

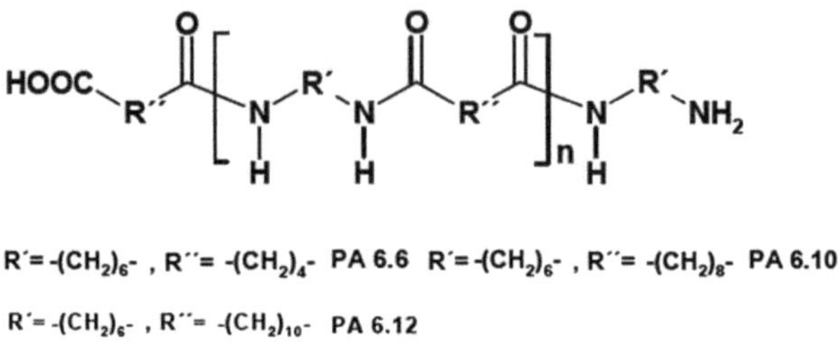

Fig. 13.4 Aliphatic polyamide from two monomers

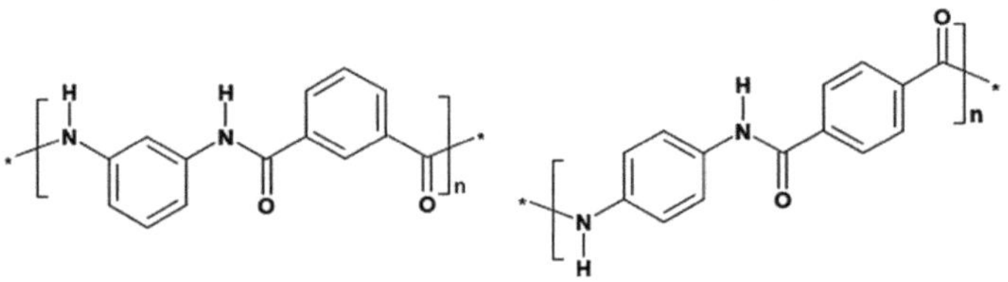

Fig. 13.5 Structure unit of poly (m-phenylene terephthalamide) (*left*); poly (p-phenylene isophathaldiamide) (*right*)

13.2.4.4 Fibers from Polyethylene Terephthalate (PET)

PET is the globally most common polymer material for the production of textile fibers [17] and belongs to the class of aromatic aliphatic polyesters. PET is composed from ethylene terephthalate units (Fig. 13.6) and is resistant to most solvents and many chemical influences.

Textile fibers made from PET have very hydrophobic surfaces causing poor wettability. This area can be activated and modified by an oxidative attack on the aliphatic $–CH_2–CH_2–$ units and by controlled ester hydrolysis (see Sect. 13.5.2.1) in order to form reactive centers and anchoring points.

13.2.4.5 Glass Fibers

Depending on the purpose of application, glass fibers are available in a large variety of material compositions (see Sect. 3.3.2), all of which consist predominantly of silicates. Here, the condensed silica structures form three-dimensional $(SiO_2)_x$ networks (see Fig. 13.7) with uncondensed free silanol functions at their phase

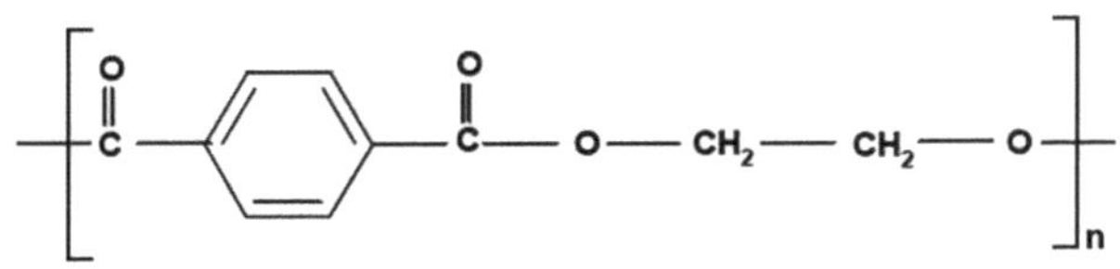

Fig. 13.6 Structure of polyethylene terephthalate

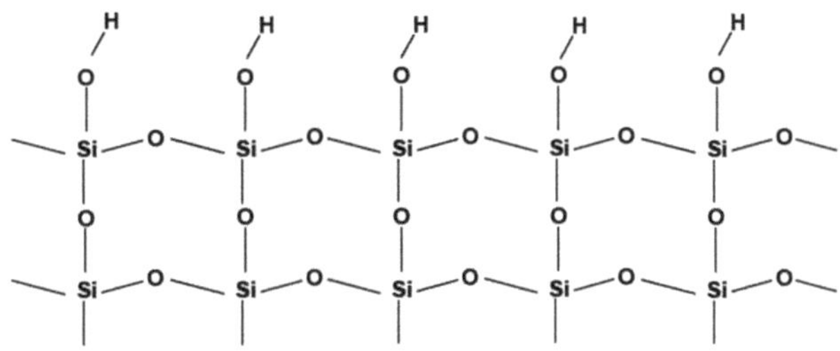

Fig. 13.7 Silicate structures at the surface of glass fibers (schematic)

boundaries [18]. These chemically reactive hydroxyl groups allow the covalent fixation of a variety of alkoxy silanes, which can form chemical bonds or permanent physical bonding with matrix materials (see Sect. 13.3.2) [19].

13.2.4.6 Carbon Fibers

Depending on the manufacturing process (see Sect. 3.3.3), carbon fibers are composed, to varying degrees, from pure, hexagonally arranged carbon structures. The ideal structure is illustrated in Fig. 13.8.

The carbon fiber can be modified in the region of surface layers using chemical oxidation, which results in physically active and chemically reactive centers at their surface structure. Interestingly, a significant increase in fiber strength and stiffness can be observed in the case of reactive gas treatment aimed at increasing surface polarity [20, 21].

13.2.4.7 Ceramic Fibers

Fibers from oxidative and non-oxidative ceramics, like aluminum oxide or silicon carbide, have high surface energy and can be wetted very well. Their surfaces are largely chemically stable and can be decomposed only under very drastic conditions [22].

13.2.4.8 Natural Fibers (Flax, Linen)

The natural flax fibers and the fiber material manufactured from them are composed of cellulosic structures (Fig. 13.9). They have reactive hydroxyl groups and are capable of forming hydrogen bonds.

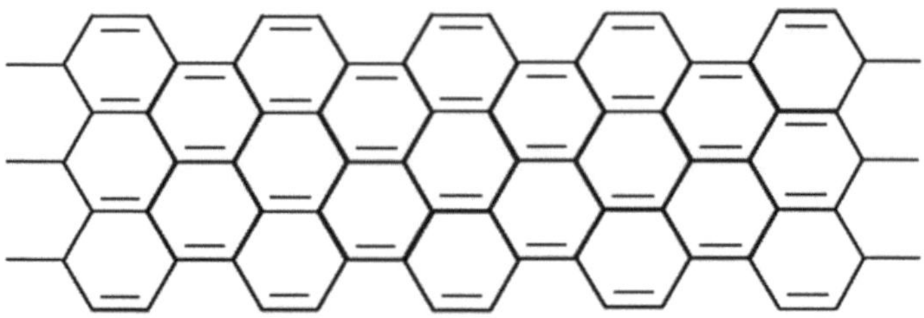

Fig. 13.8 Ideal structural segment of carbon fibers

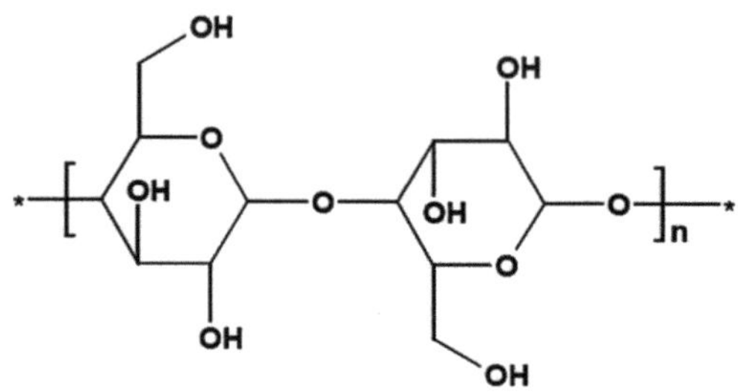

Fig. 13.9 Structural unit of cellulose

The OH groups of cellulose allow for a wide range of chemically utilizable reactions, such as addition, condensation or substitution, where the primary hydroxyl functions exhibits much higher disposition to reactions than the two secondary hydroxyl functions. The oxygen bond between the rings is very sensitive to acids and can easily be separated by them. The fiber material is very hydrophilic, contains a porous structure and offers excellent wetting properties for liquid phases.

13.2.4.9 Other Fiber Materials

Other textile technically interesting fiber materials are polyphenylene sulfide (PPS), polyether ether ketone (PEEK), polybenzimide (PBI) and polyphenylenebenzbisoxazol (PBO). They all share a high chemical stability and inert fiber surfaces.

13.2.5 Chemical Properties of Spin Finishes and Sizing Materials

13.2.5.1 Spin Finish for Polymeric Synthetic Fibers

Synthetic fibers are finished using spin finishes (with a weight addition of 0.5–2 %) immediately after their spinning in order to provide the properties required for further processing. Outstanding running and sliding properties are the main requirements to ensure an optimum frictional behavior for the reduction of electrostatic charge accumulation and to guarantee internal cohesion of individual filaments in a bundle. Depending on the substrate and spinning technique, the chemical products used to this end usually consist of the following substances [23]:

- anti-friction agent (40–60 %),
- emulsifying agents,
- anti-electrostatics, and
- additives (wetting agent, biocides, corrosion inhibitors, anti oxidants).

Anti-friction agents are based on fatty acid tri-glycerides (vegetable oil), esterified oleic acids such as octyl- and tridecyl stearates, trimethylolpropane trinonanoates and ethylene oxide/propylene oxide adducts. Fatty alcohol and fatty acid ethoxylate, partial glyceride, triethoxy glyceride as well as anionic detergents with antistatic effects, such as sulfonated or sulfated vegetable oils are applied as emulsifiers. All these substances have relatively low molecular weight, consisting to a great extent of aliphatic hydrocarbon structures and do not possess any chemically reactive centers. In a fiber-matrix interface they are a disturbing element and counter the required adhesion forces of a composite. Therefore, it is recommended to remove these spin finishes by means of wet chemical, plasma technology or reactive gas based treatment (see Sect. 13.5.2).

13.2.5.2 Finishes for Glass and Carbon Fibers

In contrast to polymeric synthetic fibers, the finishing of glass and carbon fibers is carried out with preparations (sizings), which usually possess interface-active substances. In this case, sizing materials have been developed, which assume the mechanical and anti-static protective functions described in the previous section and also interact with the matrix component.

13.2.5.3 Sizing for Glass Fibers

Conventional textile sizing materials for glass consist of a mixture of starch or dextrin with oil and fats [1]. Such preparations are applied especially for textile-reinforced composites and must be removed before further processing, i.e. impregnation or coating, as they counteract adhesion. After desizing, which is preferably executed thermally, the reinforcement structure is dip-coated with adhesion promoter. Immediately after their manufacturing, glass fibers are often sized with so-called silane coupling agents. These are composed of organic functional trialkoxysilanes and able to form a condensation product (reaction with water) with free hydroxyl groups of the glass fiber surface after hydrolysis (see Sect. 13.2.4) and can thus find a permanent anchoring (Fig. 13.10).

The organic functionality of condensed silane can be used for the chemical bonding of matrix polymers. A wide range of functionalities is available [24], some of which need to be considered here.

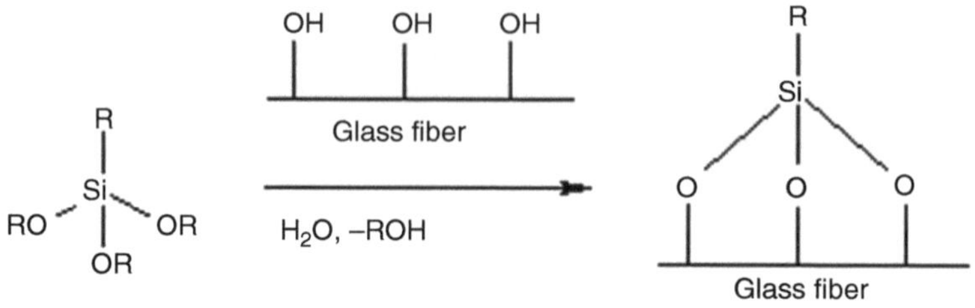

Fig. 13.10 Covalent bond of trialkoxysilanes to the surface of the glass fiber (R = functional)

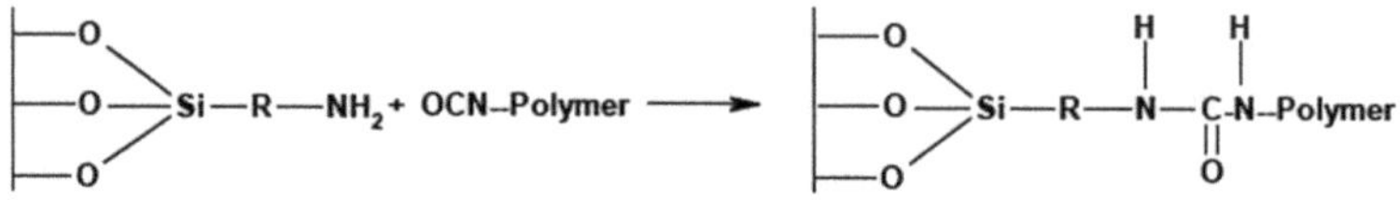

Fig. 13.11 Addition of anchored amino silanes to PU-prepolymers ($R = C_3H_6$)

Silanes containing amino groups are versatile. For instance, they can be added to isocyanat functions of polyurethane (PU) pre-polymers (Fig. 13.11).

The addition of amino-functionalized silanes to epoxy group containing epoxy resin (EP) can also be carried out easily (Fig. 13.12).

Silanes carrying epoxy functions are available for the addition of hydroxyl groups. Such hydroxyl groups are present, for example, in phenol resin or thermoplastic polyesters (Fig. 13.13).

Further alkoxysilane bonds carry functional groups with vinyl, acrylic or aliphatic functional structures. Vinyl silanes and acrylic silanes can react with unsaturated polymers and aliphatically modified silanes serve the adhesion with nonpolar matrix materials such as polyolefins [24].

13.2.5.4 Sizing for Carbon Fibers

Carbon fibers are provided with application-adapted sizings by the manufacturers [25]. For this purpose, mainly sizings containing epoxy (EP) and polyurethane are used, which are suitable for the processing of fiber materials in EP-resin or in thermoplastic, as phases with approximately identical surface energies come to meet with each other (see Sect. 13.3). However, it is also possible to increase the polarity of sized surfaces of carbon fibers using different methods (see Sect. 13.5.2), so that active interfaces can be obtained.

13.2.5.5 Other Sizing Materials

Generally, all textile yarns and the fabrics produced from them contain sizings in order to provide the mechanical and antistatic properties, allowing high-speed

Fig. 13.12 Addition of anchored amino silanes to EP-resin (R = C_3H_6)

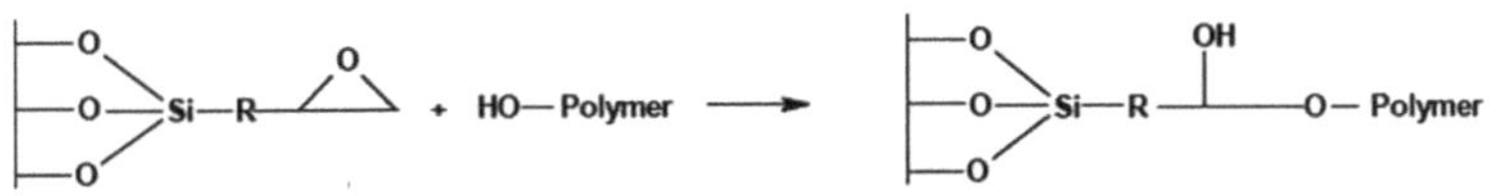

Fig. 13.13 Addition of anchored amino silanes to hydroxyl groups

processing. Mineral oil products, aqueous polymer dispersions such as polyacrylate or polyvinyl alcohols, and formulations based on natural materials are applied. As these sizing materials often exhibit very poor interaction potential and block the access to the fiber interface, they must be removed. This applies especially for sized aramid fibers [1, 26], which show reduced adhesion properties if applied without removal of the mentioned auxiliary agents.

13.2.6 *Chemistry of the Matrix Interface*

Thermoset, thermoplastic or inorganic material system (e.g. concrete) are used as matrix materials as well as coating agents depending on the requirements and application. The limits of application of fiber-reinforced composites or textile-based membranes are set by the choice of matrix or coating systems. Apart from the mechanical properties of the matrix, its suitability for chemical bonding to pretreated and finished fiber materials is of special interest.

13.2.6.1 Reactive Cross-Linking Matrix Resins

Reactive resin systems are low-molecular pre-polymer substances available for processing in liquid form [1]. This facilitates wetting of textile surfaces and offers possibilities for chemical bonding between fiber and matrix, as a wide range of reactive groups are available before the resin is cured. The most common thermoset resin systems are described briefly below:

Epoxy Resins

Epoxy resins (EP) have reactive centers which are developed as epoxy functions (Fig. 13.14). Hydroxyl and amino substituted molecules or structural elements can be added easily to these chemically active groups, which are used for the curing of

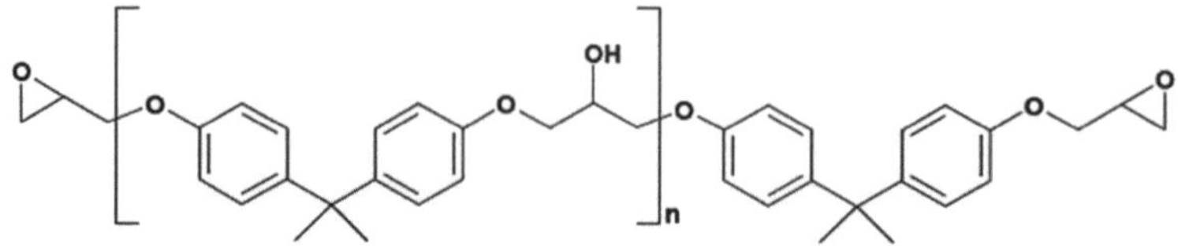

Fig. 13.14 Structural element of an epoxy resin in its uncured state, with reactive epoxy functionalities as end groups

the resin. Such an addition reaction is also possible with appropriate chemical structures offered at the fiber surface.

Vinyl Ester Resins

Vinyl ester resins are acrylic or methacrylic esters of epoxy resin. They have reactive C–C double bonds (Fig. 13.15), are dissolved in styrene and cured by radical polymerization. The vinyl functional groups anchored on the fiber can participate in this curing process and thus form a chemical bond with the matrix.

Vinyl Ester–Urethane Hybrid Resin

Vinyl ester urethane hybrid resins consist of long-chain diisocyanates in addition to the vinyl ester resin, which serve to cross-link over secondary OH-groups of the vinyl esters. The isocyanate functional groups are also suitable to create bonds with the fiber, if amino or hydroxyl groups are present there. As a result of the reaction urethane or urea structures are formed (Fig. 13.16).

Unsaturated Polyester Resins

Similar to vinyl ester resins, unsaturated polyester resins have reactive double bonds which can be cross-linked by styrene. Again, it is also possible to incorporate vinyl groups anchored in the fiber interface in the radical polymerization (Fig. 13.17).

Phenol Resins

Phenol resins are pre-polymeric condensed products from phenol and aldehydes, in which formaldehyde is used preferentially. They have reactive methylol groups and release formaldehyde during their curing. The methyl functional groups as well as free formaldehyde can serve the bonding between matrix and fiber when hydroxyl and amino functional groups are present from the side of the fiber (Fig. 13.18).

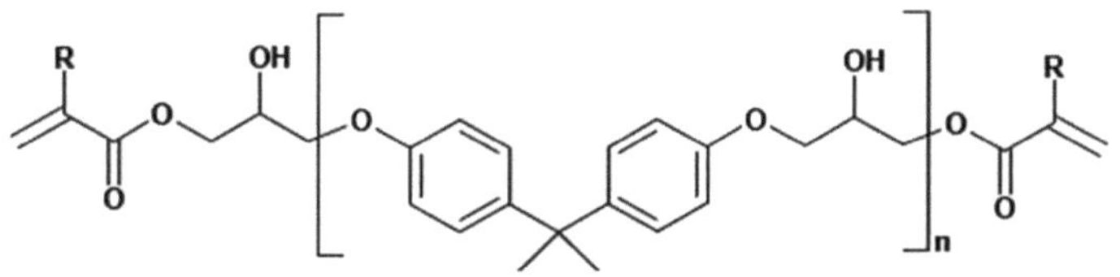

Fig. 13.15 Vinyl ester resin of pre-polymer with polymerizable vinyl functional groups (molecule edges on *left* and *right*, R = H, CH_3)

Fig. 13.16 Reactions of isocyanates with hydroxyl (*top*) and amino (*bottom*) functional groups on a fiber surface

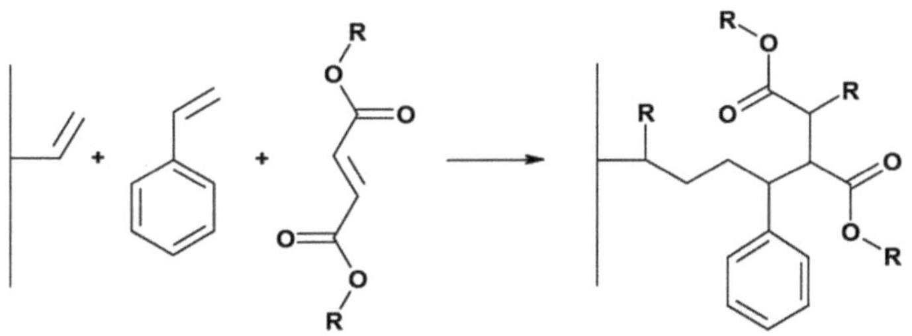

Fig. 13.17 Reaction between interface with vinyl functional groups, styrene, and unsaturated polyesters (R = polymer functional groups)

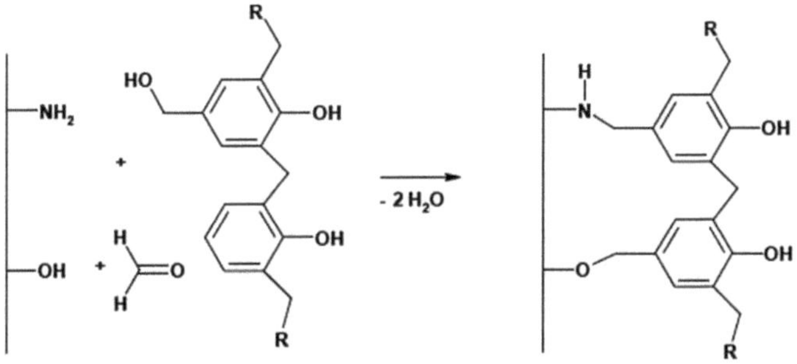

Fig. 13.18 Reaction possibility between amino and hydroxyl functional groups from the side of the fiber, formaldehyde, and phenol resin (R = polymer functional groups)

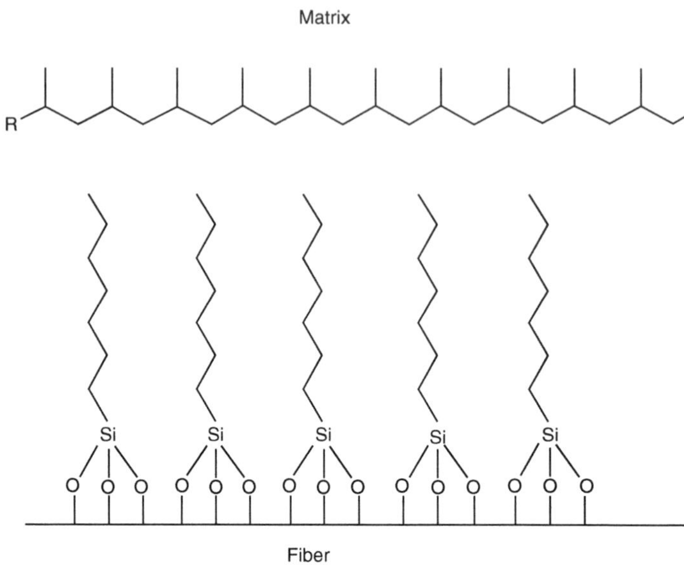

Fig. 13.19 Modification of glass fibers with aliphatic structures

13.2.6.2 Thermoplastic Matrices

A wide range of materials are available for the application of thermoplastic matrices [1]. Their application is realized by melting, distribution, and pressing. If the thermoplastic materials have no chemically reactive center and show relatively low surface energy, then the polar structured surface of reinforcement fibers can be assimilated to the matrix. This can be achieved by the application of alkoxysilanes with distinctive aliphatic structural elements on the glass fiber (see Fig. 13.19).

13.2.6.3 Mineral Matrices

In the last few years, textile reinforcement structures have found applications as reinforcements of concrete structures for the restoration and maintenance of civil constructions (see Sect. 16.4) due to their performance capability. The composite material used for this purpose is concrete, which is a mixture of cement, aggregates, and water. As the binder material, cement is composed mainly of calcium oxide (CaO), silicon dioxide (SiO_2), aluminium oxide (Al_2O_3) and ferric oxide (Fe_2O_3) [27], representing a very complex chemical system which reacts with water to form stable and insoluble bondings. As a result calcium silicate hydrates are formed, which give strength to the building material. The frictional coupling of the inorganic matrix with the textile reinforcement structure generally produced from the coated glass fiber and/or carbon fiber bundle, can occur only via a mechanical dovetailing and polar interactions. Good composite properties are achieved by fiber materials coated with sulfonated or carboxylated cross-linkable styrene-butadien copolymers [28].

13.2.6.4 Coating Materials

The most commonly used materials for the coating of textile reinforcement structures, such as polyvinyl chloride [29], polyurethane, polyacrylic ester, polytetrafluoroethylene, polysiloxane, polychloropropene, and natural rubber are explained in Sects. 13.5.3 and 16.5.

13.2.7 Topography of Textile Surfaces

Surface topography is another influencing factor for a good wetting and adhesion behavior of fibers. It deals with the height deviations of the actual boundary surface from the ideal average smooth boundary plane. As the reasons for adhesion, the mechanical dovetailing and chemical bonding are discussed [25, 30]. The contribution of mechanical interlocking for adhesion can be estimated. It definitely increases with the increase of roughness of the fiber surface. This roughness implicates an enlarged surface, which facilitates chemical interactions. Optimum mechanical properties of the composite can be achieved if the adhesion energy in the boundary surface between fiber surface and matrix exceeds the cohesion energy of the polymer matrix [31–35]. The fiber roughness plays a decisive part, considering the mechanical adhesion theory (see also Sect. 3.2.2.5) for the improvement of adhesion properties, which contributes to the improvement of bonding [36, 37].

13.3 Material Combination and Compatibility

For the evaluation of the compatibility of materials for composites, the physical and chemical properties of their interface should be considered, from which a conclusion can be drawn regarding the wettability and the adhesion properties.

13.3.1 Physical Compatibility

Generally, it is apparent that a solid surface with higher surface energy containing distinct polar parts exhibits good wetting properties. At the same time, it is also known that the wettability of a liquid phase, e.g. of a polymer melt, coating paste or reactive resin formulation, increases with the decrease of its surface tension. This means that solid phases (fiber materials) with higher surface energy (σ_s) and liquid phases (melts, pastes, resins) with lower surface tension (σ_l) are advantageous for a good wetting of textile reinforcement structure with polymer matrices during the production of composites for lightweight construction. This can be formulated as follows Eq. (13.2):

$$\sigma s > \sigma l \tag{13.2}$$

The work of adhesion (W_A), which is used for the separation of two phases (with surface energies σ_1 and σ_2) from each other, is described by DUPRÉ in Eq. (13.3).

$$WA = \sigma_1 + \sigma_2 - \sigma_{12} \tag{13.3}$$

Equation (13.1) by OWENS and WENDT described in Sect. 13.2.2 is applicable for the calculation of the interface energy from the surface energies of individual components of a material combination, such as fiber and matrix. An inclusion of this concept in the DUPRÉ equation (13.3), yields the following equation [38]:

$$W_A = \sigma_1 + \sigma_2 - \sigma_1 - \sigma_2 + 2\left(\sqrt{\sigma_1^D \times \sigma_2^D} + \sqrt{\sigma_1^P \times \sigma_2^P}\right) \tag{13.4}$$

As is apparent from Eq. (13.4), the maximum possible value for the work of adhesion W_A is obtained in the disperse and polar parts in case of exactly identical surface energies of the partners. Thus, the material combinations should be selected in such a way that all materials have similar characteristics in their surface energy and also exhibit good wetting behavior. The relationship is illustrated schematically in Fig. 13.20.

The values given in tables can be used to estimate compatibility of material combinations regarding their surface energies. A small selection of relevant data is given in Table 13.1. For common textile materials, the specific values are available in tables. However, they can deviate significantly from existing real textile materials. On the one hand, this is due to the fact that generally copolymers and various additive mixtures rather than pure material systems are used for the production of textile yarns and threads. On the other hand, textile surfaces are provided with sizings and spin finishes, which cause a significant change in the interface characteristics compared to desized raw materials. Furthermore, it is also possible to change the surface energy qualitatively and quantitatively by chemical-physical treatments and finishing processes as well as through functional coating (see Sect. 13.5.2). Thus, an increase of compatibility of a given fiber-matrix system can be achieved. For the determination of conditions prevailing in the material, direct measurements are advantageous. The methods applied for this purpose are discussed in detail in Sect. 13.4.

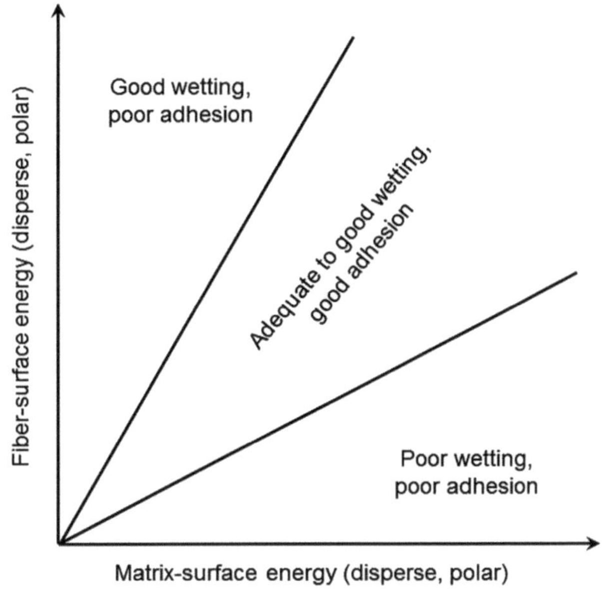

Fig. 13.20 Qualitative relationship between surface energies, wetting behavior, and adhesion properties of fiber-matrix systems

Table 13.1 Surface energies of selected materials [39]

	Surface energy (mN/m)		
Material	Total	Disperse	Polar
Polyethylene	35.1		
Polypropylene	31.2	30.5	0.7
Polyamide	40.5	33.7	6.8
Aramid (desized)	30.8	10.3	20.5
Polyester (PET)	44.0	43.0	1.0
Glass (desized)	73.3	29.4	43.9
Glass (sized)	30.2	2.2	18.0
Carbon	34.8	33.0	1.8
Carbon (sized)	22.2	16.2	6.0
Basalt (sized)	39.2	19.7	19.5
Steel	34.4	34.0	0.4
Water	72.8	51.0	21.8
SBR dispersion	22.2	16.3	5.9
PU resin	43	Not available	Not available
PMMA resin	45.8	39.2	6.6
Epoxy resin	36	Not available	Not available
Phenol resin	42	Not available	Not available
Polyester resin	41	Not available	Not available

13.3.2 Chemical Compatibility

If the textile fibers possess reactive groups in their surface structures, these can serve as the binding points for the covalent bonds with reactive matrix materials. Basically, three different types of reactions take place here, which can lead to the following combinations [40].

13.3.2.1 Addition Reactions

Fiber materials with amino (–NH_2) and hydroxyl (–OH) functional groups are readily available for the addition reaction with epoxy groups containing epoxy resin and isocyanato (–NCO) groups of pre-polymer polyurethanes. If the textile components have epoxy or isocyanate functionalities, then matrix systems with amino and hydroxyl groups can be added.

13.3.2.2 Condensation Reaction

Condensation reactions are only possible in interfaces of material combinations when amino and carboxyl (–COOH) functional groups are available alternatively, which form an amide bond after secession.

13.3.2.3 Polymerization Reactions

With the availability of carbon-carbon double bonds (π bonds) in the phase boundaries of fiber and matrix, polymerizations can take place. The material combinations can include a fiber modified by finishing and an unsaturated polyester resin or polymethacrylic resin.

In most cases the aforementioned functional groups at the fiber surface required for the chemical bonding must be developed by direct surface modification by means of functional sizing and coating as well as adhesion promoters. Processes and methods of textile finishing are suitable for this purpose (see Sect. 13.5).

13.4 Experimental Determination of Physical and Chemical Characteristics of Interfaces

The characteristics of surface properties of materials are the result of their chemical and physical structuring at micro, meso and macro levels. Starting from the atomic composition of macro molecules, to their arrangement in polymer and the formation of phase boundaries (surface), there are a number of structural elements contributing to the interface phenomenon. Under these circumstances, it is favorable to represent the different influencing factors in a brief description. Such a representation is based on the knowledge of the surface energy of textile materials, by which the interaction potential of the surface is described in numerical values. These values describe the wetting behaviors, the physical/chemical adhesion capability, and thus the compatibility with given matrix materials. The determination of this parameter is realized by means of the measurements of liquid contact angels directly at or on the surface of the material to be considered in fabric and as yarn form [41]. Since the determined values for surface energy, subdivided into disperse and polar part, represent the physically and chemically active structural elements as a whole, it can be useful to test the chemical functions separately.

13.4.1 Investigations on Surface Energy of Textile Materials

The surface energy of solids can be determined horizontally in a plane surface or in a vertically hanging form (e.g. as fiber). In the simplest case, so-called test inks can be used, which enables only an estimation of the total energy. However, this is often sufficient, particularly when it concerns the testing of finishing effects of a treated material. Exact information can be gathered by the application of a contact angle measurement device (for the analysis of a plane surface) and a tensiometer (for the perpendicular measurement of a specimen).

13.4.1.1 Test Ink

Test ink represents a simple and cheap means for the testing of surface energy of solid materials. Inks are colored liquid mixtures with a precisely defined surface tension. The liquids, whose composition is standardized according to DIN ISO 8296, are available in the range from 30 to 72 mN/m with an adjustment of 2 mN/m [42]. They are applied directly to the surface, and wetting or lack thereof is assessed immediately. The value of the test ink inducing a stable wetting over a period of time corresponds to the surface energy of the specimen under consideration. Matrix materials to be evaluated can be formed into a film in order to be subjected to the test. The testing of individual fibers requires some skill but can be performed with the help of a light microscope. Test inks are particularly suitable for the testing of a successful chemical and physical treatment to increase interface activity.

13.4.1.2 Contact Angle Analysis of a Plane Surface

The contact angle measurement for the calculation of surface energy differentiated into disperse and polar parts (see Sect. 13.2.1) of solid plane surfaces is performed by means of devices which consist of an illuminated specimen table, a dropper

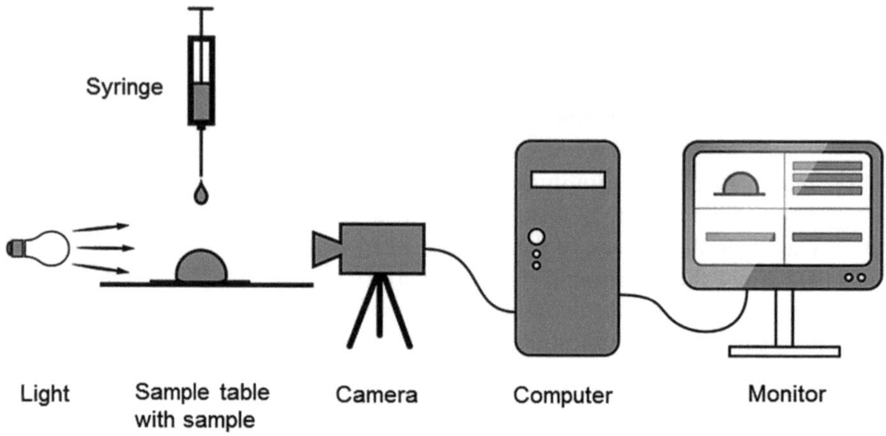

Fig. 13.21 Schematic set-up of a contact angle measurement system

(syringe and needle) for the liquid measurement, a camera, and a computer with image processing and calculation programs (Fig. 13.21).

On the surface, different test fluids are applied as drops and their contours are transferred to the image processing program using the camera. The program is able to measure the contact angle between the specimen surface and the drop of fluid. The measured contact angle serves the program as the basis for the calculation of surface energy values, which should subsequently consider the method of OWENS, WENDT, RABEL and KAELBLE [43]. The determination of the contact angle is carried out through tangents applied to the drop contours at the contact point of liquid/solid (see Fig. 13.22). The value used for the calculation is the average of individual values measures on the left-hand (Θ_l) and right-hand (Θ_r) sides.

For the measurement of contact angles in practice, minimum two liquids are used, whose surface tensions in the disperse and polar part are known. Furthermore, the liquids must exhibit clear differences in their properties, i.e. a polar and a nonpolar test liquid should be applied. One suitable liquid pair is consisting of water and diiodomethane. The values of this system for the surface tension are given in Table 13.2.

The method for the determination of surface energy described above is based on the fundamentals of the relationship between the interfacial tension at a point of a 3-phase contact line (Fig. 13.23), formulated by YOUNG in 1805.

The surface tensions of the two condensed phases are described by σ_s and σ_l, where s and l represent solid and liquid respectively. The interfacial tension between these two phases is designated by σ_{sl}, and Θ is the contact angle, which corresponds to the angle between the vectors σl and σ_{sl} [41]. The relationship between these values is designated as Young's equation

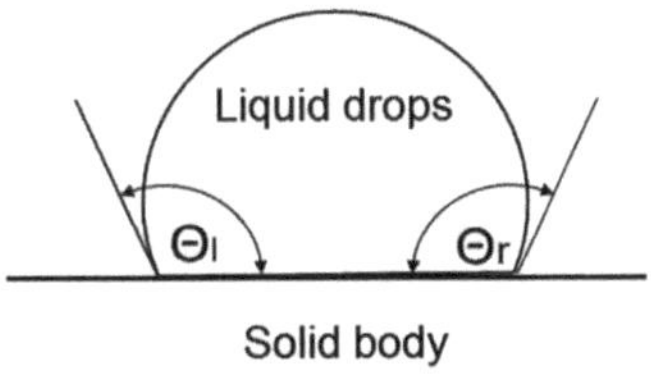

Fig. 13.22 *Left* (Θ_l) and *right* (Θ_r) side contact angle of a drop of liquid on a solid surface under the application of tangents [39] (*left*), video image of a drop on a material surface (*right*)

Table 13.2 Surface energy of test liquids [39] in mN/m

Liquid	σ_s	σ_l^D	σ_l^P
Water (H_20)	72.8	21.8	51.0
Diiodomethane (CH_2I_2)	50.8	50.8	0.0

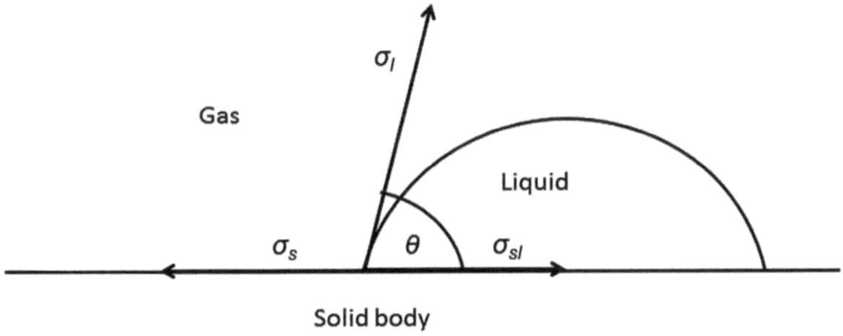

Fig. 13.23 Contact angle and interface tensions on a wetted surface, according to YOUNG

$$\sigma s = \sigma sl + \sigma l \times \cos \theta \tag{13.5}$$

The combination of the equation of interfacial tension, formulated by OWENS and WENDT [see Eq. (13.1), Sect. 13.2.2], with the Young's equation

$$\sigma_{sl} = \sigma_s + \sigma_l - 2\left(\sqrt{\sigma_s^D \times \sigma_l^D} + \sqrt{\sigma_s^P \times \sigma_l^P}\right) \tag{13.6}$$

leads to a general linear equation according to the work of KAELBLE and RABEL:

$$y = mx + b \tag{13.7}$$

which becomes clear in the following form:

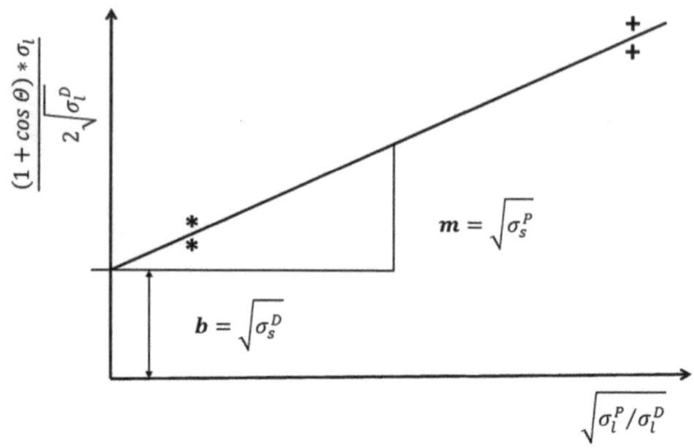

Fig. 13.24 Determination of disperse and polar parts of surface energy of a solid according to RABEL (*: measurement with fluid 1, +: measurement with fluid 2)

$$\frac{(1+\cos\theta)\times\sigma_l}{2\sqrt{\sigma_l^D}} = \sqrt{\sigma_s^P}\times\sqrt{\frac{\sigma_l^P}{\sigma_l^D}} + \sqrt{\sigma_s^D} \tag{13.8}$$

Equation 13.8 allows the calculation of the surface energy of a solid with the help of measured contact angle of two liquids with known surface tension. For visualization, this is shown in Fig. 13.24.

With the aforementioned method for the measurement of contact angle and the calculated surface energy, it is possible to characterize plane surfaces such as e.g. matrix films, dense woven fabrics (treated and untreated), coatings and coarse monofilaments (in the presence of a good measuring device) regarding their potential for boundary layer interactions.

13.4.1.3 Determination of Surface Energy of Fibers and Filaments

The surface energy of fiber and filaments with diameters of minimum 5 μm can be determined by a tensiometer. Such a tensiometer is equipped with a vertically movable table to accommodate the test fluids, and a sample holder with a highly sensitive fine scale (Fig. 13.25). The data acquired by the measurements are transferred to the computer, which then calculates the contact angle and the surface energy according to Eqs. (13.8) and (13.9) respectively.

Apart from the characterization of fiber surfaces with the help of test fluids, it is also possible to determine the wetting angle of a given textile fiber with matrix or coating materials with the described test setup. The measurement principle follows the WILHELMY methods [41], where the solid body (fiber) is brought into contact with the liquid, which exerts a force on it (Fig. 13.26).

This Wilhelmy force (F) is measured and the contact angle Θ is calculated from it.

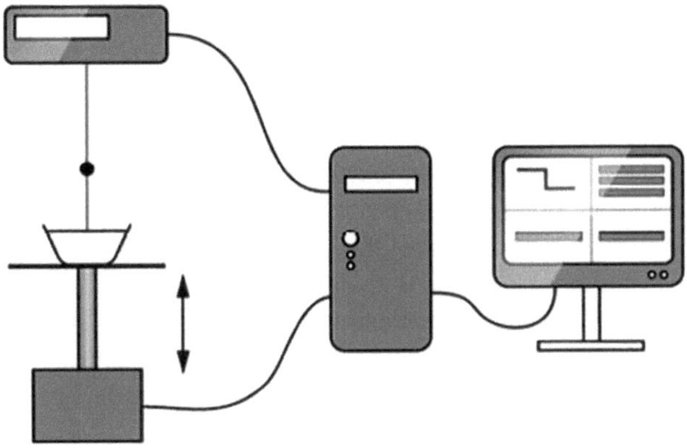

Fig. 13.25 Schematic representation of a single fiber tensiometer

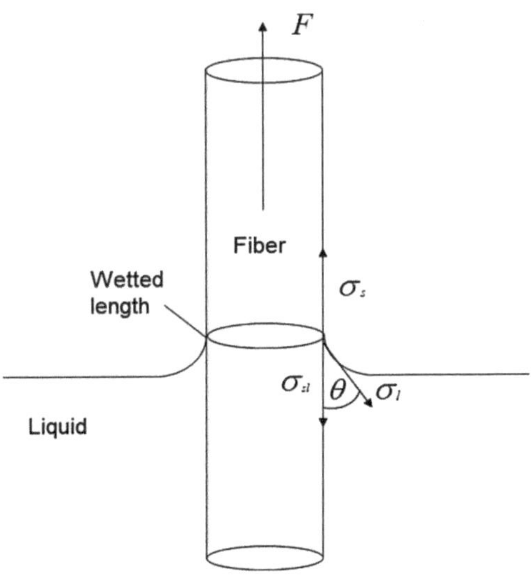

Fig. 13.26 Wetting of single fiber in single fiber tensiometer

$$\cos\theta = \frac{F}{\sigma_l \times l} \quad (13.9)$$

F (N)	measured Wilhelmy force
l (m)	wetted length
σ_l (N/m)	surface tension of the test fluid

If two different test fluids are used, such as water and diiodomethane, then the surface energy can determined in a further calculation using Eq. 13.8 developed by OWENS, WENDT, RABEL, and KAELBLE.

13.4.2 Investigations of the Chemistry of the Surfaces of Textile Materials

Apart from the determination of physical/chemical surface parameters, the chemical characterization of textile materials can be useful in order to directly detect the active chemically reactive centers in the boundary layer of fibers. This is very important to evaluate the effectiveness of methods and processes after the pretreatment of textile materials for their activation and functionalization. The required investigations can be performed with laboratory-based instrumental analytical equipment, such as infrared spectrometer with surface measurement units (IR-ATR) or with X-ray photoelectron spectroscopy. They can also be performed with simple methods.

13.4.2.1 Instrumental Chemical Analytics

From the instrumental analysis methods used for the characterizations, two methods will briefly be discussed. They are suitable for the representation of chemical characteristics of textile surfaces. For a detail description, relevant text books can be referred to [44, 45].

Infrared Spectroscopy (IR) Investigations

In the study of solid, liquid or gaseous substances by means of infrared spectroscopy, the absorption spectrum in medium infrared range (MIR) of the electromagnetic spectrum is obtained. With the obtained absorption signals (absorption bands), the molecular substructures can be designated based on the involved atoms and their bonding characteristics. The atoms present in the molecule are irradiated by electro-magnetic waves in IR-range, depending on the type and bonding of atoms. As a result, the atoms are stimulated through the absorption of defined wave lengths (=energies) to valence (ν) and deformation vibration (δ). The relevant wave length range is 2.5–25 μm, which is typically indicated in wave number per centimeter. This corresponds to wave numbers between 400 and 4,000 cm^{-1}. For the spectroscopic inspection of surfaces like those of textile fiber materials or fabrics, the IR-spectrometer must feature a suitable surface measuring device working according to the method of attenuated total reflection (ATR).

The information documented by an IR-spectrometer are mainly of qualitative nature and enable statements regarding the presence of functional groups (e.g. –NH_2, –OH, –COOH, –NCO, –CN), the existence of particular bonds (e.g. amide, urethane, ester, Si–O and other bonds), and the type of hydrocarbon (aliphatic, vinyl or aromatic) or carbon-carbon bonds (single, double or triple bonds). The evaluation of observed signals in a measured spectrum is done over reference spectra of

Table 13.3 Vibration data of important molecule groups

Molecular group	Name	Wave number (cm^{-1})
=NH	Imino	3,100–3,500
$-NH_2$	Amino	3,200–3,400
=C=O	Carbonyl	1,700
–C≡N	Cyano	2,200–2,260
–C=C–	Vinyl	1,650
–OH	Hydroxyl	2,500–3,000
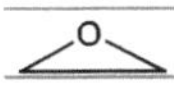	Epoxy	840–900
–COOH	Carboxyl	1,200–1,400
–NCO	Isocyanato	2,250–2,300

known structures and table data. A large number of text books are available on this subject [45]. Some exemplary data is given here (see Table 13.3).

Investigations Through X-ray Photoelectron Spectroscopy

X-ray photoelectron spectroscopy, also known as ESCA (*Electron Spectroscopy for Chemical Analysis*), is a strongly surface-sensitive analysis method. Using energetically defined X-rays, the photoelectrons are knocked out of the specimen interface, and can be detected by means of an electron analyzer [46]. Depending on the bonding energy of the photoelectrons and intensity of the resulting photoelectron signals, it is possible to recognize the existing elements, grouping of atoms, and structural components as well as their relative mass fraction. For the characterization of textile material surfaces, it can be applied in number of ways: for the proof of usable functionalities such as hydroxyl, carboxyl or amino groups, segregated substances, e.g. modified silica layers, or for the detection of repelling substituents containing fluorine or chlorine. For detailed information on the analysis methods, text books can be referred to [45, 47, 48].

13.4.2.2 Wet Chemical Dyeing

Of the different ways of wet-chemical analysis of active groups, such as by means of acid-base and redox titrations or specific chemical reactions and derivative synthesis, the dyeability test can be emphasized here, because it enables the functionality of textile surfaces to be made visible. To perform this test, acidic and basic dyes are required. Detailed investigations on the dyeability reactions and practical references are illustrated in [49].

Dyeing with Acid Dyes

Acid dyes like e.g. Acid Blue 83, enter electrostatic interactions with protonated amino functional groups on the surface of textile materials and are bonded permanently [49]. Here, a dyeing of surfaces takes place, whose regularity and intensity can be evaluated visually or colorimetrically (see Fig. 13.27). This practical application is very simple: the specimens are dipped into an acetic dye bath for 30 min where they are swirled occasionally. It is important to clean the textile surface so that only the bonded dyes remain on the surface. Since the bonding of the dyes is reversible under alkaline conditions, it is possible to measure bonded dye molecules quantitatively and hence the available amino groups. For this purpose, the concentration of detached dyes in the decoloration solution must be determined photometrically.

This dyeing technique is very suitable to obtain very fast information about the effectiveness of a textile pretreatment or finishing process. The other uncomplicated and qualitative method is the wetting with a colorless ninhydrin solution. This reagent frequently used in bio chemistry results in clear deep violet coloration in the presence of $-NH_2$ groups (Fig. 13.28).

Dyeing with Acid Dyes

Basic dyes possess a positive charge in the molecule and are suitable to form an ionic bond with negatively charged groups on the fiber surface. This can be used to visually verify functionalities like carboxylic acid or deprotonated hydroxyl groups.

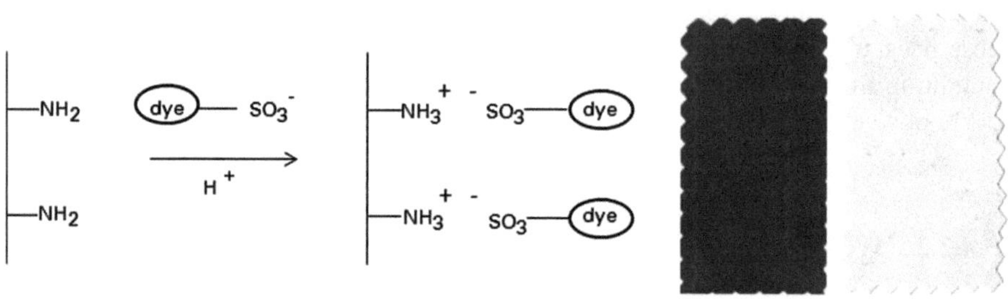

Fig. 13.27 Schematic illustration of bonding of a dye to the amino functional groups of a finished PET woven fabric (*left*), dyeing results for functionalized polyester and base material (*right*)

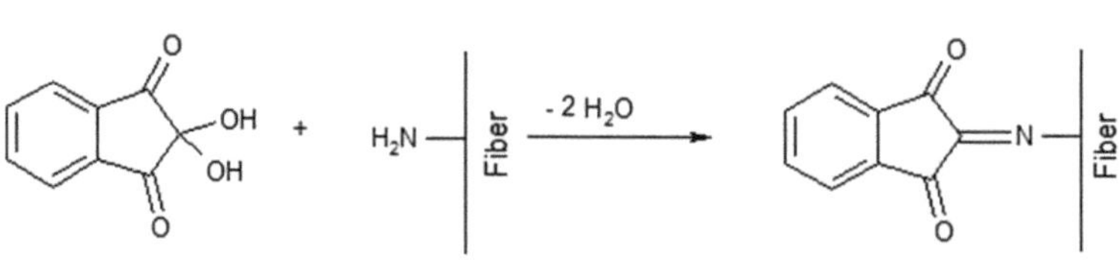

Fig. 13.28 Reaction of ninhydrin with amino-functionalized fibers (in simple form) [12]

In practice it is used for instance for quick the evaluation of oxidative treatments of textile structures, preferably with methyl blue dye [13]. This dye can be removed from the textile under an acidic condition and be quantified.

Further Dyeability Test for the Verification of Reactive Groups on Fiber Surfaces

Dyes can be added to the textile boundary layers, which after functionalization or finishing exhibit epoxy or isocyanate groups with free amino functionalities (Fig. 13.29). For this, small molecules from dispersion dyes are suitable, which do not have any other reactive centers. The release of the bonded dyes from the textile surface is not easily realizable. During the dyeability test, a blank sample without functionalization must be subjected to the same test conditions in order to eliminate any chance of misinterpretations.

13.5 Finishing Processes, Methods, and Technologies

The surface activation, functionalization, and finishing of textile materials can be performed in all steps of textile process chain. Different processes and technologies are applied depending on whether a textile fabric needs to be activated by a corona before a coating or a special functional yarn to be processed into a fabric surface. The processes, methods, and technologies which can be applied for this purpose and how they can be used will be discussed in this section. Special attention will be paid to the pre-treatment and finishing of textile fibers, yarns, and fabrics specified for use in lightweight construction and as technical textiles.

13.5.1 Spinning-Integrated Finishing of Fibers

The application-oriented finishing of synthetic fibers can be performed during their manufacturing stage in order to realize different additional functions beyond the

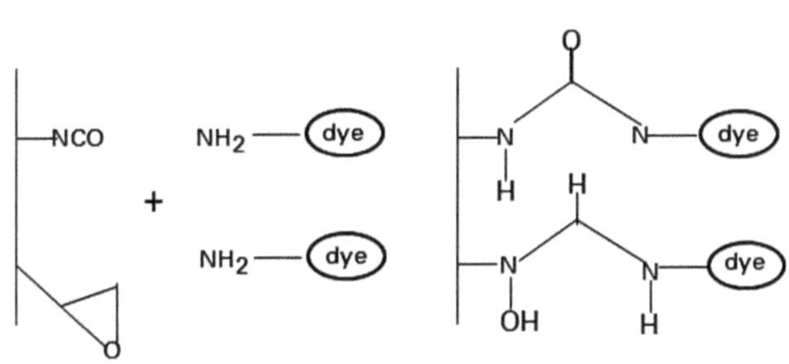

Fig. 13.29 Addition of dyes to isocyanate and epoxy functional groups

original material characteristics, which enables the development of functional groups, especially in the fiber interface. This can be done by the incorporation of any required active substances (additives) by homogeneous mixing in the spin mass before or during the spinning or on the other hand can be applied on the fiber surface after the fiber formation.

The direct incorporation of additives (internal additives) is usually performed by means of mixing in the form of a master batch [50] and this leads to a homogenous distribution of effective substances in fiber material [51]. However, additives active in the marginal layers of the fiber are only partially utilized. This drawback can be overcome by the application of functional molecules capable to migrate, since they concentrate at the periphery of fiber. As a result a small amount of them is required. Additives accumulating at the fiber surface for the finishing of textiles made of polyamide are a good example of this [52].

The external application of additives is usually performed simultaneously during the sizing or finishing, whose primary function is to ensure the textile processability of fibers. Through the appropriate composition and formulation, it is possible to add functions beyond the mechanical protection of fibers, e.g. for good adhesion properties (see Sect. 13.2.5), antistatic properties, or UV-protection [53]. The application of carbon nanotubes (CNT) is particularly interesting, as they can be mixed in portions with sizing and preparations. Thus, electro-conductivity can be imparted to fiber materials by finishing and can be adjusted selectively by the concentration of CNT [54]. Furthermore, due to their nano-scale dimension and enormous strength, the CNT are able to contribute to defect repair at the fiber surface and to increase the load-bearing capacity of their interface, thus increasing fracture energy [55]. For the application of CNT containing sizing, it has been shown that the CNT at the interface of fibers influences not only the breaking properties, but also causes a characteristic transcrystallinity. With the application of a very little amount of CNT, the tensile strength perpendicular to the fiber axis and compression shear strength can be increased by 10 %. With a good dispersion of CNTs, it is possible to achieve the percolation threshold of a system with a low amount of CNT, allowing the development of an interface sensor [56]. CNT-based interface sensors can be used in real composite components and also for different sizing systems in glass fiber-based composites in order to identify different interface damages.

13.5.2 Pretreatment of Textile Yarns and Fabrics in the Finishing Process

The pretreatment of yarns and textile fabrics serves the purpose of preparation with regards to their further processing steps such as finishing, coating, or application in composites. In many cases, the removal of spin preparations or sizes is one of the first tasks, as they hamper the further processing. Sizings showing functionality or compatibility apart from their good mechanical processability and being applied to glass and carbon fibers can be exceptions to this rule (see Sect. 13.2.5). For aramid fiber materials, it is necessary to remove the sizing to improve the surface properties [1]. This removal of preparations and sizing is ensured by separate washing with appropriate detergents or in the course of further pretreatment measures. This process can be divided as follows:

- wet chemical pretreatment
- pretreatment by means of corona/plasma, and
- gas phase treatment with a mix of fluorine/air

With all the mentioned processes, an activation or functionalization of textile surfaces is attained. The possibilities range from etching the material boundary the creation of cavities and roughnesses, over the realization of polarity, and the preservation of reactive groups at the surface. To perform wet chemical operations, the original machines and devices of textile finishing are suitable such as winch, jet or high temperature (HT) appliances, while special processing plants are required for corona/plasma and fluorination.

13.5.2.1 Wet Chemical Pretreatment of Textile Materials

Four of the many kinds of wet chemical pretreatments will be described below. These include the alkali treatment of aromatic/aliphatic polyester fibers from polyethylene terephthalate (PET), the biocatalytic pretreatment of polyamide (PA) fibers, the oxidation of carbon fibers, HT processing of fibers from polypropylene (PP) and high strength polyethylene (UHMWPE).

Activation of Polyethylene Terephthalate by Means of Alkali Treatment

PET is a hydrophobic material with a relatively low surface energy (see Table 13.1), especially in its polar parts. It does not have any directly utilizable chemical functions and can therefore be dyed only with dispersion dyes, using the HT method [57]. However, the ester groups typically found in these polymers are susceptible to alkali hydrolysis (saponification) and amines, which result in aminolysis. For the activation of the PET interfaces, only saponification is considered as it easier to control than aminolysis. By adjusting alkali concentration, temperature, and time, the alkali treatment can be controlled in a way that minimizes material and strength loss. As a result, fiber surfaces with higher surface energies, hole-like dents (cavities) as well as a higher roughness are attained. By this, the capacity of the interaction property of polyester fiber increases, which improves wetting and adhesion properties of the material. This type of pretreatment allows the realization of a permanently anchored SiO_2 and functionalized SiO_2 layers on the PET [59] as well as a successful metallization of such surfaces [60] by the application of wet chemical methods. PET fibers are shown exemplarily before and after alkali treatment in Fig. 13.30.

An exact view of the treated surface is provided by from Fig. 13.31, where almost no differences in micro-roughness are found.

Bio-catalytic Surface Treatment of Aliphatic Polyamide

The enzyme treatment is a very elegant and environmentally friendly pretreatment method to increase the boundary layer activity of polyamides [20]. Here, a significant increase of surface energy of circa 50 % is observed.

Oxidation of Carbon Fibers with Nitric Acid

Oxidative treatments are often applied to carbon fibers immediately [61] after their manufacturing and before their sizing (see Sect. 13.2.5). These sizings are capable

Fig. 13.30 PET fibers with 5,000-fold magnification (*left*: untreated, *middle*: lightly treated, *right*: strongly treated with alkali)

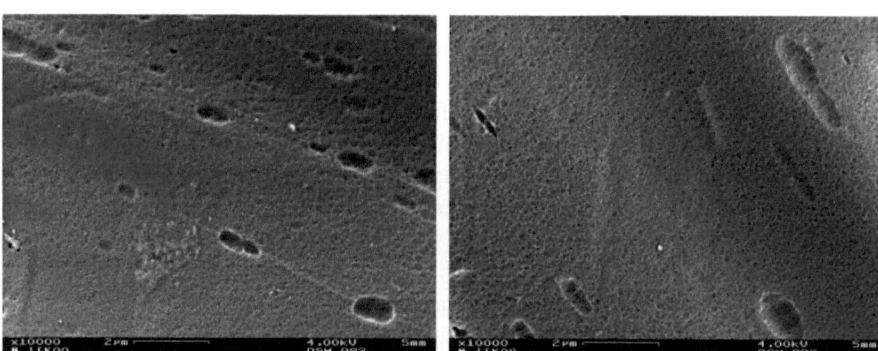

Fig. 13.31 Surface structure of lightly (*left*) and strongly (*right*) alkalized PET fibers (10,000-fold magnification)

of forming permanent bonding with matrix or coating materials of the same chemical structure. The measurable surface energies of individual filaments fall into the range of 22 mN/m where polar parts of 6 mN/m can be observed [20]. By means of a treatment with strongly oxidizing nitric acid (HNO_3), these values can be increased to about 40 mN/m or 18 mN/m respectively, which results in a considerably more active fiber surface, especially regarding the polar properties. This increase in polarity leads to higher wettability of the thousands of individual filaments in a carbon fiber bundle (roving) and thus facilitates their impregnation with applied matrices or coating materials. As a result, higher composite strength can be expected. As a side-effect of the treatment of carbon fibers with HNO_3, significantly higher values for the strength of individual filaments have been observed [20].

Wet Chemical Functionalization of Fibers from PP and High-Strength PE

Textile materials from aliphatic PP and PE polymers possess good to excellent textile-physical properties. However, these materials with a pure hydrocarbon structure (see Sect. 13.2.4) do not have any chemically and physically utilizable molecular element, so that these can hardly be applied in composites. These materials can also be activated oxidatively, which will be discussed later on (see Sects. 13.5.2.2 and 13.5.2.3). At this point, a method is needed that can successfully be realized with relatively simple means, i.e. with common machines and devices of textile finishing. This is done by adding reactive amino groups to polymer surfaces, anchoring them physically/mechanically in the base polymer. This method relies on a conventional process of textile finishing, which is applied for the dyeing of hydrophobic fibers with disperses dyes [57]. Instead of color-producing structures, long aliphatic amines with their hydrophobic molecule part are anchored in the fiber surface, so that free, reactive amino functional groups are established at the boundary surface of the treated fibers (see Fig. 13.32).

This methods of finishing of polyolefine work without problem for PP [12] and UHMWPE [13] and produce high reactive fiber interfaces, which avails themselves for further functionalization or direct application in composites. The amino-functionalities can be proved very easily (see Sect. 13.4.2) and can be used diversely. An example here is the covalent bonding of nano and sub micro particles, which occur with reactive epoxy groups. One part of these chemically active groups serve to anchor on the fiber surface (see Figs. 13.33 and 13.34), while the rest can be applied to create bonds to matrices for the adjustment of defined adhesive forces or further highly functional finishing such as the deposition of catalytic precious metal nanoparticles [62].

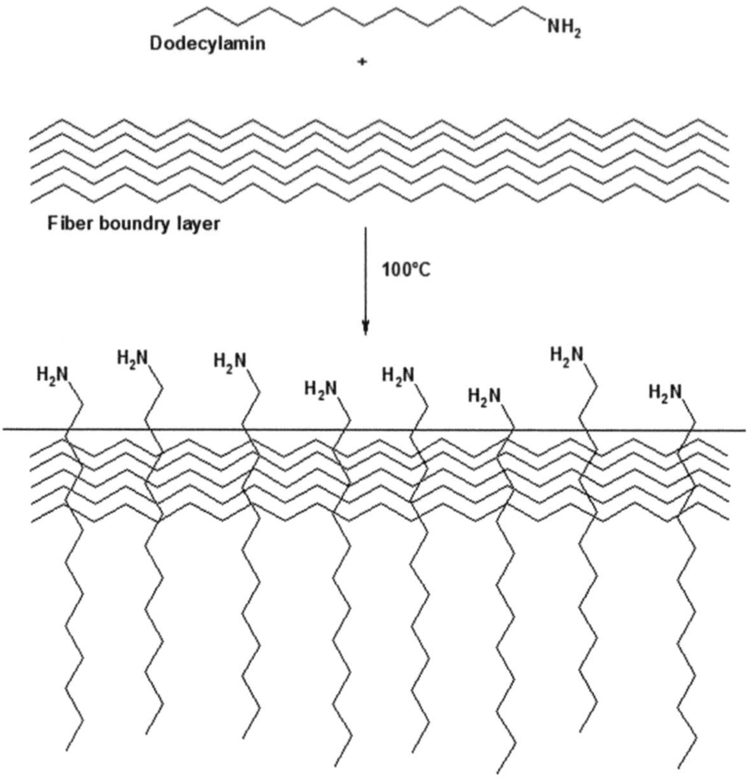

Fig. 13.32 Schematic representation of functionalization of PE (*left*) and PP (*right*) with physically anchored aliphatic amine

13.5.2.2 Pretreatment of Textile Interfaces by Means of Corona/Plasma Technology

A very effective pretreatment of textile materials can be performed by treatment plants generating generate cold plasma under atmospheric pressure on the basis of electric barrier discharge (DBD) [62]. In the simplest case, this can be used for the activation of different textile materials such as polyolefine, polyester, polyamide, or carbon fibers. This is carried out to increase the surface energy and to find possibilities for the chemical modification of the fiber surface. The changes in surfaces caused by the reactions of material interfaces with atmospheric oxygen lead to the formation of polar groups. By the application of gases such as nitrogen (N_2), nitrous oxide (N_2O), carbon monoxide (CO), carbon dioxide (CO_2), or low molecular organic bonding, defined functionalized surfaces are feasible. If a treatment plant with an atomizing device (Fig. 13.35) is available to finely blend the

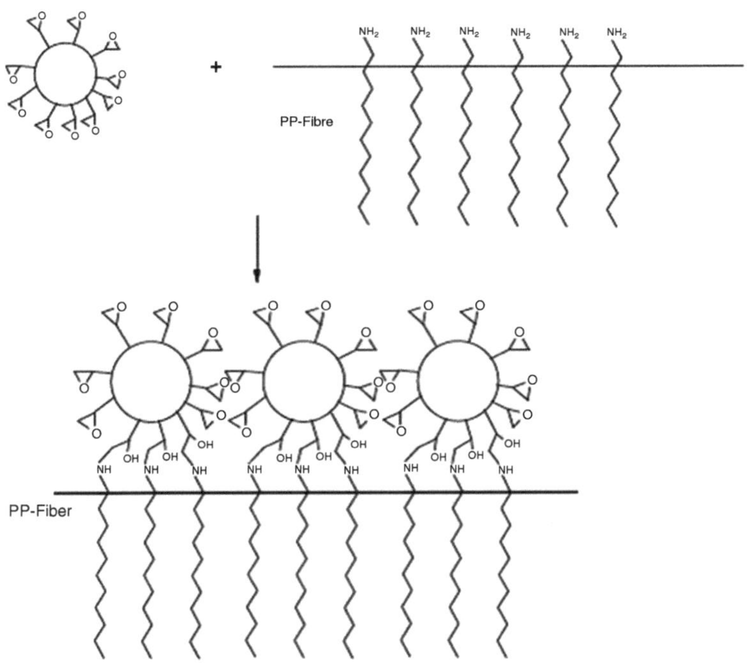

Fig. 13.33 Schematic illustration of the bonding of nano and sub-micro particles on an amino-functionalized PP fiber surface

Fig. 13.34 Reactive sub-micro particles anchored on PP fibers in 1,000-fold (*left*), 5,000-fold (*middle*) and 20,000-fold (*right*) magnification

liquid substance mixture with the plasma, further surface modifications are achievable (Fig. 13.36).

The pretreatment or functionalization of textile materials through corona/plasma processing at atmospheric pressure requires a double pass to achieve the required finishing effect on both sides.

The results of the treatment for the processing of a UHMWPE woven fabric under pure atmospheric condition are shown in Fig. 13.37. A roughening on the micro level can be clearly recognized.

Fig. 13.35 Corona/plasma unit by Ahlbrandt, with gas input and atomizing system for pretreatment of textile materials under atmospheric pressure

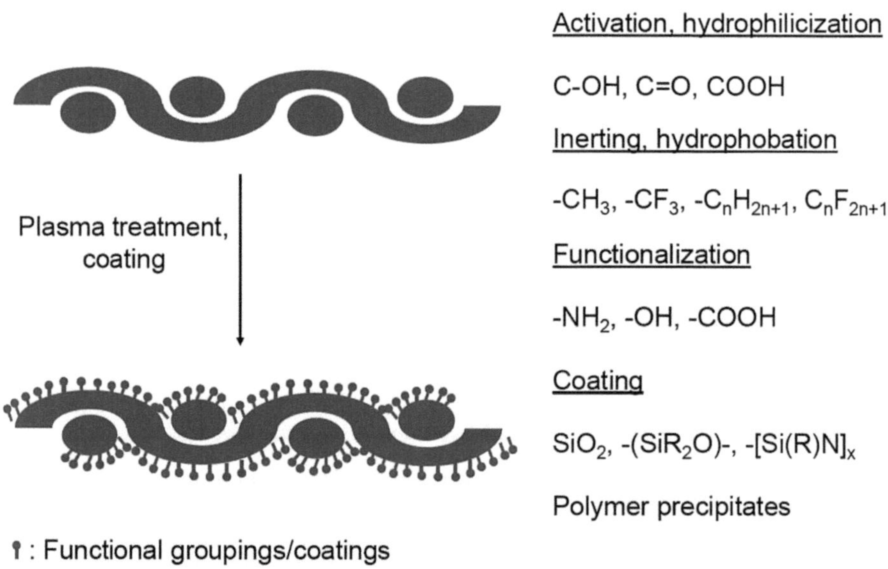

Fig. 13.36 Activation and functionalization of textile materials through DBD generated atmospheric pressure plasma

13.5.2.3 Pretreatment of Fibers, Yarns, and Fabrics in Gas Phase by Means of Fluorine/Air Mixtures

Another very elegant and extremely effective approach to the pretreatment or post-treatment of textile materials in all types of forms is the gas phase treatment by means of fluorine. The reactive substances are introduced for the modification of

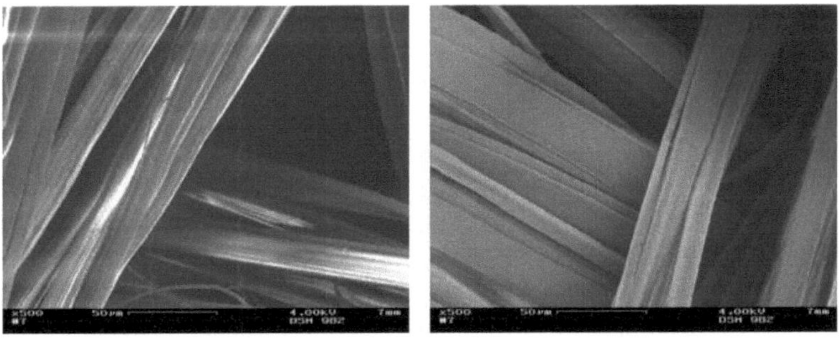

Fig. 13.37 UHMWPE woven fabric, untreated (*left*) and treated with DBD air plasma (*right*)

Fig. 13.38 Fluorine treatment in continuous processing (*Source*: Fluor Technik System GmbH)

fiber surfaces as well as their boundary layers in a gaseous aggregate state as individual molecules. This guarantees the coverage of the entire fiber sheath without any shadowing effect and allows a quick implementation due to high diffusion speeds occurring in the gas. Continuous as well as discontinuous processes are applied for this purpose, where the inline process seems most interesting with which textile goods of greater length can be processed continuously. But also the operation of a reactor in offline mode is highly useful for the treatment of textile preforms. A continuously working treatment plant is shown in Fig. 13.38.

For the gas phase fluorination, fluorine/nitrogen mixtures (1:9 ratio) are used. They are applied either pure or together with atmospheric air (oxi-fluorination) to carry out the treatment. The principle of the process is shown Fig. 13.39.

The spectrum of achievable effects and functionalities covers the removal of sizing for textile processing, the design of repellent material surface, and the development of highly active and reactive interfaces with very good composite properties. The process can be applied for all textile materials, but must be adjusted to the material by initial investigations in order to avoid unacceptable damages. In case of oxi-fluorination of carbon fibers, an increase of strength and E-modulus can

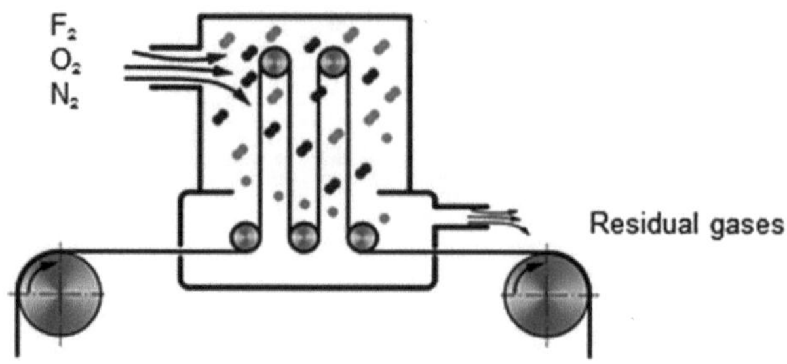

Fig. 13.39 Basis representation of fluorine gas treatment in a continuous process

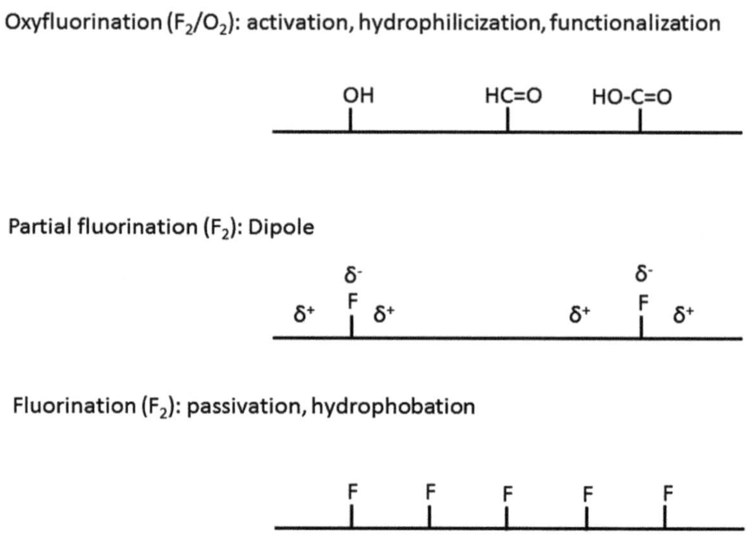

Fig. 13.40 Possibility of activation, hydrophilation, and functionalization of textile interfaces by means of oxyfluorination and fluorination

also be observed [20, 21] apart from the increase of surface energy, leading to the improvement of wetting and adhesion properties.

From Fig. 13.40, it is apparent which possibilities of activation, hydrophilation, and functionalization are offered by fluorination and oxyfluorination.

The morphological changes of fiber surfaces by oxyfluorination are illustrated in Fig. 13.41.

13.5.2.4 Further Methods of Pretreatment of Textile Materials

As further methods of pretreatment for the activation of textile material surfaces, irradiation with high energy electrons [63] (electron radiation treatment) and with light of the ultra violet range of the electromagnetic spectrum [64] (UV-treatment)

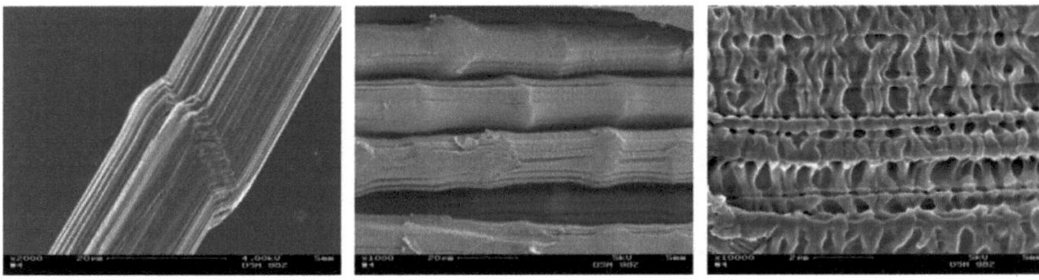

Fig. 13.41 High-performance PE fiber, untreated (*left*), oxyfluorination in 1,000-fold (*middle*) and 10,000-fold (*right*) magnification

can be mentioned. Both methods cause topographical changes of the interface by means of material degradation and provide them with polar molecular structures, caused by the atmospheric oxygen present during the irradiation. These very interesting methods were forgotten in past years and are recently gaining attention again.

13.5.3 Finishing of Textile Materials

The finishing of textile materials must involve the appropriate selection of process and methods to ensure a product that meets the requirements. Finishing processes in aqueous media (bath) and a variety of coating methods or lamination process [65] are available for this purpose. When used on their own, these textile processing steps can serve a specific purpose of finishing, e.g. to make the textiles hydrophobic or dimensionally stable. They can also be part of successful combinations, as in the aqueous application of an adhesion promoter followed by a coating. In many cases, a pretreatment is necessary for textile finishing, which can range from cleaning or removing the spinning preparations to a highly functional design of the fiber interfaces (see Sect. 13.5.2).

13.5.3.1 Textile Finishing by Means of Aqueous Baths

Imparting defined material and performance characteristics to textile materials by means of aqueous baths with relevant active ingredients such as antistatic agents, textile auxiliaries for softening, or high grade finishing agent for dimensional stabilization is one of the most common tasks of textile finishing. The manufacturing of highly demanding technical textiles can also benefit from this so-called wet finishing. This can for instance include flame-retardant properties, resistance against climatic and biological influences as well as an increased reactivity of fiber surfaces (see Sect. 13.5.2.1). For the application from aqueous liquid, two procedures are available: the discontinuous exhaust process and the continuously running impregnation process.

Exhaust Process

If soluble or dispersed substances and active agents (finishing auxiliaries) are absorbed by textile fibers in aqueous liquid because of their fiber affinity, i.e. exhausted from the liquid, this type of application is referred to as exhaust process. Exhaust processes often work with a high (long) liquid ratio, measured by the quotient of the mass of the textile in kg to liquid volume in l, which is circa 1:10. New finishing devices offer results at significantly lower (shorter) liquid ratios of 1:1 and even lower. For all types of textile products, such as fibers, yarns, or fabrics, finishing machines and devices are available, which makes it possible to integrate the finishing variably in the textile process chain [23].

The conventional objectives of the wet finishing of textiles in exhaust processes are a more evenly completed dyeing and its post-treatment [23]. It has also been shown that this procedure allows the surface modification of textile materials. This applies to the finishing of polyolefine with freely available amino groups [12, 13] (see Sect. 13.5.2.1) as well as metallization of fiber interfaces in yarns and surfaces [59]. Furthermore, with the help of these finishing techniques, the immobilization of biocatalysts (enzymes) by means of their bonding at textile surfaces [66, 67] and other modifications are realizable.

The equipment assembly for a wet treatment of textile materials often consists of machines and devices, whose mode of operation has proven its worth for decades. For the processing of wrapped surfaces and spooled yarns, devices which allow a very homogenous finishing result by means of their flexibly selectable flow of the bath fluid are appropriate, also making them especially suitable for high temperature processes. For textiles in hank form, textile machines further developed to work with low bath ratios should be used. These treatment machines, known as jet, over flow, and air flow are also suitable for HT, and guarantee good results.

Impregnation Process

The impregnation process is characterized by the application of highly concentrated treatment liquid, a very good impregnation of the material, and its continuous working procedure. By virtue of their precise adjustability of the dipping time of the textile materials and of the agent concentrations in the finishing liquid or dispersions, these processes help achieve defined results as well as resource-saving working procedures.

The aims of textile finishing by means of impregnation largely match those of exhaust processes, in which considerably higher concentrations of finishing agents are used. Furthermore, finishing agents and finishes can be applied to textile sheets. The impregnation process is suitable for open surface coating of textile structures in

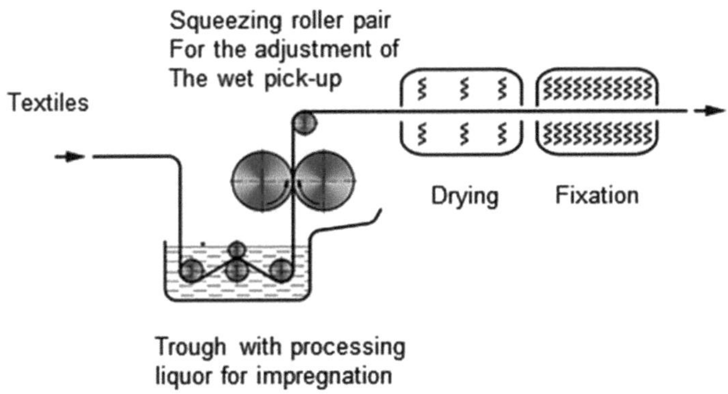

Fig. 13.42 Schematic diagram of padding

order to impart stability, strength, and adhesion capabilities to these fabrics. Particularly in the case of an open-meshed fabric structure manufactured from multifilament yarns or rovings, optimum results are made possible by the high rate of impregnation of coating formulations.

Pretreatment, application, and coating in impregnation processes are performed by means of a padding-squeezing device, known as a padder (see Fig. 13.42). The goods (substrate) are passed through a tank in their spread state. The tank contains the treatment liquid (e.g. finishes, coatings, or finishing liquid). Then the impregnated substrate is passed through a pair of rollers in order to squeeze the absorbed liquid in a defined manner and to transport the goods onward. Depending on the aim of the finishing, this is followed by a thermal drying and fixation, a hot vapor treatment, or in many cases the immediate storage of the impregnated textile materials to perform the required reaction of finishing agents. Except for the cases of coating and finishing, the treated textile structures are washed and subsequently dried.

13.5.3.2 Coating of Textile Materials

A continuous application (coating) of a textile substrate with polymeric materials results in new composites [29] with entirely new characteristics. They represent not only the simple combination of individual materials, but also possess new properties in themselves. The mechanical loading capacity of the new composite is determined mainly by the textile components, while the surface properties are attributed to the coating agents. The suitable selection of textile reinforcements, coating materials, and their formulation as well as application process (see Sect. 13.5.3.3) allow the manufacturing of composites, which can be optimized for the individual purpose of application. The application areas for coated textiles are diversified and applied in many areas such as in membrane, automobile, and aircraft construction, in civil engineering, in agriculture, and geo industries, medicine, clothing as well as in the areas of room and building design. A number of other technical applications are possible with coated textile reinforcements, where functional textiles occupy being the most important one. Here, textile systems for the generation of electricity by means of fuel or solar cells and for sensoric and actoric tasks are to be mentioned.

Coating Materials

Coating materials consist of film-forming synthetic polymers and solvents or water as their second component. The dispersions or solutions that can be manufactured from them are based on vinyl chlorides, urethanes, acrylic ester, and other raw materials. As the only natural product natural rubber latex is used [53]. Some polymers applied for coating will be discussed in this section.

Polyvinyl Chloride (PVC) Coating

Polyvinyl chloride (PVC) is the single commonly used polymer for coating of textiles. PVC is manufactured by the radical polymerization of vinyl chloride, where a brittle and hard polymer (Fig. 13.43) is obtained, which is referred to as unplasticized PVC.

This product is not suitable for the coating of flexible textile materials and must be transferred into a so-called plastisol by an addition of softeners. These softeners, which consist mainly of phthalate ester [68], can reach a content of up to 40 %. The clear film formed by soft PVC exhibits a high abrasion resistance, as well as low permeability. By means of a mixing of different additives, different properties such as dyeability, flame retardance, weather resistance, or dirt repellence can be imparted to soft-PVC coatings. These coatings are resistant to acids and bases. However, organic solvents can cause the extraction of softeners, which may make the film brittle and susceptible to break [69].

Polyurethane (PU) Coating

Polyurethanes (PU) result from the addition reaction of diisocyanate with dioles (Fig. 13.44). The resulting polymer can have varying consistencies depending on the type of raw materials and their cross-linking ability. Thus, depending on the

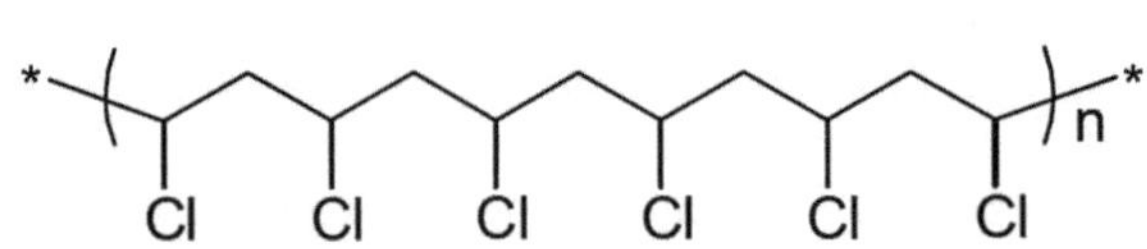

Fig. 13.43 Structural element of polyvinyl chloride

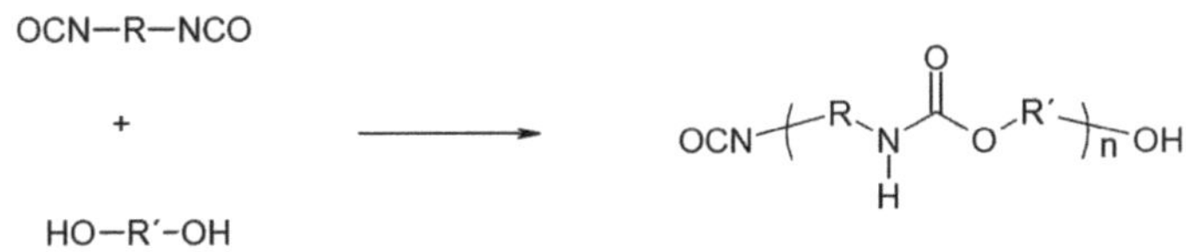

Fig. 13.44 Reaction principle of diisocyanates with dioles to polyurethane (R = $(CH_2)_6$, $C_6H_3(CH_3)$, C_6H_4–CH_2–C_6H_4; Ŕ = Polyester, Polyether)

purpose of application, the properties, such as flexibility, softness, elasticity or also vapor permeability of materials coated with them can be adjusted.

The coating is carried out by means of aqueous dispersion of polymers or by mixing pre-polymers, which react under the influence of thermal energy [70]. PU coating is characterized by the high tensile strength, tear strength, ductility, and soft handling [53].

Polyacrylate (PAA) Coatings

The basic elements of polyacrylic esters (polyacrylates) are either esters of acrylic acid or the methyl-substituted methacrylic acid (Fig. 13.45). By polymerization, polyacrylates are formed, which can exhibit different characteristics depending on the type of monomers and their esterification. Coatings with polyacrylic ester are rather soft and sticky, while those with polymethacrylic ester are hard and brittle. These materials are characterized by high resistance to UV-light, heat, ozone, chemicals, water, ageing, and solvents [53].

Polytetrafluoroethylene (PTFE) Coatings

Polytetrafluoroethylene (PTFE), a polymerization product of tetrafluoroethylenes, possesses an extremely low surface energy, so that coated textiles exhibit strong water and oil-repellent properties (Fig. 13.46). Coatings with PTFE can be applied in temperatures up to 250 °C and are resistant against most solvents and chemicals. PTFE surfaces are etchable with strong alkali substances, which can improve the adhesion properties of the materials.

Furthermore, PTFE is an excellent material for the manufacture of climate membranes, for example GoreTex® membrane. The water vapor permeable pores of these materials are created from by biaxially stretched thin PTFE-films [71, 72].

Elastic Materials for the Coatings

If coatings with elastic properties are required, a number of other materials apart from elastically adjustable polyurethanes are available. Some of these materials will be described below.

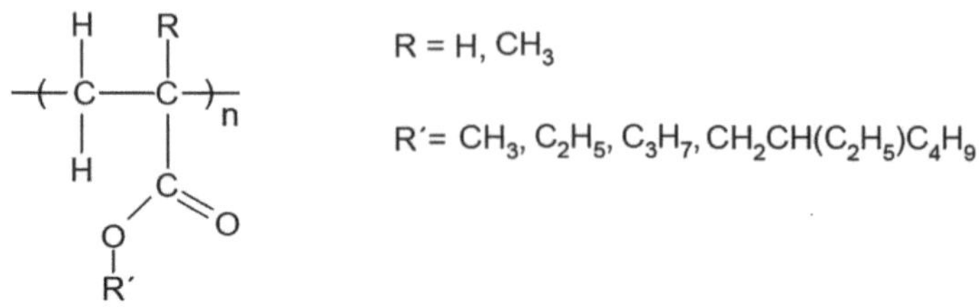

Fig. 13.45 Structure of polyacrylates from acrylic acid and methacrylate ester

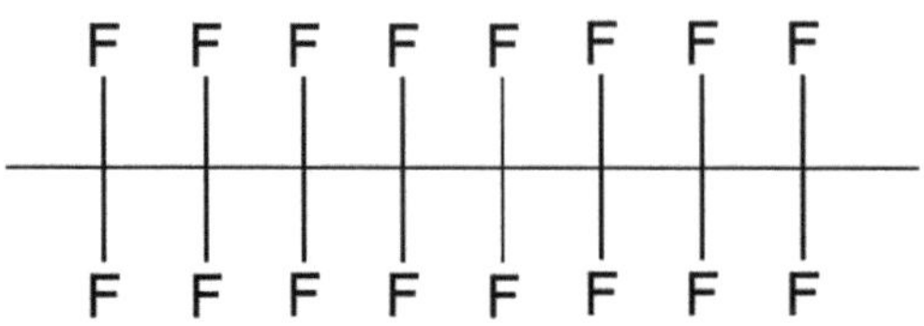

Fig. 13.46 Structural element of Polytetrafluoroethylene

Natural Rubber

Natural rubber coatings are based on naturally obtained emulsions (latex), which are applied directly for processing, or on solid materials obtained from latex, which are applied by the addition of appropriate fillers. The polymer contained in it is linear, built from isoprene-monomers (Fig. 13.47) and results in high-strength materials. The double bonds available in the macro molecule allows the cross-linking (vulcanization) of polymeric materials. Furthermore, natural gum rarely shows ageing phenomena and possesses excellent adhesion properties, making it ideal for the manufacturing of tires.

Polyoxane Coating

Polyoxanes can form as molecule chain or cross-links (Fig. 13.48), where alternating silicum-oxygen units form the frame, which is occupied by non-reactive, organic alkyl or aryl functional groups. The silicon rubbers required for coating are masses that can be transferred into a rubbery-elastic state, whose polysiloxane molecules contain cross-linkable functional groups such as oxygen atoms, hydroxyl or vinyl groups. Depending on their structure at different temperatures, they react when mixed with initiators. The resulting elasticity does not change at a temperature range from −70 to 270 °C. Polysiloxane coatings are resistant to many chemicals, oils, acids, and gases. However, they are sensitive to ozone, UV, and other environmental influences [53].

Fig. 13.47 Formation of polyisoprene from methyl butadien (isoprene)

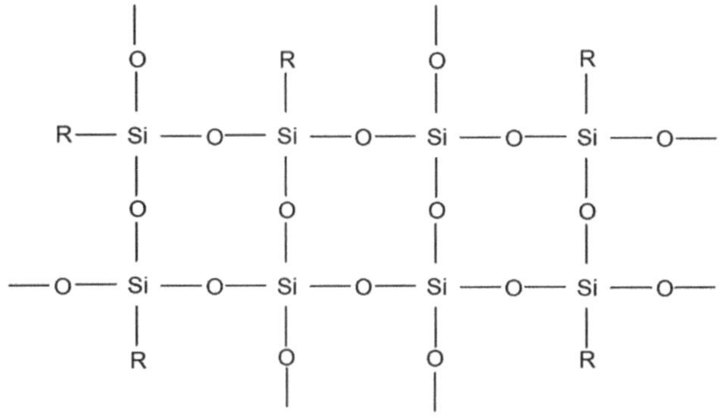

Fig. 13.48 Structure of cross-linked polysiloxane ($R = CH_3$, C_2H_5, C_6H_5)

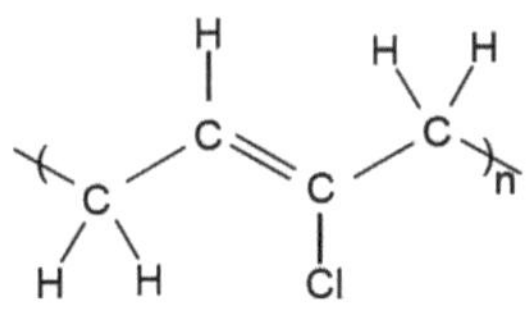

Fig. 13.49 Structure of polychloroprene (neoprene)

Polychloroprene (CR) Coatings

Polychloroprene (CR), also known as neoprene, is a synthetic, elastic material with very balanced properties, even though they are not as excellent as those of natural rubber (Fig. 13.49). It displays good mechanical stability, resistance to ozone, weather as well as ageing, an ability to adhere to many substrates, and low flammability.

Styrene-Butadien (SBR) Coatings

Styrene-butadiene (SBR) (styrene-butadiene rubber) is manufactured by emulsion polymerization of styrene and butadiene, creating an irregular copolymer (Fig. 13.50). The formulation and application of coating materials is similar to the processing of natural rubber. In comparison to the latter, SBR coatings have low

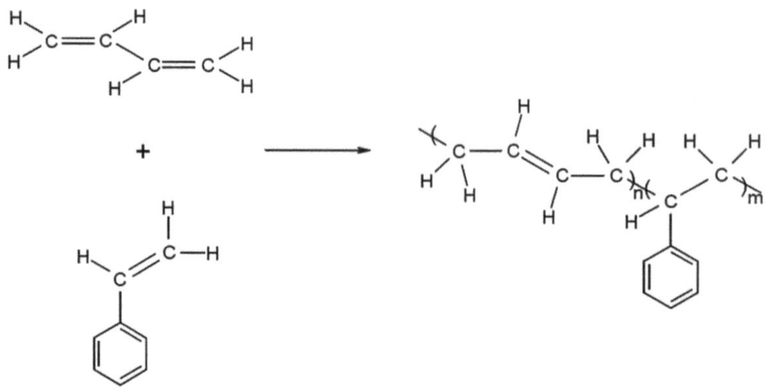

Fig. 13.50 Reaction of styrene and butadiene to styrene-butadien (SBR)

elasticity and cause a higher heat development by dynamic loading [53]. The coating of textiles can be carried out easily and results in weather-stable and ozone-resistant products.

13.5.3.3 Application of Coating by Means of Rollers (Knife Coater)

Knife Coater

For plane and flexible substrates, coater blades or knife coaters are suitable for the application of coatings. They are positioned over the whole width of the goods and work in the opposite direction of the roller, spreading the coating paste on the substrate. A differentiation can be made between roller knife, air knife, and rubber belt system. Post knives, table knives, commabar systems and spiral knives are seldom used. Detailed descriptions of the different systems are given in the literature [65]. Thin layers in the range of a few micrometers can be realized with knife coating.

Roller Knife

The roller knife works against a rubber, chrome-plated, or chilled cast iron roller (see Fig. 13.51). Rubber rollers are frequently used. Chrome-plated steel rollers or chilled cast iron roller are used for uneven substrates. The coating thickness is determined by different parameters, the important ones being the type of substrate, the viscosity, the speed of application as well as the respective placement of the knife to the rollers. Coatings between 10 and 1,250 g/m^2 are possible with this process.

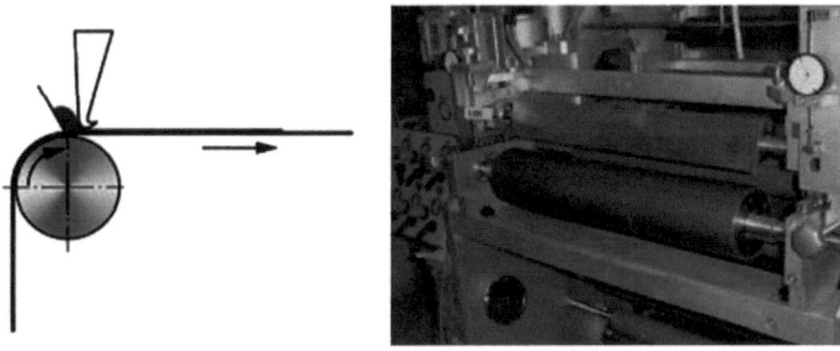

Fig. 13.51 Principle of roller knife (*left*) and photo of a Coatema plant (*right*)

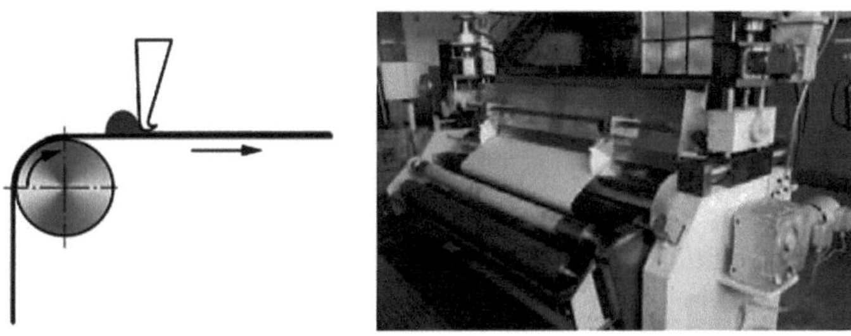

Fig. 13.52 Principle of air knife (*left*) and photo of a Coatema plant (*right*)

Air Knife

On air knife systems, air under the stretched goods acts as the opposing pole (Fig. 13.52). The amount of coating to be applied is determined by the tension of the goods. Air knives are used for less extensive coatings (5–80 g/m^2).

The coating knife can be positioned by a linear guide over or behind the coating roller.

The application of different technologies depends on the rheology of the coating paste and substrate, and the functionality to be obtained by the coating.

Drying and Hardening of Coatings

The process of coating is followed by drying and curing. Drying means that the applied solvents are removed and a film is formed. During the subsequent curing (cross-linking) a functional layer is formed by the chemical reactions (different polymerization reactions (see Chap. 3) of two or more components of the matrix. The required heat can be transferred or applied by contact and convection. In contact methods, the goods are passed over hot cylinders, while convection processes rely hot air circulation. In the application of an infrared dryer, the required energy for drying is applied through the absorption of IR radiation. In the case of coil coating, short-wavelength UV radiation is utilized. Induced high-frequency eddy currents are also used [65].

13.5.3.4 Other Coating Processes and Methods

Apart from the described coating systems, other roller systems are used for the application of coating pastes. The principle of kiss coaters and reverse roll coaters are described as representative examples below.

Kiss Coater

Kiss coaters are suitable for low to medium viscosity coatings. Pastes are applied to the substrate from a container by means of an application roller (between two guiding roller). A rotation in the same or contrary to the direction of the fabric run is possible. Fabric tension, as well as the speed of the rollers proportional to the process speed, are the important parameters for a homogenous coating. The application of very thin coatings is possible, for instance for surface functionalization. Kiss coaters are also suitable for the hydrophobic treatment of nonwovens in medical applications.

Reverse Roll Coaters

In reverse roll coaters, the rollers are arranged in counter-rotating directions. This is suitable for the processing of easily flowing, low viscosity pastes. Highly viscous and dilatant pastes, whose viscosity increases with higher shearing force cannot be processed (Figs. 13.53 and 13.54).

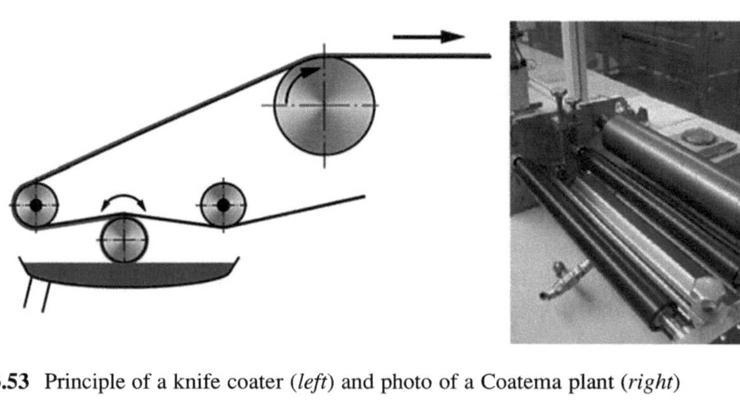

Fig. 13.53 Principle of a knife coater (*left*) and photo of a Coatema plant (*right*)

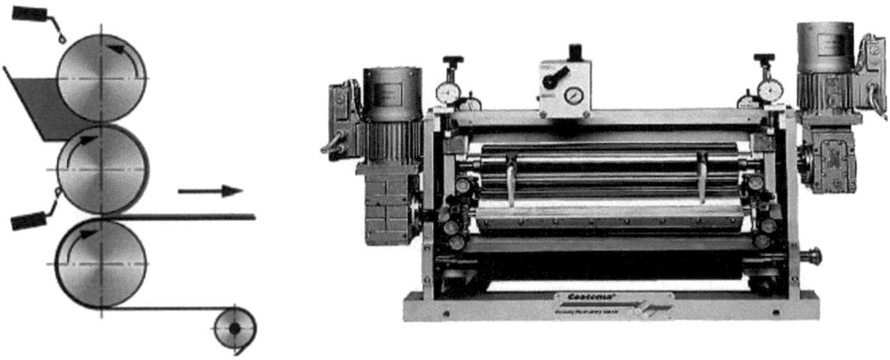

Fig. 13.54 Principle of a reverse roll coater (*left*) and photo of a Coatema plant (*right*)

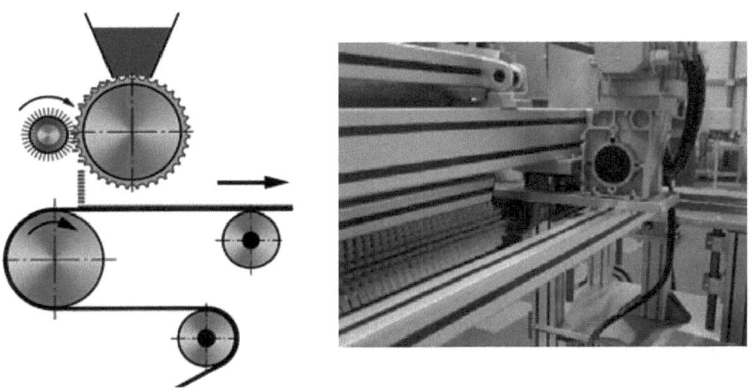

Fig. 13.55 Principle of powder coating (*left*) and photo of a Coatema plant (*right*)

Powder Coating

For specific applications, the use of powdery materials is necessary. A powder scattering unit always consists of a gridded roller. The type of grating and the rotational speed during coating define the amount of applied powder (Fig. 13.55).

Hotmelt Coating

Polymers which soften and harden again at a temperature between 80 and 220 °C are referred to as hotmelt. They can be dosed easily, applied in liquid state, and require only short coating distances and cross linking time. By means of a continuous coating process, plane and closed films can be applied. In combination with a proper "mask", the coating weight and width can be defined during the application of a closed film >10 g/m^2 or of a partly porous film <10 g/m^2 (Fig. 13.56).

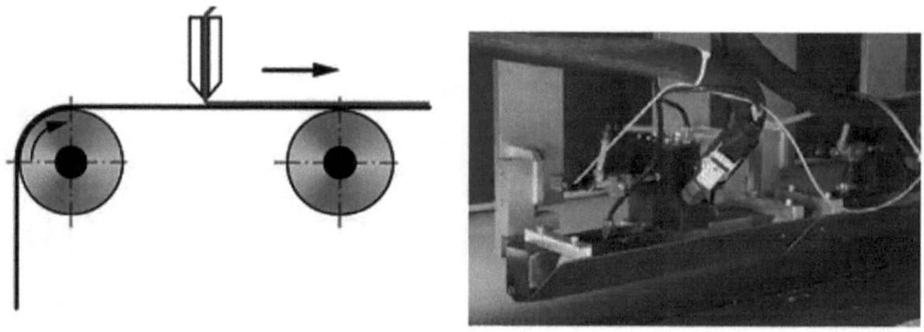

Fig. 13.56 Principle of hotmelt coating (*left*) and photo of a Coatema plant (*right*)

Apart from the aforementioned processes, which are applied in practice for a wide range of applications, there are a number of special application technologies. Among others, chemical vapor deposition (CVD) [73, 74] and physical vapor deposition (PVD) [75] are used.

References

1. EHRENSTEIN, G. W.: *Faserverbundkunststoffe: Werkstoffe, Verarbeitung, Eigenschaften.* München, Wien : Carl Hanser Verlag, 2006
2. FLEMMING, M.; ZIEGMANN, G.; ROTH, S.: *Faserverbundweisen: Fasern Matrices.* Berlin, Heidelberg : Springer Verlag, 1995
3. ACHEREINER, F.: *Verbesserung von Adhäsionseigenschaften verschiedener Polymerwerkstoffe durch Gasphasenfluorierung.* Erlangen-Nürnberg, Technische Fakultät der Universität Erlangen-Nürnberg, Dissertation, 2009
4. NEITZEL, M.; MITSCHANG, P.: *Handbuch der Verbundwerkstoffe (Werkstoffe, Verarbeitung, Anwendung).* München, Wien : Carl Hanser Verlag, 2004
5. BISCHOF, C.; POSSART, W.: *Adhäsion - Theoretische und experimentelle Grundlagen.* Berlin: Akademie Verlag, 1983
6. JAKUBASCH, H. J.: *Oberflächenchemie faserbildender Polymere.* Berlin : Akademie Verlag, 1984
7. LONDON, F.; EISENSCHITZ, R.: On the Ratio of the van der Waals Forces and the Homopolar Binding Forces. In: *Z. Physik* 60 (1930), pp. 491
8. NOLLER, H.; PAULING, L.: *Die Natur der chemischen Bindung.* Weinheim : Verlag Chemie GmbH, 1968
9. ERBIL, H. Y.: *Surface Chemistry of Solid and Liquid Interfaces.* Oxford : Blackwell Publishing Ltd., 2006
10. ERHARD, G: *Konstruieren mit Kunststoff.* München, Wien : Carl Hanser Verlag, 2008
11. OWENS, D. K.; WENDT, R. C.: Estimation of the Surface Free Energy of Polymers. In: *Journal of Applied Polymer Science* 13 (1969), pp. 1741–1747
12. PRINZ, K.; HUND, R.-D.; CHERIF, Ch.: Gezielte Eigenschaftsmodifizierung hoch inerter textiler Polypropylenfasern mittels chemisch reaktiver Gruppen. In: *Proceedings. 48. Chemiefasertagung Dornbirn.* Dornbirn, Austria, 2009
13. BARTUSCH, M.: *Gezieltes Oberflächendesign textiler Polymerwerkstoffe zur Erzielung nutzbarer physikalisch/chemisch/biologisch aktiver Grenzschichten.* Dresden, Technische Universität Dresden, Master thesis, 2010

14. BOTTENBRUCH, L.; BINSACK, R.: *Polyamide, Kunststoffhandbuch Bd.3/4: Technische Thermoplaste*. München, Wien : Carl Hanser Verlag, 1998
15. BEYER, H.; WALTER, W.: *Lehrbuch der Organischen Chemie*. Stuttgart : Hirzel, 2004
16. RAHM, J.: *Beitrag zur Herstellung Langfaserverstärkter Aluminium-Matrix Verbundwerkstoffe durch Anwendung der Prepregtechnik*. Chemnitz, TU Chemnitz, Fakultät Maschinenbau, Dissertation, 2008
17. Business Analytic Center: Polyethylenterephthlat (PET) Konjunkturaussichten und Prognosen. 2011
18. BOBETH, W.; JACOBASCH, H. J.: Zur Bedeutung der physikalisch-chemischen Oberflächenparameter von Chemiefaserstoffen für deren Färbe-, Veredlungs- und Gebrauchseigenschaften. In: *Lenzinger Berichte* 33 (1972), pp. 33–42
19. SCHÜRMANN, H.: *Konstruieren mit Faser-Kunststoff-Verbunden*. Berlin, Heidelberg : Springer Verlag, 2005
20. ELAHI, F.: *Surface activation of carbon fibers using gas-fluorination, plasma treatment and chemical procedures*. Dresden, Technische Universität Dresden, Master thesis, 2008
21. MATHUR, R. B.; GUPTA, V.; BAHL, O. P.; TRESSAUD, A.; FLANDROIS, S.: Improvement in the mechanical properties of polyacrylonitrile PAN/ -based carbon fibers after fluorination. In: *Synthetic Metals* 114 (2000), pp. 197–200
22. BALDUS, H. P. et al.: Properties of Amorphous SiBNC-Ceramic Fibres. In: *Key Engineering Materials* 177 (1997), pp. 127–131
23. ROUETTE, H.-K.: *Enzyklopädie Textilveredlung - Band 4*. Frankfurt am Main : Deutscher Fachverlag GmbH, 2009
24. ARCLES, B.; LARSON, G.: *Silicon Compounds: Silanes Silicones*. Morrisville, PA: Gelest, Inc., 2004
25. CHUNG, D. D. L.: *Carbon Fiber Composites*. Boston, London, Oxford, Singapore, Sydney : Butterworth-Heinemann, 1995
26. EYERER, P.; ELSNER, P.; VOIT, B.; DUNGERN, A. v.: Verfahrenstechnische Grundlagen für verbesserte Faserverbundwerkstoffe durch online Herstellung (AiF 118 ZBG) / Leibniz-Institut für Polymerforschung e.V. Dresden. Dresden, 2002. – Final report
27. KUCH, H.; SCHWABE, J. H.; PALZER, U.: *Herstellung von Betonwaren und Betonfertigteilen, Edition Beton*. Düsseldorf : VBT Verlag Bau und Technik, 2009
28. CURBACH, M.: Sonderforschungsbereich 528 (2008/2-2011/1) Textile Bewehrungen zur bautechnische Verstärkung und Instandsetzung / Technische Universität Dresden. Dresden, 2008. – Research application
29. WERNER, A.: *Encyclopedia of PVC New York*. New York : Litton Educational Publishing, 1986
30. HÜTTINGER, K. J.: The Fundamentals of Chemical Interactions in Composite Interfaces. In: FIGUEIREDO, J. L.; BERNARDO, L. A.; BAUER, R. T. K.; HÜ TTINGER, K. J. (Hrsg.): *Carbonfibres filaments and composites*. Dordrecht, Boston, London : Kluwer Academic Publishers, 1990, pp. 245–261
31. DONNET, J.-B.; WANG, H-T.; PENG, J. C. M.; REBOUILLAT, S.: *Carbon Fibers*. New York, Basel, Hong Kong : Marcel Dekker Inc., 1998
32. HÜTTINGER, K. J.; ZIELKE, U.; HOFFMAN, W. P.: Surface-oxidized carbon fibers: III. Characterization of carbon fiber surfaces by the work of adhesion/pH diagram. In: *Carbon* 34 (1996), pp. 1007–1013
33. SHEN, W.; LI, Z.; LIU, Y.: Surface Chemical Functional Groups Modification of Porous Carbon. In: *Chemical Engineering* 1 (2008), pp. 27–40
34. ZIELKE, U.: *Untersuchungen zur Aktivierung einer Ultra-Hochmodul-Kohlenstofffaser durch eisenkatalysierte Oxidation*. Karlsruhe, Universität Karlsruhe, Institut für Chemische Technik, Dissertation, 1992
35. KREKEL, G.: *Works of Adhesion at the Carbon fibres*. Karlsruhe, Universität Karlsruhe, Institut für Chemische Technik, Dissertation, 1995

36. SEIDEL, C.: *Verbunde aus Hochtemperaturthermoplasten und Kupfer für flexible Schaltungsträger*. Erlangen-Nürnberg, Friedrich-Alexander-Universität Erlangen-Nürnberg, Dissertation, 2007
37. SEIDEL, C.; DAMM, C; MUENSTEDT, H.: Surface modification of films of various high temperature resistant thermoplastics. In: *Journal of Adhesion Science and Technology* 21 (2007), pp. 423–439
38. ATKINS, P. W.: *Physikalische Chemie*. 2. Auflage. Weinheim : VCH Verlagsgesellschaft mbH, 1996
39. KRÜSS GmbH: *KRÜSS Informationsdatenbank (KIDB)*. http://www.kruss.de/de/informationsdatenbank.html (05-04-2010)
40. BREITMAIER, E.; JUNG, G.: *Organische Chemie*. 5. Auflage. Stuttgart : Georg Thieme Verlag, 2005
41. KRÜSS GmbH: *Theorie und Methoden*. http://www.kruss.de (05-04-2010)
42. Ahlbrandt System GmbH: *Testtinte*. http://neu.ahlbrandt.de/index.php?page=teststiftetesttinten (05-04-2010)
43. OWENS, D. K.; WENDT, R. C.: Estimation of the Surface Free Energy of Polymers. In: *Journal of Applied Polymer Science* 13 (1969), pp. 1741–1747
44. GÜNZLER, H.; GREMLICH, H.-U.: *IR-Spektroskopie: Eine Einführung*. 4. Auflage. Weinheim: Wiley-VCH, 2003
45. HESSE, M.; MEIER, H.; ZEEH, B.: *Spektroskopische Methoden in der organischen Chemie*. 6. Auflage. Stuttgart : Thieme, 2002
46. ERTL, G.; KÜPPERS, J.: *Low Energy Electrons and Surface Chemistry*. Weinheim : Wiley-VCH, 1985
47. HÜFNER, S.: *Photoelectron spectroscopy, principles and applications*. Berlin, Heidelberg, New York : Springer Verlag, 1996
48. SCHINDLER, K.-M.: Photoelektronenbeugung. In: *Chemie in unserer Zeit* 30 (1996), No. 1, pp. 32–38
49. HARTWIG, A.; ALBINSKY, K.: Qualitätssicherung der Oberflächenvorbehandlung von Kunststoffen in der Fertigung durch selektive Farbreaktionen (AiF-No. 10900 N/1) / Fraunhofer-Institut für Fertigungstechnik und Angewandte Materialforschung. Bremen. – Final report
50. HUND, M. C.; GRÜNEWALD, N.: Additives and Masterbatches. In: *Kunststoffe* 93 (2003), No. 7, pp. 84–85
51. WOLF, R.; LAL KAUL, B.: *Ullmanns Encyclopedia of Industrial Chemistry Vol. A*. Weinheim: Wiley-VCH, 1992
52. STRUBEL, R.: *Oberflächenanreicherbare Additive für die Permanentausrüstung von Polyamid 6-Textilien*. Dresden, Technische Universität Dresden, Dissertation, 2007
53. ALAGIRUSAMY, R.; DAS, A.: *Technical Textile Yarns*. Oxford, Cambridge, New Dehli : Woodhead Publish Ing., 2010
54. RAUSCH, J.; MÄDER, E.: Health monotoring in continous glass fibre reinforced thermoplastics: Manufacturing and application of interphase sensors based on carbon nanotubes. In: *Composites Science and Technology* 70 (2010), No. 11, pp. 1589–1596
55. GAO, S. L.; MÄ DER, E.; PLONKA, R.: Surface Defects Repairing by Polymer Coating with Low Fraction of Nano-reinforcments. In: *Proceedings. The Fifth Asian-Australian Conference on Composite Materials (ACCM-5)*. Hong Kong, China, 2006
56. RAUSCH, J.: *Grenzflächenmodifizierung von Glasfaserverstärktem Polypropylen durch Einsatz von Carbon Nanotubes*. Dresden, Technische Universität Dresden, Dissertation, 2010
57. ROUETTE, H. K.: *Handbuch der Textilveredlung, Farbgebung*. Frankfurt am Main : Deutscher Fachverlag, 2009
58. ROUETTE, H. K.: *Handbuch der Textilveredlung, Ausrüstung*. Frankfurt am Main : Deutscher Fachverlag, 2009
59. FATEMA, U. K.: *Verankerung chemischer Funktionen auf inerten textilen Fasermaterialien aus Polyester und Polyolefinen mittels Methoden der Textilveredlung*. Dresden, Technische Universität Dresden, Master thesis, 2006

60. ONGGAR, T.; HUND, R.-D.; HUND, H.; CHERIF, Ch.: Surface Functionalization and Silvering of inert Polyethylen Terephthalate Textil Materials. In: *Material Technology* 25 (2010), No. 2, pp. 106–111
61. ALMANSA, E.; HEUMANN, S.; EBERL, A.; KAUFMANN, F.; CAVACO-PAULO, A.; GUBITZ, G. M.: Surface hydrolysis of polamide with a new polyamidase from Beauveria brongniartii. In: *Biocatalysis and Biotransformation* 26 (2008), No. 5, pp. 371–377
62. HEGEMANN, D.; BALAZS, D.J.: Nanoscaled treatment of textiles using plasma technology. In: SHISHOO, R. (Ed..): *Plasma Technologies for Textiles*. Cambridge : Woodhead Publishers, 2007
63. CHARLES, A.: *Atomic Radiation and Polymers*. Oxford : Pergamon Press, 1960
64. FOZZA, A. C.; KLEMBERG-SAPHIA, J. E.; WERTHEIMER, M.R.: Vacuum Ultraviolet Irradiation of Polymers. In: *Plasma and Polymers* 4 (1999), No. 213, pp. 183–206
65. GIESSMANN, A.: *Substrat- und Textilbeschichtung*. Heidelberg, Dordrecht, London, New York : Springer Verlag, 2010
66. STRUCH, M.: *Immobilisierung der Laccase aus Trametes versicolor an Chitosan und Chitosanfasern*. Dresden, Technische Universität Dresden, Bachelo Thesis, 2009
67. KÜCHLER, K.: *Immobilisierung der Laccase aus Trametes versicolor an Chitosanfasern*, Technische Universität Dresden, Diploma thesis, 2010
68. MAITHEN, B.; JANDEL, A. S.: *JOT Fachbuch*. Wiesbaden : Friedrich Vieweg Sohn/ GWV Fachverlag GmbH, 2005
69. HALL, M. E.: Coating of technical textiles. In: HORROCKS, A. R. (Hrsg.); ANAND, S. C. (Ed.): *Handbook of Technical Textiles*. Cambridge : Woodhead, 2000, pp. 173–186
70. ROUETTE, H. K.: *Handbuch der Textilveredlung, Beschichtung*. Frankfurt am Main : Deutscher Fachverlag, 2009
71. PAINTER, C. J.: Waterproof, breathable fabric laminates. A perspective from film to market place. In: *Journal of Coated Fabrics* 26 (1997), pp. 107
72. TRÄUBEL, H.: *New Materials Permeable to Water Vapor*. Berlin, Heidelberg, New York : Springer, 1999
73. ALLENDORF, M.: From bunsen to VLSI: 150 years of growth in chemical vapor deposition technology. In: *The Electrochemical Society Interface* 7 (1998), No. 1, pp. 36–39
74. ALLENDORF, M.: On-line Deposition of Oxides on Flat Glass. In: *The Electrochemical Society Interface* 10 (2001), No. 2
75. SREE HARSHA, K. S.: *Principles of Physical Vapor Deposition of Thin Films*. Elsevier – Science Technology, 2006

Additional Literature

AHN, H. W.; PARK, C. H.; CHUNG, S.E.: Waterproof and breathable properties of nanoweb applied clothing. In: *Textile Research Journal, 81* (2011), No. 14, pp. 1438-1447

CLOUGH, R.L.: Nuclear Instruments and Methods in Physics Research Section B: Beam Interactions with Materials and Atoms. In: *High-energy radiation and polymers: A review of commercial processes and emerging applications* 185 (2001), No. 1–4, pp. 8–33

HÜFNER, S.: *Photoelectron Spectroscopy: Principles and Applications*, Berlin, Heidelberg. New York: Springer Verlag, 2003

NASKAR, A. K.: Carbon Fiber Composites, Presentation to the Bridging the Gap Conference at ORNL. *Materials Science and Technology Division*, March 5, 2013

SARKAR, S.: *The fabrication of polymer-derived SiCN/SiBCN ceramic nanostructures and investigations of their structure-property relationship*. Calcutta, University of Calcutta, Dissertation, 2010

Chapter 14
Textile Testing Methods

Thomas Pusch

This chapter explains basic aspects and methods for the physical characterization of technical textiles and the fiber-reinforced composites manufactured from them. Testing methods from all steps of the supply chain will be considered, from the filament, the yarn, fabrics, preforms, to the final composite. Commercially available testing device exist for this, realizing standardized test environments and procedures. A representative selection of standardized testing methods will be outlined. The focus will be on testing methods providing information on the mechanical properties, in particular on the strength, of the textile structures and the resulting fiber-reinforced composites.

14.1 Introduction

To ensure the serviceability of fiber-reinforced composites, one has to have detailed knowledge of these structures' relevant properties, which are determined by means of physical testing methods. Due to the advanced state of plastics engineering and fiber-reinforced plastics, a great number of testing methods exists for this material group. The following chapter will compile the most important methods. For newer developments, e.g. textile-based membranes or textile-reinforced concrete, however, there are few obligatory and generally accepted testing methods. Some aspects of the testing of these material classes can only be touched on at the end of the chapter.

The property profile of fiber-reinforced composites is extraordinarily complex. An effective development of new structures can only be expedited with profound

T. Pusch (✉)
Institute of Textile Machinery and High Performance Material Technology, TU Dresden, Dresden, Germany
e-mail: thomas.pusch@tu-dresden.de

C. Cherif (ed.), *Textile Materials for Lightweight Constructions*,
DOI 10.1007/978-3-662-46341-3_14

knowledge of the basic materials' and the intermediate products' parameters. It is therefore necessary to accompany the entire supply chain of composites by test, which is why the following chapter will include testing methods for filaments, yarns and textile fabrics/prepregs.

In practical applications as well as in research and development, standardized testing methods are used for the physical characterization of the structures. The testing standards are developed by a variety of standardization institutions and exist parallel, which can result in a variety of definitions and testing standards for the same material property. The standards are neither consistent in terms of the designation of the parameters to be tested, nor are they fixed—instead, they are subject to frequent alterations and amendments. This complicates a systematic portrayal of this subject, which can only be presented with a focus on a limited number of points due to its extent.

Another difficulty is presented by the fact that the testing environment influences the results while different materials require different testing environments. Therefore, material or material group-based standards exist for individual textile-physical parameters. Drafting such standards only becomes necessary after the materials or material groups have found widespread use in practice. From among the materials treated in this chapter, this applies to carbon, glass, and aramid. Therefore, the following will be concerned primarily with testing methods for textile structures produced from these materials.

14.2 Laboratory and Test Engineering Basics

With an eye to the setting of priorities of this chapter, an overview of some technical test of engineering basics is appropriate. The aspects of this sub-chapter are equally relevant to the deliberations following it, and also apply to the testing methods described in subsequent chapters.

14.2.1 Measuring Versus Testing Technology

The test engineering-based characterization of textile structures is based on established and modern methods of metrology—in this regard, textile test engineering is based on primarily metrological aspects. The substantial metrological terminology, definitions, and procedures are specified and described in detail in [1–3].

Metrology's perspective is based on the assumption that the measurand has a constant value, and that all scatterings of a measurement given by a measurement chain are an imperfection of the measuring chain. The *true value* x_w of a measurand is overlaid by a recognizable and a non-recognizable complex of influences. The recognizable complex is the *systematic error* e_s. It "systematically" influences the

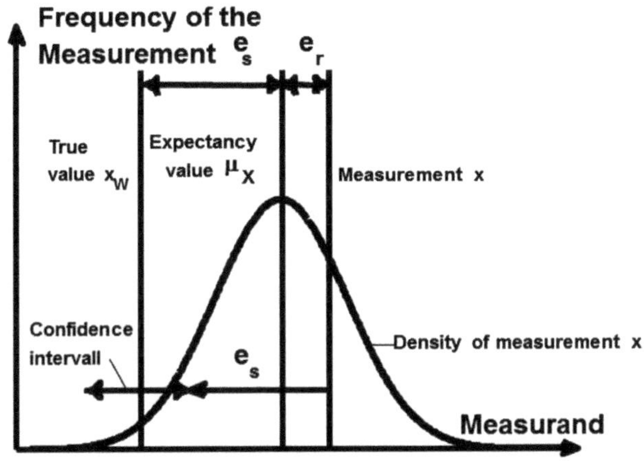

Fig. 14.1 Measurements scatter due to imperfections (errors) of the measuring method

measurand x in one direction. Unpredictable fluctuations in the readings of a measurement cause non-unilaterally directed scatterings of the individual measurements of a measurand around the mean or expectancy value $\mu_x = x_w + e_s$, and are labeled as *random errors* e_r.

The measured value x can be amended with the value from e_s. Based on the random errors e_r, sufficient numbers of repeat measurements will yield information on the scatter range of the measurements. While these are not suitable to amend the measurement x, they allow a statement regarding which value interval $m \pm u$ the (constant) true value x_w is located in with a predetermined probability (usually 95 %). Figure 14.1 clarifies these relations.

The value interval $m \pm u$ is called the *confidence interval*, u is the measurement uncertainty. Methods for the calculation of the measurement uncertainty u are detailed in [4–7]. They are generally accepted and form the basis for the conventional depiction of measuring results.

Textile tests are measurements with an exactly stipulated procedure. This procedure includes, at the minimum, sampling, specimen size, specimen preparation and processing, environmental conditions, design and properties of the testing device, as well as interpretation and statement of the results. This is set down in the test standards. A presentation and assessment of systematic and random errors is usually omitted. Due to the established measuring and environmental conditions it can be assumed that systematic errors are similar in laboratories worldwide, and that a correction of the test results is rather a hindrance to the comparability of the data gathered in different testing laboratories. Furthermore, it is assumed that the testing devices and procedures are designed to keep random errors significantly smaller than the scatter of test results of a specimen property. Scatters of the test values during repeat measurements are therefore—this being a marked difference to the approach in metrology—exclusively regarded as scatters of the material parameters. This means that the confidence interval (calculated analogously to metrology) defines the value range in which the true mean value of the tested material parameters is located in with a predetermined probability (confidence level) of

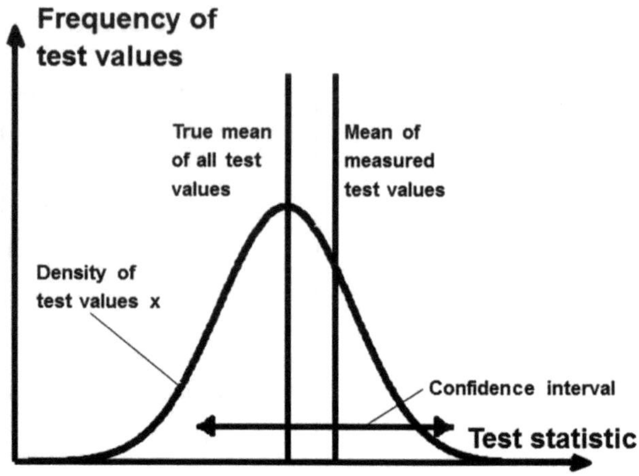

Fig. 14.2 Testing—test values scatter due to (location-dependent) scatterings of the tested parameters

usually 95 %. The proportion of random errors in the scattering of test values is usually omitted from the assessment. Figure 14.2 shows how this approach differs from that of metrology.

14.2.2 Sensors

Tests of the mechanical properties of textiles frequently include examinations of the reactions of a material to be tested, i.e. a specimen, to loads typically occurring during use in practice, and at maximum loads. A special role is played by the defined deformation of the specimen under simultaneous measuring of the force reaction. Length and force measurement systems will therefore be subjected to a closer inspection.

14.2.2.1 Extensometers

Lengths and length variations are fundamental variables in many textile-physical examinations. An absolute determination of length at high precision is rarely required. It only applies to the determination of dimensions of textile fabrics and components, which is performed in accordance with metrological principles.

A test engineering assessment of textile and textile-based structures usually requires knowledge of the state of deformation of the component to be tested, making a detection of length and position variations necessary. Indirect and direct measuring methods are employed for their determination. The direct measuring methods can work either with or without contact with the component.

14.2.2.2 Indirect Length Measurement

The testing device is set to defined positions, i.e. lengths and elongations, and used to impart these onto the specimen. The typical case is that of a rigid transfer of defined crosshead movements to fixing clamps and their nip lines, and from them to the specimen. This makes a length determination on the specimen itself unnecessary. Crossheads are moved by high-precision spindles, and the angle of rotation becomes easily and accurately ascertainable by means of incremental rotary encoders. With them, the crosshead position can be derived accurately from 1 to 10 μm. Strains and deflections on components of the testing device and specimen holders, particularly under load, can cause systematic errors an order of magnitude above the adjustment precision of the crosshead position. Therefore, the manufacturers of testing devices offer corrective algorithms aimed at eliminating the measurement deviations from the measuring sequence.

14.2.2.3 Direct Length Measurement

A direct measurement of length on the specimen is to be preferred from a metrological perspective. However, it requires greater effort, because a length measuring device has to be provided. Usually, extensometers serve to measure variations in span on specimens. They are used whenever the variation in span of the fixing clamps is not representative of the mechanical deformation of the specimen, as it is the case, for instance, when using rope-grip clamps for yarn tests.

There are contact-type extensometers, called *clip-on* extensometers, featuring two pairs of knife-edges which can be clipped-on to the specimen. The measuring spans of the knife-edges range from a few millimeters to centimeters. Variations in span between the knife-edges can be detected with errors on the scale of 1 μm. Using this type of device is only sensible if it neither influences force measuring nor damages the specimen. Especially tests on thin specimens and yarns are at risk of being influenced. It is equally important to assess the situation for breaking tests, as the devices are at risk of being destroyed by the elastic resilience of the specimen after breaking, or by splicing specimen components. Figure 14.3a shows an example of the use of a clip-on extensometer in a tensile test.

Optical length measuring systems or *optical extensometers* are contact-free, do not influence the force measurement, and are not at risk of being damaged when the specimen breaks. Optical measuring systems for variations in length detect specimen deformations by tracking optical characteristics on the surface of the specimen. These characteristics are usually provided by self-adhesive reflective gages. Figure 14.3b shows an example of an optical extensometer. In most cases, CCD cameras are used for determining the gages their variations in position. Evaluation of the pictures can be performed using a number of methods, with correlation analyses being the current method of choice.

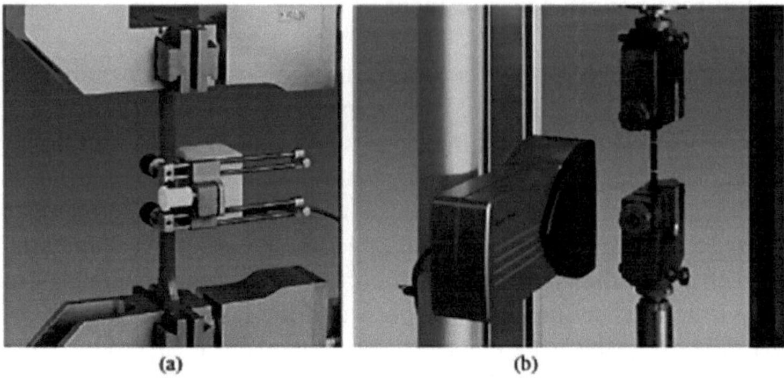

Fig. 14.3 (a) Clip-on extensometer, (b) optical length measuring by means of a camera and reflective gage marks (Zwick GmbH & Co. KG, Ulm, Germany)

Depending on their calibration, extensometers can detect span variations between the optical gages with measuring accuracy of 1–10 μm. Greater changes in span are detected with two CCD cameras following exactly the movement of the gages. Changes in position of the camera can also be detected with an accuracy of 1–10 μm. In the near future, the development of suitable software for image analysis will make it possible to video-optically record the specimen surface during deformation by testing devices, and calculate the strain profile from these records. With a grid applied to the specimen surface, this is already feasible.

The use of strain-gauges on the specimen surface is another manner of establishing strain. However, it relies on thin contact wires to connect the gauges and the specimen, and is therefore not contact-free.

14.2.2.4 Force Transducer

Force transducers for static or quasi-static force measurement are often called load cells, as a tribute to their design. They are usually equally suitable for tensile force and compression force measurements. The most important parameter is the nominal force, which marks the limit of manufacturer warranty for compliance with maximum values for measuring errors set down in the accuracy class. As force is basically only perceptible by its effects, force transducers contain loaded members, and their deformation proportional to the force is usually measured by means of strain-gauges. Testing devices for technical textiles and fiber composite materials are fitted with standard load cells with typical nominal forces ranging from 0.1 to 1,000 kN, and measuring errors, related to the nominal value, on a scale of 0.1–1 %. It is important that the transducers are sufficiently insensitive to transverse loading. Figure 14.4a shows an example of a typical load cell with strain-gauges for tensile testers.

For measuring extremely fast variations in force, as they occur in high-speed tensile tests or impact and crash tests, these load cells are unsuitable, because they

(a) (b)

Fig. 14.4 (**a**) Load cell with strain-gauge (Hottinger-Baldwin Messtechnik, Darmstadt), (**b**) piezoelectric load cell (Kistler Instrumente AG, Winterthur, Switzerland)

have relatively low resonant frequencies and cannot correctly register fast variations in force on the specimen. For such tests, dynamic force transducers with a piezoelectric sensor element (quartz) are used. These are characterized by their high rigidity and small mass, which allows them to achieve resonant frequencies of up to 100 kHz in the relevant force range. This type of transducer is only partly suitable for quasi-static force measurements, and wholly unsuitable for static ones. Regarding their dimensions, piezoelectric force transducers are very similar to load cells with strain-gauges, as seen in Fig. 14.4b, which gives an example of this type of load transducer.

14.2.3 Test Conditions

To achieve reproducible results, tests have to be performed in defined environmental conditions, i.e. in standard atmosphere, as the material parameters often depend on temperature and humidity in the surrounding air. Therefore, an air conditioning of testing laboratories in accordance with DIN EN ISO 139 ("textile atmosphere", "alternative textile atmosphere") [8] or DIN EN ISO 291 ("plastics atmosphere") [9] is required (Table 14.1).

Tests on technical yarns and the fabrics produced with them are performed in "20/65" conditions in accordance with DIN EN ISO 139, or in "23/50" atmosphere under DIN EN ISO 291, always depending on the type of yarn. Here, the requirements of the respective test standards have to be fulfilled. Fiber-reinforced plastic composites and semi-finished products are tested in "23/50" atmospheres following DIN EN ISO 291, and have to fulfill the requirements of a tolerance class depending on the test. In any case, it is important to correctly condition the specimen to the standard atmosphere, which requires conditioning times ranging from a few hours to several days. Criteria for the correct conditioning are given in the test standards.

Table 14.1 Standard atmospheres for conditioning and testing

Standard	Designation	Temperature (°C)	Relative humidity (%)
DIN EN ISO 139	"20/65"	20 ± 2	65 ± 4
	"23/50—alternative"	23 ± 2	50 ± 4
DIN EN ISO 291	"23/50 K1"	23 ± 1	50 ± 5
	"23/50 K2"	23 ± 2	50 ± 10

14.2.4 Mechanical Displacement

Composite components are produced to accommodate and to absorb mechanical loads in practice. Knowing the characteristics of these materials and their precursors (fibers, filaments, yarns, fabrics, prepregs) under mechanical loads is therefore crucial. Mechanical exposure is performed at

- normal-use load cases, in which the specimen is not damaged
- extreme load cases causing failure of the specimen

In the first case, mechanical parameters of the structure can be established, based on which the force-deformation behavior of any structural geometry can be calculated. The second case offers information on the limits of the material's suitability for its intended purpose, i.e. its strength. Both aspects are often considered in one test, by exposing the specimen to the load (linear variation of a spatial coordinate) from zero to failure, and simultaneously measuring the force reaction.

These tests primarily measure force and displacement characteristics. If possible, the force is depicted as (mechanical) tension related to the cross-sectional area, while the displacement is always converted into elongation. The general stress-strain behavior of a structure in the elastic range (i.e. if there is a clear, reversibly linear connection) can be expressed as

$$\underline{\varepsilon} = \underline{S} \times \underline{\sigma} \tag{14.1}$$

where $\underline{S}$ is the compliance tensor, while σ and ε are the tension and elongation tensors. In order to be able to handle this equation, specimen size and load application during the tests are designed to zero as many elements of the Eq. (14.1) or render them negligible. One typical case for this are biaxially reinforced structures of small specimen thickness, which are displaced in fiber directions 1–2. In that case, there is a planar state of stress with only normal stresses σ_1 and σ_2, and shear stress τ_{12}. All stresses acting perpendicular to the specimen plane (direction 3, perpendicular to the fiber directions) are zero or can be neglected. This is a case of orthotropy, i.e. normal stresses do not cause shear distortions, and tangential stresses cause pure shear distortions. Equation (14.1) can be given in simplified matrix form as

$$\begin{pmatrix} \varepsilon_1 \\ \varepsilon_2 \\ \gamma_{12} \end{pmatrix} = \begin{pmatrix} \frac{1}{E_1} & -\frac{\nu_{21}}{E_2} & 0 \\ -\frac{\nu_{12}}{E_1} & \frac{1}{E_2} & 0 \\ 0 & 0 & \frac{1}{G_{12}} \end{pmatrix} \times \begin{pmatrix} \sigma_1 \\ \sigma_2 \\ \tau_{12} \end{pmatrix} \quad (14.2)$$

Mechanical parameters can only be depicted in an elasticity matrix for tests with slow displacements (quasi-static tests). For monaxial loading of a thin specimen in direction i,

$$E_i = \frac{\sigma_i}{\varepsilon_i} \quad (14.3)$$

applies. E is the Young's modulus in the linear part of the stress-stain curve ("Hooke's law"). It is a gauge of the resistance of the specimen to forced deformation in load direction, and thus a gauge of its rigidity. Tensile and compression loads cause a contraction/elongation of the specimen perpendicular to the load direction. The *lateral contraction ratio* ν characterizes this change in elongation. It indicates to what degree the specimen is influenced in transverse direction under load in a principal direction. For this,

$$\nu_{ij} = -\frac{\varepsilon_j}{\varepsilon_i} \quad bei \ \sigma_i \neq 0, \quad \sigma_j = 0 \quad (14.4)$$

applies. In tensile tests, the lateral contraction ratio is referred to as *Poisson's ratio* μ.

The shear modulus G characterizes the shear deformation resulting from shear loading. The shear deformation is expressed by the shear angle γ. The shear modulus is the gradient of the graph in the linear region of the shear stress-shear angle diagram.

$$G_{ij} = \frac{\tau_{ij}}{\tan \gamma_{ij}} \overset{\gamma \approx 0}{\rightarrow} G_{ij} = \frac{\tau_{ij}}{\gamma_{ij}} \quad (14.5)$$

The assumption of orthotropy offers the advantage that there is no coupling of elongation and shear distortion. However, this means that the Young's modulus E and shear modulus G have to be determined by parallel, independent tests. If the determination of the shear modulus is to be based on elongation measurements, orthotropy has to be offset. This is usually done with specimens in which the reinforcement directions are at a ±45° angle to the longitudinal direction of the specimen.

Apart from the parameters quoted above, which apply to the low load range (reversible deformation), the strength values are important for the assessment of the utility fitness (irreversible deformation). The tests are performed as a deformation

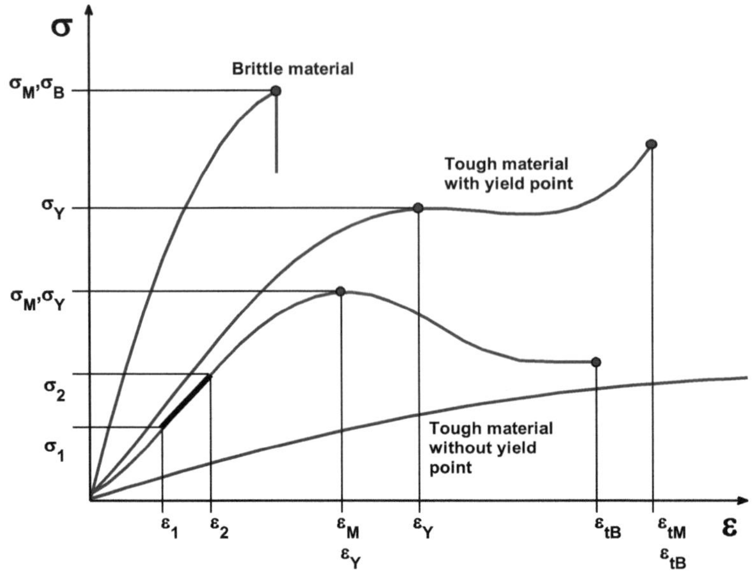

Fig. 14.5 Stress-strain curves according to DIN EN ISO 527-1

on a principal axis at constant deformation rate, with simultaneous registration of the force reaction of the specimen. These experiments result in stress-strain curve diagrams. Typical stress-strain diagrams, as well as the relevant general definitions, especially for the tensile test, are given in DIN EN ISO 527-1 [10]. Figure 14.5 gives relevant stress-strain curves and the corresponding definitions.

When stating values for stresses and strains in the elastic region of the specimen, the formula symbols σ and ε are used. Beyond the yield points σ_Y and ε_Y, indexing is performed as σ_t and ε_t. The nominal strain ε_t represents the relative elongation which takes place along the total length of the specimen, i.e. it is derived from the increase in length of the distance of the grips. The parameters relevant for the specimen are derived from the diagrams. Table 14.2 summarizes these parameters.

The stated definitions of parameters are used analogously in other load situations—e.g. compression, shearing, bending.

If the specimen does not have a defined cross-sectional area in the direction of load exposure, no statement can be made regarding cross-sectional force. Therefore, it will not be subjected to tensile testing. Instead, the (non-standardized) reaction force during deformation is measured. Comparability of the test data is ensured by established specimen dimensions.

Linear specimens, such as fibers, filaments, and yarns, have only a single load direction, and Eq. (14.1) is reduced to Eq. (14.3). For technical filaments/yarns, a cross-section/total cross-section can often be specified, which makes it possible to ascertain a stress-strain behavior as in Fig. 14.5. If cross-sectional tensile testing are not possible or undesired, these materials can also be tested for their tenacity (tension per unit of linear density).

Table 14.2 Parameters according to DIN EN ISO 527-1

Parameter	Definition	Unit
Tensile strength σ_M	Maximum stress value	MPa
Tensile stress at break σ_B	Stress at specimen failure	MPa
Tensile stress at yield σ_Y	Stress value at which elongation increase occurs without increase in stress	MPa
Strain at tensile strength ε_M	Strain at maximum stress	%
Nominal strain at tensile strength ε_{tM}	Nominal strain at maximum stress, if maximum stress occurs after yield point	%
Tensile strain at failure ε_B	Strain at specimen failure, if failure occurs before the yield point	%
Nominal tensile strain at break ε_{tB}	Nominal strain at specimen failure, if failure occurs beyond the yield point	%
Tensile strain at yield ε_Y	Strain at tensile stress at yield	%
Young's modulus E	(Initial) increase of the $\sigma(\varepsilon)$ graph in the 0.05 % ... 0.25 % elongation interval	MPa
Poisson's ratio μ	Negative ratio of transverse strain and extension	–

14.2.5 Presentation of Test Results

The presentation of test results is specified by standards, and usually takes the form of graphical depictions of relevant parameters in relation to the selected exposure of the specimen and/or other parameters. Graphical presentations, while cumbersome, contain the most information and details. They are preferred in the research and development. Specific values are convenient, easy to convey, but often unable to reflect details. They are usually used in commercial fields, for the specification of the properties of the structure at hand.

To ensure the reliability of individual parameters, textile tests always rely on the preparation and testing of several specimens under identical conditions. Deviations between the individual results are exclusively regarded as scatterings of the relevant parameter across the entirety of the test material, as laid out in Sect. 14.2.1. As a rule, the individual results are mathematically processed into statistical parameters. There are estimators for the test values measured during a general testing of the test material. These are

- arithmetic mean—as estimator for the average of the population
- standard deviation—as estimator for the standard deviation of the population
- coefficient of variation—as reference for the relative scattering of the parameter
- confidence interval—as the value range in which the mean of the population is located at a pre-specified probability (confidence level, nearly always 95 %)

These four statistical parameters are usually given in any test protocol.

The following chapters will use the designations drawn from the cited standards to specify tested parameters. Therefore, it has to be noted beforehand that the remarks concerning terminology and symbols cannot always be consistent. This is due to the fact that a large number of authors, the very complex situation of the field, and frequent changes to terminology and formula symbols have prevented the standards from being designed consistently over the past decades.

14.3 Tests on Fibers and Filaments

Fibers and filaments are characterized by:

- material (type, composition, internal structure/homogeneity)
- dimensions, shape (diameter, length, crimp, uniformity)
- mass (longitudinally related mass/linear density)
- mechanical parameters (stress-strain behavior, strength, flexural and torsion stiffness, retardation, friction coefficient)
- physical properties of the material (thermal conductivity, thermal capacity, electric conductivity, thermal expansion, dielectric behavior)
- chemical properties of the material (reactivity, biochemical compatibility, wettability), and
- durability (stability of the parameters over time, creeping, influence of radiation, humidity, temperature)

In-depth knowledge of the filament properties is crucial, as it allows a formulation of the limit properties of the yarns produced from the filament. The real yarn parameters are considerably influenced by the yarn construction and cannot be calculated with sufficient accuracy, even with detailed knowledge of the filament properties.

The properties of the fibers and filaments are only rarely tested during the production of composite structures, due to the fact that the creation of value usually starts with the purchased yarns, as their properties are the decisive factor. Therefore, the remarks in this chapter will be kept brief and limited to some of the parameters of technical fibers and filaments. Furthermore, the selection is largely limited to the standards for fibers and filaments from glass, aramid, and carbon. These can analogously be used for other materials, for which there are normally no special test standards (Table 14.3).

14.3.1 Diameter

The diameter of a filament is measured after removal of the sizing. For the measurements, optical methods are available, as specified, for instance, in DIN 65571 for diameter measurement by means of:

- light microscope and micrometer eyepieces (span measurement cylinder diameter)

Table 14.3 Testing methods for fibers and filaments

Parameter	Unit	Material	DIN/ISO standard
Diameter	µm	Glass Glass, aramid, carbon	ISO 1888 DIN 65571
Linear density	dtex	Glass, aramid, carbon	DIN EN ISO 1973
Tensile strength Young's modulus Elongation at break	MPa MPa %	Glass, aramid, carbon	(DIN EN ISO 5079)

- light microscope and digital camera, with subsequent planimetry (transverse microsection measurement)
- laser interferometry, and
- projection

Determining the filament diameter is of importance to the manufacturers of technical yarns, including the developers of filaments based on new materials. Users of technical yarns are primarily interested in knowing the distribution of the filaments of a yarn used in a textile fabric. For this, transverse microsections are required. For planimetry, specialized software is available for both ascertaining the diameter and detecting the distribution of the filaments in the cross-section. The software is often offered in a full package with microscope and digital camera.

14.3.2 Linear Density

Linear density is the mass per unit length of a filament. Filaments for technical yarns have a defined cross-section area A. Combined with the mass density ρ of the material, the length-related mass can be calculated, which is the linear density T_t, in the form of

$$T_t = \frac{m}{L} = \rho A \tag{14.6}$$

According to ISO 1144, and the analogous DIN 60905-1 standard, the linear density is given in tex, where the following applies:

$$1\ \text{tex} = \frac{1\ \text{g}}{1000\ \text{m}} = \frac{1\ \text{mg}}{1\ \text{m}} \tag{14.7}$$

The linear density of filaments is often given in *dtex*.

As set down in DIN EN ISO 1973, filament linear density can be determined gravimetrically by cutting a specified number of fibers to a given length. From the total length and total mass of the filament bundle, the filament linear density is calculated, using Eq. (14.6). Handling filaments is complicated, so it is easier to establish filament linear density using the vibroscope method, which is also contained in the standard. For this, the filament is arrayed between to two fixed points (edges) set at a span L between them. Under a pre-tension F_v, the filament is set to oscillating at a variable frequency while the amplitude of the transversal oscillation of the filament is simultaneously determined optically. The highest oscillation amplitude will be measured at the resonant frequency f_r of the filament, from which the filament linear density T_t can be calculated according to

$$T_t = \frac{F_v}{4 \times f_r^2 \times L^2} \tag{14.8}$$

Equation (14.8) is the well-known physical equation for a vibrating string.

14.3.3 Tensile Strength, Young's Modulus

There are few general standards for measuring the load-deflection behavior and fatigue parameters of technical filaments. Some aspects of the tests can be drawn from DIN EN ISO 5079, which was developed for spun fibers. Tensile testers are required to allow a clamping of the filaments at a nip-to-nip of the clamp jaws of 10 or 20 mm, and to enable measurements of small forces in a range from 0 to ca. 100 cN. Deformation speed is set to ensure a strain rate of the specimen of 50 % min^{-1}. The clamping of the filament requires special clamp surfaces to prevent the filaments from slippage between the clamps and being damaged. The measured force, or the tension calculated from it, is recorded over the elongation. This results in typical characteristics, as shown in Fig. 14.5 in the curve "brittle material". The maximum values of tension and elongation of the filament occur at its break. Test protocols give these values as tensile strength σ_M and ultimate elongation ε_M. The Young's modulus is then determined based on the slope of the measured stress-strain plot.

As shown above, the testing of filaments allows the determination of essential parameters on the properties of the yarns to be produced. Particularly, the theoretical yarn strength cannot exceed the sum of all individual strengths of the filaments contained in the yarn. The degree to which the real yarn strength differs from this limit value due to yarn construction and filament damages gives crucial hints regarding the quality of the yarn and its production.

14.4 Yarn Testing

Technical yarns are characterized by the properties described for filaments at the beginning of Sect. 14.3. Additional characteristics describe

- yarn construction (number of filaments, twist, material ratios, and mixture in hybrid yarns)
- coating (sizing/preparation, and their moisture content)
- yarn defects (filament breakage, protruding filaments, lint), and
- roving construction (deviation from cylinder shape, fallen roving)

The relevant test methods and standards are listed in

- DIN EN ISO 14020-2 for textile glass rovings
- DIN EN 13003-2 for para-aramid fiber yarns, and
- DIN EN 13002-2 for carbon filament yarns

Table 14.4 summarizes a number of essential standards. Here, too, the deliberations will be limited to glass, aramid and carbon fiber yarns, for the same reasons as in the sections concerning filaments.

14.4.1 Linear Density

The linear density of a yarn is its mass per unit length, with or without sizing/preparation. The mass m of a piece of yarn of a known length L is determined, and its length-related mass is then calculated. The method is defined in DIN EN ISO 1889. The linear density is stated, as described previously in Sect. 14.3.2, in accordance with the international tex system found in ISO 1144.

The yarn length for the determination of length-related mass can range from 5 to 500 m and is set to yield a yarn piece of 3–10 g of mass. The yarn length is provided by winding the yarn on a wheel with a circumference of 1 m. For a consistent winding, a traversing yarn guiding system is provided.

The following methods are currently used to remove the sizing, which would otherwise contribute a few percent to linear density:

- Glass Calcination at 625 °C,
- Aramid Extraction by means of a Soxhlet extractor at 105 °C, subsequent drying,
- Carbon Extraction by means of a Soxhlet extractor at 105 °C, or thermal decomposition at 450 °C in a nitrogen atmosphere

14.4.2 Twist

In accordance with DIN EN ISO 1890, twist is determined using a twist testing device consisting of two horizontally yarn clamps aligned with each other and situated at a span of 500 mm. One of the clamps has to be rotating, and the number of turns of the clamp is given by a counter. The yarn is stretched between the clamps with a pre-force (typically 0.25 cN/tex) and untwisted until the filaments are oriented parallel to each other. This parallel alignment is determined by means of

Table 14.4 Test methods for yarns

Parameter	Unit	Material	DIN/ISO standard
Linear density	tex	Glass, aramid, carbon	DIN EN ISO 1889
Twist	1/m	Glass, aramid, carbon	DIN EN ISO 1890
Tensile strength Maximum tensile strength/tensile strain at break Young's modulus	MPa % MPa	Glass Aramid Carbon	ISO 3341 DIN EN ISO 9163 DIN 65382 DIN EN 12562 DIN EN ISO 2062 DIN 65382 DIN EN ISO 10618 DIN 65382
Sizing content Preparation mass portion	% %	Glass Aramid, carbon	ISO 1887 or ISO 15039 DIN EN ISO 10548
Moisture content	%	Glass, aramid, carbon	DIN EN ISO 3344

a dissecting needle inserted between the clamps and through the entire filament group. The twist T of the yarn is the number of turns N of the yarn at a yarn length L of 1 m

$$T = \frac{N}{L} \tag{14.9}$$

The yarn has to be taken off from the bobbin tangentially to avoid an additional imparting of twist by an over-end take-off of the yarn from the bobbin. It is also common to state the twist direction of the yarn, which is symbolized by the letters S or Z. When a yarn is held perpendicular, the windings of the filaments around their axis incline in the direction of the central portion of the letters S or Z, respectively. The statement of a twist factor often found for conventional spun yarns is uncommon for technical yarns, as their number of turns is very small.

14.4.3 Tensile Strength, Young's Modulus

Yarn strength and the elastic behavior of yarns are determined in tensile tests. Force/tension and elongation are recorded simultaneously, and the desired parameters are derived from the result. Generally, tests with untreated yarns and yarns treated with plastic matrices have to be distinguished. Tests on untreated yarns require much less effort.

The impregnation and lamination of filament yarns with polyester matrices improves the cohesion between the individual fibers. Thus, the load introduction

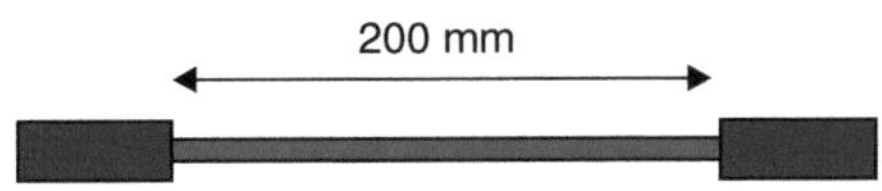

Fig. 14.6 Example of load insertion by means of end tabs ("bonded end tabs") as stated in DIN 65382

can be distributed more evenly to all filaments of a yarn. Dispersions based on acrylate, polychloroprene, and polyurethane are used for matrices, as are epoxy resins. The suitable polymers differ regarding their mechanical properties. The polymer properties are used to adjust the properties of the impregnated yarn region. Therefore, for them to be used in specimen preparation in yarn tensile tests, a high reproducibility of the matrix properties has to be ensured. This also applies whenever only the load introduction sections of the yarn specimens are impregnated with a polymer matrix.

For tensile tests of glass, aramid and carbon yarn/rovings, DIN 65382 can be used. Paper or plastic end tabs are attached to the ends of the impregnated specimens and serve as load insertion elements. This prevents damages to the filaments in the clamping area. Figure 14.6 gives a schematic example for the test design of yarn specimens.

The test is started with a pre-tension of 2 cN/tex. The initial span between grips depends on the specimen dimensions, is 200 mm, but at least 60 mm. Deformation speed is to be slower than 10 mm/min. The use of external elongation sensors applied to the center of the specimen allows a precise measurement of the deformation rate of the yarn specimen.

DIN EN ISO 9163 contains principles of specimen preparation and impregnation, and of attaching epoxy tabs for load introduction. Examples of yarn test specimen designs are given in Figs. 14.7a, b.

The advantage provided by this design is the improved load introduction from the clamps to the end tabs, as the load insertion elements do not have to be glued on in an extra process step, but can be attached to the specimen during production. Sampling and test procedure are very similar to those given in DIN 65382.

For carbon fibers and carbon filament yarns, DIN EN ISO 10618 applies. The variants of load insertion elements made from resin, reaction resin or paper, are specified in this standard. Specimen length is either 200 or 150 mm, deformation speed is not to exceed 250 mm/min. Figures 14.8a–d show examples for the design of yarn specimens. It is clear that the load insertion elements from the various standards do not differ in principle.

To ensure that as many filaments as possible are strained uniformly during tensile tests, the filaments within the yarn have to be aligned properly before impregnation. This filament arrangement differs from that of the base material, and influences the tensile-mechanical properties of the yarns. Usually, increased tension absorption can be observed.

Yarn impregnation is extremely time-consuming. The preparation of the yarn specimens requires an additional mold that needs to be properly tempered before

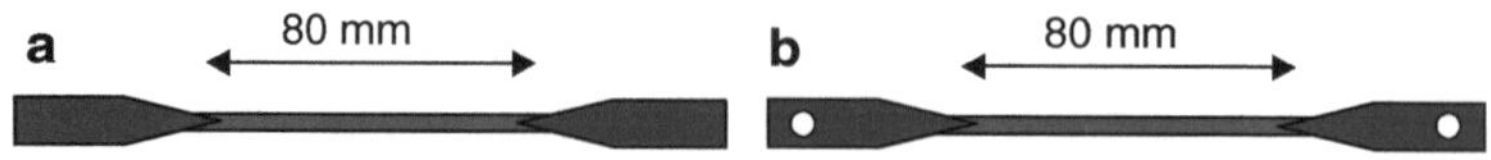

Fig. 14.7 Example of load introduction by means of end tabs ("bonded end tabs") as stated in DIN EN ISO 9163

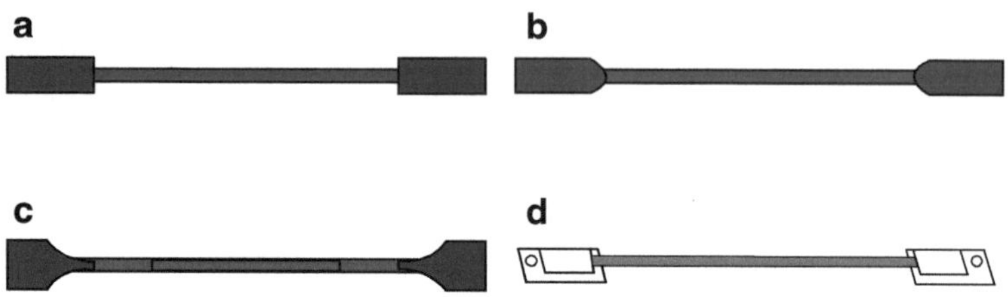

Fig. 14.8 Example of load introduction by means of end tabs according to DIN EN ISO 10618

the molding process. Furthermore, consistent properties of the impregnation agent have to be guaranteed at all times. Figure 14.9 shows an example of a device for the preparation of yarn specimens.

To prevent slippage of the laminated yarn ends between the clamps, and to ensure an even load introduction, additional clamp surfaces are often required.

Specimen preparation and testing is uncomplicated if the yarns are not impregnated at the time of testing. For aramid yarn in particular, DIN EN 12562 applies here, as does ISO 334 for glass rovings. DIN EN 12562 closely follows the standard used for textile yarns, DIN EN ISO 2062, which means that the initial spans between grips are 500 or 250 mm, the deformation speeds are set to 100 % of the clamping length per minute (500 or 250 mm/min). The pre-tension is 2 cN/tex. A predefined twist has to be added to untwisted aramid multifilament yarns. As per ISO 3341, glass rovings are also to be tested with initial spans between grips of 500 or 250 mm, while the deformation speed is to be 200 mm/min and the pre-tension is 0.5 cN/tex.

The proper selection of specimen holders and clamp surfaces is crucial for the prevention of test failures, such as jaw break (breaking of the yarn between clamps due to compressive loads) or jaw slippage (slippage of the yarn between the clamps due to insufficient frictional locking). Straight specimen holders (clamps) with a defined nip are suitable for impregnated yarns with load insertion elements. For yarns without impregnation, the tensile tests have to be performed with rope-grip clamps, in which the tensile force is reduced by friction through looping. Figure 14.10 gives the schematic of both clamping options for yarn specimens in a tensile tester.

When using rope-grip clamps, the yarn elongation cannot be derived from the crosshead motion of the tensile tester, as the yarn slips in the forward end of the yarn looping on the rope-grip clamp. Therefore, an elongation measurement directly on the yarn is required, which is performed by means of an optical length measurement system. For this, the optical extensometer described in Sect. 14.2.2

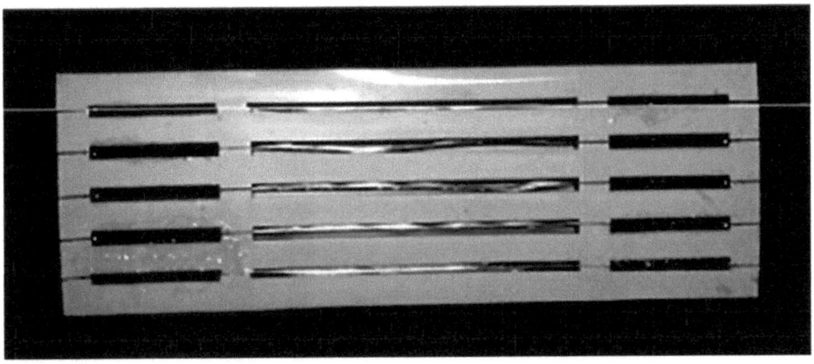

Fig. 14.9 Embedding form for the tensioning of the yarns and application of the load insertion elements

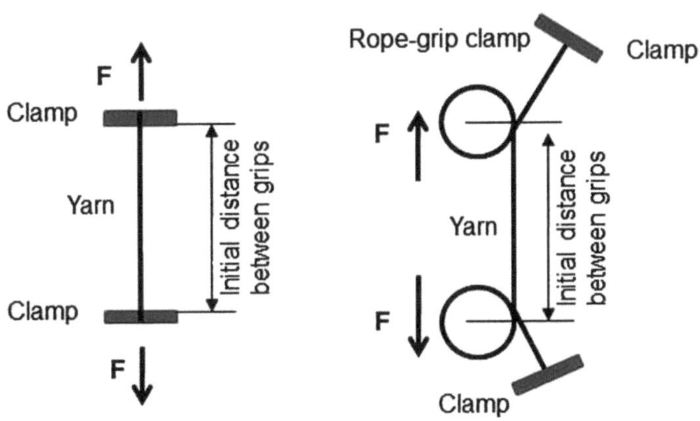

Fig. 14.10 Options for specimen clamping in tensile tests on yarns; left: clamps for impregnated yarn specimens with load insertion elements, right: "rope-grip clamps" for untreated yarn specimens

can be used. Figure 14.11 shows one such optical elongation measuring system, in which two cameras record the change of position of reflective foil attached to the yarn specimen, and derive the elongation of the yarn.

The actual test of the yarn is fundamentally similar in all test variations. The yarn specimen is tensioned in a suitable device in the tensile tester and loaded to pre-tension. Then, the specimen is loaded at constant deformation speed until the yarn breaks, while variations in length and tensile force are recorded simultaneously. As a rule, the tensile forces are represented in relation to the cross-section (tension σ), while length variations are represented length-related (elongation ε), resulting in some form of the σ-ε characteristics shown in Fig. 14.5. Tensile strength σ_M and Young's modulus can be determined as shown in Fig. 14.5, as can tension at break σ_B, and the respective elongations ε_M and ε_B.

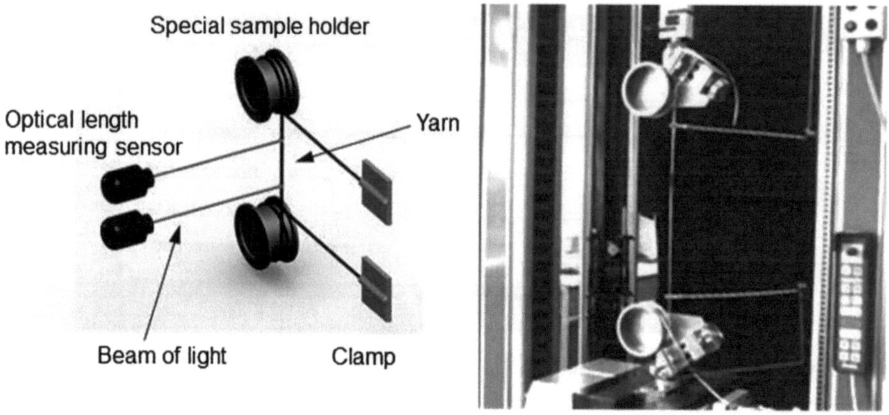

Fig. 14.11 Optical elongation measurement with two CCD cameras on a yarn specimen with two reflective foils; *left*: schematic of the test principle, *right*: rope-grip clamps on a tensile tester

If the yarn cross-section is not known at all or not in detail, the tensile force can also be related to the easily determined linear density of the yarn (tenacity, given in N/tex).

The tensile tests described above are performed at low deformation speeds under standard atmosphere, according to DIN EN ISO 291 or DIN EN ISO 139. With awareness of the end use of the yarns, especially in impact applications, tensile tests with high deformation speeds and high temperatures are becoming a requirement. Such tests at high strain rates are still being developed at the moment. Generally accepted standards for these cases have not been drawn up yet.

14.4.4 Sizing and Preparation Content

The sizing or preparation content *SC* of yarns characterizes the mass fraction of the sizing/preparation in the total mass of the yarn. It is defined as the percental ratio of the loss of mass Δm of a piece of yarn by removal of the sizing/preparation in relation to the total mass m of the yarn. Mass is always measured on dried specimens.

$$SC = \frac{\Delta m}{m} \times 100\,\% \tag{14.10}$$

The sizing/preparation is removed in a manner similar to that described for the determination of yarn linear density. Glass rovings are dried at 105 °C, after which their mass is measured. The sizing is removed by at least one hour of calcination in a muffle furnace, preferably at 625 °C. The subsequent cooling has to be performed in a desiccator, to prevent moisture absorption. The loss of mass ("calcinations loss") is determined, and the sizing content calculated by means of Eq. (14.10). The removal of the sizing can also be performed by extraction with a Soxhlet extractor.

A Soxhlet extractor is also used for aramid yarns, and for carbon yarns with soluble preparation. If the latter is not the case, wet-chemical oxidation (sulfuric acid/hydrogen peroxide mixture) or a pyrolysis under nitrogen atmosphere can be used.

14.4.5 Moisture Content

The moisture content H of yarns characterizes the mass fraction of water in the total mass of the yarn. Analogously to the sizing and preparation content, it is defined as the percental ratio of the loss of mass Δm of a piece of yarn by dehumidification, related to the total mass of the yarn.

$$H = \frac{\Delta m}{m} \times 100\% \qquad (14.11)$$

The test is standardized in DIN EN ISO 3344. Initially, the specimen is to be conditioned under a standard atmosphere for 6 h, as specified in DIN EN ISO 291. This is followed by the determination of its mass. The specimen is dried in a drying oven at a temperature, at which the sizing/preparation do not escape (usually 105 °C). A lower temperature can be selected in case of sizing/preparation losses, but it must be at least 50 °C. Afterwards, the specimen is cooled down in a desiccator, before the loss of mass is determined.

14.4.6 Other Test Methods

The mentioned test methods are largely sufficient for the characterization of the technical yarn with regard to its suitability for the desired product. In commercial trade, additional quality assurance tests are common. This concerns tests of the state of yarn (protruding filaments, lint, contaminations) and the roving (cylindrical shape, fallen roving). These tests are crucial for the unimpeded processability of the yarns on fabric-forming machines.

The advantageous properties of textile-based composite structures open up a growing number of applications. The result is an increased demand of tests for yarn under extreme

- load situations and
- environmental conditions

Extreme load situations are primarily simulated in high-speed tests. High-speed tensile tests or crash tests are already common for composite structures (see

Sect. 14.6.3), but are increasingly in demand for yarns. The realization of such tests has proven difficult. Commercially available high-speed tensile testers achieve speeds of 20–25 m/s and yarn strain rates of a few $10^2\ s^{-1}$—these strain rates will not suffice for future requirements. Knowledge of the deformation and failure behavior of the yarns at extremely high strain rates is necessary to estimate and specifically influence the impact behavior of the composites produced from the yarns. The measuring devices for force and length/elongation have to meet extreme measuring dynamic requirements, in particular for carbon yarns, as the low ultimate elongation of carbon results in a time interval from zero load to failure of the yarn of 100 μs or less.

Extreme environmental conditions include high and low temperatures as well as influence from surrounding media. Exposure to temperature is only sensible in tests of tensile strength and elasticity, but is gaining in importance, as the applications have expanded, and composites are used in a wide temperature range these days. The tests require additions to the test devices, among them climatic chambers and IR heaters. These additions are either offered by the manufacturers of the test devices or developed and provided by the user.

Determining the influence of surrounding media on tensile strength is also becoming more and more important as the range of applications of technical yarns is growing. One example is the use of yarns for the reinforcement of concrete, for which the properties of the yarns have to be determined after exposures typically found in practice, such as storage in an alkaline environment or simulated pore water solutions.

14.5 Tests on Textile Fabrics

Textile yarns are processed into textile fabrics with textile-technological methods. Fabrics produced from continuous fibers and pre-impregnated with an uncured thermoset plastic matrix in an additional process step are called prepregs. Textile fabrics, including prepregs, are primarily characterized by information on

- material (yarn),
- manner of fabric formation (woven, weft-knitted, non-crimp fabric; yarn and mesh densities; defects in the fabric),
- dimensions (length, width, thickness),
- mass per unit area,
- strength, elasticity,
- flexural stiffness, drapability, and
- sizing and matrix contents

Material and textile construction type are given factors only tested subsequently in exceptional cases. The lateral dimensions (length and width) are also primarily of an informative nature. Some relevant physical properties and corresponding test standards are given in Table 14.5.

Table 14.5 Test methods for textile fabrics

Parameter	Unit	Material	DIN/ISO
Thickness	mm	Glass, aramid, carbon	DIN EN ISO 5084
Mass per unit area	g/m^2	Glass, aramid, carbon Glass prepregs Carbon prepregs Glass, aramid, carbon prepregs	DIN EN 12127 DIN EN 2329 DIN EN 2557 DIN EN ISO 10352
Tensile strength Maximum tensile force/ultimate elongation	N, MPa %	Glass, aramid, carbon	DIN EN ISO 12934-1 (DIN EN ISO 527-1)
Flexural stiffness	Nm^2	Glass, aramid, carbon	DIN 53362
Resin/fiber content	%	Glass prepregs Glass prepregs Carbon prepregs Glass, carbon prepregs	DIN EN ISO 1172 DIN EN 2331 DIN EN 2559 DIN EN ISO 11667

14.5.1 Thickness

Thickness is defined as the perpendicular span between the top side and the bottom side of a textile fabric. A textile fabric usually has a distinct surface profile, so assigning the thickness parameter requires precise determinations, which can be gathered from the test standards. According to DIN EN ISO 5084, the textile is placed on an even (polished) reference plate. A second, parallel plate, which is designed as a round presser-foot and coupled with a length measuring system, is applied to the specimen with a defined pressure. Then, the span between the presser-foot and the reference plate is determined and quoted as the thickness of the fabric. Preferred values for pressure and time, as well as for the area of the presser foot are given in the standard.

Fabrics from technical yarns are subject to special standards, such as ISO 4603 for woven fabrics. They are not fundamentally different from the stated standard.

14.5.2 Mass per Unit Area

The parameter represents the mass in grams of 1 m^2 of the fabric. The most common standard for determining the mass per unit area is DIN EN ISO 10352. The standards for glass prepregs (DIN EN 2329), and carbon prepregs (DIN EN 2557) differ only in details from the general standard DIN EN ISO 10352 (sampling).

Determining the mass per unit area is relatively effortless: specimens sized $A = 0.1 \times 0.1\ m^2$ or $A = 0.2 \times 0.2\ m^2$, depending on the standard, are taken from the fabric, and their mass is measured. The mass per unit area M results from

$$M = \frac{m}{A} \qquad (14.12)$$

Particularities of the individual standards concern the specific sampling conditions and the proper procedure of the tests. For prepregs, the (partly volatile) preparations have to be taken into account in determining the mass per unit area, while any separating foils can be disregarded. The areic fiber mass is determined after removal of all preparations and resin contents.

14.5.3 Tensile Strength, Young's Modulus

Tensile strength is expressed by the parameters maximum tensile load F_M (maximum force under tensile strain) or breaking load F_B (force at the tearing or breaking of the specimen). These parameters help assess the capability of the fabric to achieve the ultimate strength resulting from the fabric construction and yarn strengths. The corresponding elongation values of the specimen are the strain at maximum load ε_M, or tensile elongation at break ε_B.

14.5.3.1 Monaxial Tensile Tests

Tensile strength and elasticity are determined by tensile tests. These are monaxial tensile tests, i.e. a loading of the textile fabric in one (preferred) direction, e.g. in weft or warp direction of woven fabrics. The preference for monaxial tensile tests is due to the relatively easy feasibility of the required test machines—as opposed to the realization of defined biaxial load exposures by test engineering means. Therefore, the term "tensile test" has become generally synonymous with monaxial tensile tests in general terminology and technical representations—a convention that holds true for this chapter.

Tensile tests primarily performed as simple tensile tests, i.e. *plain tensile tests*, and are detailed in DIN EN ISO 13934-1. For these tests, strip-shaped specimens are clamped in a tensile tester and loaded with a constant deformation rate of 20–100 mm/min until failure (break, rupture). During this loading, both the elongation and force reaction of the specimen are recorded. The initial span between grips of the specimens in the tester is specified as 200 or 100 mm, and their width as 50 mm. The specimen preparation has to provide sufficient length for clamping between parallel clamps and, if applicable, a defined specimen width by removing yarns from each of the long edges by unraveling. Fabrics from technical yarns are often prepared differently than stated in the standard. If the sum of the linear densities of

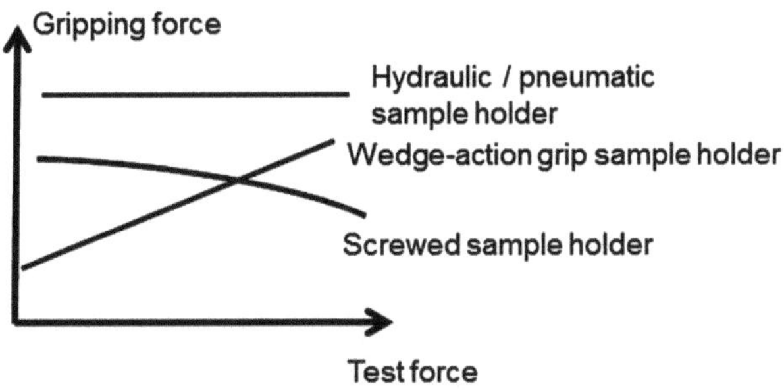

Fig. 14.12 Gripping force of various specimen holders depending on test force

the yarns running in load direction is known, it is sensible to establish the relation of the measured forces to the linear density, for the sake of comparability of specially shaped specimens. If the sum of yarn or filament cross-section surface areas is known, a relation to the cross-section is to be preferred, and the parameters given in Fig. 14.5 apply.

Correct clamping of textile fabrics made from technical yarns is often problematic. Usually, specimen size only allows clamping between parallel clamps. Gripping force acting perpendicularly to the specimen plane, is created by mechanical fastening or hydraulic/pneumatic cylinders. Figure 14.12 is a schematic representation of the manner in which the gripping force is subject to possible changes, depending on the test force of the tensile tester. The most advantageous characteristics are exhibited by the wedge-action grip-specimen holders, which exerts a higher gripping force on the specimen with increasing test force.

Sufficient gripping force is required to prevent a slippage of the specimen between the clamps. However, a very high gripping force can cause mechanical damage to the specimen between clamps, maybe even cause jaw break of the specimen. To simultaneously preempt clamp slippage and jaw break is a feat for experienced testing staff, which can use variations of clamp type, gripping force, and clamp surfaces. Corrugated or rough profiles are much more common than smooth ones. The steel clamps are often fitted with hard-elastic surfaces, which ensure the required gripping force but allow a certain degree of form-fit grip on the textile surface. For example, smooth or corrugated surfaces made from polyester-urethane rubber (Vulkollan™) are often used.

The elasticity properties of the textile fabrics can be derived from the determination of tensile strength by means of the strip method, if the stress-strain behavior in the lower elastic range of loading is considered. The notion of elasticity that applies to classic textiles, according to DIN 53835-13 or DIN EN 14704-2, is concerned with the relation of elastic and residual elongation after exposure to loads, and can therefore not be applied to textile fabrics produced from technical yarns. Analogously, DIN EN ISO 527-1 can be used, if the yarns of the fabric are aligned straightly in load direction, and if no length variations due to geometrical stretching of the yarns occur, which is virtually never the case. The complete lack of generally acknowledged and used standards for the elasticity or the Young's modulus of textile fabrics made from technical yarns is a manifestation of that fact. Such tests only gain significance after lamination of the textile fabrics.

14.5.3.2 Biaxial Tensile Tests

As it is important to closely describe the complex material behavior of textile fabrics, biaxial tests are a necessary addition to the test range methods. Biaxial tensile testers fundamentally consist of two perpendicularly oriented monaxial tensile testers. Two load axes allow simultaneous exposure to loads in longitudinal and transverse direction. Usually, the specimen is fixed in place by four specimen holders, in most cases consisting of parallel clamps. The motion of each individual specimen holder can be controlled separately. This makes complex load situations realizable in the experiment, in which loading in one direction always influences the mechanical state of the other directions. The biaxial tensile tester—as opposed to the monaxial tensile tester, for which one load cell suffices—requires four load cells to record the force reactions of the specimen in the four load directions.

Biaxial tensile tests are performed to test the suitability of the tested materials for their planned application in practice. The specimens are conditioned to typical practical defaults and then exposed to loads until failure occurs. Elongation and force reaction of the specimen are detected simultaneously for all four load directions. Regarding the clamping of the specimen, the same aspects as for single-axis tensile tests apply. Rounded compression dies arranged perpendicular to the textile fabric enable the user to test the suitability of preforms for curved composite components. Specimen size, length as well as type of nips during clamping, and above all the elective exposure to loads in four directions, which can be strain- or stress-controlled, result in unique test regimens and results. As the standardization of the biaxial tensile test is therefore problematic, there are no generally accepted standards and parameters.

14.5.4 Flexural Rigidity

The *flexural rigidity* of textile fabrics gives indications regarding their drapability and is therefore important for their assessment of suitability as three-dimensional semi-finished product. Using the definition "stiffness in bending" from engineering mechanics, where it denotes the quotient of bending moment and curvature, is only partly useful, since the material is not displaced elastically, but instead the filaments glide alongside each other. The most important standardized method for ascertaining stiffness in bending is the cantilever method from DIN 53362 (ASTM D1388). In this test, shown in Fig. 14.13, a strip-shaped specimen is slid over an edge, bending downward under its own weight.

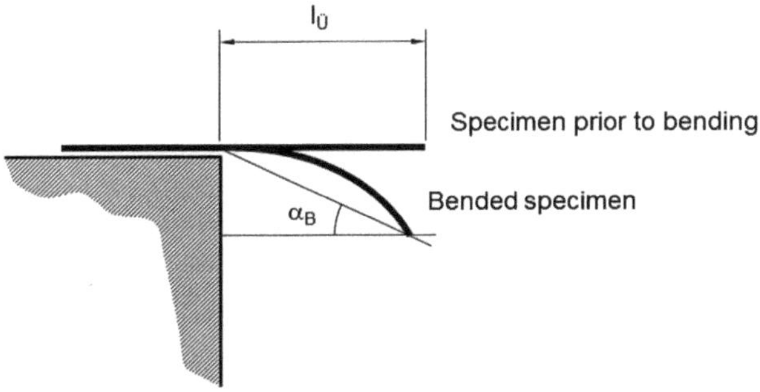

Fig. 14.13 Schematic of the Cantilever method

At a predefined bending angle $\alpha_b = 41.5°$, the overhang length $l_Ü$ is measured, and the stiffness in bending B is then formally calculated according to the equations of engineering mechanics:

$$B = F_L \times \left(\frac{l_Ü}{2}\right)^3 \quad (14.13)$$

F_L is the length-related weight force of the specimen derived from the mass per unit area described in Sect. 14.5.2. The bending angle $\alpha_B = 41.5°$ is convenient because the specimen assumes a representative bending at this angle, and the equation for the stiffness in bending is simplified.

The cantilever method is complemented in practice by other special (often in-house) methods simulating the respective draping process. Fundamentally, the material is subjected to a defined flexural deformation, while its force reaction is being measured. The reproducibility of the flexural deformation is of utmost importance, as the force reaction of the specimen caused by relaxation of the yarn/filament shifts is time-dependent.

14.5.5 Resin and Fiber Content

The filler content ratio and the fiber content or content ratio of reinforcement fibers φ_m are relations between the fiber mass m_f and the total mass m_v of a prepreg.

$$\varphi_m = \frac{m_F}{m_V} \times 100\,\% \quad (14.14)$$

They are calculated by determining the total mass of the prepreg and the mass after removal of the matrix. The matrix is removed by

- calcination, i.e. heating in a muffle furnace (glass fiber) and

- chemical extraction (glass fiber: hydrochloric acid, methyl ethyl ketone; carbon fiber: sulphuric acid used with hydrogen peroxide solution, dichloromethane, acetone, methyl ethyl ketone)

  A Soxhlet extractor is the preferred tool for extraction.

14.5.6 Additional Test Methods

The tests described above are typical for textile fabrics. Beyond them, there is a variety of other standardized test methods, all of them adapted to the respective type of material, for example the groups of standards for nonwovens or geotextiles. Aspects in further test methods for textile fabrics include:

- deformability (draping),
- deformation resistance (elastic recovery, retardation, relaxation),
- sewability,
- insulation properties (heat, water, air/gas, sound), and
- resistance to surrounding media, heat/flames

Apart from standardized test methods, companies employ their own unique, non-standardized test methods to test special properties under special conditions. Often, these tests are concerned with deformability and deformation resistance. One example is given in Fig. 14.14. The test device applies shear deformations or combined shear- und flexural deformations in order to enable an assessment of the drapability of textile fabrics.

14.6 Tests on Fiber-Reinforced Plastic Composites

Tests on fiber-reinforced plastic composites are a complex field, including the extensive characterization of these structures. Due to the great variety of test methods, not the entire range of methods will be portrayed in this chapter.

Among the tests on composite structures, the tests depicted in the previous chapters, beginning with the filament, can be included. At the other end of the validation chain are methods with defined special-shape specimens, i.e. specimens of standardized dimensions, from whose properties conclusions can be drawn as to the behavior in use of the finished composite structures. A selection of these methods will be detailed. For this, the following classification is practical:

- tests without mechanical loads,
- tests at low deformation rate (including static load), and
- tests at high deformation rate

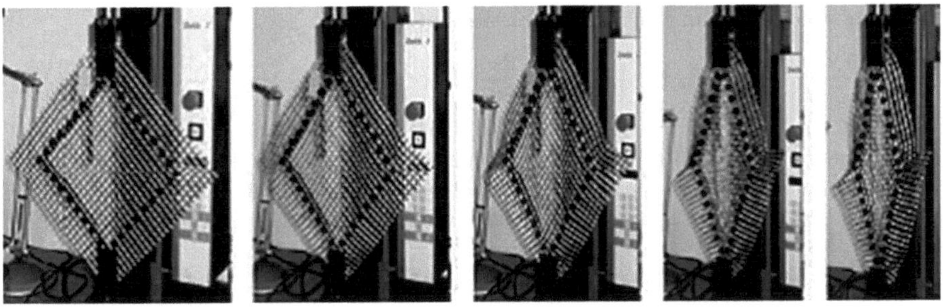

Fig. 14.14 Shear of a textile fabric for drapability assessment

Individual test methods exist for the separate tests. From the test results of one category, the properties of another category cannot be derived with precision (or at all). Often, this also applies to the assessment of the practical suitability of complex composite components, based on test values gathered with standardized specimen dimensions. In this case, component-specific test methods have to be developed, which naturally are not part of the following descriptions.

The importance of defined specimen production and preparation and of the test conditions has to be repeated here. The standard tests are always conducted in the conditioned state, i.e. specimens are leveled to the test atmosphere described in DIN EN ISO 291 and tested in that atmosphere.

14.6.1 Tests Without Mechanical Loads

General tests determining load-independent parameters often only serve informational purposes, but can be of importance in commercial communication. Test methods are concerned primarily with

- material (type, ratios of different components),
- construction (setup and orientation of reinforcement layers),
- dimensions,
- mass, amount (density, mass per unit area, fiber volume content),
- physical properties (thermal expansion, electric conductivity, dielectric properties, resistance to temperature and radiation), and
- surface properties (adhesion of filament and matrix)

Two test methods often applied in practice are given in Table 14.6 and will be described in the following sections.

The fabric mass or mass per unit area is determined in the usual way, which is already described in Sect. 14.5.2, as the quotient of mass and surface area of the specimen. The specimens are cut from the laboratory sample with a cutting device, and weighed. The preferable specimen dimensions are 0.20×0.20 m^2 or

Table 14.6 General test methods without mechanical loads

Test method	Parameter	DIN/ISO standard
Determination of mass	Mass per unit area Fabric mass	DIN EN ISO 10352
Determination of composition	Fiber content Filler content Resin content	DIN EN ISO 11667 DIN EN ISO 1172 DIN EN 2564

$0.20 \times 0.10\ m^2$. The methods are inspired by the approach to the determination of the mass per unit area of textile fabrics.

The fiber volume count φ_v results from Eq. (14.15), as the quotient between fiber volume v_f and the total volume v_v of the fiber-reinforced plastic composite.

$$\varphi_V = \frac{V_F}{V_V} \times 100\,\% = \frac{A_F}{A_A} \times 100\,\% \tag{14.15}$$

To determine the fiber volume count, a microsection of a specimen is produced, from which the total area A_f of the fibers can be determined by counting filaments and multiplying their number with the known cross-section area of a filament. This is then related to the total area A_A. Figure 14.15 clarifies this situation.

Filler material content or fiber content, or the content of reinforcement fibers by mass φ_F are the relation between fiber mass m_F and total mass m_V. The content of reinforcement fibers by volume φ_V is defined analogously. The fiber content is usually written as a percentage of the total mass.

$$\varphi_F = \frac{m_F}{m_V} \times 100\,\% \tag{14.16}$$

They are calculated by determining the total mass and mass after removal of the matrix. Depending on the type of the fiber the matrix is removed by

- calcinations
- chemical extraction

as described in Sect. 14.5.5. The relation of the mass-related ratios φ_F is of limited informational value. It is more important to know which volume content ratio φ_V is taken up by fibers. For this reason—and with knowledge of the mass densities of the fiber ρ_{fiber} and the matrix ρ_{matrix}, or after determining the mass density $\rho_{composite}$ of the composite—the mass-related value is converted to the volume-related value by means of Eq. (14.17).

$$\varphi_v = \varphi_F \times \frac{\rho_{composite}}{\rho_{fiber}} = \frac{\frac{m_{fiber}}{\rho_{fiber}}}{\frac{m_{fiber}}{\rho_{fiber}} + \frac{m_{matrix}}{\rho_{matrix}}} \times 100\,\% \tag{14.17}$$

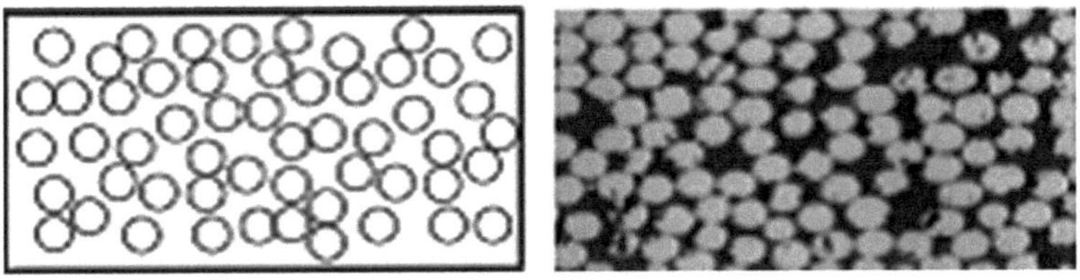

Fig. 14.15 Cross-sections of filaments in a composite; *left*: schematic, *right*: microsection

Established test methods give information on the composite "as a whole". The properties of individual components and their interactions are included integrally in the test results. The development of technologies for fiber-reinforced plastic composites shows that the achievable limit values for relevant properties of these composites, such as their strength, can be influenced by an ideal design of the interactions between the individual components. This pertains primarily to the adhesion of filament and matrix, and the integration and distribution of the filaments in the composite. However, generally accepted standards for the testing of these properties (such as pull-out-tests of yarns from matrixes) do not exist yet.

14.6.2 Tests at Low Deformation Rates

Tests at low deformation rates are the most important test methods for the characterization of fiber-reinforced plastic composites. They mostly serve to determine

- material parameters, according to the remarks in Sect. 14.2.4,
- strengths, and
- service life times

For these tests, specimens produced in a defined manner are exposed to a defined deformation—usually along the main axes—while the force reaction is measured. The advantage of defined load cases in the direction of the main axes at low deformation rates is the relatively small scatter range of the test value. The matrix elements of the Eqs. (14.1) and (14.2) can be determined in this manner. If the deformations are followed through to the destruction of the specimen, information is gathered regarding its strengths.

Complex load situations are more common in practice. The respective tests often require special test devices and regimens. As a rule, test values from differing methods are not comparable, since the influence of the test method and the specimen dimensions on the test result is too significant. Only tendencies of the material parameters—i.e. parameters increasing or decreasing—could be stated with results from differing test methods. Suitable and widespread methods are given in Table 14.7 and will be detailed in the following sections.

Table 14.7 Test methods with low deformation rate

Test method	Parameter	DIN/ISO standard
Tensile test	Parameters as in Table 14.2	DIN EN ISO 527-1,-4,-5
Compression test Flexural test	Parameters analogous to Table 14.2 "pressure" instead of "tensile" "compression" instead of "strain" Flexural modulus, flexural strength	DIN EN ISO 14126 DIN EN ISO 14125
Shear strength test Shear test	Shear strain/shear stress response Shear modulus, shear strength Apparent interlaminar shear strength	DIN EN ISO 14129 DIN EN ISO 14130
Creep test	Tensile-/flexural-creep strain Tensile-/flexural-creep modulus	DIN EN ISO 899-1,-2

14.6.2.1 Tensile Tests

Tensile tests are concerned with determining the Young's modulus E_i, the lateral contraction coefficients ν_{ij} (or the Poisson's ratio μ_{ij}), and the strengths σ_B. This is made possible by loading the specimen from zero to failure, which covers the low and high tension areas in a single test. There are several international standards for tensile tests, such as DIN EN 2561 or ASTM D3039/D3039M, but in a European context, DIN EN ISO 527-4 and -5 should be applied.

DIN EN ISO 527-5 describes the conditions for tensile tests on unidirectionally reinforced fiber-reinforced plastic composites in longitudinal and transversal directions. The necessary specimen shape is given in Fig. 14.16a.

Load insertion is ensured by means of load lead-in elements called bonded end tabs, with a mandatory length of at least 50 mm. Their thickness can vary between 0.5 and 2 mm, and they are made from GRP laminate whose fibers are oriented at ±45° to the load direction. These are attached to the specimen with a highly elastic adhesive. The initial span between the bonded end tabs is 150 mm, the specimens are 15 or 25 mm wide, and the test rate is set to 2 or 1 mm/min. Further details are included in DIN EN ISO 527-5.

DIN EN ISO 527-4 lists requirements to be met by tensile tests on isotropically and anisotropically (more accurately: orthotropically) reinforced fiber-plastic composites. The specimen dimensions and test conditions differ from those designated in DIN EN ISO 527-5. Three variants are set down; one of the most widely used is shown in Fig. 14.16b. To guarantee a central clamping, centering pins can be used.

The clamping of the specimen into the tensile tester can be effected by means of a variety of specimen holders, and the aspects detailed in Sect. 14.4.3 also apply to tensile tests on fiber-reinforced plastic composites. The clamping must not result in pre-tension in deformation direction being imprinted into the specimen. Wedge-action grip specimen holders, exerting a higher tensioning on the specimen with increasing test force, usually display the most advantageous properties. For force and length measurements, the deliberations in Sect. 14.2.2 apply. Use of the crosshead travel position for length measurements is to be avoided. Extensometers or strain-gauges applied to the specimens are more advantageous. The

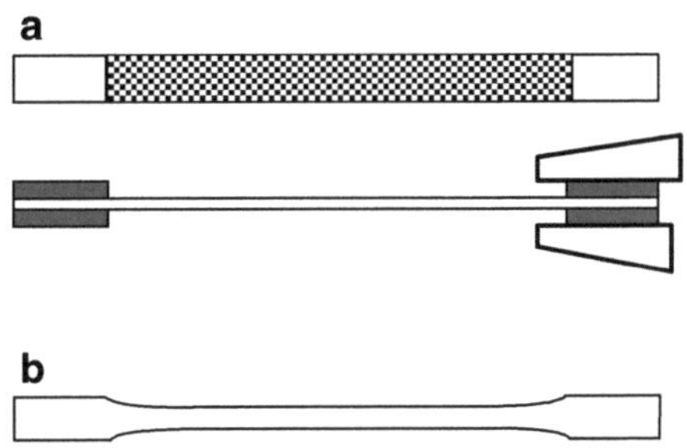

Fig. 14.16 Examples of specimen shape for tensile tests; (**a**) according to DIN EN ISO 527-5, (**b**) according to DIN EN ISO 527-4

characteristics are always represented as σ-ε diagrams in the vein of Fig. 14.5; the parameters are calculated according to Table 14.2. Often, an additional assessment of the type of failure, i.e. a statement regarding the fracture pattern, can be useful.

Modern test laboratories use biaxial tensile testers allowing a biaxial load introduction in 0° and 90° direction simultaneously. However, no generally accepted standards exist concerning specimen shape, test regimen, or evaluation.

14.6.2.2 Compression Test

In compression tests, the axial compression properties are determined. The compression test is a distinct test, not a simple inversion of the tensile test because composite structures display failure types such as buckling or bulging, or split vertically at relatively small compressive stresses. Therefore, compressive strengths are usually lower than tensile strengths.

Due to the complex mechanical failure mechanisms, compression test are problematic. There are many test standards resulting in different values because of different specimen dimensions and specimen fixtures. Reproducible results are difficult to obtain, even if the respective standard is consequently adhered to. Specimen preparation is once more of crucial importance here.

DIN EN ISO 14126 attempts to unify as many aspects of other standards as possible in one. It standardizes two types of specimens, one of which is to be used preferably for unidirectional materials, the other for "other materials". Conventional specimen holders/clamps are allowed, as long as the deformation by bending or buckling at failure of the specimen does not exceed 10 %. With this, an axial loading during the compression test can be assumed.

The specimens prepared in accordance with DIN EN ISO 14126 are sized 110×10 mm^2 or 125×25 mm^2, and the largest part of the specimen serves as clamp area. The free specimen area for compression tests is 10×10 mm^2, or 25×25 mm^2. The specimens can be fitted with load lead-in elements (bonded end tabs), for which the same aspects apply as for tensile tests. Figure 14.7a shows the schematic of a specimen with bonded end tabs.

For load insertion, two principles are provided by the standard, and are shown in Figs. 14.17b, c. A lateral compressive force introduction essentially causes a shear

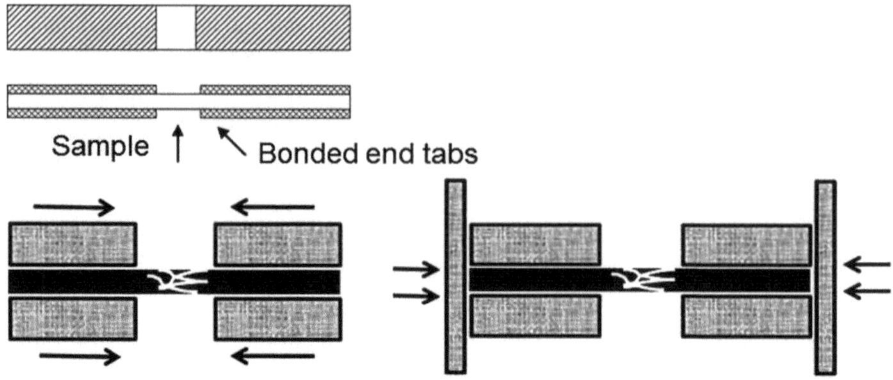

Fig. 14.17 Schematic of a compression test in accordance with DIN EN ISO 14126; (**a**) specimen, (**b**) load introduction via lateral surfaces (shear loading), (**c**) load introduction via ends (end loading or mixed loading)

loading of the specimen, while compressive force introduction via the ends provides end loading or mixed loading.

For the compression test, tensile testers with a directional inversion of the crosshead motion can be used to realize a compressive loading of the specimen. The clamping of the specimens into the tester is secured with compression test fixtures (trapezoids), which are adaptions of the globally common Celanese and IITRI compression test fixtures used in ASTM D3410 and DIN EN ISO 14126.

The compression of the specimen is to be measured preferably by strain-gauges applied to the free specimen surface with an adhesive. The crosshead rate is set to 1 mm/min, and the specimen is loaded until failure. The σ-ε characteristics are recorded. The test report contains the parameters compressive failure stress, compressive failure strain, and modulus of elasticity in compression, analogous to Table 14.2. Furthermore, the type of failure (shear, breaking, or vertical split/delamination) is to be recorded in the test report.

Apart from DIN EN ISO 14126, other test methods can be applied, all of which can differ with regards to specimens, test conditions, and primarily specimen holders/compression test fixtures. Examples are given in the normative references at the end of this chapter.

14.6.2.3 Flexural Tests

Flexural tests serve to determine the *flexural modulus* (modulus of elasticity in flexure) and the *flexural strength*. Near exclusively, three-point and four-point flexural tests are used to establish these parameters. Several standards exist for this purpose, which share the same principle but differ in terms of flexural support/test fixture, specimen dimensions, and experimental parameters. The flexural support is always designed symmetrically, so as to ensure a symmetrical load insertion

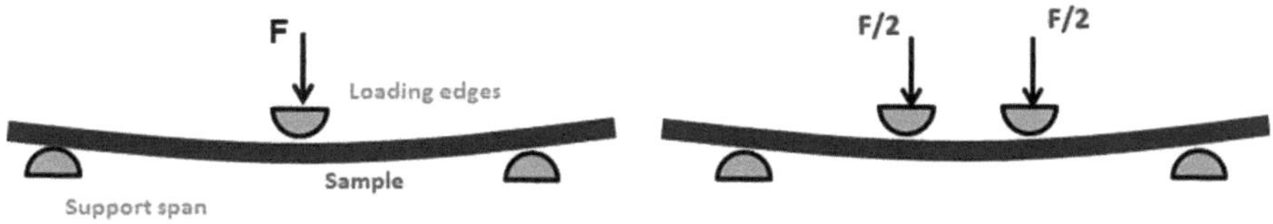

Fig. 14.18 Schematic of a flexural test arrangement according to DIN EN ISO 14125; (**a**) three-point flexural test, (**b**) four-point flexural test

for bending the specimen. Bending creates a multiaxial state of stress in the specimen, resulting in tensile, compressive, and shear stresses. For this reason, a variety of failure types can occur. Shear stresses during flexural tests are kept small by a sufficiently large ratio of specimen length and specimen thickness. Principal test assemblies following DIN EN ISO 14125 are shown in Fig. 14.18. The advantage of the four-point flexural test is the constant bending moment created between the two symmetrically arranged loading edges.

Specimen dimensions differ greatly, depending on the individual standards. Specimen lengths usually range from 50 to 100 mm, specimen width from 10 to 25 mm, and specimen thicknesses from 2 to 4 mm. For three-point flexural tests, the supporting span L of the two supports is 40–80 mm. In four-point-flexural tests the support span is 45–81 mm, while the loading edges are spaced from 15 to 27 mm.

In the flexural test, the loading edges apply a constantly paced deflection to the specimen to the point of failure. The speed to be set for the loading edges depends on the force response due to corresponding deflection; the DIN EN ISO 14125 standard demands values in the range from 0.5 to 500 mm/min. The deflections of the top or bottom surfaces of the specimen in the middle of the support span, and the total force F applied by the loading edges are recorded. From these measured values, flexural stress and flexural strain are derived. Flexural stress σ_f and flexural strain ε_f designate the tensions and elongations on the exterior surfaces of the specimen in the midspan of the loading edges L. No differentiation is made between the compressive and tensile stresses on the top and bottom surfaces of the specimen, as they amount to equal values for reasons of symmetry. The characteristics of σ_f and ε_f match the graphs in Fig. 14.5, if the diagram uses flexural strain as abscissa, and flexural stress as ordinate, according to

$$\sigma_f = \frac{3FL}{2bh^2} \quad and \quad \varepsilon_f = \frac{6sh}{L^2} \tag{14.18}$$

for the three-point bending test or

$$\sigma_f = \frac{FL}{bh^2} \quad and \quad \varepsilon_f = \frac{4,7sh}{L^2} \tag{14.19}$$

for the four-point bending test, b and h being the width and thickness of the specimen. Equations (14.18) and (14.19) apply to small deflections ($s < 0.1\,L$), higher values require modifications of the equations. A direct measurement of the flexural strain ε_f by means of strain-gauges applied to the specimen is an alternative to the relations stated in Eqs. (14.17) and (14.18). Analogously to the representations in Fig. 14.5, the flexural modulus E_f is the slope of the σ_f-ε_f characteristic in the flexural strain range of $0.0005 \leq \varepsilon_f \leq 0.0025$.

Flexural tests are always performed until failure of the specimen. Apart from the σ_f-ε_f characteristic, the peak values of σ_f and ε_f, and the type of failure have to be stated in the test report. Acceptable failure types in flexural tests are fractures, which occur at a distance from the supports and are caused by either tensile or compressive strains. Failure caused by interlaminar shear is not accepted, and related results are dismissed.

14.6.2.4 Shear Test

Shear loading in the laminate level often occurs in the everyday applications of fiber-reinforced plastic composites, introducing opposing shear forces into the component. In test engineering and in test standards, these cases are referred to as shear stresses, especially when the failure of the specimen is the focus of the test. Strictly speaking, a shear stress requires opposing shear forces acting on the exact same line of action and causing a "shearing off" of the material. This case occurs inside the component and results in its shear failure if the external shear loading is high enough. The methods detailed in this section apply mainly to low shear loading, in order to characterize the elastic range of the shear deformation. Assertions regarding the shear failure are made possible by continuing the shear test to the failure of the specimen.

A shear test is performed to characterize the behavior of a material under shear loading. It serves to determine the shear modulus G_{ij} in the compliance tensor in Eq. (14.1), according to Eq. (14.5). A test engineering replication of this situation has proven difficult, partly due to the technical feasibility of defined homogeneous shear deformations G_{ij} and the simultaneous measurement of resulting shear stresses γ_{ij}, and partly due to the complex reaction of the fiber-reinforced plastic composites characterized by more or less inhomogeneous interior states of stress and non-linearity. The result—similar to compression test and as opposed to tensile and flexural tests—is a multitude of individual methods for the assessment of the shear behavior and the shear failure. The methods described in the following are the most common, although their results differ on account of the different experimental parameters.

±45° Tension Test (*In-plane Shear Method*)

The ±45° tension test, usually referred to as "in-plane shear method" is described in DIN EN ISO 14129, and is based on ASTM D3518. The method is attractive from a test engineering point of view, as no special preparations for the clamping of the specimen are required. A flat specimen of rectangular dimensions and a fiber orientation of ±45° to the longitudinal axis is subjected to a simple tensile test. The specimens measure $250 \times 25\ mm^2$, the active specimen length being 150 mm.

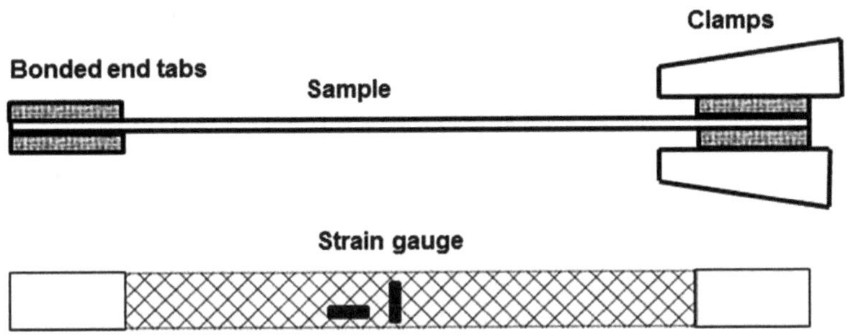

Fig. 14.19 Specimen shape for shear test according to DIN EN ISO 14129

The specimens have a thickness of 2 mm. Strain-gauges are applied parallel and perpendicular to the longitudinal axis of the specimen (Fig. 14.19).

A tensile strain of the specimen in longitudinal directions, performed at a test rate of 2 mm/min, causes extensions ε_x and transverse strain ε_y of the specimen, which are measured by means of strain-gauges. In addition, the tensile force F_x is measured in 0° direction. The Eqs. (14.20) and (14.21)

$$\gamma_{12} = \varepsilon_x - \varepsilon_y \quad (\varepsilon_y < 0) \tag{14.20}$$

$$\tau_{12} = \frac{F_x}{2hb} \tag{14.21}$$

allow the calculation of shear deformation γ_{12} and shear stress τ_{12}, with b being the width, and h being the thickness of the specimen. The in-plane shear modulus G_{12} is determined from the shear stress-shear deformation curve in shear deformations of $\tau'_{12} = 0.001$ and $\tau''_{12} = 0.005$, using Eq. 14.5, and becomes

$$G_{12} = \frac{\tau''_{12} - \tau'_{12}}{\gamma''_{12} - \gamma'_{12}} \tag{14.22}$$

The test ends upon reaching a value of $\gamma_{12} = 0.05$, or at the failure of the specimen, the corresponding value τ_{12} is referred to as "in-plane shear strength". This test primarily serves the determination of the shear modulus, not the assessment of the shear failure. In principle, the test can also be performed with a fiber orientation deviating from 45°.

Rail Shear Tests (*Rail Shear Methods*)

The rail shear method is standardized in ASTM D4255. The test requires special rail-shaped specimen holders with which a shear loading of the specimen is realized. In the simplest case, the two-rail shear test, the specimen is screwed onto two rails. The specimen holder can be exposed to pressure or tension by means of a tensile tester, and the rails create shear stresses within the specimen. The shear force is derived from the measurement of the force F on the tensile tester,

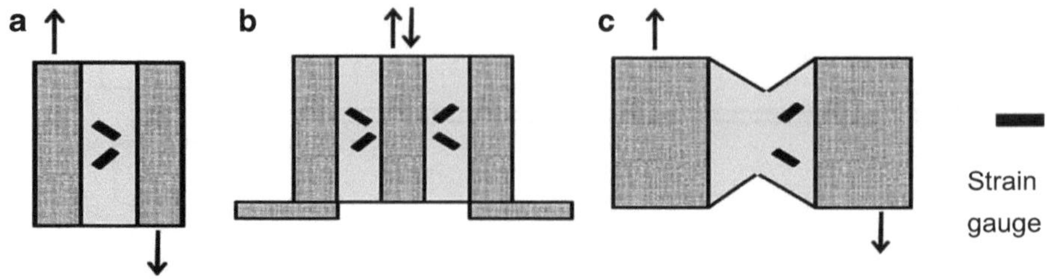

Fig. 14.20 Schematic of shear test arrangements; (**a**) two -rail shear test, (**b**) three -rail shear test, (**c**) shear test with V-notches

the shear deformations γ_{12} are those deformations $\varepsilon_{45°}$ on the specimen which are directly measured with a strain-gauge.

An extension of the previous method is constituted by the three-rail shear test, which requires three eponymous rails for holding the specimen. The specimen holder consists of two outer rails firmly connected to a base plate. Mandatory boreholes in the specimen enable the screw fixing of the specimen to the rails. A central rail, guided by the specimen holder serves for load insertion, causes a shear loading of the specimen. The central, symmetrical loading of the specimen ensures that no additional shear forces can affect the specimen. A homogeneous specimen will display a homogeneous displacement field. The shear modulus is calculated in a manner analogous to the two-rail shear test. Figures 14.20a, b show the schematic of the two test variations.

Iosipescu Shear Test (*V-notched Beam Method*)
For this method, which is named for its original developer, shear loading is created by V-shaped notches, which are cut into the specimen to a depth of up to 20 % of the total width of the specimen. The method was initially developed for metals, but ASTM D5379 is an interpretation of the original method for composite materials.

V-notched Rail Shear Test (*V-notched Rail Shear Method*)
This method combines the advantages of the two-rail method and the Iosipescu method standardized in ASTM D7078. The improvements on the Iosipescu shear test apply mainly to the increased span between the horizontal central baselines of the V-notches. The result is a greater space for the application of the strain-gauges. In comparison to the two-rail method, the specimen holding was improved, and no boreholes are necessary, because the specimen can simply be clamped. Furthermore, the total surface area of the specimen is reduced. Both improvements positively affect the test methods efficiency. The performance of the test and the determination of the shear modulus are performed in a similar manner as in the Iosipescu test method. Figure 14.20 c shows the schematic of the shear tests with V-notches. Both variants described above differ with regard to the shape and dimensions of the specimen and the specimen holder, particularly the proportions of specimen length and specimen width.

Pivotable Plate ("*Plate Twist Method*")
The test is described in DIN EN ISO 15310 and requires square specimens. The test device has two supports located in a diagonal of the specimen. The load introduction is performed symmetrically in the other diagonal of the specimen. The shear modulus G_{12} can be derived from the stress-deflection-plot, the necessary relations are part of the cited standard. This method is rarely used.

Large-panel Shear Test
This test method is detailed in DIN 53399 and requires specially shaped, square specimens. The load introduction area of the specimen is incised and bored diagonally, and fixed in a shear frame with screws. The shear test requires a tensile tester in which the shear frame is stretched at predefined rates. The elongations have to be recorded right on the specimen in various directions. The shear modulus G_{12} is derived from this data.

14.6.2.5 Tests of Interlaminar Shear Strength

Tests of interlaminar shear strength as well as the flexural and shear tests described above are in some manner related, which would allow a different structure than the one chosen here. Despite the similarities, *interlaminar shear strength* is a parameter of its own, especially so because it cannot be derived from either other parameters or flexural and shear tests. The result of the test is not an absolute value. Therefore, the correct, standardized term is apparent interlaminar shear strength. The term used here corresponds to common usage.

In principle, it relies on a short-beam shear test in the form of three-point method, analogous to the representation in Fig. 14.18a. The specimen of rectangular cross-section is placed on two supports and loaded by a centrally located loading edge until failure by interlaminar shear occurs. This can be achieved because the span L between the two supports is no greater than the quintuple specimen thickness (for example, L = 10 mm at a typical specimen thickness of 2 mm).

Interlaminar shear strength is the value of interlaminar shear stresses in the neutral plane of the specimen, which occurs at the failure of the specimen.

The specimen is arranged symmetrically over the supports and loaded at a speed of 1 mm/min by the loading edges. The force F is recorded. In accordance with DIN EN ISO 14130 the apparent interlaminar shear strength τ_{12} is calculated from the force value F, using the Eq. (14.23)

$$\tau_{12} = \frac{3}{4} \times \frac{F}{bh} \tag{14.23}$$

Where b and h are the width and thickness of the specimen.

It is important that the mode of specimen failure is caused by interlaminar shear between two or more layers (single shear or multiple shear). Specimens with other types of failure are not included in the test analysis.

Further test standards are included in the normative references at the end of this chapter. These methods are also based on the short-beam shear method. Specimens and test devices are designed differently from DIN EN ISO 14130.

14.6.2.6 Tests of Creep Properties

The creep of a material denotes its change of shape under constant or cascadingly altering load. Quantitatively, the creep parameters are determined by means of a creep test of at least 1,000 h duration. The specimen is exposed to tensile, bending and pressure loads, and its deformation is registered across an extended interval of time. The result of the creep test provides information regarding the viscoelastic behavior of the material. This knowledge is important, as composite components in practical use are subjected to strains over long periods of time and have to retain their serviceability.

Creep tests have been common in metallurgy, construction, and plastics engineering for a long time. The standards developed for the bending and tensile loading of plastics can also be applied to fiber-reinforced plastic composites. For tests concerning the creep behavior, the DIN EN ISO 899-1 and -2 standards should be consulted. In accordance with these, creep is the increase in elongation of the specimen over time under tensile (Part 1) or bending (Part 2) loads.

The creep test according to DIN EN ISO 899-1 is a special tensile test. Aspects of specimen dimensions and the performance of tensile tests are provided by the respective sections of DIN EN ISO 527, which are detailed in Sect. 14.6.2.1, and can be used for the creep test. The creep test in the manner prescribed by DIN EN ISO 899-2 is a special type of three-point test, as represented under different aspects, in Sects. 14.6.2.4 and 14.6.2.5. The radii of supports and loading edge, as well as the support span depend on a variety of aspects. Their dimensions are set to produce a bending load while preventing interlaminar shear. The support span has to amount to at least the 16-fold value of the specimen thickness. The load is to be set at ca. 50 % of the tensile strength or the shear failure.

Increase in strain or deflection are preferably to be measured in a time interval ranging from of 1 min to 1,000 h, where the time intervals increase exponentially, which is the reason for a logarithmic representation of the time axis in the graphical depiction of the time dependence of the creep parameters. For reasons of the—quite desirable—small deformations during the creep test, an adherence to a constant test environment is crucial. For loading the specimens, the use of defined weights, possibly in combination with a lever system, should be considered. To measure the deformation, commonly used length measuring systems are sufficiently stable over long periods of time. Employing long-term stable mechanical dial gages with a length resolution of 5–10 μm is often sensible. This way, the test is largely safe from outages in electrical supply of the test equipment.

The tensile creep test uses the increase in length between the gauge marks $(\Delta L)_t$ as well as the initial gauge length without load L_0 to determine the tensile creep strain under tensile load ε_t, and the tensile-creep modulus E_t, following Eqs. (14.24 and 14.25)

$$\varepsilon_t = \frac{(\Delta L)_t}{L_0} \tag{14.24}$$

$$E_t = \sigma \times \frac{1}{\varepsilon_t} = \frac{F}{bh} \times \frac{L_0}{(\Delta L)_t} \tag{14.25}$$

Where F is the introduced force, b is the width, and h is the thickness of the specimen.

The flexural creep test measures the deflection S_t centrally between the supports, at the position of the loading edge. This, and knowledge of the span between the supports L, allows the calculation of the flexural creep strain ε_t—i.e. the elongation of the top or bottom surface of the specimen, using the following Eq. (14.26).

$$\varepsilon_t = \frac{6S_t h}{L^2} \tag{14.26}$$

The flexural creep modulus E_t is the quotient of the tension σ on a specimen surface in the center of the support span and the strain ε_t at the same position. It results from using Eq. (14.18) according to Eq. (14.27) as

$$E_t = \sigma \times \frac{1}{\varepsilon_t} = \frac{3FL}{2bh^2} \times \frac{L^2}{6S_t h} = \frac{L^3 F}{4bh^3 S_t} \tag{14.27}$$

Where F is the introduced force, and b and h are width and thickness of the specimen. Depending on the top and bottom surfaces of the specimen, the tension σ of the specimen is either a compression stress or an equal tensile tension. The results of the creep tests are graphically represented as

- creep curves and
- creep-modulus curves

They show the characteristics over time $\varepsilon_t(t)$ and $E_t(t)$ with logarithmic time axes, often with different load levels. Almost without exception all characteristics at any load level display an increase of the elongation ε_t and a decrease of the modulus $E_t(t)$ over time. If several load levels are used, and if the creep test is performed to failure of the specimen, the results can also be represented as

- isochronous stress-strain plots
- time-to-rupture curves

For this, the stress-strain value pairs determined in the individual measuring times are entered into a σ-ε_t diagram and connected to form a curve.

Creating the time-to-rupture curves requires higher loads to ensure that failure of the specimen occurs during the test, i.e. during the loading time of typically 1,000 h.

The time-to-rupture curve is the graphic representation of the tension σ over the length of time until failure. To gather conclusive data, ten tests per load level are recommended.

Creep tests are very time-consuming, but entirely indispensible. The value of creep tests and their graphical representations is the possibility to extrapolate creep parameters and make informed statements on the serviceability of the tested structures for the entire duration of their use in the field—usually 10–30 years.

14.6.3 Tests at High Deformation Rates

In practice, fiber-reinforced plastic composites are usually exposed to high deformation rates. However, the behavior of the structures under high deformation rates, such as crashes, cannot be securely derived from the test parameters for low-speed tests, or from the knowledge of material parameters according to Eq. (14.1). Attempts to recreate typical situations found in practical use have inspired a number of test methods, many of them having been developed independently from one another. All of them aim to gather information on

- material laws (constitutive equations) at high deformation speeds,
- material damage under impact loading, and
- material fatigue

Table 14.8 compiles a few of the test methods.

14.6.3.1 High-Speed Tensile Tests

The failure behavior of fiber-reinforced plastic composites depends on the deformation rate or loading rate, respectively. Apart from deformation-speed-dependent viscoelastic behavior of the materials, local interior warming in the specimen or resonances may occur, and further influence the failure behavior.

To derive the material laws, the high loading rates encountered in practical application have to be reproduced in high-speed tensile tests. The crosshead movement of conventional tensile testers is screw- or spindle-driven, which keeps deformation rates low. For high-speed tensile tests, an entirely distinct machine type with an auxiliary hydraulics unit exists. This hydraulics unit accelerates a piston rod to speeds of 20–25 m/s. At that speed, the piston rod couples with the specimen holding device, causing a deformation. Depending on the dimensions, strain rates of up $10^2\ s^{-1}$ are achievable. Due to the required dynamics, force measurements are performed with piezoelectric load cells, and an inspection of the specimen with a high-speed camera is sensible. One advantage of the method is the suitability of the piston rod equipment to also test at considerably lower speeds, which allows examinations of the failure behavior at a large interval of deformation rates, including tensile tests and compression tests.

Table 14.8 High deformation rate test methods

Test method	Parameter	DIN/ISO test standard
High-speed tensile test	Maximum tensile force/strength F_m, σ_m Stress at break σ_m	– –
Impact test	Impact-failure energy E_{50} Impact energy E_P Charpy impact strength a_{CU} Izod impact strength a_{IU}	DIN EN ISO 6603-1 DIN EN ISO 6603-2 DIN EN ISO 179-1,-2 DIN EN ISO 180-1,-2
CAI test	Residual compressive strength σ_{dBR}	DIN 65561 ISO 18352
Fatigue test	S-N curve	DIN 50100

Analyzing the measured force characteristics requires experience, as they are also influenced by deformations of components in the test equipment and by the dynamic properties of the load cell. The design of the specimens, their mounting, and the test regimen are usually adapted to the respective object of investigation, special test standards for simple high-speed tensile tests have not been designed yet.

Often, the strain rates achieved by high-speed tensile testers do not suffice for real, highly dynamic load situations. In these cases, the clamp speed has to be increased by using a drop tower, in which a defined mass is accelerated by free fall and makes a high-speed impact on the device in which the specimen is mounted. Depending on the height of fall and the specimen dimensions, strain rates (tensile or compression deformation) of the order of $10^3\ s^{-1}$ are achievable.

Extremely high strain rates under compression load are created with *Split Hopkinson pressure bars*. The test equipment consists of the bars, a projectile, and cylindrical incident as well as transmitter bars aligned in a straight line with the specimen between them. The specimen is thin compared to the length of the bars, and the specimen cross-section is smaller than the cross-section area of the bars. The elongation of the incident and transmitter bars is measured with strain gages. The projectile is accelerated with compressed air, and pushed onto the incident bar at high speeds (10–50 m/s), triggering an elastic shock wave on the incident bar. When the shock wave hits the specimen, part of it is reflected, while the remainder goes through the specimen and deforms it due to its smaller cross-section. Within the specimen, this enables elongation speeds of up to $10^4\ s^{-1}$. Analyses of the strain gage signals allow assessments of the viscoelastic behavior of the specimen.

14.6.3.2 Crash Testing

Generally, the high-speed test method described above is the most suitable for deriving material laws (constitutive equations) at high deformation rates. Application-relevant statements regarding damage tolerance for shock stresses are provided by crash tests, which are also known as impact tests. They aim to damage or destroy a component with predefined shock stresses. The relations are highly complex, which minimizes comparability of the results of various methods of impact testing, several of which have been standardized in the past. The ones described in the following have become the most widespread.

Impact Test (*Drop Weight Impact*)

The impact test adhering to DIN EN ISO 6603 was developed for plastics, but can also be used for fiber-reinforced plastic composites. It serves to determine the impact behavior of the materials. For this, specimens are fixed to a support device with a clamping ring. In the impact test standardized in the first part of the cited standard (which also matches ASTM D5628), a lubricated striker or dart, formed as a ball burst device with a hemispherical, polished and hardened impact surface is dropped onto the center of the specimen from a predefined height, damaging it.

After impact, the specimen is assessed with regard to the damage feature defined beforehand, e.g. crack, break, penetration, shattering. If the feature is missing, the impact energy is increased for the following test by incrementing the mass of the striker. If damage occurs, the mass has to be reduced for the next round of testing. Using a method for adjusting the mass detailed in the standard ("staircase method"), and energy value at which 50 % of the specimens fail with the predefined damage feature is derived. This value is to be interpreted as the impact-failure energy E_{50}.

DIN EN ISO 6603-2 standardizes the instrumented impact test. Specimen support and clamping device are to be designed analogously to part 1 of the standard. The impact has a preferable drop height of 1 m, its mass has to be large enough to ensure a fracture of the specimen at near-constant speed. The force exerted on the specimen by the striker is measured using a piezoelectric load cell and recorded as a function of time. With knowledge of the striker speed being 4.4 m/s, the force- deflection diagram, the impact or puncture energy E_p, and other parameters can be derived.

Charpy Impact Test

The Charpy impact test detailed in DIN EN ISO 179 was also developed for plastics, and has been widely used in plastics testing. The method can be applied for fiber-reinforced plastic composites. The test equipment consists of a pendulum impact tester into which the specimens mounted touching two supports ("horizontal beam"). The span of the supports has to maintain a certain ratio to the thickness of the specimen. At this point, the standard allows various variations. The pendulum has a striking edge impacting the center of the specimen. For this, kinetic energy is available, equal in amount to the potential energy of the pendulum at the beginning of the test. Impact on the specimen can hit either the narrow or broad longitudinal surface of the specimen (edgewise or flatwise impact). Furthermore, the impact direction can be parallel or normal to the reinforcement plane.

Part 1 of the cited standard describes the non-instrumented impact test. After breaking of the specimen, the pendulum reaches a certain height corresponding to a remaining potential energy resulting from the initial energy minus the absorbed impact energy E_c, i.e. the energy required for the failure. The pendulum impact testers are constructed to allow easy determination of the potential energy of the pendulum after impact, preferably with a slave pointer on an appropriately calibrated energy scale. The Charpy impact strength a_{CU} is then calculated as the quotient of absorbed impact energy E_C and the specimen cross-section. The second index U denotes the un-notched specimens, as opposed to the index N for notched specimens, which is redundant here.

DIN EN ISO 179-2 describes the instrumented impact strength test in which impact force and deflection of the specimen are recorded. In this case, the striking edge has to be fitted with a force sensor such as a piezoelectric load cell. The deflection can be derived with an angle sensor at the fulcrum of the pendulum. In this case, the absorbed impact strength of the specimen required for determining the Charpy impact strength a_{CU} is calculated from the integral of the force over the deflection. The biggest advantage of the instrumented impact strength test is the possibility to derive statements on the type of failure—yielding followed by plastic deformation, tough break, brittle break, splintering break—from the measured force characteristics. The respective failure type has to be documented in the test report.

Izod Impact Strength

The Izod impact test is described in DIN EN ISO 180. In principle, it is comparable to the non-instrumented Charpy test. The crucial difference is the manner of specimen mounting. While the specimen is positioned centrally on two supports for the Charpy method, it is held on one side only in the Izod variant. The Izod impact strength a_{IU}, analogously to its Charpy counterpart a_{CU}, is the quotient of impact energy E_c and the specimen cross-section.

Compression After Impact (CAI)

The previously described impact tests determine the energy needed to completely destroy the specimen in a defined crash. To properly assess the practicability of composite components, it is often more important to know their residual strength after a "small" impact strain. The fiber-reinforced plastic composites are damaged in a strictly defined manner, and the resulting property changes are examined by test engineering means. Suitable methods for this were first developed in aeronautics.

Today's standard method for this kind of tests adds a compression test after the crash or impact pre-damaging, and then determines the residual compressive strength. These tests are known as "compression after impact" tests, usually abbreviated to CAI. Several standards exist for CAI, varying slightly with regard to specimen design and test conditions. The Normative References at the end of the chapter include examples of the standards usually used. The specimen is a flat rectangular plate of standardized dimensions (e.g. 150×100 mm^2 according to DIN 65561, or 127×76.2 mm^2 (5×3 in.) for the Boeing specification BSS-7260). The specimen is fixed in a stable frame and, as in impact tests, hit and locally damaged by a striker in a drop device. Details are given in the standards included in the normative references. It is important to construct the test devices in such a way as to ensure that the specimen is subjected to only a single impact and to prevent bouncing.

The specimen is then inserted into a compression test fixture allowing an even compressive strain on the ends of the specimen in direction of its longitudinal axis.

To determine the compressive strength, the specimen is compressed at a constant test speed, i.e. 1 mm/min according to DIN 65561, and stressed to specimen failure. The compressive stress at failure of the damaged specimen is called residual compressive strength. The damage resistance of the specimen is assessed in a plot in which the residual compressive strength σ_{dbr} is entered as a function of the impact energy. In addition, the compression at failure of the damaged specimen can be determined with strain gages applied to the specimen.

14.6.3.3 Fatigue Test/Fatigue Behavior

Fatigue test complete the list of common exposures in practical use of fiber-reinforced composites. They have been in use for the assessment of the long-term behavior of metals for some time. While the creep test described in Sect. 14.6.2.6 ascertains the deformations of a specimen under constant loads, the fatigue test exposes specimens to alternating loads until failure. With randomly changing loads, information on the structural durability of the materials can be obtained. On the other hand, sinusoidal loads offer information on the *fatigue behavior*. Due to the easier realization in experiments, practicality demands the fatigue of a component to be determined by long-term tests with single-axis sinusoidal loading of the specimens.

Terminology and parameters of the long-term fatigue tests are detailed in DIN 50100. Mechanical exposure is described by the cyclic stress range $\sigma_a = (\sigma_O - \sigma_u)$ of the maximum stress σ_O and the minimum stress σ_u amplitudes alternating symmetrically by a mean value σ_m. The specimens can be subjected to pure compression-compression loads ($\sigma_O < 0$) or pure tension-tension loading ($\sigma_u > 0$). Tension-compression loading occurs if both compressive and tensile loads ($\sigma_O > 0$, $\sigma_u < 0$) occur in a cycle.

The fatigue of the specimens can be caused by tensile/compression loading, or by bending and torsion. The frequency of mechanical oscillations, depending on peak amplitudes, ranges from ca. 0.1 Hz to a few 10 Hz. Dynamic test devices are used in order to realize the variety of load profiles. In particular, the loading levels can be kept constant (single-step test), or changed incrementally (multi-step test) or close to conditions in the field (fatigue test).

Most commonly, the Wöhler test method is used, which is a single-step test in which the specimen is strained to by a suitable sinusoidal load until failure after 10^4–10^5 cycles. The type of failure is defined with a view to the planned field of application of the specimen material, e.g. breaking, cracking, or similar cases. Afterwards, the test is repeated with decreased loads. For this, the peak amplitude of a load cycle σ_a can be decreased incrementally at constant mean stress σ_m. Other variations of load reduction are also possible. Load reduction is continued until no failure occurs after 2×10^6–10^7 cycles. The loading parameter—usually some type of mechanical tension—is plotted over the number of load cycles at failure of the specimen, resulting in the *S-N curve*. The S-N curve is divided into three sections. Low-cycle fatigue (“K”) exists if the failure of the specimen occurs at less than 10^4 cycles. The section of high-cycle fatigue (“Z”) designates strains in which failure occurs only after 10^4–2×10^6 cycles. For strains which the specimen can withstand for 2×10^6–10^7 cycles, fatigue strength (“D”) is achieved, i.e. the specimen is

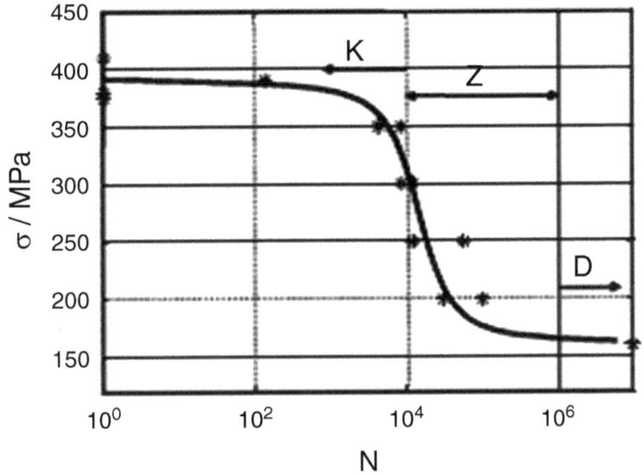

Fig. 14.21 S-N curve

theoretically capable of withstanding this exposure to loads for an infinite amount of time. Figure 14.21 shows the diagrammatic representation of an S-N curve.

For the Wöhler test, one load parameter—mean stress σ_m, or the stress amplitude of a cycle σ_a—has to be kept constant. If both parameters are altered systematically, a set of S-N curves results, from which the respective fatigue strengths can be derived. The graphical representation of fatigue strengths as a function of the various types of strain provides fatigue strength plots. There is a variety of possibilities, all of which are described in DIN 50100.

14.6.4 Other Test Methods

Apart from the abovementioned test methods, a variety of other methods exists. They are standardized and provide information on fiber-reinforced plastic composites at typical states in practical use, e.g. tests of notched compression, bearing strength, stiffness in torsion, and crack growth, to name only a few. A complete account of this rapidly changing field cannot be given here.

Newly developed textile structures, semi-finished products or components often display properties which are beyond the capabilities of establish test methods and equipment. One increasingly important branch, already discussed above, is testing under exposed conditions (temperature, radiation, and environmental media). These methods are exclusively published in expert literature, if at all. The know-how contained in them is bound to influence test standards in the future. Non-destructive test methods are also becoming increasingly important, as they are the only methods for controlling the fiber-reinforced plastic composites over their entire life cycle. Examples of non-destructive test methods, cited from [11], are

- radiography tests (identification of cavities, porosities, cracks),
- confocal laser-scanning microscopy (photoelastic examinations),
- speckle interferometry (deformation measurement, resolution 0.05 μm),
- thermography based on optical, acoustic or electric excitation (identification of anisotropy, impact damages, delamination, cracks),
- ultrasound, mechanical vibration (identification of defects), and
- vibrometry, vibration analysis (identification of defects)

Detailed information concerning these and other methods can be obtained from the respective technical literature.

14.7 Tests on Other Fiber-Based Composite Materials

Previous remarks were concerned with the physical characterization of those textile materials and structures playing a role in the value chain from the filament to the fiber-reinforced plastic composite component. Beyond those, there are other fiber-based composite materials. The methods and aspects discussed in Sects. 14.1–14.5 also apply to these materials. However, the methods detailed in Sect. 14.6 have to be modified in most cases. Often, the properties and dimensions of these structures require the development of independent test methods. Examples of such structures include

- textile-based membranes and
- textile-reinforced concrete

The list is incomplete. Worldwide activity in this scientifically and economically interesting field will make other types of composite materials play a role, all of which will require their own test methods. Exemplarily, this can be shown for textile membranes and textile-reinforced concrete.

14.7.1 Textile Membranes

Textile membranes are characterized by mass per unit area, tensile strength, cutting resistance, tear strength, strength of welded joints, coating adhesion, and low flammability. The most important standards are given at the end of the chapter. A detailed assessment of textile membranes for lightweight construction (see Sect. 16.5), for example, is based on the results of biaxial tensile tests. More information on biaxial tensile tests on textile fabrics is included in Sect. 14.5.3, and is also applicable to tests on textile membranes. Textile carriers for membranes are commonly glass, PES, or aramid woven fabrics, but can also be bi- and multiaxial non-crimp fabrics. To assess the stress-strain behavior of the textile membranes produced with them, a biaxial load introduction is practical. Load cases in the specimen are defined primarily by

- specimen dimensions, specimen shape,
- mounting of the specimen (clamp segments, total length of the nip), and
- load introduction (symmetrical/asymmetrical, strain- or stress-controlled)

European standards for the characterization of textile membranes are in preparation.

For biaxial tensile tests on membranes, square specimens are well-suited. These specimens can have rounded-off corners and/or yarn-parallel slits. An undefined load introduction by shear forces at the nips of the specimens is prevented by transversely mobile clamp segments. One preferred solution for membranes are specimens with cross-shaped incisions, stripes of at least 30 cm length and 20 cm width, and a test area (20×20 cm^2) in the center of the cross [12]. Furthermore, equal travelling spans of the specimen holder in both load axes are favorable. Generally, a largely homogenous load introduction into the measuring field can be achieved in this way, allowing a relatively simple analysis of the test data. Any deviation from these conditions results in special load exposures, rendering the results useless for anything except the assessment of this exact load case. However, they are extremely important whenever a situation common in practical use—for instance on fastening elements, on seams with overlapping areas, on tensile bars or compression bars, etc.—has to be assessed. It is much more complicated to derive verified statements on the

- properties of large surface areas (roof constructions), based on test results determined for small test specimens, and
- practicability over the entire period of use, which can exceed 25 years (roof constructions) [13], based on long-term tests spanning no more than a few weeks or months

As the viscoelastic (and the elastic) properties of these materials play an important part in assessing their performance characteristics, biaxial tensile tests are also crucial in the low and medium load range. The elastic and viscoelastic behavior of the membranes is established by tests under cyclic loads. Regarding the elastic properties, the same aspects as monaxial tests apply. The viscoelastic behavior is determined from the residual elongations of the membranes after the individual load cycles. Tests featuring constant elongations along the load axes in the measuring field are preferred. Generally, the specimen surface is furnished with a drawn grid, whose distortion is recorded video-optically and converted to elongation profiles by means of image processing software. Further details can be found in Sect. 16.5.

14.7.2 Textile-Reinforced Concrete

Textile-reinforced concrete is a term for concrete in which the usual steel reinforcement is substituted by textile-based reinforcement structures, such as fabrics made of glass or carbon filament yarns. The composite behavior between textile reinforcement and concrete matrix determines important properties of the textile-reinforced concrete, such as durability and load capacity. Durability describes the temporal changes of relevant parameters of textile-reinforced concrete. Furthermore, the load capacity, i.e. the capability of withstanding a diversity of different load situations, is of crucial importance. To assess the properties of textile-reinforced concrete, special test methods are available. Broad use is being made of

- SIC test,
- pull-out test,
- tensile and flexural test

14.7.2.1 SIC Test

Fiber materials are exposed to an alkaline environment in the concrete matrix, which makes them susceptible to damage. The use of alkaline-resistant yarns, such as AR glass filament yarns, is therefore necessary. However, load capacity and strength losses can occur, caused by pore water. The SIC (strand in concrete) test, which is standardized in DIN EN 14649, is used to quantitatively describe such situations. The embedding of the yarns in concrete specimen of 30 mm length and $10 \times 10\ mm^2$ cross-section creates a situation which is very close to real practical conditions. The yarn lengths have to be set as to ensure protruding ends on either side, for mounting in a tensile tester. The specimens are aged at 80 °C over several days or weeks, exposed to high humidity, alkaline pore water solutions or, in accordance with DIN EN 14649, deionized water (for 96 h). It is important to protect the protruding yarn ends from the exposure to avoid strength losses. After exposure, a tensile test concerning the aspects detailed in Sect. 14.4.3 is performed. A comparison of the measured breaking strengths ("residual strengths") to the breaking strengths of undamaged yarns allows an assessment of the selected reinforcement. Apart from chemical damages, the SIC tests also serve to examine other influences on the composite properties, such as mechanical damages to the yarns caused by the concrete texture or additional transverse compression caused by the formation of hydration products [14].

14.7.2.2 Pull-out Tests

The pull-out test primarily gives information on the composite behavior between the yarn selected for the reinforcement and the concrete matrix. From this, estimates can be derived for the failure behavior, and thus, for the durability and load capacity. In the simplest version, the yarn is inserted into a concrete matrix and pulled out under predefined test conditions. The pull-out force and span are recorded. In a two-sided pull-out test, the yarn is placed centrally in a prismatic specimen, which features a saw cut forming a predetermined breaking notch in its center, running transversely to the yarn axis [15]. The specimen is tensile-loaded at low deformation speed. In the crack cross-section, only the yarn absorbs the tensile load. By recording the tensile load and deformation span, the force value is determined at which the adhesion bond between concrete and yarn fails.

14.7.2.3 Tensile and Flexural Tests

The load capacity of textile-reinforced concrete can be assessed by tensile and flexural tests. Principal aspects of these tests are detailed in Sects. 14.6.2.1 and 14.6.2.3, although the test regimens described there have to adapted in order to apply to textile-reinforced concrete [15, 16].

Tensile tests on textile-reinforced concreted requires relatively large specimen, for example of 100 cm in length and 10 cm in width. Mounting these specimens in a tensile tester is not done with the usual clamping between clamps, but sheaths embedded in concrete, which are connected to the crosshead of the tensile tester with a hinge [15]. The range of loading of the specimen has a tapering design ("waisted specimen"), and predetermined breaking notches can be applied. The exposure to load is performed at very low elongation speeds, typical values are around $1\,‰\,\mathrm{min}^{-1}$. The tensile test is continued until specimen failure. The standard load cell of the tensile tester is used to measure the breaking load, from which the tensile stress at break can be calculated. Local elongation measurements on the specimen are taken with applied strain gages. Crack initiation and crack propagation are analyzed by means of optical/photogrammetric methods.

The properties of textile-reinforced concrete under bending loads are preferably determined in a four-point flexural test. The specimens are 100 cm long and designed as beams with various profiles (Π, T, or I profiles). The span of the support and span between the loading edges can be designed variably. Load exposures, flexural force, and deflection measurements are performed analogous to tensile tests, as are analyses of crack initiation and crack propagation. To assess the textile-reinforced concrete, specimen areas in which tensile stresses occurred are used for reference. Further details are laid out in Sect. 16.5.

References

1. Standard DIN 1319-1: Fundamentals of metrology - Part 1: Basic terminology (1995-01)
2. Standard DIN 1319-2: Fundamentals of metrology - Part 2: Terminology related to measuring equipment (2005-10)
3. Standard DIN 1319-3: Fundamentals of metrology - Part 3: Evaluation of measurements of a single measurand, measurement uncertainty (1996-05)
4. Standard DIN 1319-4: Fundamentals of metrology - Part 4: Evaluation of measurements; uncertainty of measurement (1999-02)
5. KRYSTEK, M.: Berechnung der Messunsicherheit - Grundlagen und Anleitung für die praktische Anwendung (Expression of Uncertainty in Measurement – Principles and Methods for Practical Applications). Berlin: Beuth Verlag GmbH, 2012

6. Standard DIN V ENV 13005: Guide to the expression of uncertainty in measurement; German version ENV 13005:1999
7. INTERNATIONAL ORGANIZATION FOR STANDARDIZATION (Publ.): ISO/IEC Guide 98-3:2008 Uncertainty of measurement - Part 3: Guide to the expression of uncertainty in measurement, 2008
8. Standard DIN EN ISO 139: Textiles - Standard atmospheres for conditioning and testing (2011-10)
9. Standard DIN EN ISO 291: Plastics - Standard atmospheres for conditioning and testing (2008-08)
10. Standard DIN EN ISO 527-1: Plastics - Determination of tensile properties - Part 1: General principles (2012-06)
11. Aktivitäten der Abteilung Zerstörungsfreie Prüfung in Forschung und Anwendung (Activities of the Department for Polymer Engineering in Research and Application). Special Print of the Institut für Kunststofftechnik (IKT) Stuttgart, 2010
12. SAXE, K.: Ein biaxiales Prüfsystem (A biaxial testing system). In: Proceedings, Symposium Membrankonstruktionen, Fachbereich Bauwesen der Universität-GH Essen; Essen 1997
13. FITZ, J.: Neuartige Architekturgewebe aus Fluorpolymeren (Innovative fabrics based on fluoropolymers for textile architecture). Lecture on Kooperationsforum Bayern Innovativ 8.7.2010, Miesbach, 2010
14. SCHORN, H.; SCHIEKEL, M.; HEMPEL, R.: Dauerhaftigkeit von textilen Glasfaserbewehrungen im Beton (Durability of glass fiber reinforcement of concrete). Bauingenieur, Vol. 79, pp. 86-94, Springer-VDI-Verlag, 2004
15. MOLTER, M.: Zum Tragverhalten von textilbewehrtem Beton (Load-bearing capacity of textile-reinforced concrete). Doctoral thesis, RWTH Aachen, Fak. Bauingenieurwesen, Aachen 2005
16. JESSE, F.: Tragverhalten von Filamentgarnen in zementgebundener Matrix (Load capacity of filament yarns in cement-bonded matrices). Doctoral thesis, TU Dresden, Fak. Bauingenieurwesen, Dresden 2004

Additional Literature

AGRAWAL, S.; SINGH, K. K.; SAKAR, P. K.: Impact damage on fibre-reinforced polymer matrix composite - A review. Journal of Composite Materials, Vol. 48 (2014), Issue 3, pp. 317-332

CHERIF, C.; YOUNES, A.; NOCKE, A.: Development of a novel test stand for the characterization of tensile mechanical material behavior of high performance fibers under impact stress. Proceedings 13th. AUTEX World Textile Conference (2013) Dresden, Technische Universität Dresden

LORENZ, E.; SCHÜTZE, E. SCHLADITZ, F., CURBACH, M.: Textilbeton. Grundlegende Untersuchungen im Überblick / Textile Reinforced Concrete. Overview of standard test methods. Beton- und Stahlbetonbau, Vol. 108 (2013), Issue 10, pp. 711-722

Normative References: Fibres/Filaments

ISO 1888 Textile glass - Staple fibres or filaments - Determination of average diameter (2006-07)

DIN 65571-1 Aerospace - Reinforcement fibres - Determination of filament diameter and cross-sectional area of filament yarns - calculation method (1992-11)

DIN 60905-1 Tex-System - Principles (1985-12)

ISO 1144 Textiles - Universal system for designating linear density (1973-12)

DIN EN ISO 1973 Textiles fibres - Determination of linear density - Gravimetric method and vibroscope method (1995-12)
DIN EN ISO 5079 Textiles - Fibres - Determination of breaking force and elongation at break of individual fibres (1996-029

Normative References: Yarns

DIN EN 14020-2 Reinforcements - Specification for textile glass rovings - Methods of test and general requirements (2003-03)
DIN EN 13003-2 Para-aramid fibre filament yarns - Methods of test and general specifications (1999-06)
DIN EN 13002-2 Carbon fiber yarns - Test methods and general specifications (1999-06)
DIN EN ISO 1889 Reinforcement yarns - Determination of linear density (2009-10)
DIN EN ISO 1890 Reinforcement yarns - Determination of twist (2009-10)
DIN 65382 Aerospace - reinforcement fibres for plastics - tensile test of impregnated yarn test specimens (1988-12)
DIN EN ISO 9163 Textile glass - Rovings - Manufacture of test specimens and determination of tensile strength of impregnated rovings (2005-07)
DIN EN ISO 10618 Carbon fibre - Determination of tensile properties of resin-impregnated yarn (2004-11)
DIN EN 12562 Textiles - Para-aramid multifilament yarns - Test methods (1999-10)
ISO 3341 Textile Glass -Yarns - Determination of breaking force and breaking elongation (2000-05)
DIN EN ISO 2062 Textiles - Yarns from packages - Determination of single-end breaking force and elongation at break using constant rate of extension (CRE) tester (2010-04)
ISO 1887 Textile glass - Determination of combustible-matter content (1995-05)
DIN EN ISO 10548 Carbon fibre - Determination of size content (2003-12)
DIN EN ISO 3344 Reinforcement products - Determination of moisture content (1997-08)

Normative References: Textile Fabrics and Prepregs

DIN EN ISO 5084 Textiles - Determination of thickness of textiles and textile products (1996-10)
DIN EN 12127 Textiles - Fabrics - Determination of mass per unit area using small samples (1997-12)
DIN EN ISO 10352 Fibre-reinforced plastics - Moulding compounds and prepregs - Determination of mass per unit area (2011-04)
DIN EN 2329 Aerospace series; textile glass fibre preimpregnates; test method for the determination of mass per unit area (1993-04)
DIN EN 2557 Aerospace series - Carbon fibre preimpregnates - Determination of mass per unit area (1997-05)
DIN 53835-13 Testing of textiles; determination of the elastic behaviour of textile fabrics by a single application of tensile load between constant extension limits (1983-11)
DIN EN 14704-1 Determination of the elasticity of fabrics - Part 1: Strip tests (2005-07)
DIN 53362 Testing of plastics films and textile fabrics (excluding nonwovens), coated or not coated fabrics - Determination of stiffness in bending - Method according to Cantilever (2003-10)
ASTM D1388 Standard Test Method for Stiffness of Fabrics (2008)

DIN EN ISO 1172 Textile-glass-reinforced plastics - Prepregs, moulding compounds and laminates - Determination of the textile-glass and mineral-filler content - calcination methods (1998-12)
DIN EN 2331 Aerospace series; textile glass fibre preimpregnates; test method for the determination of the resin and fibre content and mass of fibre per unit area (1993-04)
DIN EN 2559 Aerospace series - Carbon fibre preimpregnates - Determination of the resin and fibre content and the mass of fibre per unit area (1997-05)
DIN EN ISO 11667 Fibre-reinforced plastics - Moulding compounds and prepregs - Determination of resin, reinforced-fibre and mineral-filler content; Dissolution methods (1999-10)

Normative References: Fibre-Reinforced Plastic Composites - No Mechanical Load

DIN EN ISO 10352 Fibre-reinforced plastics - Moulding compounds and prepregs - Determination of mass per unit area (2011-04)
DIN EN ISO 1172 Textile-glass-reinforced plastics - Prepregs, moulding compounds and laminates - Determination of the textile-glass and mineral-filler content; calcination methods (1998-12)
DIN EN 2564 Aerospace series - Carbon fibre laminates - Determination of the fibre-, resin- and void contents (1998-08)

Normative References: Fibre-Reinforced Plastic Composites - Low Rate of Deformation - Tensile Tests

DIN EN ISO 527-1 Plastics - Determination of tensile properties - Part 1: General principles (2012-06)
DIN EN ISO 527-4 Plastics - Determination of tensile properties - Part 4: Test conditions for isotropic and anisotropic fibre-reinforced plastic composites (1997-07)
DIN EN ISO 527-5 Plastics - Determination of tensile properties - Part 5: Test conditions for unidirectional fibre-reinforced plastic composites (2010-01)
DIN EN 2561 Aerospace series - Carbon fibre reinforced plastics - Unidirectional laminates - Tensile test parallel to the fibre direction (1995-11)
ASTM D3039 Standard Test Method for Tensile Properties of Polymer Matrix Composite Materials (2008)

Normative References: Fibre-Reinforced Plastic Composites - Low Rate of Deformation - Compression Tests

DIN EN ISO 14126 Fibre-reinforced plastic composites - Determination of compressive properties in the in-plane direction (2000-12)
DIN 65375 Aerospace; fiber reinforced plastics; testing of unidirectional laminates; compression test transverse to fiber direction (1989-11)

DIN V 65380 Aerospace; fiber-reinforced plastics; testing of unidirectional laminates; compression test, parallel and transverse to fiber direction (1987-04)
ASTM D3410 Standard Test Method for Compressive Properties of Polymer Matrix Composite Materials with Unsupported Gage Section by Shear Loading ("Celanese") (2003)
ASTM D695 Standard Test Method for Compressive Properties of Rigid Plastics (2010)
ASTM D6641 Standard Test Method for Compressive Properties of Polymer Matrix Composite Materials Using a Combined Loading Compression (CLC) Test Fixture (2009)

Normative References: Fibre-Reinforced Plastic Composites - Low Rate of Deformation - Bending Tests

DIN EN ISO 178 Plastics - Determination of flexural properties (2013-09)
DIN EN ISO 14125 Fibre-reinforced plastic composites - Determination of flexural properties (2011-05)
DIN EN 2562 Aerospace series - Carbon fibre reinforced plastics - Unidirectional laminates; flexural test parallel to the fibre direction (1997-05)
DIN EN 2746 Aerospace series - Glass fibre reinforced plastics - Flexural test, three point bend method (1998-10)
ASTM D790 Standard Test Methods for Flexural Properties of Unreinforced and Reinforced Plastics and Electrical Insulating Materials (2010)
ASTM D7264 Standard Test Method for Flexural Properties of Polymer Matrix Composite Materials (2007)
ASTM D6272 Standard Test Method for Flexural Properties of Unreinforced and Reinforced Plastics and Electrical Insulating Materials by Four-Point Bending (2010)

Normative References: Fibre-Reinforced Plastic Composites - Low Rate of Deformation - Shear Tests

DIN EN ISO 14129 Fibre-reinforced plastic composites - Determination of the in-plane shear stress/shear strain reponse, including the in-plane schear modulus and strength, by $\pm 45°$ tension test method (1998-02)
ASTM D3518 Standard Test Method for In-Plane Shear Response of Polymer Matrix Composite Materials by Tensile Test of a $\pm 45°$ Laminate (2013)
DIN 53399-2 Testing of reinforced plastics; shear test on plane specimens (1982-11)
ASTM D4255 Standard Test Method for In-Plane Shear Properties of Polymer Matrix Composite Materials by the Rail Shear Method (2001)
ASTM D5379 Standard Test Method for Shear Properties of Composite Materials by V-Notched Beam Method, ("Iosipescu-Method") (2012)
ASTM D7078 Standard Test Method for Shear Properties of Composite Materials by V-Notched Rail Shear Method (2012)
ASTM D5448 Standard Test Method for Inplane Shear Properties of Hoop Wound Polymer Matrix Composite Cylinders (2011)
DIN EN ISO 15310 Fibre-reinforced plastic composites - Determination of the in-plane shear modulus by the plate twist method (2005-10)

Normative References: Fibre-Reinforced Plastic Composites - Low Rate of Deformation - Interlaminar Shear Strength Tests

DIN EN ISO 14130 Fibre reinforced plastic composites - Determination of apparent interlaminar shear strength by short beam-method (1998-02)
DIN EN 2563 Aerospace series - Carbon fibre reinforced plastics - Unidirectional laminates; determination of apparent interlaminar shear strength (1997-03)
DIN EN 2377 Aerospace series; glass fibre reinforced plastics; test method; determination of apparent interlaminar shear strength (1989-10)
ASTM D2344 Standard Test Method for Short-Beam Strength of Polymer Matrix Composite Materials and Their Laminates (2013)

Normative References: Fibre-Reinforced Plastic Composites - Low Rate of Deformation - Creep Tests

DIN EN ISO 899-1 Plastics - Determination of creep behaviour - Part 1: Tensile creep (2003-10)
DIN EN ISO 899-2 Plastics - Determination of creep behaviour - Part 2: Flexural creep by three-point loading (2003-10)

Normative References: Fibre-Reinforced Plastic Composites - High Rate of Deformation - Crash Tests

DIN EN ISO 6603-1 Plastics - Determination of puncture impact behaviour of rigid plastics - Part 1: Non-instrumented impact testing (2000-10)
DIN EN ISO 6603-2 Plastics - Determination of puncture impact behaviour of rigid plastics - Part 2: Instrumented puncture test (2002–04)
DIN EN ISO 180 Plastics - Determination of Izod impact strength (2013-08)
ASTM D3763 Standard Test Method for High Speed Puncture Properties of Plastics Using Load and Displacement Sensors (2010)
ASTM D5420 Standard Test Method for Impact Resistance of Flat, Rigid Plastic Specimen by Means of a Striker Impacted by a Falling Weight (Gardner Impact) (2010)
ASTM D5628 Standard Test Method for Impact Resistance of Flat, Rigid Plastic Specimens by Means of a Falling Dart (Tup or Falling Mass) (2010)

Normative References: Fibre-Reinforced Plastic Composites - High Rate of Deformation - Compression After Impact

ISO 18352 Carbon-fibre-reinforced plastics - Determination of compression-after-impact properties at a specified impact-energy level (2009-08)
DIN 65561 Aerospace; fibre reinforced plastics; testing of multidirectional laminates; determination of compressive strength after impact test (1991-05)

ASTM D7136 Standard Test Method for Measuring the Damage Resistance of a Fiber-Reinforced Polymer Matrix Composite to a Drop-Weight Impact Event (2012)
ASTM D7137 Standard Test Method for Compressive Residual Strength Properties of Damaged Polymer Matrix Composite Plates (2012)
Airbus AITM 1-0010 Determination of Compression Strength After Impact-Stress
Boeing BSS-7260 Boeing Company Specification BSS-7260

Normative References: Fibre-Reinforced Plastic Composites - High Rate of Deformation - Fatique Tests

DIN 50100 Testing of Materials; Continuous Vibration Test; Definitions, Symbols, Procedure, Evaluation (1978-02)
DIN 53442 Flexural fatique testing of plastics using flat specimens (1990-09)
ISO 13003 Fibre-reinforced plastics - Determination of fatigue properties under cyclic loading conditions (2003-12)
ASTM D3479 Standard Test Method for Tension-Tension Fatigue of Polymer Matrix Composite Materials (2012)

Normative References: Membranes

DIN EN ISO 527-3 Plastics - Determination of tensile properties - Part 3: Test conditions for films and sheets (2003-07)
DIN EN ISO 1421 Rubber- or plastics-coated fabrics - Determination of tensile strength and elongation at break (1998-08)
DIN EN ISO 2411 Rubber- or plastics-coated fabrics - Determination of coating adhesion-Beschichtungen (2000-08)
DIN 53363 Testing of plastic films - Tear test using trapezoidal test specimen with incision (2003-10)
DIN 4102 Fire behaviour of building materials and building components - Part 1: Building materials; concepts, requirements and tests (1998-05)

Normative References: Textile Based Concrete

DIN EN 14649 Precast concrete products - Test method for strength retention of glass fibres in cement and concrete (SIC TEST) (2005-07)

Chapter 15
Modeling and Simulation

Lina Girdauskaite, Georg Haasemann, and Sybille Krzywinski

This chapter will describe the fundamental aspects and methods for the modeling and simulation of textile reinforcement structures and fiber-reinforced plastic composites (FRPCs). Due to the anisotropic material properties, the simulation of the deformation behavior of textile reinforcement structures is a complex matter. Various approaches will be introduced and simulation solutions based on kinematic models will be discussed in detail. The focus of this chapter is to assist designers and engineers in the design of preforms for complex FRPC components. To correctly configure the composite material according to the expected strains by means of Finite Element Models (FEM), extensive experimental tests for the quantification of composite characteristics are necessary. This contribution will therefore also address modeling and simulation methods based on multi-scale approaches to the determination of special mechanical values of materials.

15.1 Introduction

The use of computer-assisted methods for the assessment of a component design and its constructive realization has become crucial to achieve ever shorter product development cycles. Apart from that, generating geometry models to describe the product dimensions requires the characterization of material behavior of the textile

L. Girdauskaite • S. Krzywinski (✉)
Institute of Textile Machinery and High Performance Material Technology, TU Dresden, Dresden, Germany
e-mail: lina.girdauskaite@tu-dresden.de; sybille.krzywinski@tu-dresden.de

G. Haasemann (✉)
Formerly: Institute of Solid Mechanics, TU Dresden, Dresden, Germany
e-mail: i.textilmaschinen@tu-dresden.de

C. Cherif (ed.), *Textile Materials for Lightweight Constructions*,
DOI 10.1007/978-3-662-46341-3_15

composite reinforcements for purposes of modeling. Ensuring a wrinkle-free shaping of textile reinforcement structures into strongly curved shapes, sometimes even double-curved spatial contours, and realizing a load-adapted orientation of the reinforcement yarns are essential criteria in the designing of FRPC components.

The mechanical behavior of textile reinforcement structures differs considerably from that of monolithic materials. Due to the inhomogeneous structure made from fibers and yarns, locally varying material properties are common. Globally varying material properties arise from the production of textile reinforcements made from approximately similar fibers and yarns but manufactured by means of different fabric formation technologies. Due to the orientation of the reinforcement yarns, the properties of the reinforcement structures are anisotropic.

The flexible, anisotropic behavior of textile reinforcements makes the modeling of such structures for simulation processes for the purpose of specific designing and construction quite complex. Established methods can be classified according to the depth of analysis into the micro (fiber) level, the meso (yarn) level, and the macro (fabric) level. While the structure in the micro and meso scales is inhomogeneous, the macro scale allows a homogeneous approach.

The load-adapted orientation of the textile reinforcement structures of complex-shape components imposes high engineering requirements not only on the computation and simulation software to be used but also on the machines and methods suitable for production. If reinforcement structures are selected specifically and provided efficient manufacture, fiber-reinforced plastic composites allow a more cost-efficient component production in comparison to the metal materials. This requires an adaption of the geometry and designing/constructive aspects to the processed material. If this is taken into account, composite materials can be used to realize complex component geometries which could (easily) be produced with metal materials [1–6].

The current technologies used to manufacture textile preforms and precisely position them in the mold for shaping/consolidation are not sufficiently economical outside of aeronautics applications, regardless of component manufacturing method and used matrix material [7–9]. Preform production costs constitute a significant part of the total production costs.

Currently, the desired reproducibility of the preform quality and the resulting structure-mechanical component properties are not achieved due to the significant geometry alterations during the handling of textiles required in building the stack of layers and in shaping and positioning in the consolidation mold. During shaping into complex components, the load-adapted yarn orientation is partially lost, making an increased use of material necessary to ensure the safety required by the design. This is done at the expense of the lightweight construction effect and increases costs due to over-dimensioning. Improving simulation tools for the characterization of the deformation behavior of dry textile structures and fiber-reinforced plastic composites is therefore a priority. For this, the material parameters needed for the simulation have to be determined.

15.2 Deformation Behavior of Textile Semi-finished Products

15.2.1 General Discussion

For the application of textile semi-finished products made from high-performance fibers in FRPCs, it is crucial to realize the reinforcement of complex, sometimes strongly curved geometries without the formation of wrinkles. To perform this task without an excessive number of iterations in the future, a simulation of the deformation is currently a priority goal. For this, parameters characterizing the deformation behavior are required.

Tensile, flexural, and *shear parameters* as well as *torsion rigidities* [10–12] have to be taken into account in the mechanical assessment of the deformation of textile fabrics. A measuring method for the determination of torsion rigidities in textile fabrics has been developed and tested for classical textile structures, and it is also suitable for textiles made from high performance fibers. The currently available experimental examinations will have to be extended in the future to specify exact test conditions [13].

15.2.2 Tensile Values

The deformation behavior of textile fabrics under tensile loads results from the mechanical properties of the processed fibers, yarns, and the construction design [12]. Changes in length, related to the initial length of the sample, are referred to as elongations [14].

Determining the tensile values by means of tensile tests on strips according to DIN EN ISO 13934 T1 is explained in detail in Sect. 14.5.3.

Stress-strain behavior is tested using tensile tests on strips, preferably in two directions usually corresponding with the direction of the reinforcement yarns. To minimize the effects of transverse deformation the test can also be performed on biaxial tensile testers (manufacturers include Zwick GmbH & Co. KG [15], Kato Iron Works Co. Ltd [16]). The measuring sample is held in place by several clamps located on all four sides. The clamps are arranged perpendicular to the load direction or move proportional to the current change in length, controlled by computers. Tensile force is measured in both directions. The biaxial tensile test offers the possibility to test in an approximately load-adapted manner [17]. For the deformation simulation, transverse normal stiffnesses can be measured in addition to longitudinal normal stiffnesses. To gather test results with minimal spread, the clamping variations and sample shape have to be defined precisely and observed accurately. At the time of writing, no standards, but numerous recommendations on biaxial tensile tests exist. To avoid transverse deformation and the influence of the material clamping, the literature mainly discusses application-related test forms and force introductions [18–21].

15.2.3 Flexural Values

The flexural rigidity B is a measure of the resistance with which a textile fabric counters the bending moment of a defined bending change [10]. Flexural rigidity in textile fabrics depends on the fiber material used and the yarn construction. Both the rigidity of the fibers in the yarn and the construction design of the fabric significantly influence this specific value. The flexural rigidity of a fabric is commonly measured using a cantilever process, which is detailed in Sect. 14.5.4.

Extensive research at the Institute of Textile Machinery and High Performance Material Technology (ITM) at TU Dresden have shown that the currently used devices for flexural testing with cantilever methods do not offer reproducible results due to manual operation and the corresponding subjective influences. The visually determined overhanging length is included in the calculation of flexural rigidity in cubic power. Thus, even slight inaccuracies in reading the results and the insufficiently accurate scale marking can cause calculation errors of the flexural rigidity. Therefore, a new flexural rigidity measuring system (ACPM 200) was developed at the ITM [22]. The high degree of test process automation and analysis compensates the mentioned disadvantages of previous devices. Furthermore, the influence of local differences in rigidity across the sample width can be quantified. The developed device can be used for flexural rigidity tests of fabrics made from high-performance fibers.

A provision of the measured values in the form of a moment-curvature curve is required for the simulation of flexural behavior. This curve is not recorded in flexural tests on the ACMP 200. EISCHEN et al. [23] gives an indirect method to determine the moment-curvature curve, based on the cantilever method. In the first step, the X-Y coordinates along the bending curve are recorded from a side view, using a digital camera (Fig. 15.1).

The data is computer-processed to determine the value pairs of the moments and curvatures along the overhang from the free endpoint to a locally defined endpoint. The zero point of the coordinate system is located at the free end of the sample. Programming the algorithm for the ascertainment of the bending moment—curvature curve (Fig. 15.2) can be performed with MATLAB.

15.2.4 Shear Values

Shearing is the change of the angle of the crossing yarn systems caused by shear loads. The yarn intersection points form a center of rotation in which the distance of the points remains constant. Four intersection points are considered in Fig. 15.3.

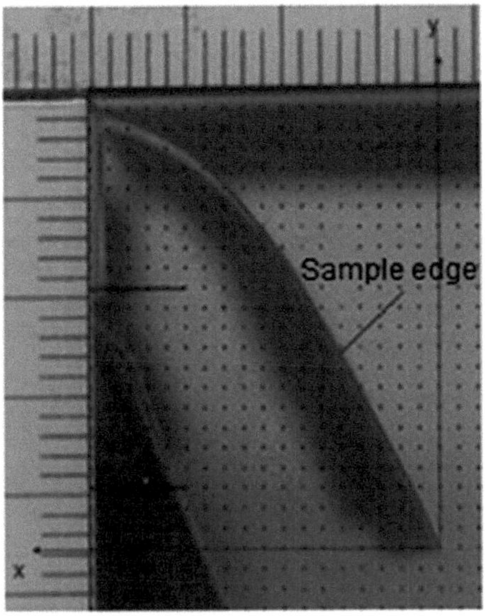

Fig. 15.1 Experimentally determined bending curve

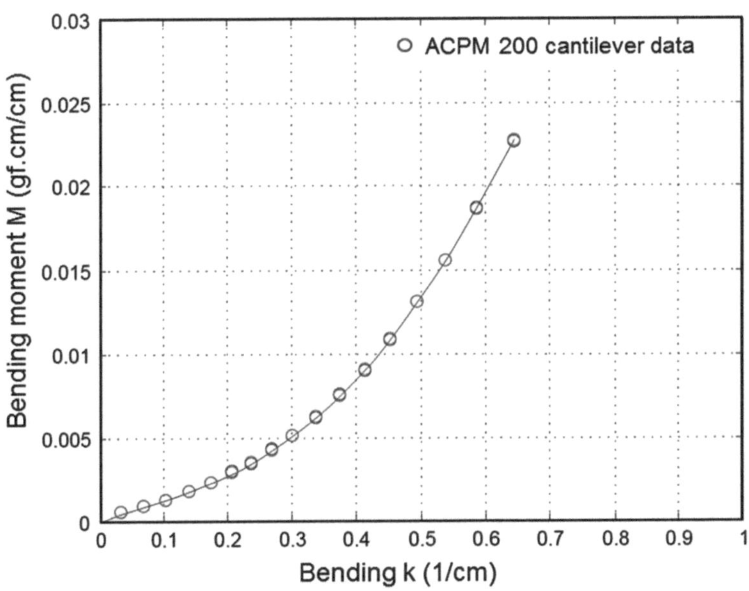

Fig. 15.2 Bending curve calculated with ACPM 200 data

The originally rectangular arrangement of the points is distorted into a rhombus under shear strain. The resulting angle φ is referred to as the shear angle. When the deformation results in a maximum yarn compression, the *shear angle* reaches the

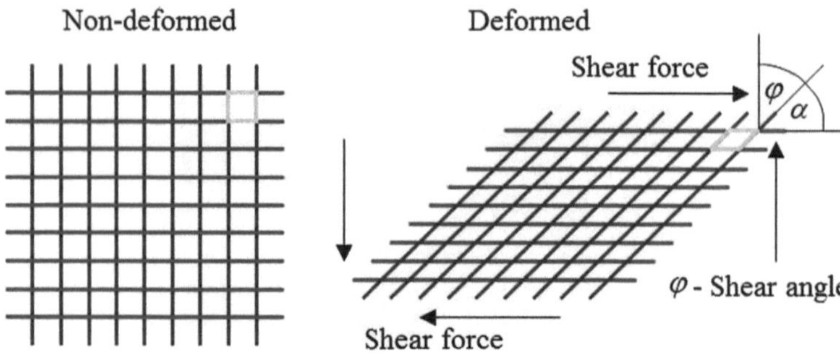

Fig. 15.3 Shear angles according to [12, 24, 25]

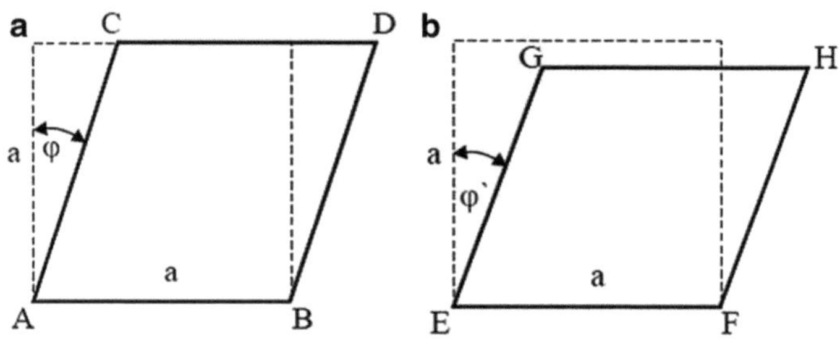

Fig. 15.4 Schematic representation of shear principles [26]: (**a**) simple shear, (**b**) pure shear

critical value φ_{kr}. Further yarn compressions cannot take place in the plane. This necessarily results in the creation of wrinkles in the fabrics.

The shear of textile fabrics can be distinguished into two essential types of shear: simple and pure shear [26]. Figure 15.4 represents the different shear principles.

Numerous methods for the characterization of the shear behavior of fabrics have been tested. However, not all of them are suitable for tests on reinforcement textiles made from high-performance fibers.

In simple shear (Fig. 15.4a), shear loads cause not only angular changes between the yarns, but also tensile strain caused by yarn torsion at the crossing points. The distance between the two clamping lines remains constant during the entire shear process, causing a change of the length of the unclamped sample edge. If a tensile force is overlaid on shear deformation across the entire load cycle, a premature buckling of the fabric can be prevented in classical textiles. Usually, classical textiles tend to form wrinkles at low shear angles. As a shear angle of up to 8° is not sufficient for tests on textile composite reinforcements, this test method is not recommended. At greater angles, simple shear testing results in tensile deformations, which can only be realized by very high forces in textile composite reinforcements and are therefore irrelevant for the application of preforming.

Therefore, pure shear testing is recommended for tests on textile reinforcements (Fig. 15.4b). In pure shear, only the angle between the yarns is changed without yarn elongation. The distance between the clamping lines remains constant, and the

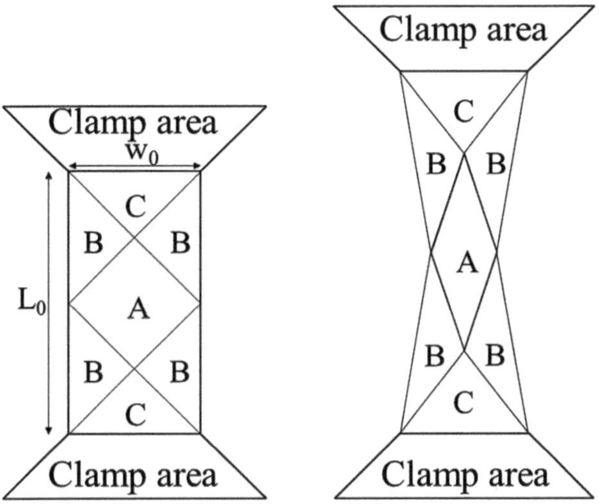

Fig. 15.5 Schematic of the bias-extension test [27]

sample edges do not change. This test principle is realized in bias extension tests and shear frame tests. The realization of shear in *bias-extension tests* can be performed on monaxial as well as biaxial tensile testers. They are tensile tests using a sample cut at a 45° angle to the yarn directions [27, 28]. In section A (Fig. 15.5), pure shear is measured. In section B, half the value of the shear measured in section A is registered. Section C does not contribute to shear force. During testing, the shear force—shear displacement diagram is recorded.

The NAISS Company developed the TEXPROOF textile testing machine, which includes a shear test module (Fig. 15.6) [29]. The sample is affixed between two clamps, one of which is fixed, while the other can travel on a curved course. This course realizes the "Trellis" effect. The effect is triggered when the directions of introduced tensile forces do not correspond to the main directions of the reinforcement yarns. Shear with a change of the angle occurs until the reinforcement yarns are oriented in force-direction or until a maximum angle is achieved, which depends of the reinforcement geometry [29]. Independent of the sample thickness, light barriers register the shear angle at a crease height of 3 mm perpendicular to the plane.

In the *shear frame test* [26–28, 30–37], pure shear is realized by fixing a square sample in a shear frame, clamped in a monaxial tensile tester at opposing corners, and deformed into a rhombus to a preset deformation distance (Fig. 15.7). Meanwhile, the force plot is recorded along the entire deformation distance. The shear angle can be calculated from the change in length of the rhombus diagonal. Clamping influences on the shear result are minimized by fixing the sample on needle bars [26]. For this test, the samples (300 mm × 300 mm) are needled onto the shear frame without tension, and secured.

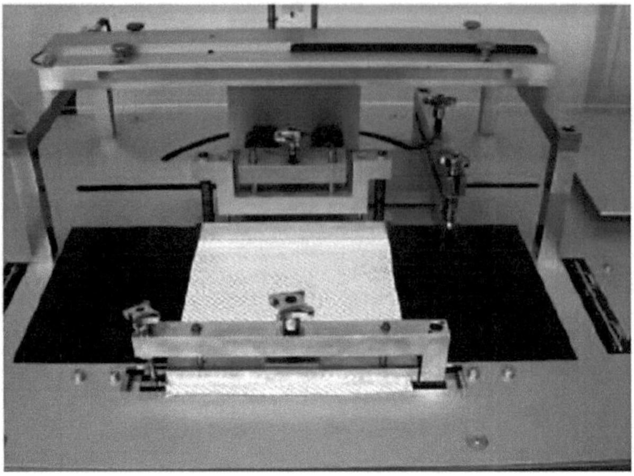

Fig. 15.6 Shear test module TEXPROOF [29]

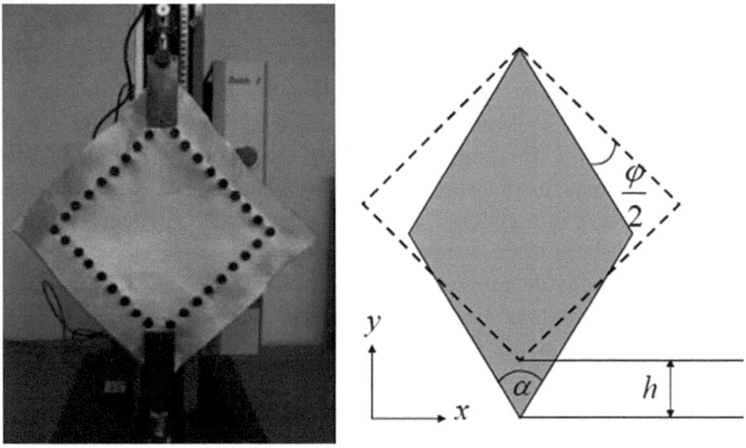

Fig. 15.7 Test set-up and sample deformation principle

The deformation distance of the shear frame in these tests ranges from 60 mm (for multiaxial non-crimp fabrics) to a maximum of 100 mm (for woven fabrics). This is equivalent to shear angles of $\varphi = 0°$ to $\varphi = 28°$, or $\varphi = 0°$ to $\varphi = 56.3°$ respectively. At the ITM, test speed is set at $v_{test} = 200$ mm/min. During the shear test, the shear force is recorded along the traveling distance and then depicted graphically.

The shear angle can be calculated according to Eq. 15.1

$$\varphi = 90 - 2\arccos\left(\frac{1}{\sqrt{2}} + \frac{h}{2a}\right) \tag{15.1}$$

Here, φ is the shear angle in °, h is the deformation distance of the shear frame in mm, and a is the side length of the shear frame in mm (200 mm in the test case).

At the beginning of the measurement series, reference measurements are taken with the frame without inserted sample, determining the friction forces of the shear frame during the test. The friction force is subtracted from all measurement curves as a reference curve. The deformation of the samples is also recorded with a camera during testing, and the wrinkling in Z direction is optically captured by means of the shade. The camera is positioned facing the surface normal in the center of the sample. At least three photographs per second are recommended to photographically record the sample over the entire test duration.

From the calculated shear force—shear angle diagrams, statements regarding the expectable deformability can be derived. To analyze the shear behavior, critical shear angles and the limiting angle are used, as they can be stored in the simulation tools.

To determine the limiting angle, the shear force—shear angle plot is separated into a linear and a non-linear zone as suggested by [38] (Fig. 15.8). The warning angle φ_{lim} is defined at the transition from linear to non-linear zone.

Due to the force increase at the beginning of the shear test, the measured values at the beginning of the curve are not considered. After this increase of force, the curve transitions into a nearly linear section. This is chosen as the initial value of the regression analysis. Usually, these are measured values in a range from 5° and 10°. In the test example, the results from the difference between the measured curve and the line are characterized by fluctuations under 3 %. In the transition from the linear to the non-linear section, the deviation is significantly higher. The end of the linear section is defined at a deviation of 5 % [26].

The critical shear angle of a fabric can be determined optically, as described above. The first bulking of the sample is often subjectively detected visually from the recorded images. It is however recommended to recognize wrinkles

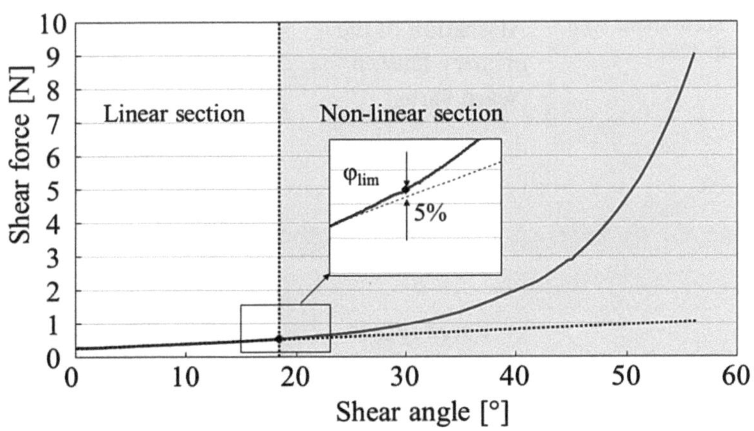

Fig. 15.8 Shear force—shear angle diagram with φ_{lim} for a woven fabric

instrumentally. Grey value image analysis or optical 3D deformation measurement systems are suitable options to detect the formation of wrinkles [26, 31].

Currently, preforms for complex FRPC components are largely manufactured manually. The textile composite reinforcements are shaped from the plane into the desired component geometry without tension. Recent developments are concerned with the automation of preform structuring. So-called "guidance systems" enable an automated laying of the reinforcement structures. These create a primary tension in the textile, which can influence wrinkling during deformation. To analyze this influence, the test device described in the following section was developed [28].

The shear measuring device built in [28] consists of a shear frame equipped with force sensors to determine primary tension and tensile force in warp and weft direction (Fig. 15.9). This allows a measuring of the tensile forces occurring in warp and weft direction during the test, and an allocation of the primary tensions to the shear deformation.

To date, a variety of research efforts to determine suitable shear test methods and to apply experimental results in modeling approaches related to the deformation behavior have been conducted in parallel, but without co-ordination between individual research groups. Therefore, academic and industrial researchers founded an international group for developing and implementing a benchmarking in 2003 [31]. Shear frame and bias-extension tests for textile composite reinforcements with symmetrical or asymmetrical reinforcement yarn arrangements were performed and compared. The tests were realized on three identical woven fabrics by seven international research institutions: (NU) Northwestern University, USA; (UT) University of Twente, Netherlands; (LMSP) Laboratoire de Mécanique des Systèmes et des Procédés, France; INSA-Lyon, France; (UML) University of Massachusetts Lowell, USA; (UN) University of Nottingham, United Kingdom; (KUL) Katholieke Universiteit Leuven, Belgium; and (HKUST) Hong Kong University of Science and Technology, China. Six of the research groups delivered

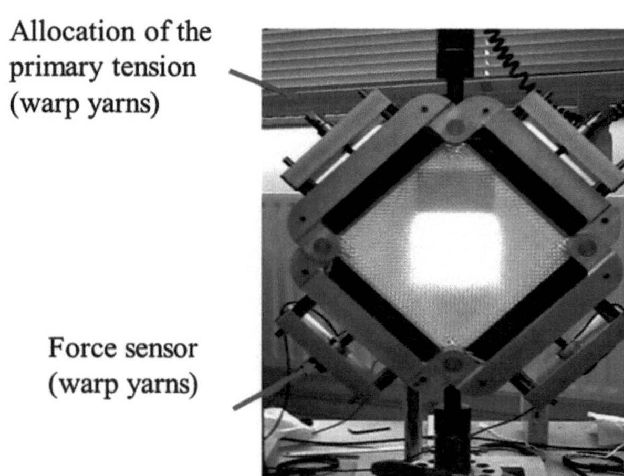

Fig. 15.9 Shear frame with force sensors [28]

results based on the use of a shear frame, and four of the groups submitted findings based on bias-extension tests.

The efforts of the abovementioned [31] and other research works aim to provide recommendation of suitable testing technology and a standardized test method. In [31] no significant limitations were given regarding the conduct of the test.

The samples are square and adapted to the respective frame dimensions (145–250 mm). More details regarding sample preparation are given below. Woven fabrics are selected as test material, meaning that the reinforcement yarns are oriented in 0°/90° direction to one another. The tests are performed at different velocities ranging from 10 to 1,000 mm/min. The shear frame constructions are not identical (Fig. 15.10), but all the frames share certain characteristics. The textile reinforcements are held by the shear frame clamping mechanisms in a fashion that prevents slipping during the test. The friction between reinforcement structure and clamping is not considered. To eliminate the share of the force resulting from the shear of the selvedge reinforcement yarns, HKUST removed all clamped selvedge yarns running parallel to the clamping direction. Likewise, UT eliminated all reinforcement yarns arranged at a defined distance parallel to clamping direction, in order to prevent premature crease formation.

For bias-extension tests, samples of different sizes are used as well, and tested at different machine speeds.

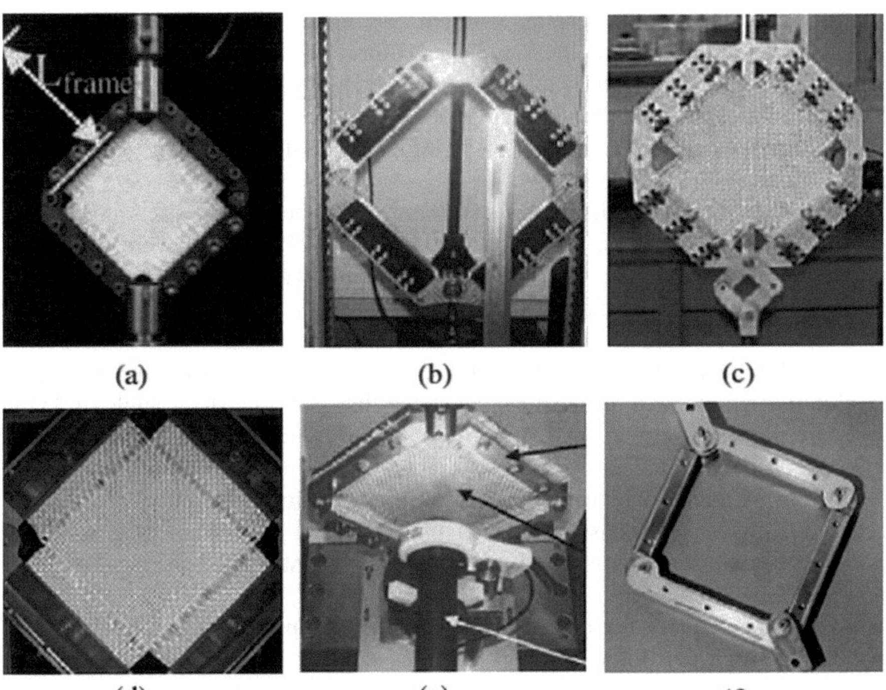

Fig. 15.10 Shear frames used [31]: (**a**) HKUST, (**b**) KUL, (**c**) UML, (**d**) UT, (**e**) LMSP, (**f**) UN

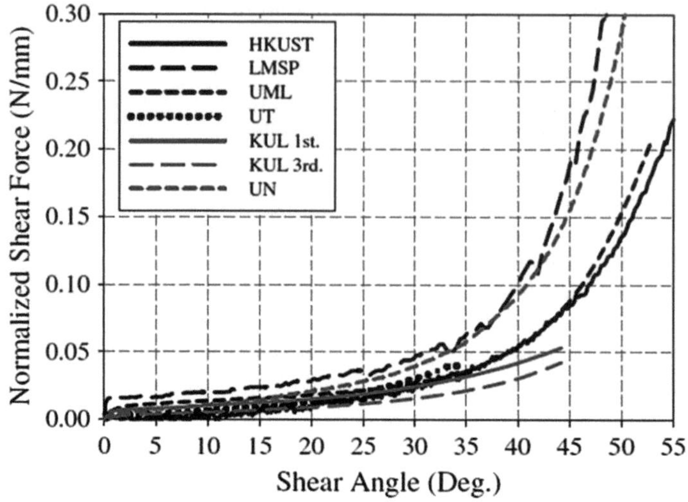

Fig. 15.11 Shear force—shear angle curves [31]

The results of the study have shown that it is possible to use shear frames to obtain valuable experimental data for the characterization of the shear behavior of reinforcement structures, despite different test engineering preconditions. The mechanical conditioning of the sample can improve repeatability, as shown by the results of UML and KUL. All samples tested by UML were mechanically conditioned and gave results showing lower fluctuations. Using mechanical conditioning, the tensions remaining in the reinforcement structure from weaving are compensated (Fig. 15.11).

In order to compare different shear frame constructions, shear frame and/or sample sizes, and designs, normalization methods [27, 31, 39] were developed and presented (Fig. 15.12). After applying the normalization methods described in the study, test results are similar in the range up to 35°, which is relevant for the deformation of textile reinforcements (Fig. 15.12).

Parallel to calculating the shear angle, optical methods are used to record wrinkle formation. As a result, it can be stated that—up to a shear angle of 35° in shear frame tests, and up to 30° in bias-extension tests—plain-woven fabrics display consistent matches with the shear angle determination based on the traversal path. At greater angles, optical wrinkling detection is recommended.

So far, this chapter has established the lack of a suitable testing standard. Currently, global research is being conducted to understand the effects of primary tension [28, 31].

As shown above, a textile-physical characterization of textile reinforcements made from high-performance fibers requires adjustments of the textile testing technology, as these textile reinforcements differ considerably from classic fabrics with regard to geometrical and structural construction.

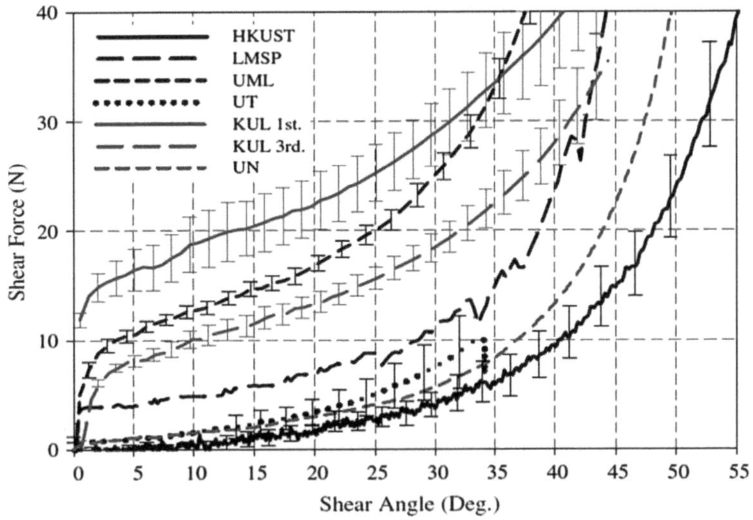

Fig. 15.12 Shear force—shear angle curves after normalization based on frame length [31]

In summary, it can be said that the deformation behavior of fabrics made from yarns produced from continuous high-performance fibers consist in shares of yarn elongation, yarn extension, yarn displacement (Fig. 15.13), and surface shearing [12, 24, 25, 38, 40]. Shear, which is the single most important specific value for the description of the deformation behavior of textile reinforcements, has been characterized in details. The other deformation mechanisms are described in the following sections.

15.2.4.1 Yarn Elongation

Woven fabrics, knitted fabrics with biaxial or multiaxial reinforcement yarns, or non-crimp fabrics made from high-performance yarns are characterized by a minimal elongation at tensions below 100 N (which is the value required for draping). Therefore, elongation has only a miniscule share in the deformation of the textile fabric [12, 25, 28].

15.2.4.2 Yarn Extension

In fabrics produced by the interlacing (e.g. woven fabrics, braided fabrics) or interlooping (knitted fabrics), the yarns are not oriented straight and stretched, but in waves, sinuses, or loops. Due to the change of curvature radii under loads, the yarns of the semi-finished product are stretched [25, 26]. Figure 15.13 shows the yarn stretching of a woven fabric.

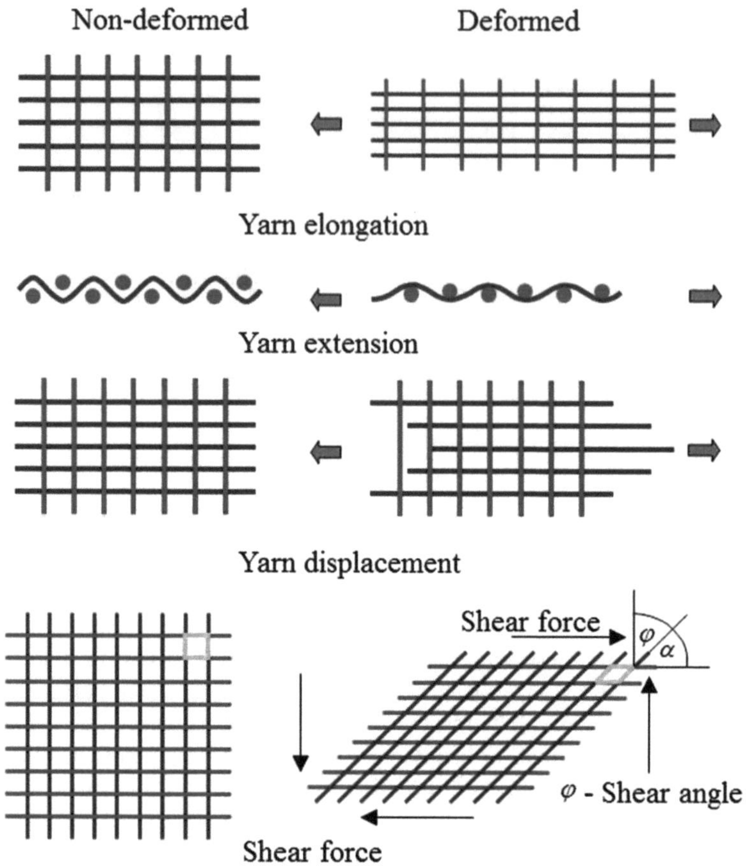

Fig. 15.13 Basic deforming mechanisms of textile reinforcement structures [24, 25]

15.2.4.3 Yarn Displacement

Yarn displacement often occurs in fabrics made from high-performance fiber materials at low yarn-yarn friction. Such fabrics include multiaxial non-crimp fabrics and fabrics with long floating loops (twill and atlas-woven fabrics).

The shear behavior is therefore the most important specific deformation value for modeling and simulating the shaping process of textile reinforcement structures made from high-performance fibers. The aim is to support the constructive component design in high-performance fiber composite application with an informed selection of reinforcement structures. Here, precise previous knowledge of the degree of crease-free shaping of dry reinforcement structures is crucial.

Model examination for a comparison of the deformation behavior of dry reinforcement structures can be performed by means of push-out tests.

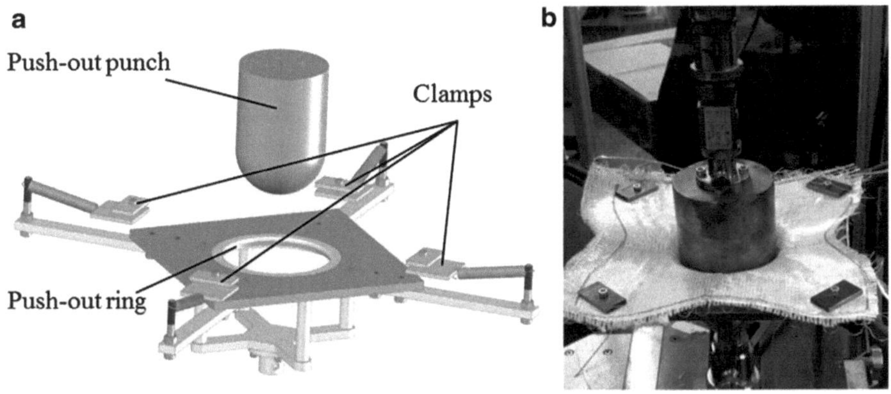

Fig. 15.14 (**a**) 3D model of the push-out test setup, (**b**) push-out test performed on a multilayered knitted fabric [8]

A test setup constructed especially for the *push-out test* is integrated into a tensile tester (Fig. 15.14). In this setup, a cylindrically tapered hemisphere (diameter: 100 mm) pushes the textile to be tested through a ring. This ring is exchangeable, but has an internal diameter of 120 mm, and an edge radius of 2 mm in the case of the performed tests [8]. The squarely cut textile to be tested is fixed by spring-attached clamps at four points and repositioned by them in a defined manner during the test. This clamping mechanism only slightly impedes the deformation of the textile semi-finished product and does not significantly influence the course of the measurement curve. The push-out punch and ring with its bracket are clamped respectively in the bottom and top clamping device of a tensile tester (e.g. the Zwick GmbH Z100). The test records and analyses the push-out force in relation to the traverse path.

15.3 Computer-Assisted Simulation of the Deformation Behavior of Textile Composite Reinforcements

15.3.1 Models for Simulating the Deformation Behavior Simulation

The achievable draping behavior for a wrinkle-free shaping in complex components is of equal importance as the load-adapted reinforcement by means of single- or multilayer structures. Here, the reinforcement-adjusted yarn orientation is to be retained after shaping into the desired component contour in order to avoid a deterioration of the mechanical properties of the component by an undefined placement of the reinforcement textile.

In general, two variations of shaping are distinguished. In the first case, a rectangular cutting of the textile reinforcement is draped into a mold that matches the desired component geometry. In combination with yarn extension, and depending on the curvature, this can cause sometimes considerable yarn displacement and surface shearing, affecting the desired yarn orientation. In areas where wrinkles occur, the reinforcement structure is incised and extensively overlapped without a joining process, resulting in according with thickenings and higher component masses.

In the second case, ready-made technology is used to project the complex geometry in the plane as undistorted as possible by means of a partition into several smaller cuts. To produce a preform largely matching the component contour, the aim is to approximately project the component geometry by as few partial cuts as possible. However, further draping is required for the shaping. Angle and distance alterations of the yarns are inevitable in the process. To achieve the desired 3D component shape in a load-adapted design without later additional process steps, the smaller cuts are designed directly on the virtual geometry model [12].

The individual smaller cuts are placed in the mold either manually or by robots, or are joined into near-net shape preforms. Usually, the joining sections change the wall thickness in the joint zone.

According to [38, 41, 42] the following approximate classification can be derived for the modeling of the draping behavior of bidirectional textiles:

- the kinematic model,
- the elasticity model, and
- the particle model

The *kinematic model* depicts the meso structure (yarn level) of the textile by a geometric pattern with regard to the geometric boundary conditions on a surface.

In the *elasticity model* the textile semi-finished products, which are largely represented anisotropically and partly orthotropically, are discretized in order to determine the deformation and the yarn orientation by means of the Finite Element Method.

The *particle model* represents microscopic interactions describing the properties of the macroscopic system. Each yarn intersection point is represented by a particle possessing the physical characteristics of the reinforcement structure. The most probable arrangement of the intersection points is determined by establishing the minimum particle energy. This approach does not require a flat placement of the textile, allowing a simulation of the free deformation of textiles under the influence of gravity. The fiber bending is factored in.

CHERIF [38] develops a significant contribution to the drape-ability simulation of woven fabrics and multiaxial non-crimp fabrics on the basis of the Finite Element Method. Apart from the anisotropic material behavior, production-technological constrains are considered as well. To describe the mechanical material behavior, a four-noded shell element is used. The macromechanical model allows the consideration of high and non-linear shear deformation degrees at negligible elongations. Extensive experiments serve to determine the required specific values. Furthermore, geometrical non-linearities caused by the contact of reinforcement structure and shaping mold, as well as boundary conditions varying with the proceeding draping process can be simulated. To omit a repetition of any detailed deliberations on the subject, the respective literature is recommended for in-depth information.

Beyond this, the textile structure can be depicted as precisely as possible by means of a unit cell (RVE—Representative Volume Element) [43–48]. Determining the modeling parameters can be an elaborate process, but given sufficient parametrization, can be used to represent a variety of structures of the same fabric formation process. The modeling of the entire structure to simulate draping behavior on different component geometries still requires enormous computing power.

Kinematic models for the simulation of the deformation behavior of reinforcement structures have disadvantages. Even so, short computing times and the precision of calculation results make them increasingly relevant for practical applications due. Both factors may vary depending on component complexity. The disadvantages include [38]:

- Only single-layer structures are simulated. Frictions between the layers during simultaneous draping of several layers are not considered.
- The effect of any drape effectors or other shaping molds on the draping result cannot be simulated.
- The influence of a guidance system for a reproducible, preferably wrinkle-free shaping is disregarded.

Despite worldwide research efforts for the description of draping processes by means of FEM, the currently commercially available simulation tools require extensive computing times and considerable experimental effort for the determination of the required specific values.

15.3.2 Kinematic Modeling of the Deformation Behavior

As established above, shearing behavior is the most important specific value to be considered in the simulation of the deformation behavior of dry reinforcement structures. When describing the deformation, the effects of yarn elongation and yarn extension are usually omitted. In the kinematic model, only deformations based on shear are considered. The placement of the bidirectional reinforcement structure on the surface of the form is simulated. The model delivers geometrical information on the shear angles required for the deformation. The mechanical properties of the reinforcement structure are not entered in the modeling [42, 49, 50].

The crossing points of the yarn axes between the yarns are modeled as joints. Between these intersection points, the yarns behave as beams with constant length. A geometric algorithm allows the determination of the crossing points of the reinforcement structure, if the location of the point on the shape geometry and the fiber orientation direction in the point are known. The yarns between the joins are geodesically placed on the geometry. The resulting angles in the joints match the shear angles of the reinforcement structure. Shear limitations for the realization of a wrinkle-free draping may prevent the placement of the intersection points. This is always the case if the critical shear angle stored in the simulation tool is exceeded [42, 49].

Using a comparison of the calculated local shear angles to the critical shear angles of the textile semi-finished products to be used (which can differ greatly between multilayered knitted fabrics, woven fabrics, and multiaxial non-crimp fabrics), a sensible default is set for the cut parts. Usually, these algorithms do not consider loads or friction effects.

FiberSIM® [51], *DesignConcept 3D* (*DC3D*) [52], *PAM Quickform* [53], *Composite Part Design* (*CPD*) [54] are examples of commercial software packages on a geometrical basis, and contain interfaces with FE-calculation programs such as *ANSYS* [55] or *MSC-Patran* and *MSC-Nastran®* [56]. With these, the yarn orientation can be simulated and represented on the basis of a geometrically sufficiently defined reinforcement structure. The result can be transferred into FE-calculation programs and is used to recalculate load cases.

The *FiberSIM®* and *DC3D* software solutions are examined with regard to deformation simulation and cut generation in [57]. The simulations are based on the kinematic model.

The textile preform has to match the desired component geometry as precisely as possible. The desired component thickness is an exception, as it is only achieved after the consolidation process. The aim is to design the cutting of the reinforcement textile in a manner that creates minimum material compressions or extensions, adheres to the set reinforcement direction, and ensures a wrinkle-free deformation. The 3D data required for cut generation of the projected components can be created with commercially available 3D-CAD software solutions (such as *CATIA* [53], *SolidWorks* [58]) or implemented into the simulation software via neutral interface formats (e.g. IGES and STEP).

Using the *DC3D* software solution [52], the user can create virtual 3D geometry models and conduct feasibility analyses based on an automated cutting generation. The cutting contours for the realization of the desired component shape are based on the 3D geometry. To this end, cutting boundaries are set on the surface. In order to be able to represent the cutting designed in 3D in the plane, the surfaces are triangulated. *Triangulation* refers to the partition of an area into triangles. Triangulation can be performed curvature-dependently or uniformly.

The simulation criterion consists of minimizing changes to the edge length of the triangles, the angles in the triangle, and the areas of the triangles [12, 57]. The flattening result is influenced significantly by two boundary conditions: the *starting point* and the *fiber orientation direction* [40]. In assembling, the term *flattening* is used for the representation of the 3D-developed cut parts in the plane.

As reinforcement textiles can have an anisotropic stress-strain behavior, fiber orientations play a crucial part. Textile-reinforced plastic components are

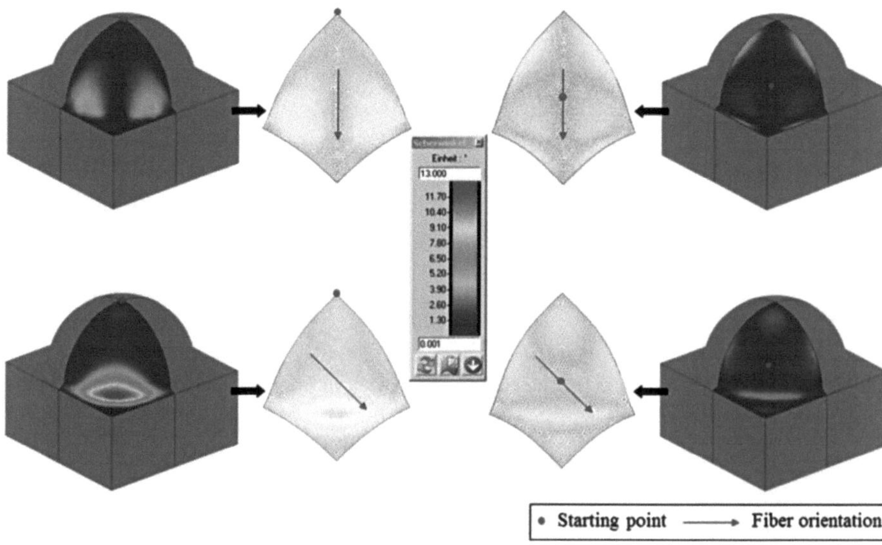

Fig. 15.15 Cuttings under variation of fiber orientation and simulations starting point [57]

differently loaded in practice, and therefore the reinforcement yarns have to be oriented according to the structure-mechanical design.

Fiber orientation directions and starting point of the simulation are varied for a reference geometry (hemisphere with flat contact surface) (Fig. 15.15).

The flattening analysis shows which shear angles are necessary to drape the cutting onto the shaping mold. By comparing the experimentally established critical shear angles of the textile reinforcements, the designer can judge whether the selected cutting determination is suitable for the component geometry to be realized. The designer has to decide whether the shear-induced local displacements of the reinforcement yarns are permissible for the load-adapted component design.

To achieve the desired wall thickness of the component, several layers of reinforcement structure are often required. The component geometry therefore has to be adjusted according to layer thickness in the compacted state, in order to ensure accurate cutting generation.

The *FiberSIM®* software is currently in use by leading manufacturers in aeronautics to develop fiber-reinforced plastic composite constructions.

The cutting contours are derived from the 3D geometry. As mentioned above, the cut boundaries are defined on the surface for this purpose. *FiberSIM®* works directly on the CAD representation of the component, evaluating the native geometry without transformation and approximation [51]. After the simulation of the reinforcement structure placement on the respective geometry, information concerning the necessary shear angles becomes available [42, 49]. The simulation results are represented on the surface of the geometry by means of a set of curves known as "fiber trajectories", which reflect the final state of the semi-finished

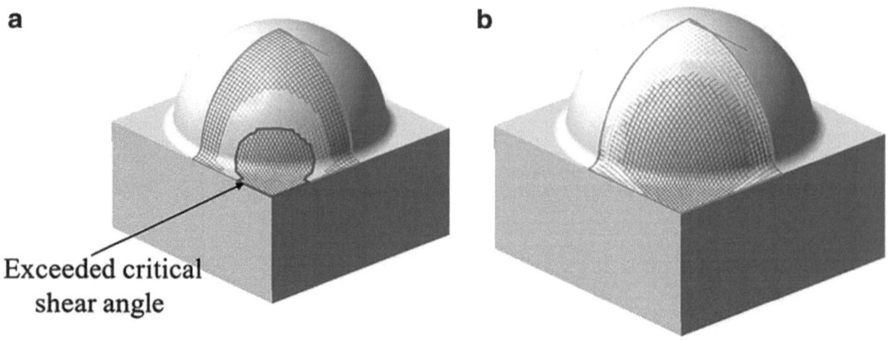

Fig. 15.16 (**a**) Shear deformation on the cutting, (**b**) Deviation from the defined orientation of the reinforcement yarns [57]

product after draping. For biaxially reinforced structures, the result is a grid on the model surface.

FiberSIM® contains a material database in which the basic specific values and characteristic values for a description of the shear behavior of reinforcement textiles can be stored. The calculation results are compared to the limiting angle and critical angle values under shear loads, stored in the database. The "fiber trajectories" are shown color-scaled. Areas visualized in blue display low strain, a yellow color signalizes that the limiting shear angle has already been reached, while a red color indicates that no wrinkle-free draping is possible. Figure 15.16a gives an example of this.

By means of the analysis of deviations from the selected fiber orientation direction, *FiberSIM®* offers additional control to avoid multiple iterations during preform development. One example is given in Fig. 15.16b.

As the surface area of the planar cutting deviates greatly from the surface area of the counterpart in 3D, problems during draping can be identified early on (Fig. 15.17). This proves that variations of the starting point can contribute to an optimization of the flattening result.

The deformation analysis results of both software solutions are similar (Fig. 15.18). As can be seen from Fig. 15.19 and Table 15.1, the calculated cuttings of both software solutions can differ significantly.

The examined software solutions are based on different algorithms (*Fishnet—FiberSIM®*, *Mosaic—DC3D*), which are explained and discussed below [42, 49, 59–62].

Both algorithms use the basis algorithm of the kinematic model. All points $\vec{x}$ on a double-curved geometry surface can be represented parametrically with surface coordinates u_i [42]

$$\vec{x} = \vec{x}(u_1, u_2). \tag{15.2}$$

The elementary length dS of a surface segment between two closely located points is given by the first surface fundamental form

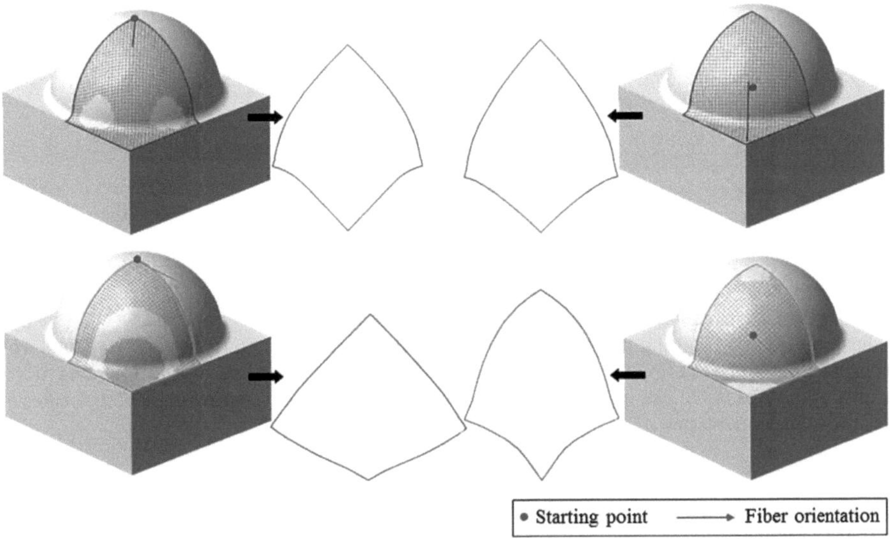

Fig. 15.17 Cuttings under variation of the fiber orientation and the simulation starting point [57]

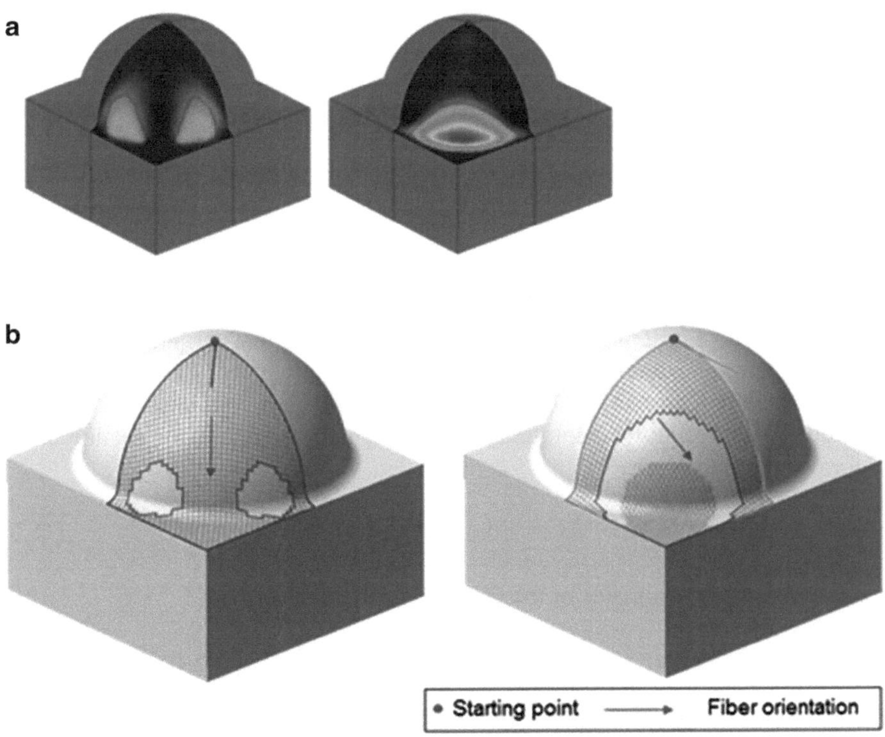

Fig. 15.18 Deformation analysis [57]: (**a**) *DC3D*, (**b**) *FiberSIM*®

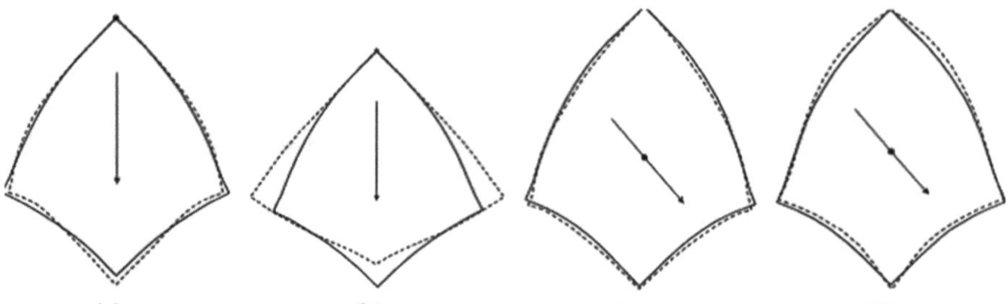

Fig. 15.19 Comparison of cuttings developed with *FiberSIM®* (*dotted line*) and DC3D (*continuous line*) [57]: (**a**) starting point at the cut tip (apex), fiber orientation 0° to warp direction, (**b**) starting point at the cut tip (apex), fiber orientation 45° to warp direction, (**c**) starting point at the center of the area of the cut, fiber orientation 0° to warp direction, (**d**) starting point at the center of the area, fiber orientation 45° to warp direction

Table 15.1 Surface area of the cuts shown in Fig. 15.19

	Cutting area size (mm^2)		
Cutting	DC3D	FiberSIM®	Area difference (%)
(a)	63111.5	63980.5	1.4
(b)	63111.5	69097.5	9.5
(c)	63598.6	64379.5	1.2
(d)	63598.6	67261.0	5.7

$$dS^2 = G_{ij} du_i du_j \tag{15.3}$$

with the coefficients

$$G_{ij} = \frac{\partial \vec{x}}{\partial u_i} \cdot \frac{\partial \vec{x}}{\partial u_j}. \tag{15.4}$$

In Eq. 15.3, the Einstein notation is used, which states that all indices occurring twice in a term are automatically summed up.

The biaxially reinforced textile structure is described by coordinates v_i along the directions of the reinforcement yarns. The elementary length ds of a section of the deformed textile is given by

$$ds^2 = (\delta_{ij} + 2E_{ij}) dv_i dv_j. \tag{15.5}$$

E_{ij} describes the coordinates of the Green-Lagrange tensor with

$$E_{11} = 0, E_{22} = 0 \text{ (no fiber elongation) and } 2E_{12} = \cos\alpha, \tag{15.6}$$

i.e. it is assumed that the deformation of the biaxially reinforced textile structure under shear is caused by pure shear at a fiber angle α. When placing the textile on the surface, the length of one surface segment is equal to that of the corresponding textile section, therefore

$$dS = ds \tag{15.7}$$

applies, from which follows

$$G_{ij} du_i du_j = \left(\delta_{ij} + 2E_{ij}\right) dv_i dv_j. \tag{15.8}$$

Using the Einstein notation, and insertion into Eq. (15.8) result in

$$G_{11} du_1^2 + 2G_{12} du_1 du_2 + G_{22} du_2^2 = dv_1^2 + 2\cos\alpha dv_1 dv_2 + dv_2^2. \tag{15.9}$$

The draping of a biaxially reinforced textile structure onto the surface is described by the equation

$$u_i = u_i(v_1, v_2). \tag{15.10}$$

Inserting

$$\begin{aligned} du_1 - \frac{\partial u_1}{\partial v_1} dv_1 + \frac{\partial u_1}{\partial v_2} dv_2 \quad \text{and} \\ du_2 - \frac{\partial u_2}{\partial v_1} dv_1 + \frac{\partial u_2}{\partial v_2} dv_2 \end{aligned} \tag{15.11}$$

into Eq. (15.9) results in

$$\begin{aligned} ds^2 = & \left[G_{11} \left(\tfrac{\partial u_1}{\partial v_1}\right)^2 + 2G_{12} \frac{\partial u_1}{\partial v_1} \frac{\partial u_2}{\partial v_1} + G_{22} \left(\frac{\partial u_2}{\partial v_1}\right)^2 \right] dv_1^2 + \cdots \\ & \cdots + \left[G_{11} \left(\tfrac{\partial u_1}{\partial v_2}\right)^2 + 2G_{12} \frac{\partial u_1}{\partial v_2} \frac{\partial u_2}{\partial v_2} + G_{22} \left(\frac{\partial u_2}{\partial v_2}\right)^2 \right] dv_2^2 + \cdots \\ & \cdots + 2\left[G_{11} \frac{\partial u_1}{\partial v_1} \frac{\partial u_1}{\partial v_2} + G_{12} \left(\frac{\partial u_1}{\partial v_1} \frac{\partial u_2}{\partial v_2} + \frac{\partial u_1}{\partial v_2} \frac{\partial u_2}{\partial v_1} \right) + G_{22} \frac{\partial u_2}{\partial v_1} \frac{\partial u_2}{\partial v_2} \right] dv_1 dv_2 = \cdots \\ & \ldots = dv_1^2 + 2\cos\alpha dv_1 dv_2 + dv_2^2. \end{aligned} \tag{15.12}$$

A comparison of the coefficients gives

$$\begin{aligned} G_{11} \left(\tfrac{\partial u_1}{\partial v_1}\right)^2 + 2G_{12} \frac{\partial u_1}{\partial v_1} \frac{\partial u_2}{\partial v_1} + G_{22} \left(\frac{\partial u_2}{\partial v_1}\right)^2 &= 1 \\ G_{11} \left(\tfrac{\partial u_1}{\partial v_2}\right)^2 + 2G_{12} \frac{\partial u_1}{\partial v_2} \frac{\partial u_2}{\partial v_2} + G_{22} \left(\frac{\partial u_2}{\partial v_2}\right)^2 &= 1 \\ G_{11} \frac{\partial u_1}{\partial v_1} \frac{\partial u_1}{\partial v_2} + G_{12} \left(\frac{\partial u_1}{\partial v_1} \frac{\partial u_2}{\partial v_2} + \frac{\partial u_1}{\partial v_2} \frac{\partial u_2}{\partial v_1} \right) + G_{22} \frac{\partial u_2}{\partial v_1} \frac{\partial u_2}{\partial v_2} &= \cos\alpha. \end{aligned} \tag{15.13}$$

The initial and boundary conditions are necessary to solve the equations of the kinematic model numerically. As boundary conditions for

$$v_1 = 0 \quad \text{and} \quad v_2 = 0 \tag{15.14}$$

the yarns are to be placed geodesically. To numerically solve the set of non-linear equations (15.13), the biaxially reinforced textile structure is discretized in a mesh of edge length d, so that the node (i, j) has the coordinates

$$\begin{aligned} v_1 &= id \text{ and} \\ v_2 &= jd. \end{aligned} \tag{15.15}$$

The surface coordinates or the spatial coordinates can then be calculated for each node, from which the angle α can be determined [42, 49].

The draping is simulated in five steps:

1. The starting point for the placement of the reinforcement structure on the surface of the geometry is selected
2. The draping direction for the yarns is determined as $v_2 = 0$,
3. The yarns with $v_2 = 0$ are placed on the surface geometry along a geodesic curve,
4. Steps 2 and 3 are repeated for the yarns with $v_1 = 0$,
5. All nodes (i, j) are traversed, and the conditions set by Eq. (15.13) will be complied for each cell with nodes (i, j) and $(i - 1, j - 1)$

Figure 15.20 illustrates the kinematic model for reinforcement yarns during the draping of a bidirectional reinforcement structure. For the illustration of u_1 (v_1, v_2), different approaches can be used [40, 49], which are described below.

In the fishnet algorithm, the bidirectionally reinforced textile structure is represented by a web of intersecting yarns arranged on the surface along geodesic curves. For this, a freely selected cell will be considered below (Fig. 15.20). Initially it is assumed that the length of the edges originating in the left intersection

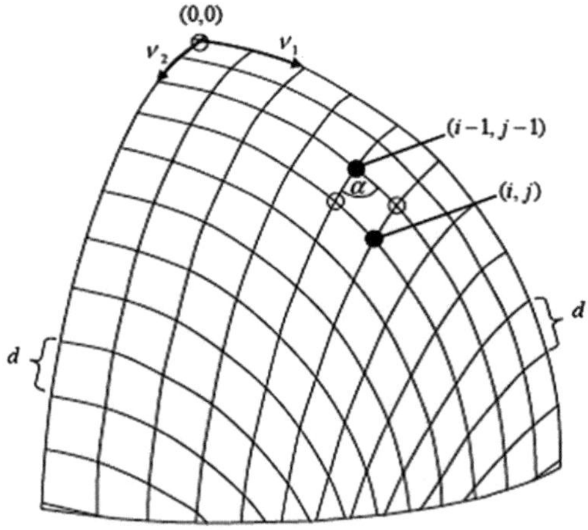

Fig. 15.20 Kinematic simulation of draping on a spherical surface, discretization of the bidirectional textile [42, 49, 57]

point is known. The initial direction of the geodesic edges beginning in the upper and lower intersection point are determined, whose terminal points have to meet. The calculation rule is framed as an optimization problem of the distance between two terminal points.

As the analysis of the objective function includes the integration of a differential for geodesic curves of given length d, the problem is not solvable analytically. Therefore, Gradient and Hessian matrix are calculated numerically by the finite difference method [49]. A geodesic edge is determined by parametrically defining the reinforcement yarn on the surface by Eq. (15.2), using

$$\begin{aligned} u_1 &= u_1(v) \\ u_2 &= u_2(v) \end{aligned} \tag{15.16}$$

This designates the arc length v. If

$$\frac{d\vec{x}}{dv} \cdot \frac{d\vec{x}}{dv} = 1 \tag{15.17}$$

and

$$\frac{d\vec{x}}{dv} = \frac{\partial \vec{x}}{\partial u_1} u_1' + \frac{\partial \vec{x}}{\partial u_2} u_2', \tag{15.18}$$

the reinforcement yarn is inextensible.

The yarn is referred to as geodesic if the position-dependent normal to the curve and the normal to the surface coincide [49]. This can be phrased as follows:

$$\frac{d^2\vec{x}}{dv^2} = \frac{1}{\rho}\overline{n}. \tag{15.19}$$

This describes the curvature 1/ρ. With the equation

$$\overline{n} = \frac{\overline{J}}{|\overline{J}|} \tag{15.20}$$

the normal to the surface is calculated, where

$$\overline{J} = \frac{\partial \vec{x}}{\partial u_1} \cdot \frac{\partial \vec{x}}{\partial u_2}. \tag{15.21}$$

The differentiation of the condition (15.17) results in

$$\left(\frac{\partial \vec{x}}{\partial u_k} u_k^{\prime}\right) \left(\frac{\partial^2 \vec{x}}{\partial u_i \partial u_j} u_i^{\prime} u_j^{\prime} + \frac{\partial \vec{x}}{\partial u_i} u_i^{\prime\prime}\right) = 0. \tag{15.22}$$

Considering Eq. (15.19),

$$\begin{Bmatrix} u_1^{\prime\prime} \\ u_2^{\prime\prime} \\ 1/\rho \end{Bmatrix} = -\left[\frac{\partial \vec{x}}{\partial u_1} \frac{\partial \vec{x}}{\partial u_2} - \overline{n}\right]^{-1} \left\{\frac{\partial^2 \vec{x}}{\partial u_i \partial u_j} u_i^{\prime} u_j^{\prime}\right\} \tag{15.23}$$

follows.

The first two rows are differentials of the second order for curvilineal coordinates $u_{k'}$, while the third row gives the curvature. The initial conditions for Eq. (15.23) in the starting point are u_1 (0) and u_2 (0), where du_2/du_1 are the adequate values for the initial orientation direction. The values u'_1 (0) and u'_2 (0) include the normalization conditions (15.17) [49].

If the surface of a model geometry is described with planar triangles, the *mosaic algorithm* [49, 62] is used. For complex surfaces, a multitude of elements are required for a precise approximation of a curvature. A geodesic curve is reduced to a zigzag line by means of such mosaic surfaces (Fig. 15.21).

If the starting point and fiber orientation are given, consecutive node points on the edges of the mosaic can be defined. The fiber direction at the beginning of the consequent triangle results from the observation that the angle between the yarn and the edge does not change between triangles ($\alpha = \beta$ in Fig. 15.21). If the trajectories AB and AC in any given cell ABCD are known, the initial directions of the trajectories BD and CD have to be defined in a way that ensures that the end points coincide at D. In turn, this is equivalent to an optimization problem, which has previously been explained for the fishnet algorithm [49]. The quality of the mosaic algorithm is impaired by a constant error caused by the discretization of the surface.

In summary it can be stated that the determination of the optimum between structure-mechanical requirements and shaping possibilities (separation into individual cuttings) has to be made accurately for each individual case. For this

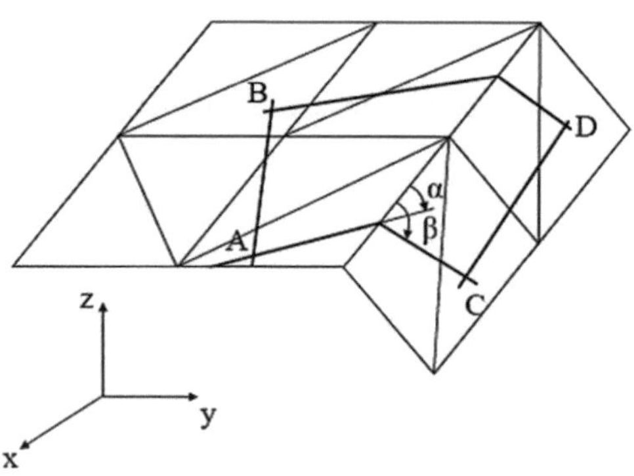

Fig. 15.21 Changes of fiber orientation in kink points [49, 57]

purpose, the abovementioned and tested simulation tools can be very helpful and reduce required development times.

15.3.3 Local Structural Fixations for the Defined Draping of Textile Structures on Strongly Curved Surfaces

During the shaping and placing of textile reinforcements for the construction of a preform, deformability plays a crucial role. As the flexible textile fabrics are very sensitive, the reinforcement yarns are at risk of being displaced in an undefined manner during handling. In addition, selvedge yarns (especially in acute-angled outlines) can become disengaged or displaced. To counter these effects, fixation agents in the form of binders can be applied to the textile reinforcements. If the binder is applied to the entire surface, the form stability of the textile is increased significantly, which usually limits the wrinkle-free deformation capacity. For this reason, only sections of the pre-assembled textile structures are fixated [63–65].

For this purpose, a method that can contribute greatly to an improvement of the preforming process will be portrayed. The developed and patented method for the structure fixation [66] realizes tailor-made local fixations with due regard to the 3D component geometry. Thus, it can solve current, sometimes substantial problems of cutting, handling, and shaping along the preform production chain. The following features characterize this software-based method, according to [66]:

- adjustment of the structural fixation according to the calculated cutting geometries,
- identification of shear deformation during spatial arrangement of the cutting into the preform geometry,
- fixation of the sections exposed to small strains/displacement by deformation,
- application of the fixation agents in a grid-like pattern or intermittently (the manner of fixation depends of the respective fabric structures and the fiber materials to be processed),
- minimization of the amount of the fixation agent with regard to the usually porous structure of the textile structure, continuous or discontinuous performance of the structural fixation in the preform production chain,
- contour-adapted stacking of fixated cuttings with due regard to the sectional drawing, and
- observance of the matrix compatibility of the fixation agent.

Sections exposed only to small displacements by the deformation are suitable as local fixation areas. These sections can be identified by calculation and flattening analyses (see Sect. 15.3.2).

The fixated areas can be arranged over lines or over areas. For the hemispherical reference geometry, the structural fixation is adjusted over lines, as shown in Fig. 15.22. The fixation lines run over the cutting in crossed and longitudinal

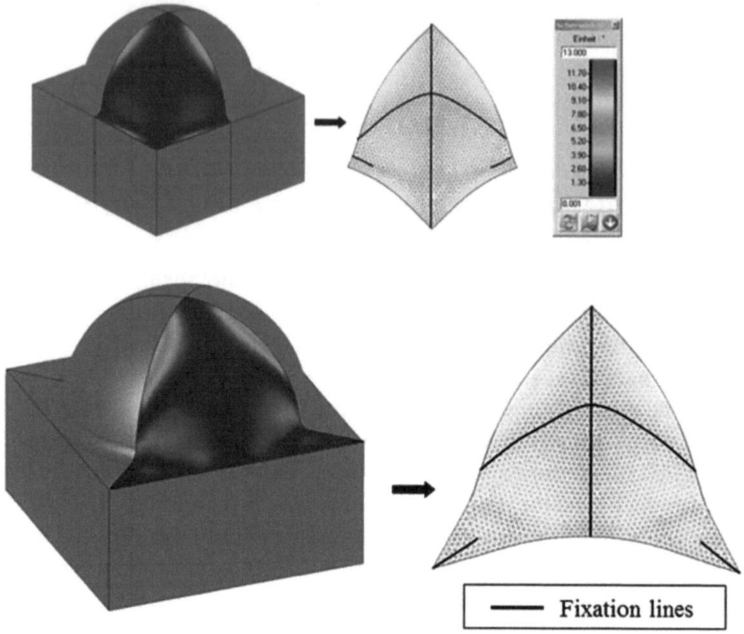

Fig. 15.22 Determination of the fixation lines of different cuttings [57]

lines, fixating all yarns running transversely and lengthwise respectively. This prevents the yarns from falling apart without fixation the edges of the cutting [64]. The amount of fixation agent is minimized in this method to achieve good deformability of the reinforcement structures into the preform and avoid negative influences on the specific mechanical values of the composite.

Due to the significantly different fabric structures, any measures of structural fixation have to be adjusted to the textiles in question. For this purpose, they are modeled with sufficient accuracy (Fig. 15.23a), based on microscopic experiments. The illustration uses a woven reinforcement fabric as an example.

The following parameters are commonly required for a 3D fabric modeling:

- width of the warp and weft yarns,
- fabrics thickness,
- distance between yarn centers normal to the woven fabric plane, and
- distance between yarn centers in the woven fabric plane

The modeled woven reinforcement textiles are shown in Fig. 15.23b.

By modeling the woven fabrics, the cutting contours and structures of the textiles can be “coupled” with each other, in order to adjust the zones of local fixation to the structure (Fig. 15.24). The applied method of structural modeling is expensive and therefore rarely realizable for the multitude of used reinforcement structures in practice. Therefore, another method allowing the virtual control of the reinforcement yarn position is being developed. A grid representing the yarn course of the

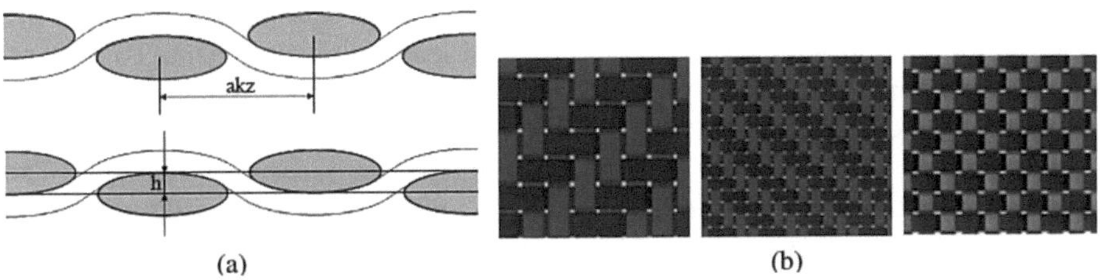

Fig. 15.23 (**a**) Distances between yarn centers: (*top*) normal distance to woven fabric plane, (*bottom*) in the woven fabric plane, (**b**) 3D woven fabric models (*top view*)

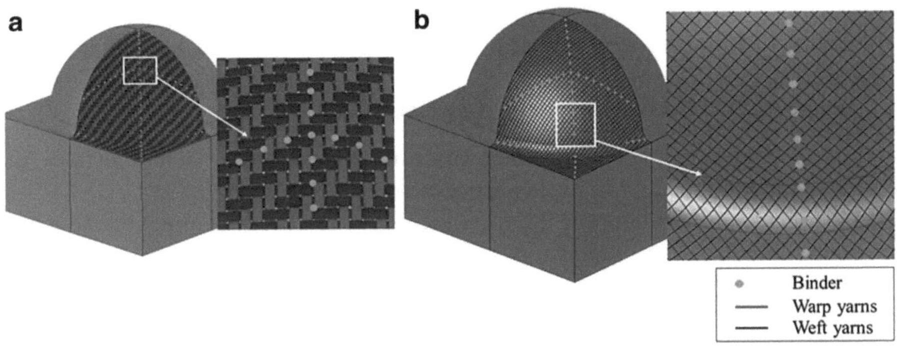

Fig. 15.24 (**a**) Matching the woven fabric to the reference geometry, binder application, (**b**) virtual control of the reinforcement yarns, binder application

reinforcement structure is drawn on the 2D cutting calculated in *DC3D*. Adjusted for strain, this is then projected onto the virtual shaping tool (Fig. 15.24b).

The structural modeling indicates the positions of reinforcement structure where binder has to be applied in order to realize the determined fixation patterns and fixate all yarns of the cutting.

15.4 Composite Material Modeling

The mechanical behavior of fiber-reinforced composite materials differs fundamentally from that of classical, monolithic materials. The significant distinguishing features are:

- locally varying material properties due to inhomogeneous composite structure,
- globally varying material properties due to changing composite configuration, and
- anisotropic material properties caused by the orientation of the reinforcement fibers

The major challenge for the designer is to find the optimum configuration of the composite material with regard to the expected load on a component. To analyze the selected component design concerning criteria such as maximum deformation or strain, FE calculation are used. For the FE models, the geometry and the constitutive laws have to be given, according to the respective material. The adaptivity and resulting multitude of possible configurations of textile composite materials and the complex material behavior require extensive experimental efforts to quantify composite properties, especially when considering the physically nonlinear behavior. Alternatively, both mechanical and other specific material values can be determined much more efficiently with modeling and simulation methods based on a multi-scale approach.

The hierarchical structure of composite materials, as shown in Fig. 15.25, allows a clear distinction between different scales, which are defined by characteristic dimensions.

Accordingly, a distinction will be drawn here between:

- *Micromechanics* (<0.1 mm), which describes e.g. the influence of interfaces, or the interaction of fiber filaments and matrix,
- *Mesomechanics* (0.1 mm–1.0 cm), which captures the properties of the yarn/matrix bundles, their interactions, and matrix cracks, and
- *Macromechanics* (>1 cm) for the determination of component behavior under defined external loads

The bracketed dimensions match characteristic dimensions for the respective level of observation.

On the one hand, a multi-scale simulation is based on the constitutive relations of composite constituents, which are often characterized by a simple formulation and well-known quantification. On the other hand, the modeling, e.g. based on the FEM, represents the composite architecture. Homogenization then refers to the transition

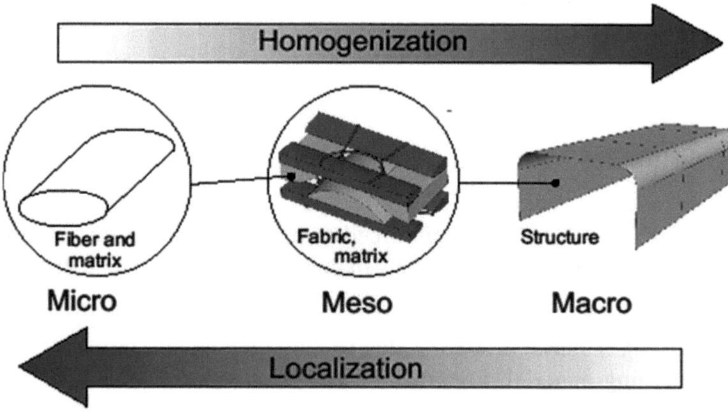

Fig. 15.25 Definition of micro, meso, and macro scale, based on the hierarchical structure of composite materials

from one scale to the next coarser scale, i.e. from micro to meso, or from meso to macro. The reversal, which is the transition from one scale into the next finer scale is equivalent to a localization. With this approach and the corresponding methods, variations of the specific geometrical or material values in the micro and meso scales allow a targeted adjustment of the macroscopic properties of the composite without experiments.

In the following, the modeling of textile-reinforced composite materials on the micro and meso level will be described in greater detail. This includes the geometrical analysis of the composite on the respective observation level as well as the application of special modeling methods to efficiently generate FE models for the representation of complex reinforcement architectures.

15.4.1 Modeling the Fiber/Matrix Composite (Micro-level)

To simulate the material behavior on the micro-level and calculate the effective properties of the meso-level by means of FE-based homogenization methods, the composite (which consist of high-performance filaments and surrounding plastics) has to be modeled. The assumptions and prerequisites made for that purpose are described below.

The impregnation of the meshed non-crimp fabric structure with matrix material depends on the technological parameters of the production method. Here, it is assumed that the space between filaments is completely filled with matrix material.

As shown by the CT scan in Fig. 15.26, the filaments are arranged parallel in a roving, without local irregularities.

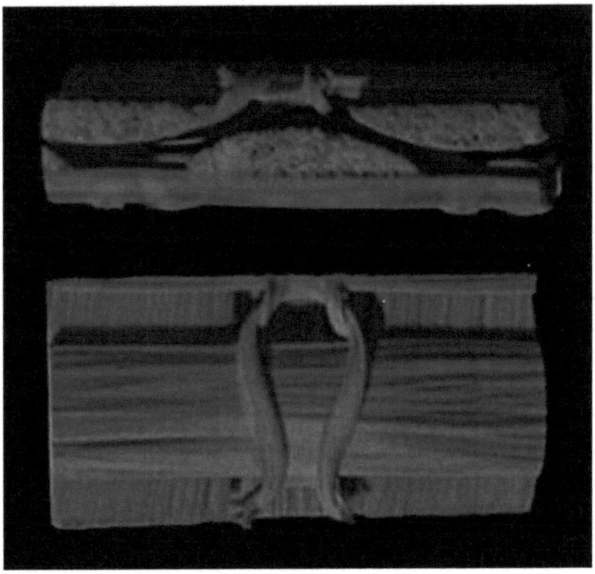

Fig. 15.26 CT scan of the composite

Thus, for the microscopic modeling of the representative volume element (RVE), the fiber/matrix area is considered as a UD composite.

The irregular arrangement of the filaments in the cross-section of the roving can be seen in Fig. 15.28, showing a polished composite section. Due to the high number of filaments and their statistical distribution the effective mechanical properties perpendicular to the roving alignment are independent of the orientation. Therefore, a transversely isotropic material law can be used to describe the mechanical behavior. To avoid an elaborate modeling of the statistical distribution, an equivalent, idealized UD composite with regularly arranged filaments, for which a unit cell can be defined as RVE, is considered. One prerequisite for the transversely isotropic material is an equally spaced positioning of neighboring filaments. Several possibilities are available to determine the unit cell area, where the case of an oblique parallelepiped is shown in Fig. 15.29.

Due to the periodic structure, the exterior dimensions of the model in filament direction have no influence on the effective properties and can therefore be selected freely. The acute angle enclosed by the sidelines of the cross-section is 60°. The length a of these lines can be calculated by

$$a = \sqrt{\frac{\pi}{2\sqrt{3}\varphi}} d_f \tag{15.24}$$

if the filament diameter d_f and the fiber volume content φ are known.

15.4.2 Modeling of the Textile in the Composite (Meso-level)

15.4.2.1 Geometrical Description of the Multilayer Knitted Fabric Reinforcement (MLG Reinforcement)

Textile-technical and manufacturing engineering parameters are the basis for a description of the geometry and of the determination of independent geometry values. Furthermore, as shown in Fig. 15.30, digital shots of the textile and composite can be made with scanners or transmitted-light microscopes, and analyzed regarding their planar dimensions. The geometry values in thickness direction can be determined using CT scans, as shown in Fig. 15.26. As CT scans are expensive, the equations for the determination of spatial reinforcement geometries derived in the following section will rely solely on specific textile-technical values, and dimensions which can be established from optical images in the textile plane.

For the modeling based on the FEM for the experiments of composite behavior on the meso-level, the orientation and course of the reinforcement yarns has to be known. In order to simplify the geometrical description and thus the modeling, only the basic geometrical shapes are used, such as lines and circle segments. By using basic geometrical shapes for the yarn cross-sections, the positions of the center of gravity can be determined. The path of this center of gravity along the respective reinforcement yarn has to be known for the Binary Model, described in Sect. 15.4.2.2.

The CT scans show that an ellipse can be assumed as an approximate cross-section for the weft yarn, while a circular segment is a suitable cross-section approximation for the warp yarn. In comparison to warp and weft yarns, the loop yarn has a smaller cross-section. For simplicity, it is therefore regarded as circular.

By means of these geometrical abstractions, the reinforcement architecture of a composite with a textile reinforcement consisting of two conversely placed MLGs can be represented as shown in Fig. 15.27. For the quantitative description of the orientation, a coordinate system matching the three views in Fig. 15.31 is defined.

In the definition of geometry parameters, it is assumed that variables with subscript We, Wa, and L refer to values in relation to weft, warp, or loop yarns.

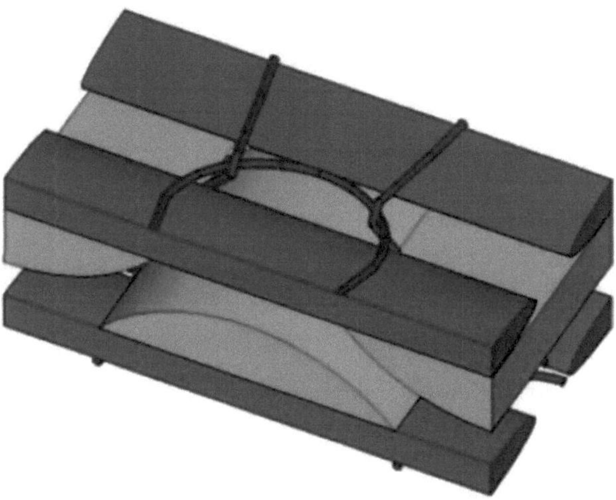

Fig. 15.27 Volume model of a multilayer non-crimp fabric

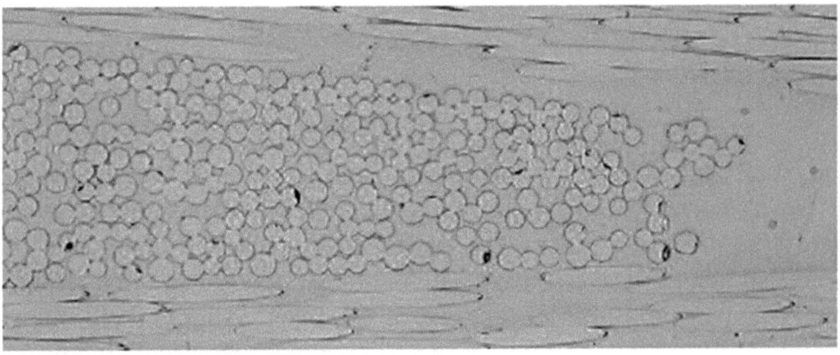

Fig. 15.28 Polished section of the warp yarn area with irregular filament arrangement

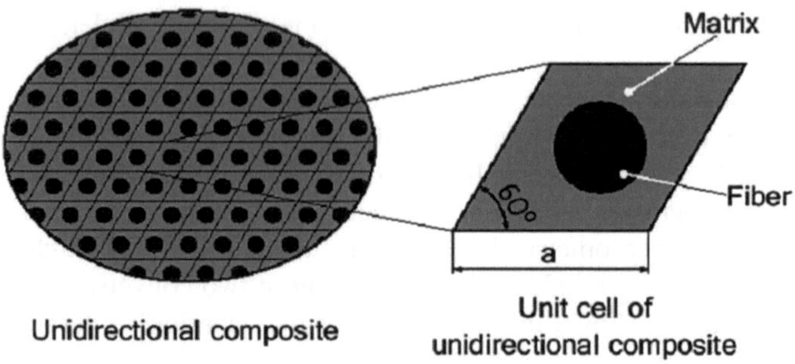

Fig. 15.29 Model of a unit cell in a UD composite

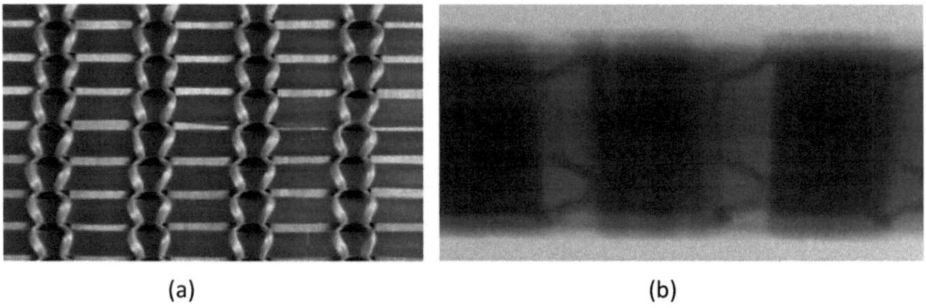

Fig. 15.30 Optical images of the MLG and the composite. (**a**) Scan of the MLG 3a. (**b**) Microscopic view of the composite

Geometry of the Biaxial Reinforcement

The distance between the neighboring warp and weft yarns can be determined from the warp and weft yarn densities η_{Wa} and η_{We} where $l_{Wa} = \frac{1}{\eta We}$ and $l_{We} = \frac{1}{\eta Wa}$, respectively. The area of the cross-sections A_{Wa} and A_{We} can be computed based on the assumptions regarding the geometrical shape, and the dimensions D_{Wa}, d_{Wa}, D_{We}, and d_{we}, which are defined in Fig. 15.31 by

$$A_K = \frac{2}{3} D_K d_K + \frac{d_K^3}{2D_K} \qquad \text{and} \qquad A_S = \frac{\pi}{4} D_S d_S \tag{15.25}$$

Here, the equation for A_{Wa} conforms to an approximation of the area of a circular section. The independent values D_{Wa} and D_{We} are easily determined from the scans and microscopic images, since they are dimensions in the composite plane. In contrast to that, d_{Wa} and d_{We} as dimensions in Z direction can only be determined by means of expansive polished sections or CT scans. Alternatively, these values can be estimated from the relations of the mass content ratio $\overline{m}_{Wa}$ and $\overline{m}_{We}$ of warp and weft yarns, which are known as textile-technical parameters. Based on the

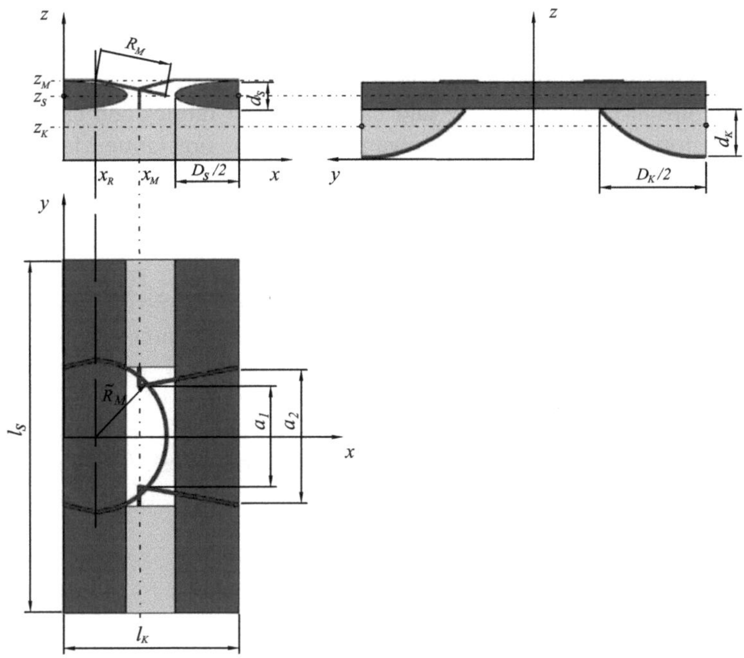

Fig. 15.31 Graphic definition of the MLG geometry parameters

assumption that the mass densities of the rovings are approximately identical, the resulting relation is

$$\frac{\overline{m}_S}{\overline{m}_K} = \frac{A_S l_S n_S}{A_K l_K n_K}, \tag{15.26}$$

where n_{Wa} and n_{ew} are the respective number of warp and weft yarn systems in the MLG. Substituting the areas A_{Wa} and A_{We} with the Eq. (15.25), neglecting the cubic part of d_K and with the definition of the ratio

$$rd := \frac{d_S}{d_K} = \frac{8\overline{m}_S D_K n_K \eta_K}{3\pi \overline{m}_K D_S n_S \eta_S}$$

the following equations can be derived for the computation of the warp and weft yarn thicknesses, respectively.

$$\begin{aligned} d_K &= \frac{d_{MLG}}{(n_S rd + n_K)\xi_K} \quad \text{with} \quad d_{MLG} := d/n_{MLG} \quad \text{and} \\ d_S &= \frac{\xi_K}{\xi_S} rd \;\; d_K \quad \text{with} \quad \frac{1}{2} \leq \xi_{K/S} \leq 1 \end{aligned} \tag{15.27}$$

Here, d_{MLG} is the thickness of an MLG in the composite. Furthermore, it is assumed that the contribution of the loop yarn to the composite thickness is negligible. The factors ξ_{We} and ξ_{Wa} describe the superposition of yarn layers by interloping. This occurs, for instance, in alternately laid biaxial knitted fabrics and can clearly be seen in the top of Fig. 15.26 for the warp yarn layer. For knitted fabrics with a double loop yarn system, the structure of the textile prevents superposition, from which follows that $\xi_{We} = \xi_{Wa} = 1$. The factors ξ_{We} and ξ_{Wa} depend extensively on the compression of the knitted fabric layers in thickness direction. For this reason, analytical solutions on a purely geometrical basis do not give satisfying results.

For an analytical approximation of the effective mechanical properties of the fiber/matrix areas in the composite, the respective fiber volume fraction has to be known. It can be determined by $\varphi := A^f/A$, if this domain is considered as a UD composite. The total cross-section area A can be determined from the Eq. (15.25) for the warp and weft yarns. The fiber cross-section area A^f is calculated from the yarn fineness T_t and the thickness of the textile material ρ, as $A^f = T_t/\rho$.

The coordinate Z_K of the warp yarn axis results from the equation

$$z_K = \frac{\left(\frac{D_K}{2}\right)^2 + d_K^2}{2d_K} - \frac{D_K^3}{12A_K},$$

derived from the circle radius und the distance of the centroid to the center of the circle. With the thicknesses d_{Wa} and d_{We}, all other Z coordinates of the reinforcement layers can be determined. For this example considered here it follow,

$$z_S = d_K + \frac{1}{2}d_S.$$

Loop Yarn System Geometry

The loop yarn is represented strongly abstracted, with due regard to the geometrical resolution on the meso-level, and as shown in Fig. 15.31. The actual geometry of the cross-section area of the loop yarns is very variable locally. Due to the relatively small diameter, the cross-section can be simplified into a circle. To calculate the area A_L, it is also assumed that the fiber volume fraction of the loop yarn $\varphi_L = A_L^f / A_L$ matches the averaged value of warp and weft yarns. Therefore, the diameter can be determined by $d_L = 2\sqrt{\frac{AL}{\pi}}$. With the yarn diameter, the y and z coordinates of the loop yarn in the area of the warp yarn enclosure are known.

The geometry of the top area is largely determined by the radius R_L of the stitch loop. This radius can be measured in its projection onto the x-y plane as R_L in the microscopic and scan views. Since the stitch, as visible in the CT scans, is inclined downwards, it is assumed that the center is placed at height of the weft yarn axis z_{We}. The coordinate of the loop section in thickness direction (which is regarded as plane) can be determined with $z_L = z_{We} + \frac{1}{2}d_{We}$. Thus, the stitch radius can be calculated from the projection radius and the z coordinates by means of

$$R_M = \sqrt{\widetilde{R}_M^2 + (z_M - z_S)^2}$$

As the center of the stitch loops (as in Fig. 15.26) is touching the weft yarn,

$$x_R = l_K - \frac{1}{2}\left(D_S + d_M + 2\widetilde{R}_M\right).$$

applies to the x coordinate of the center of the circular arc. Value a_1 describes the y distance of the contact point between the downward running loop yarns and the stitch loop (Fig. 15.31). With due regard to the yarn layer orientation, this distance can be estimated with $a_1 = l_{We} - D_{Wa} - 2d_L$. From the intersection point of the stitch loop arc with the coordinate $y = a_1/2$ follows the equation

$$x_M = x_R + \sqrt{\widetilde{R}^2 - \left(\frac{a_1}{2}\right)^2}$$

With the help of the theorem of intersecting lines, the equation for the calculation of the distance a_2 can be derived:

$$a_2 = \frac{l_K - x_M}{l_K - x_M + x_R}\left(2\widetilde{R} - a_1\right) + a_1.$$

With these previously mentioned equations, the spatial geometry of the loop yarn can be described entirely.

The mentioned analyses for the determination of specific geometrical values of the MLG can be transferred analogically to other MLGs, if attention is paid to possibly varying number of warp and weft yarn systems, for instance. Furthermore, the equations can be adjusted to determine the geometry of similar textiles such as woven multiaxial non-crimp or knitted fabrics.

15.4.2.2 Finite Element Modeling with the Binary Model

Concept and Numerical Implementation of the Binary Model

The Binary Model was developed for the efficient modeling of textile-reinforced composites [67–69], and is used in a multitude of simulations for static, thermal, and dynamic problems [70, 71].

The central characteristic of the binary model is the separation of the mechanical composite properties. This principle is based on great differences in stiffness between the reinforcement fibers and the matrix material. In the case of the composite examined here, the Young's modulus of the glass fibers is 25 times higher than that of the resin.

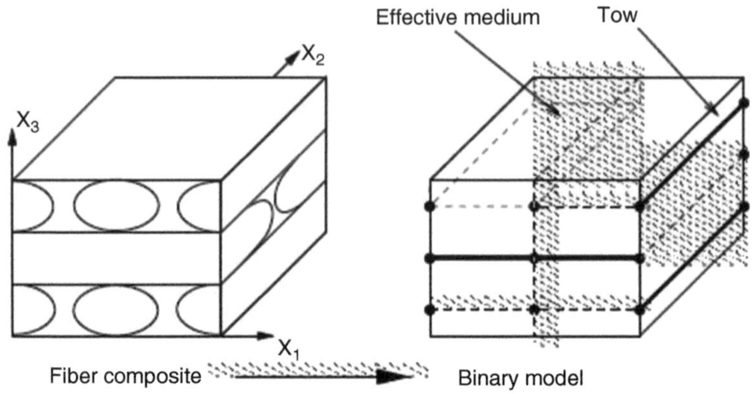

Fig. 15.32 Binary model of a biaxial fiber composite

Figure 15.32 gives a schematic representation of the Binary Model for a biaxial fiber composite regarding the meso-level.

Here, the axial stiffnesses of the fiber/matrix bundles are replaced with tows. The so-called effective medium represents all remaining properties of matrix and yarn material. In the case of a purely mechanical model, these include the Poisson's effect as well as shear and normal stiffness of the matrix material.

With the transfer of this observation to the FE modeling, the axial stiffnesses are replaced with two-noded line elements, and the effective medium with eight-noded volume elements. The latter take up the entire space of the considered component domain. The element geometry does not have to be adjusted to the interface between matrix and yarn, as is required for a conventional FE-mesh. Therefore the shape of the elements can always match that of a cuboid, which benefits convergence and precision of the solutions.

The locations of the line elements therefore represent the center of gravity line of the respective yarn cross-section. Suitable constraints are to be used to ensure continuity of the displacement field of the superimposed volume and line elements. For this, the two following possibilities have to be distinguished.

If a node of the line elements occupies the same place as a node of the volume element, both can be replaced by a shared node.

If, as in Fig. 15.32, the direction and location of the line elements are identical to the edges of the volume elements, the node positions of line and volume elements can always be correlated.

For irregular geometries as those of the loop yarn, extensive efforts would be required to match the positions of the respective volume element with those of line elements. Therefore, a second case will be taken into account, in which the line element nodes, as shown in Fig. 15.33, can be placed in any desired location within the volume elements. For this, the continuity of the displacement field is ensured by eliminating the degrees of freedom of the node of the beam element *i*, here designated generally as $\{\hat{\mathbf{u}}\}^{(T)}$. The displacement constraints between $\{\hat{\mathbf{u}}\}^{(T)}$ and

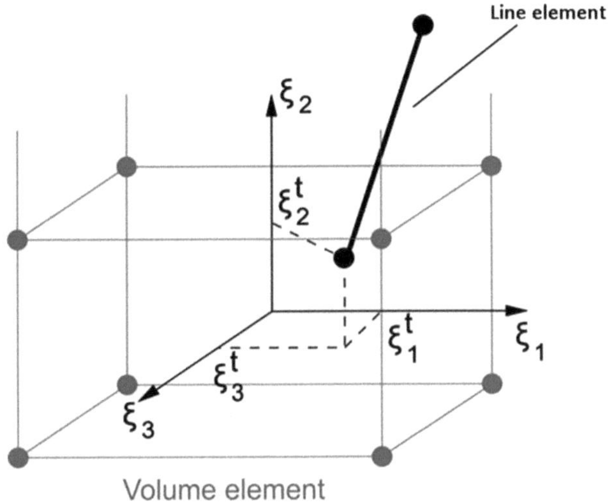

Fig. 15.33 Local volume element coordinates of a line element node

the degrees of freedom of the volume elements $\{\hat{\mathbf{u}}\}^{(\mathrm{EM})}$ can be described in the equations

$$\{\hat{\mathbf{u}}^i\}^{(T)} = \sum_{I=1}^{8} \begin{bmatrix} N_I(\xi^t) & 0 & 0 \\ 0 & N_I(\xi^t) & 0 \\ 0 & 0 & N_I(\xi^t) \end{bmatrix} \{\hat{\mathbf{u}}^I\}^{(EM)} =: \left[\mathbf{T}^i\right] \{\hat{\mathbf{u}}\}^{(EM)} \qquad (15.28)$$

where N_I are the eight shape functions of the volume element, and ξ^t are the local coordinates of the volume element of the line node (Fig. 15.33).

With the matrix $[\mathbf{T}^i]$ thus determined, the constraints can be implemented as described in [72]. The number of degrees of freedom of the FE Binary Model is therefore exclusively determined by the number of volume element nodes. The FE mesh of the volume elements and the structure of the line elements of the RVE with two alternatively placed biaxial weft-knitted fabrics are given as an example of the Binary Model in Fig. 15.34.

The evaluated biaxial weft-knitted fabric and the employed geometry match the example analyzed in Sect. 15.4. This FE model requires 3,927 degrees of freedom. A similar conventional FE model describing just the structure of the warp and weft yarns with volume elements only needs circa 18,200 degrees of freedom. This comparison clarifies the efficiency of the Binary Model concerning the numerical efforts. The meshing of the loop structure based on volume elements also entails considerable modeling efforts. For this reason, the mesh has to be refined significantly, which in turn further increases the number of degrees of freedom of the equation system.

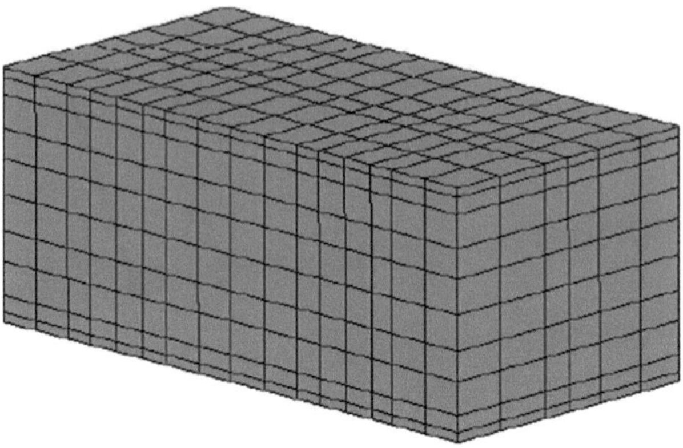
(a)

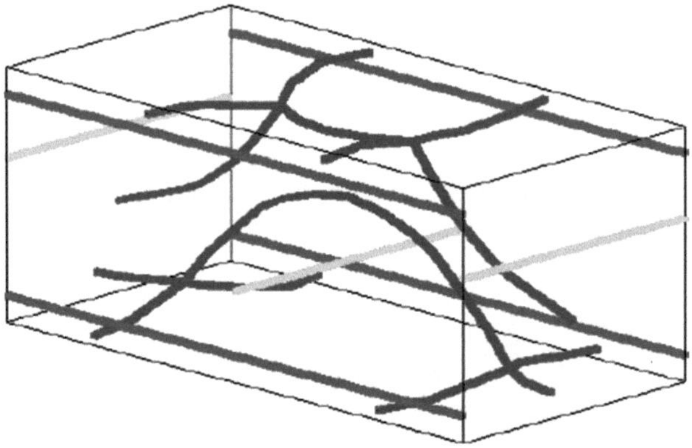
(b)

Fig. 15.34 Binary model of the RVE with MLG reinforcement, (**a**) FE mesh of the volume elements, (**b**) Structure of the line elements

The geometrical abstraction of the binary model requires a derivation of the constitutive equations which are then assigned to the line and volume elements. According to the statements of XU et al. [68], an elastic material behavior is detailed below.

Constitutive Equations of the Effective Medium

The mechanical properties of the effective medium for a biaxial reinforcement can be derived from the effective properties of the UD composite. These can be calculated by means of homogenizing a UD model according to Sect. 15.4.1 or by analytical approximation (see [73]). For reasons of differentiation, all values of the UD composite or of the effective medium will be designated by a superscript (UD) or (EM) respectively. Beyond that, the fiber volume fraction $\varphi^{(EM)}$ is defined as the relation of the volumes of fiber reinforcement and RVE. The fiber orientation in the UD composite matches the x_1 direction. The following indices of the values of the effective medium are based on the coordinate system given in Fig. 15.32.

When applying specific UD values to the characteristics of the effective medium, it is assumed that the transverse stiffness of the fiber matrix bundle significantly influences the Young's modulus $E_3^{(EM)}$ in thickness direction. For lateral contractions and shear moduli of the effective medium, an estimation with $\nu_{12}^{(UD)}$ or $G_{12}^{(UD)}$ respectively, is the obvious choice [68]. The remaining Young's moduli regarding the composite planes $E_1^{(EM)}$ and $E_2^{(EM)}$ can be calculated from shear modulus and lateral contraction, based on a transversal isotropy where fibers are aligned with the x_3 axis. The material properties of the linear-elastic effective medium can be summarized as:

$$\begin{aligned} G_{23}^{(EM)} &= G_{13}^{(EM)} = G_{12}^{(EM)} = G_{12}^{(UD)}, \\ \nu_{23}^{(EM)} &= \nu_{13}^{(EM)} = \nu_{12}^{(EM)} = \nu_{12}^{(UD)}, \\ E_3^{(EM)} &= E_2^{(UD)} \quad \text{and} \\ E_1^{(EM)} &= E_2^{(EM)} = 2\left(1 + \nu_{12}^{(EM)}\right) G_{12}^{(EM)}. \end{aligned} \tag{15.29}$$

This elastic behavior only deviates from an isotropic material because of the different Young's moduli. As stiffnesses in textile composite are primarily determined by the reinforcement yarns, the identities of $E_1^{(EM)} = E_2^{(EM)} = E_2^{(UD)}$ can be simplifyingly assumed, as in [68]. From this follows an isotropic behavior of the effective medium.

The reinforcement architecture of MLG composites differs from that of the previously considered biaxial composite in terms of the loop structure. As the loop structure only makes up 4–14 % of the total textile mass, the structural influence on the properties of the effective medium can be neglected. The volume of the loop yarn is only taken into account for the calculation of $\varphi^{(EM)}$.

Constitutive Equations of the Lines

As described in this section, the line elements in the Binary Model represent the axial stiffnesses of the filaments consolidated with matrix material in the composite material. The rovings used for warp and weft yarns do not contain twisted filament fibers. As visible from Fig. 15.26, the respective orientations of the filaments and the rovings match well, which allows the treatment of these fiber/matrix areas as a UD composite.

If the loop yarn consists of a twisted glass yarn, there are limitations to treating it as UD. But as the fiber volume content ratio of the loop yarn is relatively low in the composite, the mechanical influence of the twisted filament arrangement can be neglected. Therefore, the area of the loop yarn in the consolidated MLG can be simplistically modeled as a UD composite.

For the following derivations for a description of effective properties of the line elements, the components of the UD composites (in this case glass and epoxy resin) are treated as continua.

The elastic stiffness of the line elements can be determined analogously to the effective medium, using an analytical UD model. In consequence, the resulting Young's modulus in orientation of the line element matches the value $E_1^{(UD)}(\varphi)$ calculated from the fiber volume content ratio φ of the fiber-matrix area and the material properties of fiber and matrix.

In the FE model, line and volume elements are superimposed. This corresponds to a parallel arrangement of the axial stiffnesses of both elements. To prevent a multiple consideration of the contribution of the elastic stiffness of the matrix material, the Young's modulus of the line element $E^{(T)}$ is set according to the equation

$$E_\beta^{(T)} = E_1^{(UD)}(\varphi_\beta) = E_{axial}^{(EM)} \quad \text{with} \quad \beta \in \{\mathrm{K,S,M}\} \tag{15.30}$$

The index β describes the relation to the respective warp, weft, and loop yarn, while $E_{axial}^{(EM)}$ matches the stiffness of the effective medium in the orientation of the line element.

To calculate $E_{axial}^{(EM)}$, the strain state of the line element is transferred on the allocated volume element. With this, the desired value can be numerically calculated using

$$E_{axial}^{(EM)} = \{\varepsilon^{-t}\}^T \left[\mathbf{C}^{(EM)}\right] \{\varepsilon^{-t}\} \tag{15.31}$$

where the vector $\{\varepsilon^{-\mathrm{t}}\}$ corresponds to a unit strain in orientation with the line, and the matrix $[\mathbf{C}^{(\mathrm{EM})}]$ matches the material stiffness of the effective medium. In the case of an FE simulation with an elastic-plastic material law for the effective medium, the matrix in Eq. (15.31) is to be replaced with the respective consistent tangent stiffness.

15.5 Material Properties of Composite Materials, Exemplified by Multi-layered Weft-Knitted Fabrics

15.5.1 Experimental Examinations

The aim of experimental examinations of composite materials with textile reinforcement usually consists of the general study of material behavior with due regard to various material behaviors at different length scales and the quantification of effective in-plane material properties. These can be used directly, or in the calculation of structural models from the corresponding composite material, or as the basis for validation of simulation and modeling methods. Furthermore, the evaluation of tests performed on pure-resin test specimens can provide the material properties for the micro and meso models of Sect. 15.4.2.

In contrast to many monolithic materials, the properties of continuous-fiber-reinforced composites are strongly anisotropic and dependent on the type of load. In comparison to the first mentioned group of materials, this requires a significantly larger test effort for a thorough and complete experimental analysis. In general, the anisotropic properties can be examined in tensile, compression, flexural, and shear tests. In the following, these tests and the related evaluations will be considered in greater detail.

15.5.1.1 Tensile and Compression Tests

As described in Sect. 14.6.2.1, tensile and compression tests are used to examine the behavior of a material under monaxial tensile and compressive loads. The anisotropic material behavior requires an evaluation of data sets of specimens with different textile orientations. For an orthotropic behavior, tests with three different directions, e.g. 0°, 45°, and 90°, are required to determine the specific elastic in-plane values. Examinations of specimens with additional textile orientations allow a validation of the assumption of orthotropic behavior.

The experimental test method for the compression behavior is to be selected under consideration of the composite thickness d and of the regarded load spectrum. In case of risk of buckling on the specimen, a buckle support (as shown in Fig. 15.35a) needs to be used.

(a) (b)

Fig. 15.35 Experimental devices for compression testing, (**a**) buckle support for thin specimens, (**b**) compression die for thick specimens

As discussed in Sect. 14.6.2.2, sufficiently thick composite materials can be clamped directly into the clamping jaws of the testing machine, or fixated between two compression dies, as shown in Fig. 15.35.

Measured data are generally recorded by the control computer of the testing machine. The force and traverse path can be measured by dedicated devices integrated in the tester. Strains on the specimen surface can be measured by means of strain gages. In comparison to monolithic material, strain gages have to be much longer for measurements on composite materials [74]. Unfortunately, the use of these longer strain gages is much more costly. Laser extensometers are an alternative for the determination of transverse strain. For this, two measuring markers are affixed to the specimen, and their distance is registered by a laser beam during the entire test. The control program uses these measurements to calculate the nominal strain, which is to be interpreted as an integral quantity over the area between the measuring markers. Another measurement possibility is simultaneously determining the deformation field of the opposite specimen surface with ARAMIS (by GOM GmbH, Braunschweig, Germany). This allows a simultaneous measurement of longitudinal and transverse strains, and the examination of bending deformations. A corresponding test setup is shown in Fig. 15.36.

For exemplary purposes, Fig. 15.37 compares three typical stress-strain plots from tensile test in 0°, 45°, and 90° direction with specimens made from a biaxial MLG composite material.

It is to be expected from the reinforcement architecture that the composite displays a substantially higher stiffness in the direction of the reinforcement fibers as compared to the 45° direction. The tests also prove the curve section, which can be correlated with the linear-elastic area, to be greater in the principal material axis. Furthermore, the curve section delineates itself from the subsequent inelastic

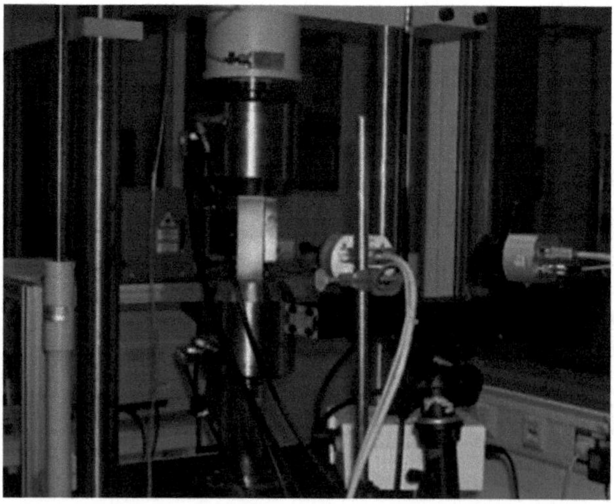

Fig. 15.36 Tensile test with parallel measuring by ARAMIS and laser extensometer

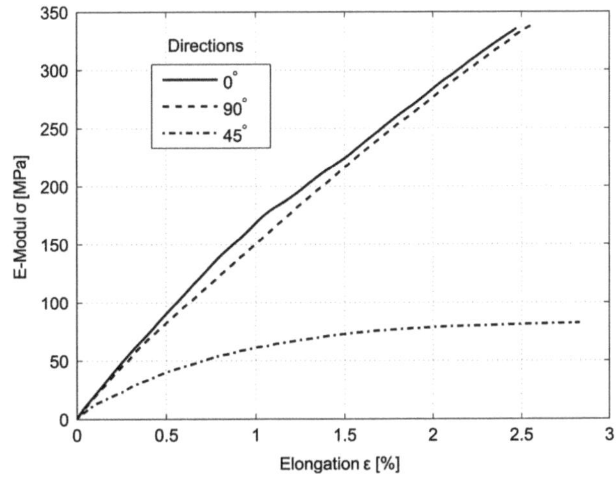

Fig. 15.37 Stress-strain diagram of tensile tests in 0°, 45°, and 90° directions on an MLG composite material

behavior by the dropping of the plot. From the initially linear slope of the stress-strain curves, the Young's modulus $E^{(\alpha)}$ is to be determined for each test with textile orientation α. The Poisson's ratio $\nu^{(\alpha)}$ is determined from the slope $\varepsilon_q\ (\varepsilon_l)$, where ε_q and ε_l are the transverse and longitudinal strains. From these engineering properties, the components of the compliance tensor are calculated by means of

$$S_{11}^{(\alpha)} = \frac{1}{E^{(\alpha)}} \quad \text{and} \quad S_{12}^{(\alpha)} = \frac{\nu^{(\alpha)}}{E^{(\alpha)}} \tag{15.32}$$

For q experimental tests on specimens with different angles α_i $(i=1, \ldots q)$ of textile orientation, these results can be related to the components of the compliance tensor $S_{xx,}$ S_{yy}, and S_{ss} in the main axis system x-y, where Sss designates the shear compliance. Assuming orthotropic material behavior, the following system of equations results:

$$\begin{bmatrix} m_1^4 & n_1^4 & 2m_1^2n_1^2 & m_1^2n_1^2 \\ m_2^4 & n_2^4 & 2m_2^2n_2^2 & m_2^2n_2^2 \\ \vdots & \vdots & \vdots & \vdots \\ m_q^4 & n_q^4 & 2m_q^2n_q^2 & m_q^2n_q^2 \\ m_1^2n_1^2 & m_1^2n_1^2 & m_1^4+n_1^4 & -m_1^2n_1^2 \\ \vdots & \vdots & \vdots & \vdots \\ m_q^2n_q^2 & m_q^2n_q^2 & m_q^4+n_q^4 & -m_q^2n_q^2 \end{bmatrix} \begin{pmatrix} S_{xx} \\ S_{yy} \\ S_{xy} \\ S_{ss} \end{pmatrix} = \begin{pmatrix} S_{11}^{(\alpha_1)} \\ S_{11}^{(\alpha_2)} \\ \vdots \\ S_{11}^{(\alpha_q)} \\ S_{12}^{(\alpha_1)} \\ \vdots \\ S_{12}^{(\alpha_q)} \end{pmatrix}, \tag{15.33}$$

Here, $m_i := \cos(\alpha_i)$ and $n_i := \sin(\alpha_i)$. If the number of tests and therefore of results to be analyzed is more than four, the equation system (15.33) is overdetermined, and the solution of the unknown components of the compliance tensor can be determined by means of a regression calculation, for instance in MATLAB.

15.5.1.2 Flexural Tests

The behavior of composite materials under flexural loads can deviate significantly from tensile or compressive loads. This is due to the structural design of the composite in thickness direction [50]. Therefore, an experimental examination of the flexural stiffness is necessary, especially for thinner composite plates.

A great number of different construction methods are available for the flexural test device. In addition to the deliberation regarding tests based on standards in Sect. 14.6.2.3, a distinction can be drawn between devices with horizontal or vertical specimen arrangement and devices with the possibility to apply alternating flexural loads. To examine the behavior of materials under alternative flexural loads, the Institute of Solid Mechanics at TU Dresden developed a horizontal device with four-point bearing. As shown in Fig. 15.38, the specimen is mounted between two fixed and two mobile rollers.

The carriers of the roller pairs have an internal distance l_b, and are pivoted in order to avoid a prevention of flexural deformation. Mobile rollers, which are fixed to the shafts with bearing pins at a distance $L_b > l_b$, expose the specimen to the load. The construction of this device allows a variable adjustment of the geometry values l_b and L_b.

Suitable sensors on the test station record data of the path and force measurement from the control computer. Additionally, the spatial displacement of the specimen surface is determined by the ARAMIS measuring system. Using the displacement field shown for 0° and 45° directions in Fig. 15.39, the lateral contraction of the composite material can be determined [50].

Fig. 15.38 Horizontal alternating flexural load device

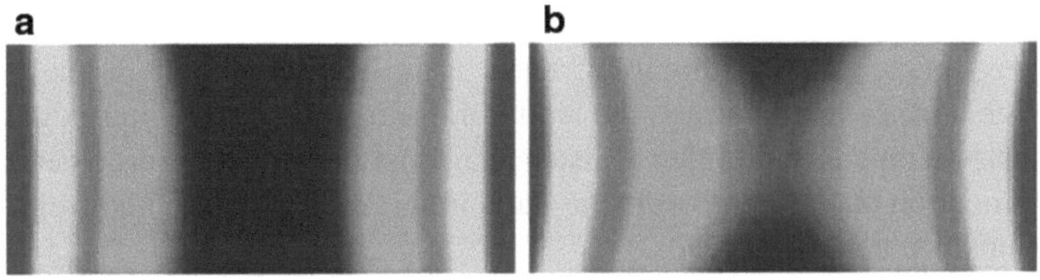

Fig. 15.39 Color-coded representation of the displacement u_z of the specimen surface of a composite with MLG reinforcement, (**a**) textile orientation 0°, (**b**) textile orientation 45°

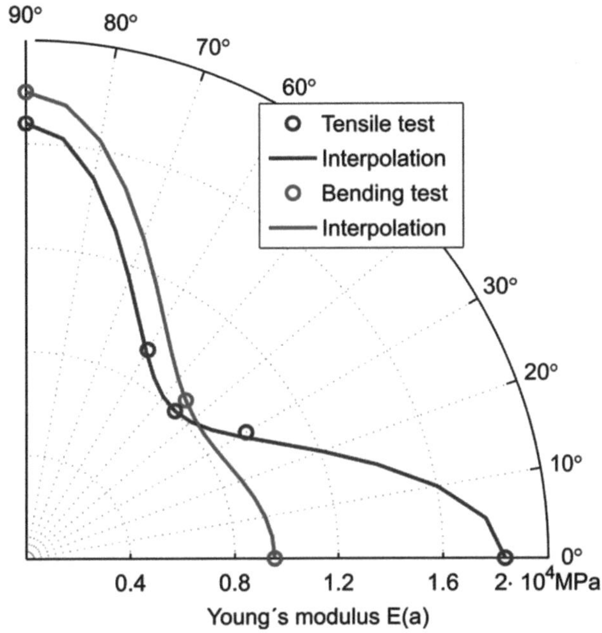

Fig. 15.40 Young's modulus in the composite plane, from tensile and flexural test on a MLG composite

Assuming a linear stress plot across the thickness of the composite (regarded as homogeneous), the measured strains ε_b on the surface allow a determination of the plots $\sigma_b\,(\varepsilon_b)$ of the flexural stress. The slope of the initially linear curve is referred to as flexural Young's modulus $E_b^{(\alpha)}$ for the textile orientation α. A comparison of the tensile and flexural Young's modulus in the composite plane of an MLG composite is shown in the polar diagram of Fig. 15.40.

An excellent match of the results in 45° direction is evident. In comparison, the Young's modulus at 0° is higher (and lower at 90°, respectively) than the values from the tensile tests. This effect is caused by an inhomogeneous loading under flexure, and by the construction of the reinforcement structure. For instance, the specimens with a 90° orientation are exposed to a flexural load around 0°. Here, the absolute flexural stress increases towards the specimen surface starting at the mid-surface plane. In this direction, there is no reinforcement in the sections close to the surface. By contrast, the weft yarns situated on the outer edges of the layer construction cause a higher resistance against the deformation when bent by 90°. The location of the reinforcement yarns in thickness direction is irrelevant for stiffness under pure tensile loads.

15.5.1.3 Shear Tests

A variety of different testing and loading devices for an experimental examination of the composite behavior under pure shear loads are available (see Sect. 14.6.2.4). As described in [75], testing with a shear frame is distinguished by a number of advantages over alternative methods. Figure 15.41 shows one such device, dimensioned based on DIN 53399-2. By means of applied cap strips, a homogeneous load introduction is ensured. Analogous to the tensile tests, the laser extensometer or ARAMIS can be used for measuring the deformation on the respectively opposite side of the specimen.

The macroscopic shear stress τ is calculated with the equation

$$\tau = \frac{\sqrt{2}}{2ld} F, \tag{15.34}$$

where l is the edge length of the measured surface, and F designates the outer tensile force on the test device. The result of the ARAMS measurements is determining the strain plots in the diagonal directions of the square measured surface. Due to the parallel determination of the vertical strain by means of the laser extensometer, a possible flexure of the specimen can be determined.

Fig. 15.41 Test setup and measuring devices of the shear test

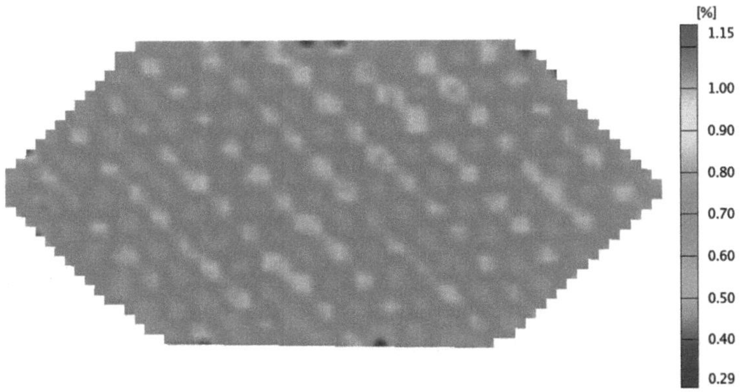

Fig. 15.42 Distribution of the major strain in the shear test on a MLG composite

The test setup in Fig. 15.41 correlates to the vertical strain of the major strain ε_1 and the horizontal strain of the minor strain ε_2. With due regard to an anisotropic deformation, the shear strain γ results from [76]

$$\gamma = \frac{\varepsilon_1 - \varepsilon_2}{1 + \varepsilon_1 + \varepsilon_2}. \tag{15.35}$$

Figure 15.42 shows the distribution of the major strain determined by ARAMIS. Except for local inhomogeneities, the homogeneous strain distribution is not disturbed by boundary effects caused by the load transmission into the specimen.

The plane shear modulus G_{xy} of the tested composite can be determined from the initial slope of the curve plot $\tau(\gamma)$. A comparison with the results of the tensile test allows the verification of the consistency of the results of both experiments.

15.5.2 Homogenization on the Basis of the Energy Criterion

15.5.2.1 Basics of Homogenization

For efficiency reasons, the modeling of the meso and micro structures is not a sensible option in the computer-assisted simulation of load and deformation states of macroscopic components. Therefore, the heterogeneous material for these calculations is created by a homogeneous continuum. The material parameters in the constitutive equations of the substitute continuum, which are here also referred to as effective material properties, can generally be determined by experiments. Due to the macroscopically anisotropic material behavior and the resulting multitude of specific values, this requires extensive efforts. In addition, every alteration of the reinforcement architecture or the material of the composite components makes new experiments necessary. To avoid this effort, the effective properties can be calculated using homogenization methods.

In this section, the heterogeneous micro- as well as the homogeneous macro-level are considered exemplarily for a multi-scale analysis at different length scales. On the macro-level, the characteristic length L can be defined as the maximum distance of two points in a homogeneous body Ω with

$$L = \max_{\mathbf{X}_1, \mathbf{X}_2 \in \Omega} |\mathbf{X}_1 - \mathbf{X}_2| \tag{15.36}$$

Analogously, l describes a characteristic length of the micro structure. One essential basis of many homogenization methods is the concept of a representative volume element (RVE). For this, a partial area $Y = \left\{\mathbf{y} = y^i \mathbf{e}_i, |y^i| < \frac{a}{2}\right\}$ on the micro-level is selected, which completely represents the macro-level characteristics of the material from a statistical point of view. This partial area is referred to as a representative volume element. For composite materials with a periodic micro-structure the RVE can also be defined as a unit cell, as shown in Fig. 15.43.

Using the previously introduced characteristic lengths, general conditions for using homogenization methods and definitions of the RVE can be stated. If both the RVE and the substitute continuum are described within in a CAUCHY continuum, the condition

$$L \gg a \gg l \tag{15.37}$$

must be met [50]. If this cannot be ensured, an extended theory, such as the COSSERAT continuum, must be used to consider the macroscopic body.

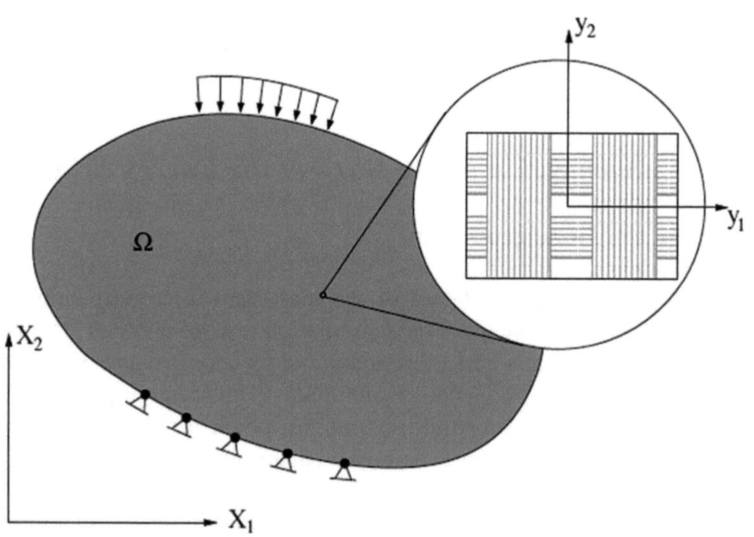

Fig. 15.43 Macro- and meso-level

Otherwise, the prerequisites for the consideration of the heterogeneous body using a homogeneous continuum are not met, making it incorrect.

In the following, it is agreed that all variables with capital letters are defined in relation with the macro-continuum, while lowercase letters refer to the micro-continuum. The basic element of homogenization within the scope of the energy criterion consists of a coupling between micro- and meso-level. The static equilibrium on micro-level, neglecting volume forces, can be stated with the equations

$$\nabla \times \sigma = 0 \quad \text{and} \quad \sigma = \sigma^T \text{ in } Y \tag{15.38}$$

where ∇ is the gradient operator, and σ is the CAUCHY-stress tensor. The constitutive equations

$$\sigma^n(\tau) = \mathrm{F}^n_\sigma(\varepsilon(t), t \in [0, \tau]) \quad \forall \quad n \in \{1 \ldots N\} \tag{15.39}$$

each describe respective current stress strain relations at the time t for the material component n in a RVE with N materials. The strains ε

$$\varepsilon(\mathbf{y}) = sym(\mathbf{u} \otimes \nabla) \tag{15.40}$$

are calculated from the displacement filed $\mathbf{u}(\mathbf{y})$ of the micro-level.

By means of the volume average, the specific values from micro- and macro-level are coupled, resulting in the equations

$$\sum := \frac{1}{|Y|} \int_{\partial Y} \mathbf{y} \otimes \mathbf{t} dA \quad \text{with} \quad \mathbf{t} : \sigma^T \cdot \mathbf{n} \tag{15.41}$$

for the macroscopic tension Σ and

$$\mathbf{E} : \frac{1}{|Y|} \int_{\partial Y} sym(\mathbf{u} \otimes \mathbf{n})\, dA \tag{15.42}$$

for the macroscopic strain $\mathbf{E}$. Equations (15.41) and (15.42) show that macroscopic state variables are described unambiguously by the stress or strain vector $\mathbf{t}$ and $\mathbf{u}$ on the surface ∂Y of the RVE.

A calculation of effective stresses and strains based on Eqs. (15.38)–(15.42) is referred to as *homogenization*. The macroscopic material properties $\mathbf{C}$ then describe the linear transformation between these fields by

$$\langle \sigma(\mathbf{y}) \rangle = \mathbf{C} : \langle \varepsilon(\mathbf{y}) \rangle \quad \text{with} \quad \langle \cdots \rangle := \frac{1}{|Y|} \int_Y (\cdots) d\mathbf{y}\,. \tag{15.43}$$

The inverse problem, i.e. the determination of the microscopic variables with defined macroscopic variables, is referred to as *localization*. As there are not boundary conditions for this case, the problem is yet only incompletely formulated. The definition of these boundary conditions aims to reproduce the state of the material within the respective area as precisely as possible. Below, a summary is given of the derivation of three types of boundary conditions, based on an energy criterion.

The energy theorem according to [77], also known as HILL-MANDEL condition, requires energy equivalence of the heterogeneous micro-continuum and the homogeneous equivalent continuum. From this follow the equations

$$\begin{aligned} \langle \varepsilon(\mathbf{y}) : \mathbf{c}(\mathbf{y}) : \delta\varepsilon(\mathbf{y}) \rangle &= \langle \varepsilon(\mathbf{y}) \rangle : \mathbf{C} : \langle \delta\varepsilon(\mathbf{y}) \rangle \\ \frac{1}{|Y|} \int_{\partial Y} \mathbf{t} \cdot \delta\mathbf{u} \; dA &= \sum : \delta\mathbf{E}, \end{aligned} \tag{15.44}$$

where δ is the variation operator.

If the strain fluctuation is defined as the deviation of the field from the mean value,

$$\breve{\varepsilon}(\mathbf{y}) := \varepsilon(\mathbf{y}) - \langle \varepsilon(\mathbf{y}) \rangle.$$

applies.

The positive definition of the elasticity tensor $\mathbf{c}$ is ensured if

$$\breve{\varepsilon}(\mathbf{y}) := \mathbf{c}(\mathbf{y}) : \breve{\varepsilon}(\mathbf{y}) \geq 0 \tag{15.45}$$

applies to the quadratic form. This relationship can be combined with Eqs. (15.43) and (15.44) as

$$\langle \varepsilon(\mathbf{y}) \rangle : (\langle \mathbf{c}(\mathbf{y}) \rangle - \mathbf{C}) : \langle \varepsilon(\mathbf{y}) \rangle \geq 0 \tag{15.46}$$

The quadratic form of the average volume of $\mathbf{c}$ is therefore higher than that of the effective elasticity tensor $\mathbf{C}$ and represents an upper bound, also referred to as VOIGT bound. Analogously, these considerations, in combination with complementary energy, i.e. the quadratic form of the compliance tensor $\mathbf{s} := \mathbf{c}^{-1}$ lead to the REUSS bound, which correlates to a bottom bound.

$$\langle \varepsilon(\mathbf{y}) \rangle : (\langle \mathbf{c}(\mathbf{y}) \rangle - \mathbf{C}) : \langle \varepsilon(\mathbf{y}) \rangle \geq 0$$

To determine these bounds, the microscopic strain or stress fields are constantly defined by

$$\varepsilon(\mathbf{y}) := \langle \varepsilon \rangle / \sigma(\mathbf{y}) := \langle \sigma \rangle$$

from which follow the definitions

$$\mathbf{C}^V := \langle \mathbf{c} \rangle \quad \text{and} \tag{15.47a}$$

$$\mathbf{C}^R := \langle \mathbf{c}^{-1} \rangle^{-1} \tag{15.47b}$$

as VOIGT and REUSS bounds for the effective elasticity tensor $\mathbf{C}$.

By specifying a constant strain or stress field respectively, the mechanical equilibrium or the compatibility conditions is generally infringed. As the microscopic stresses and strains in Eqs. (15.41) and (15.42) are unambiguously defined by the corresponding boundary conditions, the question of boundary conditions fulfilling the HILL-MANDEL condition arises Eq. (15.44). The three following statements fulfill this condition:

1. The specification of $\mathbf{u}$ on the surface of the RVE by

$$\mathbf{u}(\mathbf{y}) = \mathbf{E} \cdot \mathbf{y} \quad \forall \quad \mathbf{y} \in \partial Y \tag{15.48}$$

matches displacements which are linear in $\mathbf{y}$.

2. Analogously, constant surface stresses can be defined with

$$\mathbf{t}(\mathbf{y}) = \sum \cdot \mathbf{n}(\mathbf{y}) \quad \forall \quad \mathbf{y} \in \partial Y \tag{15.49}$$

3. The validity of periodic boundary displacements and anti-periodic stress vectors

$$\mathrm{u}(\mathrm{y}^+) - \mathrm{u}(\mathrm{y}^-) = \mathrm{E} \cdot (\mathrm{y}^+ - \mathrm{y}^-) \tag{15.50a}$$

$$\mathrm{t}(\mathrm{y}^+) = -\mathrm{t}(\mathrm{y}^-) \tag{15.50b}$$

is based on a periodic microstructure in the unit cell.

$\mathbf{y}^+ \in \partial Y^+$ and $\boldsymbol{y}^- \in \partial Y^-$ define the coordinates of two points, which are associated according to periodic geometry on the opposite surfaces ∂Y^- and ∂Y^+ of the RVE (Fig. 15.44).

In a heterogeneous continuum, the three types of boundary conditions have different results. In general, and comparable to the VOIGT and REUSS approaches, the behavior of an RVE is stiffer for linear boundary displacements, and more flexible for constant boundary stresses in comparison to periodic boundary

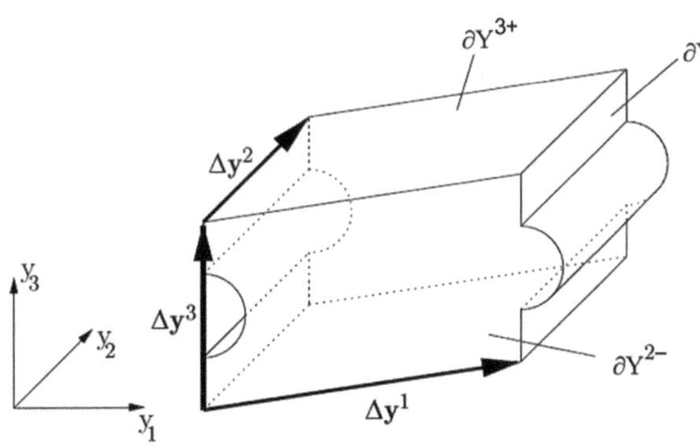

Fig. 15.44 RVE with three associated surface pairs

displacements. For many materials, it has been shown that periodic boundary conditions create the best results by comparison [78]. Therefore, the following considerations will be reduced to this form.

15.5.2.2 Introduction of Generalized Variables

To transfer the periodic displacement boundary conditions (15.50a) to an FE model representing the unit cell, the following generalized variables will be introduced below. They are also prerequisites for an efficient evaluation of the simulation results, e.g. for the calculation of the effective properties or tangent stiffnesses.

When considering any unit cell, for instance as shown in Fig. 15.44, the surface can be segmented into respectively associated lateral surfaces $\alpha \in \{1, 2, 3\}$. These surfaces $dY^{\alpha+}$ and $dY^{\alpha-}$ have to be compatible for the purpose of periodicity. Thus, the difference of the spatial coordinates in Eq. (15.50a) is constant for each surface pair α and can be expressed with the definition

$$\Delta \mathbf{y}^{\alpha} := \mathbf{y}^{\alpha+} - \mathbf{y}^{\alpha-} \quad \forall \quad \alpha \in \{1,2,3\} \tag{15.51}$$

With this equation and the introduction of a generalized displacement increment $\Delta \mathbf{u}^{\alpha}$, Eq. (15.50a) can be described by

$$\Delta \mathbf{u}^{\alpha} := \mathbf{u}(\mathbf{y}^{\alpha+}) - \mathbf{u}(\mathbf{y}^{\alpha-}) = \mathbf{E} : \Delta \mathbf{y}^{\alpha} \tag{15.52}$$

The complete formulation of a spatial or plane boundary value problem (BVP) therefore requires the specification of nine or four generalized displacement increments respectively.

Analogously to the displacements, generalized forces in the form of

$$\mathbf{F}^{\alpha} := \int_{\partial Y^{\alpha+}} \mathbf{t}^{+} dA \tag{15.53}$$

are added, making the homogenized stress Eq. (15.41) calculable with

$$\sum = \frac{1}{|Y|} \sum_{\alpha} \Delta \mathbf{y}^{\alpha} \otimes \mathbf{F}^{\alpha} \tag{15.54}$$

A comparison of this equation to the previous integral equation (15.41) clarifies the significantly reduced effort regarding a numerical realization made possible by the introduction of generalized variables.

To determine the effective linear-elastic stiffness tensor **C**, six different macroscopically homogeneous deformation states have to be considered, for which the strain tensor **E** is specified by

Table 15.2 Assignment for the Indices I and J in Eq. (15.55)

BVP no.	1	2	3	4	5	6
IJ for $E_{IJ} = 1$	11	22	33	23	13	12

$$E_{ij} = E_{ji} = \begin{cases} 1 & \text{for} \quad i = I, j = J \\ 0 & \text{otherwise} \end{cases} \tag{15.55}$$

Table 15.2 contains a possible assignment for the indices I and J.

With the displacement boundary conditions (15.52) and the solution of the BVP, the generalized forces $\mathbf{F}^{\alpha}$ (E_{IJ}) at averaged strain can be calculated. Thus, the macroscopic stresses in Eq. (15.54) are equivalent to the material stiffnesses induced by the deformation, and

$$\sum\nolimits_{kl}(E_{IJ}) = C_{klIJ}E_{(IJ)} = \frac{1}{|Y|}\sum_{\alpha} \Delta y_k^{\alpha} F_l^{\alpha}(E_{IJ}) \tag{15.56}$$

applies, where no summation convention is applied to bracketed indices.

15.5.2.3 Homogenization in the Macroscopic In-plane or Monaxial Stress State

From a macroscopic perspective, any heterogenic materials, such as the textile composite considered here, can be regarded as shell structures within the scope of a plane stress condition (PSC), giving rise to the question of a homogenization method with which the effective mechanical properties of this structural observation can be determined directly. Due to the distinct three-dimensional architecture in the textile composite, homogenization is required to ensure the transfer from a spatially heterogeneous RVE onto a homogeneous, plane structure.

In the PSC, $\Sigma_{i3} = 0$ applies, with the coordinate direction X_3 concurs with the normal vector of the shell mid-surface. Therefore, constant surface tensions are specified on the shells surface according to the abovementioned second kind of boundary conditions. From this follows

$$\mathrm{t(y)} = 0 \quad \forall \quad \mathrm{y} \in \partial Y^3 . \tag{15.57a}$$

The transfer of planar macroscopic strains $\{E_{\beta\gamma} \mid \beta, \gamma \in \{1, 2\}\}$ onto the RVE is then performed by the periodic displacement boundary conditions on the remaining surfaces ∂Y^1 and ∂Y^2, whose normal vectors are tangentially in the shell plane. With Eq. (15.52), these boundary conditions can be described by

$$\Delta\mathbf{u}^{\alpha} = \mathbf{E} \cdot \Delta\mathbf{y}^{\alpha} \quad \forall \quad \alpha \in \{1, 2\} \tag{15.57b}$$

As $E13 = E23 = 0$ applies for macroscopic strains in the context of the PSC, and the vectors Δy^1 and Δy^2 are located in the $y_1 - y_2$ plane, the displacement increments are Δu_3^1 and $\Delta u_3^2 = 0$.

As the result of the homogenization, the planar stresses $\Sigma_{\beta\gamma}$ calculated with Eq. (15.54), where summation convention is applied over $\alpha \in \{1, 2\}$. Additionally, the macroscopic strain E_{33} can be calculated with Eq. 15.42.

As a consistent continuation of the consideration regarding the PSC, the boundary conditions for a monaxial stress state can be derived. With the specification of Σ_{11} as the macroscopic stress (different from zero), the stress vectors

$$\mathbf{t}(\mathbf{y}) = \mathbf{0} \quad \forall \quad \mathbf{y} \in \partial Y^2 \cup Y^3 \tag{15.58a}$$

and the displacement increments are

$$\Delta\mathbf{u}^1 = E_{11}\Delta\mathbf{y}^1 \tag{15.58b}$$

to be specified on the surface of the RVE. To ensure a completed formulation of boundary conditions in this case, it is important to prevent rigid body translation and rotation along the tensile axis.

15.5.2.4 Example for the Homogenization of a MLG Composite

In the following, an example for the application of the homogenization method applied to a composite with glass fiber MLG and epoxy resin will be considered. Here, the textile reinforcement in the composite consists of two alternatively laid knitted fabrics, whose geometry is analyzed in Sect. 15.4.2.1.

Based on the multi-scale approach, calculating the effective specific material values of a UD composite of the micro-level is the initial step. The corresponding deformation of the unit cell described Sect. 15.4.1 at unit strain $E_{11} = 1$ and unit shear stress $\Gamma_{12} = 1$ are shown in Fig. 15.45a, b.

With these calculated specific effective elastic values, the constitutive equations of the binary model of the meso-level can be specified according to the description in Sect. 15.4.2.2.

The FE model of the meso-level of this composite is shown in Fig. 15.34 as a binary model. The homogenization within the scope of the plane stress state requires three simulations with specified macroscopic strain. The deformation state $E_{xx} = 1$ and $\Gamma_{xy} = 1$ are shown with the calculated local strain distribution ε_{xx} and γ_{xy} respectively in Fig. 15.46.

The color coding of both illustrations shows that strain or shear stress = 1 for large areas of the RVE, corresponding to the macroscopically specified deformation. The local fluctuation occurs only because of the line elements.

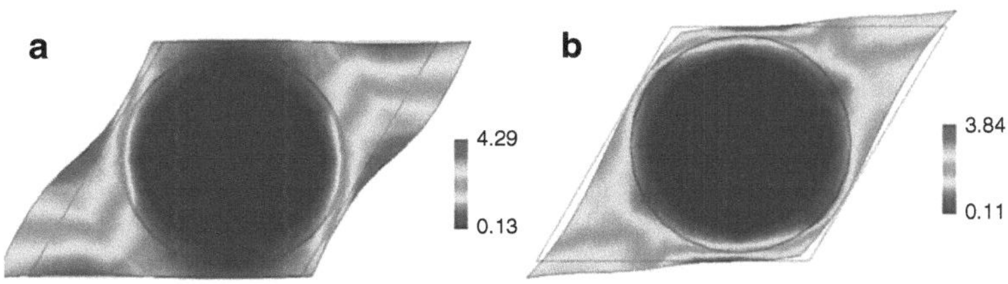

Fig. 15.45 Deformation of the UD unit cell, (**a**) macro-deformation E_{11} with color-coded ε_{11}, (**b**) macro-deformation Γ_{12} with color-coded γ_{12}

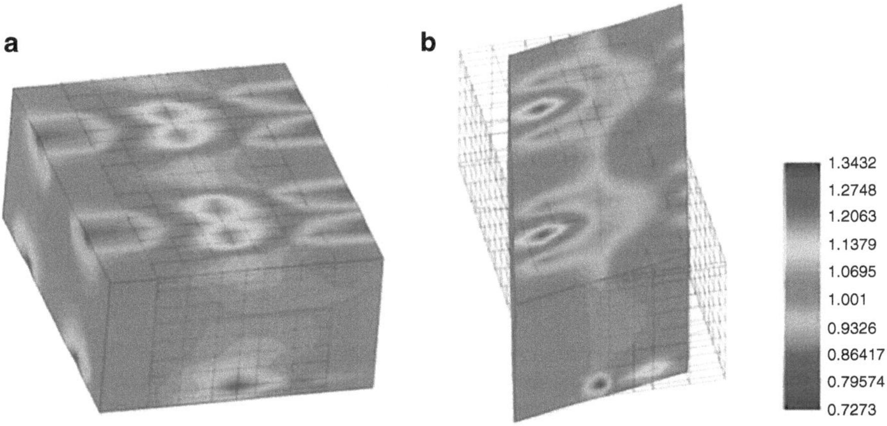

Fig. 15.46 Deformation of the composite unit cell, (**a**) macrodeformation $E_{xx} = 1$ with color-coded ε_{xx} (x, y), (**b**) macrodeformation $\Gamma_{xy} = 1$ with color-coded γ_{xy} (x, y)

The verification of simulation results is performed with experimental data based on engineering constants. For instance, the polar diagram of Fig. 15.47 shows the Young's modulus of the $x - y$ plane dependent on the angle α.

The lines oriented in radial direction mark the scatter ranges of the experimentally determined Young's moduli. The curve plot of the numerically calculated Young's moduli is very good prediction of the experimentally determined values. The little differences between simulation and experiment can often be traced to technological causes. For example, slight deformations of the textile during the consolidation process create local variations in the reinforcement architecture. This leads to a minor deviation between the modeled and the real composite geometries on the meso-level.

In summary, it can be stated that the linear-elastic effective properties of the MLG composite can be estimated very well by means of the homogenization method in combination with the binary model. Therefore, this modeling concept is much more efficient in comparison to conventional volume meshing and of equal value with regard to mechanical evaluation criteria.

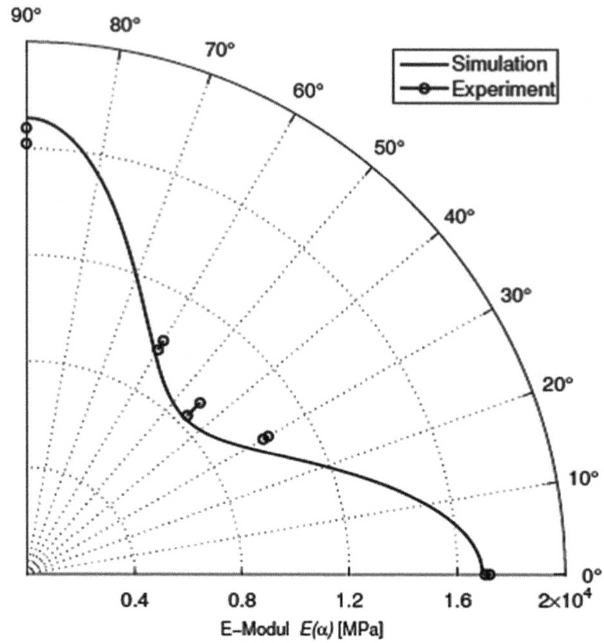

Fig. 15.47 Polar diagram with experimentally and numerically determined in-plane Young's modulus

References

1. www.tu-dresden.de/mw/ilk/sfb639 (22-03-2011)
2. DRECHSLER, K.: *Textiltechnik und Fahrzeugbau.* http://www.unistuttgart.de/wechselwirkungen/ww2004/K.20Drechsler.pdf (21-02-2011)
3. DRECHSLER, K: Latest developments in stitching and braiding technologies for textile preforming. In: *Proceedings. 60. International SAMPE.* Los Angeles, USA, 2004
4. HUFENBACH, W.: Funktionsintegrativer Leichtbau im globalen Spannungsfeld - Vorwort. In: *Proceddings. 10. Dresdner Leichtbausymposium.* Dresden, Germany, 2006
5. STAUBER, R.: Kunststoffe und Verbundwerkstoffe - Anwendungen und Trends. In: *Proceedings. Symposium Polymere im Automobilbau, Bayer Innovativ.* München, Germany, 2006
6. STAUBER, R.: Verbundwerkstoffe im Automobilbau - Anforderungen und Trends. In: *Proceedings 2. Aachen-Dresden International Textile Conference.* Dresden, Deutschland, 2008
7. CHERIF, Ch.: Textilbasierte Leichtbaustrukturen für den Fahrzeug- und Maschinenbau. In: *Proceedings. 11. Dresdner Leichtbausymposium"Materialeffizienz durch Systemleichtbau - den Fortschritt nachhaltig gestalten".* Dresden, Germany, 2006
8. CHERIF, Ch.; DIESTEL, O: Hochdrapierbare Mehrlagengestricke für Schutzhelme (AiFProjekt-No. 15153 BR) / Technische Universität Dresden, Institut für Textilmaschinen und Textile Hochleistungswerkstofftechnik der Technische Universität Dresden. Dresden, 2009. – Final report
9. MITSCHANG, P.: Kontinuierlich faserverstärkte Thermoplaste – NeueWerkstoff- und Prozessoptionen. In: *Proceedings. 10. Europäische Automobil-Konferenz Vision Kunststoff-Automobil 2015.* Bad Nauheim, Deutschland, 2006
10. FISCHER, P.: *Ermittlung mechanischer Kenngrößen textiler Flächen zur Modelierung des Fallverhaltens unter Berücksichtigung konstruktiver, faserstoffbedingter und technologischer*

Abhängigkeiten. Dresden, Technische Universität Dresden, Fakultät Maschinenwesen, Dissertation,1997

11. SCHENK, A.: *Berechnung des Faltenwurfs textiler Flächengebilde*. Dresden, Technische Universität Dresden, Fakultät Maschinenwesen, Dissertation, 1996
12. KRZYWINSKI, S.: *Verbindung von Design und Konstruktion in der textilen Konfektion unter Anwendung von CAE*. Dresden : TUDpress (Dresdner Forschungen, Maschinenwesen Bd. 19), Habilitation, 2005
13. KRZYWINSKI, S.; SCHENK, A.; HAASE, E.: Berücksichtigung der Materialeigenschaften textiler Mehrschichtstrukturen und Nähte in der Simulation und virtuellen Passformkontrolle von Bekleidungstextilien (DFG RO 1303/13-2) / Technische Universität Dresden, Institut für Textilmaschinen und Textile Hochleistungswerkstofftechnik. Dresden, 2010. – Final report
14. REUMANN, R.-D.: *Prüfverfahren in der Textil- und Bekleidungstechnik*. Berlin, Heidelberg, New York : Springer Verlag, 2000
15. http://www.zwick.de (03-02-2010)
16. KAWABATA, S.; NIVA, M.; KAWAI, H.: 3-the finite deformation theory of plain-weave fabrics Part I: the biaxial deformation theory. In: *The Journal of the Textile Institute* 64 (1973), pp. 21–46
17. REICHARDT, C. H.; WOO, H. K.; MONTGOMERY, D. J.: A Two-Dimensional Load-Extension Tester for Woven Fabrics. In: *Textile Research Journal* 23 (1953), No. 6, pp. 424
18. BALLHAUSE, D.: *Diskrete Modellierung des Verformungs-und Versagensverhaltens von Gewebemembranen*. Stuttgart, Universität Stuttgart, Institut für Statik und Dynamik der Luft- und Raumfahrtkonstruktionen, Dissertation, 2007
19. BÖGNER, H.: *Vorgespannte Konstruktionen aus beschichteten Geweben und die Rolle des Schubverhaltens bei der Bildung von zweifach gekrümmten Flächen aus ebenen Streifen*, Universität Stuttgart, Institut für Werkstoffe im Bauwesen, Dissertation, 2004
20. MINAMI, H.: A Multi-Step Linear Approximation Method for Nonlinear Analysis of Stress and Deformation of Coated Plain-Weave Fabric. In: *Journal of Textile Engineering* 52 (2006), No. 5, pp. 189–195
21. BIGAUD, D.; SZOSTKIEWICZ, C.; HAMELIN, P.: Tearing analysis for textile reinforced soft composites under mono-axial and bi-axial tensile stresses. In: *Composite Structures* 62 (2003), No. 2, pp. 129–137
22. SEIF, M.: *Bereitstellung von Materialkennwerten für die Simulation von Bekleidungsprodukten*. Dresden, Technische Universität Dresden, Fakultät Maschinenwesen, Dissertation, 2007
23. EISCHEN, J. W.; CLAPP, T. G.; PENG, H; GHOSH, T. K.: Indirect Measurement of the Moment-Curvature Relationship for Fabrics. In: *Textile Research Journal* 60 (1990), No. 9, pp. 525–533
24. HÖRSTING, K.: *Rationalisierung der Fertigung langfaserverstärkter Verbundwerkstoffe durch den Einsatz multiaxialer Gelege*. Aachen, RWTH Aachen, Fakultät Maschinenwesen, Dissertation, 1994
25. KÖRWIEN, T.: *Konfektionstechnisches Verfahren zur Herstellung von endkonturnahen textilen Vorformlingen zur Versteifung von Schalensegmenten*. Bremen, Universität Bremen, Dissertation, 2003
26. ORAWATTANASRIKUL, S.: *Experimentelle Analyse der Scherdeformation biaxial verstärkter Mehrlagengestricke*. Dresden, Technische Universität Dresden, Fakultät Maschinenwesen, Dissertation, 2006
27. HARRISON, P.; CLIFFORD, M. J.; LONG, A. C.: Shear characterisation of viscous woven textile composites: a comparison between picture frame and bias extension experiments. In: *Composites Science and Technology* 64 (2004), No. 10-11, pp. 1453–1465
28. LAUNAY, J.; HIVET, G.; DUONG, A. V.; BOISSE, P.: Experimental analysis of the influence of tensions on in plane shear behaviour of woven composite reinforcements. In: *Composites Science and Technology* 68 (2008), pp. 506–515
29. http://www.naiss.de (22-10-2006)

30. BOISSE, P.; GASSER, A.; HIVET, G.: Analyses of fabric tensile behaviour: determination of the biaxial tension-strain surfaces and their use in forming simulations. In: *Composites : Part A* 32 (2001), pp. 1395–1414
31. CAO, J.; AKKERMAN, R.; BOISSE, P.; CHEN, J.; CHENG, H. S.; DE GRAAF, E. F.; GORCZYCA, J. L.; HARRISON, P.; HIVET, G.; LAUNAY, J.; LEE, W.; LIU, L.; LOMOV, S. V.; LONS, A.; DE LUYCKER, E.; MORESTIN, F.; PADVOISKIS, J.; PENG, X. Q.; SHERWOOD, J.; STOILOVA, Tz.; TAO, X. M.; VERPOEST, I.; WILLEMS, A.; WIGGERS, J.; YU, T. X.; ZHU, B.: Characterization of mechanical behavior of woven fabrix: Experimental methods and benchmark results. In: *Composites: Part A* 39 (2008), pp. 1037–1053
32. LOMOV, S. V.; VERPOEST, I.: Model of shear of woven fabric and parametric description of shear resistance of glass woven reinforcements. In: *Composites Science and Technology* 66 (2006), No. 7-8, pp. 919–933
33. MORNER, B.; EEG-OLOFSSON, Z.: Measurement of shearing properties of fabric. In: *Textile Research Journal* 27 (1957), pp. 611
34. POTLURI, P.; CIUREZU, D.A. P.; RAMGULAM, R.B.: Measurement of meso-scale shear deformations for modelling textile composites. In: *Composites Part A* 37 (2006), pp. 303–314. DOI 10.1016/j.compositesa.2005.03.032
35. ZHU, B.; YU, TX; TAO, XM: An experimental study of in-plane large shear deformation of woven fabric composite. In: *Composites Science and Technology* 67 (2007), No. 2, pp. 252-261
36. STUMPF, H.: *Study on the manufacture of thermoplastic composites from new textile preforms*. Hamburg, TU-Hamburg-Harburg, Dissertation, 1998
37. LEBRUN, G.; BUREAU, M. N.; DENAULT, J.: Evaluation of bias-extension and picture-frame test methods for the measurement of intraply shear properties of PP/glass commingled fabrics. In: *Composite structures* 61 (2003), No. 4, pp. 341–352
38. CHERIF, Ch.: *Drapierbarkeitssimulation von Verstärkungstextilien für den Einsatz in Faserverbundwerkstoffen mit der Finite-Element-Methode*. Aachen : Shaker Verlag, 1999
39. PENG, X. J.; CAO, J.; CHEN, P.; XUE, P.; LUSSIER, D. S.; LIU, L: Experimental and numerical analysis on nomalization of picture frame tests for composite materials. In: *Composites Science and Technology* 64 (2004), pp. 11–21
40. HANCOCK, S. G.; POTTER, K. D.: Inverse drape modelling-an investigation of the set of shapes that can be formed from continuous aligned woven fibre reinforcements. In: *Composites: Part A* 36 (2005), pp. 947–953
41. BOGDANOVICH, A.: Three-dimensional continuum micro-, meso- and macromechanics of textile composites. In: *Proceedings. TEXCOMP-8*. Nottingham, England, 2006
42. ERMANNI, P.; ENDRUWEIT, A.: Textile Halbzeuge. In: ERMANI, P. (Hrsg.): *Composites Technologien*. Zürich : Eidengenössische Technische Hochschule Zürich, 2007, pp. 1–45
43. http://www.nottingham.ac.uk/emxmns/texgen.htm (14-01-2008)
44. http://www.mtm.kuleuven.be/Research/C2/poly/software.html (14-01-2008)
45. LOMOV, pp. V. et al.: Mathematical modelling of internal geometry and deformability of woven preforms. In: *Int. J. of Forming Processes* 61 (2003), No. 3/4, pp. 413–442
46. KOISSIN, V. E.; IVANOV, D. S.; LOMOV, S. V.; VERPOEST, I.: Fibre distributions inside yarns of textile composite: geometrical and FE modelling. In: *Proceedings. TEXCOMP-8*. Nottingham, England, 2006
47. CROOKSTON, J. J.; KARI, S.;WARRIOR, N. A.; JONES, I. A.; LONG, A. C.: 3D textile composite mechanical properties prediction using automated FEA of the unit cell. In: *Proceedings. 16th Int. Conf. on Composite Materials (ICCM-16)*. Kyoto, Japan, 2007
48. LOMOV, S. V.; TRUEVTZEV, A. V.; CASSIDY, C.: A predictive model for the fabric-to-yarn bending stiffness ratio of a plain-woven set fabric. In: *Textile Research J.* 70 (2000), No. 12, pp. 1088–1096
49. WEEËN, F. van d.: Algorithms for Draping Fabrics on Doubly-Curved Surfaces. In: *International Journal for Numerical Methods in Engineering* 31 (1991), pp. 1415–1426

50. HAASEMANN, G.: *Effektive mechanische Eigenschaften von Verbundwerkstoffen mit Biaxialgestrickverstärkung*. Dresden : TUDpress, Dissertation, 2008
51. http://www.vistagy.com (22-02-2011)
52. http://www.lectra.com (22-02-2011)
53. http://www.esi-group.com (22-02-2011)
54. http://www.3ds.com (22-02-2011)
55. http://www.ansys.com (22-02-2011)
56. http://www.mscsoftware.com (22-02-2011)
57. GIRDAUSKAITE, L.: *Lokale Strukturfixierung im Preformherstellungsprozess für komplex gekrümmte Faserkunststoffverbundbauteile*. Dresden, Technische Universität Dresden, Fakultät Maschinenwesen, Dissertation, 2011
58. http://www.solidworks.com (22-02-2011)
59. BYUN, J. H.; CHOU, T. W.: Modelling and characterization of textile structural composites: A review. In: *The Journal of Strain Analysis for Engineering Design* 24 (1989), No. 4, pp. 253–262
60. WHITNEY, T. J.; CHOU, T.-W.: Modeling of 3-D Angle-Interlock Textile Structural Composites. In: *Journal of Composite Materials* 23 (1989), No. 9, pp. 890–911
61. PASTORE, C. M.; CAI, Y. J.: Applications of computer aided geometric modelling for textile structural composites. In: WILDE, W. P. D. (Hrsg.); BLAIN, W. R. (Hrsg.): *Composite Materials Design and Analysis*. Berlin : Springer Verlag, 1990, pp. 127–142
62. LOMOV, S. V.; IVANOV, D. S.; VERPOEST, I.; ZAKO, M.; KURASHIKI, T.; NAKAI, H.; HIROSAWA,S.: Meso-FE modelling of textile composites: Road map, data flow and algorithms. In: *Composites Science and Technology* 67 (2007), No. 9, pp. 1870–1891
63. GIRDAUSKAITE, L.; KRZYWINSKI, S.; RÖDEL, H.; BÖ HME, R.; JANSEN, I.: Trockene Prefroms für komplexe Faserverbundkunststoffbauteile. In: *Technische Textilien* 52 (2009), No. 6, pp. 280–281
64. GIRDAUSKAITE, L.; KRZYWINSKI, S.; RÖDEL, H.; WILDASIN-WERNER, A.; BÖHME, R.; JANSEN, I.: Local Structure Fixation in the Composite Manufacturing Chain. In: *Applied Composite Materials* 17 (2010), No. 6, pp. 597–608
65. BÖHME, R.; GIRDAUSKAITE, L.; JANSEN, I.; KRZYWINSKI, S.; RÖDEL, H.: Reproduzierbare Preformfertigung für textilverstärkte Kunststoffe. In: *Lightweightdesign* (2009), No. 5
66. Protective right DE102007032904 (27th November 2008).
67. CARTER, W. C.; COX, B. N.; FLECK, N. A.: A binary model of textile composites - I. Formulation. In: *Acta Metallurgica et Materialia* 42 (1994), No. 10, pp. 3463–3479. DOI 10.1016/0956–7151(94)90479–0
68. XU, J.; COX, B. N.; MCGLOCKTON, M. A.; CARTER, W.C.: A binary model of textile composites–II. The elastic regime. In: *Acta Metallurgica et Materialia* 43 (1995), No. 9, pp. 3511–3524
69. MCGLOCKTON, M. A.; COX, B. N.; MCMEEKING, R. M.: A Binary Model of textile composites: III. High failure strain and work of fracture in 3D weaves. In: *J. Mech. Phys. Solids*. 51 (2003), No. 8, pp. 1573–1600
70. HAASEMANN, G.: An application of the Binary Model to dynamic finite element analysis. In: *Proc. Appl. Math. Mech.* 3 (2003), No. 1, pp. 176–177
71. HAASEMANN, G.; ULBRICHT, V.; BRUMMUND, J.: Modelling the mechanical properties of biaxial weft-knitted fabric reinforced composites. In: *Proc. Appl. Math. Mech.* 4 (2004), No. 1, pp. 193–194
72. COOK, R. D.; MALKUS, D. S.; PLESHA, M. E.: *Concepts and applications of finite element analysis*. 3. Auflage. New York, USA : John Wiley Sons, 1989
73. ALTENBACH, H.; ALTENBACH, J.; RIKARDS, R.: *Einführung in die Mechanik der Laminat und Sandwichtragwerke*. Stuttgart : Deutscher Verlag für Grundstoffindustrie, 1996
74. TUTTLE, M. E.; BRINSON, H. F.: Resistance-foil strain-gage technology as applied to composite materials. In: *Experimental Mechanics* 24 (1984), No. 1, pp. 54–65

75. SKUDRA, A. M.; BULAVS, F. J.; ROCENS, K. A.: *Kriechen und Zeitstandverhalten verstärkter Plaste*. Leipzig : VEB Deutscher Verlag für Grundstoffindustrie, 1975
76. WELZ, M.; GÄDKE, M.: Versuche zur Bestimmung des Schubmoduls glasfasermattenverstärkter Kunststoffe (FD-30) / DLR Braunschweig. Braunschweig, 1971. – Department report
77. HILL, R.: On constitutive macro-variables for heterogeneous solids at finite strain. In: Proc. R. Soc. Lond. A 326 (1972), No. 1565, pp. 131–147
78. HOLLISTER, S. J.; KIKUCHI, N.: A comparison of homogenization and standard mechanics analyses for periodic porous composites. In: Computational Mechanics 10 (1992), No. 2, pp. 73–95

Additional Literature

ABDIN, Y.; TAHA, I.; EL-SABBAGH, A.; EBEID, S.: Description of draping behaviour of woven fabrics over single curvatures by image processing and simulation techniques. In: Composites: Part B 45 (2013), pp. 792-799

DÖBRICH, O.; GEREKE, Th.; CHERIF, Ch.; KRZYWINSKI, S.: Analysis and finite element simulation of the draping process of multilayer knit structures and the effects of a localized fixation. In: Advanced Composite Materials 22 (2013), pp. 175-189. DOI: 10.1080/09243046.2013.791239

GEREKE, Th.; DÖBRICH, O.; HÜBNER, M.; CHERIF, Ch.: Experimental and computational composite textile reinforcement forming: A review. In: Composites: Part A 43 (2013), pp. 1-10

HÜBNER, M.; DÖBRICH, O.; DIESTEL, O.; KRZYWINSKI, S.; GEREKE, Th.; CHERIF, Ch.: Wrinkle free draping. In: Kunststoffe International 9 (2013), pp. 99-101

HÜBNER, M.; DÖBRICH, O.; GEREKE, Th.; KRZYWINSKI, S.; CHERIF, Ch.: Drape simulation of textile reinforcement structures. In: tts6-4-1100-matthias-huebner.pdf. Techtextil Symposium, Frankfurt, 11th - 13th June 2013

KAUFMANN, M.: DeMaCo – Design for Manufacture of Composites. Sirris, Public report, 2013

ZHANG, Y.; SUN, F.; WANG. Y.; CHEN, L.; PAN, N.: Study of intra/inter-ply shear deformation of three dimensional woven preforms for composite materials. In: Materials and Design 49 (2013), pp. 151-159

Chapter 16
Processing Aspects and Application Examples

Chokri Cherif, Olaf Diestel, Thomas Engler, Evelin Hufnagl, and Silvio Weiland

The global energy and climate situation requires future activities to reduce energy consumption as much as possible, not only in transport technology and the construction sector, but also in all other industries. Lightweight construction with textile-reinforced composite materials offers fascinating possibilities in developing energy-efficient and function-integrated structural components, especially when compared to their conventional metallic counterparts. The combination of two or more different types of material results in novel composites, whose performance exceeds the sum of properties of the individual components.

This chapter presents some selected aspects of processing and use of textile semi-finished products for lightweight applications in the fields of fiber-reinforced plastics, textile concrete, and textile membranes. The performances of textile materials for lightweight construction as well as their practical feasibility in large-scale applications are demonstrated. In addition, the manufacturing technology associated with these lightweight applications will be discussed.

C. Cherif (✉) • O. Diestel
Institute of Textile Machinery and High Performance Material Technology, TU Dresden, Dresden, Germany
e-mail: chokri.cherif@tu-dresden.de; olaf.diestel@tu-dresen.de

T. Engler • E. Hufnagl
Formerly: Institute of Textile Machinery and High Performance Material Technology, TU Dresden, Dresden, Germany
e-mail: i.textilmaschinen@tu-dresden.de; i.textilmaschinen@tu-dresden.de

S. Weiland
Tudalit Markenverband e.V., Dresden, Germany
e-mail: info@tudalit.de

C. Cherif (ed.), *Textile Materials for Lightweight Constructions*,
DOI 10.1007/978-3-662-46341-3_16

16.1 Introduction

Lightweight construction with textile-reinforced composites offers extensive possibilities with advantages over conventional metallic construction in the development of materials and energy-efficient structural components. Through the targeted combination of two or more different materials, composites with a performance far exceeding the sum of capacities of the individual components can be manufactured. Composite materials with a multilayered arrangement of individual components can often be seen in nature and are already found in many areas of applications today. The exceptionally high design flexibility resulting from adjustable anisotropic properties under engineering and design aspects offers ample opportunities for achieving lightweight benefits based on textile high-performance materials. This makes the new group of continuous-fiber reinforcements all the more lucrative in fiber-reinforced plastics (FRP), textile-reinforced concrete (TRC), and fiber-based membrane technology for different application fields.

Textile materials and semi-finished products exhibit an extraordinarily diverse and freely adjustable property profile with the existing design and manufacturing opportunities, including their functionalization. This opens up a particularly high potential for their large-scale usage in complex and heavy duty lightweight components in traffic engineering and mechanical engineering. Furthermore, applications are opened where reinforcements are required in slender and filigree concrete components, in the maintenance and repair of existing buildings, and for textile membranes. Process-optimized and load-optimized design of textile materials and semi-finished products makes a significant contribution to resource conservation by reducing the mass of moving components and use of materials, as well as saving energy. These materials are characterized by the flexible adaptability of material structure and meet the structural requirements by a controlled adjustment of material and anisotropic properties. The individually configurable fiber production and textile technology provide an ideal foundation for a targeted development and making the naturally found ideal lightweight construction suitable in all its breadth and complexity for a comprehensive industrial usage. The potential of textile semi-finished products as a powerful lightweight material is made possible by a careful selection and combination of textile materials and processes as well as through a custom load-path-aligned arrangement of rovings in the textile structures, making a wide variety of property profiles and design variations to function-integrated near-net shape components possible. In general, the textile materials offer a wide range of variations and enormous possibilities to meet design requirements and construction specifications of load-bearing structures in terms of strength, stiffness, and energy absorption.

For future large-scale applications in the fields of traffic and mechanical engineering a multi-material design system is currently being followed to achieve high efficiency in resource utilization. Continuous-fiber-reinforced plastic composites are experiencing a steady growth in the various application areas because they have a particularly high potential for series usage in complex, heavy duty lightweight components with large weight-saving benefit, thereby effectively contributing to energy conservation (see Sect. 16.3).

The current climatic, resource-relevant, social, and economic conditions have increasingly led to an exploitation of all opportunities for reducing energy consumption not only in transportation and mechanical engineering, but also in the construction sector. This has led researchers to intensively pursue the development of resource-efficient technologies required in the construction industry for both new construction and renovation (see Sect. 16.4).

Textile membranes as thin and primarily tensile-stressed materials serve as innovative composites for industrial, building, and architectural purposes and can be customized according to the required specifications by suitable material selection (reinforcement material, coating system, and binders), relevant textile designs, and coating technologies. The achievable flexible constructions allow a variety of almost all free-form objects and lightweight solutions that are not feasible using conventional metallic materials (see Sect. 16.5).

The existing lightweight applications in the fields of fiber-reinforced plastics, textile-reinforced concrete, and textile membranes demonstrate the impressive performance of textile materials and semi-finished products for reinforcing and strengthening tasks as well as their practical suitability. The range of applications is growing steadily and will certainly continue to significantly increase further with the current development trends resulting in a promising economic prospect for many industries. Selected developments and technologies are discussed in detail in Sects. 16.3–16.5.

16.2 Composite Material Structure

The fiber-reinforced composite materials generally consist of at least three components. Figure 16.1 illustrates the basic principle of composite structures and presents the key functions of the two main components, namely reinforcement fibers and matrix, their interaction in the composite formation, and a subsequent building of a three-dimensional boundary layer in the composite.

The fiber component assumes a reinforcing and load-bearing function. The fibers used for this purpose are reinforcing materials with very high mechanical and possibly further functional properties in comparison to the matrix component. Polymer-, mineral- and metal- based high performance filament yarns or rovings are of extreme relevance to the composite industry. These high performance fiber materials are marked by very high tensile strengths and moduli of elasticity due to their molecular and super-molecular structure that mostly exceeds the traditional materials as well as plastics or mineral materials in terms of material density.

The matrix as a second component in the fiber-reinforced composite sheaths the reinforcing material until all the individual filaments are preferably fully enclosed. Thereby it is supposed to realize an efficient composite in conjunction with the formation of the fiber-matrix interface, and lead to a load-conforming fixation of the

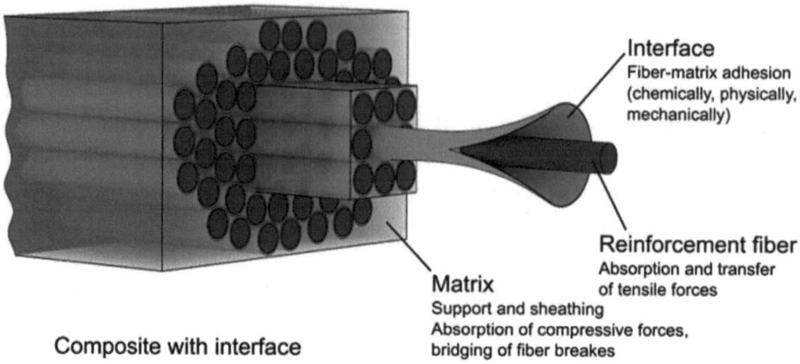

Fig. 16.1 Structure of a fiber-reinforced composite

reinforcing fibers in the desired spatial orientation. The pressure-absorbing matrix must contribute significantly to the discharge and distribution of the externally occurring forces and momenta to the reinforcing fiber. Furthermore, the matrix must assume the function of protecting the reinforcement fiber against applied external mechanical, thermal and chemical factors. Depending on the field of application, the matrices can be made of thermoplastic, thermoset, elastomer, mineral, metallic, or ceramic materials. In this chapter, materials used for coatings are also regarded as matrices.

Fiber-matrix phase boundaries are a third important component that is not clearly evident but which significantly determines the properties and the performance of the composite component. These boundary layers, composed of the boundary surfaces of the reinforcing fiber and matrix as well as the transition region between them, are relevant to the interaction and adhesion capacity of the constituent material components. These properties have a direct influence on the load deflection into the fibers as well as on the impact behavior of the composites.

In addition to the mechanical properties of the individual components, the load transfer between the two components fiber and matrix as well as the growth of existing cracks are decisive, as they are dominated by the adhesive strength of the composites. The strength and toughness of a fiber composite can be significantly modified by the fiber-matrix interface and thereby customized according to the requirements.

The distance between the fiber-matrix phase boundaries can extend down to the molecular level, making direct interactions between the components possible. Furthermore, the introduction of other active mediating substances in the boundary layer is possible in order to equip them to sufficiently meet the requirements of the final composite. The upper or boundary surfaces of the textile reinforcement structures can be equipped with additional sensor and actuator functions integrated into composite components, so as to achieve a continuous structure monitoring, self-diagnosis and control. The design possibilities of interfaces and layers as well as the required consequent surface modifications for different material combinations are discussed in Chap. 13.

16.3 Fiber-Reinforced Composites

16.3.1 General Remarks

The interest in the industrial applications of composite materials as an alternative to metallic components in the field of automotive, machine and construction, sports equipment and medical technology is growing steadily with an increasing focus on the weight requirements. A high lightweight potential of the components is of great importance particularly when large masses must be accelerated or transported. Due to their high lightweight potential, fiber-reinforced plastics (FRP) can make a significant global contribution in terms of energy and resource conservation as well as towards environmental protection efforts.

The three main principles of lightweight constructions are:

- Lightweight material and substances
- Lightweight design and form
- Functional lightweight construction (functional integration, assembly step reduction etc.)

High-performance fibers or rovings made of glass or carbon are ideal for lightweight construction due to their very high specific tensile stiffness and strength (see Sect. 16.3.3). However, their limp textile characteristics and their sensitivity to stresses perpendicular to the fiber axis restrict their usage as solid materials components in the construction of vehicle, machinery or plant parts. Hence, the principle of fiber-reinforced composites is based on the load-bearing reinforcement fibers or yarns with adjustable anisotropic or form-conforming material targeted to specific requirements so that a solid component can be formed in combination with a matrix material (see Fig. 16.1).

All fiber composites based on polymeric matrix systems are referred to as fiber-reinforced Plastics (FRP). Generally, every plastic can be used for this purpose. The matrix systems can be divided into three basic groups: thermosets, thermoplastics and elastomers. In addition to plastics, metals or ceramics can also be used as matrix systems in conjunction with selected reinforcement fiber types. Here, different production techniques are used for FRPs. The following section focuses exclusively on FRP.

The relative significance of the two input components, i.e. reinforcing fibers and matrix (see Fig. 16.1), for the important mechanical and physical properties of the composite are estimated in the literature as shown in Table 16.1. This illustrates that the mechanical composite properties are largely determined by the reinforcing fibers. The decisive factor is the proportion of the reinforcing fibers in the composite, normally specified as a fiber mass or fiber volume ratio, which in turn depends on the textile reinforcement structure and the respective production process as well. The matrix supports and protects the fibers. The breaking strength of the matrix should be well above the reinforcing fibers.

Table 16.1 Relative importance of fibers and matrix for the composite properties [1, 2]

	Fiber	Matrix
Mechanical properties		
Stiffness	3	1
Strength	3	1
Fatigue	3	1
Damage tolerance	1	3
Impact behavior	3	1
Thermo-mechanical properties	3	1
Fiber-matrix adhesion	2	2
Physical properties		
Corrosion behavior	1	3
Temperature resistance	0	4
Chemical resistance	0	4
Electrical properties	2	2
Processing properties	0	4

0: 0 %, 1: 25 %, 2: 50 %, 3: 75 %, 4: 100 %

The quality of the interface/boundary layer during the composite formation is crucial for the interaction between the reinforcing fibers and the matrix and is responsible for the actual properties of the resulting FRP (see Sect. 16.2 and Fig. 16.1). The high load transmission potential required in FRP can be only achieved by a high fiber-matrix adhesion in conjunction with a complete wetting of the reinforcing fibers during the composite formation. This depends on various process parameters and also on the matrix viscosity during the impregnation of the reinforcing structure as well as on the surface structure of the fibers. Adhesion-compliant matrix systems, functionalization and design of reinforcing fibers with suitable sizing, binders or surface activation methods are crucial for a functional interface/boundary layer. Many manufacturer-customized material options are available, and used industrially for many of the standard FRP material combinations. However, the fiber manufacturers usually do not publish information on the composition and effect of the sizing or binding agents. Hence, further improvement in bonding through the fiber-matrix design interface is now often the subject of research activities [3–6]. FRPs have particularly high potential for the consequential combination of the aforementioned lightweight construction principles and thus on achieving high lightweight construction potential in the final component. This is due to the fact that they can be designed in a requirement-adapted way using various degrees of freedom and produced with anisotropic properties by various textile- and plastics-engineering production methods. Figure 16.2 illustrates the high lightweight construction potential of this material group, based on the specific stiffness and strength in comparison to metals. FRPs in specific combination with other material groups offer an excellent approach for intelligent material combinations in realizing integrated components with hybrid design or components with multi-material design, which combines the advantages of all corresponding material groups [7, 8].

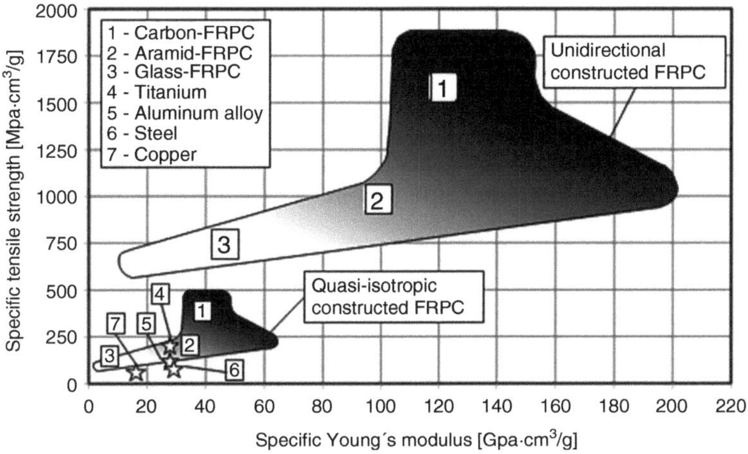

Fig. 16.2 Comparison of the specific material properties of FRPs and metals, according to [9]

Compared to metallic materials, FRP also exhibit a number of other properties that are of interest for a potential industrial utilization [9, 10]:

- High form stability, including adjustable thermal expansion,
- Good damping behavior,
- Low thermal conductivity,
- High corrosion resistance,
- Extensive possibilities for integral construction, and
- High design flexibility

Reinforcing fiber types used in FRP include glass, carbon, aramid, polyethylene, ceramic or basalt, as well as natural fibers, such as flax, hemp, jute or sisal. The forms and characteristics of these reinforcing fiber types are provided in Sect. 3.3.

The polymeric matrix systems must be in a liquid state in order to ensure a full impregnation of the reinforcing fibers.

Curing of thermoset matrix systems occurs by chemical cross-linking reactions based on polymerization, polyaddition, or polycondensation. In elastomeric matrix systems such as silicone or rubber, the degree of cross-linking is much weaker, so that these materials have elastic properties. Neither system is fusible. Usually, these matrix systems are mixed with additional materials such as fillers, additives, flame retardants, or colorants in order to achieve the required processing performance and the required structural characteristics.

Thermoplastic matrix systems must first be converted into a molten state by heating. After impregnation, the necessary transition of the polymer melt into a solid state occurs through cooling and solidification. Thermoplastics soften or melt even after the composite formation under the influence of temperature and solidify again after cooling. Hence thermoplastic composites are re-formable and also re-weldable.

The following sections are limited to the production of FRPs using thermoset or thermoplastic matrix systems. The production of prepregs (pre-impregnated with matrix materials) and their further processing can take place independently in a separate process and has already been discussed in Sects. 11.2 and 11.3. Therefore, the focus of this chapter is placed on the production process in which the textile reinforcement structures are impregnated or infiltrated directly during component production with the respective matrix system by injection or infusion.

16.3.2 Reinforcing Structures

The possibilities to form the reinforcing yarn construction and the textile reinforcement structure are wide-ranging as described in Chaps. 4–10, taking into account the diverse textile manufacturing processes and the numerous plastic-technical process available for the application of reinforcing fibers or filaments in the composite. Figure 16.3 shows a typical example of reinforcing fibers and their individual layer arrangements in various textile fabrics. The fiber length and orientation significantly determine the achievable fiber proportion and mechanical properties in the composites.

Figure 16.3a, c show non-oriented short or continuous fiber reinforcements that are mostly processed as chemically or mechanically bonded nonwovens (so-called chopped mats or continuous mats) (see Sect. 9.4.3). These usually exhibit no marked preferential orientation of the reinforcing fibers. They allow only a comparatively low fiber content of up to 30 % by volume and have low mechanical composite properties equally in all directional planes, i.e. quasi-isotropic, and are mostly used in components with standard requirements.

Certain measures in non-woven production also allow the generation of short- and continuous-fiber reinforcements with a certain fiber orientation (Fig. 16.3b), which has a corresponding influence on the achievable fiber volume and the mechanical properties in the composite.

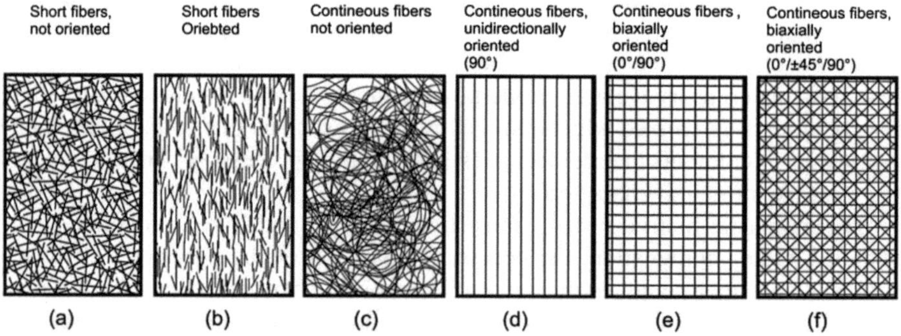

Fig. 16.3 Typical reinforcement fiber or reinforcement yarn arrangements in the individual layers or textile fabrics, depending on fiber length and orientation

Unidirectional (UD) reinforcements with monaxial orientation of continuous reinforcing fibers (Fig. 16.3d) usually come as a thermoset or thermoplastic prepregs (see Sect. 11.2) for processing as they cannot be handled without fixation and would fall apart. Besides, in practice, structures like woven fabrics, having a secondary yarn system with uniaxial reinforcing fiber orientation, are also referred to as UD-reinforcements. This secondary yarn system fixates the monaxially aligned reinforcing fibers and makes no significant contribution to the reinforcing effect in the composite. UD- reinforcements allow highest fiber volume fractions of up to 80 % in the reinforcing direction and the best mechanical properties. However, the reinforcement effect in transverse direction is very low (see Fig. 16.4) since it is mainly determined by the matrix properties and the interfacial properties between fiber and matrix. UD reinforcements are mainly used for components with high mechanical requirements.

Continuous fiber based multiaxial reinforcements with appropriate arrangement of the reinforcing yarns in the plane (Fig. 16.3e, f) are produced using various yarn processing and surface building technique, such as weaving, multilayer knitting, multiaxial warp knitting or braiding (see Chaps. 5–8). Reinforcement fibers arranged in different orientations can be interlaced with each other (woven, braided). Alternatively, the reinforcement structure can be comprised of many stacked parallel fiber layers in different orientations that are fixed together by a loop or binding yarn (multiaxial warp knitted fabric, multilayer knitted fabric, multilayer woven fabric). With the specific arrangement of the reinforcement textile, the mechanical composite properties in the corresponding layer can be set ranging from strongly anisotropic to "quasi-isotropic" (see Table 16.2). However,

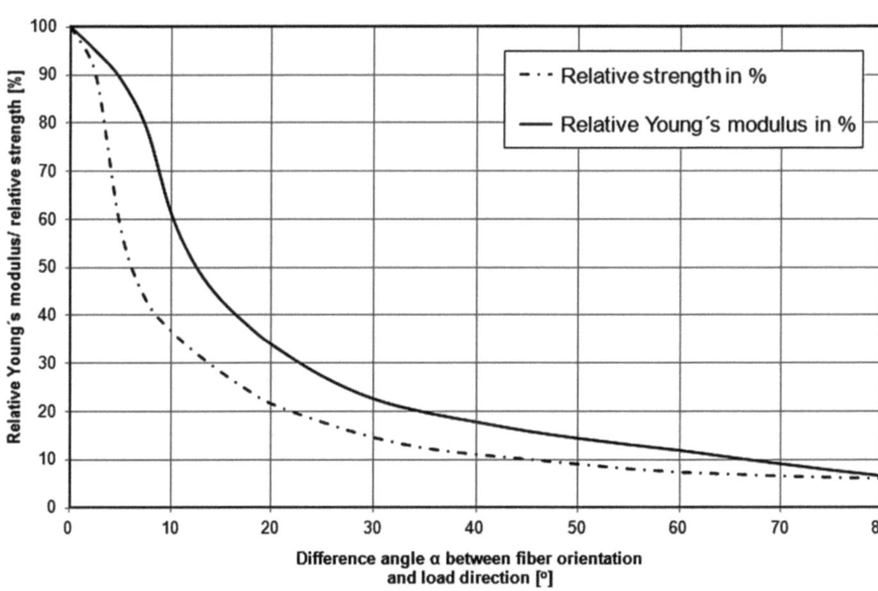

Fig. 16.4 Direction dependence of Young's modulus and strength in UD-FRPs, according to [9]

Table 16.2 Typical reinforcement fiber/reinforcement yarn arrangements from continuous-fiber-reinforced prepregs or textile semi-finished products

	Reinforcing filament arrangement in the composite			
Reinforcement arrangement in individual layers	Unidirectional	Multiaxial, biaxial	Multiaxial, quasi-isotropic	Multiaxial, three-dimensional
Unidirectional	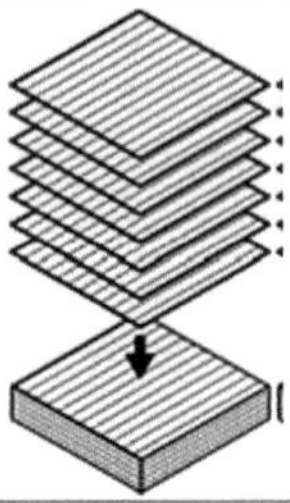	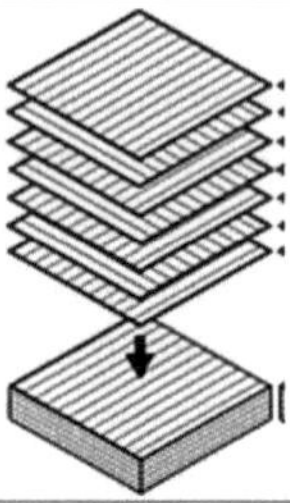	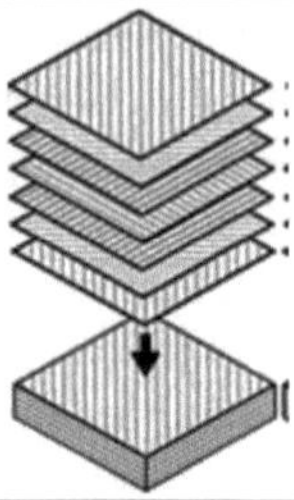	
Bi or multiaxial		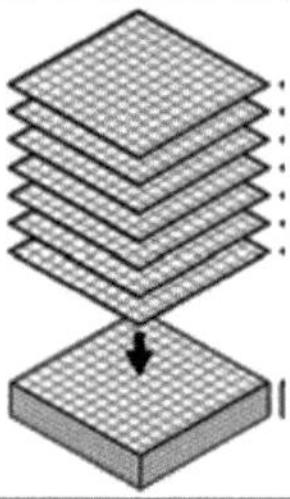	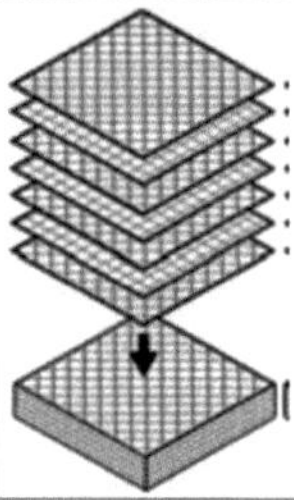	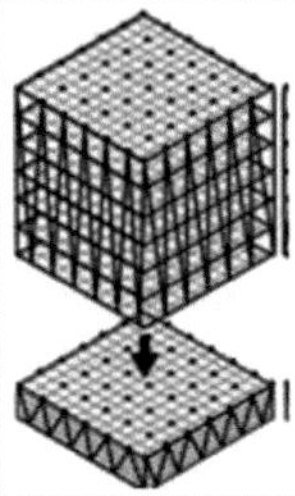

the achievable fiber volume proportion in the composite is dependent on the specific semi-finished textile construction and the realizable composite strength and stiffness of reinforcing fibers under load in the respective textile manufacturing process or from the partially stretched arrangement of the reinforcing fibers or yarns in the textile structure. They lie within a range of up to 65 % and are thus lower than those of FRPs made from UD reinforcements.

Figure 16.4 shows the high directional dependence in the mechanical properties of FRP using a sample tensile strength and modulus of elasticity of a UD composite in terms of the angle difference between the reinforcing fiber orientation and the loading direction. A 5 % deviation in the fiber orientation from the load direction of UD reinforcement would consequently result in a tensile strength reduction of at least 30 %. Hence it is required that the permissible deviation of the reinforcing fiber orientation occurring during component manufacturing variations must be kept as small as possible in order to exploit the performance potential of the reinforcing fibers in the FRP.

Since real loading resulting from monaxial component stresses occurs only in individual cases, FRP components are usually designed and produced to have a reinforcing structure (laminate construction) of several reinforcing layers arranged and aligned over each other in the direction of its thickness according to the operational needs. In order to avoid shrinkage-related residual stresses and the resulting unwanted distortions in the consolidated composites, it is important that the arrangement of these layers is made as symmetrical as possible in the thickness direction of the reinforcing structure.

Table 16.2 presents different possibilities for the combination of continuous-fiber reinforcements according to Fig. 16.3d–f in isotropic laminate structures with unidirectional and multiaxial reinforcement.

The illustrations in the right column represent complex textile semi-finished products with a high proportion in z-direction, i.e. reinforcement fibers or filaments arranged perpendicular to the fabric plane, for example, by 3D weaving, multi-layer weaving, or multi-ply knitting (cf. Chaps. 5 and 6). Such z-reinforcements reduce the interlaminary delamination, i.e. the bonding failure between the reinforcement layers in the composite components, ensuring a damage-tolerant component and having positive effects on the crash and impact behavior of the composites.

It should also be noted that the z-reinforcement, for example, causes a partial reduction in tensile and bending properties in-plane, as arranged reinforcement fibers perpendicular to the composite plane cannot absorb the occurring tensile forces. In addition, the z-reinforcements may create to undesirable resin-rich zones in the respective yarn sections.

Figure 16.5 shows a polar graph, schematically representing the qualitative distribution of the mechanical composite properties in relation to the load angle for a unidirectional reinforcement assembly in the 0°—direction, as well as for two multiaxial reinforcement assemblies (0°/90° or 0°/+45°/−45°/90°).

The impregnation of the reinforcing fiber assembly with the matrix, especially for injection or infusion processes in which the matrix has to flow through the textile structure, is largely determined by its permeability. This provides an important material parameter for these processes and in turn depends on the textile reinforcement structures and the degree of compaction during the component

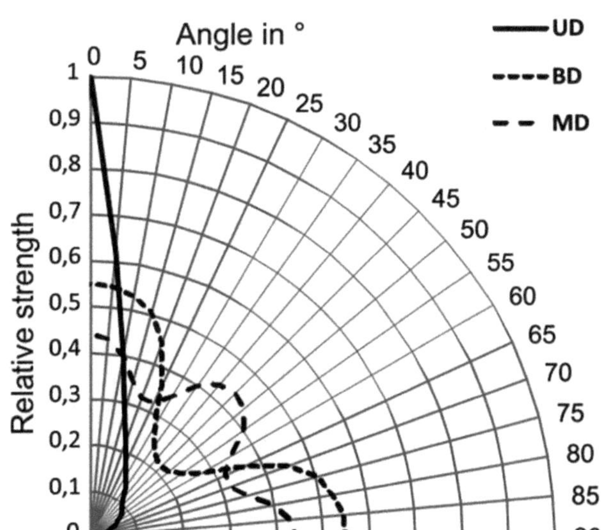

Fig. 16.5 Direction dependence of tensile strength in unidirectionally (UD), bidirectionally (BD), and multi-axially quasi-isotropically reinforced FRPC, in relation to the tensile strength of the UD reinforcement in reinforcement direction

production. The speed of the impregnation process can be influenced mainly through the injection or infusion pressure and the matrix viscosity, depending on the processing method and the use of additional flow additives (see Sect. 6.3.4). Since the textile reinforcing structure is defined mainly on the basis of the final component requirements and not by the impregnation behavior, it is important to be able to recognize and predict these in advance. The determination of the permeability has not yet been standardized. A number of measurement procedures have been developed and are being currently used. They vary significantly in terms of technical complexity and reliability [11–15].

In case of injection or infusion processes, the porosity of the textile reinforcement structure plays a decisive role, as it is important not to filter out the additional materials like fillers and flame retardants that determine the functional characteristics.

16.3.3 Matrix Systems

In principle, all plastics can be used as matrix material, offering near-innumerous possibilities. The most important thermosetting materials are unsaturated polyester resins (UP), vinyl ester resins (VE), epoxy resins (EP) and phenolic resins (PF). The spectrum of thermoplastics is broad and ranges from standard thermoplastics, such as polypropylene (PP) or polyamide (PA), to those for high-performance applications, such as polyphenylene sulfide (PPS), polyetherimide (PEI), polyethersulfone (PES) or polyether ether ketone (PEEK).

The influence of the matrix on the production and properties of FRP and their potential fields of application should not be underestimated. In addition to the formation of an adhesion-friendly interface with the reinforcing fiber, it is crucial to damage tolerance, corrosion, temperature, and fire or chemical resistances of the composites (see Table 16.1).

The thermal properties of the matrix determine the temperature range for the application of the FRP as their thermal or temperature resistances, with few exceptions (polymeric reinforcement fibers) lie well below those of the reinforcing fibers. The elongation properties of the matrix together with the interface layers is decisive, especially for the compressive stresses perpendicular to the reinforcing fiber orientation as well as for the long-term behavior of the FRPs and their behavior under dynamic loads [1, 16]. Figure 16.6 gives a general overview of the areas for the maximum continuous operating temperature and the breaking strength of the abovementioned matrix groups as stated in the literature.

Apart from that, the usable methods and process parameters for composite manufacturing and possibilities of subsequent processing for component finishing are significantly dependent on the matrix properties (e.g. viscosity in the processing state, the nature and quantity of the additional materials). In the case of the injection method, the thermoset matrix systems at their processing temperature behave approximately as Newtonian fluids and have comparatively low viscosities between

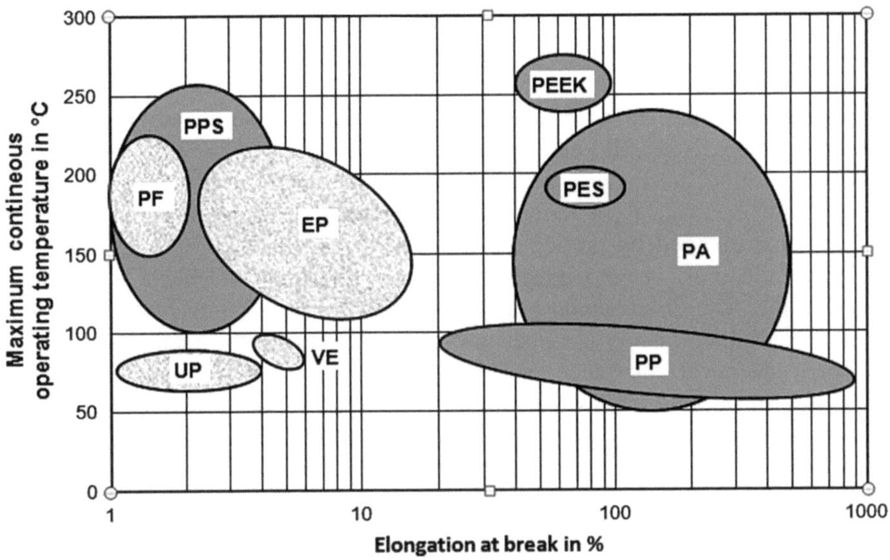

Fig. 16.6 Maximum long-term service temperature and maximum elongation at break of typical plastic matrix systems, based on [9]

20 and 500 mPas. In contrast, the typical polymer melts of thermoplastic composites at processing temperature cannot be regarded as Newtonian fluids. Also, ensuring the flow of the matrix in the range of 105–106 mPas during the impregnation of textile reinforcement structures is much more difficult [3, 16].

The matrix viscosity is highly dependent on temperature and can be reduced by increasing the processing temperature. However, the temperature cannot be arbitrarily chosen, as at higher temperatures polymer degradation occurs, causing decreased matrix properties. In case of thermosetting resin systems, an increased viscosity occurs during the reaction time, caused by the cross-linking or curing process which limits the processing time and is generally termed as "pot life". It should be noted here that a reduction in matrix viscosity through an increase in temperature leads to an improved curing rate and a decreased pot life.

Since the most important traditional thermoplastics matrix component in question are already described in Sect. 3.4.2, the following section is limited to few selected thermoset matrix systems.

Unsaturated polyester resins (UP) and vinyl ester resins (VE) are cured by a radical polymerization. This process occurs very fast and the cross-linking is very intense.

Unsaturated polyester resins are very easy to use and are wide spread in practical application. There exist numerous types of resins for example exhibit properties like high level of thermal stability, increased toughness and chemical resistivity, often at the expense of other properties. The disadvantages are an only moderate resistance to weathering due to the existing double bonds in the cured state, high shrinkage (reduction in matrix volume due to the chemical cross-linking) and styrene emission. The significantly more reactive vinyl ester resins in comparison are characterized by better mechanical properties, lower mold shrinkage and high thermal as well as chemical resistance [1, 3].

In contrast to the curing of thermosets by radical polymerization, the cross-linking in epoxy resins (EP) and phenol resin (PF) occurs through polyaddition or polycondensation, a stepwise reaction of the functional groups. Epoxy resins exhibit better mechanical properties when compared to the other three resin systems and are characterized by particularly high heat resistance. However, the resin/hardener ratio must be very precisely set here in order to achieve a complete cross-linking. The disadvantages are their price and higher moisture absorption. Phenolic resins are mainly used because of their excellent fire-retardant behavior and high chemical resistance, such as in vehicle interior applications. The processing of phenolic resins leads to an emission of toxic formaldehyde, which requires appropriate safety measures or a closed-device fabrication process [1, 3].

In Table 16.3 important parameters for selected thermoset matrix materials are summarized.

16.3.4 Component Production and Applications

16.3.4.1 Overview

As mentioned in Sect. 16.3.1, the component manufacturing processes discussed here are limited to methods based on the impregnation or infiltration of dry textile reinforcements with a liquid matrix system. The range of processable textile preforms begins with yarns or rovings to fabric formation including complex-shape preforms. The actual component manufacturing typically involves the sub-processes of impregnation or infiltration of the reinforcing structure, shaping, and consolidation. If required, the processing steps can also integrate cores based on foams, balsa wood or honeycomb structures into the composite components and thus produce sandwich structures. This will not be discussed here in detail. In order to achieve good mechanical properties, thermal treatment of the components by annealing is required.

Essential criteria for the use of different component manufacturing processes and process variants include [1, 9]:

- Component size and complexity,
- Mechanical requirements of the components
- Base materials (reinforcing structures, matrix system),
- Required fiber volume,
- Required surface quality,
- Quantity,
- Costs
- Availability of the technology

Table 16.3 Typical properties of thermoset matrix systems [3, 9]

Property	UP	EP	PF	VE
Density g/cm^3	1.17 … 1.30	1.10 … 1.25	1.25 … 1.30	1.16 … 1.25
Tensile strength in MPa	40 … 75	45 … 100	50 … 100	50 … 80
Tensile modulus of elasticity in GPa	2.8 … 3.5	2.8 … 3.4	5.0 … 6.0	2.9 … 3.1
Processability at room temperature	Yes	Yes	No	Yes
Maximum processing temperature in °C	180	170	190	175
Glass transition temperature in °C	90 … 115	110 … 210		90 … 190
Molding shrinkage in %	6.0 … 10.0	1.0 … 3.0	0.5 … 1.5	0.1 … 1.0

The typical processing methods for the component manufacturing made of textile fabrics or continuous reinforcing filaments with similar functionalities are classified according to [17] in the following categories A to E. This list is enhanced with categories F and G:

A. Manual lamination, vacuum infusion, autoclave and tube blow molding methods
B. Resin-Transfer-Molding (RTM) method
C. Compression molding
D. Sleeve and round braiding methods
E. Pultrusion methods
F. Fiber-resin injection and fiber centrifuging
G. Injection molding and method combinations with injection molding.

Since the autoclave and molding process are already covered by Sects. 11.2 or 11.3, these will not be further discussed here. The methods in categories A and B are based on the processing of textile reinforcement fabrics or prepared preforms (see Chap. 11). The categories D and E include methods preferably based on the processing of reinforcing filaments or narrow textile structures. In addition, the methods of the category F are widespread in the industry where continuous reinforcing fibers, usually made of GF rovings, are separated into non-continuous fibers and directly processed with the resin system into components with isotropic short fiber reinforcements. Since injection molding in combination with continuous-fiber-based textile reinforcements is becoming increasingly important, it is included as category G.

The selected methods are briefly explained and then compared based on certain criteria. Due to the considerable wide range of possibilities for component applications with specific requirements and features in the FRP area, the application-related inferences are here confined to the realizable component geometries and the reinforcing structures.

In addition, important application areas are presented. For more information, based on specific examples of applications, relevant technical literature can be referred to.

16.3.4.2 Manual Lamination-, Vacuum Infusion and Tube Blow Molding Processing

The manual lamination and vacuum infusion processes (category A) are based on the production of simple or complex, shell-shaped components with low-cost open tools. The contour of the mold corresponds to the final smooth visible front side of the component. The back side of the component usually exhibits a rough surface due to process-inherent reasons. The textile reinforcing fabrics (e.g. nonwovens, woven or non-crimp fabrics) are cut according to the required global and local reinforcing layers and the desired wall thickness of the respective component. They are then manually positioned or draped in the mold. Hence, a reproducible production is limited. The impregnation with the reactive resin system differs in these processes. After curing of the component, it is demolded and post-processed into the final component.

In the manual lamination process the layup and subsequent impregnation of the textile reinforcement with the resin system is realized layer wise. The incorporation of resin as well as the compaction and deaeration of the laminate structure are performed with simple tools such as brush and pressure rollers. The lamination time and the pot life of the resin system must be made compatible with each other according to the part size and complexity. In principle, the uncured laminate construction can also be packed in a vacuum using a vacuum foil, and cured by applying low pressures. Since manual lamination is an open component manufacturing process in which the processing of the resin system is connected with the emission of harmful substances (i.e. phenol resin), appropriate safety measures have to be complied with. Manual lamination is used in the production of rotor blades for wind turbines, in boat building and commercial vehicle construction [18–20].

In the vacuum infusion process, also known as resin infusion (RI), the complete reinforcing structure or the preform including various production aids is vacuum-tightly packed, fitted with a sprue and suction, finally compressed by applying a vacuum, and impregnated with the resin system (see Fig. 16.7a). In this case, the low pressure (<1 bar) in the storage container aids the resin flow through multiple rings and the reinforcing textiles to the suction point lying under an ambient pressure. The film serves as a flexible upper tool. Alternatively, an additional counter-tool can be incorporated into the vacuum setup making smooth surfaces on both sides possible. Due to the limited pressure difference, the matrix flow paths and the component sizes are currently limited. This limitation can be countered by various measures, such as selective tempering, use of flow additives or distribution medium, or appropriate sprue designs [2, 3, 20, 21].

Figure 16.7b illustrates a typical vacuum infusion construction with textile reinforcement structure (preform), release fabric, distribution medium and vacuum foil. The separating or peel ply allows a subsequent separation of the production aids from the component surface. The so-called distribution medium (e.g. 3D warp-knitted or non-crimp fabric) must have a high permeability and compressive

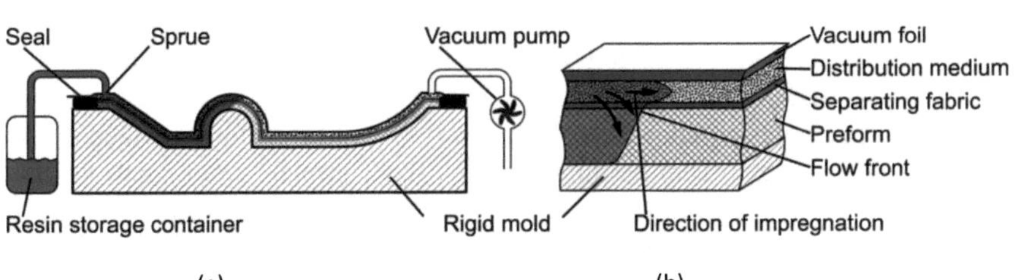

Fig. 16.7 Vacuum creation for vacuum infusion, based on [7]. (**a**) Vacuum infusion. (**b**) Flow front progression during vacuum infusion

strength to ensure fast and secure impregnation of the reinforcing textiles. The flow channel of the distribution medium allows an even distribution of the resin system after compression due to the low pressure. In this case, the actual impregnation of the fiber structure preferably takes place in the thickness direction, and thus over a short distance (Fig. 16.7b). The cycle time depends on the component and can be up to 60 min [3, 7, 17, 22].

Thin-walled hollow structures, such as load-bearing frames, can be produced by the tube blow molding processes. Here two mold halves with the textile reinforcement structure and a tubular vacuum foil are combined, sealed, and the fabric is then impregnated by vacuum infusion. The tubular vacuum foil forms the cavity of the component [17].

The methods described under category A are used for cost-effective small scale production of small to large parts due to the usage of inexpensive molds. The disadvantage here is the waste created due to release fabric, distribution medium, and vacuum foils containing the resins. Typical components include vacuum-infused machine components (panels, structures), boat and ship components and rotor blades of wind turbines [1, 9, 17–19]. In addition, the use of the vacuum infusion process for the aerospace industry is of growing relevance (e.g. wing boxes of aircraft or booster cases of launch vehicles) [23, 24].

16.3.4.3 Resin-Transfer-Molding Processing Method

In the RTM process (Category B), the impregnation of the textile reinforcement structures or near-net-shape preforms occur by injection of low-viscosity reactive resin systems under high pressure in rigid, multi-part and closed molds with a defined cavity corresponding to the component geometry. The components have smooth surfaces on both sides and can have a very complex shape, but this requires highly complex components. Figure 16.8 shows the main process steps for the component manufacturing in the RTM process.

Initially, a preform is produced and inserted into the cavity of the open mold. Alternatively, the textile reinforcement can also be shaped directly in the mold. The surface quality of the component or the UV-resistivity are improved by applying gel coats (thin layers of pure resin) up to the mold cavity surface before the lay-up, or alternatively by using non-wovens. After closing the mold, the compression of the preform occurs through tempering via built-in heating ducts and heating of the preform. Impregnation of the preform is carried out by injection of the liquid resin system under pressure so that it flows from the sprue in the preform surface through the textile to the risers. Tempering simplifies the complete impregnation of the preform, as it allows a decrease in the viscosity of the resin system during the injection. After curing the component under pressure to reduce the size of pores or sink marks caused by shrinkage, the component is removed from the mold and post-processed into the final component [1, 3, 9, 25].

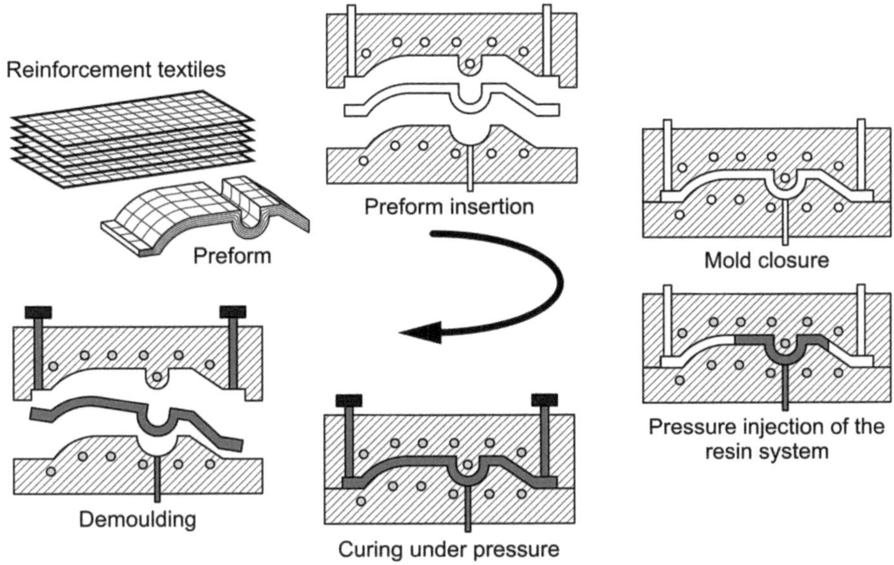

Fig. 16.8 Typical process sequence of component production in RTM processes

Typical cycle times for the structural components ranges from approximately 5 to 25 min [8, 22]. The uniform heating and fast curing of the resin system can be achieved by the use of microwave technology (300 MHz–300 GHz) [8].

There are numerous known modifications of the RTM process which are aimed at improving or shortening the injection process, enhancing the processable matrix systems and producible component spectrum with high quality. A few examples are listed below [1, 3, 26–28]:

- VARTM method (Vacuum-Assisted Resin Transfer Molding); deaeration before injection,
- ARTM method (Advanced Resin Transfer-Molding or gap impregnation method); injection into partially closed tool followed by closure of the tool to final dimensions; preform impregnation by the shortest flow path in thickness direction and
- SRIM method (Structural Reaction Injection Molding); injection of highly reactive low-viscosity polymer blends at high pressure (>20 bar).

A recent process development presents the component production based on prepregs produced by resin transfer prepregging. These are first impregnated in a deep-frozen mold by injecting a thermosetting resin system without compressing the preform, using cascaded injection units with needle valve nozzles so that excess resin is formed. The bonding reaction of the matrix is delayed by the cooling of the preform, resulting in usable preforms. In a further processing step, the compression, molding, and bonding take place under the influence of temperature and pressure. Excess resin is directed into the overflow cavity of the mold so as to ensure correct fiber volume content in the final component. Current works show that the first processing step can be completed after approx. 6 min and the second step can be completed after about 10 min [28, 29].

The RTM process and related variants allow a reproducible production of components with a weight ranging from a few grams to up to about 50 kg, of reproducible quality with smooth surfaces on both sides and a high reinforcing fiber content. More extensive components require an elaborate molding due to the high injection pressures and are limited by the processing time of the resin systems [17]. The components usually require little post-processing. As this involves a closed process, harmful emissions are largely avoided in the manufacture of RTM components. The process can be easily automated and is economically feasible for medium to large production [3, 17]. However, it is characterized by significantly higher mold cost due to the required high pressure and is also very demanding in terms of mastering the process.

In addition to the processing of thermosetting matrix materials, it is also possible to carry out the RTM process with thermoplastics, for example by using additives for improving the flow behavior of the melt [22]. Alternatively, the RTM process is connected directly to the in-situ polymerization, for instance, e-caprolactam in the presence of textile reinforcement structures inside the cavity. The preparation of such textile-reinforced polyamide six components requires a cycle time of 3 min [8].

Typical RTM components are available for example in the automotive industry (e.g. spoilers, roof superstructures, tailgates, cladding), as high quality sports equipment (e.g. bicycle frames), in mechanical engineering (e.g. components, cladding) and in the aircraft industry (e.g. connection fittings in the spoiler area) [1, 8, 17, 25].

16.3.4.4 Winding Processes

The winding process (category D) is based on the layer-by-layer wrapping of reusable or e.g. washable lost form cores with reinforcing fibers or rovings or band-shaped reinforcement fabrics that are then directly impregnated with reactive resin in the winding system (see Fig. 16.9, roller impregnation).

An alternative to the open resin bath for an improved impregnation and the reduction of plant contamination is offered by siphon impregnation, in which the rovings are encapsulated in tubes until the drop-off point [22]. The compression of the laminate structure is achieved by the yarn tension. Excess resin can be removed, for example, by means of absorption fleece/fabric. Reusable cores, if used, must be removed after curing by means of suitable doffing equipment due to the shrinkage of the matrix material. Alternatively, divisible cores can be used [1, 3, 10].

Complex component geometries and winding patterns are carried out by elaborate CNC-controlled winding devices with four or more axes or adapted according to winding pattern in combination with robotic technology for the positioning of the winding core [1].

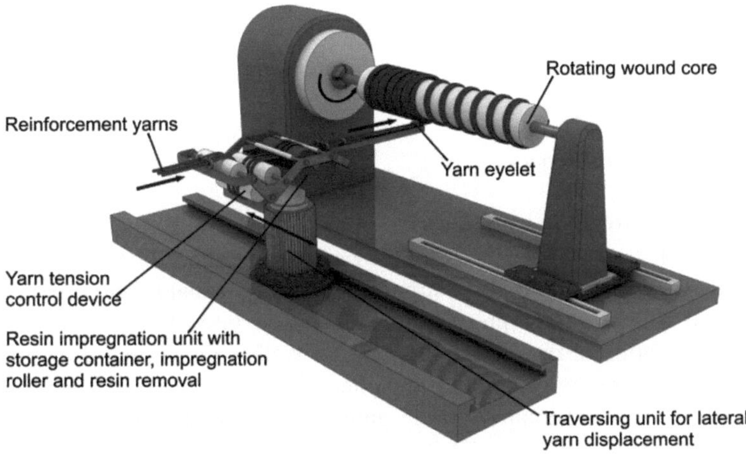

Fig. 16.9 Schematic representation of a (cross-) winding plant

The winding process is similar to the production of yarn bobbins where the pattern is realized through a relative movement between the rotating hub and the lateral displacement of the yarn eye by circumferential, parallel or crossed windings (Fig. 16.9). In addition, polar windings are used for ellipsoidal or cylindrical components with curved surfaces, which permit a small yarn placement angle relative to the longitudinal axis component. The yarn placement is possible over an angle range of $\pm 10°$ to $\pm 89°$ [3, 17, 30].

The winding method is characterized by a high degree of automation in combination with good reproducibility of the fiber layer, better mechanical properties of the component and high fiber volume content of up to 65 %, which may also be significantly higher in individual cases. The process-related industrial application however largely focuses on the production of cylindrical, conical or ellipsoidal components with or without curved surfaces (e.g. pipes, drive shafts, axles, rollers, masts, leaf springs, containers or tanks) [3, 10, 17, 31]. Rotor blade components for wind turbines are also wrapped using this process [8, 19]. The production of smooth wound surfaces or concave components is not feasible without costly additional measures. Depending on the design of the winding system, small to very large components such as up to 42.5 m long tubes [31] or a highly integrated structure wagon for railway vehicles [32] in laboratory conditions can be produced. For the load application in wound rod or tubular structures, specific technologies have been developed, such as the formation of integrative internal pipe yarn using multipart winding cores, or the technical integration of loaded pins with metallic flanges [31–34]. The high performance capability of the winding technology can be seen in the 52 m high self-supporting sculpture “Mae West” in Munich, which is made from 32 to 40 m long tubes wound with carbon fiber having a diameter ranging from 220 to 280 mm [31].

The processing of thermoset or thermoplastic prepreg tapes (see Sects. 11.2.2 or 11.3.2) in a winding process is also possible and requires a corresponding consolidation, for example in an autoclave or by laser, hot air, or heated pressure rollers [3, 22].

16.3.4.5 Pultrusion Process

As shown in Fig. 16.10, in the pultrusion process (category E) the reinforcing yarn or roving bundle are infiltrated with a low-viscosity thermosetting matrix in a resin impregnation unit and pulled into a strand with unidirectional filament arrangement by means of a mixture-compressing funnel-shaped inlet through the pultrusion mold. The formation of the profile geometry as well as the curing of the matrix at an increased temperature generally occurs here. In the last mold section, the profile in accordance to the calibrated profile geometry is actively cooled. The doffing force is transferred to the rovings through an active doffing unit containing the hardened profile. These doffing units may be designed as a band or caterpillar and alternatively may also consist of two alternately engaging gripper systems that alternate in the longitudinal profile axis (see Fig. 16.10). Owing to the principle, the reinforcing fibers are oriented parallel to the longitudinal direction of the profile. In order to realize other configurations transversely to the profile longitudinal axis for improving the mechanical properties, braided hoses or band-shaped reinforcement fabrics can be introduced into the profile. The cutting unit is meant for cutting the profile into manageable portions [3, 10, 17].

The pultrusion method is suitable for automated and continuous production of open or closed FRP profiles in reproducible quality with high fiber volume fractions of circa 60 % to about 80 %. With appropriate system design, profiles with complex profile cross-sections and even shell structures of great length can be produced, which can be used in many areas of mechanical and vehicle engineering (e.g., elevators, cable management systems, truck cabcots), construction (e.g., concrete reinforcement, segments used for stairs or gratings) or for sports applications. The high equipment and molding costs, and the extensive manual effort required to

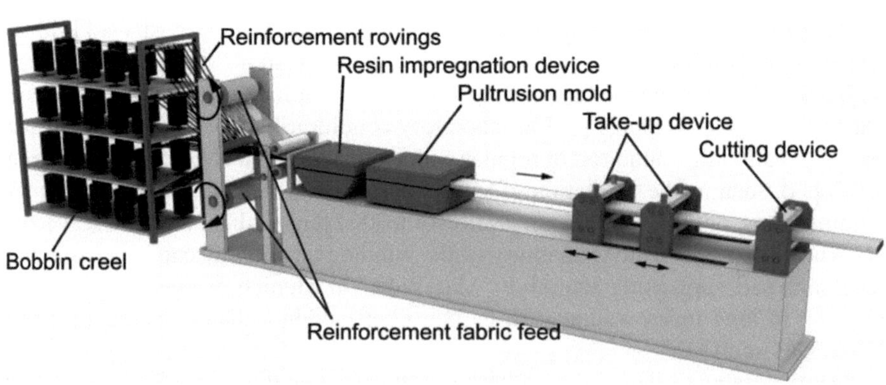

Fig. 16.10 Schematic representation of a pultrusion plant

prepare the system makes the pultrusion process economically feasible for parts of at least several 100 m [1, 17, 35–37].

More recently, the principle for pultrusion was successfully enhanced through an innovative process called Radius Pultrusion™ for the industrial production of curved profiles, which will lead to a significant expansion of the product range [38, 39].

The processing of thermoset or thermoplastic prepreg tapes (see Chap. 11) is possible, as in the winding process, even after the pultrusion. The production of fully matrix-impregnated thermoplastic profiles is not trivial due to the high viscosity of the matrix melt and therefore are hybrid yarns or strand-like braided or knitted hybrid prepregs (see Sect. 11.3.2) used as a starting material [3, 40, 41].

The so-called profile-reinforcement pultrusion combines the advantages of pultrusion and winding technology in conjunction with a thermoplastic matrix for the production of closed profiles. Here, the unidirectional reinforcing filament assembly of the pultruded tube is enhanced subsequently by directly wrapping with dry reinforcement rovings at an adjustable angle. This is followed by the impregnation of the rovings, compacting, and consolidation of the composite [42].

16.3.4.6 Fiber Resin Spraying und Fiber Centrifuging

There are numerous types of fiber injection systems for fiber-resin spraying (category F), which differ from each other with regard to the working pressure (from 2 to 100 bar) and the atomization. Figure 16.11 illustrates the functional principle. In general, the reactive matrix system (usually UP resin) is sprayed into an open mold together with the short fibers (15–50 mm in length) produced in the cutting head.

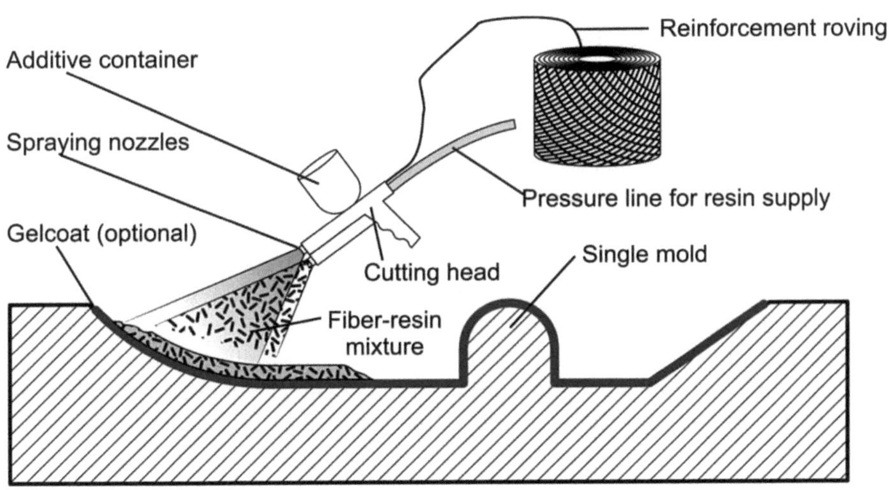

Fig. 16.11 Schematic representation of fiber resin injection

Mixing of the components can occur inside as well as outside of the spraying device, as shown in Fig. 11.16. In order to obtain a high-quality component surface, the tool surface is pre-coated with gelcoat. The compression and ventilation of the sprayed fiber-matrix mixture is manually ensured as in manual lamination, or partly mechanized. After curing and demolding, the post-processing of the component takes place, where local overstressing should be avoided due to the isotropy of the short-fiber reinforcement. The fiber volume fraction can be adjusted between 20 and 40 % by means of the mixing ratio of the starting components [1, 43].

Fiber resin spraying is a low-cost fabrication process (cost of production, cutting, and handling of flat textile structures) using moderately sophisticated molds for the processing of direct rovings which creates only minimally loadable components with anisotropic short fiber reinforcement and non-uniform wall thicknesses.

The method is mostly used for large components, such as containers, swimming pools or roof elements, in small to medium quantities [1].

According to a new and innovative method, fiber spraying is used to produce manageable thermoplastic 3D preforms with short-fiber reinforcement, which can be further processed into components using the conventional thermal pressing processes described in Sect. 11.3. For this purpose, in accordance with Fig. 16.1, 1 thermoset matrix hybrid yarns without any additives (see Sect. 4.1.3) are cut to non-continuous, short fibers about 50 mm long by means of an industrial robot, sprayed into an air-permeable mold and fixed by means of a vacuum. The fibers can be laid with a locally definable preferred orientation, where the ratio of the strength in the preferred direction and orthogonal to it can be up to 3:1. The heating of the fibers during placement leads to a slight melting of the thermoplastic hybrid yarn components, which ensures a good fixing of the preform. Another advantage here is that no trimming of the preforms is necessary and the thickness and the fiber orientation of the preform can be varied locally. Compared with composites made of continuous fiber-reinforced semi-finished products (with equal hybrid yarn and reinforcing fiber volume) the strengths of "fiber-sprayed preform"-based composites are approx. 10 % lower [44, 45].

In fiber centrifuging (category F), short-fiber/matrix mixture realized directly during the component manufacturing is processed similarly to fiber-resin spraying. However, the spraying here is done using a lance with the mixing head and the nozzle on the inner walls of a rotating centrifugal drum. The acting forces lead to the compression of the mixture and the subsequent hardening. The anisotropic short-fiber reinforcement exhibits a mass percentage of 25–45 %, which is especially higher on the outer surface due to the higher density of GF when compared to resin. The demolding is not critical due to the shrinkage of the resin system. The mechanical properties can be improved with an additional placement of reinforcement structures. The mostly cylindrical to slightly conical and thick-walled hollow body parts exhibit a smooth outer surface and are used e.g. for underground tanks, silos, masts, or pipes for transporting aggressive medium [1, 10, 43].

16.3.4.7 Injection Molding Process

Injection molding (category G) is a suitable molding method for the large-scale production of plastic parts in a single manufacturing process. Usually, the granular materials, (e.g. a thermoplastic) is plasticized in an extruder similarly to the process used in melt spinning (see Sect. 3.2.1). It is then injected into a closed multi-part form mold representing the component shape through a nozzle under a very high pressure. This process normally consists of injection, compression, and holding-pressure phase (compensation for volume shrinkage). The typical processing temperature is between 150 and 300 °C and depends on the material. The subsequent cooling process effectively determines the quality of the component in terms of texture, shape, and dimensional accuracy. After the solidification of the thermoplastic, demolding is executed, where the component is removed by the integrated ejector system. An appropriate mold quality reduces the subsequent finishing work on the component to a bare minimum. The mold costs are generally high and increase considerably with the complexity of the component shape, making this process economically viable only for large-scale production [46, 47].

Through the multicomponent injection molding based on multiple injection units like composite molding, assembly molding and the co-injection process, it is possible to combine different types of materials, e.g. hard and soft, in a single component [46–48]. Other special process include back-molding of textiles or foils, and micro-injection molding [49].

Thermosets, such as UP or PF resin systems or elastomers, such as nitrile butadiene rubber, can also be processed by injection molding, where the cross-linking reaction occurs only in a heated mold (120–200 °C) [46, 47]. However, it is not discussed in detail here.

Fiber-reinforced injection-molded components exhibit higher mechanical properties compared to unreinforced component and are produced by processing fiber filled molding compounds (fiber length from 0.2 to 0.5 mm). The use of longer non-continuous fibers (approx. 2–5 mm) to reinforce injection molded components is possible with so-called long fiber thermoplastic granules, pellets, or chips. During melting, a significant reduction of the average fiber length occurs in the extruder plasticizing unit due to bending, shear, and friction processes, and also by the injection molding process itself. The typical degradation in the average fiber length on a GF/PP long-fiber granulate having an initial length of 10 mm and a fiber volume content of 30 % is given as 6 mm in front of the nozzle and 2.4 mm in the component [50].

In order to achieve even greater fiber length, processes are being developed to reduce the distance covered by the textile reinforcement fibers in the screw. This is applicable in the case of the Injection Molding Compounder (IMC) process. Here, the reinforcement rovings are fed in the rear part of the twin-screw extruder, where the thermoplastic material is already present in a fully plasticized state [51]. However, the short screw distance makes the complete mixing and wetting of the reinforcement fibers difficult.

It is also possible to reinforce the injection-molded parts with a pre-consolidated textile reinforcement made of continuous-filament yarns, in order to achieve higher mechanical properties i.e. low creep behavior or improved crash properties. The current approaches focus on simple structures [52, 53]. Based on the LOREF process (Locally Reinforced Thermoplastics), continuous-fiber-reinforced rod elements produced by pultrusion are inserted into the injection cavity and molded. However, due to the high flow speed of the thermoplastic matrix, it is to fix these elements in position with slides, clamps, or spacers, as applicable [53].

Alternatively, pre-consolidated grid structures made of stretched GF/PP hybrid yarns with a fiber volume of 60 %, and biaxial and multiaxial orientation suited for full-surface or local reinforcements of injection-molded parts have also been developed. These grid structures are produced with a modified warp knitting (stitch-bonding) technology, where the pre-consolidation by infrared radiation is performed directly after the fabric formation while the yarns are held in tension by the transport chains. This method of consolidation leads to stabilization of the structure without any geometrical changes and allows better handling and further processing in the component production. The grid structures can be made with symmetrical fiber layers in thickness direction, load-conforming warp yarn course and integrated functional elements. The experimental results demonstrate good processability of the pre-consolidated grid structures during the injection molding or LFI (Long Fiber Injection) processes [54, 55]. Figure 16.12 shows a corresponding preform with integrated load application elements as well as the final component.

With the aim to achieve a large-scale and economically feasible production of components in short cycle times, various process combinations for the hybrid FRP-composite production have been developed recently, where the forming of the preheated organic sheets (see Sect. 11.3.2.2) is integrated in the molding cycle. As a short reference to this case, In-mold forming (IMF) and Fluid Injection Technique (FIT) are briefly illustrated. Both methods allow short manufacturing processes and prevent more energy-intensive heating of organic sheets [56].

In in-mold forming the forming process takes place separately by tool closure, for instance, with infrared heated organic sheets fixated with special clamping and

Fig. 16.12 Pre-consolidated textile grid preform with load introduction element (*left*), and demonstrator component (*right*) (*Source*: Institute of Lightweight Engineering and Polymer Technology, TU Dresden)

gripper technology. This is followed directly by the injection molding process, where the residual heat of the placed preform results in a good material connection between the two matrix components. The die-cutting of the redundant materials at the edges is performed in the same tool, so that no further finishing is required. In this way, it is possible to produce highly stable molded rib-based shell structures like steering column links or side impact protection structures for automobiles [56–59].

With the so-called FIT hybrid method, the production of heavy-duty tubular structures is achieved through the forming of two preheated organic sheets placed inside the mold by injection molding, where the thermoplastic material is applied between the two sheets. The final forming occurs by the generation of gas pressure. This is achieved by the induction of gas which either inflates the thermoplastic material that is laid in the form of string or drives the material into the secondary cavity, or by the use of gas-loaded cast [56, 60].

16.3.4.8 Comparison of the Component Manufacturing Methods and Relevant Trends

As illustrated in this section, a number of component-relevant and economic relevant factors are decisive in the selection of the manufacturing process of specific FRP components. Table 16.4 shows a comparison of selected methods in terms of the textile reinforcements, achievable fiber volume proportions and typical producible quantity ranges for an economic component manufacturing.

It is clear that the methods belonging to category A, and fiber-spraying are economically most suited and productive for prototypes and small-scale manufacturing. For the medium and large scale production of components, the pressing methods based on the short-fiber-reinforced flowable thermoset or thermoplastic prepregs (see Chap. 11) are suitable, as are RTM and pultrusion methods (approx. 106 parts per year). Heavy-duty profile and cylindrical structures with load compliant continuous-fiber reinforcements are predominantly produced by pultrusion and winding processes. For the manufacturing of shell-shaped high-performance structures of varying complexity, vacuum infusion or RTM processes based on the use of processable and load-compliant textile preforms are suited depending on the production scale. In addition, the explained methods based on the autoclave technology in Sect. 11.2.2 are also suited, preferably for unidirectional reinforced prepregs.

The component manufacturing methods exhibit further process- and application-specific parameters such as the achievable component geometries and the mechanical properties. Figure 16.13 shows the results of a portfolio analysis of FRP components [3]. Here, the typical component groups are arranged in relation to their size and complexity. The size and complexity of the component part is considered decisive for the selection of the manufacturing process, while the component loading is the most important criterion for the selection of materials [3, 61].

Table 16.4 Comparison of selected component production methods [1, 3, 9, 17, 22, 43]

Category, method	Output reinforcing material	Achievable fiber volume fraction in%	Economic numbers per year
A, Hand lay-up	Fabrics	15 ... 30 (mat) 40 ... 50 (woven fabrics)	1 ... 150
A, Vacuum infusion process	Fabrics, preforms		10 ... 100 (2,000)
B, Resin injection method	Fabrics, preforms		100 ... 10,000 (50,000)
D, Winding method	Rovings, band shaped fabrics		1 ... 1,000
E, Pultrusion method	Rovings, band shaped fabrics		1,000 ... 100 0,000
F, Fiber-resin spraying	Rovings (cut)		1 ... 400
G, Injection molding	Short fibers Rovings (cut) Fabrics	Up to 40 Up to 10	>1,000 >1,000

In general, it can be seen that these trends and developments in the production of heavy-duty FRP components are significantly driven by their high lightweight potential and their high benefit of application especially in the field of mechanical and automotive engineering. These require economically feasible manufacturing methods and process chains especially for medium- and large-scale production. Developments in the part manufacturing methods mainly focus on the reproducibility of production processes for components with high to Class-A surface quality, high degree of automation, multiple process-integrated methods, and the development as well as use of highly reactive resin systems in order to achieve cycle times of less than 1 min. Besides, current research and development works seek not only to develop various energy-saving production methods, but also to economize the use of still quite expensive input materials for high-performance FRP components.

16.4 Textile-Reinforced Concrete

16.4.1 Special Features of the Concrete Matrix

Concrete is an artificial rock created by mixing cement, stone aggregates, and water. As a cheap mass building material it is characterized by a high compressive strength of about 60 N/mm^2 and above in high-strength concrete. Its very low tensile strength, which is negligible in most engineering-based modeling, is compensated for by steel reinforcements in conventional concrete. For the sake of simplification, all tensile forces occurring in the concrete are considered to be absorbed by the reinforcement.

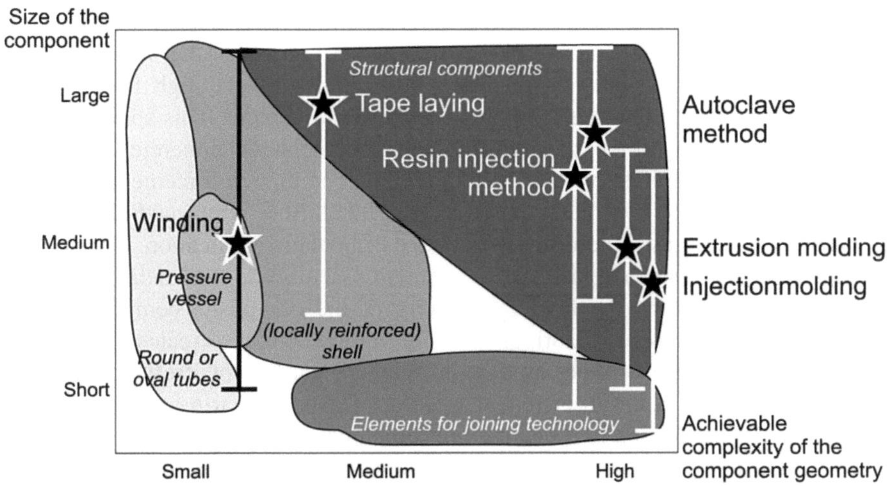

Fig. 16.13 Classification of the component production methods regarding achievable component geometry, according to RIEBER [61]

In an uncracked state (state I), the composite referred to as steel-reinforced concrete displays approximately linear-elastic behavior. Under an increasing load, the absorbable concrete stresses in the tension zone are locally exceeded. This results in an initial crack, and the subsequent crack-bridging action of the reinforcement causes the formation of the final crack pattern (state II a). As soon as the tensile strength of the concrete between two adjacent cracks is no longer exceeded, the crack pattern is completed. Further loading of the composite material leads to an opening of the cracks (state II b). The crack opening is limited by the increasingly stretched steel reinforcement. Only in this state can the compression zone of the concrete be utilized effectively, while their usability in the construction can be ensured by defining the maximum possible crack widths. Reliable measurements normally indicate the failure by visible deformation and increased crack width, caused by the plasticization of the reinforcing steel (state III).

The states I, II a, and II b can also be used to characterize the stress-strain relationship of textile-reinforced concretes (see Sect. 16.4.4, Fig. 16.24). State III is not applicable due to the material used. Nonetheless, sufficient predictability of ductile loading and deformation behavior of the textile-reinforced concrete, in contrast to short-fiber-reinforced concretes, allows as the creation of load-bearing components. The positive influence of short fibers on the crack formation and distribution can only increase the suitability of concrete structures, as well as imparting increased impact and abrasion resistances. The load-bearing capacity of the composite is less strongly influenced due to the process-limited fiber volume content of about 10 %, and the arbitrary orientation of the short fibers.

The acquired knowledge and experience in the formation of fiber-reinforced plastic composites (see Sect. 16.3) can hardly be transferred to the merging of inhomogeneous, brittle concrete matrices and ductile steel or quasi-ductile continuous fiber reinforcement into a composite. On the one hand, the latter combine ductile plastics with usually brittle fibers and require a very high fiber volume content of up to 75 % in order to obtain a good reinforcement. This approximately corresponds to ten times the reinforcement content of concrete structures. Depending on the application, the share of textile concrete reinforcements however makes up only 1–7 % by volume. On the other hand, fiber-reinforced plastics find themselves in an elastic deformation state during their application. An angular deviation of just 3° of the plastic reinforcement assembly to the loading direction has a serious effect on the loading capacity of the composite component. The permissible pseudo-plastic deformation of reinforced concrete creates more favorable conditions, since a fiber-based reinforcement can orient itself between the edges of the crack in the direction of loading. The load-bearing capacity of the continuous filaments used for concrete reinforcement decreases linearly to approximately 50 % for an angular deviation of 35° in comparison to the load-conforming orientation [62]. This experimentally determined dependency already takes into account the deflection forces at the crack edges, which cause a local transverse-compressive stress of the continuous fibers.

Conventionally designed construction concrete is not suited for the use as continuous fiber grid-like textile reinforcements. Recipes for textile-reinforced concrete require an adapted grain composition for the concrete matrix, as is the case in fine-grain concrete. The largest grains used have diameters of 1–2 mm and require high cement paste content due to the large wetting surface (Fig. 16.14). The mechanical properties of such concrete formulations correspond to those of high-performance concretes.

The concrete matrix must satisfy the requirements of chemical compatibility with the used fibers and their coatings. In case of textile reinforcement made from alkali resistant glass (AR glass), the high alkalinity of the concrete has an unfavorable effect on the long-lasting decisive bonding behavior, although it provides exceptional protection against corrosion of steel reinforcements. Therefore, composite and blast furnace cements with addition of pozzolanic contents are used preferentially. The latter also improve the grain structure as well as the processing

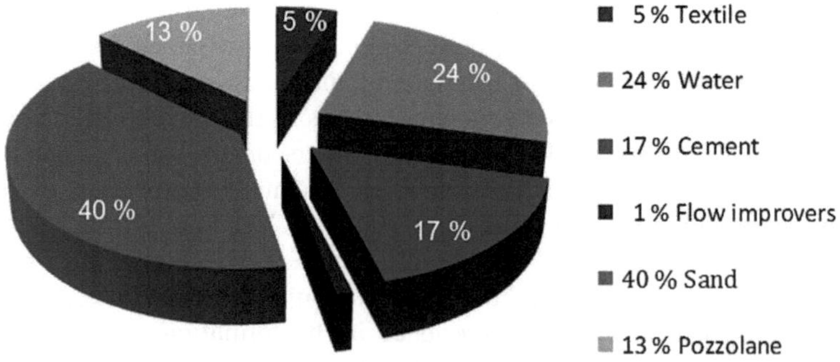

Fig. 16.14 Composition of textile-reinforced fine-grain concrete in volume percentage

performance of the concrete matrix during lamination, spraying, or injections. For these applications, the consistency of the fresh concrete can be suitably adjusted by various additional concrete mixtures.

16.4.2 Suitable Textile Fiber Materials

All continuous fibers are generally suited for the textile reinforcement in concrete, as they are characterized by a very good bonding behavior with concrete, high strength, and low fracture elongation. The Young's modulus of the fiber material should be well above the concrete matrix otherwise the strength of the component is significantly reduced when cracks develop. In addition, the fibers must be permanently compatible with the mineral matrix system both chemically (alkali resistance) and physically.

These requirements correspond best to the fibers made of AR glass and carbon. The development and production of textile concrete reinforcements is therefore primarily carried out on the basis of these two fibers. Basalt fibers can also be used, taking into account possible property fluctuations. Fibers made of polypropylene, polyvinyl alcohol, and polyacrylonitrile are used as short-fiber reinforcements in concrete [63]. Above all, short steel fibers have proven useful in special steel-fiber-reinforced concrete for various building construction applications. Typical properties of the relevant fibers are compared in Sect. 3.3.

The AR-glass and carbon fibers are available in the form of filament yarns or rovings and exhibit low material strain under tensile load. In order to ensure even distribution of load on all the filaments of the yarn in the final composite component, stretched and parallel filament orientation is necessary [64]. Therefore, yarns that are directly produced should be used for textile concrete reinforcement instead of unassembled yarns or direct rovings. In the textile manufacturing process, AR glass filaments are drawn from the interior of coreless bobbins. Therefore, the yarn undergoes a rotation around the yarn axis for each drawing cycle. On the other hand cheaper cross-wound bobbins with cylindrical coil core ensure a rotation-free drawing of the yarn material in the case of tangential drawing. These bobbins are available for carbon filament yarns in particular.

Hybrid yarns (friction-spun yarns, co-extruded yarn) that are specially designed for textile-reinforced concrete applications could not yet be used in textile concrete reinforcements. Other types of yarn, such as foil strips made of polypropylene, are used in applications where the fiber material has largely constructive functions, e.g. in the production or further processing of textile reinforcements. This allows a reduction of the use of highly expensive high-performance yarns can be reduced, which can account for circa 75 % of the material cost in the composite.

The use of AR glass and carbon does not require a minimum concrete cover for corrosion protection as is necessary in steel-reinforced concrete, since these fibers do not corrode under normal ambient conditions. Thus, very thin textile concrete layers are achievable. In comparison to steel bars the use of delicate textile

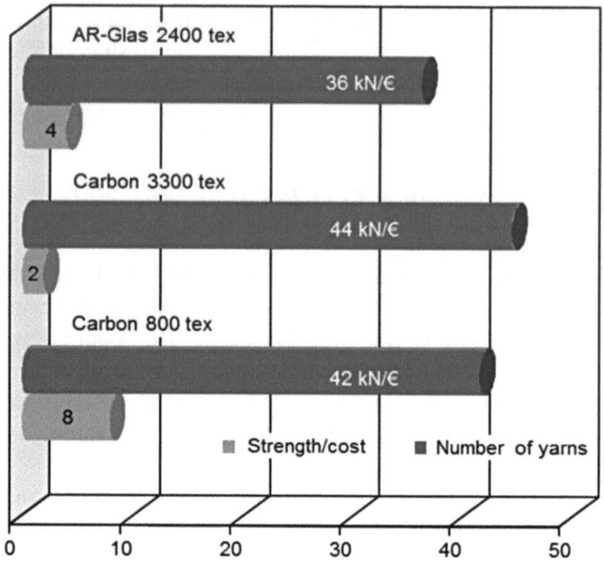

Fig. 16.15 Strength/cost relation and number of yarns to be inserted in a single concrete tensile specimen (diameter: 1 cm^2, length: 100 cm; reinforcement content in load direction: 3.5 volume percent)

reinforcement allows short anchorage lengths and very fine crack distribution because high bonding forces can be transmitted due to the large surface area of the yarns. Finally, the textile reinforcements made from AR glass or carbon with tensile strengths up to 2,500 N/mm^2 in concrete are significantly stronger than conventional reinforcing steel.

From a construction technological and economic point of view, it is advantageous if only few layers of textile reinforcement are introduced during concreting. Since the achievable reinforcement content or fiber volume fraction is limited by the building materials, the payload-related costs primarily determine the cost effectiveness of the reinforcement (Fig. 16.15). However, slight differences in fiber-related textile manufacturing costs are not taken into account here. Although the practice-relevant filament yarns with a count of more than 1,000 tex allow for a high fiber volume content, an increasing yarn count also leads to deterioration of the composite properties in the concrete. Furthermore, the more complex need for the secondary coatings (see Sect. 16.4.4) and the required enlargement of the load transfer lengths for the practical construction implementation of the textile-reinforced concrete should not be considered solely in terms of costs alone [65].

16.4.3 Basics of Production Technology for Textile Concrete Reinforcement

The continuous fiber-based concrete reinforcement in the form of textiles is suitable for both flat and curved structures. It allows for an effective implementation of varying reinforcement proportions in different loading directions under complex loads. The reinforcement can be inserted directly and oriented specifically in the tension zone of the components. From a technological point of view, the structural-property requirements of the reinforcement textile semi-finished products are characterized mainly by the feasibility of open, grid-like structures with sufficient strength and good drapability. The grid reinforcement structure is required for an adequate concrete flow capability and to ensure complete enclosure of the yarns with concrete. Hence, the grid openings must be three to four times larger than the diameter of the largest grain of the concrete matrix.

The advantageous tensile properties of the filament yarns must not be lost either as a result of mechanical processing or through inaccurate layer assembly in the textile. The comparison of stress-strain relationships between filaments, yarns, and textiles illustrated in Fig. 16.16 highlights the objectives associated with this requirement, namely the maximum utilization of the potential strength of the monofilaments in the reinforcement structure. The individual filaments fail successively during the tensile test of a yarn. Since the tensile strength for stress determination is based on the cross-sectional area of the entire filament bundle, the strength of the entire bundle is expected to be smaller than the average strength of the filaments. The yarn strength itself depends on the statistical fracture distribution, the stress distribution in the filament bundle, the rearrangement of existing as well as preliminary damages. The processing of the yarn into the textile structure (e.g. in stitch-bonding processes) results in filament breakage or yarn abrasion, yarn broaching, or lateral pressure in the yarn leading to a further loss in strength. The stress-strain curves of filaments, yarn, and textile also differ further due to non-ideal, parallel, and stretched filaments in the yarns and the textile structure.

In addition to the structural properties, the achievable productivity and the required production costs are further selection criteria for the fabric construction. Thus, only a few industrial processes qualify for the production of concrete

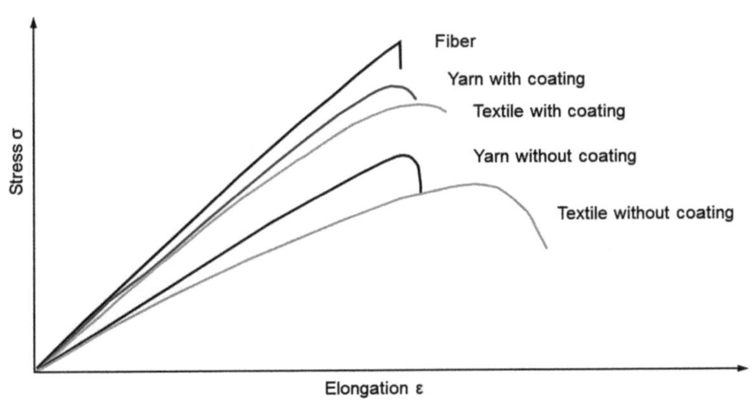

Fig. 16.16 Qualitative stress-strain plots of filament, yarn, and textile at comparable test conditions

reinforcements [66]. Currently established non-crimp textile production or knitting techniques are preferred.

Highest stiffness and tensile strengths in the textile reinforcement can be achieved with an exact and extended arrangement of high-performance yarns in warp knitted fabrics. Such structures are composed of several layers of biaxial to multiaxial oriented yarns or yarn groups, which are connected by a knitting yarn (Fig. 16.17). This reinforcement fiber group can have one of many designs with regard to angle and sequence of individual layers. The character of the grid structures can easily be adapted to the application by the variation of arbitrary fiber material layers and surface area as well as through the choice of bonding (see Sect. 7.4.2).

Alternatively, grid structures can be produced with the help of modern weaving process. Flat fabrics in the form of full or half leno fabrics are suitable for the reinforcement of concrete as open lattice structures need an increase in stability by having a fixed design. However, the major disadvantage of conventional leno fabrics is the principle-related undulation of the reinforcement yarns. A process enhancement of leno weaving (e.g. EasyLeno® 2 T, see Sect. 5.5.4) allows the formation of actual woven structure as delicate yarn systems which can be processed as undulation-free reinforcement yarn.

Three-dimensional warp-knitted spacer fabrics with two lattice surfaces provide a special type of textile concrete reinforcement (see Sect. 7.5). The outer surfaces of the spacer fabric are reinforced by means of biaxial reinforcement yarn insertion. In concrete applications the gaps between the outer surfaces range from 15 to 100 mm. The spacer warp-knitted fabric technology enables the production of continuous structures with varying surface gaps. In addition, the two outer surfaces can be made of different materials with varying reinforcing fiber density.

Stitch-bonded fabrics have a two-dimensional character. They can be used as additional reinforcement of existing concrete structures and for the implementation of textile-reinforced concrete components. The production of three-dimensional

(a) Biaxial non-crimp fabric (b) Leno fabric

Fig. 16.17 Biaxial grid structures manufactured on the basis of (**a**) extended stitch-bonding, and (**b**) weaving processes textile structures with integrated surface reinforcement is of importance for only a few special applications.

16.4.4 Composite Behavior and Load-Bearing Capacity of Textile Structures in Concrete

In concrete, the filaments along the yarn and within the yarn cross-section are of varying quality. Depending on their location in a given cross-section, the filament comes into contact with the fine-grain concrete to different degrees, drawing a distinction between inner and outer composite, or edge and core filaments. The outer matrix composite zone is featuring with matrix-enclosed filaments at the edges. The inner regions are only occasionally penetrated by hydration products of the concrete and hence participate in the load transfer only by means of a low friction in the composite. The resulting ideal strains in the edge and core filaments are illustrated in Fig. 16.18.

The individual outer filaments in the embedded state are only in partial contact with the concrete matrix and not with their entire surface. While the characteristics of these adhesion bridges depends essentially on the composition of the concrete, the displacement of the core filaments against each other, and the composite bonding significantly determine the activation for load transfer. This bonding in the boundary layer filament is generally insufficient and can be selectively adjusted by an additional coating of the textile reinforcement. Therefore, the core filaments in Fig. 16.19 exhibit different strain levels across the yarn cross-section, which is then balanced out with an increased coating level on the outer filaments. The influence of the coating on the properties of the interface filament/matrix improves the structural behavior of the composite significantly [68].

The load-bearing reserves activated by means of an additional coating for the textile concrete reinforcement are shown qualitatively in Fig. 16.16, as part of Sect. 16.4.3. In addition to the strengthening of the filament yarns, the coating also assumes the following functions:

- Additional stability of the grid structure in terms of the required processing behavior,
- Improved bonding in the concrete matrix,
- Reduction of capillary suction and of gas transport through the textile reinforcement, and
- Improving the long-term behavior of textile and textile concrete composites.

Aqueous polymer dispersions are used as coatings. These are based on self-cross-linking, carboxylated styrene-butadiene copolymers and epoxy resins with high molecular weight [69]. Likewise, resin impregnation is also suitable for the homogenization of the stress distribution in the yarn as well as for the composite bonding improvement in the concrete [70]. The coating formulations must be

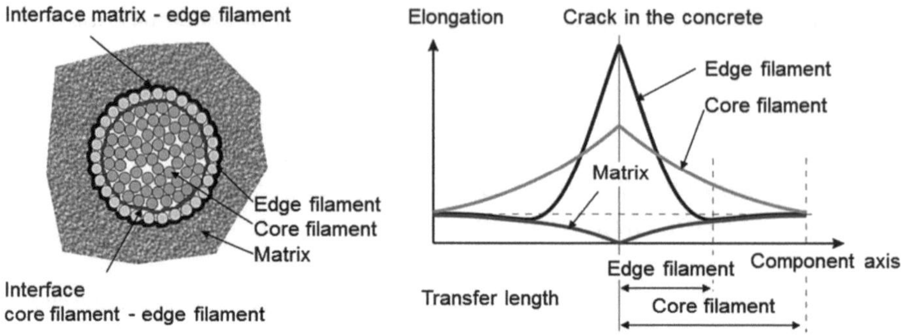

Fig. 16.18 Composite model and qualitative elongation behavior of outer filaments, core filaments, and matrix [67]

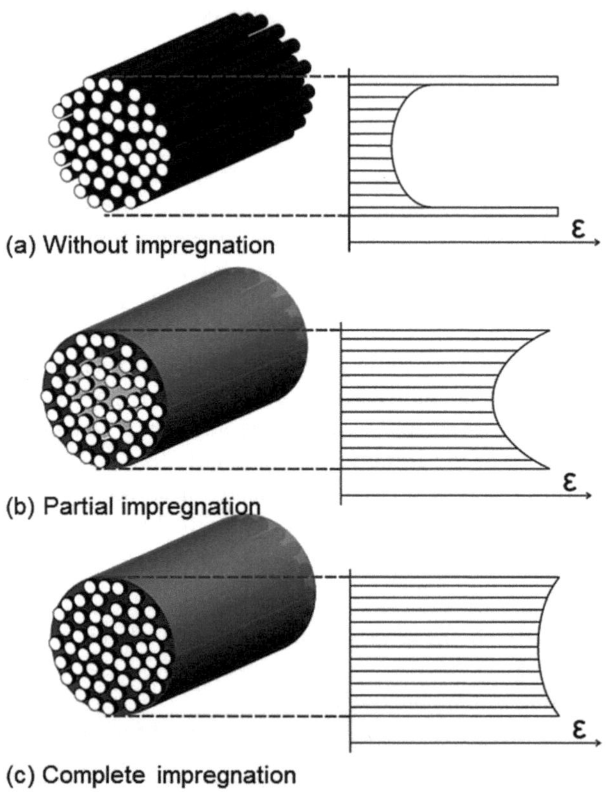

Fig. 16.19 Coating-dependent strain distribution in the filament yarn

selected according to both the primary coating and sizing of the yarn material as well as the cement or concrete binder used (see Sect. 13.3).

The best results for coating are obtained when yarns are optimally aligned, i.e., the fixation should preferably occur during the fabric formation process. For the

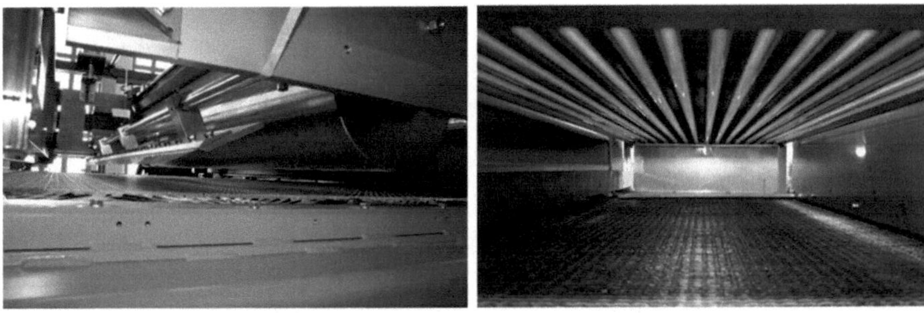

(a) (b)

Fig. 16.20 "Kiss Coater" roller application system, and carbon round rube infrared radiators. (**a**) Roller application. (**b**) Infrared radiators

application of appropriate fixatives, suitable coating systems with one or two rollers are available (Fig. 16.20). During the drying of the wet-coated grid structures, short heating and cooling times are required due to the dynamic nature of the textile manufacturing process and the high production speeds. Therefore, medium waved carbon round tube infrared heaters are particularly suitable for integration into the textile manufacturing process and can be custom-arranged unilaterally in combination with radiation converters for drying open textile structures [71].

The coating of the textile reinforcement in a separate process is technologically more elaborate and unfavorable, especially with regard to the required stretched alignment of the filament yarns. The sufficiently long designable hot air zone also allows very precise adjustment of reproducible drying conditions in an external coating process. The hot air zone, which can be designed to be of sufficient length in an external coating process, allows an accurate setting of reproducible drying conditions.

16.4.5 Influence of Textile-Technological Parameters of Stitch-Bonding (Multiaxial Warp Knitting) Technology on Composites

The reinforcement effect of open-grid fabrics or multiaxial warp-knitted structures in concrete is mainly due to the material properties of the reinforcement yarns used, and the stabilization as well as the bonding of the applied coating. The latter greatly influences the handling of the fabrics as well as their possible arrangement in the component, while these characteristics sometimes are partly contradictory to each other.

However, the behavior of the component under load depends on structure-related textile properties related to the stitch-bonding process:

- Machine gauge and layup gauge as well yarn feed,
- Lay-up arrangement and orientation,

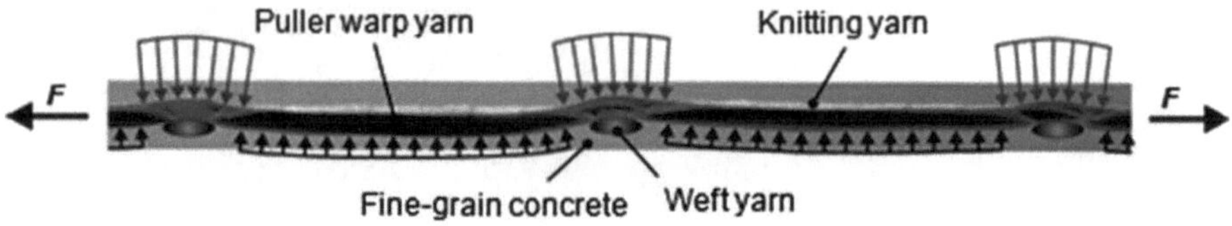

Fig. 16.21 Transverse tensile forces in concrete caused by deflections of the reinforcement yarns, at the risk of delaminations and concrete spalling [65]

- Course-oriented or non-course-oriented bonding of the reinforcing yarns,
- Tension of the knitting (bonding) yarn,
- Bonding type, and
- Stitch length.

The type of interlooping (e.g. right/left tricot), the binding of the reinforcement yarns (course-oriented and non-course-oriented), and the extension of the bonding process with a needle bar displacement connected to the two former textile-technological influencing factors (Sect. 7.2) are of extreme significance.

Technically, conventional stitch-bonding leads to a relatively small, but unavoidable ripple of the 0° reinforcing yarns (warp) in the intersection area with the 90° reinforcing yarns (weft). This results in cracks along the reinforcement yarns, and in concrete spalling under loads, eventually causing component failure (Fig. 16.21).

The bonding construction in the extended stitch-bonding [65] process allows a clearly more uniform warp yarn course (Fig. 16.22). The resulting dual-side bonding of the warp yarns leads to a compact yarn cross-section and creates larger grid openings. In addition, the distribution of the warp yarn material into an upper and a lower layer as well as a symmetrical arrangement of the layers of the yarns is possible. The ripple effect in yarns can be reduced to a negligible scale.

The reinforced concrete components with such textile structures are characterized by an increased stiffness and durability. The coating of textile reinforcement improves the bonding between the warp yarn and concrete resulting in a further enhancement of the composite properties.

Another way for the bonding to influence the handling and processing of the textile reinforcement and the load-bearing capacity of the composite is the course-oriented feed-in of the weft yarns in the stitch-bonding machines. In its basic design, these enhanced stitch-bonded fabrics are similar to the described existing fabrics and only differ in the course-oriented interlooping of the weft yarns. A piercing of the yarns is eliminated, so that the damage to the reinforcement material is much smaller. The producible structures are drapeable due to their course-oriented bonding, but also exhibit lower displacement stability. The stress-strain behavior in the warp and weft direction can be designed independently of the direction in comparison with the conventional bonding.

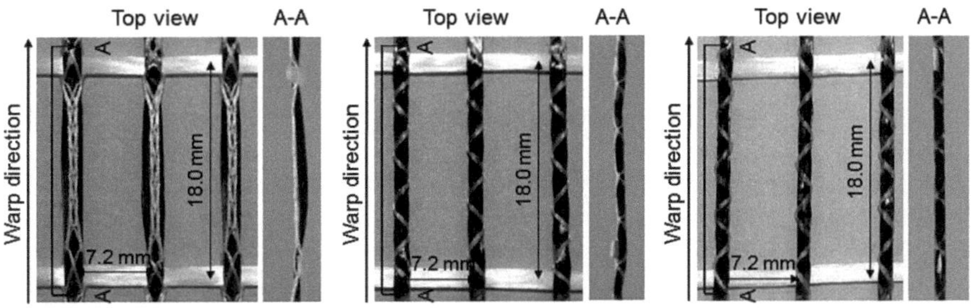

0°/90° Reinforcement yarns with conventional double tricot construction

0°/90° Reinforcement yarns with needle offset tricot, basic offset

Layer-symmetrical arrangement (0°/90°/0°) of reinforcement yarns with needle offset tricot construction, basic offset

Fig. 16.22 Constructive development of yarn layered stitch-bonding fabrics for concrete reinforcement (0° reinforcement yarns in warp directions)

16.4.6 Relevant Characteristics of Textile-Reinforced Concrete

Usually, it is not possible to conclude the final structural behavior of the composite material directly from the known properties of the individual raw materials of textile concrete. Hence, they will be described using experimentally determined parameters. These can be determined by means of monoaxial tensile tests on yarns or textile-reinforced fine concrete layers, as well as in yarn pull-out test (see Sect. 14.7.2).

The recorded data of elongation tests yields the average expansion of the textile concrete layer related to the tensile forces. The deformation is measured across several cracks along the direction of force. The forces are related to the expanding body or concrete cross-section A_c and interpreted as material strength $\sigma_{c,}$ or are considered in relation to the reinforcement area A_t, in order to obtain the tension σ_t of the reinforcement textile or the textile strength in the composite material:

$$\sigma c = F/Ac \tag{16.1}$$

or

$$\sigma t = F/At \tag{16.2}$$

The resulting stress-strain relationships (Fig. 16.23) are an essential prerequisite for the design and dimensioning of the composite material as well as for the derivation of their characteristic parameters, scatter, and partial safety factors.

The reinforcement surface area required for the evaluation of yarn in the elongation test $A_{t,Yarn}$ can be determined by customary conventions of the textile

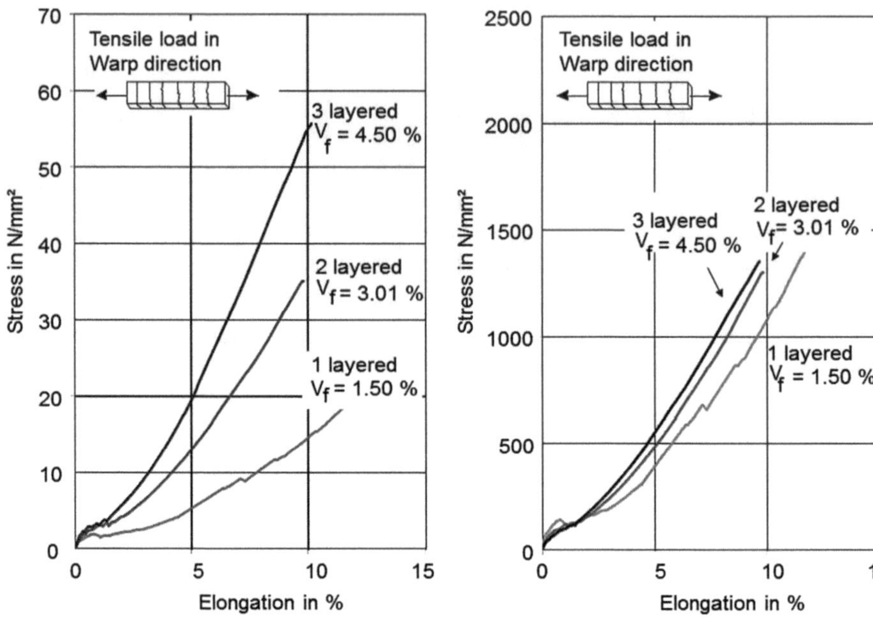

Fig. 16.23 Stress-strain relationships of expansion elements with textile reinforcement made from carbon filament yarns of 3,300 tex fineness (warp yarn spacing: 10.8 mm, weft yarn spacing: 18.0 mm, mass per unit area of the textile: 609 g/m^2 including 15 mass percent for polymer coating)

industry, such as yarn fineness or linear density, using the values of the uncoated yarn:

$$At, Yarn = \frac{Yarn\, fineness}{Fiber\ material\ density} = \frac{Tt\ Yarn}{\rho\, Fiber} \tag{16.3}$$

The total cross-sectional area of the textile reinforcement A_t results from the number of layers and the axial spacing of the reinforcement yarns:

$$At = Number\ of\ layers \times \frac{Component\ width}{Yarn\ distance} \times At, Yarn \tag{16.4}$$

Alternatively a reinforcement area per meter can be defined in the longitudinal and transverse directions for a textile reinforcement structure in the component.

Instead of theoretical values, such as the stress and strain values, input values (e.g. Young’s moduli and strength of the textile reinforcement) can be derived from experimental tests on composites and these can be used for the design.

In addition, a trilinear approximation of the stress-strain relationships (Fig. 16.24) with the characteristic states I, II a and II b is possible (see

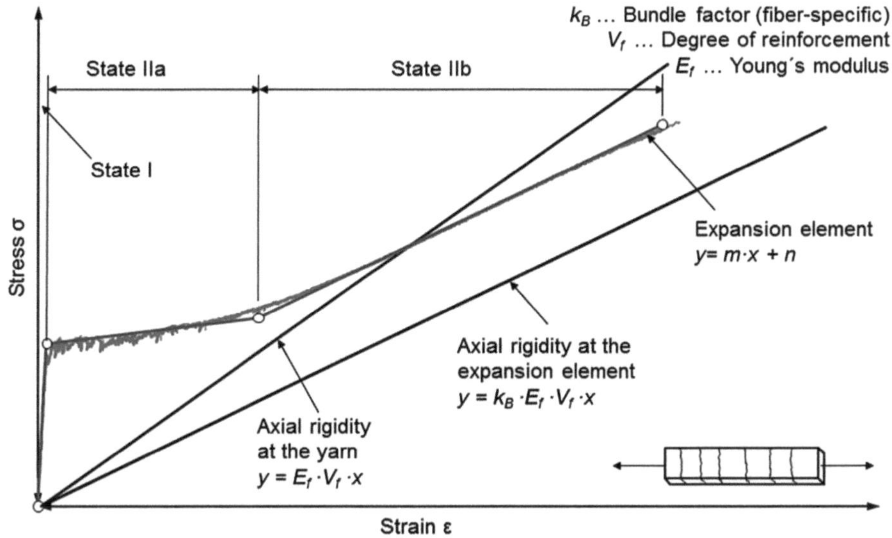

Fig. 16.24 Approximation of the stress-strain relationship of a monaxially tensile-loaded expansion element made from textile concrete [67]

Sect. 16.4.1) for a specific configuration of textile reinforcement and fine concrete. The iterative measuring of textile-reinforced concrete layers can be performed by establishing the approximated, linearized relationship between textile breaking strength and medium textile deformation of the composite material [72].

Apart from the experimental characterization, a variety of mechanical, analytical, and numerical models are available for the description of the load bearing capability of textile-reinforced concrete. Different structural levels are selected depending on the model concept, observation scale, and descriptive parameters (e.g. filament or yarn properties, composite properties). Models at the macro level consider the composite as a homogeneous replacement material in which damage and failure mechanisms are mapped only on a small number of model parameters. Mesomechanical modeling considers concrete matrixes, composite and reinforcing elements, which can also be classified into groups. At the micro level it is necessary to formulate the basic mechanisms between filaments and the constituents of the concrete. Currently, universal multi-scale models for the highly complex structural behavior of textile-reinforced concrete, considering various factors und their mutual influence, are prepared. A related approach is a micro-meso-macro forecasting model based on the simulation of the microstructure behavior, predicting the macroscopic material behavior of the composite material [73]. Thus the mechanical parameters characterizing the textile-reinforced concrete (Fig. 16.25) can be derived from the experimentally determined force-deformation dependencies.

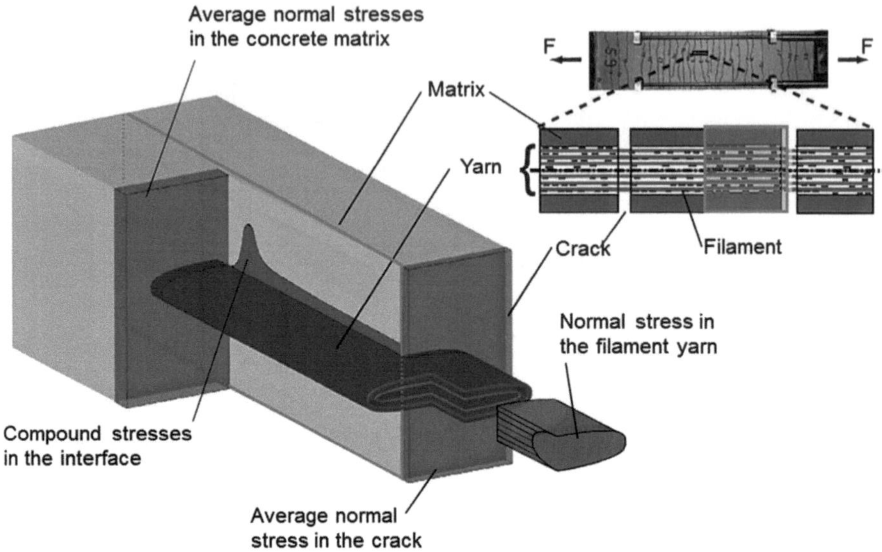

Fig. 16.25 Stress distribution and compound force transfer in the yarn-matrix model [73]

16.4.7 Repair and Strengthening of Existing Buildings/ Constructions

The strengthening or repair of steel-reinforced concrete structures is carried out locally or for the entire building. Shotcretes with reinforcement or bonded reinforcement made of steel as well as fiber plastic composites are used for this purpose. The use of textile concrete reinforcements (Fig. 16.26) combines the advantages of shotcrete with the advantages of adhesive reinforcements [72]. Material-related disadvantages such as the high weight of shotcrete reinforcements or moisture sensitivity, the lack of fire resistance, and high implementation cost of adhesive reinforcements are compensated. This makes reinforcing textile concrete a promising alternative and supplement to conventional reinforcement methods and increases the repair options for steel-reinforced concrete.

The corrosion resistivity of the textile reinforcement materials allows for a thin reinforcement layer. Four layers of textile reinforcement and a maximum grain size of 1 mm result in a reinforcement thickness of only 12–18 mm in the concrete. The weight of the old component is only slightly increased. Textile reinforcements are also easy to shape and can be adapted to almost any component geometry. Thus, profiled cross-sections, columns, or shell-like components can be repaired. Textile-reinforced concrete as reinforcements is an obvious choice only when an enhancement of the load-bearing area with reinforcement leads to an increased load bearing capacity (Fig. 16.27). The degree of reinforcement can be selected by varying the reinforcement content through the number of layers and the fiber cross-sectional area per layer.

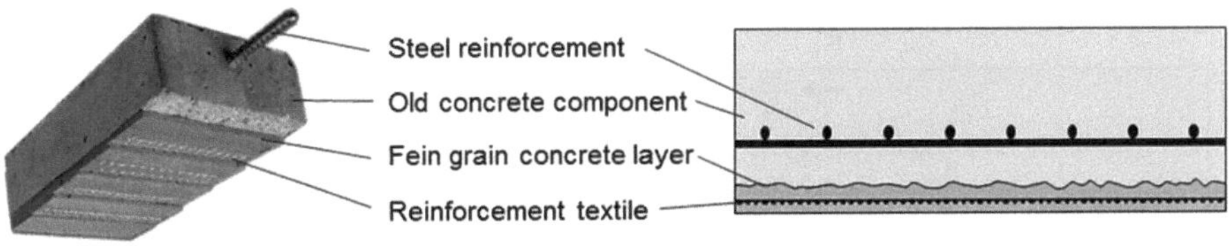

Fig. 16.26 Structure of a component refurbished with textile concrete

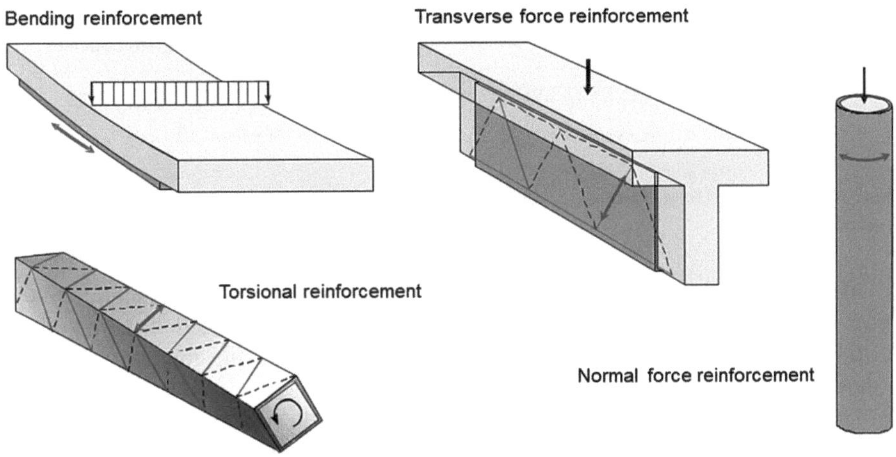

Fig. 16.27 Possibilities of textile-reinforced concrete reinforcements

Textile-reinforced steel concrete structures can fail in various ways depending on the geometry, loading, and reinforcement content. The components are often designed specifically for a bending failure by appropriate design measures. This form of failure in comparison with the other failure mechanisms allows a ductile component behavior and thus exhibits an early failure indication by cracks and deformations. The textile reinforcement structure and the reinforcing steel in combination act as mixed concrete reinforcement in the textile-reinforced steel concrete components subjected to bending stresses. Generally, it is possible to apply the classical reinforced steel and already established design methods during the interaction of textile and reinforcement steel (Fig. 16.28). However, the various lever arms of the reinforcement must be considered [72]. Apart from increasing the bearing capacity, further repair tasks, such as the production of a new or additional dense concrete cover, the design of surfaces, the stiffening of structures, or the influence of crack formation can be realized.

For the application of the individual textile-reinforced concrete layers into a single reinforcement layer, simple and reliable application methods can be used. Another advantage over lamellar reinforcements lies in the achievable planar force transmission. The risk of failure due to delamination is thereby significantly reduced.

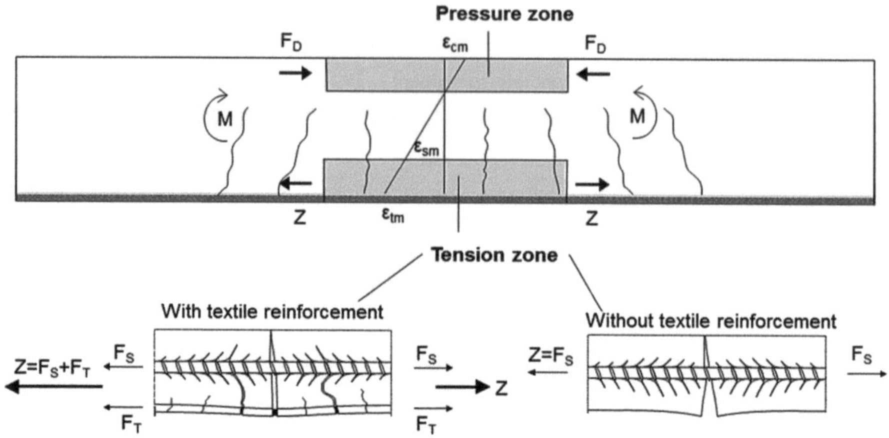

Fig. 16.28 Bending load bearing capacity and tension rod analogy of textile-concrete reinforced components [72]

October 2006 saw the first refurbishment with textile-reinforced concrete made from carbon filament yarn in three layers (Fig. 16.29), when a shell structure of the lecture hall building at FH Schweinfurt was refurbished using this material. The renovation of the structure was necessary because the steel-reinforced concrete structure of only 8 in. thickness showed overloading in the upper steel reinforcement layer of the projecting areas of the shell. Since the textile-reinforced concrete is not yet a standardized construction material, an individual approval was required for the use of the textile reinforcement.

Two years after this successful application, another textile concrete reinforcement could be executed in Zwickau. A barrel-shaped roof structure of reinforced steel concrete was constructed in 1903. Since the viability of this design is not verifiable according to current standards, additional reinforcement was required for the conversion and continued use of this building. The use of a pre-mixed fine concrete mix in large quantities enormously reduced the cost of dosing and mixing on-site. The installation of the textile concrete reinforcement layer in an overhead position (Fig. 16.30) was another special feature of this application. This specially developed application technology was also used to upgrade several 1,000 m^2 of ceiling area in an office and retail building in Prague as well as in a production building in Koblenz with two to four layers of fabric made from carbon heavy tows.

16.4.8 Formation of Textile-Reinforced Individual Components

Due to the corrosion-resistant textile reinforcement and the associated low concrete cover requirements, the production of very thinly walled concrete structures is

Fig. 16.29 Reinforcement of the hypar shell structure at FH Schweinfurt

Fig. 16.30 Installation of the textile concrete reinforcement layer

possible. Therefore, TRC is ideal for lightweight, yet highly durable concrete structures, e.g. as facade, roof, and balcony elements, noise barriers, city furniture, or for tank and pipeline construction.

In addition, spatial shapes can also be realized. Different application examples with a general building authority or equivalent approvals in certain individual cases are available here.

The first building approval for a component made of textile-reinforced concrete was issued for a textile-reinforced, thin-walled facade panel (Fig. 16.31) known as betoShell®. The required certification was carried out at the TU Dresden. The precast elements have a thickness of just 20 mm. In addition to significant weight conservation and resulting ease of structure mounting, commonly found surface structures of conventional reinforced concrete facade panels (washed, acidified, blasted or polished surfaces) can be colorized easily, at smaller cost and high diversity due to lower requirements of color pigments. In addition, the use of fine-grain concretes with a maximum grain size of 1–2 mm makes it possible to create very smooth and dense structural surfaces (smooth exposed concrete) as well as sharp-edged profiles and joints, leading to a completely new appearance of concrete surfaces.

Fig. 16.31 betoShell® textile concrete facade elements (*Source*: Ulrich van Stipriaan)

The dimensions of the textile-reinforced concrete facade elements are restricted to 2.40 m × 1.20 m by the technical certification (DIBt Z-33.1-843). Larger elements can be used after approval according to each individual case. Such large-scale, self-supporting elements from textile-reinforced concrete were installed for instance at RWTH Aachen in the form of a sandwich facade system of approximately 590 m^2. 15 mm thin textile concrete layers were applied to both sides of a load-bearing rigid foam core (Fig. 16.31). The element dimensions are 3.45 m × 1.0 m × 0.18 m.

The world's first bridge made of textile-reinforced concrete was created in 2005 for the State Horticultural Show in Oschatz, Germany. In the fall of 2007, a second 17 m-long pedestrian and bicycle-bridge, strong enough to carry a bulldozer was presented to the public in Kempten, Germany (Fig. 16.32). Both bridges were initially developed at the TU Dresden and consist of U-shaped textile-reinforced concrete segments joined by tendons to form a supporting structure. The weight of the textile concrete structures is only one-third of the weight of conventional steel or pre-stressed concrete bridges. Due to the light weight, the entire superstructure of the bridge can be assembled and transported to the construction site with a single truck and lifted by mobile crane.

The longest textile-reinforced concrete bridge so far, at circa 100 m in length, was built in the summer of 2010 in Albstadt, Germany as a six-span continuous beam with pre-stressed seven-web tee beam cross-section (Fig. 16.33). The six precast elements have a length of up to 17.2 m and spans of up to 15.05 m. Textile fabric reinforcements, GRP reinforcing bars, and monostrand braces are used for the linear load transfer in the bridge structure [74].

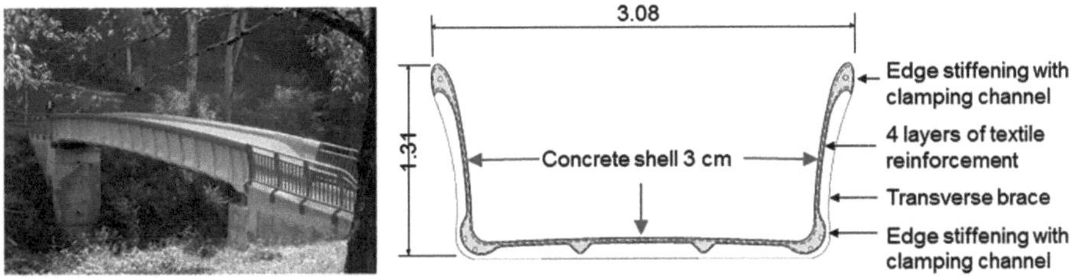

Fig. 16.32 Cyclist and pedestrian bridge in Kempten, Germany

Fig. 16.33 Pedestrian bridge in Albstadt, Germany (*Source*: Groz-Beckert KG)

16.4.9 Development Trends for Textile-Reinforced Concrete

Textile-reinforced concrete for the reinforcement of steel concrete structures bridges the gap between the commonly used carbon fiber adhesive reinforcements and conventional short-fiber-reinforced shotcrete. The feasibility of thinly layered textile-reinforced concrete and their free shapeability in the manufacturing process enable the production of finer and lighter, textile-reinforced precast concrete elements with varied design. The required technology for the production of textile concrete reinforcement and the basis for a material-specific application of TRC are available. Currently, the existing application difficulties are caused by the open questions concerning the interaction between concrete steel and textile reinforcement as well as the lack of building inspectorate regulations. Thus, the task is no longer a question of just developing a suitable design model for textile-reinforced concrete, but also involves a quantification of the relevant input parameters taking into account safety aspects by means of testing. Here, the definition of quality characteristics for the entire textile manufacturing process chain of (filament, yarn, textile, and coating) as well as the test methods for the determination of yarn parameters in conjunction with fine concrete matrix, as described in Sect. 14.7.2, are of great significance.

A subsequent coating of the filament leads to a more uniform utilization of all filaments. Particularly in case of AR-glass the yarns are also protected against the alkaline environment and a poor morphology of the boundary surface with the concrete. In order to achieve a practical application in construction, material-specific adjustments to the coating formulations, e.g. for carbon, basalt, and steel fibers must be made, whose primary coatings are currently being optimized for use in fiber plastic composites. Furthermore, it is necessary to carry out standardization of suitable textile designs and to specify textile fine-grain concrete matrix combinations as well as to detail the requirements for their conformability (for example, limits of the crack width and spacing) and largely integrate them into the already developed assessment bases.

The conventional manufacturing technologies for producing grid-like reinforcing textiles are expanded radically through the introduction of new textile machinery concepts for the stretched and symmetrical yarn placement with integrated coating and downstream contouring. With newly available binding variants, especially for the textile concrete, advantageous arrangements of reinforcement yarns as well as a more favorable bond behavior and significantly improved quality of the textile reinforcement can be achieved.

For the application of TRC, both European as well as country-specific and regional approval regulations for construction products have to be considered.

16.5 Textile Membranes for Lightweight Constructions

16.5.1 Definition and Applications

Membranes for lightweight construction provide an innovative, generally thin, flexible, and loadable material that can be designed by choice of material and construction to fulfill many different functions, such as cutting, filtering, wrapping, load bearing and distribution as well as weather, sound, and heat protection. This results in a wide range of applications, from membranes for textile construction to sails for yachting, including sun protective textiles, billboards, tents, swimming pool and pond liners, containers, conveyors, oil barriers, bellows, geotextiles, truck tarpaulins, convertible tops, airbags as well as inflatable boats and life rafts. Figure 16.34 shows various roofing solutions for the use of membranes.

From the various operating conditions, specific requirements are derived which can often only be met if the can act as strength carriers in the form of a textile surface—in contrast to film structures, such as pneumatically preloaded air cushions made from ethylene tetrafluoroethylene copolymer (ETFE) film [75]. These strength carriers are surface treated according to the application requirements, for example by coating or lamination. Such fabric structures, which will be referred to as textile membranes below, form the basis for a variety of novel technical lightweight solutions, especially for high-performance application. These mainly include the application of textile architecture in construction and also the realization of mass-reducing construction structures for vehicle construction as well as for the sports and leisure sectors.

Fig. 16.34 Application examples for membranes in roofing (*Source*: Mehler Texnologies GmbH)

Thus, textile membranes can be used as architectural roofing solutions for outdoor use (Fig. 16.34). These can be configured in a multitude of ways. Examples include permanent and semi-permanent roofs, movable roof surfaces, subsequent adaptations of existing buildings such as canopies or walkways, and also temporary, reusable designs for exhibition stands, exhibition pavilions, tents, protective systems against hurricanes etc. [76, 77].

The construction of such membranes is increasingly popular with architects and planners, as indicated by a host of impressive large objects. Among these are the "Grand Pavilion Showgrounds" exhibition center in Melbourne (Fig. 16.35, 10,000 m^2 of built-up membrane surface), the international Airport in Bangkok (100,000 m^2 of membrane surface), three stadiums for the Football World Cup 2010 in South Africa (Green Point Stadium Cape Town, Moses Mabhida Stadium Durban, Nelson Mandela Bay Arena in Port Elizabeth) or the Haj Terminal of the international Airport in Jeddah, Saudi Arabia (440,000 m^2 covered surface) [77, 78].

Higher rates of innovation and growth opportunities are predicted for this type of building in the future [79, 80], due to the fact that these extremely filigree textile membrane structures or substructures [81] can help achieve considerable mass reductions, for example in the field of roofing and facades, which allow the realization of large spans of more than 200 m with. In addition, they offer attractive detail solutions and creative effects, such as dyeing or printing of materials [82]. Furthermore, new approaches for the implementation of free, organic forms

Fig. 16.35 Membrane edifice "Grand Pavilion Showgrounds", Melbourne (*Source*: Mehler Texnologies GmbH)

instead of such a simple geometry convey further the advancements in membrane structure [77].

In addition to outdoor use, textile membranes have experienced widespread distribution in interior design in recent years due to their multi-faceted design options. Here, the acoustics and lighting effects are most important, e.g. for ceiling, wall or room dividing elements, where they minimize noise pollution and provide a pleasant lighting [82, 83].

Examples of lightweight solutions in the automotive, sports, and leisure sectors are sliding roofs for trailers, sail cloths for water vehicles, hot air balloons, gliders, and kites [84, 85].

The current wide range of applications allows is indicative of the further potential of textile membranes, which has to be tapped by further developments in the textile strength carriers, including surface functionalization. Application-specific requirements always create new requirements for the membranes to fulfill such as climate and acoustics control [86]. The requirements on such structures and their implementation and installation will be illustrated using the example of membranes as a construction material, referred to as construction or engineering membranes.

16.5.2 Requirements for Construction Membranes

An evaluation of the technical literature lists the essential requirements of construction membranes relating to protection, permanent resistance, processability and mechanical properties, as follows [87]:

- High tensile strength at low elongation and surface mass [80], high tear and cut resistance [80, 88],

- Low flammability, achieving fire protection class A2, B1, or B2 according to DIN 4102 [82],
- High weather resistance including resistance to hail or sandstorms and UV radiation [80, 82],
- High biopersistence (mold, microbes) [87],
- High fatigue resistance to aging, wear and tear, dirt, failure [87],
- High water and gas tightness [87],
- Excellent low and high thermal stability [80],
- High resistance to buckling [87],
- Good welding, sewing capability, printability, ability to connect to solid components [87, 88],
- Self-cleaning surface properties [80],
- Optical properties, such as adjustable light transmittance, low tendency to yellowing, and color fastness [87],
- Recyclability [88], and
- Noise and thermal insulation properties [89].

It should be noted that the requirements of the construction membranes increase or decrease depending on the application area and can vary in importance. As an example, the requirements of membrane structures for sports facilities can be considered, which significantly differ from those of facade applications.

16.5.3 Properties of Construction Membranes

16.5.3.1 Basic Structure

Construction membranes are usually made of textile reinforcements coated on both sides. They can exhibit additional surface coatings depending on the requirements (Fig. 16.36).

The bond between textile carrier and coating (main coat), and between coating and surface finish is supported by an adhesion promoter. These contribute to the full utilization of load bearing capacity of the support structure. Both the coating and the special surface coating protect the load-carrying member and provide the membrane with the properties needed from the application perspective. The coating primarily contains such additives, as among them UV stabilizers, agents for increasing fire resistivity, coloring agents and fungicides while the surface coating facilitates cleaning of the membrane [88].

16.5.3.2 Raw Materials, Textile Designs and Composite Formation

The primary raw fiber materials for load-bearing textile surfaces of the membrane are glass, polyester, polyamide, aramid (Nomex) and polyethylene (Dyneema)

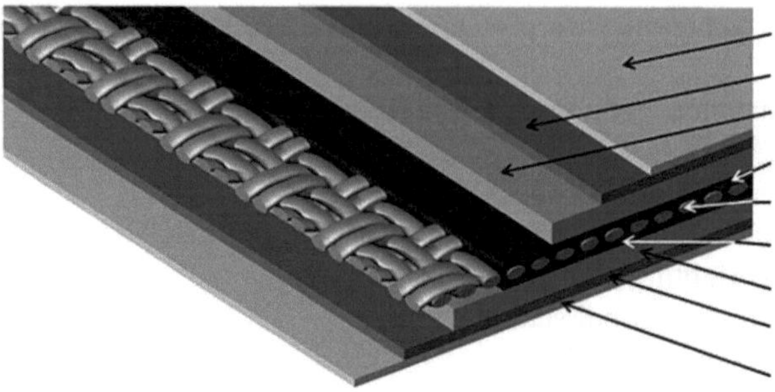

Fig. 16.36 Principal structure of a construction membrane using a carrier fabric (*Source*: Mehler Texnologies GmbH)

fibers, which are used in the form of rovings or filament yarns with different porosities [88, 90–93]. For example, typical polyester filament fineness varies from 110 tex to 220 tex.

In the coating or lamination of the textile strength carrier, polyvinyl chloride (PVC), polyethylene (PE), fluoroplastics, such as polytetrafluoroethylene (PTFE) and polyvinylidene fluoride (PVDF), silicone, polyurethane (PU), as well as natural and synthetic rubber in paste, dispersion, or film form are normally used (see Sect. 13.5.3.2). The additionally applied protective coatings, for example, consist of acrylic or polyvinyl fluoride (PVF) [80, 88, 93]. Various machines are available for the realization of the coating or lamination, discussed in detail in Sects. 13.5.3.3 and 13.5.3.4. For the application of a PVC coating in the form of a paste, the coating machine shown in Fig. 16.37 is used. For the application of varnishes, a coating module can be combined with a coating machine (Fig. 16.38).

The stress-strain behavior of the finished membrane depends on the design and production of the textile load-carrying member using the coating or laminating machine. The clamping of the fabric in the edge region ensures that approximately identical stress distribution are achieved in the longitudinal and transverse directions during processing, resulting in an almost homogeneous stress-strain behavior of the end membrane. Without the lateral guidance, later application of forces leads to differences between the longitudinal and transverse strains, which must considered in the implementation of membrane structures.

Woven fabrics have so far dominated (see Chap. 5), mainly in plain and panama weaves. The membranes produced from these fabrics have found widespread application with the following material combinations and their characteristic properties:

PVC-coated closed or open polyester, polyamide and aramid are particularly suitable for convertible and mobile structures due to their good breaking resistance [77, 80, 82], sometimes with lacquered or laminated outer surfaces. They are used as weathering, soiling, and embrittlement protection [88, 90, 92, 94–97]. Polyester fiber fabrics with PVC coating still form the bulk of textile canopies.

PTFE-coated or laminated glass, aramid, and polyester fiber fabrics have anti-adhesive and self-cleaning properties. The also have a very high durability, are

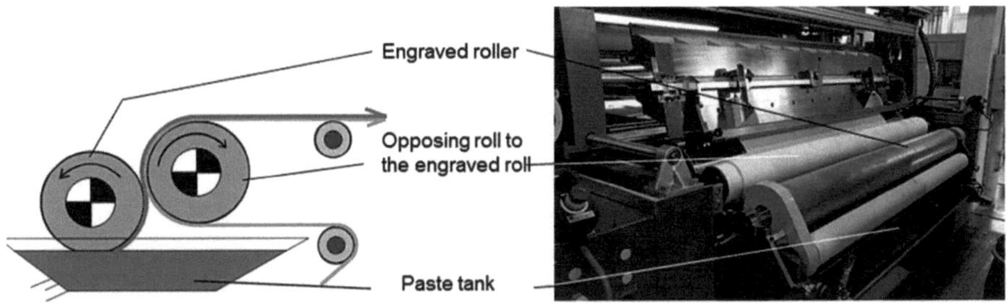

Fig. 16.37 Coating machine (*Source*: Coatema Coating Machinery GmbH)

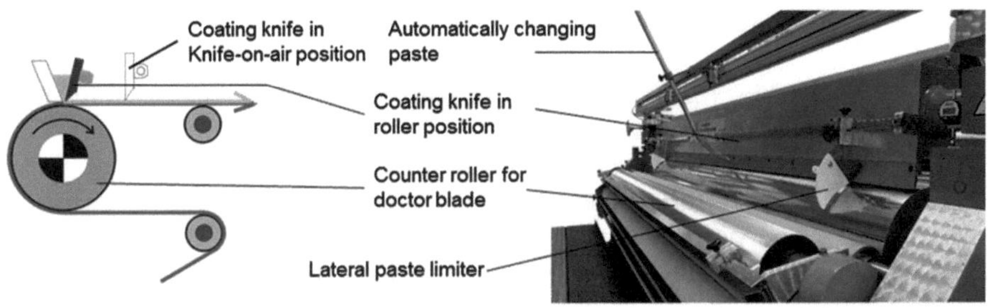

Fig. 16.38 Coating module in combination with coating machine (*Source*: Coatema Coating Machinery GmbH)

relatively hard, show good shear strength and are kink-sensitive [80, 82, 90, 94]. The coating is often executed on both sides to achieve water tightness, high aging resistance, and tensile strength [81, 91]. Such membranes very often have translucent properties [96].

Tetrafluoroethylene-hexafluoropropylene vinylidene fluoride (THV) coated polyester fibers and ETFE fabrics, despite their high cost, are the optimal solution for convertible structures, on account of their extreme kink resistance, dirt-repellent properties, and high abrasion resistance [80, 94].

Fluoropolymer film laminated fiberglass meshes have particularly high light transmittance [77].

Silicone coated fiberglass fabrics with varying fiber densities [77, 80, 82, 93], are characterized by a high degree of translucency, good weather resistance, and varied colored designability [96].

Polyurethane-coated polyester and glass fiber fabrics, sometimes with surface sealers made of acrylic, PVDF, and tetrafluoroethylene-hexafluoropropylene (Teflon FEP), are extremely flexible and last longer [92, 95].

Further developments of the carrier fabric include the processing of aramid filaments and PTFE yarns in order to increase tear resistance, in addition to the crack blocking [98]. These are integrated into the woven fabrics as linear, exposed filament yarn stretches (floats, see Sect. 5.6.3) and, if necessary, can be made visible

Fig. 16.39 Examples of woven fabrics with integrated floats

on one side only (Fig. 16.39). Although the tear strength can be increased fivefold by this, the mass per unit area of the membrane increases only slightly.

Sound insulation is improved in combination with so-called acoustic nonwovens [99]. Further ongoing efforts focus on the application of a hybrid fabric consisting of PTFE yarns, and various mono or multi filaments, such as PVDF, ETFE, FEP, THV, and woven fabrics from new fluorpolymer filaments, which are coated with the same polymer [100]. The latter are much easier to recycle.

As an alternative to conventional fabrics, open or closed leno fabrics, for example made from polyester filaments yarns are used. In these, the warp and weft yarns are connected together by leno yarns [101], achieving a largely straight/linear warp and weft yarn paths. Furthermore, high-strength warp-knitted fabrics made of polyester filament yarns with weft insertion in the longitudinal, transverse, and diagonal directions, e.g. the so-called DOS structures (Directly Oriented Structures) are used in various applications [102]. Open grid-like multiaxial warp knitted fabrics made of aramid and polyester filament yarns, produced according to the stitch-bonding technology (see Sect. 7.4.2), are also used as reinforcements [84]. Non-crimp fabrics are also used as carrier structures for membranes consisting of unbonded yarns laid over the other [87, 95, 103]. In this context, it should be noted that the high strengths allowed by the stretched filament arrangement can be used to achieve open-grid structures. A coating with PVC is one typical possibility for such textile carriers. Nonwovens as well as woven and knitted spacer fabrics are possible textile carriers, for instance for rubber coatings. However, they offer less favorable conditions in comparison to previously described textile structures [99, 104], i.e. they cannot achieve maximum strengths while maintaining small masses per unit area. They are rather suited as sound-insulating middle layer in multi-layered membrane structures.

16.5.3.3 Characteristics of Construction Membranes

In order to quantify the demands placed on the characteristic parameters of construction membranes (see Sect. 14.7.1), the following ranges or limits are usually used:

- Thickness: 1–4 mm [82],
- Surface mass: 300–2,000 g/m^2 [82, 88–90, 92, 93, 104],
- Maximum tensile strength: 200 N/5 cm–20,000 N/5 cm, with different values are present in the warp and weft direction (Figures are due to the small thickness across the width in N/5 cm.) [88, 90, 92, 93, 99, 104],
- Elongation: 2–35 % [93, 99],
- Tear force: 150–2,000 N [88, 90, 93],
- Quantity for coating: 10–1,500 g/m^2 [80],
- Adhesive force: 75–150 N/5 cm [93, 98],
- Buckling behavior: $\geq$50,000–100,000 breaks [90, 92],
- Resistance to cold: -20 to -50 °C [90, 93],
- Thermal resistivity: 70–180 °C [90, 93],
- Light transmission: 2–50 % [88, 93, 105], and
- Fire protection corresponding to fire classification for building A2, B1, B2 [93].

The extremely broad parameter ranges are a result of the wide variety of applications, which clearly indicates the multitude of realized construction membranes. This is an impressive demonstration of the extensive range of application-ready materials available to membrane construction businesses. In addition, the efforts of the membrane manufacturers are focusing on offering structures according to the application requirements.

16.5.4 Realization of Membrane Structures

Textile membranes for architectural applications can be realized as stressed, frame-supported, or suspended constructions [88]. Figures 16.40 and 16.41 show selected sample applications.

Based on the construction design, distinguished distinction is made between single- and multi-layered constructions, with the former being suitable only for structures without stringent construction-physical requirements, and the latter being able to fulfill functional requirements like insulation [82].

In preparation of the assembly of the individual membranes, the individual sections are transferred into a three-dimensional seam pattern using special computer-based sewing methods. Taking into account the required level of pre-stresses as well as the additional components like welds and edge details, the determination of the final part size and the cutting is performed by means of a digitally controlled cutter. For the welding under tension, high frequency or thermal

Fig. 16.40 Mechanically tensioned structure (*Source*: Ceno Tec GmbH)

Fig. 16.41 Suspended construction (*Source*: Ceno Tec GmbH)

welding are used, depending on the membrane material and application, with only the coating being welded [82]. The pattern pieces can also be joined by sewing or glueing.

The connections of construction membranes to the supporting structure are created by linear clamping of welded edge welts using clamp profiles, or by means of cables or high-strength belts [82]. The support structures used for the application are, for example, structures made of steel, filigree steel structures from trusses, curved ring carriers, internal poles, and external tensile and compression bars with rear tensioning, as well as cable structures [78, 90, 106].

In order to reduce the amount of effort in planning and constructive implementation of membrane structures, efforts are being made to standardize the membrane modules, thereby creating modular membrane structures and adaptive building systems [81, 94]. A special emphasis in their development is placed on equipping the membranes as multifunctional room shells [94]. So-called low-E coatings (low emissive coatings) on the membranes, for examples, reduce heat radiation and minimize heat losses [77, 94]. In order to improve the thermal buffering, latent heat storage based on phase change materials (PCM) is used [86]. Furthermore, a combined use of membrane, isolation, and reflecting layers reduces the heat conduction and convection as well as heat radiation [105]. With respect to energy generation, the membranes can be expanded into collector surfaces or flexible photovoltaic elements by means of capillary systems [99]. With the woven insertion of electroluminescent fibers, the application of pigmented particles changing color under UV radiation, or the lamination of flexible OLED surfaces, new lighting functions can be realized in the future. These heated, shape-changing, or self-repairing materials will be used in further applications [94]. These trends reflect the general tendency toward the functionalization of technical textiles.

A completely new approach when building with membranes is to develop modular multi-layered textile building shells with adaptive properties as a variation of modern shell structures [107]. Such systems must have the highest possible adaptability, e.g. to sunlight, humidity, and wind loads, while at the same time retaining an ability to produce and save energy. Reaching a favorable recycling behavior with minimum material consumption is another objective. Such adaptive buildings provide a fundamental contribution to the advancement of construction technology [108].

References

1. ANONYMOUS: *Handbuch Faserverbundkunststoffe*. Dresden, 2009
2. ERMANNI, P.: *Composites Technologien (Skript zur ETH-Vorlesung 151-0307-00 L, Version 4.0)*. Zürich, August 2007
3. NEITZEL, M.; MITSCHANG, P.: *Handbuch Verbundwerkstoffe*. München, Wien : Carl Hanser Verlag, 2004
4. MÄDER, E.: *Grenzflächen, Grenzschichten und mechanische Eigenschaften faserverstärkter Polymerstoffe*. Dresden, Technische Universität Dresden, Fakultät Maschinenwesen, Habilitation,2001
5. BURGERT, I.; SCHLAAD, H.; BERTIN, A.; MILWICH, M.; SPECK, T.; FERY, A.: Optimierung der Faser-Matrix Grenzfläche von Faserverbundwerkstoffen nach dem Vorbild der Natur. In: *Proceedings. Denkendorfer Bionik-Kolloquium"Bio-Inspired Textile Materials"*. Denkendorf, Deutschland, 2008
6. SIDDIQUI, N. A.; LI, E. L.; SHAM, M.-L.; TANG, B.-Z.; GAO, S.-L.; MÄDER, E.; KIM, J.-K.: Tensile strength of glass fibres with carbon nanotube-epoxy nanocomposite coating: Effects of CNT morphology and dispersion state. In: *Composites Part A: Applied Science and Manufacturing* 41 (2010), No. 4, pp. 539–548. DOI 10.1016/j.compositesa.2009.12.011
7. HUFENBACH, W. (Hrsg.) et al.: *Textile Verbundbauweisen und Fertigungstechnologien für Leichtbaustrukturen des Maschinen- und Fahrzeugbaus*. Dresden : SDV - Die Medien AG, 2007
8. HENNING, F.; GEIGER, O.: Großserienfähige Faserverbundtechnologien für den Automobilbau. In: *Proceedings. Symposium Automobil Innovativ"Faserverbundwerkstoffe und High-Tech-Metalle für Automobil- und Maschinenbau"*. Augsburg, Deutschland, 2010
9. ANONYMOUS: *Leitfaden zur Realisierung von Bauteilen aus faserverstärkten Kunststoffen im Textilmaschinenbau*. Aachen, Frankfurt : VDMA, 2008

10. EHRENSTEIN, G.: *Faserverbund-Kunststoffe, Werkstoffe - Verarbeitung - Eigenschaften*. München, Wien : Carl Hanser Verlag, 2006
11. ENDRUWEIT, E. et al.: Experimental determination of the permeability of textiles: A benchmark exercise. In: *Composites Part A: Applied Science and Manufacturing* (2011). DOI doi:10.1016/j.compositesa.2011.04.021
12. KLUNKER, F.: *Aspekte zur Modellierung und Simulation des Vacuum Assisted Resin Infusion*. Clausthal, Technische Universität Clausthal, Dissertation, 2008
13. RIEBER, G.; MITSCHANG, P.: 2D Permeability changes due to stitching seams. In: *Composites: Part A* 41 (2010), No. 1, pp. 2–7. DOI 10.1016/j.compositesa.2009.09.006
14. VERLEYE, B.; CROCE, R.; GRIEBEL, M.; KLITZ, M.; LOMOV, S. V.; MORREN, G.; SOL, H.; VERPOEST, I.; ROOSE, D.: Permeability of textile reinforcements: Simulation, influence of shear and validation. In: *Composites Science and Technology* 68 (2008), S. 2804–2810. DOI 10.1016/j.compscitech.2008.06.010
15. LOUIS, B. M.; FRATTA, C. di; DANZI, M.; ZOGG, M.; ERMANNI, P.: Improving time effective and robuste techniques for measuring in-plane permeability of fibre preforms for LCM processing. In: *Proceedings. SEICO 11 - SAMPE EUROPE 32nd International Technical Conference"New Material Characteristics to cover New Application needs"*. Paris, France, 2011, pp. 204–211
16. FLEMMING, M.; ZIEGMANN, G.; ROTH, S.: *Faserverbundbauweisen - Fasern und Matrices*. Berlin, Heidelberg : Springer Verlag, 1995
17. HUFENBACH, W.; HELMS, O.: *Leitfaden zum Methodischen Konstruieren mit Faser-Kunststoff-Verbunden*. Dresden, 2010
18. DAUN, G.: Schaltbare Härter verbessern Rotor-Fertigung. In: *Kunststoffe* 99 (2009), No. 7, pp. 74–76
19. HAU, E.: *Windkraftanlagen - Grundlagen, Technik, Einsatz, Wirtschaftlichkeit*. 3. Auflage. Berlin, Heidelberg, New York : Springer Verlag, 2003
20. HENNE, M.; MARTI, A.; STAUFACHER, S.: Boot im Gleitflug. In: *Kunststoffe* 96 (2006), No. 10, pp. 72–76
21. LONG, A. C.: *Design and manufacture of textile composites*. Abington Cambridge : Woodhead Publishing Ltd., 2005
22. SCHLEDJEWSKI, R.: Fortschritte bei der Verarbeitung faserverstärkter Kunststoffe. In: *Kunststoffe* 96 (2006), No. 10, pp. 182–188
23. ANONYMOUS: FACC entwickelt Flügelkasten aus Composite. In: *Takeoff - Das FACC Informationsmagazin* (2010), No. 26, pp. 4–7
24. ANONYMOUS: *CFK-Verbundtechnologie - Technische und wirtschaftliche Vorteile."Carbon - Schwabens Schwarzes Gold"*. Augsburg, 2009
25. ANONYMOUS: Schlüsseltechnologie: Eigene RTM Fertigung bei FACC. In: *Takeoff - Das FACC Informationsmagazin* (2010), No. 25, pp. 14–15
26. MICHAELI, W.; FISCHER, K.: Untersuchungen zur Fertigung von flächigen Bauteilen aus faserverstärkten Kunststoffen mithilfe des Spältimprägnierverfahrens. In: *Kunststofftechnik* 4 (2008), No. 4, pp. 1–29
27. BHAT, P.; MEROTTE, J.; SIMACEK, P.; G., ADVANI S.: Process analysis of compression resin transfer molding. In: *Composites: Part A* 40 (2009), pp. 431–441. DOI 10.1016/j.compositesa.2009.01.006
28. HABERSTROH, E. et al.: Aus der Technologieschmiede Aachen. In: *Kunststoffe* 96 (2006), No. 6, pp. 76–80
29. FISCHER, K.; MICHAELI, W.: Durch neue Prozessketten zur FVK-Großserie. In: *Proceedings. 17. Nationales Symposium SAMPE Deutschland e.V."Faserverbundwerkstoffe - Hochleistung und Großserie"*. Aachen, Germany, 2011
30. FLEMMING, M.; ZIEGMANN, G.; ROTH, S.: *Faserverbundbauweisen - Fertigungsverfahren mit duroplastischer Matrix*. Berlin, Heidelberg : Springer Verlag, 1999
31. DAWSON, D. K.: Mae West: Pipe Dream in Munich. In: *High-Performance Composites* (2011), No. 3, pp. 46–53

32. ZIEGMANN, G.: Faserverbunde im Schienenfahrzeugbau - Materialien und Technologien. In: *Kunststoffe* 87 (1997), No. 9, pp. 1142–1146
33. HUFENBACH, W.; KROLL, L.; GUDE, M.; HELMS, O.; ULBRICHT, A.; GROTHAUS, R.: Integrative Rohrgewinde in Wickeltechnik für hochbeanspruchte Verbindungen. In: *VDIBerichte No. 1903*. Düsseldorf : VDI-Verlag GmbH, 2005, pp. 301–316
34. ANONYMOUS: *Metallanbindung*. http://www.carbon-grossbauteile.com/Beispiele.Metallanbindung.aspx (16-05-2011)
35. KIPF, O.: Pultrusion von Rundprofilen in einem industriellen Produktionsprozess. In: *Proceedings.euroLITE 2008*. Salzburg, Austria, 2008
36. MIRAVETE, A.: *New Materials and New Technologies Applied to Elevators*. Mobile, USA : Elevator World, 2002
37. JANSEN, K.; WEIDLER, D.: Click-clac-Snap - Verbindungstechnologie für pultrudierte Profile im Automobilinnenraum. In: H.-P., Degischer; (Hrsg.): *Proceedings. Verbundwerkstoffe 14. Symposium Verbundwerkstoffe und Werkstoffverbunde*. Weinheim, Deutschland
38. WILEY-VCH Verlag, 2003, pp. 890–895 *Radius-PultrusionTM*. http://www.thomas-technik.de/pdf/Radius*Pultrusion:pdf* (16-05-2011)
39. ANONYMOUS: Krümmungen endlos produziert. In: *Kunststoffe* 99 (2009), No. 11, pp. 77
40. MILWICH, M.: Forschung und Entwicklung von Faserverbundwerkstoffen am ITV Denkendorf vom Rohstoff bis zum Bauteil. In: *Proceedings. Workshop des Carbon Composites e.V."Garne und Textilien für den Faserverbundleichtbau"*. Denkendorf, Germany, 2010
41. OFFERMANN, P.; DIESTEL, O.; FUCHS, H.; HUFNAGL, E.; ARNOLD, R.: Strangförmige Textilarmierungen für thermoplastische Kunststoffprofile. In: *Technische Textilien* 42 (1999), No. 2, pp. 143–147
42. SCHOLL, S.; SCHÜRMANN, H.: Endlosprofile mit gezielter Orientierung. In: *Kunststoffe* 98 (2008), No. 7, pp. 84–87
43. MURPHY, J.: *The Reinforced Plastik Handbook*. 2. Auflage. Oxford, New York : Elsevier Advanced Technology, 1998
44. MICHAELI, W.; PÖHLER, M.: 3D-Faserspritzen mit Faserorientierung. In: *Lightweight design* 3 (2010), No. 6, pp. 57–62
45. MICHAELI, W.; PÖHLER, M.: 3D-Faserspritzen - Komplexe Preforms mit lokal einstellbaren Eigenschaftsprofilen. In: *Proceedings. 25. Internationales Kunststofftechnisches Kolloquium des IKV, Session 12:"Verarbeitung lang- und endlosfaserverstärkter Thermoplaste"*. Aachen, Germany, 2010, pp. 4–8
46. SCHWARZ, O.; EBELING, F.-W.; LÜPKE, G.: *Kunststoffverarbeitung*. 6. Auflage. Würzburg : Vogel Buchverlag, 1991
47. MICHAELI, W.; GREIF, H.; KRETZSCHMAR, G.; KAUFMANN, H.; VOSSEBÜRGER, F.-J.: *Technologie der Kunststoffe*. München, Wien : Carl Hanser Verlag, 1992
48. EGGER, P.; KRALICEK, M.; ZEIDLHOFER, H.: Starke Partner. In: *Kunststoffe* 99 (2009), No. 11, pp. 26–30
49. HELLRICH, W.; HARSCH, G.; HAENLE, S.: *Werkstoff-Führer Kunststoffe - Eigenschaften, Prüfungen, Kennwerte*. München, Wien : Carl Hanser Verlag, 2004
50. SCHEMME, M.: Langfaserverstärkte Thermoplaste - Status und Perspektiven. In: DRUMMER, D. (Hrsg.): *Thermoplastische Faserverbundkunststoffe - Werkstoff/Verarbeitung/Simulation/Anwendung*. Erlangen : Lehrstuhl für Kunststofftechnik, Universität Erlangen-Nürnberg, 2011, pp. 1–33
51. SPÖRRER, A.; ALTSTÄDT, V. et al.: Verarbeitung von langfaserverstärkten, hochgefüllten Thermoplasten im einstufigen Spritzgießprozess. In: *Proceedings. 8. Internationale AVK-TV Tagung*. Baden-Baden, Germany, 2005
52. SCHEMME, M.: Langfaserverstärkte Thermoplaste (LFT) - Entwicklungsstand und Perspektiven. In: *Proceedings. Fachtagung Technische Kunststoffe*. Würzburg, Germany, 2006

53. KOCH, T.; SCHÜRMANN, H.: Spritzgussbauteile lokal verstärken. In: *Kunststoffe* 96 (2006), No. 1, pp. 55–58
54. HUFNAGL, E.; HUFENBACH, W.: Weiterentwicklung und Anwendung thermoplastischer endlosfaserverstärkter mehraxialer Gitterstrukturen als Funktionselemente (IGF 282 ZBR) Technische Universität Dresden, Institut für Textilmaschinen und Textile Hochleistungswerkstofftechnik, Institut für Leichtbau und Kunststofftechnik. Dresden, 2010. – Final report
55. CHERIF, Ch.; FRANZKE, G.; HUFNAGL, E.; ERTH, H.; HELBIG, R.; HUFENBACH, W.; BÖHM, R.; KUPFER, R.: Thermoplastische endlosfaserverstärkte Spritzgussbauteile durch Einsatz textiler Gitter. In: *Kunststofftechnik/Journal of Plastics Technology* 4 (2008), No. 4
56. MÜLLER, T.; DRUMMER, D.: Neue Prozessstrategien für Hybridstrukturen auf Basis faserverstärkter Thermoplaste. In: DRUMMER, D. (Ed.): *Thermoplastische Faserverbundkunststoffe Werkstoff/Verarbeitung/Simulation/Anwendung*. Erlangen : Lehrstuhl für Kunststofftechnik, Universität Erlangen-Nürnberg, 2010, pp. 123–141
57. ANONYMOUS: Werkzeugtechnologie für gewebeverstärkte Leichtbauteile. In: *Gummi. Fasern, Kunststoffe* 64 (2011), No. 3, pp. 142
58. MITZLER, J.; RENKL, J.; WÜRTELE, M.: Hoch beanspruchte Strukturbauteile in Serie. In: *Kunststoffe* 101 (2011), No. 3, pp. 36–40
59. SCHMACHTENBERG, E.; AL-SHEYYAB, A.: Kurzer Prozess bei hybriden Strukturen. In: *Kunststoffe* 97 (2007), No. 12, pp. 120–124
60. HOFFMANN, L.; RENN, M.; DRUMMER, D.; MÜLLER, T.: FIT-Hybrid - Hochbelastbare Faserverbundbauteile großserientauglich hergestellt. In: *Lightweightdesign* 4 (2011), No. 2, pp. 38–43
61. RIEBER, G.: Die Rolle der Permeabilität für Injektionsprozesse. In: *Proceedings. Workshop des Carbon Composites e.V."Garne und Textilien für den Faserverbundleichtbau"*. Denkendorf, Germany, 2010
62. MOLTER, M.: *Zum Tragverhalten von textilbewehrtem Beton*. Aachen, RWTH Aachen, Fakultät Bauingenieurwesen, Dissertation, 2005
63. NAAMAN, A. E.: *Ferrocement Laminated Cementitious Composites*. Michigan, USA : Techno Press 3000, 2000
64. ABDKADER, A.; GRAF, W.; MÖ LLER, B.; OFFERMANN, P.; SICKERT, J.-U.: Fuzzy-Stochastic Evaluation of Uncertainties in Material Parameters of textiles. In: *AUTEX Research Journal* 2 (2002), No. 3, pp. 115–125
65. HAUSDING, J.; LORENZ, E.; ORTLEPP, R.; LUNDAHL, A.; CHERIF, Ch.: Application of stitch-bonded multi-plies made by using the extended warp knitting process: reinforcements with symmetrical layer arrangement for concrete. In: *The Journal of the Textile Institute* 102 (2011), No. 8, pp. 726–738. DOI 10.1080/00405000.2010.515729
66. GRIES, T.; ROYE, A.; OFFERMANN, P.; ENGLER, T.; PELED, A.: Textiles for the Reinforcement of Concrete. In: BRAMESHUBER, W. (Ed.): *RILEM Report 36, State-of-the-Art Report of RILEM Technical Committee TC 201-TRC'Textile Reinforced Concrete'*. Bagneux, France : RILEM Publications s.a.r.l., 2006
67. JESSE, F.; CURBACH, M.: Verstärken mit Textilbeton. In: BERGMEISTER, K. (Ed.); FINGERLOOS, F. (Ed.); WÖRNER, J.-D. (Ed.): *Beton-Kalender 2010, Teil I*. Berlin : Ernst Sohn, 2009, pp. 457–565
68. BUTLER, M.: *Zur Dauerhaftigkeit von Verbundwerkstoffen aus zementgebundenen Matrices und alkaliresistenten Glasfaser-Multifilamentgarnen*. Dresden, Technische Universität Dresden, Fakultät Bauingenieurwesen, Dissertation, 2008
69. GAO, S.-L.; MÄDER, E.; PLONKA, R.: Nanostructured coatings of glass fibers: Improvement of alkali resistance and mechanical propertiers. In: *Acta Materialia* 55 (2007), pp. 1043–1052
70. KRÜGER, M.: *Vorgespannter textilbewehrter Beton*. Stuttgart, Universität Stuttgart, Fakultät Bau- und Umweltingenieurwissenschaften, Dissertation, 2004

71. KÖCKRITZ, U.: *In-Situ Polymerbeschichtung zur Strukturstabilisierung offener nähgewirkter Gelege*. Dresden, Technische Universität Dresden, Fakultät Maschinenwesen, Dissertation, 2007
72. WEILAND, S.: *Interaktion von Betonstahl und textiler Bewehrung bei der Biegeverstärkung mit textilbewehrtem Beton*. Dresden, Technische Universität Dresden, Fakultät Bauingenieurwesen, Dissertation, 2009
73. LEPENIES, I. G.: *Zur hierarchischen und simultanen Multi-Skalen-Analyse von Textilbeton*. Dresden, Technische Universität Dresden, Fakultät Bauingenieurwesen, Dissertation, 2007
74. HEGGER, J.; KULAS, C.; SCHNEIDER, H. N.; BRAMESHUBER, W.; HINZEN, M.; RAUPACH, M.; BÜTTNER, T.: TRC Pedestrian Bridge - Design, Load-bearing Behavior and Production Process of a Slender and Light-weight Construction. In: BRAMSHUBER,W. (Hrsg.): *International RILEM Conference on Material Science - Volume I*. Bagneux, France : RILEM Publications s.a.r.l., 2010, pp. 353–364
75. ANONYMOUS: *ETFE-Folien - das"flexible Glas"*. http://www.sattler-ag.com/sattler-web/de/produkte/338.htm (03-03-2011)
76. ANONYMOUS: Das"fünfte Element". Textile Membranen in der Architektur. In: *architektur* 7 (2000), No. 2, pp. 60–68
77. KOCH, K.-M. (Hrsg.): *Bauen mit Membranen*. München : Prestel Verlag, 2004
78. NIENHOFF, H.: Solitäre im urbanen Zusammenhang. Drei Stadien für die Fußballweltmeisterschaft in Südafrika. In: *[Umrisse] Zeitschrift für Baukultur* 9 (2009), No. 2, pp. 10–17
79. ANONYMOUS: *BAUGENIAL erwartet deutlichen Wachstumsschub im Leichtbau*. http://www.rigips.com/web/at/press/pressdetail.php?SubCoIID=4216 (19-11-2009)
80. GLAWE, A.; GIESSMANN, H.: Production of roof membranes and coated textiles for textile constructions. In: *Technische Textilien* 49 (2006), No. 4, pp. 239–243
81. ALTEVOLMER, Ch.: Modulare Gebäudehülle für Verkehrsknotenpunkte. In: *[Umrisse] Zeitschrift für Baukultur* 9 (2009), No. 2, pp. 40–45
82. ZETTLITZER, W.: *Der konstruktive Membranbau. Transparentes Bauen am Beispiel der Allianz-Arena München*. www.covertex.com/version01/de/presse/pdf/konstruktivemembranbau.pdf (23-11-2009)
83. ANONYMOUS: *Koch Membranen: Der Spezialist für Textile Architektur-Textiles Bauen-Technische Konfektion-Licht und Akustik-Verdunkelung*. http://www.kochmembranen.com/de/produkte/produkteuebersicht.php (19-04-2010)
84. TE RIELE, D.: Carapax®: Twaron Multi-axial Composite Trailer Roof. In: *Proceedings. 49. Chemiefasertagung Dornbirn*. Dornbirn, Austria, 2010
85. MÜLLER, J.: Theorie und Praxis der Textilbeschichtung - Funktionale Technische Textilien durch Beschichtung mit Silicon, Anforderung an Beschichtungstechnologie, Anwendungsgebiete.In: *Proceedings. Praxisseminar Textilbeschichtung*. Mönchengladbach und Dormagen, Germany, 2008
86. HOLZBACH, M.: *Adaptive und konditionierende textile Gebäudehüllen auf Basis hochintegrativer Bauteile*. Stuttgart, Universität Stuttgart, Fakultät Architektur und Stadtplanung, Dissertation, 2008
87. HAMSEN, K.-H.: Theorie und Praxis der Textilbeschichtung - Polyvinylchlorid als Beschichtungsrohstoff. In: *Proceedings. Praxisseminar Textilbeschichtung*. Mönchengladbach und Dormagen, Germany, 2008
88. MEHLER TEXNOLOGIES: *Technical guideline to permanent tensile architectures (2009)*. –Hückelhoven. – Company document
89. SEDLBAUER, K.: Textilien im Bau - bauphysikalische Anforderungen an den konstruktiven Membranbau. In: *Proceedings. Bayern Innovativ Kooperationsforum mit Fachausstellung" - Textilien für Bau und Architektur"*. Miesbach, Germany, 2010
90. ANONYMOUS: *Textile Architecture*. http://www.sattler-ag.com/sattler-web/de/produkts/138.htm (23-04-2010)

91. ANONYMOUS: *PTFE (Teflon®) Coated Glass Fabrics*. http://www.fiberflon.de/PTFE_coated_glass_fabrics/EN/5/fiberflon.html (07-05-2010)
92. JULIUS HEYWINKEL GMBH: *Beschichtete Gewebe*. Bramsche, Germany, 2009 – Company document
93. ANONYMOUS: *Precontraint-Technologie*. http://www.ferarri-architecture.com/de/precontraint.php (28-09-2009)
94. MÜLLER, J.: Membran-Bau-Material. In: *[Umrisse] Zeitschrift für Baukultur* 9 (2009), No. 2, pp. 34 – 36
95. LANDSKROON: *Technical Textiles*. Apeldoorn, Netherlands, 2009 – Company document
96. JANSSEN, E.: Theorie und Praxis der Textilbeschichtung - Beschichtung ohne Grenzen. Die Vielfalt beschichteter Erzeugnisse für die Herstellung funktioneller technischerTextilien. In: *Proceedings. Praxisseminar Textilbeschichtung*. Mönchengladbach und Dormagen, Germany, 2008
97. SYNTEEN LÜCKENHAUS TEXTIL-TECHNOLOGIE: Technische Textilien für Dichtungsbahnen. Klettgau-Erzingen, Germany, 2009 – Company document
98. HOFFMANN, G.; CHERIF, Ch.; TROMMER, K.; BÖHME, Y.; STOLL, M.; WIRTH, F.: Gezielter Einsatz von Hochleistungsgarnen und Beschichtungen zur Steigerung der Weiterreißfestigkeit von Planenmaterialien. In: *Proceedings. 48. Chemiefasertagung Dornbirn*. Dornbirn, Austria, 2009
99. WAGNER, R.: Textile Architecture Future in Membrane Construction. In: *Proceedings. 15. Techtextil-Symposium*. Frankfurt/M., Germany, 2009
100. FITZ, J.: Neuartige Architekturgewebe aus Fluorpolymeren. In: Proceedings. Bayern Innovativ Kooperationsforum mit Fachausstellung"Textilien für Bau und Architektur". Miesbach, Germany, 2010
101. HÄNSCH, F. S.: Flexible Structure for new Possibilities in Textile Building. In: Proceedings. 15. Techtextil-Symposium. Frankfurt/M., Germany, 2009
102. ENGTEX AB: Reinforcement for coating. Mullsjö, Sweden, 2009 – Firmenschrift
103. THEODOR PREUSS: Technische Gewirke. Ubstadt-Weiher, Deutschland, 2009 – Company document
104. ANONYMOUS: Properties and Applications of Industrial Materials. Highly resistant composite materials from more than 100 substrates and 300 elastomer coatings. http://www.contitech.de/pages/produkte/gewebe/stoffe/technische_stoffe_en.html (04-03-2011)
105. STEGMAIER, T.; ABELE, H.; RIETHMÜLLER, C.; SCHWEINS, M.; PLANCK, H.: Lichtmanagement für Textile Architektur. In: Proceedings. Bayern Innovativ Kooperationsforum mit Fachausstellung"Textilien für Bau und Architektur", Miesbach, Germany, 2010
106. ANONYMOUS: Das etwas andere Dach. In: [Umrisse] Zeitschrift für Baukultur 9 (2009), No. 2, p. 47
107. HAASE, W.: Modulare, mehrlagige, textile Gebäudehülle mit adaptiven Eigenschaften. http://www.irb.fraunhofer.de/bauforschung/projekte.jsp?p=20088034119 (04-04-2011)
108. HAASE, W.; SEDLBAUER, K.; KLAUS, Th.; SOBEK, W.; SCHMID, F.; SYNOLD, M.; SCHMIDT, T.: Adaptive textile und folienbasierte Gebäudehüllen. In: Bautechnik 88 (2011), No. 2, pp. 69–75

Additional Literature

de LLORENS, J. I.: Report Latin American Symposium of Tensile Structures TENS-SCL 2012. Tensinews Nr. 24 - April 2013, p. 12-15

GEREKE, T.; HUFNAGL, E.; DÖBRICH, O.; CHERIF, Ch.: Concepts for adaptive textile membrane constructions. In: *Proceedings. TensiNet Symposium 2013*, Istanbul, (TR), May 8-10, 2013

HUFNAGL, E.; WALDMANN, M.; HUND, H.; KRZYWINSKI, F.; CHERIF, Ch.: Brave new worlds built from textile membranes - Load-bearing textiles for membranes based on multiaxial warp knitting technology. In: *Kettenwirk-Praxis* 1/2014, pp. 20-22

SIEMENS, P. M.; BOSCHMANN, D.; MURATOVIC, D.; LEUER, H.; DEVECIOGLU, M.: Functionalized surfaces in textile architecture. In: *CD-Rom and Book of Abstracts. 6. Aachen-Dresden International Textile Conference,* Dresden, November 29-30, 2012, p. 171

SYNOLD, M.; SOBEK, W.; van der ROHE, M.: Future building envelopes - A chance for multifunctional membranes? In: *CD-Rom and Book of Abstracts. 6. Aachen-Dresden International Textile Conference,* Dresden, November 29-30, 2012, p. 168

Formula Symbols

α_{tex}	Twist factor (–)
Γ	Fiber mass content (%)
γ	Gradient angle of the fibers to the yarn axis (°)
γ	Shear deformation (%)
δ	Deformation vibration (–)
η	Dynamic viscosity of fluids (Pa·s)
ε	Strain (%)
ε_B	Tensile strain at failure (%)
ε_M	Strain at tensile strength (%)
ε_Y	Tensile strain at yield (%)
Θ	Contact angle (°)
Θ_l	Contact angle (left) (°)
Θ_r	Contact angle (right) (°)
μ	Average (mean) value (–)
μ	Poisson's ratio (–)
ν	Capillary flow speed (m/s)
ν	Stretching vibration (–)
ν	Lateral contraction coefficient (–)
ν_L	Proportion of enclosed air (%)
ρ	Fiber density (g/m^3)
ρF	Fiber material density (kg/m^3)
ρ_{Fiber}	Reinforcement fiber density (g/m^3)
ρ_{Yarn}	Packing density of the yarn (g/m^3)
ρ_{Matrix}	Matrix fiber density (g/m^3)
ρ_V	Nonwoven material raw density (kg/m^3)
$\rho_{Composite}$	Composite density (g/m^3)
σ	Mechanical tension (Pa)
σ^D	Disperse part of the surface energy (mN/m)
$\sigma_1{}^D$	Disperse part of the surface energy Phase 1 (mN/m)

C. Cherif (ed.), *Textile Materials for Lightweight Constructions*,
DOI 10.1007/978-3-662-46341-3

σ_2^D	Disperse part of the surface energy Phase 2 (mN/m)
σ^P	Polar part of the surface energy (mN/m)
σ_{12}	Interface energy Phase 1/Phase 2 (mN/m)
σ_1	Surface energy Phase 1 (mN/m)
σ_1	Surface tension of wetting fluid (mN/m)
σ_1^P	Polar part of the surface energy Phase 1 (mN/m)
σ_2	Surface energy Phase 2 (mN/m)
σ_2^P	Polar part of surface energy Phase 2 (mN/m)
σ_B	Tensile stress at break (Pa)
σ_f	Flexural stress (Pa)
σ_l	Surface tension of liquid phases (mN/m)
σ_M	Tensile strength (Pa)
σ_s	Surface energy of solid phases (mN/m)
σ_V	Tensile stress at yield (Pa)
τ	Shear stress (Pa)
τ	Tortuosity factor (–)
φ	Shear angle (°)
φ	Fiber volume content (%)
A	Area (m^2)
A_{Pore}	Percental proportion of cavities (pores) (%)
B	Stiffness in bending (flexural rigidity) (Nm^2)
b	Sample width (m)
d	Thickness of nonwoven material (m)
d	Diameter (wa ... warp, we ... weft) (mm)
$d_{p,\, max}$	Maximum pore diameter (m)
e	Twisting (%)
e_r	Systematic measurement deviation (–)
e_s	Systematic measurement deviation (–)
F	Force (N)
f	Number of individual filaments within the yarn (–)
f	Frequency (Hz)
f_r	Resonant frequency (Hz)
F_V	Pre-tension (N)
G	Shear modulus (Pa)
H	Moisture content (%)
K	Dimensionless adjustment factor (–)
l	Length (m)
l_0	Yarn length of original yarns (Single yarn or initial twisted yarn) (m)
L_F	Loft factor according to Jordan (–)
l_P	Length of a pore canal (μm)
m	Mass (g)
M	Areic mass (kg/m^2)
M	Molar mass of the polymer (g/mol)
M0	Molar mass of the monomer (g/mol)

mA	Areic mass of the nonwoven material (kg/m^2)
m_{Fiber}	Reinforcement fiber mass (kg)
m_i	Total mass of all molecules of the molecular fraction (g)
m_{Matrix}	Matrix fiber mass (kg)
$m_{Composite}$	Total mass of the composite material (kg)
M_n	Number average of the molar mass averages (g/mol)
M_w	Mass average of the molar mass averages (g/mol)
M_i	Relative molar mass of a narrow molecular fraction (g/mol)
n	Yarn spacing (wa ... warp, we ... weft) (mm)
n_{Fiber}	Number of reinforcement fiber yarns (–)
n_i	Number of molecules of the molecular fraction (mol)
n_{Matrix}	Number of matrix fiber yarns (–)
n_{Spi}	Spindle speed (revolutions per minute) (min^{-1})
p	Hydrostatic pressure (Pa)
P	Degree of Polymerization (–)
P	Porosity (%)
p	Weave factor (weave pattern-dependent) (–)
p_B	Wetting pressure (Pa)
r	Pore diameter (m)
SC	Sizing/preparation content (%)
T	Number of twists per m (–)
T_t	Fineness in tex (tex)
T_{tFiber}	Reinforcement fiber fineness (tex)
$T_{tMatrix}$	Matrix fiber fineness (tex)
U	Molecular inconsistency (–)
V	Volume (m^3)
V_F	Fiber volume (m^3)
V_{Fiber}	Reinforcement fiber volume (m^3)
V_L	Volume of enclosed air (m^3)
V_{Matrix}	Matrix fiber volume (m^3)
$V_{Composite}$	Composite material volume (m^3)
v_L	Delivery speed (m/min)
V_V	Total volume nonwoven (m^3)
Vol.%	Fiber volume content (%)
W	Number of twists per m in a co-wrapping ply yarn (–)
W_A	Work of adhesion related to area (N/m)
x_w	True value of a measured measurand (–)

Abbreviations

1D	One-dimensional
2D	Two-dimensional
3D	Three-dimensional
A, A1, A2	Clamping points in the descriptions of twist insertion
AR	Aramide fiber material
ART	Advanced Resin Transfer Moulding
AS4	Carbon fiber type
ATR	Attenuated Total Reflection
B, B1, B2	Clamping point in twist insertion
C	Clamping point in the descriptions of twist insertion
CA	Acetate
CAD	Computer Aided Design
CAI	Compression After Impact
CCD	Charge-coupled device
CD	Cross direction (perpendicular to production/machine direction)
CD tow	Continuous discontinuous tow
CF	Carbon fiber material
CFRP	Carbon fiber-reinforced plastic
CLF	Polyvinyl chloride fiber material
CNC	Computer Numerical Control
CNT	Carbon nanotubes
COM	*In-situ* commingled
co-PBT	Polybutylene terephthalate
CR	Chloroprene rubber
CUP	Cuprammonium rayon
CV	Viscose fiber material
CVD	Chemical Vapor Deposition
DBD	Dielectric Barrier Discharge
DFG	Deutsche Forschungsgemeinschaft (*German Research Foundation*)
DFG-FOR	DFG-Forschergruppe (*Research Group*)

C. Cherif (ed.), *Textile Materials for Lightweight Constructions*,
DOI 10.1007/978-3-662-46341-3

DG	Fabric density
DIN	Deutsches Institut für Normung (*German Institute for Standardization*)
DMS	Strain gauge
ECPE	Extended Chain polyethylene fibers
EDV	Elektronische Datenverarbeitung (*Electronic Data Processing*)
EL	Elastane
E-Modul	Elastic modulus
EN	Europäische Norm (*European Standard*)
EP	Epoxy resin
ESCA	Electron Spectroscopy for Chemical Analysis
ESZ	Ebener Spannungszustand (*2-dimensional stress state*)
Fbst.	Farbstoff (Dye)
FRPC	Fiber-reinforced plastic composite
FOY	Fully oriented yarn
FCM	Fiber composite material
GF	Glass fiber
GFRP	Glass fiber-reinforced Plastic
GMT	Glass mat-reinforced thermoplastic
hf	hochfest (high strength)
HM	High-modulus type (for AR), High modulus (for CF)
HMS	High modulus/high strength
HOY	Highly oriented yarn
HST	High Strain and Tenacity
HT	High temperature
HT	High tensile
IGES	Initial Graphics Exchange Specification
IM	Intermediate Modulus
IR	Infrared
ISO	International Organization for Standardization
ITA	Institut für Textiltechnik at RWTH Aachen
ITM	Institute for Textile Machinery and High Performance Material Technology at TU Dresden
K	1,000 filaments
LC	Liquid Crystal
LCP	Liquid Crystal Polymers
LFT	Long-fiber-reinforced thermoplastic
LI	Flax
LL	Left-left
LOY	Low oriented yarn
MAG	Multiaxial stitch-bonded fabric
m-AR	meta-Aramide
MD	Machine direction (stitch direction, production direction)
MIR	Medium infrared range

ML	Stitch length
MLG	Mehrlagengestrick (Multi-layered knitted fabric)
MOY	Medium oriented yarn
MPIA	Poly(m-phenylene isophthalamide)
MPP	Mesophase pitch
MR	Stitch course
MRR	Stitch course direction
MS	Wale
MSR	Wale direction
MTF	Metal fibers
N	Normal type
NeB	English cotton yarn number
Nm	Metric number—Length numbering
NMP	N-Methylpyryolidone
OE	Open end (open yarn end)
OT	Top reversal point
PA	Polyamide
PAA	Polyacrylic acid
p-AR	para-Aramide
PBI	Polybenzimidazole
PBI_M	Polybenzimide for matrix fibers
PBO	Polyphenylene-benzbisoxazole
PBO_M	Polybenzoxazolene for matrix fibers
PC	Polycarbonate
PDM	Product data management system
PE	Polyethylene
PEEK	Polyether ether ketone
PEI	Polyetherimide
PES	Polyester
PET	Polyethylene terephthalate
PF	Phenolic resin
POY	Pre-oriented yarn, partially oriented yarn
PP	Polypropylene
PPD	p-Phenylenediamine
PPS	Polyphenylene sulfide
PPTA	Poly-paraphenylene terephthalamide
PSU	Polyethersulfone
PTFE	Polytetrafluoroethylene
PU	Polyurethane
PVA	Polyvinyl alcohol fibers
PVC	Polyvinyl chloride
PVD	Physical Vapor Deposition
R	Organic Group
SEM	Scanning electron microscope

RFID	Radio Frequency Identification
RL	Right-left
RR	Right-right
RRG	Right-right crossed
RTM	Resin Transfer Molding
RWP	Boundary value problem
S	Twist in S-direction
SBCR	Stretch-broken carbon fiber
SBR	Styrene-butadiene rubber
SBS	Side-by-side
SE	Silk
Si	Sisal
SIC	Strand-in-Concrete
SMA	Shape Memory Alloys
SMC	Sheet molding compound
SME	Shape Memory Effect
SRIM	Structural Reaction Injection molding
STEP	Standard for Exchange of Product model data
T300	Type of Carbon fibers
TDC	Terephthaloyl chloride
tex	Fineness—weight numbering
Td	Titer/Denier system
TFP	Tailored fiber p
UD	unidirectional
UHM	Ultrahigh Mode
UHMWPE	Ultrahigh molecular weight polyethylene
UP	Unsaturated polyester resin
UT	Bottom reversal point
UV	Ultraviolet
VARTM	Vacuum Assisted Resign Transfer Molding
VE	Vinyl resin
VI	Vacuum Injection
x	Designation of shedding
XPS	X-ray photoemission spectroscopy
Z	Twist in Z direction

Index

C. Cherif (ed.), *Textile Materials for Lightweight Constructions*,
DOI 10.1007/978-3-662-46341-3

MIX
Papier aus verantwortungsvollen Quellen
Paper from responsible sources
FSC® C105338

If you have any concerns about our products,
you can contact us on
ProductSafety@springernature.com

In case Publisher is established outside the EU,
the EU authorized representative is:
Springer Nature Customer Service Center GmbH
Europaplatz 3, 69115 Heidelberg, Germany

Printed by Libri Plureos GmbH
in Hamburg, Germany